Fungal Diseases of Amenity Turf Grasses

To the late Drs F.T. Bennett and J.H. Western, successively Provincial Advisory Mycologists, Northern Province, at King's College, Newcastle upon Tyne, University of Durham (now University of Newcastle upon Tyne), to J. Monteith Jr and A.S. Dahl, lately of the US Golf Association and Dr F.L. Howard, Professor Emeritus of the University of Rhode Island, for their important contributions to research on the fungal diseases of turf grasses.

Fungal Diseases of Amenity Turf Grasses

THIRD EDITION

J.D. Smith MSc, CBiol, MIBiol, PAg

Department of Horticulture Science
University of Saskatchewan, Saskatoon, Canada

N. Jackson PhD

Department of Plant Sciences
University of Kingston, Rhode Island, USA

A.R. Woolhouse BSc

Formerly at the Sports Turf Research Institute, Bingley,
West Yorkshire, UK

LONDON
NEW YORK
E. & F.N. SPON

First published in 1959 by
The Sports Turf Research Institute
Second edition 1965
This edition first published in 1989 by
E. & F.N. Spon Ltd
11 New Fetter Lane, London EC4P 4EE
Published in the USA by
E. & F.N. Spon
29 West 35th Street, New York NY 10001

Phototypeset in 9/10 pt Times by
Thomson Press (India) Ltd, New Delhi
Printed in Great Britain by
Richard Clay Ltd, Bungay, Suffolk

ISBN 0 419 11530 7

British Library Cataloguing in Publication Data

Smith, J. Drew
Fungal diseases of amenity turf grasses.
——3rd ed.
1. Turfgrasses——diseases 2. Fungal diseases of plants
I. Title II. Jackson, N
III. Woolhouse, A.R. IV. Smith, J. Drew.
Fungal diseases of turf grasses
635.9'642 SB608.T87

ISBN 0-419-11530-7

Contents

Preface to the third edition xi

Acknowledgements xii

PART ONE—Principles of disease control 1

1 Environmental effects 3
1.1 Water 3
1.2 Temperature 3
1.3 Soil texture 4
1.4 Soil reaction 4
1.5 Soil organic matter 4
1.6 Plant nutrients 5
1.7 References 6

2 The effects of cultural practices on turf grass diseases 7
2.1 The establishment of turf by seeding 7
2.2 The establishment of turf by vegetative means 7
2.3 Mowing 8
2.4 Top dressing 8
2.5 Liming and turf acidity control 9
2.6 Fertilization of established turf 9
2.7 Pesticides 10
2.8 Soil compaction 11
2.9 Scarification 11
2.10 Switching, poling and hose dragging 11
2.11 Partial sterilization of top dressing (aerification) 11
2.12 Aerations 11
2.13 Irrigation 12
2.14 Temporary turf covers 13
2.15 References 13

3 Species and cultivars of turf grasses in relation to disease management 15
3.1 References 16

4 Turf disease control with fungicides 18
4.1 History 18
4.2 Fungicide formulations 22
4.3 Fungicide applications 22
4.4 Fumigant fungicides 24
4.5 References 25

PART TWO—Seed and seedling diseases (including some minor root diseases) 27

5 Seed rots, pre- and post-emergence blights 29
5.1 Symptoms 29

5.2 Causal fungi 30
5.3 Pre- and post-emergence seedling blights caused by '*Fusarium*' species 30
5.4 Pythium seedling diseases and root rots 38
5.5 Turf grass seedling diseases caused by *Rhizoctonia* spp. 40
5.6 Minor seedling or root pathogens and parasitic root fungi 43
5.7 Control of seed rots and seedling diseases 49
5.8 References 50

PART THREE—Low-temperature fungal diseases of amenity turf grasses 57

6 Low-temperature fungal diseases 59
6.1 Microdochium patch and pink snow mould 59
6.2 Cottony snow mould—low-temperature basidiomycete (LTB) and sclerotial low-temperature basidiomycete (SLTB) phases of *Coprinus psychromorbidus* 67
6.3 Frost scorch, leaf rot, tip blight, or string of pearls disease 74
6.4 Grey or speckled mould or typhula blight 75
6.5 Sclerotinia snow mould 88
6.6 Other low-temperature-tolerant turf grass pathogens and antagonists 92
6.7 Snow mould complexes and competition 95
6.8 Summary of snow moulds of turf grasses 98
6.9 References 103

PART FOUR—Root diseases 115

7 Fusarium blight (fusarium foliar blight, crown and root rot) 117
7.1 Symptoms 118
7.2 The causal fungi 119
7.3 Host range 120
7.4 Disease development 120
7.5 Control 121
7.6 References 122

8 Spring dead spot 124
8.1 Symptoms 124
8.2 Causes 125
8.3 Host range 126
8.4 Disease development 126
8.5 Control 127
8.6 References 128

9 Necrotic ring spot 129
9.1 Symptoms 129
9.2 The causal fungus 129
9.3 Host range 131
9.4 Disease development 131
9.5 Control 131
9.6 References 132

10 Summer patch 133
10.1 Symptoms 133
10.2 The causal fungus 133
10.3 Host range 134

10.4 Disease development 135
10.5 Control 135
10.6 References 135

11 Take-all patch (formerly ophiobolus patch) 137
11.1 Symptoms 137
11.2 The causal fungus 137
11.3 Host range and susceptibility 139
11.4 Epidemiology 140
11.5 Control 143
11.6 References 144

12 Mycorrhizae 147
12.1 Symptoms 147
12.2 The causal fungi 147
12.3 Mycorrhizal development 148
12.4 Effects of mycorrhizae on grasses 148
12.5 Control 149
12.6 References 149

13 Fungal endophytes 151
13.1 References 153

PART FIVE—Rusts **155**

14 Rusts 157
14.1 Introduction 157
14.2 Crown rust—*Puccinia coronata* 162
14.3 Cynodon rust—*Puccinia cynodontis* 168
14.4 Rust of fine-leaved fescues—*Puccinia festucae* 168
14.5 Black or stem rust—*Puccinia graminis* 168
14.6 Brown fleck rust—*Puccinia poae-nemoralis* 170
14.7 Orange stripe rust—*Puccinia poarum* 171
14.8 Brown or leaf rust—*Puccinia recondita* 171
14.9 Yellow stripe rust—*Puccinia striiformis* 172
14.10 Zoysia rust—*Puccinia zoysiae* 173
14.11 Uromyces leaf rust—*Uromyces dactylidis* 173
14.12 Some other rusts of amenity turf grasses 173
14.13 Some useful diagnostic features of common, cool-season rust species on turf grasses 174
14.14 Control of rusts on amenity turf grasses 174
14.15 References 175

PART SIX—Smuts **181**

15 Smut diseases 183
15.1 Leaf smuts 183
15.2 Sheath smut 191
15.3 Head smuts 192
15.4 References 196

PART SEVEN—Inflorescence diseases (other than smuts) **199**

16 Some important inflorescence diseases (other than smuts) 201
16.1 Blind seed 201

16.2 Choke or cat's tail disease 203
16.3 Ergot 205
16.4 Silvertop, whiteheads or white-ears 207
16.5 Stem eyespot of fescues 208
16.6 Twist 210
16.7 References 211

PART EIGHT—Foliar blights and patches **217**

17 Anthracnose and colletotrichum basal rot 219
17.1 Symptoms 220
17.2 The causal fungus 222
17.3 Pathogenicity tests 222
17.4 Host range 223
17.5 Disease development 224
17.6 Species and cultivar resistance 225
17.7 Control 225
17.8 References 225

18 Copper spot disease 227
18.1 Symptoms 227
18.2 The causal fungus 227
18.3 Host range 227
18.4 Disease development 228
18.5 Control 228
18.6 References 229

19 Dollar spot disease 230
19.1 Symptoms 230
19.2 Host range 231
19.3 The causal fungus 231
19.4 Disease development 232
19.5 Control 235
19.6 References 236

20 Downy mildew or yellow tuft disease 238
20.1 Symptoms 238
20.2 The causal fungus 239
20.3 Host range 240
20.4 Disease development 240
20.5 Control 242
20.6 References 242

21 Grey leaf spot 244
21.1 Symptoms 244
21.2 The causal fungus 244
21.3 Host range 245
21.4 Disease development 245
21.5 Control 246
21.6 References 246

22 Powdery mildew 247
22.1 Symptoms 247
22.2 The causal fungus 247

22.3 Host range 247
22.4 Disease development 248
22.5 Control 249
22.6 References 250

23 Pythium foliar blights (spot blight, grease spot, cottony blight) 252
23.1 Symptoms 252
23.2 The causal fungi 253
23.3 Host range 253
23.4 Disease development 254
23.5 Control 256
23.6 References 257

24 Red thread and pink patch diseases 259
24.1 Symptoms 259
24.2 The causal fungi 260
24.3 Host range 262
24.4 Disease development 262
24.5 The control of red thread and pink patch diseases 263
24.6 References 264

25 Rhizoctonia foliar diseases of established turf 266
25.1 Brown patch 266
25.2 Yellow patch 272
25.3 References 274

26 Southern blight or southern sclerotium blight 276
26.1 Symptoms 276
26.2 The causal fungus 276
26.3 Host range 277
26.4 Disease development 277
26.5 Control 278
26.6 References 278

PART NINE—Leaf spots, leaf blights and leaf streaks 281

27 Leaf spots, leaf blights, crown rots and seedling blights of turf grasses caused by *Drechslera, 'Helminthosporium'* and *Curvularia* species 283
27.1 Introduction 283
27.2 General taxonomy of the pathogens 283
27.3 *Bipolaris* species 285
27.4 *Cochliobolus* species 285
27.5 *Curvularia* species 291
27.6 *Drechslera* species 293
27.7 *Pyrenophora* species 301
27.8 *Setosphaeria rostrata* 305
27.9 References 305

28 Some leaf spots associated with pycnidial fungi 312
28.1 *Ascochyta* 312
28.2 *Phyllosticta* 314
28.3 *Pseudoseptoria* 314
28.4 *Selenophoma* 315
28.5 *Septoria* 315

28.6 *Stagonospora* 317
28.7 *'Hendersonia'* 318
28.8 *Phaeoseptoria* 319
28.9 *Wojnowicia* 319
28.10 Disease development and control of pycnidial leaf spots 319
28.11 References 320

29 Some other leaf spots, leaf blights and leaf streaks 321
29.1 Leaf spots caused by *Cercospora* and *Pseudocercosporella* 321
29.2 *Cercosporidium graminis*—brown stripe or leaf streak 323
29.3 *Cladosporium phlei*—timothy eyespot 324
29.4 *Hadrotrichum* leaf spot 325
29.5 *Leptosphaerulina* leaf blight 325
29.6 *Mastigosporium* spp.—leaf fleck of bentgrasses and timothy 326
29.7 *Phyllachora* spp.—black leaf spot or tar spot of turf grasses 327
29.8 *Physoderma* leaf spot and leaf streak 329
29.9 *Ramularia* and *Ramulaspera* leaf spots 330
29.10 *Rhynchosporium* spp.—leaf blotch or scald 332
29.11 *Spermospora* leaf spot, scald and leaf blight 333
29.12 *Cheilaria agrostis*–Char spot 334
29.13 References 335

PART TEN—Fairy rings **339**

30 Fairy rings 341
30.1 Causal fungi 341
30.2 The rate of growth of fairy rings 342
30.3 The effects of fairy rings on the soil 343
30.4 Fairy rings caused by *Marasmius oreades* (Bolt ex Fr.) Fr. 344
30.5 Suppression and elimination of fairy rings 347
30.6 Superficial fairy rings 350
30.7 References 351

PART ELEVEN—Slime moulds and lichens **355**

31 Slime moulds 357
31.1 Causal organisms and their biology 357
31.2 Effects on turf 357
31.3 Control 358
31.4 References 358

32 Lichens 359
32.1 Control 359
32.2 References 359

PART TWELVE—Summary **361**

33 Key to common turf diseases caused by fungi 363
34 Summary of pathogens, hosts and controls 367
35 Common and chemical names of commonly used turf fungicides 384

Index **387**

COLOUR PLATES (Figures 1–65) between pages **20–21**

Preface to the third edition

Fungal diseases continue as a major cause of deterioration in the quality of amenity turf although other disease incitants are receiving increasing attention.

Viruses, bacteria, rickettsias, mycoplasmas, nematodes, algae, moss and insects may cause varying degrees of turf damage. Injuries result from chemical excesses and deficiencies, climatic factors relating to temperature and moisture extremes, soil conditions and the activities of small and larger mammals including man! These deserve separate treatment and are outside the scope of this text. However, there is an excellent introduction to the whole range of turf grass diseases in 'Compendium of turf grass diseases' prepared by Dr R.W. Smiley, published by the American Phytopathological Society in 1983.

We have extensively revised the second edition of 'Fungal diseases of turf grasses', published in 1965 by the Sports Turf Research Institute, and now out of print. Whereas the last edition was concerned mainly with some of the important turf grass diseases in cool temperate climates, in this third edition we have widened the scope to include major diseases of warm-season grasses. We have also expanded the coverage of overwintering diseases and those of seed crops of turf grasses.

Grasses for turf purposes have improved markedly through selection and breeding during the past 25 years and numerous cultivars are now available specifically for turf usage. Although many of them show greater tolerance to the stresses of wear, drought, shade, heat or cold, persistence under mowing and are leafier, more attractive in colour and more resistant to specific diseases than their predecessors, they may still be damaged by fungi if overstressed. Resistance to all major diseases in a cultivar is probably unattainable, given pathogen mutability. Fortunately, another great advance in turf grass disease management during this same period has been the development and adoption of new fungicides, especially systemic materials, that afford excellent control of previously intractable fungal disease problems. Already, however, widespread pathogen resistance to some of these fungicides has been observed.

The development of good management strategies against turf disease depends upon an accurate diagnosis of their cause and an understanding of their development. It is hoped that this text will assist the professional turf manager, turf grass agronomist, breeder and plant pathologist in identification and control of many of the diseases likely to be met.

Although the text and number of illustrations have been considerably expanded, the layout is similar to that of the previous edition. The main literature search was completed on 31 December 1983, with some updating to 31 December 1987.

J. Drew Smith
Department of Horticulture Science, University of Saskatchewan, Saskatoon, Canada S7N 0W0

N. Jackson
Department of Plant Sciences University of Kingston, Rhode Island, USA

A.R. Woolhouse
Formerly at the Sports Turf Research Institute, West Yorkshire, UK
1988

Acknowledgements

We are indebted to R.E. Underwood, Photographer, Saskatoon Research Station, Research Branch, Agriculture Canada, for assistance in many ways with illustrations used in this text; to Dr Neil A. Baldwin Plant Pathologist of the Sports Turf Research Institute for assistance in the final preparation of the text for publication; to colleagues in several countries who contributed information and illustrations which are credited appropriately.

To the following scientists of the Biosystematics Research Centre, Research Branch, Agriculture Canada, Ottawa: Dr R.A. Shoemaker for review of the sections on *'Helminthosporium'* and *Curvularia* and seedling diseases caused by *Fusarium* species; Dr D.J.S. Barr for review of *Pythium* seedling diseases, minor seedling or root rot pathogens and parasitic root fungi; Dr J. Bissett for review of the section on leaf spot diseases; Dr S. Redhead for comments and information on cottony snow mould; Dr D.B.O. Savile for review of the chapter on rusts. H.B. Gjaerum of the Plant Protection Research Institute, Ås-NLH, Norway, also reviewed the rust chapter and Dr R. Koske, Department of Botany, University of Rhode Island, the chapter on mycorrhizae.

We are most grateful for this assistance, but accept full responsibility for errors, inaccuracies and opinions expressed in the text.

We also thank colleagues of our respective organizations for information and interest in the undertaking. The work would not have been possible without the support of the Directors of our respective institutions: Dr J.R. Hay of Agriculture Canada Research Station, Saskatoon; Dr G.A. Donovan, Director of Rhode Island Agricultural Experiment Station, Kingston, Rhode Island; Dr P. Hayes of The Sports Turf Research Institute, Bingley; Mr J.R. Escritt, former Director of the Sports Turf Research Institute, under whose aegis the work was started.

Chapter 6 on 'Low temperature fungal diseases of amenity turf grasses' has been adapted from 'Winter hardiness and overwintering diseases of amenity turf grasses with special reference to the Canadian prairies' by J. Drew Smith (the latter is published as Technical Bulletin 1987–12E by Research Branch Agriculture Canada, Ottawa).

Part one

Principles of disease control

1 *Environmental effects*

1.1 WATER

Most of the fungi concerned with diseases of aerial parts of turf grasses require either free water or a very high atmospheric humidity for infection to proceed rapidly. These moisture conditions are supplied by rain, dew, fog, guttation drops and by the humid microclimate which develops around the base of plants in a dense sward growing on a moist or wet soil.

When the soil is moist a good dense turf is more suitable for the development and spread of turf diseases than is a sparse one. In a dense turf the tendency is for the crowded plants to shade each other and to hold moisture by preventing air movement and reducing solar drying. A dense sward nevertheless is the ideal one for most sports (except perhaps in the case of cricket tables). Grass plants are continually transpiring water particularly if growing rapidly. With high soil moisture, high atmospheric humidity and restricted transpiration, water appears at the tips of the leaves as guttation drops that provide a suitable environment for the germination of spores and the development of fungal mycelium. Guttation fluid may exude from the mown ends of plant parts where damaged or dead tissues provide a relatively easy point of entry for spore germ-tubes or mycelium.

The humidity of the microclimate within a dense turf is governed principally by the moisture content of the soil and the relative humidity of the atmosphere above the ground. These factors, plus wind, govern evaporation of moisture from the soil. Grass plants make little growth in dry soil and plants that are not growing freely cannot combat an established infection by regenerating injured parts sufficiently rapidly to keep pace with disease. Some fungi grow actively at moisture tensions far lower than those conducive to grass growth and can take advantage of drought-stressed turf (e.g. *Fusarium* spp.). Infection and decay of plant parts is constantly taking place even in an apparently healthy sward.

In soil which contains too much moisture, i.e. waterlogged, root systems cannot function satisfactorily. Their cellular respiration is impaired due to the soil air being replaced in the pore spaces of the soil by soil water. Some fungal root diseases caused by Phycomycetes, e.g. *Pythium* spp., that spread by means of motile zoospores, are favoured by an abundance of soil-water and water films on the surface of the soil and lower plant surfaces.

Water can be an important agent in the release and dispersal of fungal spores from aerial parts of grasses in the case of some diseases of amenity turf, e.g. those caused by rusts, *Cochliobolus* or *Drechslera* spp. Air shock waves produced by raindrops impacting on dry, spore-bearing tissues may detach the spores and disperse them locally. Spores may also be carried in or on water droplets splashed to infect new hosts. Intermittent wetting and drying of infected plant debris may promote sporulation of some leaf spot pathogens (e.g. *Drechslera* spp.).

With well-regulated soil moisture, plants in turf are supplied with sufficient water for good growth and yet root aeration is satisfactory. The amount of soil drainage should be suited to the type of soil available. On heavy or peaty soils in an area of high precipitation there will be more soil water to be controlled. Conversely, in drier areas and on light soils it may be necessary to add materials such as organic matter which will retain water in the soil during dry periods.

1.2 TEMPERATURE

Each fungus species and strain has a minimum, an optimum and a maximum temperature for growth. These so-called 'cardinal' points may be determined by growing the fungus in pure culture on artificial media. The optimum temperature for the growth of the fungus is not necessarily the optimum temperature for the onset and progress of the disease. In a fungal disease there are two organisms concerned, the fungus and the plant. The optimum temperature for plant growth may coincide with that for the fungus, in which case, although fungal growth may proceed rapidly, plant growth also being rapid may keep pace with the onslaught of the fungus. The temperature at which a disease progresses most rapidly is the one which lowers the vitality of the plant more than that of the fungus. A complicating factor in soil-borne diseases is that soil organisms antagonistic to the pathogen may be relatively more effective in checking it at its optimum temperature than at a higher or lower temperature.

Extremes of temperature may injure both plant and fungus, but rarely are the air and soil temperatures encountered in temperate climates sufficiently extreme to kill out turf disease fungi completely. They may, however, prevent the further development of disease. In foliar diseases, such as those caused by *Rhizoctonia solani, Microdochium nivale, Limonomyces roseipellis, Laetisaria fuciformis* or *Sclerotinia homoeocarpa*, where further leaf involvement results from the growth of aerial mycelium, it is the lower atmospheric humidity which often accompanies the higher temperature, as much as the high temperature alone which reduces the rate of attack. Dry conditions desiccate the aerial mycelium and the tissues on and in which the fungi are living and prevent their further progress (Lucas and Vargas, 1978).

Damage to turf grasses occurs when there is a sudden increase or decline in temperature during periods when growth is proceeding rapidly. Grass blades and shoots may be injured and cells killed, especially near their tips and the dead tissues form suitable infection points for pathogenic fungi. During these periods, play or management operations such as mowing, may aggravate the situation. In diseases of the 'damping-off' type, germination and growth at low temperatures takes place so slowly that the young shoot and root systems are in a susceptible state during periods when pathogens such as *Pythium* or *Fusarium* spp. are quite active. Severe heat stress may damage turf directly or predispose the weakened grasses to various fungal pathogens, e.g. *Colletotrichum graminicola, Fusarium* spp. and *Sclerotinia homoeocarpa*.

1.3 SOIL TEXTURE

Few soil-borne diseases appear to be directly favoured by soils of heavy texture but heavy soils because of impeded drainage may favour some diseases as they promote a higher relative humidity round stem bases just above soil level. Most fungi, whether soil-borne or otherwise, are strong aerobes and most pathogenic soil fungi are favoured by soils of light texture which allow the natural diffusion of soil gases.

1.4 SOIL REACTION

While there is a considerable body of information on the effect of pH on the growth of turf grass pathogens *in vitro* (see individual diseases), there is not much available on the effect of soil pH on fungal diseases of turf grasses.

In interpreting the direct effects of soil reaction on turf grass diseases the observer may be misled, as soil acidity or alkalinity is often related to the availability of nutrients or the occurrence of toxic materials in the soil.

Couch and Bloom (1958) found that susceptibility of *Agrostis stolonifera* to *Rhizoctonia solani* was not affected by pH in the range 4.0 to 10.0 under low nitrogen nutrition. When nitrogen was moderately high the plants were less susceptible under alkaline conditions.

The prevalence of take-all disease in cereals caused by *Gaeumannomyces graminis*, which is also important as the cause of take-all patch in turf grasses, was correlated with soil reaction by Garrett (1937). He found that there was more rapid growth of the fungus along roots in alkaline soils than in acid soils. In soils of heavy texture and in acid soil, fungal growth is checked by the accumulation of respiratory carbon dioxide. It was suggested that alkaline soils increased the fungal growth rate by acting as acceptors of carbon dioxide. During pathogenesis, the runner hyphae of *G. graminis* are mainly exogenous and probably influenced more by soil reaction (and antagonists) than in the case of an endogenous pathogen.

In take-all patch disease of turf, increased incidence of the disease has been correlated with an increase in turf surface pH following liming and decreased incidence with applications of sulphur and acid-tending fertilizers (see Chapter 11). Similary the patch disease caused by *Microdochium nivale* is increased in severity after liming.

Copper spot caused by *Gloeocercospora sorghi* is most destructive to *Agrostis canina* turf growing in soils with pH between 4.5 and 5.5. Liming to reduce the acidity of the soil will reduce the incidence of copper spot (Howard, Rowell and Keil, 1951).

1.5 SOIL ORGANIC MATTER

There is little information on the direct effects of soil organic matter on turf grass diseases, but its role in the maintenance of plant vigour and hence in minimizing disease damage is well established (Beard, 1973). It is concerned with soil structure, water relations, nutrient retention and release, aeration and root respiration, survival of pathogens and their antagonists. If organic matter is lacking, particularly on light-textured soils, drought resistance may be poor, due to inadequate root systems and inadequate moisture retention; soluble nutrients, such as nitrogen and potassium may leach rapidly in wet weather or under heavy irrigation. The resulting lack of plant vigour may lead to greater susceptibility to diseases such as dollar spot (*Sclerotinia homoeocarpa*), pink patch (*Limonomyces roseipellis*), red thread (*Laetisaria fuciformis*) and the rusts. The addition of organic matter to soil sometimes has the effect of reducing parasitic activities of soil-borne pathogens such as *Gaeumannomyces graminis* and *Rhizoctonia solani*. Excessive amounts of organic matter incorporated into soil increases water retention, impedes drainage and favours root diseases such as those caused by Phycomycetes. Heavy surface applications of organic top dressings have a smothering effect, increase turf surface humidity and favour the incidence of disease caused by *Microdochium nivale*. However, in establishing turf from seed, especially on light-textured soils or sand, incorporation of organic matter can assist rapid germination and seedling establishment by holding moisture at stages of growth when seeds or seedlings are very disease susceptible (see Part Two).

1.6 PLANT NUTRIENTS

The amendment of soil before seeding with suitable fertilizers containing nitrogen, phosphorus and potassium is a common practice and may be expected to result in better seedling establishment than in nutrient-deficient, unamended soil (Madden, 1935). The early stages of germination of grass seeds are controlled by major factors of water supply, temperature, oxygen and in some cases, by light. If these are suboptimal seed germination may be delayed and the risk of seed rotting increased. However, as endosperm nutrient reserves are depleted the young seedling becomes dependent on external sources of nutrients for photosynthesis and growth to carry it past the early stages of development. Post-emergence seedling diseases may still beset it at this time (see Part Two).

Nitrogen level and its balance with other soil nutrients, especially phosphate and potassium are major factors that determine proneness to particular diseases and the rate of recovery from disease attacks. Generally, nitrogen deficiency is shown by low vigour, poor sward colour and density. More specifically, grasses become more susceptible to 'low fertility' diseases such as those caused by *Sclerotinia homoeocarpa, Laetisaria fuciformis, Limonomyces roseipellis*, *Puccinia* spp. and *Gloeotinia granigena*. In excess, nitrogen encourages excessively rapid, soft growth, delays maturation of tissues, increases guttation (Healy, 1967) and results in cell walls being thin and easily penetrated by fungal hyphae. Lesions caused by leaf-spotting fungi elongate rapidly where too much nitrogen has been applied (see *Cochliobolus sativus*, p. 286). However, nitrogen also plays a major role in recovery from disease by enabling the grass to produce more herbage than is being destroyed. Excessive or untimely usage of nitrogenous fertilizer increases susceptibility to *Cochliobolus* and *Drechslera* spp., *Microdochium nivale*, *Leptosphaeria korrae*, *Gloeocercospora sorghi*, *Magnaporthe grisea*, *Rhizoctonia solani*, *Sclerophthora macrospora* and the snow moulds generally. The quantification of the term 'excessive' is related to management level. Amenity turf, especially that on golf greens, is more intensively managed in North America with higher nitrogen usage and concomitant greater need for fungicidal control measures than in western Europe. Financial strictures in recent years have resulted in a trend towards 'low-maintenance turf' where possible. A reduction in luxury use of nitrogen and less irrigation may herald an increase in 'low-fertility' diseases in North America.

Phosphorus in adequate amounts encourages the formation of vigorous roots and reduces the severity of seedling diseases. If it is deficient in turf, rooting is poor and swards will lack density. Deficiencies of phosphate increase susceptibility to *Gaeumannomyces graminis, Myriosclerotinia borealis* and *Typhula* snow moulds and interfere with acclimation. In snow mould caused by *Coprinus psychromorbidus*, phosphate applied without nitrogen significantly increases disease severity, but not when applied with nitrogen. This suggests a direct effect on the pathogen or its antagonists rather than on the grasses. Liming may increase phosphate availability.

Potassium effects on disease susceptibility are often more apparent than those of phosphorus, but as with the latter nutrient, adequate supply and NPK balance are important in maintaining plant health. Susceptibility to *Erysiphe graminis* and *'Helminthosporium'* diseases (Evans, Rouse and Gudauskas, 1964; Holben, 1950) is greater when potassium is deficient and increasing the supply has been reported to reduce susceptibility to *Leptosphaeria korrae* and *Sclerotinia homoeocarpa*. *Cladosporium phlei* causes more severe injury where potassium is deficient, but NPK balance is also important in reducing disease damage. Potassium applications have a greater effect than those of phosphorus in reducing susceptibility to *Microdochium nivale*. Applications of potassium and phosphate suppress *Gaeumannomyces graminis*. Potassium may become deficient in light soils or turf formed on sand because its salts are soluble and readily leach out.

Calcium is applied to turf in the form of calcium hydroxide or carbonate, usually for its effect on soil pH rather than as a plant nutrient. However, calcium as a nutrient, when deficient increased the susceptibility of *Agrostis tenuis* to *Pythium ultimum* (Moore, Couch and Bloom, 1963) and of *Festuca rubra* ssp. *rubra* to 'Corticium' red thread disease (Muse and Couch, 1965) in pot experiments.

Magnesium deficiency, shown by poor growth and extreme chlorosis, may occur occasionally on some turf grasses, but particularly *Poa annua*. Symptoms may appear in spring as grass growth becomes active and often they are aggravated temporarily by the spring fertilizer dressing.

Iron is not likely to become deficient in turf maintained under neutral to acid conditions. As calcined ferrous sulphate, iron is applied as a top dressing in low dosage and gives a dark-green colour to grasses, 'hardening' and 'toning' the plants. It reduces the incidence of *Microdochium nivale* patch, but the mode of action is uncertain. Excessive use of calcined ferrous sulphate results in the reduction of availability of phosphate as shown by soil analysis.

Sulphur deficiency reduces turf grass vigour and induces chlorosis and marginal necrosis of grass leaf blades (Love, 1962). The severity of the deficiency is increased by high nitrogen fertilization (Goss, 1974). Escritt and Legg (1969) reported that sulphur, applied as ammonium sulphate reduced the severity of *Microdochium nivale* patch. As elemental sulphur, ammonium sulphate or sulphur-containing fungicide it was reported to inhibit the development of *Gaeumannomyces graminis* patch disease when used with balanced NPK fertilization (Davidson and Goss, 1972). When applied as elemental sulphur at high dosage (224 kg/ha) it inhibited the development of *M. nivale* regardless of level of nitrogen application (Goss, Brauen and Olsen, 1975). Since sulphur is an essential nutrient, a fungicide, and at high dosage depressed surface pH considerably, its mode of action in reducing the incidence of the diseases is complex (Goss, 1974).

Copper and manganese have been reported to increase resistance to *Microdochium nivale* in *Lolium perenne*.

Few instances of the effects of minor nutrients on turf have been recorded but it is not usual for deficiencies of these materials to occur in turf. Excess of certain elements, including trace elements, in concentrations harmful to turf grasses may occur through the misuse of some chemical aids to turf management (herbicides, vermicides and fungicides) or the indiscriminate use of some industrial waste materials in fertilizer and top dressing mixtures. The effects of these materials on turf grass diseases is generally not known.

1.7 REFERENCES

Beard, J.B. (1973) *Turfgrass: Science and Culture*. Prentice-Hall, New Jersey, 658 pp.

Couch, H.B. and Bloom, J.R. (1958) Influence of soil moisture, pH and nutrition on the alteration of disease proneness in plants. *Trans. N.Y. Acad. Sci.*, **20**, 432–7.

Davidson, R.M. and Goss, R.L. (1972) Effects of P, S, N, lime, chlordane and fungicides on *Ophiobolus* patch disease of turf. *Plant Dis. Rept.*, **56**, 565–7.

Escritt, J.R. and Legg, D.C. (1969) Fertilizer trials at Bingley. *Proc. 1st Int. Turfgrass Res. Conf.*, Harrogate, England, pp. 185–190.

Evans, E.M., Rouse, R.D. and Gudauskas, R.T. (1964) How soil potassium sets up coastal bermuda for leaf spot disease. *Highlights Agric. Res.*, **11** (2), 14.

Garrett, S.D. (1937) Soil conditions and the take-all disease of wheat. II. The relation between soil reaction and soil aeration. *Ann. Appl. Biol.*, **24**, 747–51.

Goss, R.L. (1974) Effects of variable rates of sulphur on the quality of putting green bentgrass. In *Proc. 2nd Int. Turfgrass Res. Conf.*, Blacksburg, Va. 1973, (ed. E.C. Roberts), pp. 172–5.

Goss, R.L. Brauen, S.E. and Olsen, S.P. (1975) Effects of N, P, K, and S on *Poa annua* in bentgrass putting turf. *J. Sports Turf Res. Inst.*, **51**, 74–82.

Healy, M.J. (1967) Factors affecting the pathogenicity of selected fungi isolated from putting green turf. Ph.D. Thesis, University of Illinois, 50 pp.

Holben, F.J. (1950) Nitrogen/potash relationships as they affect growth and diseases of turf grasses. *Penn. State Coll. 19th Ann. Turf Conf. Proc.*, pp 74–93.

Howard, F.L., Rowell, J.B. and Keil, H.B. (1951) Fungus diseases of turf grasses. *Rhode Island Exp. Sta. Bull.*, **308**, 56 pp.

Love, J.R. (1962) Mineral deficiency symptoms on turf grass. 1. Major and secondary nutrient elements. *Wisc. Acad. Sci. Arts Lett.*, **51**, 135–40.

Lucas, L.T. and Vargas, J.M. Jr (1978) The effects of weather on disease. *Golf Superintendent*, **46**, 31–5, 38.

Madden, E.A. (1935) Garden lawns and playing fields. Establishment and maintenance. *N.Z. Dept. Agric. Bull.*, **165**. (Reprint from *N.Z. J. Agric.* 1935).

Moore, L.D., Couch, H.B. and Bloom, J.R. (1963) Influence of environment on diseases of turf grasses. 3. Effect of nutrition, pH, soil temperature and soil moisture on *Pythium* blight of Highland bentgrass. *Phytopathology*, **53**, 53–7.

Muse, R.R. and Couch, H.B. (1965) Influence of environment on diseases of turf grasses. IV Effect of nutrition and soil mositure on Corticium red thread of creeping red fescue. *Phytopathology*, **55**, 507–10.

2 *The effects of cultural practices on turf grass diseases*

2.1 THE ESTABLISHMENT OF TURF BY SEEDING

Operations which encourage the rapid germination and emergence of grass seeds directly or indirectly reduce the severity of attack by fungi causing pre- and post-emergence damping-off. The soil should be fallowed and cultivated or chemically sterilized to reduce the number of weeds and eliminate competition. For the successful establishment of seedlings the seed bed should be well prepared, firm on the surface, sufficiently moist and not too acid. Conditions which delay germination such as low temperature, excessive soil moisture (which reduces seed respiration), too little soil moisture (which slows down imbibition of water) should be avoided. Chemicals in the seed bed such as residual herbicides, excessive amounts of copper, arsenic and some organo-mercury compounds will reduce the vigour of germination, or may completely inhibit it. Fertilizer applications, best made before sowing, encourage the rapid establishment of grass seeds (Beard, 1973; Madden, 1935). Where sowings are to be made on land which carried grasses in the recent past, the use of a fungicide seed dressing (See pp. 49–50) is an insurance against poor and/or patchy stands caused by pre- and post-emergence diseases. This is especially the case when small-seeded grasses are being sown, but a fungicidal seed dressing may also improve establishment of larger-seeded species such as *Lolium perenne*.

Amenity turf is often established in industrial areas on land which would be considered unfit for agricultural grassland. The site is often levelled by the 'cut and fill' method and the subsoil and topsoil may not be replaced in their correct relative positions. Heavy earth-moving equipment which is commonly used compresses and compacts the soil unevenly. The site may be a refuse dump or levelled pit or shale heap with a few inches of imported soil perhaps of heavy texture. In many of these cases excess water troubles occur, even when tile – or mole – drainage systems have been installed. It may take many years, particularly in clay soils, for a suitable structure with natural drainage channels to re-form in soil which has been extensively disturbed. Where soil moisture is excessive, newly sown turf may suffer greatly from diseases of the damping-off type. These seedling diseases are aggravated by the over-abundant seeding rates frequently used in the renovation of worn turf at the end of the playing season. Crowding grass seedlings further disposes them to seed-rotting and damping-off diseases (Madison, 1966a; Smith, 1957). After emergence, unless seedlings are topped at the appropriate height they soon become too tall and encourage humid conditions at soil level conducive to disease development.

2.2 THE ESTABLISHMENT OF TURF BY VEGETATIVE MEANS

To obtain a rapid ground cover turf may be established by planting stolons, sprigs or plugs of grasses such as *Agrostis stolonifera*, *A. canina*, *Cynodon* spp., *Stenotaphrum secundatum*, *Eremochloa ophiuroides* or *Zoysia* spp. or by transplanting sod of many cool- and warm-season grasses grown in turf nurseries or commercially on sod farms.

In stolonizing, sprigging or plugging the main problems, other than those of preparing propagative material of suitable grass strains, are those of establishing good cultural conditions and maintaining adequate soil moisture in the initial stages of stolon growth. However, disease outbreaks may be severe in established monostands unless a disease-resistant strain has been selected.

In sod growing the aim is to produce a uniform, attractive, dense turf which can be lifted and re-established as quickly as possible. Species and cultivars are selected with this in mind and disease problems are uncommon in the early growth stages, but are just as likely to develop with age as in sown turf. Sod can be severely damaged by heating and secondary rot fungi if the stacking of lifted sod is protracted.

The use of pasture sod to establish lawn or playing surfaces is much less common nowadays than previously (Dawson, 1959). The quality of the material varies greatly. In many cases fine turf of inland origin is too fibrous to allow rapid rooting and establishment while coarse grass species present often make it unsuitable for fine turf areas. In the British Isles, it is still common practice to harvest sea-marsh turf from a semi-littoral environment and use it to form bowling greens and lawns inland. Sooner or later this sod loses its original botanical composition, depending on the management imposed. Intensive mowing reduces the root systems of the component plants (sea-marsh fescues and sea-marsh bent *Festuca rubra* L. ssp. *litoralis* and *Agrostis stolonifera* L. var. *compacta* Hartm.) which in turn reduces the shoot vigour. In an attempt to maintain the original dominant fescue, generous fertilizer treatment is often discouraged on the grounds that this promotes the invasion of *Poa annua*. However, experience has shown this to be unsound practice since the lowered fertility permits attacks of dollar spot and (to a lesser extent) red thread disease to develop, which thin out the susceptible sea-marsh fescue and permit the invasion of the turf by other plant species. If, on the other hand, too much nitrogen is used, both the fescue and bent become susceptible to *Microdochium nivale*. Scars due to this disease fill up with *Poa annua* and other weed species and the original botanical composition is lost. Sea-marsh turf is expensive and extremely difficult to manage and its use can be justified only because it provides a quick cover and, largely because of its silt layer, it rolls out evenly to provide a true surface for lawn bowling.

2.3 MOWING

The tolerance of different turf grass species and cultivars to cutting height varies from very close (5–13 mm) in fine *Agrostis* spp. and *Poa annua*, through moderate height (25–50 mm) for most *Poa pratensis*, *Lolium perenne* and *Festuca rubra* cultivars, to high (40–75 mm) in *Agropyron cristatum* and *Festuca arundinacea*. If the mowing is lower than that normally tolerated by a particular species or cultivar plant vigour is reduced. Repeated foliage removal decreases carbohydrate reserves which is reflected in decreased root growth. Consequently, vigorous aerial growth and ability to recover from disease declines.

Close mowing may increase damage from the foot rot phase of *Drechslera poae* on *Poa pratensis* (p. 297), grass rusts generally (p. 155), *Rhizoctonia solani* (Madison, 1966b; Rowell, 1951), *Erysiphe graminis* (p. 247), *Sclerotinia homoeocarpa* (p. 230), red thread (p. 259) and *Leptosphaeria korrae* (p. 129). Some fungi find the infection courts of dead and damaged shoot tissues a ready (or the main) means of access to the host tissues.

Frequent mowing removes invaded leaf tips of some of the leaf spotting fungi such as *Drechslera, Ascochyta, Septoria* and *Stagonospora* spp, before tissues below the invaded areas have been colonized. If these tissues are collected during mowing, inoculum for later attacks is reduced, e.g. in *Puccinia zoysiae* and for those pathogens which are poor competitors in tissues in a later stage of degradation in the thatch, e.g. *Cochliobolus sativus* (see p. 286). Unmown, neglected turf in late autumn and early winter may become matted, trapping tree leaves, which maintains a high humidity, favouring outbreaks of patch disease caused by *Microdochium nivale*. *Sclerotinia* or *Typhula* spp. also find conditions for germination more suitable in long turf at this time. Nevertheless, where low-temperature grass injury is a problem the crowns of closely mown grass are less well-insulated than in longer turf.

Mowing is a potent means of dispersal of the spores of fungi, especially dry-spored types such as *Drechslera poae* (see p. 298) and some of these are deposited on freshly cut leaf blades. Where a mowing machine is poorly adjusted or the cutters are blunt leaf tissues may be shredded, particularly in some cultivars of *Festuca arundinacea* and *Lolium perenne* providing greater amounts of substrate for saprobic development of some facultative pathogens. Mowing also assists in the spread of infected leaf fragments in diseases such as dollar spot (Smith, 1955) and red thread (Smith, 1954).

The repeated collection of grass clippings when mowing removes large amounts of major nutrients (especially nitrogen) and soil fertility is reduced. If these are not replaced grasses lose vigour and become more prone to damage by 'low fertility' diseases. If, on the other hand, clippings are returned. soil fertility is maintained, but a spongy, moisture-retentive turf surface often will develop. If *Poa annua* seed is returned in the clippings its germination and establishment is favoured, especially under alkaline to slightly acid soil conditions, increasing also the risk of attack by *Microdochium nivale*.

2.4 TOP DRESSING

The top dressing of turf (mainly of the finer lawns, putting and lawn bowling greens) with bulky materials such as peat, rotted animal or vegetable wastes, soil, sand or mixtures of these is done to control thatch (Adams and Saxon, 1979; Eggens, 1980), improve levels, surface drainage, rooting (when combined with coring), fertility (to a lesser extent), to afford winter protection, to repair injuries and to act as a carrier to ensure the even application of seeds, fertilizers and pesticides. In some instances, microorganisms present in the top dressing antagonize and may suppress soil-borne pathogens (Schneider, 1982; Wildermuth, 1982 and take-all patch-p. 142). Unless sterilized, some of the organic components may be sources of weed seeds of disease-susceptible grasses (e.g. *P. annua*) or the propagules of persistent fungal pathogens (sclerotia, oospores and chlamydospores especially).

The physical condition of the top dressing is all-important. It should not be too heavy in texture (cricket tables are an exception) or it will seal the surface and it should be in a friable, fairly dry condition to facilitate its even distribution and to allow it to be worked into the turf with a brush or drag mat. The quantity applied should be such that it can be accommodated by the sward, without smothering it. If the rate of application is too high there is a risk that diseases such as microdochium patch will be encouraged to develop. In general, the composition of the top dressing should be similar to the medium on which the turf was produced. If it is too dissimilar from the underlying soil, stratifications in the profile may develop with root restriction or even root breaks. Frequent light top dressing with straight sand is currently in vogue in North America but the practice may generate problems when improperly used on old-established, thatchy turf. High sand media may also be associated with *Pythium* spp. root infections.

2.5 LIMING AND TURF ACIDITY CONTROL

Some effects of soil reaction on turf grass diseases have been noted in Chapter 1 (p. 4).

In temperate climates turf for soccer, rugger, cricket outfields, field hockey and for golf fairways is produced mainly from mixtures which are predominantly composed of *Lolium perenne* or *Poa pratensis*. For this type of turf a pH of 6.0 or slightly above will permit good growth. This level is also suitable for *Cynodon* or *Zoysia* turf. Finer turf composed of *Festuca rubra* ssp. *rubra*, and *commutata*, or bentgrasses (e.g. *Agrostis stolonifera*, *A. canina*, *A. tenuis* or *A. castellana*) will grow satisfactorily at pH 5.0–6.0 and the maintenance of a moderate acidity level assists to a large extent in keeping turf free from worm casts, some weeds and some diseases (Dawson, 1959; Smith, 1965; STRI, 1980).

The leaching out of bases, especially calcium, in areas of high precipitation, the effects of acid fumes in polluted air, or the over-use of acidifying materials such as ammonium sulphate results in turf becoming more acid. It may become too acid even for the vigorous growth of acid-tolerant grasses such as some *Festuca* and *Agrostis* spp. Under such circumstances little plant response can be expected from fertilizer applications without first raising the soil pH. However, the correction of over-acid conditions must be done with great circumspection on the basis of a soil test. A sudden increase in the turf surface pH as the result of even light liming may encourage the onset of diseases such as take-all patch and microdochium patch. In both of these diseases most of their activity occurs near the turf surface. Availability of the lime used to correct the acidity influences the severity of take-all patch. Coarser grades of calcium carbonate do not alter turf surface pH as rapidly as fine material (see take-all patch, pp. 137–143).

Improvements in fungicides, and selective herbicides have made it possible to maintain fine turf at pH levels near the optimum for the species concerned and have reduced the tendency for thatch accumulation which builds up more rapidly under acid conditions. Abrupt changes in pH and microbial balance such as occur near the turf surface after liming may be avoided. Especially on smaller turf areas, such as golf and bowling greens and lawns, a pH balance may be struck by alternate use of acid- and alkaline-tending fertilizers and top dressings applied in split applications without the sudden lift in pH which results from heavy liming.

2.6 FERTILIZATION OF ESTABLISHED TURF

The role of plant nutrients was considered in Chapter 1.

Root initiation of grasses takes place mainly in late summer and autumn but most active root growth apparently occurs during the spring, reaching a maximum several weeks before maximum top growth and then declining rapidly as the season progresses. The demand for nutrients by the grass plant is greatest during the active growth period and the provision of adequate amounts of nitrogen, phosphate and potash at this time is essential. During the summer nitrogen is the major requirement to maintain plant vigour. An initial heavy dressing of a slow release nitrogenous fertilizer theoretically should fulfil this summer requirement, but in practice more precise control of growth is obtained by several small dressings of quicker acting, but less persistent, forms of nitrogen. Lack of nitrogen during the summer months leads to reduced vigour and predisposes the turf grasses to attack by diseases such as dollar spot, red thread and rusts.

Particularly in temperate regions with mild winter climates care must be taken to avoid applying too much nitrogen in late summer and autumn as this may lead to a flush of soft, succulent growth which is very susceptible to snow moulds generally. At one time, in Britain it was a common practice on fine turf to apply ferrous sulphate with the autumn fertilizer to 'tone up and harden' the grass lessening the risk of autumn diseases. In climates like those found in the Great Plains of North America where turf grass growth goes quiescent during late autumn and remains so until spring and the ground carries a snow cover for most of this period, dormant fertilization in late autumn is practicable and often advantageous. There is little loss of soluble nutrients and they are in position in spring to assist in early recovery from winter injury and snow mould diseases (see cottony snow mould and myriosclerotinia snow mould– p. 67 and 88).

If the soil is too dry or too cold little immediate response results from the application of inorganic fertilizers. Even less response is to be expected from organic fertilizers applied under the same conditions, for these require to be broken down to simpler materials before they can be absorbed by the grass roots. (Dried blood, is however, an example of an organic nitrogenous fertilizer which is almost as rapid in action as inorganic materials.)

Synthetic fertilizers with controlled nitrogen release such as urea-formaldehyde, sulphur-coated urea, crotonylidene urea, and isobutylidene urea have received acceptance in turf grass management practice in several parts of the world, but not yet generally in the British Isles (Woolhouse, 1983). They are used in an attempt to supply nitrogen at a rate equal to the demand of the growing plant. Nitrogen sources with the rapid action of ammonium nitrate, ammonium sulphate and urea produce disease-susceptible flushes of growth and need to be applied several times during the growing season if control of fertility is to be maintained. The breakdown of the slow-acting organic fertilizers such as hoof and horn meal, fish meal and bone meal which were traditionally used with the faster-acting nitrogen fertilizers to prolong growth stimulation, is often not predictable. The organic materials are now expensive or in short supply.

Organic fertilizers add organic matter to the soil as well as plant nutrients. Organic matter is important in the maintenance of an active soil microflora and good soil structure but the use of organic fertilizers to replenish soil organic matter is a slow and expensive procedure. It can be supplied quickly and cheaply to turf by top dressing with a compost high in organic matter content.

The even application of the fertilizers selected is of great importance or uneven growth and, in extreme instances, damage to the turf or disease may result. It has frequently been noted that even with the 'brush and endless belt' type of distributor, overlapping applications of nitrogenous fertilizer will result in strips and lines of microdochium patch disease where the grass has been over-stimulated by the heavier application. Areas where fertilizer application has been missed may show 'low fertility' diseases. The crushing action of the wheels of any distributor on the grass combined with the fertilizer applied may also result in scorched lines.

2.7 PESTICIDES

Several different categories of biocides are used extensively in amenity turf grass management. These include selective and non-selective herbicides, both pre-emergence and post-emergence kinds; insecticides, nematicides, bactericides and fungicides. Most of these affect both the target and non-target organism (Pennypacker *et al.*, 1982; Smiley, 1981). Their actions may be direct on the grass pathogen, reducing or enhancing mycelial growth, viability or longevity. They may influence spore or sclerotium survival or germination (Altman and Campbell, 1977; Katan and Eshel, 1973). Many of the post-emergence herbicides are synthetic auxins which even at dosages tolerated by the grass without obvious symptoms may be expected to cause physiological changes. Some of these may result in increasing or decreasing the susceptibility to a pathogen or its ability to outgrow the disease (Altman and Campbell, 1977; Rodrigez-Kabana and Curl, 1980). This is an indirect effect on the pathogen.

The susceptibility of *Poa pratensis* cv. 'Merion' to stripe smut was shown to increase when the dithiocarbamate pre-emergence herbicide bandane was used for weed control (Turgeon *et al.*, 1974). The post-emergence herbicides 2,4-D 2,4,5-T and dicamba showed variable effects of stimulation or inhibition of sporulation, spore germination, germ tube growth and pathogenicity of *Cochliobolus sativus* on *Poa pratensis* leaves depending on herbicide concentration (Hodges, 1980). After soil of putting greens has been fumigated with methyl bromide, mainly for the purpose of weed seed control, attacks of *Gaeumannomyces graminis* var. *avenae* may be severe on the *Agrostis* turf, probably because of the elimination or suppression of antagonistic microorganisms (see take-all patch–p. 142). Red thread disease is often more severe on turf on which the growth retardant, maleic hydrazide has been used (Smith, unpublished). Thiabendazole, applied for the control of *Myriosclerotinia borealis* apparently stimulates the mycelial growth of *Typhula ishikariensis* var. *canadensis* (see p. 88). The high selectivity of quintozene applied for the control of snow mould complexes may increase the severity of *Microdochium nivale* attacks (see p. 67). The insecticide and vermicide, chlordane, has shown fungicidal activity against *G. graminis* var. *avenae* in Washington, but not in the British Isles (Nilsson and Smith, 1981). Ferrous sulphate and mercury compounds which are used in some countries for moss control show indirect or direct effects in disease control. There is an association of worm casts with disease caused by *M. nivale* and the elimination of the worms with materials which do not have fungicidal properties (e.g. lead or calcium arsenates) reduces risk of the disease.

2.8 SOIL COMPACTION

Excessive rolling, or foot traffic compacts surface layers of soil, destroys the soil structure of the surface, impairs drainage and reduces root aeration, thereby reducing root and plant vigour. It may also result in 'surface sealing' and algal colonization of the surface. Dollar spot is often more prevalent at rink ends which receive more treading than other parts of the bowling green. Basal rot of *Poa annua* (*Colletotrichum graminicola*) appears to be associated with compacted turf adjacent to foot paths or golf green aprons (see p. 225).

2.9 SCARIFICATION

Scarifying by means of rotary machines, hand rakes, brushes or harrows to remove dead and dying plant material from the sward usually results in an increase in plant vigour and is a recognized means of improving neglected turf. It exerts a beneficial influence by allowing the easier penetration of air, moisture and nutrients from fertilizers to the root systems of grass plants. Light scarification is an adjunct to mowing, lifting the foliage into a vertical position when pressed down so that it may be clipped. Deeper scarification assists in the control of thatch. However, scarification also detaches sclerotia of *Typhula* spp., *Myriosclerotinia borealis*, *Rhizoctonia solani*, stromata of *Laetisaria fuciformis* and microsclerotial flakes of *Sclerotinia homoeocarpa* and foliage infected with mycelium of many pathogens. Wet- and dry-spored organisms are disseminated also. In most cases the beneficial results outweigh the risk of spreading disease. Suction collection of sclerotia of *Myriosclerotinia borealis* following thorough scarification is suggested as a means of reducing inoculum of the pathogen (p. 92).

2.10 SWITCHING, POLING AND HOSE DRAGGING

These practices, which are usually done early in the day, disperse dew or raindrops on fine turf, grounding or reducing the water to films which evaporate more rapidly. The drying process is increased if a current of air passes over the turf surface. Dew forms when atmospheric moisture over the turf condenses on leaves during radiation heat loss from the ground, mainly at night. Clear skies, still weather and/or sheltered conditions favour dew formation. While dew is a source of some moisture for turf grasses during dry weather, it also provides sufficient water for bridging growth of the mycelium of fungal pathogens from leaf to leaf in *Microdochium nivale*, *Sclerotinia homoeocarpa* and *Rhizoctonia solani* (Howard, Rowell and Keil, 1951; Smith, 1953, 1955). In the absence of rain dew also provides suitable moisture conditions for the germination of the spores of rusts (see p. 155) and other turf grass pathogens. As with scarification, switching disseminates pathogens. Its advantages in managing turf surface humidity and in disease control probably outweigh these disadvantages.

2.11 PARTIAL STERILIZATION OF TOP DRESSINGS

This is an effective way of ensuring that top dressings containing compost or soil are free from weeds, pathogens and pests. It is not customary in North America to sterilize peat, sand or other minerals used in top dressings as they are usually substantially free from problem-causing organisms. Activated sewage sludges have already been steam-sterilized, but digested sewage sludge has not been so treated and may contain weed seeds and pathogens and requires 'sterilization' if used in top dressings. Although partial sterilization eliminates many of the unwanted organisms it may also kill beneficial antagonists (see Section 2.7). Partial sterilization of compost etc. is effected through the use of heat or chemicals such as methyl bromide, dazomet or metham-sodium.

2.12 AERATION (AERIFICATION)

Turf which receives much foot and/or machine traffic tends to become compacted. This restricts water penetration, drainage and root respiration. Less vigorous growth and slower recovery from disease and mechanical injuries are the consequence. Because of grass growth habit, a mass of roots, rhizomes or stolons in/on the upper layer of the turf (Adams and Saxon, 1979; Troughton, 1957) form thatch especially where no earthworms are present (Richardson, 1938) and where soil pH is low (Stapledon and Hanley, 1927): the former probably resulting from the latter. Different turf grass species accumulate thatch at different rates according to management, but there may be so much thatch that its hydrophobic properties when dried out may prevent or restrict water and nutrient penetration. Conversely, it is so water-retentive when wet that the turf surface becomes waterlogged and spongy and remains so too long for convenience.

The need to relieve compaction is lessened if turf is established on suitable soil, sand or sand/soil mixture using sound construction methods and proper cultivations. Pricking, slicing, slitting, solid-tine forking and the use of surfactants aid in water penetration and root ventilation in matted turf. Hand or mechanical equipment is available for this. Thatch may be tackled by hand or machine using hollow tines or spoons. The former usually remove cores up to 10 cm long, special machines remove much larger cores; the spoons remove small pieces of turf and thatch. The material which is removed may be broken up and worked back into the holes or top dressing may be used for this purpose (see Section 2.4). Other methods of turf ventilation and thatch removal with vertical mowers or groovers have been developed.

The importance of aeration operations in pathogen dispersal is uncertain, but it is probable that soil infested with fairy ring mycelium may be transported to clean turf in hollow tine fork cores (see fairy rings, p. 349). This is a risk which is outweighed by the benefits which ensue from aeration, but it could be reduced by 'sterilizing' the cores and disinfecting the equipment used.

2.13 IRRIGATION

Water is essential for the vital functions of both host plant and pathogen. In some regions of the world with temperate oceanic climates, like most of the British Isles, excess water and its disposal is often a major turf management problem, especially in wet seasons. In the interior of great continents such as North America there is insufficient precipitation or it is so unevenly distributed that artificial watering is needed if good quality turf is to be established and maintained.

Established turf should be irrigated before it dries out to the point that severe wilting occurs, especially on acid, matted turf, where one of the difficulties is of obtaining water penetration. Irregular watering based on measured (or observed) requirements, to thoroughly moisten the soil to, say, 15 cm depth is to be preferred to frequent, regular sprinkling which wets only the top 5 cm. The latter merely encourages shallow rooting and subsequent drought susceptibility and keeps the surface moist encouraging development of several diseases. These include brown patch, microdochium patch, leaf spot diseases in general and yellow turf (see pages 59, 238, 266). Pre-and post-emergence diseases of seedlings caused by Phycomycetes with motile spores such as *Pythium* spp. (see. p. 38) are generally favoured by excessive irrigation, but that caused by *Fusarium culmorum* is often more severe where soils are dry, as is post-emergence seedling blight caused by *Cochliobolus sativus* in hot weather when grass seeds are sown in sand (Fig. 1, colour plate section). Drought stress increases severity of rust diseases generally (p. 174). Irrigation influences the build up of stripe smut of *Poa pratensis*, by preventing the death of perennially infected plants which would normally die off in droughted soil (Hodges, 1967).

Excessive irrigation or flooding of turf may result in the death of deep roots by asphyxiation. Since these are required to take up and supply water during periods of drought stress such turf may be rendered more drought susceptible. Too much water, especially on sands or sandy soil, may result in leaching of the more soluble nutrients, nitrogen and potassium, increasing the incidence and severity of 'low fertility' diseases (see Section 1.5, and Section 2.6). In take-all patch disease, which is most severe (in Britain) in wet years, irrigation may increase its severity by rapidly solubilizing lime applied to correct acidity in *Agrostis* turf (p. 140). Excessive or mis-timed irrigation may wash off contact fungicides or dilute or leach systemic materials thereby reducing their effects. On the other hand correctly timed watering can effectively prevent foliar damage from fertilizers or wash fungicides applied to control pathogens in the thatch or on root systems to where they are most effective. Light irrigation or syringing is an effective means of diluting fungal stimulants in guttation drops or cooling and reducing transpiration rate of turf grasses in hot, droughty and windy weather (see *Curvularia* diseases, p. 291).

There is some controversy over the most suitable time of the day for irrigation. Experiments quoted by Monteith and Dahl (1932) indicated that when golf greens were overwatered it did not matter when this was done, but when irrigation was done correctly much less disease occurred if the water was applied in the early morning. Generally, night or early morning watering is done since this causes the least inconvenience to turf users and this also allows the surface to dry out during the day, reducing the period when moist conditions, favouring disease, prevail. The maintenance of high turf surface moisture is favourable to the growth of *Poa annua* which is a very disease-susceptible species.

Dispersal of spores, particularly of dry-spored species of fungi by raindrop 'tap-and-puff' (Hirst and Stedman, 1963) occurs at the start of irrigation. As the turf is wetted spore dispersal continues of both wet- and dry-spored types by rain splash. As water films build up on the plants, movement continues of hydrophobic dry spores in water tension films particularly on glossy, wettable leaf surfaces (Crawley, Campbell and Smith, 1962; Smith and Crawley, 1964).

2.14 TEMPORARY TURF COVERS

In temperate climates, where winter turf sports are possible, the covering of football pitches with straw has been practised to permit play for an important game when freezing temperatures are forecast. Likewise, temporary covers of canvas or synthetic materials are sometimes used on cricket squares or grass tennis courts in showery weather to minimize interruptions of games. If these covers remain on for up to a day or so they cause little apparent change in disease susceptibility. When light intensity and temperatures are low from late autumn to spring, covering of football pitches increases liability to aetiolation and more rapid grass growth due to the heat conservation affected by the cover. The softer growth and higher humidity favours the development of microdochium patch disease. There is also some loss in acclimation of the grass which increases the risk of 'winter burn' when the covers are removed.

In severe winter climate regions, such as in the interior of the North American continent, winter desiccation of the fine turf of putting and bowling greens, especially those containing *Poa annua* is common. This occurs particularly in low-snowfall regions where the turf loses its protective cover quickly, perhaps due to warm 'chinook' winds, leaving it exposed to drying winds, especially in late winter or early spring. An adequate water reserve in the soil before freeze-up is one aspect of prevention. Control of evapotranspiration is also necessary. This may be achieved by measures such as windbreaks placed to the windward side of the turf or pieces of brushwood laid on the turf to trap the snow. The turf may be covered with organic mulches, top dressing or sheets of woven fabric, netting with fine meshes, clear or coloured polyethylene or a porous synthetic material (Evans, 1975; Smith, 1987).

2.15 REFERENCES

Adams, W.A. and Saxon, C. (1979) The occurrence and control of thatch in sports turf. *Rasen-Turf-Gazon*, **10** (3), 75–83.

Altman, J. and Campbell, C, L. (1977) Effect of herbicides on plant diseases. *Ann. Rev. Phytopathol.*, **15,** 361–85.

Beard, J.B. (1973) *Turfgrass: Science and Culture.* Prentice-Hall, New Jersey, 658 pp.

Crawley, W.E., Campbell, A.G. and Smith, J.D. (1962) Movement of spores of *Pithomyces chartarum* on leaves of ryegrass. *Nature*, **193,** 295.

Dawson, R.B. (1959) *Practical Lawncraft.* 5th ed., Crosby Lockwood, London, 320 pp.

Eggens, J.L. (1980) Thatch control on creeping bentgrass turf. *Can J. Plant Sci.*, **60,** 1209–13.

Evans, G.E. (1975) Winter mulch covers, spring vigor and subsequent growth of *Agrostis*. *Agron. J.*, **67** (4), 449–54.

Hirst, J.M. and Stedman, O. J. (1963) Dry liberation of fungus spores by raindrops. *J. Gen. Microbiol.*, **33,** 335–44.

Hodges, C.F. (1967) Etiology of stripe smut, *Ustilago striiformis* (West.) Niessl, on Merion bluegrass, *Poa pratensis.* Ph.D. Thesis, University of Illinois, 64pp.

Hodges, C.F. (1980) Post-emergent herbicides and pathogenesis by *Drechslera sorokiniana* on leaves of *Poa pratensis.* In *Advances in Turfgrass Pathology.* (eds B.G. Joyner and P.O. Larsen) Harcourt Brace Jovanovich, Duluth, Minn., 197 pp.

Howard, F.L., Rowell, J.B. and Keil H.L. (1951) Fungus diseases of turf grasses. *Rhode Island Exp. Sta. Bull.*, **308,** 56pp.

Katan, J. and Eshel, Y. (1973) Interactions between herbicides and plant pathogens. *Residue Rev.*, **45,** 145–77.

Madden, E.A. (1935) Garden lawns and playing fields. Establishment and maintenance. NZ Dept. Agric. Bull., **165** (Reprint from *NZ J. Agric.*, 1935).

Madison, J.H. (1966a) Optimum rates of seeding turf grasses. *Agron. J.*, **58,** 441–3.

Madison, J.H. (1966b) Brown patch of turf grass caused by *Rhizoctonia solani.* Kuhn. *Calif. Turfgrass Cult.*, **16** (2), 9–13.

Monteith, J. Jr and Dahl, A.S. (1932) Turf diseases and their control. *Bull. US Golf Assn Green Sect.*, **12** (4), 186 pp.

Pennypacker, B.W., Sanders, P.L., Gregory, L.V., Kilbride, E.P. and Cole, H.Jr. (1982) Influence of triadimefon on the foliar growth and flowering of annual bluegrass. *Can. J. Plant Pathol.*, **4,** 259–62.

Richardson, H.L. (1938) The nitrogen cycle in grassland soils with special reference to the Rothamsted Park Grass experiment. *J. Agric Sci.*, **28,** 73–121.

Rodrigez-Kabana, R. and Curl, E. A. (1980) Non-target effects of pesticides on soil-borne pathogens and disease. *Ann. Rev. Phytopathol.*, **18,** 311–88.

Rowell, J.B. (1951) Observations on the pathogenicity of *Rhizoctonia solani* on bentgrass. *Plant Dis. Rept.*, **35,** 40–242.

Schneider, R.W. (ed.) (1982) *Suppressive Soils and Plant Disease.* American Phytopathological Society, 88pp.

Smiley, R.W. (1981) Nontarget effects of pesticides on turf grasses. *Plant Dis.*, **65**(1), 17–23.

Smith, J.D. (1953) Fungi and turf diseases 3. Fusarium patch disease. *J. Sports Turf Res. Inst.*, **8**(29), 230–52.

Smith, J.D. (1954) Fungi and turf diseases 4. Corticium disease. *J. Sports Turf Res. Inst.*, **8**(30), 365–77.

Smith, J.D. (1955) Fungi and turf diseases 5. Dollar spot disease. *J. Sports Turf Res. Inst.*, **9**(31), 35–59.

Smith, J.D. (1957) Seed dressing trial. *J. Sports Turf Res. Inst.* **9**, 369–72.

Smith, J.D. (1965) *Fungal Diseases of Turf Grasses*. 2nd ed., (revised Jackson, N. and Smith, J.D.) Sports Turf Research Institute, 97 pp.

Smith, J.D. (1976) Snow mold control in turf grasses with fungicides in Saskatchewan, 1971–74. *Can. Plant Dis. Surv.*, **56**, 1–8.

Smith, J.D. and Crawley, W.E. (1964) Disturbance of pasture herbage and spore dispersal of *Pithomyces chartarum* (Berk. & Curt.) M.B. Ellis. *NZ Agric. Res.*, **7**(3), 281–98.

Smith, J.D. (1987) Winter hardiness and overwintering diseases of amenity turf grasses with special reference to the Canadian prairies Research Branch, Agriculture Canada, Ottawa, *Tech. Bull.* **12E,** 193 pp.

Stapledon, R.H. and Hanlcy, T.A. (1927) *Grassland.* Claredon Press, Oxford.

STRI (1980) *Turfgrass Seed, 1980*. Sports Turf Research Institute, 105pp.

Troughton, A. (1957) The underground organs of herbage grasses. *Commonw. Bureau Pasture Field Crops. Bull.*, **44,** 163 pp.

Turgeon, A.J., Beard, J.B., Martin, D.P. and Meggitt, W.F. (1974) Effects of successive applications of preemergence herbicides on turf. *Weed Sci.*, **22,** 349–51.

Wildermuth, G.B. (1982) Soils suppressive to *Gaeumannomyces graminis* var. *tritici:* Effects on other fungi. *Soil Biol. Biochem.*, **14,** 561–7.

Woolhouse, A.R. (1983) An investigation of the effectiveness of IBDU as a slow release source of nitrogen. *J. Sports Turf. Res. Inst.*, **59,** 93–102.

3 *Species and cultivars of turf grasses in relation to disease management**

Species and cultivars of turf grasses vary greatly in disease reaction and selection of the most suitable genotype forms the basis for all disease management. Disease resistance of a particular cultivar may be considerably modified by different environmental conditions, management practices, changes in the spectrum of pathogens or their virulence. A grass which under prevailing environmental conditions is resistant to a particular pathogen may become susceptible in abnormal environments (e.g. waterlogged soil, persistently high atmospheric humidity or drought, a curtailed period of prehibernal acclimation or unusually long snow cover); following changes in turf management practices (e.g. excessively heavy top dressing, turf surface pH changes after liming, high nitrogenous fertilization or changes in fertilizer balance); or pathogen changes (e.g. the development of new, more virulent races of rusts).

Since the 1940s there have been great changes in the availability of grasses selected or bred specifically for amenity turf usage. In Britain, formerly, fine turf for putting or lawn bowling greens was developed by mowing (and/or sheep grazing) turf known to contain fine-leaved grasses, mainly *Festuca* and *Agrostis* spp. Coarser turf for soccer or rugby pitches or cricket outfields was often mown pasture or meadow turf. Sheep-grazed moorland or sea-marsh turf was often lifted and transported to form lawn bowling greens. Fine turf was also established from seed of land-race strains such as South German Mixed Bentgrass (*Agrostis* spp.) or Chewings fescue (*F. rubra* spp. *commutata*) and browntop (*A. tenuis*) harvested as a secondary crop from pastures in the South Island of New Zealand. With 'lean' fertilization (Dawson, 1959) and irrigation restricted to natural precipitation, diseases other than microdochium patch, red thread, dollar spot, take-all patch and fairy rings were usually not of major importance. Where coarser lawn, soccer or rugby turf was started from seed, pasture-type agricultural grasses were available and some of these (e.g. *Lolium perenne* 'S 23') were extensively used until a few years ago. The climate of the British Isles and much of western Europe is such that only cool-season grass species can be used successfully in turf. Warm-season species of *Cynodon* and *Zoysia* are suitable for use in southern Europe, extending the range of pathogens encountered. Some regions of northern Europe, Scandinavia and northern Japan require amenity grasses capable of surviving low temperatures, long snow covers and low-temperature-tolerant pathogens, a characteristic of successful northern cultivars (Jamalainen, 1974).

North America has a great diversity of climatic conditions, from boreal to sub-tropical and cool- and warm-season grasses were needed. Most of the cool-season turf grass species used were not native but were European or Eurasian introductions. The introduced warm-season species were from widespread centres of origin in Africa, Asia and South America (Ward, 1969; Hanson, 1972; Whyte, Moir and Cooper, 1959). The spectrum of fungal pathogens on turf grasses consequently was also wide, comprising practically all the fungi found on European grasses, most of those associated with the warm-season grasses and

* In general, we have tried to avoid the use of common names of turf grass species, since these may be different in different parts of the world. In the Latin names of cool-temperate grasses we have followed Hubbard, C.E. (1984) *Grasses: A guide to their structure, identification, uses and distribution in the British Isles.* 3rd edn, revised by J.C.E. Hubbard, (Pelican original) Penguin Books, Harmondsworth, England, 476 pp. and STRI (1980) *Turfgrass seed, 1980.* Sports Turf Research Institute, Bingley Yorkshire, England, 105 pp.

some which have been found, so far, only in North America (Conners, 1967; Couch, 1973; Gould, 1964; Smiley, 1982; Sprague, 1950; USDA, 1960). The progression in usage through natural, land-race and agricultural grasses was also followed in North America, but selection of strains for turf use took place much earlier than in Europe. For putting green turf much greater use was made in North America than in western Europe of vegetatively propagated clones of *Agrostis stolonifera* and *A. canina*. Many of these were selected from the time of the First World War onwards from greens sown with seed of South German Mixed Bentgrass (Hanson, 1972). Individual clones were susceptible or very susceptible to one or more of the major diseases, brown patch (*Rhizoctonia solani*), dollar spot (*Sclerotinia homoeocarpa*), snow moulds (*Microdochium nivale* and *Typhula* spp.), leaf smuts (*Ustilago* and *Urocystis* spp.) and leaf spot diseases (mainly *'Helminthosporium'* spp.) (Beard 1973; Hanson, 1972; Monteith and Dahl, 1932). Most of these clones appear to have been selected in the eastern half of North America and mainly for turf production characteristics rather than disease resistance, which was (and remains) a major problem in bentgrass putting green turf according to Monteith and Dahl (1932). Selections of bentgrass cultivars to be propagated by seed, although fewer in number were made from about the same period, and they were found susceptible to one or more major diseases (Beard, 1973; Hanson, 1972).

An increasing public interest in professional sports played on turf, such as football in its various forms, and greater public access to golf courses in the developed countries has spurred the selection and breeding of new cultivars of a wide range of grass species in North America and Europe since the Second World War, especially of *Poa pratensis* and *Lolium perenne*. In Europe most of these new grasses have been developed in the Netherlands. Most of them show greater tolerance to wear, drought, shade, heat and cold, persistence under mowing, compactness, leafiness as turf grasses than their predecessors. However, it is not surprising that it has been impossible to combine all the requisite agronomic characters in one grass, especially when economical seed production is necessary. Disease resistance to perhaps two or three major diseases is all that can be expected without losing some other suitable turf-forming characteristics. Compensation for deficiencies in resistance to less important diseases can be achieved by specific cultural measures, by using mixtures of more than one grass species, or to a lesser extent by blending two or more cultivars of the same species (Vargas and Turgeon, 1980). It is very difficult to maintain turf of a high standard when it is composed of species susceptible to many diseases (e.g. some *A. stolonifera* cultivars and *Poa annua*) without recourse to constant fungicidal treatment (Smith, 1965).

References to sources of resistance to disease will be found with particular diseases throughout the text and in the following: Beard, (1973, 1978); and Braverman (1967, 1986); Braverman and Oakes (1972); Bundessortenamt (1979); Funk (1983); Goss and Law (1977); Hanson (1972); Jamalainen (1974); Manner (1972); Meyer (1982); Pick (1978); RIVRO (1980); Smith (1965); Smith (1980a,b); STRI (1980); Turf-Seed Inc. (1983); Vargas and Turgeon (1980); Vargas *et al.* (1980).

3.1 REFERENCES

Beard, J.B. (1973) *Turfgrass: Science and Culture*. Prentice Hall, New Jersey, 658 pp.

Beard, J.B. (1978) Kentucky bluegrass cultivar update. *Grounds Maintenance*, 18 Nov., 20, 67.

Braverman, S.W. (1967) Disease resistance in forage, range and turf grasses. *Bot. Rev.* **33**, 329–78.

Braverman, S.W. (1986) Disease resistance in cool-season forage, range and turf grasses. *Bot. Rev.*, **52** (1), 1–12.

Braverman, S.W. and Oakes, A.J. (1972) Disease resistance in warm-season forage, range and turf grasses. *Bot. Rev.*, **38**, 491–544.

Bundessortenamt (1979) *Descriptive List for Turfgrasses*. Alfred Strother Verlag, Hanover, 198 pp. (in German).

Conners, I.L. (1967) An annotated index of plant diseases in Canada. *Res. Br. Can. Dept. Agric.* Queen's Printer, Ottawa. Publ. 1251, 381 pp.

Couch, H.B. (1973) *Diseases of Turfgrasses*, Krieger, New York, 348 pp.

Dawson, R.B. (1959) *Practical Lawncraft*, 5th edn., Crosby Lockwood, London, 320 pp.

Funk, C.R. (1983) Ryegrasses reviewed. *Golf Course Mgmt*, **51** (4), 81, 83, 85.

Goss, R.L. and Law, A.G. (1977) Turf varieties differ in performance. *Golf Superintendent*, **45** (7), 27–9.

Gould, C.J. (1964) Turf grass disease problems in North America. *Golf Course Rept.*, **32** (5), 36–54.

Hanson, A.A. (1972) Grass varieties in the United States. *US Dept Agric., Agric. Res. Service, Agric. Handbk.* **170**, 124 pp.

Jamalainen, E.A. (1974) Resistance in winter cereals and grasses to low-temperature parasitic fungi. *Ann. Rev. Phytopathol.*, **12**, 281–302.

Manner, R. (1972) Species questions and current breeding objective for turf grass plants in Finland. *Rasen-Turf-Gazon*, **3**, 72–4. (German, French, English).

Meyer, W.A. (1982) Breeding disease-resistant cool-season turf grass cultivars for the United States. *Plant Dis.*, **66**, 341–4.

Monteith, J.J. and Dahl, A.S. (1932) Turf diseases and their control. *US Golf Assn Green Sect. Bull.* **12**, 85–188.

Pick, M.C. (1978) Professional turf varieties for professional turf managers. *Greenmaster*, **14** (4), pp. 4, 5, 7, 8.

RIVRO (1980) 55th descriptive variety list of agricultural crops Rjksinstituut voor het Rassenonderzoek van Cultuurgewassen, Wageningen, The Netherlands. 336 pp. (Dutch).

Smiley, R.W. (1982) *Compendium of Turfgrass Diseases*. American Phytopathology Society, 102 pp.

Smith J.D. (1965) Fungal diseases turf grasses. 2nd edn. (Revised Jackson, N. and Smith, J.D.), *Sports Turf Res. Inst.*, 97 pp.

Smith, J.D. (1980a) Snow mold resistance in turf grasses and the need for regional testing. In *Proc. 3rd. Int. Turfgrass Res. Conf.* Munich. ed. J.B. Beard, American Society of Agronomy, pp. 275–82.

Smith, J.D. (1980b) *Major diseases of Turfgrasses in Western Canada*. Extension Dept. University of Sask., Saskatoon. Publ. 509, 14 pp.

Sprague, R. (1950) *Diseases of Cereals and Grasses in North America*. Ronald Press, New York, 538 pp.

STRI (1980) *Turfgrass Seed, 1980*. The Sports Turf Research Institute, 105 pp.

Turf-Seed Inc. (1983) Turf-seed variety review. *Turf Tech.* Turf Seed Inc. Hubbard, Ore.4 pp.

USDA (1960) Index of plant diseases in the United States. *United States Dept. Agric. Agric. Handbk.*, **165**, 531 pp.

Vargas, J.M., Payne, A.T., Turgeon, A.J. and Detweiler, R. (1980) Turf grass disease resistance – Selection, development and use. In *Advances in Turfgrass Pathology.* (eds B.G. Joyner and P.O. Larsen), Harcourt Brace Jovanovich, Duluth, Minn., pp. 179–82.

Vargas, J.M. and Turgeon, A.J. (1980) The principles of blending Kentucky bluegrass cultivars for disease resistance. In *Proc. 3rd Int. Turf Res. Conf.* (ed. J.B. Beard), American Society of Agronomy, pp. 45–52.

Ward, C.Y. (1969) *Climate and adaptation.* In *Turfgrass Science* (eds F.V. Juska and A.A. Hansen), Am. Soc. Agron. Monograph 14, pp. 27–79.

Whyte, R.O., Moir, T.R.G. and Cooper, J.P. (1959) Grasses in agriculture. *FAO Agric. Studies*, **42**, Rome, 417 pp.

4 *Turf disease control with fungicides*

Monteith and Dahl (1932) outlined the problem of controlling turf diseases in the following words:

> Cultural practices influence to a great degree the frequency and severity of diseases. Many non-parasitic diseases can be corrected only by proper cultural practices. These practices, however, cannot completely control fungus diseases, but they may to a great extent lessen the frequency and severity of the attacks. If by cultural practices the amount of disease can be materially diminished, the cost of controlling it with fungicides may be greatly reduced. It should be the aim of every greenkeeper to use those agricultural methods which tend to discourage fungal diseases.

These conclusions are basically correct 50 years after they were written. In natural grasslands the component species have been subjected over long periods of time to varied selection pressures, including that from fungal pathogens. Diseases occur, but suscepts and pathogens tend to achieve an equilibrium. Forage grasses which have been specially developed for grazing or conservation are more vulnerable to fungal diseases, particularly if they are mismanaged. Bred mainly for high yields, there is a tendency to push these agricultural grasses to the limit of their output by heavy fertilization, irrigation, intensive cutting or grazing. Alternatively, their needs may be neglected. The stresses resulting from these situations exacerbate disease problems. The managed amenity turf environment is even more exacting. Here the breeder is less concerned with high yield, but vigorous and sustained growth is needed to replace tissues and organs damaged or lost through foliage removal, mechanical injury, pest depredation and senescence so that the desired quality of the turf is maintained. There has been a tendency for new turf grass cultivars, developed by selection and breeding for better adaptation to the changing conditions of use, to be subjected to more intensive fertilization, mowing and watering. This is often to satisfy demands for longer periods of use, extension of season, or play under unfavourable conditions. As new cultivars resistant to attack by particular pathogens are developed, so in the natural course of evolution, pathogens develop new strains capable of causing disease in the newly developed plant genotypes. However, recent financial stringencies seem to have, to some extent, reduced this trend and resulted in more attention being paid to the breeding and utilization of 'minimum maintenance' amenity turf grasses which are subjected to less management and disease pressure.

No matter how well cultivars are selected and cultural practices planned to keep turf disease to a minimum, it is impossible to cover all eventualities of weather, soil and use and plant response to them. Errors of judgement occur with those who plan the management and those who perform the operations. The use of fungicides helps to close this gap in schemes for preventing and dealing with outbreaks of turf diseases, provided that the disease is correctly diagnosed.

4.1 HISTORY

Much of the pioneer work on the use of fungicides for turf disease control was done in the USA. In 1917 trials were started on the use of Bordeaux mixture for the control of brown patch (*Rhizoctonia solani*) on golf courses and in 1919 it was in general use for that purpose (Monteith and Dahl, 1932). It was soon found that the repeated use of Bordeaux mixture led to the accumulation of copper in the soil sometimes causing injury to the turf more severe than the disease it was intended to control. Also the fungicide did not control the important dollar spot disease caused by *Sclerotinia homoeocarpa*.

In Britain, Bennett (1933, 1935) and Smith (1953b, 1957d) found that a solution of the dye malachite green at 1:10 000, in combination with Bordeaux mixture (mainly used as a sticker) was effective in the control of microdochium (fusarium) patch (*Microdochium nivale*) and red thread disease (*Laetisaria fuciformis*). But the repeated applications needed in wet weather to maintain a protective coating on leaves also resulted in copper accumulation in the soil. As well as being fungicidal, the malachite green dye improved the colour of diseased turf cosmetically, foreshadowing some modern turf dyeing practices (Karnok, 1984). Malachite green was also used in formulations of other fungicides as a tracer and safener. It was also shown to stimulate grass growth (Smith, 1957b). Howard (1956) reported that the material was effective against the mycelial stage of *Pythium* spp.

Ferrous sulphate has been used successfully for many years as a temporary control material for microdochium patch disease. It is sometimes included in fertilizer mixtures, partly for its ability to lessen the risk of disease, but also to intensify the green colour of the turf and to control moss and algae.

Mercuric chloride, first used as a fungicide in 1890 (Frear, 1948) was employed successfully in 1920 in the Chicago area for the control of brown patch. In 1925 and 1927 several inorganic and organic mercury fungicides controlled dollar spot in experiments in Rhode Island. Monteith and Dahl (1932) considered that the efficiency of these materials was largely in proportion to the amount of mercury applied. From these experiments, fungicides based on mercuric chloride (corrosive sublimate) and mercurous chloride (calomel) were developed. The insoluble mercurous chloride was included because it gave to the mixture a longer period of disease control. Fungicides based on these inorganic mercurials are still in use, in some parts of the world, mainly for turf grass snow mould control, but in other jurisdictions they are banned as environmentally hazardous (Davidson, 1986; Fushtey, 1982; Smith, 1976; Smith and Evans, 1985).

One of the first organo-mercury compounds used against turf disease was 'chlorophenol mercury'. It was not a conspicuously successful fungicide. Like the inorganic materials its effectiveness was considered to be related to the amount of mercury applied (Monteith and Dahl, 1932). The interest in organomercurials as turf grass fungicides stems mainly from their successful use as seed dressings. Phenyl mercury compounds, in particular the acetate (PMA), have been widely used as turf fungicides. PMA has a broad range of effectiveness over common turf diseases, but it may be phytotoxic if the recommended rates are exceeded or if it is applied when air temperatures are high or the turf grass species or cultivar is sensitive to injury.

Howard (1947) reported that a fungicide based on phenylamino cadmium dilactate was effective in the control of dollar spot. Subsequently, cadmium fungicides were adopted as standard materials in the USA and elsewhere for the control of dollar spot (Ludbrook and Brockwell, 1952; Smith, 1955, 1956b, 1957b). Cadmium fungicides proved effective at low dosages of 9–11 g/100 m^2 of actual cadmium for the control of dollar spot (Howard *et al.*, 1951; Smith, 1957d). They also gave effective control of copper spot (*Gloeocercospora sorghi*) and red thread disease (Howard, Rowell and Keil, 1951; Smith 1957d). Boyce (1952) found cadmium compounds and complexes ineffective in the control of snow moulds in eastern Canada, but Meiners (1955) found cadmium succinate gave fairly effective control of these diseases in the Pacific Northwest of the USA. Smith (1957e) and Jackson (1959, 1960) reported that cadmium chloride gave good protection against microdochium patch in plot experiments in England and Adams and Howard (1963) effectively controlled grey snow mould (*Typhula incarnata*) with cadmium salts in Rhode Island. Cadmium fungicides are no longer permitted in Europe, but are still used in North America. Since the mid-1960s the causal agent of dollar spot disease (*S. homoeocarpa*) has developed a high and widespread tolerance of cadmium compounds (Cole, Taylor and Duich, 1968; Jackson, 1966).

During the widespread mercury shortage in the USA in the 1939–1945 War and with the need to find less-expensive substitutes, tetramethylthiuram disulphide (thiram or TMTD) was used for the control of turf grass diseases. For most workers, this material gave good control of brown patch, but proved either ineffective or only moderately effective in controlling dollar spot when compared with inorganic mercurials, phenylmercury or cadmium fungicides (Harrington, 1941; Howard, 1955; Smith, 1954; Vaughn, 1953; Vaughn and Klomparens, 1952; Wilson and Grau, 1952). Boyce (1952) reported thiram ineffective in the control of snow mould (cause not reported) in eastern Canada but Meiners (1955) found it effective when *T. incarnata* was the dominant snow mould in complex with *Microdochium nivale*. Thiram acts as a safener and adjuvant when used in mixture with other potentially phytotoxic fungicides, such as mercurials and cycloheximide. It has been extensively used as a seed protectant against seedling diseases of turf grasses. The use of thiram has continued in turf grass disease control with the related ethylene bisdithiocarbamate (EBDC) compounds developed in the post-war years. These have included nabam (disodium EBDC) as a preventative drench for the control of spring dead spot, *Leptosphaeria* spp. (Kozelnicky, 1974). Zineb and maneb (respectively the zinc and manganous salts) have been used for the control of leaf spots, red thread, rusts and *Pythium* diseases. Mancozeb (maneb reacted with zinc) is employed for broader spectrum turf disease control than zineb, including the control of brown patch disease (Gould, 1967).

Quintozene (pentachloronitrobenzene), introduced in the 1930s, found widespread application in the control of soil-borne diseases of many plant species, particularly those incited by *Rhizoctonia solani* including brown patch of turf grasses (Gould, 1967). At high temperatures, some cool-season turf grasses may be injured by this chemical and doubts have been cast on its environmental safety. Extensive work in Finland by Jamalainen (1960, 1974) and others showed that quintozene at 4–5 kg/ha gave good protection against the low-temperature pathogens, *Microdochium nivale, Myriosclerotinia* (*Sclerotinia*) *borealis* and *Typhula* spp. in winter cereals. Jackson (1962) found that the material effectively controlled microdochium patch disease and its effectiveness persisted under long-duration snow covers in England. It is a fungicide recommended for the control of the same snow moulds and of cottony snow mould (*Coprinus psychromorbidus*) in western Canada (Smith and Evans, 1985).

Some antibiotics have been used in the control of turf diseases. Cycloheximide was reported effective in the control of foot rot and leaf spot of *Poa pratensis* caused by *Drechslera poae* in the USA (Couch, 1957; Howard, 1953). Cycloheximide (formulated with ferrous sulphate) was only moderately effective and derivatives of the antibiotic were not effective against microdochium patch in England (Smith, 1957d). Mixtures of cycloheximide and other fungicides, e.g. thiram and quintozene, are still utilized in the USA* for leaf spot, dollar spot, brown patch and powdery mildew control in turf. Griseofulvin, found effective in preventing microdochium patch and dollar spot in small plot trials, was inadequate in practice (Smith, 1956a). Dinocap, developed during the late 1940s as a miticide, was a favoured product as a control for powdery mildew by late 1960s (Gould, 1967).

Captan, discovered in 1949, soon established a wide application as a seed protectant and was adapted as a turf foliar protectant, particularly against '*Helminthosporium*' diseases. Captan, alone or in combination with cycloheximide, was shown by Wells (1962) to be moderately effective for the control of pythium blight. Subsequent trials by Wells confirmed fenaminosulf as a more reliable, but short-lived specific for the latter disease. The use of captan is now restricted because of environmental considerations. More persistent control of *Pythium*-incited diseases in turf was achieved during the late 1960s following the introduction of ethazole and chloroneb (Freeman and Horn, 1967; Freeman and Meyers, 1969; Wells, 1969). Chloroneb was advocated also as a preventative fungicide for grey snow mould (*T. incarnata*) in Rhode Island (Jackson and Fenstermacher, 1969) and Michigan (Vargas and Beard, 1970).

Two organic contact fungicides, chlorothalonil and anilazine, found application during the 1960s against a range of turf grass diseases. Although largely ineffective against *Pythium*-incited diseases, their otherwise broad spectrum of activity, and wide margin of safety and compatibility with other turf fungicides led to their world-wide adoption as reliable components of turf disease control programmes.

Most of the chemicals previously mentioned and used prior to 1970 for turf disease control are of the protectant/contact type. They have limited residual activity (low persistence) and little, if any, therapeutic (eradicant) action. With these materials, a coating of the chemical is applied in a wet or dry formulation to the aerial parts of the turf grass (and, incidentally, to the immediate layer of the thatch) in order to inhibit propagule germination and/or mycelial growth of the fungal pathogens. The chemical coating must be in place before any ingress of the pathogen has occurred since movement of this type of fungicide within the plant is usually negligible. Provided that the concentration of the fungicide remains sufficiently high, and the barrier complete, susceptible fungi (both targeted and non-targeted) are suppressed (fungistatic effect) or killed (fungicidal effect). Frequent applications of these protective materials are required to accommodate new growth in the sward and to compensate for the continuous depletion of the active ingredients by weathering processes, irrigation and clipping removal.

From their introduction into general use in the 1970s, systemic fungicides have gained increasing acceptance in turf grass disease management. Such chemicals are absorbed through leaf, stem or root surfaces and are translocated varying distances within the plant to preserve the integrity of the host tissues or suppress established infection, mainly by acting as fungistats. For most systemics, movement occurs in the symplast in an acropetal direction via the transpiration stream. The recent development of systemics that also circulate basipetaly in the apoplast offers exciting new possibilities for the control of some intractable soil-borne root and crown diseases. Systemics have a considerable advantage over protectant/contact fungicides in that they can be used at relatively low dosages to prevent or inhibit the development of infections in tissues or organs remote from the site of application. Once within the host they are sheltered from the influences of the external environment. This affords a potential increase in persistence and a reduction in application frequency. Since most are effective at low concentrations, the

* Now discontinued.

Fig. 1 *Bipolaris sorokiniana (Cochliobolus sativus)* causing seedling blight on *Agrostis stolonifera* on a sand green.

Fig. 2 Post-emergence damping-off on bowling green turf. *Fusarium culmorum.*

Fig. 3 Post-emergence damping-off of *Festuca rubra. Pythium* sp.

Fig. 4 Microdochium patch and pink snow mould (*Microdochium nivale*) Fusarium patch. Top: early symptoms. Middle: late symptoms before snow cover. Bottom: pink snow mould – symptoms after snow melt.

Fig. 5 Severe damage by cottony snow mould (non-sclerotial *Coprinus psychromorbidus*) on *Agrostis stolonifera* turf.

Fig. 6 *Coprinus psychromorbidus*. Sporophore on stolon of *Agrostis stolonifera*.

Fig. 7 Cottony snow mould. Sclerotial phase of *Coprinus psychromorbidus* on *Poa pratensis* turf. Formation of sclerotial primordia.

Fig. 8 Sclerotia of *Typhula ishikariensis* var. *ishikariensis* (dark brown) and *T. incarnata* (light brown). (J.M. Fenstermacher)

Fig. 9 Pink sporophores of *Typhula incarnata*. (J.M. Fenstermacher)

Fig. 10 Cream to brown sporophores of *Typhula ishikariensis* var. *canadensis*.

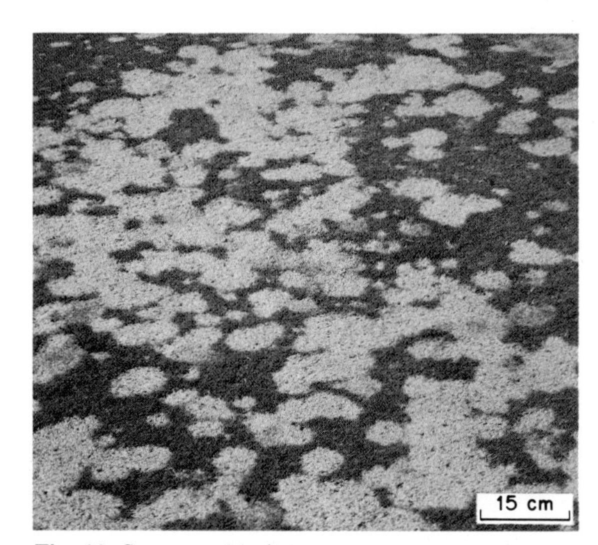

Fig. 11 Snow scald (*Myriosclerotinia borealis*) on *Agrostis stolonifera* turf.

Fig. 12 Abundant mycelium of *Typhula ishikariensis* var. *canadensis* on *Agrostis stolonifera* turf.

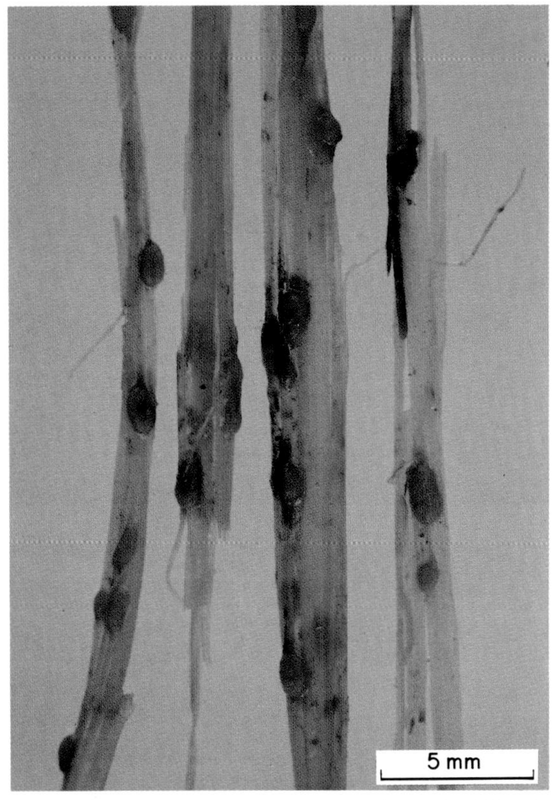

Fig. 13 Sclerotia of *Acremonium boreale* (*Nectria tuberculariformis*) on over-wintered grass.

Fig. 14 Spring dead spot on U3 *Cynodon* turf.

Fig. 15 Necrotic ring spot (*Leptosphaeria korrae*). Symptoms on *Poa pratensis*.

Fig. 16 Summer patch (*Magnaporthe poae*). Early and severe symptoms on *Poa pratensis*.

Fig. 17 Take-all patch symptoms on fine *Agrostis tenuis* turf caused by *Gaeumannomyces graminis* var. *avenae*.

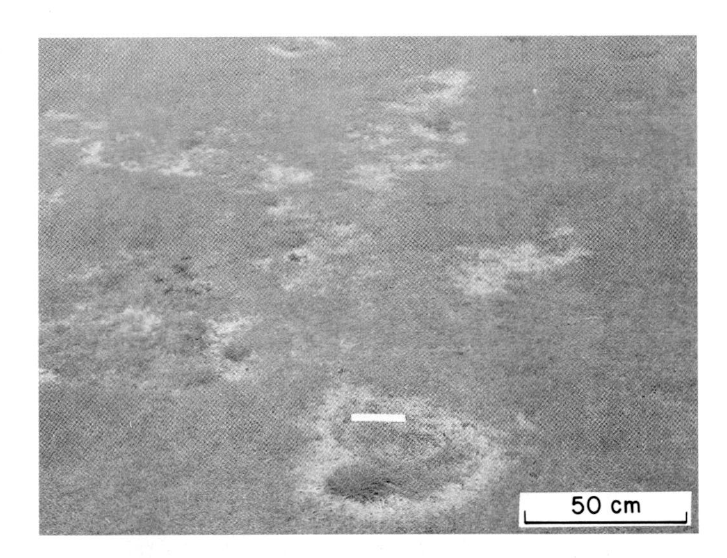

Fig. 18 Uredinia of *Puccinia graminis* ssp. *graminicola* on leaf blade of 'Merion' *Poa pratensis*.

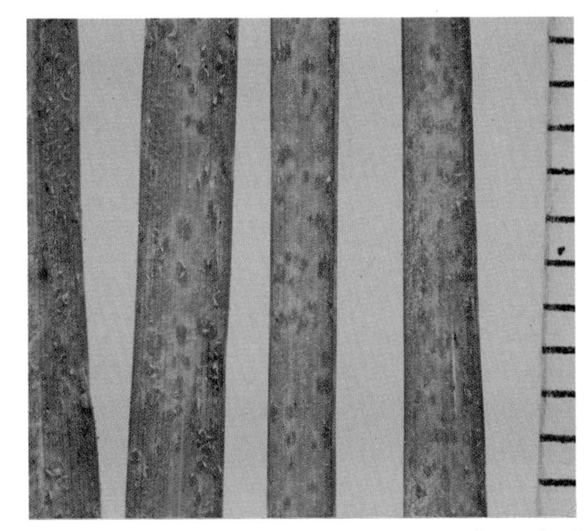

Fig. 19 Uredinia of *Puccinia poae-nemoralis* on leaf blade of *Poa pratensis*.

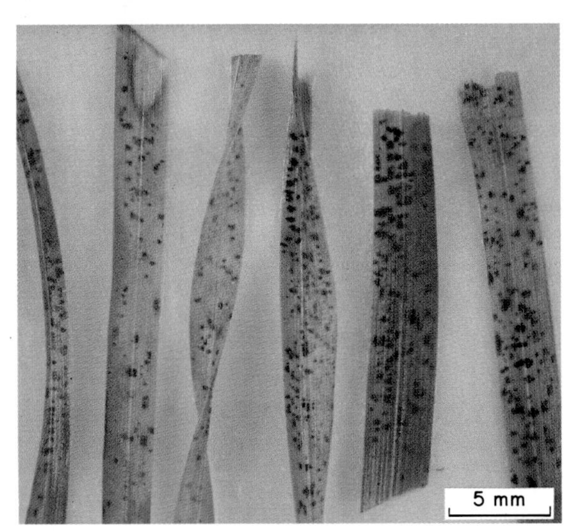

Fig. 20 Blister or spot smut. *Entyloma dactylidis* on leaves of *Poa pratensis*.

Fig. 21 Stripe smut (*Ustilago striiformis*) in leaves of *Poa pratensis*.

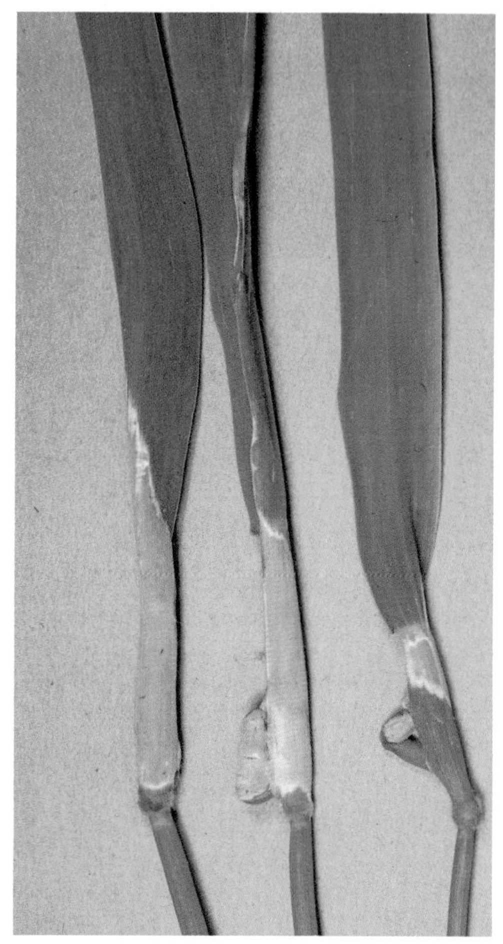

Fig. 22 Choke or cat tail. Condial stromata of *Epichloe typhina* on *Arrhenatherum elatius*.

Fig. 23 Ergots of *Claviceps purpurea* on *Agropyron cristatum*.

Fig. 24 Silvertop or whiteheads of *Poa pratensis*. Right – normal, left – bleached culms and inflorescences.

Fig. 25 Twist caused by *Dilophospora alopecuri* on *Holcus lanatus*. Emergence of inflorescences is restricted. Normal inflorescence on left.

Fig. 26 Anthracnose and basal rot. *Colletotrichum graminicola* causing patching on *Poa annua* in an *Agrostis stolonifera Poa annua* golf green.

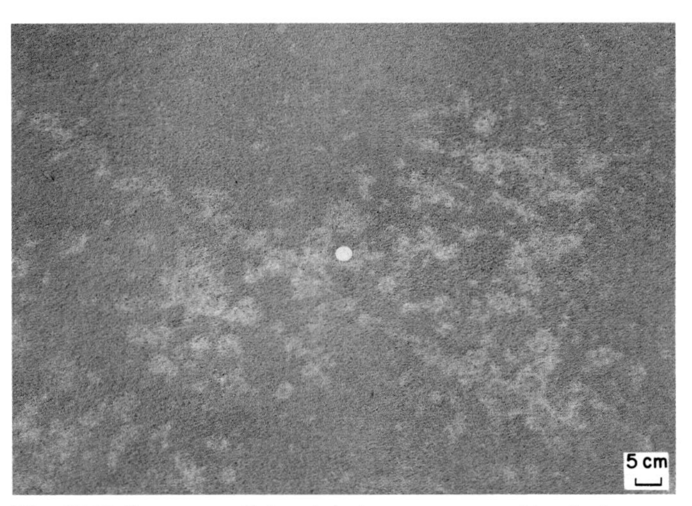

Fig. 27 Dollar spot – *Sclerotinia homoeocarpa* – bleached spots. Copper spot – *Gloeocercospora sorghi* – spots on *Agrostis canina* turf have a pink tinge.

Fig. 28 Dollar spot on *Poa pratensis*. Leaf lesion and mycelium of *Sclerotinia homoecarpa.*

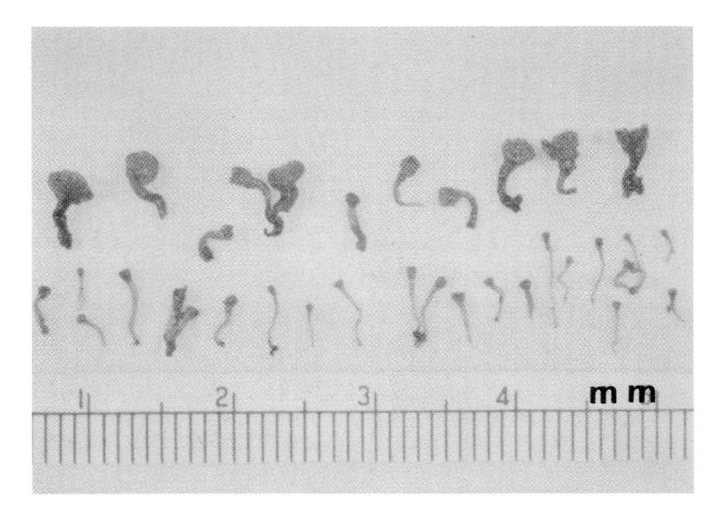

Fig. 29 Apothecia of *Sclerotinia homoeocarpa* (F.T. Bennett's material)

Fig. 30 Dollar spot on *Poa pratensis* – fertility response.

Fig. 31 Yellow turf symptoms caused by *Sclerophthora macrospora* on *Poa pratensis* tuft.

Fig. 32 Plants of *Agrostis stolonifera* infected by *Sclerophthora macrospora* causing yellow tuft.

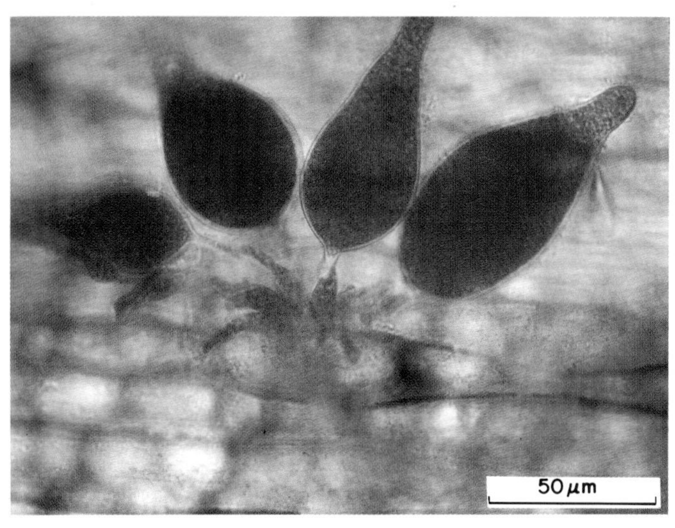

Fig. 33 *Sclerophthora macrospora* – sporangia (stained) emerging from stomata.

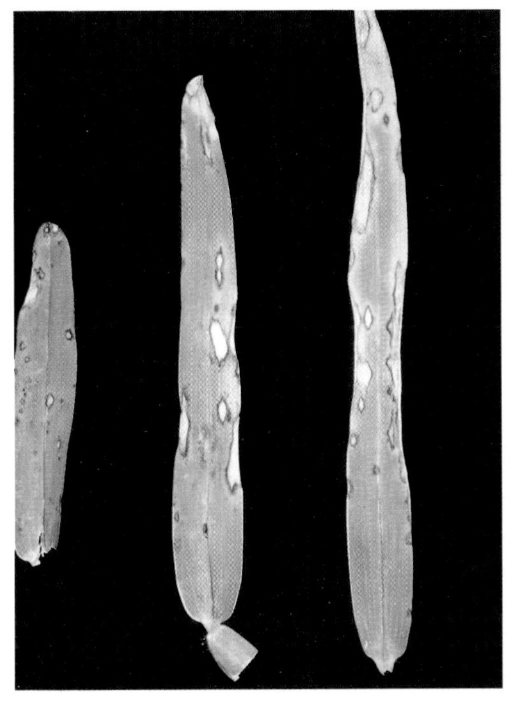

Fig. 34 *Pyricularia grisea* leaf spot on *Stenotaphrum secundatum.* (T.E. Freeman)

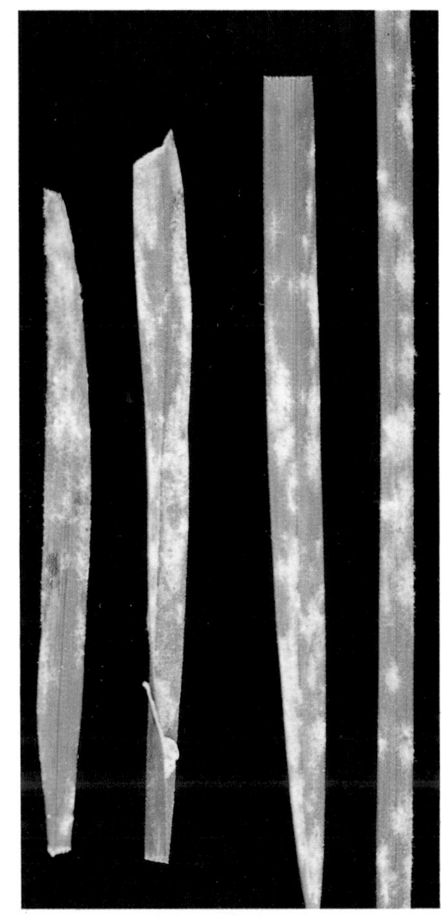

Fig. 35 Powdery mildew. *Erysiphe graminis* on *Poa pratensis.*

Fig. 36 Cottony blight. *Pythium* sp.

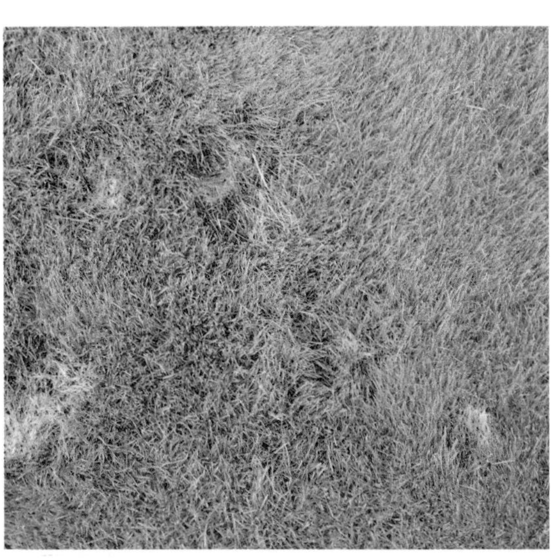

Fig. 37 Pythium spot blight on oversown *Lolium perenne.* (T.E. Freeman)

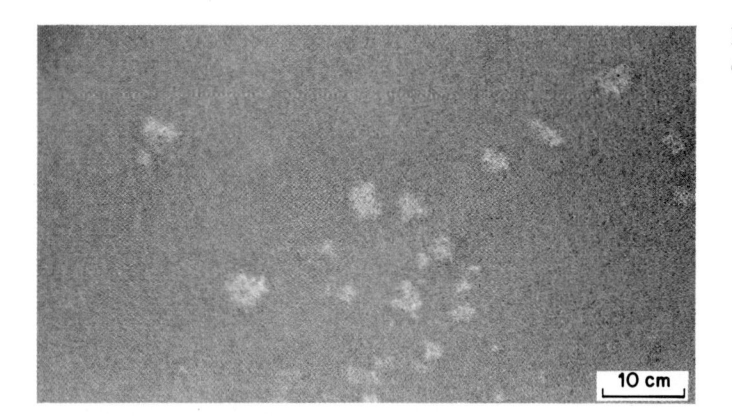

Fig. 38 Pythium grease spot. *Pythium aphanidermatum*. (E. Sharvelle)

Fig. 39 Red thread disease. Stromata of *Laetisaria fuciformis* on *Lolium perenne*.

Fig. 40 Pink patch – *Limonomyces roseipellis* – on *Lolium perenne*.

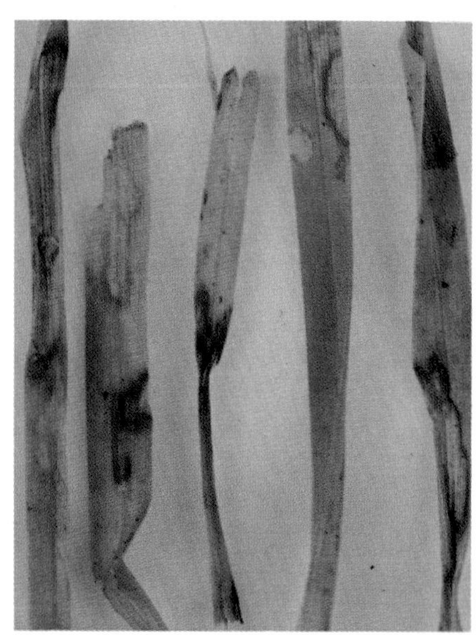

Fig. 41 Leaf lesions caused by *Rhizoctonia solani* on leaf blades of *Agrostis* sp.

Fig. 42 Brown patch with 'smoke ring' symptoms on *Agrostis* sp. turf. *Rhizoctonia solani.*

Fig. 43 Leaf lesions caused by *Rhizoctonia solani* on tall fescue, *Festuca arundinacea.*

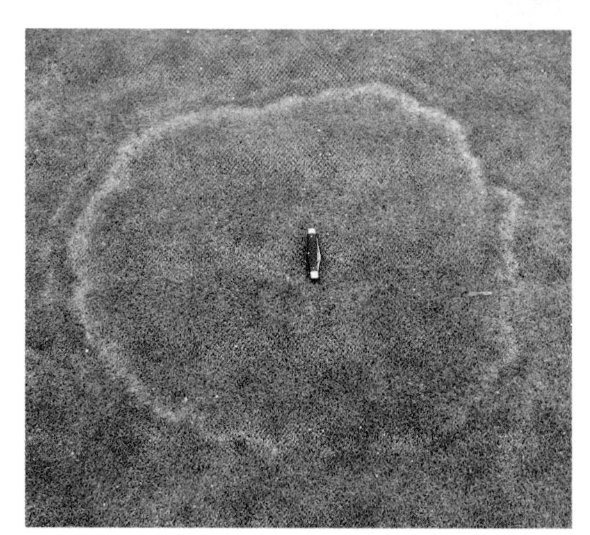

Fig. 44 Yellow patch on *Agrostis* sp. turf. *Ceratobasidium cereale* (*Rhizoctonia cerealis*).

Fig. 45 Yellow patch (*Rhizoctonia cerealis*). Leaf symptoms from yellow ring. (P.O. Larsen)

Fig. 46 Southern blight on *Agrostis* sp. turf. *Sclerotium rolfsii.* (V. Gibeault)

Fig. 47 Damage by *Drechslera gigantea* on *Agrostis canina* turf.

Fig. 48 Melting out of *Poa pratensis* – *Drechslera poae.*

Fig. 49 Brown blight of *Lolium perenne* – *Drechslera siccans.*

Fig. 50 Red leaf spot on *Agrostis* sp. – *Pyrenophora erythrospila*.

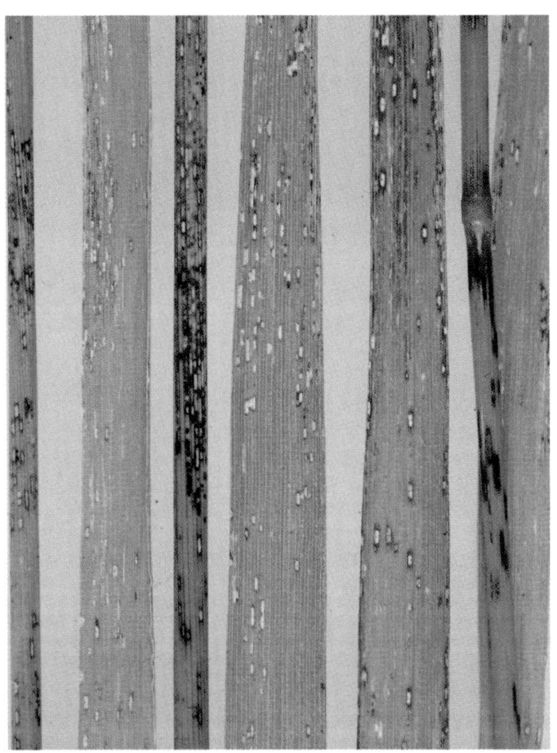

Fig. 51 Leaf and culm spot on *Agropyron* sp. caused by *Pseudoseptoria donacis*.

Fig. 52 *Ascochta phleina* causing a tip blight on 'Merion' *Poa pratensis*.

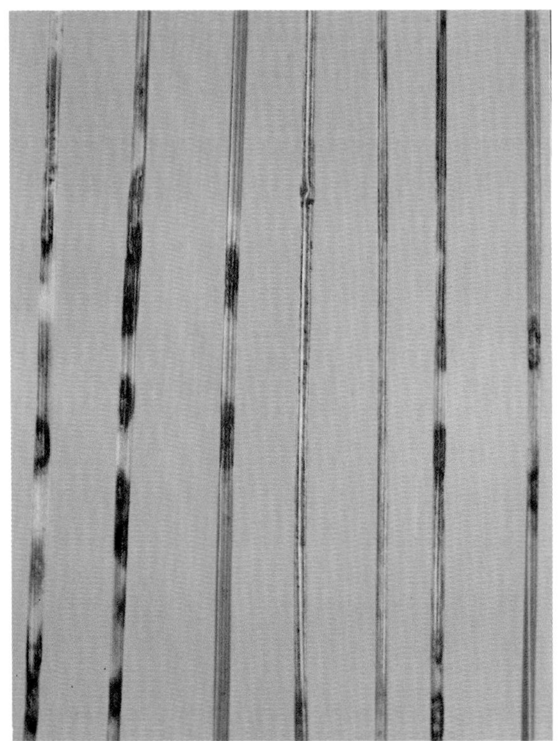

Fig. 53 Brown stripe (*Cercosporidium graminis*) on leaves of *Festuca rubra*.

Fig. 54 Purple leaf spot of *Phleum pratense* – *Cladosporium phlei.*

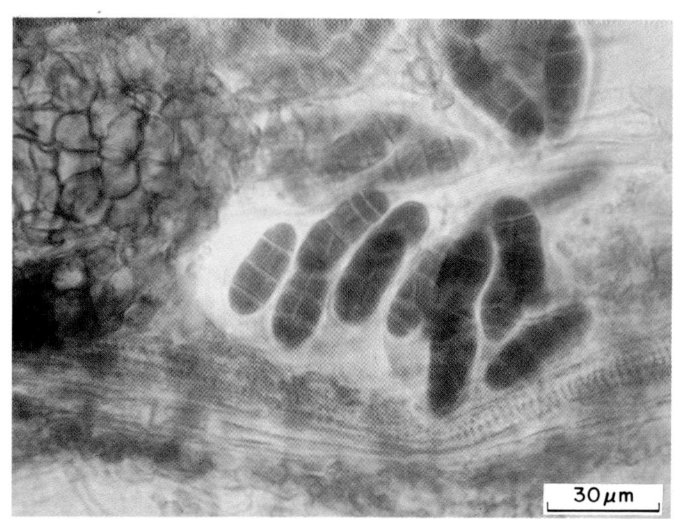

Fig. 55 *Leptosphaerulina australis* – ascus and ascospores.

Fig. 56 Scald on *Bromus inermis* – *Rhynchosporium secalis.*

Fig. 57 Spermospora leaf spot (*S. ciliata*) on *Agrostis* sp.

Fig. 59 Vertical section through the soil below a *Marasmius oreades* fairy ring with dense mycelium.

Fig. 60 *Agaricus campestris* fairy ring (Type II).

Fig. 58 Char spot (*Cheilaria agrostis*) on *Agropyron* sp.

Fig. 61 Large *Marasmius oreades* fairy rings on a golf course fairway showing stimulated and bare zones.

Fig. 62 Sporophores of *Marasmius oreades*.

Fig. 63 Mutual obliteration of two contacting fairy rings.

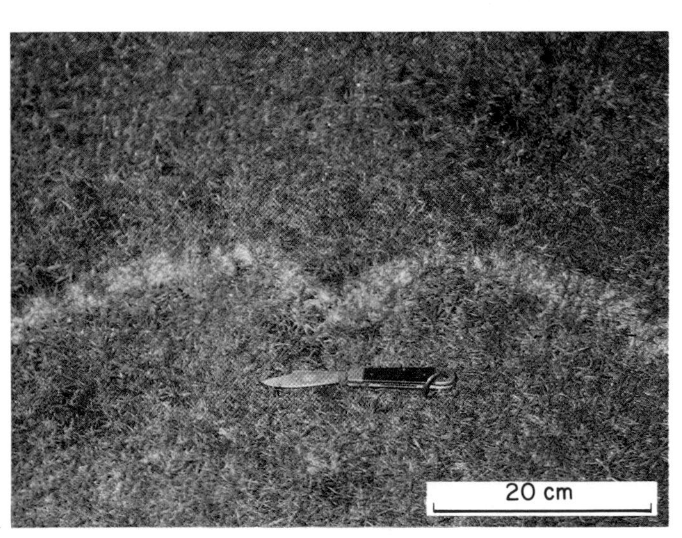

Fig. 64 Superficial fairy ring with arc of mycelium.

Fig. 65 Slime mould. Sporangia of *Physarum cinereum* on grass blades.

risk of phytotoxicity damage to the turf is small. As more products compete for turf grass applications, efficient control of disease at lower cost should result (Crowdy, 1977; Edgington *et al.*, 1980; Lucas, 1983).

The earlier systemics, such as the oxathiins and benzimidazoles, tended to be more selective in activity against certain groups of fungi than the contact fungicides (Dekker, 1976; Edgington *et al.*, 1980). Of the newer systemics in commercial use, triadimefon, fenarimol and propiconazole, function effectively against a number of turf grass pathogens but, in common with the benzimidazoles and the oxathiins, their spectrum of activity does not embrace oomycetous fungi or the important leaf spot pathogens in the genera *Drechslera* and *Bipolaris*. Use of particular systemics may, in fact, exacerbate existing leaf spot and melting out (Jackson, 1970). Similar examples of increased disease incidence and intensity in response to systemic fungicide use have been reported for superficial fairy ring (Smith, Stykes and Moore, 1970), *Pythium*-incited turf diseases (Warren *et al.*, 1976) and snow mould caused by *Typhula ishikariensis* var. *canadensis* (Smith, 1976). The dicarboximide fungicides (iprodione, vinclozolin and procymidone) are only slightly and locally systemic and may be regarded as protectant/contact fungicides. Although they are effective leaf spot fungicides, they are inadequate substitutes for the true systemics in many situations where the latter are strikingly competent, e.g. against rusts, smuts, powdery mildew and summer patch disease.

The limited number of fungicides effective against oomycete-incited diseases has been augmented with the systemics, metalaxyl, propamocarb and fosetyl aluminium, which are largely specific for the group. They move, to some extent, in the symplast and may be translocated in a basipetal direction and have demonstrated effective control of some *Pythium* diseases of turf grasses (Sanders, 1984; Sanders *et al.*, 1978; Watkins, 1983). Fungicide resistance in some oomycetes to metalaxyl has been reported (Staub and Sozzi, 1984) including resistance towards *Pythium aphanidermatum*, a causal agent of pythium blight of turf grasses (Sanders, 1984).

The mode of action of many fungicides remains obscure and the exact mechanism involved is particularly difficult to elucidate where activity of the toxicant impairs basic metabolic processes at many sites within the fungal cell. Most of the protectant/contact fungicides are multisite toxicants and, with some exceptions, their field performance has remained stable over the years (Dekker, 1976). In marked contrast, some systemic fungicides lost effectiveness shortly after their commercial release. The target fungi rapidly became resistant to particular toxicants and often demonstrated cross-resistance to structurally related chemicals or to chemicals similar in mode of action. This disturbing turn of events relates to the mode of action of some systemics as single-site toxicants in the fungal cell and the capacity of particular fungi to evade or circumvent any deleterious effects of the fungicide at the reactive site (Delp, 1980; Georgopoulos, 1977).

Unfortunately, any antiresistance strategies which might be employed, e.g. increasing the diversity of fungicides and integrated host and disease management, are often used too late – after fungicide resistance by a pathogen has already developed (Skylakakis, 1981; Staub and Sozzi, 1984).

The dollar spot fungus, *Sclerotinia homoeocarpa*, already with demonstrated resistance to some multisite category fungicides (cadmium, mercury and anilazine), rapidly developed widespread resistance to the benzimidazoles (Cole, Taylor and Duich, 1968, Cole, Warren and Sanders, 1974; Jackson, 1966; Massie, Cole and Duich, 1968; Nicholson *et al.*, 1971; Warren, Sanders and Cole, 1974). The same fungus now has resistance to dicarboximide fungicides in some situations (Pennucci and Jackson, 1983; Detweiler, Vargas and Danneberger, 1983). Instances of resistance by *Microdochium nivale* to the dicarboximides have been reported from the Pacific north west of the USA (Chastagner and Vassey, 1982). *Erysiphe graminis* and *Fusarium roseum* isolates have been reported resistant to the benzimidazoles (Smiley and Howard, 1976; Vargas, 1973).

Development of resistance in turf fungi to fungicides prompts caution in disease control programmes. No longer is it prudent to rely on one or two compounds. Tank mixes or alternating treatments with as wide a range of appropriate fungicides as possible are advocated to lessen the risk of resistance build-up. It says much for the diversity and resiliency of turf pathogens that after sustained assaults by a wide range of chemicals, there is still a need for more effective materials. There seem to be no imminent breakthroughs by plant breeders to produce turf grass cultivars with multiple disease resistance, so fungicides will remain a prime means of disease control, especially in intensively managed, high value amenity turf, even in a well-designed integrated disease and host management programme. Although industry must continue to develop new turf grass fungicides, attempts to remove arbitrarily some of the older, proven effective materials should be resisted. The range of fungicides available to combat turf diseases should be a broad one and include several alternatives for chemical control of each disease.

4.2 FUNGICIDE FORMULATIONS

Turf grass fungicides are formulated according to the intended method of application:

1. Dry powders, dusts (D) or dry granules (G) are applied to the turf with a duster or drop spreader without water, either undiluted or bulked with an inert carrier of sand, top dressing, or fertilizer.
2. Wettable powders (W or WP), dispersible granules (DG), emulsifiable concentrations (E or EC), flowable or soluble powders (F or FL or S) and liquids (L) are intended for application in water.

Couch (1973) discussed ten requirements of the ideal fungicide, most of which were still considered relevant in a recent review by Jackson (1983). (1) The first and basic requirement of the ideal turf grass fungicide is effective control of the disease. In its physical and chemical nature the formulation should show (2) early dispersal and trouble-free spraying properties, such as freedom from nozzle blockage and maintenance of droplet size characteristics, (3) the absence of objectionable visible residues and (4) compatibility with other pesticides, allowing tank-mix applications for broad spectrum turf grass treatments (Hermitage, 1982).

The formulation of commercial products has greatly improved over the years, particularly in spray dispersal, operator safety and convenience. Preweighed soluble bags or measured flowable formulations of spray materials have assisted in this respect, particularly in reducing the risks of dust inhalation. These first four points are linked with the commercial/economic requirements of fungicides: (5) convenience in packing, (6) a good storage life and (7) ready availability, (8) lowest possible cost in relation to efficacy (Gathercole, 1981). The fungicide should show (9) low phytotoxicity, so that turf grasses are not damaged, even by overdoses. Inorganic and organic mercurials, copper, cadmium, quintozene and cycloheximide, among others, have been reported to be phytotoxic under specified conditions, e.g. on particular species or cultivars and under certain climatic conditions. (10) Low mammalian toxicity is a primary concern in reducing environmental risks. Because this risk is high with some materials, some, such as mercurials and cadmium, have been banned from usage in some countries and registration procedures have been introduced in most jurisdictions regarding labelling, storage and application of pesticides to reduce environmental risks to a minimum.

4.3 FUNGICIDE APPLICATIONS

Dry powder formulations of fungicides, often inorganic or organic mercurials, bulked with sand or top dressing materials were in general use for control of disease on golf and bowling greens in Britain in the years before the Second World War. These materials were usually applied by hand or with a shovel since often no other distribution equipment was available. Later, granular formulations were developed, particularly to suit home lawn application in the USA. These materials are often applied with drop spreaders or other simple distributors. While these are convenient and easy to apply, it is difficult to get good coverage and even distribution, as also is the case with dry powders. Although professional turf managers have used the granular materials quite extensively, there is a further move to sprayable formulations as manufacturers explore ways to lessen the risks of handling dusts which may be inhaled or contaminate the skin. The effectiveness and safety of granular and spray fungicides is under scrutiny. Couch, Garber and Jones (1984) made field studies on the relative efficiency of control of *Drechslera poae* leaf spot of *Poa pratensis*, rhizoctonia blight of *Festuca arundinacea*, *Sclerotinia homoeocarpa* dollar spot on *Agrostis stolonifera*, with granular and spray formulation of the same fungicides. Granular formulations of *non-systemic* fungicides needed 2–3 times the active ingredient to give the same effectiveness of control as the spray formulations. The granular formulations were slower acting and less persistent in effect against the target fungus. Application to wet leaves improved, but irrigation after application reduced effectiveness of disease control by granular materials. The effectiveness of granular fungicides was reduced by mowing and removal of clippings immediately after application. Efficiency of disease control varied between commercial granular formulations at the same level of active ingredient.

The simplest method of application of a fungicide diluted or suspended in water is with a sprinkling (watering) can equipped with a rose. It is still a useful method where a limited area of turf is to be treated and where a soil drench is to be used to control (eradicate or suppress) a litter pathogen, a root disease or a fairy ring. Systemic fungicides are often employed in this manner with a surfactant (wetting agent) to assist penetration of thatch or soil mycelium in the case of fairy rings. Since to be effective the fungicide must reach its target on the turf grass roots in sufficient concentration, much higher dosages of active ingredient and volumes of water are required than with sprays of contact fungicides applied to control diseases of aerial parts of the plants.

The range of equipment used to spray fungicides is quite wide. There are simple pneumatic or knapsack sprayers with spray lances equipped with single or multiple nozzles, suitable for small areas. Stationary motorized pumps with hoses and single or multiple nozzles are frequently used to spray golf or bowling greens where passage of a wheeled sprayer is undesirable. Domestic lawn care companies use similar equipment on difficult to reach domestic lawns. Coverage of the turf with this type of equipment may often be uneven, in part dependent on the skill of the operator. More precision is possible with tractor-mounted or hand-propelled mobile sprayers equipped with booms and multiple nozzles. (Figs 4.1, 4.2 and 4.4.)

Figure 4.1 Motorized boom sprayer towed by a utility golf cart for larger turf areas.

Figure 4.2 Boom sprayer, p.t.o. driven, mounted on a garden tractor.

The optimum volume of spray to be applied for the most efficient disease control has been arrived at largly by rule of thumb in the case of most turf grass diseases. For most leaf diseases this ranges from 10 to 25 litres to 100 m^2 (2–5 gallons (Imp.) to 1000 ft^2). For root drenches in fine turf this volume might be increased 4-fold. For fairy ring drenching 20 litres to 1 m^2 is recommended by some fungicide manufacturers. Long grass, e.g. *Poa pratensis* mown at 75 mm, will require a greater volume of spray to protect it adequately than *Agrostis stolonifera* golf or bowling green turf mown at 5–8 mm. Couch (1984) has supplied some data from field studies on the effect of dilution rates, nozzle size and pressure on disease control in sclerotinia dollar spot, drechslera melting-out of *Poa pratensis* and brown patch with the fungicides cycloheximide, triadimefon, iprodione, chlorothalonil, anilazine and fenarimol. A boom-type precision sprayer was used. It was found that there are specific combinations of dilution rate at which the various fungicides performed most efficiently, within the range of 2–125 litres of spray per 100 m^2 of turf, using flat-fan nozzles of three different tip sizes and nozzle pressures in the range 0.68–6.1 atm (10–90 psi). Since the optimum efficiency changed with each fungicide it appears necessary to adjust the spray system with each material to reap the potential benefit. Spray nozzles should be selected according to spray rate and method of application. While several types are available, flat-fan nozzles are particularly well suited to multiple nozzle booms. Care should be taken to see that coverage is uniform with adequate overlap. This will depend on the spray angle of the tip, the nozzle spacing and the distance of the nozzle from the ground. Flat-fan nozzles are usually operated at 1–2 atm (15–30 psi). If deeper penetration in denser foliage is required, hollow-cone nozzles operated at a higher pressure (2–4 atm) may be preferred. Blocked nozzles often show as strips of uncontrolled disease. This can be prevented by ensuring sprayers are rinsed with clean water before use, the nozzles are clean and the fungicide, if a suspension in being used, is properly mixed and all pre-nozzle filters are in place. Each nozzle should be checked for uniform discharge and pattern, since tips may become corroded or worn with extensive use of some wettable powders which can be quite abrasive. Calibration can be done with the sprayer filled with water by driving at a recorded, uniform speed for a measured distance, at the same time collecting the water from a nozzle in the collecting cup. From these data the rate of spray application can be calculated (Smith, 1984).

In a few cases a single application of an appropriate fungicide or combination may be sufficient to prevent the appearance of diseases such as snow mould caused

Figure 4.3 Fungicide test on artificially inoculated bentgrass turf. Foreground – *Typhula ishikariensis* var. *canadensis*. Rear – *Myrioslerotinia (Sclerotinia) borealis*.

Figure 4.4 Motorized boom sprayer for small turf areas.

by *Typhula incarnata* (Taylor, 1974; Vargas and Beard, 1971) or stripe smut (*Ustilago striiformis*) (Jackson and Dernoeden, 1979). However, most turf grass diseases require repeated applications of fungicides to obtain adequate control. Applications are made at intervals, ranging from 4 to 28 days, dictated by the life cycle and biology of the pathogen, the nature, formulation and persistence of the fungicide, the prevailing weather conditions, the management intensity and the disease pressure. The optimum interval between applications is usually indicated on the registration label issued with the product. A useful compilation for scheduling fungicide applications for turf grass fungicides registered in Canada up to 1981 is given by Gathercole (1981).

4.4 FUMIGANT FUNGICIDES

Practically all the chemicals used for the control of soil-borne turf grass diseases by fumigation are phytotoxic and are usually applied prior to seeding. Many of them, e.g. methyl bromide, chloropicrin, carbon disulphide, methyl isothiocyanate, vapam and formaldehyde, are general biocides although they may be more toxic to particular groups of organisms than to others. Fumigant fungicides may have non-target effects on vesicular-arbuscular mycorrhizae and nematodes and fumigants applied to control nematodes may affect particular fungal species (Munneke, 1972; Rodriguez-Kabana and Curl, 1980). Soil biocides with a broad spectrum of activity will first reduce microbial numbers generally and then there will be an increase in bacterial numbers followed by the colonization of organic debris by particular groups of fast-growing fungi in the genera *Trichoderma*, *Aspergillus* and Mucorales (Rodriguez-Kabana and

Curl, 1980). Smith (1957a) noted that after injection of formalin solution into *Marasmius oreades* fairy rings abundant colonies of a *Penicillium* spp. developed down the sides of the injection hole. Methyl bromide fumigation of soil for control of weeds, insects, nematodes and seedling disease pathogens before planting bentgrass is often followed by the development of take-all (*Gaeumannomyces graminis*) in the Pacific northwest of the USA, coastal British Columbia, and has been reported from Britain and New South Wales (see p. 142). This is possibly the result of the selective reduction in numbers of competitive microorganisms.

4.5 REFERENCES

Adams, P.B. and Howard, F.L. (1963) Residual control of gray (*Typhula*) snow mould of bentgrass. *Golf Course Rept.*, **31** (5), 20–2.

Bennett, F.T. (1933) Fusarium patch of bowling and golf greens. *J. Bd. Greenkpg. Res.*, **3** (9), 79–86.

Bennett, F.T. (1935) Corticium disease of turf. *J. Bd. Greenkpg. Res.*, **4,** 32–9.

Boyce, J.H. (1952) Private communication, 21 May, and *Winter Injury and Snow Mould Problems in Canada.* Central Expt. Farm, Ottawa. 4 pp. (Mimeo.).

Chastagner, G.A. and Vassey, W.E. (1982) Occurrence of iprodione-tolerant *Fusarium nivale* under field conditions. *Plant Dis.*, **66,** 112–14.

Cole, H. Jr, Taylor, B. and Duich, J.M. (1968) Evidence of differing tolerances to fungicides among isolates of *Sclerotinia homoeocarpa. Phytopathology,* **58,** 683–86.

Cole, H. Jr, Warren, C.J. and Sanders, P.L. (1974) Fungicide tolerance – a rapidly emerging problem in turf disease control. In *Proc. 2nd Int. Turfgrass Res. Conf.* (ed. Beard J.B.), Am. Soc. Agron., Crop Sci. Soc. Am., Int. Turfgrass Soc., pp. 344–9.

Couch, H.B (1957) Melting out of Kentucky bluegrass, its cause and control. *Golf Course Rept.*, **25** (7), 5–7.

Couch, H.B. (1973) *Diseases of Turfgrasses.* 2nd edn, Krieger, New York, 348 pp.

Couch, H.B. (1984) Turfgrass fungicides: Part 2. Dilution rates, nozzle size, nozzle pressure and disease control. *Golf Course Mgmt.*, **52** (8), 73–76, 78, 80.

Couch, H.B. Garber, J. and Jones, D. (1984) Turf grass fungicides: Application methods and effectiveness. Part 1. *Golf Course Mgmt.*, **52** (7), 40, 42–43, 46, 48, 50–51, 52.

Crowdy, S.H. (1977) Translocation. In *Systemic Fungicides*, 2nd edn, (ed. R.W. Marsh). Longman, London, pp. 92–4.

Davidson, J.G.N. (1986) Chemical control and differentiation of the snow mould complex on a new bentgrass green. Abstract in *Proc. Can. Soc. Hortic. Sci. 66th Ann. Conf.* Agr. Inst. Can. Univ. Sask., Saskatoon. 6–10 July 1986. 1 p. (Duplicated).

Dekker, J. (1976) Acquired resistance to fungicides. *Ann. Rev. Phytopathol.*, **14,** 405–28.

Dekker, J. (1977) Chemotherapy. In (eds J.G. Horsfall and E.B. Cowling), Plant Disease. An Advanced Treatise. Vol. 1 Academic Press, New York.

Delp, C.J. (1980) Coping with resistance to plant disease control agents. *Plant Dis.*, **64,** 652–7.

Detweiler, A.R., Vargas, J.M. Jr and Danneberger, T.K. (1983) Resistance of *Sclerotinia homoeocarpa* to iprodione and benomyl. *Plant Dis.*, **67,** 627–30.

Edgington, L.V., Martin, R.A., Bruin, G.C. and Parsons, I.M. (1980) Systemic fungicides: A perspective after ten years. *Plant Dis.*, **64,** 19–23.

Frear, D.H.H. (1948) *Chemistry of Insecticides, Fungicides and Herbicides.* 2nd edn. D. Van Nostrand, New York, 417pp.

Freeman, T.E. and Horn, G.C. (1967) *Pythium* fungicides. *Golf Superintendent*, **35**(2), 58–60.

Freeman, T.E. and Meyers, H.G. (1969) Control of *Pythium* blight. *Golf Superintendent*, **37**(5), 24–5, 44–5.

Fushtey, S.G. (1982) The role of mercurial fungicides in turf grass disease control and the implications of such use in environmental pollution. *Greenmaster*, **18**(4).

Gathercole, W. (1981) The selection and use of fungicides. *Greenmaster*, **17**(4), 4–6.

Georgeopoulos, S.G. (1977) Pathogens become resistant to chemicals. In *Plant Disease: An Advanced Treatise*, Vol. 1 (eds J.G. Horsfall and E.B. Cowling), Academic Press, New York, pp. 327–45.

Gould, C.J. (1966–67) The use of fungicides in controling turf grass diseases. *Golf Superintendent*, **34**(9, 10), **35**(1).

Harrington, G.E. (1941) Thiuram disulphide for turf diseases. *Science* (NS), **93**(2413), 311.

Hermitage, R. (1982) Pesticide compatibility. *Greenmaster*, **18**(4), 10–12.

Howard, F.L. (1947) An organo-cadmium fungicide for turf disease. *Greenkprs. Rept.*, **15**(2), 5–9.

Howard, F.L. (1953) *Helminthosporium–Curvularia* blights of turf and their cure. *Golf Course Rept.*, **2**(12), 5–9.

Howard, F.L. (1955) What's new in turf grass diseases and their control. *Golf Course Rept.*, **21**(3), 18–24.

Howard, F.L. (1956) Quoted in South Western Turf Letter. *US Golf Assoc. Green Sect.* 1, 3.

Howard, F.L., Rowell, J.B. and Keil, H.L. (1951) Fungus diseases of turf grasses. *Rhode Island Agr. Expt. Sta. Bull.*, **308,** 56pp.

Jackson, N. (1959) Turf disease notes, 1959. *J. Sports Turf Res. Inst.*, **10**(35), 47–54.

Jackson, N. (1960) Turf disease notes, 1960. *J. Sports Turf Res. Inst.*, **10**(36), 171–6.

Jackson, N. (1962) Turf disease notes, 1962. *J. Sports Turf Res. Inst.*, **10**(38), 410–16.

Jackson, N. (1966). Dollar spot disease and its control with special reference to changes in the susceptibility of *Sclerotinia homoeocarpa* to cadmium and mercury fungicides. In *Proc. 7th Illinois Turf Conf.*, pp. 21–5.

Jackson, N. (1970) Evaluation of some chemicals for control of stripe smut in Kentucky bluegrass turf. *Plant Dis. Rept.*, **54**(2), 168–70.

Jackson, N. (1983) Turf disease control of the past and present. *Am. Lawn Applicator*, **4**(5), 14–18.

Jackson, N. and Dernoeden, P. (1979) Fall fungicide application for control of stripe smut and leaf spot. *Fungicide Nematicide Tests*, **34**(307), 143–4.

Jackson, N. and Fenstermacher, J.M. (1969) Typhula blight;

its cause, epidemiology and control. *J. Sports Turf Res. Inst.*, **45,** 67–73.
Jamalainen, E.A. (1960) Low temperatue parasitic fungi of grassland and their chemical control in Finland. In *Proc. 8th Int. Grassland Congr.*, 1960, pp. 194–6.
Jamalainen, E.A. (1974) Resistance in winter cereals and grasses to low temperature parasitic fungi. *Ann. Rev. Phytopathol.*, **12,** 281–302.
Karnok, K.J. (1984) Plant hormones, overseeding and colorants. *Am. Lawn Applicator*, **5**(8), 70–71.
Kozelnicky, G.M. (1974) Updating twenty years of research: spring dead spot on Bermuda grass. *US Golf Assoc. Green Sect. Rec.*, **12**(3), 12–15.
Lucas, L.T. (1983) Fungicide and nematicide update. Recently labelled fungicides for use on turf grasses. *Plant Dis.*, **67,** 120.
Ludbrook, W.V. and Brockwell, J. (1952) Control of dollar spot on turf in Canberra. *J. Aust. Inst. Agric. Sci.*, **18,** 39–40.
Massie, L.B. Cole, H. Jr and Duich, J. (1968) Pathogen variation in relation to disease severity and control of dollar spot disease of turf grass on fungicides. *Phytopathology*, **58,** 1616–19.
Meiners, J.P. (1955) Etiology and control of snow mould of turf in the Pacific Northwest. *Phytopathology*, **45,** 59–62.
Monteith, J. Jr and Dahl, A.S. (1932) Turf diseases and their control. *US Golf Assoc. Green Sect. Bull.*, **12**(4), 85–187.
Munneke, D.E. (1972) Factors affecting the efficacy of fungicides in the soil. *Ann. Rev. Phytopathol.*, **10,** 375–98.
Nicholson, J.F., Meyer, W.A., Sinclair, J.B. and Butler, J.D. (1971) Turf isolates of *Sclerotinia homoeocarpa* tolerant to dyrene. *Phytopathol. Z.*, **72,** 169–72.
Pennucci, A. and Jackson, N. (1983) Tolerance of *Sclerotinia homoeocarpa* to iprodione and chlorothalonil (Abstr.). *Phytopathology*, **73,** 372.
Rodriguez-Kabana, R. and Curl, E.A. (1980) Non-target effects of pesticides on soil borne pathogens and disease. *Ann. Rev. Phytopathol.*, **18,** 311–22.
Sanders, P.L. (1984) Failure of metalaxyl to control pythium blight on turf grass in Pennsylvania. *Plant Dis.*, **68**(9), 776–7.
Sanders, P.L., Burpee, L.L., Cole, H. Jr and Duich, J.M. (1978) Control of pythium blight of turf grass with CGA-48988. *Plant Dis. Rept.*, **62,** 663–667.
Skylakakis, G. (1981) Effects of alternating and mixing pesticides on the build-up of fungal resistance. *Phytopathology*, **71,** 1119–21.
Smiley, R.W. and Howard, R.J. (1976) Tolerance to benzimidazole-derivative fungicides by *Fusarium roseum* on Kentucky bluegrass turf. *Plant Dis. Rept.*, **60,** 91–4.
Smith, A.M., Stykes, B.A. and Moore, K.J. (1970) Benomyl stimulates the growth of a Basidiomycete on turf. *Plant Dis. Rept.*, **54,** 774–5.
Smith, E.S. (1984) Sprayers: Nozzle selection and calibration. *Greenmaster*, **20**(5), 19.
Smith, J.D. (1953a) Fungi and turf diseases 3. Fusarium patch disease. *J. Sports Turf Res. Inst.*, **8**(29), 230–52.
Smith, J.D. (1953b) Fungi and turf diseases. Corticium disease. *J. Sports Turf Res. Inst.*, **8**(29), 253–8.
Smith, J.D. (1954) Dollar spot disease–Fungicide trials, 1954. *J. Sports Turf Res. Inst.*, **8**(29), 439–44.
Smith, J.D. (1955) Fungi and turf diseases 5. Dollar spot disease. *J. Sports Turf Res. Inst.*, **9**(31), 35–59.
Smith, J.D (1956a) The use of griseofulvin against dollar spot and fusarium patch disease of turf. *J. Sports Turf Res. Inst.*, **9**(32), 203–9.
Smith, J.D. (1956b) Dollar spot trials, 1956. *J. Sports Turf Res. Inst.*, **9**(32), 235–43.
Smith, J.D. (1957a) Fungi and turf diseases 7. Fairy rings. *J. Sports Turf Res. Inst.*, **9**(33), 324–52.
Smith, J.D. (1957b) The control of certain diseases of sports turf grasses in the British Isles. MSc. Thesis, Univ. of Durham. 2 Vol., 266 pp.
Smith, J.D. (1957c) Dollar spot – Fungicide trials, 1956. *J. Sports Turf Res. Inst.*, **9**(33), 353–4.
Smith, J.D. (1957d) Fusarium patch disease – Fungicide trials, 1957. *J. Sports Turf Res. Inst.*, **9**(33), 360–3.
Smith, J.D. (1957e) Corticum disease – Fungicide trial, 1957. *J. Sports Turf Res. Inst.*, **9**(33), 367–8.
Smith, J.D. (1976) Snow mould control in turf grasses with fungicides in Saskatchewan, 1971–74. *Can. Plant. Dis. Surv.*, **56,** 1–8.
Smith, J.D. and Evans, I.R. (1985) Major diseases of turf grasses in Western Canada. Alberta Agric. Edmonton, Agdex. 273/636-3. 14pp.
Staub, T. and Sozzi, D. (1984) Fungicide resistance: a continuing challenge. *Plant Dis.*, **68**(12), 1026–31.
Taylor, D.K. (1974) British Columbia Turf grass Report. Proc. NW Turf grass Assn. Sun River Ore. 2 pp.
Vargas, J.M., Jr (1973) A benzimidazole resistant strain of *Erysiphe graminis. Phytopathology*, **63,** 1366–8.
Vargas, J.M. and Beard, J.B. (1970) Chloroneb, a new fungicide for the control of typhula blight. *Plant Dis. Rept.*, **54,** 1075–7.
Vargas, J.M and Beard, J.B. (1971) Comparison of application dates for the control of typhula blight. *Plant Dis. Rept.*, **55**(12), 1118–19.
Vaugn, J.R. (1953) National turf fungicide trials, 1952. *Golf Course Rept.*, **21**(3), 18–24.
Vaugn, J.R. and Klomparens, W. (1952) Drugs on the green. *Golf Course Rept.*, **20**(2), 5–7.
Warren, C.G., Sanders, P.L. and Cole, H., Jr (1974) *Sclerotinia homoeocarpa* tolerance to benzimidazole-configuration fungicides. *Phytopathology*, **64,** 1139–42.
Warren, C.G., Sanders, P.L. and Cole, H., Jr (1976) Increased severity of *Pythium* blight associated with the use of benzimidazole fungicides on creeping bentgrass. *Plant Dis. Rept.*, **60**(11), 932–5.
Watkins, J.E. (1983) New fungicides for ornamental disease control. *Plant Dis.*, **67**(4), 351–2.
Wells, H.D. (1962) Cottony blight disease. *Golf Course Rept.*, **30**(5), 33–6.
Wells, H.D. (1969) Chloroneb, a foliage fungicide for control of cottony blight of ryegrass. *Plant Dis. Rept.*, **53,** 528–9.
Wilson, C.G. and Grau, F.V. (1952) 1951 National Cooperative fungicide trials. *Golf Course Rept.*, **20**(2), 8–13.

Part two

Seed and seedling diseases (including some minor root diseases)

5 *Seed rots, pre- and post-emergence blights*

While many environmental factors influence the successful germination and establishment of grass seedlings (Beard, 1973), their failure to braird and subsequent thin or patchy stands are often the result of seed rotting or attack on the seedling in its immediate post-germination state by soil- or seed-borne fungi under sub-optimum conditions for germination. Other biotic agents such as bacteria, nematodes and small insects may be involved. Poor 'takes' are often seen in seed lots of low vigour (Perry, 1972), particularly in small-seeded, cool-season species of *Agrostis, Cynosurus, Festuca, Phleum* and *Poa* (Smith, 1965). Poor emergence, even in seed of high vigour may be noted when it is sown too deeply, in soil with poor physical condition or composition, containing toxic materials, lacking in nutrients, or with inadequate or excessive moisture. Such unfavourable factors retard early germination and rapid seedling growth and prolong the period when seedling tissues have little resistance (Garrett, 1970; Hendrix and Campbell, 1973) to ubiquitous rotting organisms and plurivorous or more specialized pathogens. If other factors are not limiting, the temperature of the soil at seeding depth determines the speed of germination. Each species has its own range of temperature for germination, with an optimum, in many cases influenced by diurnal temperature progression which would be experienced in their natural environment. With most cool-season species diurnal cyclic fluctuations of 15–30 °C promote rapid germination and with warm-season turf grasses this is increased to 20 or 25–35 °C (Anon, 1960). Even so, under optimum conditions the period from seeding to emergence can vary from about 7 days for *Lolium perenne* to over 3 weeks for many cultivars of *Poa pratensis*. Different cultivars and seed lots of the same species may have different temperature optima (Harrington, 1923) and vary in speed of emergence.

5.1 SYMPTOMS

5.1.1 Seed rot

Seedlings fail to emerge in small patches or over large areas or the stand is generally very thin. Seeds which can be found give no indication of having germinated. Caryopses are rotten and permeated with fungal mycelium and bacteria. After they are sown in moist soil, grass seeds imbibe moisture and mobilize nutrient reserves of the caryopsis. If germination is delayed by unfavourable temperatures, in particular, their nutrients provide substrates for unspecialized seed-borne and soil-borne saprobes and pathogens. Seed-borne fungi are at an advantage in this respect, being strategically located on the host tissues.

5.1.2 Pre-emergence blight

As with seed rot, seedlings fail to emerge in small or large patches or the stand is very thin. If they escape seed rotting and adverse conditions for germination persist, seedlings may be parasitized before they emerge from the soil. Seminal roots are lesioned and may collapse in decay. Pre-emergence blight may occur in cool-season grass species in cold, wet, spring or autumn weather in temperate regions or in cool-season grasses used to overseed warm-season ones in winter months in warm-humid regions (Freeman, 1980).

5.1.3 Post-emergence blight

This is often referred to as 'damping-off'. The coleoptile and/or seedling leaves appear and then collapse, rotting at soil level, often being parasitized by fungi and/or bacteria as primary or secondary invaders. The attack may have commenced before emergence, on tissues weakened by adverse conditions for germination or may develop in response to poor environmental factors above ground. Seedlings which survive may be a source of infection for disease of mature aerial tissues, e.g. in leaf spots caused by *Drechslera* spp.

5.2 CAUSAL FUNGI

Some of the commoner seed-borne fungi of turf grasses incite these diseases. These include: *Acremoniella atra* (Cda.) Sacc.; *Alternaria alternata* Kiessler; *Ascochyta* spp.; *Aureobasidium pullulans* (de Bary) Arn.; *Botrytis cinerea* Pers.; *Cephalosporium acremonium* Cda.; *Chaetomium globosum* Kze.; *Cladosporium cladosporoides* (Fr.) De Vries; *Colletotrichum graminicola* (Ces.) Wils.; *Cochliobolus geniculatus* Nelson: Boedijn; *C. sativus* (Ito and Kuribay) Drechsl. ex Dastur; *Cochliobolus* spp.; *Curvularia inaequalis* (Shear) Boedijn; *Drechslera* spp.; *Epicoccum* spp.; *Fusarium culmorum* (W.G. Smith) Sacc.; *Fusarium* spp.; *Microdochium nivale* (Fries) Samuels and Hallett; *Nigrospora* spp.; *Septoria* spp.; *Stagonospora* spp.; *Trichoderma viride* Pers. ex Fries and *Trichoderma* spp. (Cole, Braverman and Duich, 1968; Conners, 1967; Mäkelä, 1972; Richardson, 1979; Smith, 1965; Sprague, 1950).

Fusarium and *Pythium* spp. and *Rhizoctonia* spp. are probably the most common soil-borne pathogens associated with seed rots and seedling blights, often in disease complexes.

5.3 PRE- AND POST-EMERGENCE SEEDLING BLIGHTS CAUSED BY '*FUSARIUM*' SPECIES

Soil- and seed-borne fungi with anamorphs in the genus *Fusarium*, and with teleomorphs, where known usually in *Gibberella* are common, cosmopolitan and frequently associated with seedling, roots, crown, leaf and sometimes with inflorescence diseases of turf grass species (Booth, 1971; Domsch and Gams, 1972; Gordon, 1959; Gordon and Sprague, 1941; Griffiths and Siddiqui, 1958; Holmes, 1976, 1979; Holmes and Channon, 1975; Jackson, 1958; Kommedahl and Siggeirsson, 1973; Michail and Carr, 1966a; Richardson, 1979; Slykhuis, 1947; Smith, 1954, 1955, 1957a, 1965; Sprague, 1950; Waid, 1974; Warcup, 1951). *Fusarium* spp. are the dominant fungi on grass roots (Kreutzer, 1972). The low-temperature tolerant *Microdochium nivale*, synonym *Fusarium nivale*, is also soil- and seed-borne and associated with similar diseases of turf grass species (Bennett, 1933; Holmes, 1976, 1979; Holmes and Channon, 1975). The role of some of these fungi in diseases of mature turf grass is discussed under fusarium blight, microdochium patch and pink snow mould (pp. 117 and 59). However, the pathogenicity of some of the *Fusarium* species from grasses is doubtful and some are only weakly pathogenic (Bean, 1966, 1969; Cole, Braverman and Duich, 1968; Gordon and Sprague, 1941; Slykhuis, 1947; Sprague, 1950). Considerable variation in pathogenicity between isolates of the same species (Cole, Braverman and Duich, 1968; Fulton, Cole and Nelson, 1974; Holmes, 1976; Holmes and Channon, 1975) and on different turf grass species (Holmes, 1979; Smith, 1957a) has been noted. Several of the *Fusarium* spp., pathogenic when inoculated on leaves of grass seedlings in the greenhouse have failed to produce the root/crown/foliage syndrome of Fusarium blight on established turf in the field (Smiley, 1980). Nevertheless, *F. culmorum* and *M. nivale* often cause seedling blights of turf grasses, especially in cool-temperate or temperate climates, often as components of fungal complexes (Holmes, 1976, 1979; Holmes and Channon, 1975; Jackson, 1964; Slykhuis, 1947; Smith, 1954, 1955, 1956, 1957a, 1965). The importance of *F. avenaceum* (Fr.) Sacc. (teleomorph – *G. avenacea* R.J. Cook) and *F. graminearum* Schwabe (teleomorph – *G. zeae* (Schwabe) Petch) in causing pre- and post-emergence blights of turf grasses in uncertain although they are recognized cereal seedling pathogens (Bennett, 1935; Colhoun and Park, 1964).

5.3.1 Symptoms

Small patches of seedlings may fail to emerge or small patches of yellow, bronzed, collapsed or stunted seedlings with brown lesions at soil level occur from emergence to the 2–3 leaf stage. This is usually most noticeable in small-seeded *Agrostis* and *Festuca* spp. although larger-seeded turf grass species may also be affected (Figs 1, 2 and 3 colour plate section).

5.3.2 Major causal fungi

1. Anamorph — *Microdochium nivale* (Fries) Samuels and Hallett
   Teleomorph — *Monographella nivalis* (Schaffnit) E. Müller
   Synonym — *Fusarium nivale* Ces. ex Sacc.
   (see Section 6.1.2 for description)
2. Anamorph — *Fusarium culmorum* (W.G. Smith) Sacc. (Synonym–*F. roseum* Link ex Gray emend. Snyd. and Hans. 'Culmorum').
   Teleomorph — Not known.

On potato – sucrose agar the aerial mycelium of *F. culmorum* is floccose, at first carmine, then reddish-brown, spore masses golden to ochre. *Macroconidia* stout, often 5-septate, 30–60 × 4–7 μm, produced in sporodochia on secondary conidiophores (Nirenberg, 1981). *Microconida* absent, but occasionally small, oval macroconidia are formed (Booth, 1971). *Chlamydospores* intercalary or terminal, produced singly, in chains or in clumps, rough or smooth-walled, 1–2 celled, 10–14 × 9–12 μm.

5.3.3 Disease development

Sources of inoculum for infection with *M. nivale* have been considered for microdochium patch disease. Although rarely isolated from soil, the fungus is able to persist in crop residues, including root and crown tissues, in the soil as a saprobe, as a pathogen on grasses and as an endophyte. It has no special survival structures.

F. culmorum is soil-borne and seed-borne on grasses (Richardson, 1979). It is often a major component of root rot complexes associated with grass roots. With *F. oxysporum*, it is the commonest *Fusarium* sp.; their highest populations being detected in British soils under permanent pasture or under grass for at least two years after cereals (McKenzie and Taylor, 1983). Kreutzer (1972) found *F. roseum* (Link) emend. Snyder and Hansen to be the dominant fungus from the outer soil of grasses in cultivated soils and the most commonly encountered *Fusarium* sp. in the root zone of grasses generally.

F. culmorum has been regularly isolated from the roots and the soil in the root zone of *Lolium perenne* and from the stem bases and leaves of that species (Griffiths and Siddiqui, 1958; Holmes, 1976), although Bean (1969), in isolating it from crowns and leaf lesions of *Poa pratensis*, failed to do this from roots. Sometimes, in complex with *M. nivale*, it is associated with a patch disease of mature amenity turf, very similar in appearance of fusarium patch disease in Britain (Jackson, 1964; Sanders and Cole, 1981; Smith, unpublished). It is often isolated from new turf where seedlings have suffered from pre- and post-emergence damping-off. Its survival structures are chlamydospores (Garrett, 1970). The behaviour of *F. culmorum* as a relatively unspecialized, vigorous, competitive, soil-inhabiting saprobe which is a primary colonizer of fresh crop debris and its complementary pattern as a pathogen whose inoculum builds up, particularly on plant crowns, is well-known in cereals (Cook and Bruehl, 1968; Garrett, 1970). In the latter crops there are periods when the host is absent and chlamydospores are likely to be of major importance as inoculum sources. Where grass is re-established on land previously under turf, chlamydospores are the likely propagules. In perennial turf all inoculum sources may be significant. Chlamydospores of *Fusarium* spp. germinate in response to nutrient stimulus from fresh host tissue and from root exudates (Garrett, 1970). Complexes of fungi and bacteria were shown by Slykhuis (1947) to suppress the development of *F. culmorum* in the immediate vicinity of the germinating seed of grasses when added to sterilized soil and reduced the incidence of seedling blight. The severity of the disease was significantly more severe in sterilized soil, inoculated with *F. culmorum* grown in sand/cornmeal medium, near the optimum growth temperature for the pathogen (25 °C), provided that the soil infestation was not high. In unsterilized soil, greater growth of *F. culmorum* and a higher incidence of seedling blight occurred at 10–20 °C than at 25 °C which favoured the development of antagonistic fungi and bacteria. Disease was more severe in unsterilized heavy field soils, uniformly infested with *F. culmorum*, than in light soils. Griffiths and Siddiqui (1958) in surveys of soils in swards of *Lolium perenne* in Wales found that the marked seasonal variation in *F. culmorum* was related neither to soil temperature nor soil moisture separately. They showed that a bacterium isolated from root surfaces inhibited the development of the disease on seedlings when inoculated with *F. culmorum* into sterile soil sown with *L. perenne*. In soil at 40% moisture holding capacity (MHC) *F. culmorum* was dominant over *Trichoderma viride* and grew abundantly from below and above 10 °C. Weste (1975) considered *F. culmorum* tolerant to soil microorganisms and their antibiotics.

Bennett (1928) reported that disease in wheat caused by *F. culmorum* in complex with *F. avenaceum* serious in wet, cold soils even at temperatures below 10 °C. Colhoun and Park (1964) found that the incidence of seedling blight, resulting from inoculation with spores of *F. culmorum* (and *F. graminearum*), rarely occurred in wet soils at any temperature, but many wheat seedlings developed lesions or were killed in dry soils at temperatures between 16° and 23 °C. In moist soils (ca. 50% MHC), even at higher temperatures there was not much pre-emergence death, but considerable post-emergence death and lesioning of seedlings. High spore inoculum levels pemitted severe injury even when soil temperatures were sub-optimal (Colhoun, Taylor and Tomlinson, 1968), similar to the result obtained by Slykhuis (1947) using cornmeal/sand medium when considerable mycelial inoculum was applied. Soil reaction and sowing depth has less effect on disease severity than the interacting factors of

Table 5.1 Some *Pythium* species which incite seed rots, seedling and foliar blights and/or root rots of turf grasses

Species	Hosts/incidence	Diagnosis	Distinguishing features	References
P. aphanidermatum (Edson) Fitzpatrick (syn. *P. butleri* Subramaniam)	Recognized mainly as cause of leaf blight of cool-season grasses in warm weather. Most warm-season grasses except *Cynodon* are resistant. Frequent on *Agrostis* Also causes seedling and root diseases, 30–35 °C most favourable for infection.	*Sporangia* terminal, swollen, hyphal branches *Oogonia* terminal, smooth, globose *Antheridia* sac-shaped, monoclinous or diclinous mostly 1 or 2 per oogonium. *Oospores* aplerotic 20–25 μm (av. 23 μm).	Infection negligible at temperatures below 30 °C.	Freeman (1980) Freeman and Horn (1963) Hendrix, Campbell and Moncrief (1970) Saladini (1980) Sprague (1950) Van der Plaats-Niterink (1979)
P. aristosporum Vanterpool	Similar hosts to *P. graminicola* and *P. arrhenomanes*, but less common in North America. Able to incite severe root rot at quite low temperatures (8–15 °C). Possibly a snow rot species.	*Appressoria* on hyphae, common, sickle-shaped. *Sporangia* are inflated hyphae. *Oogonia* mostly terminal, on short side branches subspherical. 27–45 μm. *Antheridia* club-shaped, crook-necked, mono- or diclinous, up to 8 per oogonium, and entangled around it. *Oospores* aplerotic, smooth, subspherical, deep brown, 13–30 μm (av. 24 μm).	Similar to *P. arrhenomanes* but has fewer antheridia (<8) than *P. arrhenomanes* (20 or more) and in the latter antheridia are diclinous. Most likely to be confused with *P. graminicola* which has smaller oogonia. *P. arrhenomanes* and *P. graminicola* have higher optimum growth temperatures.	Sprague (1950) Van der Plaats-Niterink (1979) Vanterpool (1938) Waterhouse (1968)
P. arrhenomanes Drechsl	With *P. graminicola* (q.v.) is one of the most common inciters of seedling blight and brown root rot of grasses – with a very wide host range.	Profuse aerial mycelum in culture. *Sporangia* inflated; lobulate often complex. *Oogonia* subspherical mostly terminal, 24–36 μm. *Antheridia* Crook-necked, making apical contact with oogonium, often numerous (15–20) produced at the ends of branches of filaments.	Characteristically with large numbers of diclinous antheridia (see *P. aristosporum*)	Drechsler (1936) Sprague (1950) Van Luijk (1934) Van der Plaats-Niterink (1979) Vanterpool and Sprague (1942) Waller (1979)

		Oospores plerotic or nearly so, 22–33 μm (av. 27 μm), often abortive.		
P. catenulatum Mattews	From turf soil under *Agrostis, Cynodon* and *Eremochloa* spp. in southern USA in particular. A warm-temperature fungus with growth optimum 30–35 °C.	*Sporangia* swollen, branched parts of mycelium. *Hyphal swellings* (conidia not sporangia) in chains of 3–8, spherical to pear-shaped 10–20 μm diam. *Oogonia* terminal and intercalary, smooth. *Antheridia* clavate or crook-necked, point contact, 1–12 per oogonium di- or monoclinous. *Oospores* mostly plerotic 16–26 μm.	Chains of conidia distinguish *P. catenulatum* from *P. aphanidermatum*.	Hendrix, Campbell and Moncrief (1970) Matthews (1931) Saladini (1980) Van der Plaats-Niterink (1979) Waterhouse (1968)
P. debaryanum Hesse	Hendrix and Papa (1974) include this species in the *P. ultimum* complex and doubt that the species has been properly identified since its original description. Van der Plaats-Niterink (1979) does not include it. Waterhouse (1968) gives Hesse's description.			
	From a wide range of hosts. Reported very destructive on young seedlings, causing damping-off.	*Sporangia* spherical and terminal on side branches with cross wall <21 μm diam. *Conidia* (not sporangia) colourless, terminal or intercalary. Intercalary ones germinating by a germ tube after shedding. *Oogonia* terminal or intercalary, spherical, smooth, 15–25 μm diam. *Antheridia* 1 to 6 per oogonium, diclinous androgynous *Oospores* smooth 10–18 μm diam.		Krywienczyk and Dorworth (1980) Matthews (1931) Robertson (1980) Waterhouse (1968) Van der Plaats-Niterink (1979)
P. dissotocum Drechsl.	Soil under *Cynodon* spp. in south and south-east USA.	*Sporangia* inflated, dendritic, finger-like. *Oogonia* terminal, intercalary and on lateral branches.	Dendritic inflated finger-like sporangia.	Drechsler (1940) Hendrix, Campbell and Moncrief (1970) Van der Plaats-Niterink (1979)

Table 5.1 Continued

Species	Hosts/incidence	Diagnosis	Distinguishing features	References
		Antheridia sac-like, sessile or nearly so on unbranched stalks, monoclinous or diclinous, 1–3 per oogonium. *Oospores* nearly plerotic smooth mostly 15–21 μm (av. 18 μm).		
P. graminicola Subramaniam	An important seed rot, seedling blight and root rot pathogen on Gramineae generally with a very wide host range including many cool-season turf grasses. An incitant of leaf blight at high temperatures.	*Sporangia* toruloid. *Oogonia* terminal, lateral or intercalary, smooth, globose 20–25 μm *Antheridia*, crook-necked with flattened tip usually 1–3/oogonia and persistent, mostly monoclinous. *Oospores*, round, smooth, plerotic, thick-walled (3 μm) sometimes brown tinted 15–26 μm.	Unlike *P. arrhenomanes* and *P. aristosporum* which it resembles *P. graminicola* has fewer antheridia/oogonia and smaller ocgonia.	Hendrix and Papa (1974) Saladini (1980) Sprague (1950) Subramanian (1928) Van der Plaats-Niterink (1979) Waterhouse (1968)
P. intermedium de Bary	Isolated from *Agrostis, Festuca* and *Lolium* spp. in the Netherlands.	No sporangia Hyphal swellings abundant, terminal or forming chains, globose or pyriform, up to 25 μm diam. *Oogonia* and *oospores* not common. *Antheridia* diclinous, branched, stalks long.	Catenulate hyphal swellings (conidia) produced in basipetal succession.	Van Luijk (1934) Van der Plaats-Niterink (1979) Waterhouse (1968)
P. irregulare Buisman	Causes seed decay and root necrosis of a wide range of grasses.	*Sporangia* rare, globose *Hyphal* swellings (conidia) up to 25 μm diam. terminal or intercalary *Oogonia* globose, irregular or with protuberances of different sizes and shapes, usually intercalary, mostly 14–16 μm diam.	A proportion of the oogonia are irregular in shape.	Freeman (1980) Hendrix and Papa (1974) Saladini (1980) Sprague (1950) Van Luijk (1934) Van der Plaats-Niterink (1979) Waterhouse (1968)

		Antheridia usually monoclinous sometimes stalked, usually 1–2 per oogonium *Oospores* mostly aplerotic 14–16 μm diam.		
P. iwayamai **S. Ito**	Incites snow rot of cereals and grasses in Japan and cereal snow rot in north-western North America.	*Sporangia* terminal, globose to lemon-shaped 24–48 × 44 μm. *Hyphal* swellings intercalary 36–48 × 24 μm. *Oogonia* terminal or intercalary, globose, smooth, 23–29 (av. 27 μm) diam. *Antheridia* 1 to 3 per oogonium mono- or diclinous, sometimes on branched stalks. *Oospores* globose, plerotic or aplerotic, 19–24 μm.		Hirane (1960) Ito and Tokunaga (1935) Lipps and Bruehl (1978) Van der Plaats-Niterink (1979) Waterhouse (1968)
P. myriotylum **Drechsler**	Occurs in warmer regions, mainly on non-grass species but also as a seedling blight, a foliar blight and a root rot of mature grasses. Optimum growth temperature 35 °C. Damage drops off rapidly below 30 °C.	*Mycelium* produces clusters of small, globose appressoria *Sporangia* filamentous, unidfferentiated, digitate *Oogonia* subspherical, mostly terminal or intercalary 23–30 (av. 27) μm *Antheridia* terminally expanded, crook-necked or arched over the oogonium making apical contact, usually 3-6 per oogonium, mostly diclinous *Oospores* hyaline or yellow, mostly 18–24 μm aplerotic.	Digitate branched sporangia. Clusters of globose appressoria. Similar to *P. aristosporum* but lacks monoclinous antheridia. Like *P. arrhenomanes* but has diclinous antheridia and aplerotic oospores.	Drechsler (1943) Freeman (1980) McCarter and Littrell (1968) Saladini (1980) Van der Plaats-Niterink (1979) Waterhouse (1968)
P. rostratum **Butler**	Isolated from turf and soil under several grass species including *Cynodon, Agrostis* and *Eremochloa.*	*Sporangia* terminal or intercalary, spherical at first then oval, 23–34 μm, discharge tube lateral, at first with a	Slow growing with intercalary oogonia in short chains, sessile antheridia.	Hendrix, Campbell and Moncrief (1970) Matthews (1931) Middleton (1943) Saladini (1980)

Table 5.1 Continued

Species	Hosts/incidence	Diagnosis	Distinguishing features	References
		short curved beak. *Oogonia* smooth, mostly intercalary, often in short chains, about 21 μm diam., sub-globose, rostrate *Antheridia* monoclinous, often almost sessile and short, mostly 1/oogonium *Oospores* plerotic or nearly so, 12–26 μm.		
P. tardicrescens Vanterpool	One of the main causes of browning root rot in the Gramineae in Canada and occurs, in England.	*Hyphal* swellings, toruloid or lobate. *Oogonia* terminal on short side branches 17–30 (av. 20) μm. *Antheridia* clavate or crook-necked, making point contact with oogonia, usually 2 or 3 per oogonium. *Oospores* 12–26 μm, av. 20 μm aplerotic.	Differs from *P. aphanidermatum* in having terminal antheridia and from *P. aristosporum, P. myriotolum* and *P. arrhenomanes* which have more numerous antheridia. Difficult to isolate because of slow growth.	Sprague (1950) Van der Plaats-Niterink (1979) Vanterpool (1938) Waterhouse (1968)
P. torulosum Coker & Patterson	Isolated from roots of *Agrostis stolonifera* and *Festuca duriuscula* in the Netherlands and frequently from *Agrostis* and *Cynodon* spp. in south and south-east USA. Foliar blight pathogen in mid-western USA.	*Sporangia* toruloid, inflated branches of mycelium *Oogonia* terminal, intercalary or stalked, smooth, spherical, thin-walled, mostly 17–18 μm. *Antheridia* on short branches of oogonial stalk, rarely branched, tuberous, ephemeral. *Oospores* plerotic, mostly 13–17 μm diam.	Similar to *P. vanterpoolii* but this has unbranched sporangia, sometimes catenulate. *P. torulosum* grows at 35 °C whereas *P. vanterpoolii* has a maximum of 30 °C.	Coker and Patterson (1927) Hendrix, Campbell and Moncrief (1970) Kouyeas (1964) Van Luijk (1934) Muse *et al.* (1974) Van der Plaats-Niterink (1979) Saladini (1980) Vanterpool (1938) Waterhouse (1968)
P. ultimum Trow var. *ultimum* = *P. ultimum* Trow. (syn. *P. debaryanum* de Bary?)	Widespread in temperate regions with a very wide host range, a causal agent of seedling disease	*Sporangia* rarely formed *Hyphal* swellings (conidia) terminal or intercalary. *Oogonia* spherical, mostly		Couch (1973) Freeman (1980) Hendrix Campbell and Moncrief (1970)

	and root rots, but also a common cause of foliar blight of mature turf.	terminal, smooth, 20–23 μm *Antheridia* usually 1 per oogonium arising directly below oogonium, monoclinous or 2, often diclinous. *Oospores* spherical, aplerotic, 15–18 μ, thick-walled.	Hendrix and Papa (1974) Saladini (1980) Sprague (1950) Trow (1901) Van der Plaats-Niterink (1979) Waterhouse (1968)
The following species is encountered mainly as a cause of foliar blighting (see Chapter 22).			
Pythium vanterpoolii V. & H. Koueas (as *vanterpooli*)	With *P. graminicola* and *P. torulosum* causes foliar blighting of *Agrostis* spp. at high temperatures.	*Sporangia* terminal or intercalary, swollen or bulbous, and catenulate, spherical or ellipsoid, not often digitate, never branched. *Zoospores* 7.5 μm diam. after encystment, *Oogonia* terminal, rarely intercalary, globose, 18.5 μm (13.5–22.5 μm) diam., form readily in culture. *Antheridia* single, rarely doubly clavate, monoclinous, disappearing after oogonia mature. *Oospores* plerotic, wall thick, <2 μm with one fatty reserve globule and one refringent body.	Kouyeas (1963) Van der Plaats-Niterink (1979) Waterhouse (1968) Muse *et al.* (1974)

inoculum level, soil moisture and temperature (Colhoun, Taylor and Tomlinson, 1968).

Isolates of *M. nivale* from roots, stem bases and leaves of *Lolium perenne*, as mycelial macerates, varied greatly in pathogenicity towards seedlings of the grass when tested in sterilized soil. Some isolates were almost non-pathogenic. Others had a conisderable effect on seed germination, reducing emergence and vigour of surviving seedlings. All *F. culmorum* isolates had a marked effect on germination and in combination, *M. nivale* and *F. culmorum* had an additive effect in reducing emergence and seedling vigour (Holmes, 1976). The small-seeded *Phleum pratense* with its delicate, susceptible root and shoot system was more severely reduced in emergence and seedling vigour than the larger-seeded *L. perenne*. However, *P. pratense* was no more susceptible to *M. nivale* when inoculated at the seedling stage than *L. perenne. F. culmorum*, more pathogenic to germinating seeds than *M. nivale*, scarcely damaged seedlings which had become established (Holmes, 1979). Gussin and Lynch (1983a) have found that the nature of the substrate base on which *F. culmorum* was grown influenced its growth and pathogenicity. Sowing *L. perenne* into infected *L. perenne* residue led to little plant death compared with sowing it into infected wheat and barley residues.

An isolate of *F. culmorum* from *Festuca rubra* grown on cornmeal/sand medium and containing mycelium and spores of the fungus caused seed rotting, pre- and post-emergence blight in *Agrostis tenuis, Cynosurus cristatus, Festuca rubra, Phleum nodosum, Poa pratensis* and *P. trivialis. Lolium perenne* did not suffer seed rot (Smith, 1957a). Sowing at unfavourable times of the year and at increased seed rates in soil inoculated with *F. culmorum* decreased the numbers of seedlings of *F. rubra* which emerged compared with uninoculated soil (Smith, 1957b). The effects of soil inoculation in reducing establishment of *P. nodosum* and *F. rubra* persisted for a season when the soil was cultivated and left fallow (Jackson, 1958).

5.3.5 Control

See p. 43.

5.4 PYTHIUM SEEDLING DISEASES AND ROOT ROTS

Pythium spp. are cosmopolitan, ubiquitous in waters and undisturbed and cultivated soils, many of them saprobes, but some causing seed rots, seedling blights (damping-off) and root injuries of various kinds in turf grasses. Many species are associated regularly with soil under turf grasses (Conners, 1967; Couch, 1973; Drechsler, 1928, 1936; Freeman and Horn, 1963; Hendrix and Campbell, 1970, 1973; Hendrix, Campbell and Moncrief, 1970; McCarter and Littrell, 1968; Robertson, 1980; Sprague, 1950; Van Luijk, 1934; Vanterpool, 1942). Some incite leaf blights, others cause rotting at low temperatures (snow rot) These are dealt with on pages 93 and 252.

5.4.1 Taxonomy

Distinctions between species are based mainly on a few morphological characters related principally to sporangial, oogonial or conidial development and features of these may be changed by environmental conditions when they are cultured for study. Among those who have reviewed the taxonomy of the genus and/or produced diagnostic keys are Frezzi (1956), Krywienczyk and Dorworth (1980), Matthews (1931), Middleton (1943), Van der Plaats-Niterink (1979), Robertson (1980), Sprague (1950) and Waterhouse (1967, 1968). Hendrix and Papa (1974) consolidated species into complexes which although not entirely satisfactory (Schmitthenner, 1980), provides a workable basis for identification. This is to be preferred to shaky species determinations or the frequent practice of stopping *Pythium* determinations at the generic level (Freeman, 1980; Hendrix and Campbell, 1973).

In those members important to turf grasses, representatives of these species complexes are (see Table 5.1 and descriptions of pythium foliar blight pathogens on p. 253):

1. *P. dissotocum*
2. *P. aphanidermatum, P. aristosporum, P. arrhenomanes, P. catenulatum, P. graminicola, P. iwayamai, P. myriotylum, P. rostratum, P. tardiscrescens, P. torulosum, P. vanterpoolii, P. volutum*
3. *P. ultimum* (and *P. debaryanum*)
4. *P. intermedium*
5. *P. vexans*
6. *P. irregulare*

5.4.2 Isolation and culture

In some cases *Pythium* spp. can be isolated after washing or steeping roots or plant parts in sterile water, preferably with the addition of non-toxic detergents, to reduce numbers of interfering bacteria. Surface sterilization with chlorine water should not be attempted because of toxicity. The tissues are then plated on to water agar or weak nutrient agar (e.g. maizemeal) or synthetic media such as the gallic acid/rose bengal medium of Hendrix and Kuhlman (Flowers and Hendrix, 1969) or the sucrose–asparagine medium of Schmitthenner (1980). Fungicides and bactericides are often added to reduce competition from unwanted organisms and the *Pythium* sp. isolated from outgrowths of 'clean' hyphae. Suitable recipes for selective media are given by Tsao (1970), Hendrix and Campbell (1973), Schmitthenner (1980) and Robertson (1980). Generally *Pythium* spp. can be readily isolated from diseased roots because they outgrow most other root-inhabiting organisms; however, isolation of *Pythium* spp. from soil is often difficult because of the low density of the organisms and the high population of unwanted organisms, mainly bacteria. Methods using selective media, or baiting techniques with seeds or living grass seedlings have been used (Buchholz, 1949; Campbell, 1949; Hendrix and Campbell, 1970; Robertson, 1980; Schmitthenner, 1980). No one method is generally applicable to all species.

For the identification of *Pythium* species it is necessary to examine sporangia, antheridia, oogonia and oospores. Waterhouse (1967) and Schmitthenner (1980) give methods for sporangia and zoospore production using sterilized grass blades and media for the induction of sex organs. Many species will sporulate on water agar or weak vegetable and seed-extract agar whereas media rich in nutrients encourage (often unwanted) mycelial development. However, many isolates do not reproduce sexually and it is then possible only to identify to a 'group'.

5.4.3 Fungi and disease development

Some *Pythium* spp. may survive in soil for long periods in the absence of a suitable host. Many survive saprophytically on plant debris in the soil and may be opportunistic parasites or secondary invaders. Other species which are primarily parasitic with mycelium which is not long lived, and which do not compete well with other soil organisms, nevertheless are capable of surviving for long periods. Zoospores generally have short lives, the duration of which does not seem to be increased much by their encystment (Burr and Stanghellini, 1973). Sporangia, conidia or chlamydospores are usually longer-lived, but the main organs of long-term persistence, especially of those species with lobate sporangia, are oospores (Stanghellini, 1974). In some species, oospores and sporangia germinate within a few hours, in others there is a dormancy factor, or soil mycostasis delays development. Rapid germination may be ecologically or pathologically advantageous for sporangia or oospores in the presence of short-lived hosts, plant structures which have short periods of high susceptibility (seedlings, for example), or when a competitive microflora builds up. This microflora may be encouraged by the same seed or root exudates which stimulate resting structure germination (Barton, 1957; Chang-Ho, 1970; Stanghellini and Burr, 1973a).

Resting structures of *Pythium* spp. are liberated by the decay of the host tissues. They may remain dormant under adverse conditions for years in some species (Hendrix and Campbell, 1973). Germination may result in the production of zoospores and mycelium, but in some species by mycelium only. Germination and infection depend on the presence of moisture, but the levels of this vary for different species. Generally, it is assumed that increased disease incidence in *Pythium* diseases is favoured by saturated or over-saturated soil. Although this is true in most cases, it probably derives from a decrease in host vigour, an increase in the production and diffusion of stimulatory exudates from the host and the provision of water-filled soil pores suited to zoospore production and movement (Stanghellini and Burr, 1973b; Stanghellini, 1974). However, resting structures of some *Pythium* spp. are capable of germinating mycelially and growing over the whole range of moisture available to plants. Host tissue exudates may inhibit zoospore production by oospores and sporangia close to it (Stanghellini and Burr, 1973a; Stanghellini & Hancock, 1971).

With many *Pythium* spp. temperature is as important as soil moisture in disease incidence. Some species can be regarded generally as causing most seedling and root damage at low temperatures, e.g. *P. irregulare, P. ultimum* (and *P. debaryanum*?). At high temperatures, *P. myriotylum, P. aphanidermatum, P. arrhenomanes* and *P. volutum* cause more severe injury (Hendrix and Campbell, 1973; Klisiewicz, 1968). The relative growth rates of the pathogen and the velocity of emergence of the seedling determine the severity of pre-emergence infection (Leach, 1947). This is not the case with post-emergence disease because after emergence seedling tissues mature to the point where they become resistant to the pathogen and the optimum temperature for aerial tissue maturation is often quite different from that for pre-emergence growth (Garrett, 1970). General relationships between temperature optima in some *Pythium* spp. and epidemic outbreaks of *Pythium* blight have been established (Saladini, 1980 and Table 5.1).

Hydrogen-ion concentrations unfavourable for plant growth and low light intensities have been shown to increase *Pythium* seedling and root disease, probably by prolonging the period of plant susceptibility (Griffin, 1958; Klisiewicz, 1968; Roth and Riker, 1943). Inoculum levels of soil-borne chlamydospores, zoospores and oospores can be related to disease severity (Mitchell, 1978). There is little reliable information on the effect of fertilization on the severity of *Pythium* seedling diseases (Freeman, 1980). Damping-off of turf grass seedlings has been associated with low soil nitrogen and phosphate levels (Smith, 1965 – and see Section 5.7).

Zoospores and mycelia, the products of direct or indirect germination of *Pythium* resting structures, are the common agents of seed, seedling and root infection. Zoospores, although flagellate and motile, are effective in starting infection only a short distance from their origin unless transported passively by water currents or soil drainage (Hickman and Ho, 1966). However, their motility enables them to congregate in susceptible infection courts, often in the region of root elongation or the young root hair region, attracted by the nutrients diffused from the host. In the absence of a host they soon lose infectivity. At the end of a motile period the zoospores round off and encyst, become attached to the host, later putting out a germ tube and forming appressoria. Mycelia, sporangia and even oospores may form appressoria. Penetration of the host is by an infection peg or slender infection hypha. Host roots may be entered through intact cell walls or between them. Hyphal development is intra- or intercellular, the hyphae enlarging to normal size in the host tissues. Pectic enzymes and phytotoxins are produced by some *Pythium* spp. (Endo and Colt, 1974) and cellulase *in vitro* (Berner and Chapman, 1977; Park, 1977) and *in vivo* (Sharma and Wahab, 1976).

Pythium spp. are often found associated with the roots of maturing and mature stands of turf grass species. Many of these fungi are saprobes, some weakly pathogenic causing little injury, and some causing extensive damage, particularly to young rootlets. Their taxonomy, distribution and effects were extensively studied in cereals and grasses in the Great Plains of North America from the 1920s to the 1950s (Andrews, 1943; Buchholz, 1949; Sprague, 1950; Vanterpool, 1942; Vanterpool and Sprague, 1942; Vanterpool and Truscott, 1932). However, two of the species, *P. graminicola* and *P. arrhenomanes*, which are capable of causing severe damage to the roots of Poaceae have a world-wide distribution (Ledingham and Vanterpool, 1967; Sprague, 1950; Waller, 1979; Waterhouse and Waterston, 1964b) and less attention has been paid to their effects on grasses than on cereals.

Disease resulting from attacks on maturing roots of grasses is not as clear cut as that described on cereals as 'browning root rot'. Plants in the late seedling stage show firm brown lesions on the coarser roots and this injury progresses to finer rootlets in subsequent growth stages and years. Premature death of established grasses, or 'going-out' may result.

The most important incitants of *Pythium* root rot in the Great Plains of North America are *P. graminicola* and *P. arrhenomanes*, which are 'high-temperature' species. *P. aristosporum, P. tardicrescens* and *P. volutum* are sometimes associated (Hodges and Coleman, 1985; Sprague, 1950; Vanterpool, 1942). Waller (1979) found *P. graminicola* and *P. arrhenomanes* common on wheat and barley in southern England, causing a soft, pale lesioning and *Pythium* root rot, ubiquitous on grasses. Van Luijk (1938) isolated *P. volutum* from roots of *Agrostis* sp. in the Netherlands.

The effects of *Pythium* root rots are sometimes severe in warm, dry periods following wet weather in May or June in the Great Plains. Severely infected plants wither and dry up. The diseases are particularly severe on new sowings following the break up of old sod and in the first year of establishment, particularly in *Agropyron cristatum* (Vanterpool, 1942). Autumn seedings are less affected than those made in spring, since hot weather, favourable to the major pathogens, is avoided in the grasses' most susceptible growth stage. Disease severity is greatest where soil phosphate is deficient.

The appearance of *Pythium* diseases of turf grass seedlings may be an annual occurrence in the warm-humid region of the southern United States where it is a common practice to winter-overseed warm season turf grass of *Cynodon* spp. Cool-season species of *Lolium, Poa* and *Agrostis* are used as a temporary measures to improve the appearance or give a playable turf during the winter months when the warm-season grass is dormant. The warm, humid conditions favour the 'warm-temperature' *Pythium* spp. and are conducive to disease development in the susceptible seedlings of the cool-season species of grasses (Freeman, 1980).

5.4.4 Control

See p. 49.

5.5 TURF GRASS SEEDLING DISEASES CAUSED BY *RHIZOCTONIA* SPP.

There is an extensive literature on *Rhizoctonia* spp. and their teleomorphs associated with foliar diseases of turf grasses such as brown patch and yellow patch (Bell, 1967; Brien, 1935; Britton, 1969; Burpee, 1980a, 1980b; Christensen, 1979; Couch, 1973; Dahl, 1933; Joyner and Partyka, 1980; Smiley, 1982; Smith, 1965; Sprague, 1950; Traquair and Smith, 1981). There is much less information on the association of these fungi with seedlings and roots of turf grasses. An understanding of their biology and pathogenesis must be obtained mostly from extensive studies made on *Rhizoctonia* diseases of roots of forage grasses, cereals and non-grass species (Blair, 1942; Campbell *et al.*, 1982; Christou, 1962; Daniels, 1963; Flentje, 1957; Garrett, 1970; Papavisas and Davey, 1962a; Parmeter, 1970; Pitt, 1964a; Thornton, 1956, 1965; Tu and Kimborough, 1975; Waid, 1974). It is, however, apparent that *Rhizoctonia* spp. are cosmopolitan, ubiquitous, soil-borne fungi which have been reported to cause seed rot, pre- and post-emergence blight on many different species of turf grasses (Andrews, 1943; Christensen, 1979; Couch, 1973; Pitt, 1964b; Sprague, 1950; Zjawin, Halisky and Funk, 1975). They may occur in complexes with other pathogenic fungi such as *Helminthosporium, Curvularia, Pythium* and *Fusarium* spp. on grass root systems. *Rhizoctonia* spp. may also occur on grass roots as saprobes or as very weak, almost harmless parasites (Campbell *et al.*, 1982; Thornton, 1965; Waid, 1957, 1974).

5.5.1 Symptoms

Seed rotting and pre-emergence blight may occur if the soil is heavily infested. Infected, emerged seedlings show necrosis at soil level which is followed by withering and narrowing of the shoot ('wirestem') which collapses and turns pale brown. Older seedlings of some grasses may show sharp eyespot lesions similar in appearance to the disease in cereals.

5.5.2 The fungal pathogens

There has been much taxonomic confusion in the identity of the anamorphs and corresponding teleomorphs of *Rhizoctonia* spp. and many fungi producing dark, sterile mycelium on underground organs of grasses have been referred to the form-genus *Rhizoctonia* which includes basidiomycetes, ascomycetes and fungi imperfecti (Boerema and Verhoeven, 1977; Burpee, 1980a; Campbell *et al.*, 1982; Joyner and Partyka, 1980; Papavisas and Davey, 1962a; Parmeter and Whitney, 1970; Tu and Kimborough, 1975; Warcup and Talbot, 1962).

The taxonomy of the *Rhizoctonia* spp. associated with leaf and crown blights of turf grasses is outlined on pp. 266–274, but the identity of the species associated with seedling and root diseases is much less certain. However, it seems probable that the aggregate species *Thanatephorus cucumeris* (Frank) Donk (anamorph *Rhizoctonia solani* Kuhn) is widely distributed on turf grasses from boreal to tropical regions of America (Andersen, 1982; Burpee, 1980a). Isolates from turf grasses are usually associated with the anastomosis groups AG1 and AG4 (Burpee, 1980a), those in AG1 probably being of most significance as causes of seed rot and seedling diseases (Andersen, 1982). *R. cerealis* van der Hoeven, one of the causes with *R. solani* of sharp eyespot in grasses (Boerema and Verhoeven, 1977), teleomorph *Ceratobasidium cereale* Murray and Burpee, is also the cause of yellow patch of turf grass (Burpee, 1980a) and seedling diseases of grasses (Boerema and Verhoeven, 1977; Pitt, 1964b). Whereas *R. cerealis* usually causes disease at lower temperatures than *T. cucumeris* (Burpee, 1980b), some isolates of the latter have the ability to cause 'cool-season' disease of turf grass (Traquair and Smith, 1981). The distribution of *T. cucumeris* in Europe is uncertain. Although it is not recognized as a common pathogen of turf grasses (Boerema and Verhoeven, 1977; Smith, 1965) it is known to occur on the Poaceae (Richter and Schneider, 1953). *Waitea circinata* Warcup and Talbot (anamorph *R. oryzae* Ryker and Gooch and *R. zeae* Voorhees) which are grass pathogens in tropical climates may also be concerned in seedling injury.

5.5.3 Disease development

Sclerotia of *R. solani* were found mixed with seed of a *Lolium* sp. in Scotland (Dr M. Noble cited by Richardson, 1979) and proven to be transmitted with seed of *Agrostis tenuis* from Oregon although incidence was very low (Leach and Pierpoint, 1958). Although seed-borne infection is possible, the disease is primarily soil-borne. The fungus can make extensive, although sparse, mycelial growth through the soil, resisting mycostasis (Daniels, 1963; Papavisas and Davey, 1962a) and surviving on crop residues, on weed hosts and as sclerotia (Mordue, 1974). Survival in soil may be by mycelium or sclerotia (bulbils) and isolates vary considerably in survival ability (Sherwood, 1970). *T. cucumeris* contains a number of pathogenic strains, some limited to hosts of a single family and others with wider host ranges (Blair, 1942; Flentje, 1957; Richter and Schneider, 1953). Populations are composed of a number of morphologically and pathogenically different clones, aerial and subterranean types classified according to CO_2 sensitivity (Papavisas and Davey, 1962b).

Seeds may be invaded by mycelium from seed-borne propagules or from the soil, the frequency and rapidity of invasion being related to inoculum density. There is a high correlation between the competitive colonization of soil by *R. solani* and percentage pre-emergence killing at different inoculum levels (Martinson, 1963). However, it is also apparent that the size of the propagules and their nutrient reserves, and hence the ability of the fungus to grow through the soil is probably as important in pathogenesis as soil inoculum level (Bowen and Rovira, 1976; Henis and Ben Yephet, 1970). Attacked seed provides nutrients enabling the mycelium to grow to infect adjacent seeds and seedlings. Although *R. solani* is a common and widely distributed soil inhabitant, seedlings probably survive disease by disease escape through local absence of the pathogen. Optimum environmental conditions shorten the time taken by a population of susceptible seedlings to grow into a population of comparatively resistant plants (Garrett, 1970).

The method of entry of *R. solani* into the host is related to the type of surface tissue. Roots may be invaded by direct penetration of epidermal cells and root hairs by hyphae of the fungus, although this seems to be an uncommon method. More frequently infection cushions develop by the aggregation of side branches of mycelium from adjacent hyphae. Plant exudates are concerned in the formation of these structures (Flentje, Dodman and Kerr, 1963; Martinson, 1965). Infection pegs, which develop from the under surface of the cushion, may then penetrate the cuticle of the host cells, otherwise, infection hyphae may push between the epidermal cells of the host. In other instances lobate appressoria with infection pegs are the means of entry. Roots may also be invaded through cracks or wounds. In aerial tissues invasion may be through stomata and hypocotyls may be entered directly through the epidermal cells (Dodman and Flentje, 1970). Cellular collapse of host cells is usually preceded by intercellular penetration, accompanied by browning of the host plant chloroplasts and followed by proliferation of the mycelium which packs the host cells. Sclerotia may or may not be formed according to strain (Chi and Childers, 1964; Christou, 1962; Samuel and Garrett, 1932) from the proliferating hyphae. Sclerotia are of the 'loose' type without a rind, composed of loosely interwoven hyphae (Coley-Smith and Cooke, 1971; Willetts, 1972) and they are mainly confined to the cortical tissues where most fungal growth takes place during pathogenesis. These tissues generally collapse leaving narrowed stelar shanks – hence the 'wirestem' symptoms (Bateman, 1970).

When considering temperature effects on the severity of disease caused by *R. solani* many workers have found that disease incidence did not correlate closely with the growth rate of either host or pathogen at different temperatures (Baker and Martinson, 1970). Leach (1947) found that the incidence of pre-emergence disease caused by *R. solani* was inversely related to the ratio between the coefficient of velocity of seedling emergence (CVE) and the fungus growth rate – as for *Pythium ultimum*. After emergence it seems probable that the CVE is not applicable, perhaps because maturing tissues have increased physiological resistance to infection (Baker and Martinson, 1970).

It is difficult to separate the influence of the two factors of soil moisture and soil aeration in their effect on seedling diseases (Griffin, 1963). *R. solani*, as are most soil fungi, is an aerobe, although populations of the fungus have been shown to be composed of a number of morphologically and pathogenically different clones, aerial and subterranean types which can be classified according to their CO_2 sensitivity (Papavisas and Davey, 1962b). High soil moisture restricts aeration, respiration results in the accumulation of CO_2 and bicarbonates. Linear growth of *R. solani* was retarded by a concentration of 2.5% CO_2 in the atmosphere while 20–25% greatly inhibited it in culture and reduced its competitive saprobic activity (Blair, 1943; Papavisas and Davey, 1962a). Strains of *R. solani* that caused root rots (and presumably also seedling diseases) were shown to grow much better than foliar or stem-attacking isolates in the presence of high concentrations of CO_2 (Durbin, 1959). The depressing effect of organic soil amendments on the growth of *R. solani* was shown to be mainly due to CO_2 accumulation (Blair, 1943). Within the available range, moisture was shown to have no effect on the incidence of brown patch on *Agrostis stolonifera* caused by *R. solani* (Bloom and Couch, 1960), but this bears little relationship to conditions affecting disease on subterranean parts of grasses.

Growth of *R. solani* through soil was reduced by poor aeration (Blair, 1943; Das and Western, 1959), its competitive saprobic activity was reduced by poor aeration when soil moisture was high (Papavisas and Davey, 1961), and its linear growth decreased with greater moisture stress (Sanford, 1938). The optimum relative humidity (RH) for the growth of *R. solani* is 100%. Growth is checked at 99.5% RH. Since the RH of soil at permanent wilting point of 98.8%, *R. solani* is always under stress over the available moisture range for plant growth. Baker and Martinson (1970) concluded that the soil moisture condition will affect the disease mainly through its influence on the plant.

Most strains of *R. solani* show good growth within the pH limits of 5.5–7.5 at which vigorous turf grass growth may be expected (Kernekamp *et al.*, 1952; Sherwood, 1970) although individual isolates vary in pH response. A *Rhizoctonia* sp., perhaps *R. cerealis*, reported to be most severe on light soils with neutral to slightly acid pH showed no differential effects of pH in the range 5.9–8.1 when tested on wheat seedlings in sand culture (Pitt, 1964b). Any effect of pH on pathogenicity is likely to be through the nutritional response of the grass plant (Baker and Martinson, 1970) or through effects on the synergistic (Weindling and Fawcett, 1936) or antagonistic (Baker and Martinson, 1970; Hadar, Chet and Heris, 1979) microflora.

There are few data on the influence of soil nutrient status and soil amendments on the incidence of *R. solani* seedling or root disease of grasses. Potassium, nitrogen and calcium deficiencies and nitrogen excess in field soils tended to increase *R. solani* root disease in cereals and mineral imbalance tended to increase disease severity in sand cultures (Baker and Martinson, 1970). The effects of organic amendments, often added to soil in an attempt to obtain biological control are quite complex. The mineralization and nutritional balance in the soil is altered influencing the response of the host, the saprobic growth of the pathogen and the development of associated microorganisms. Kommedahl and Young (1956) found that the incorporation of maize stalks (low in nitrogen) into soil, with or without light applications of ammonium sulphate in the ratio of 100:1 reduced *R. solani* infection of wheat seedlings. However, Sanford (1952) and Davey and Papavisas (1960) found that nitrogenous substrates or nitrogen added to organic residues reduced the disease. The latter workers correlated this reduction with the lower competitive saprobic activity of *R. solani.*

5.5.4 Control

See p. 49.

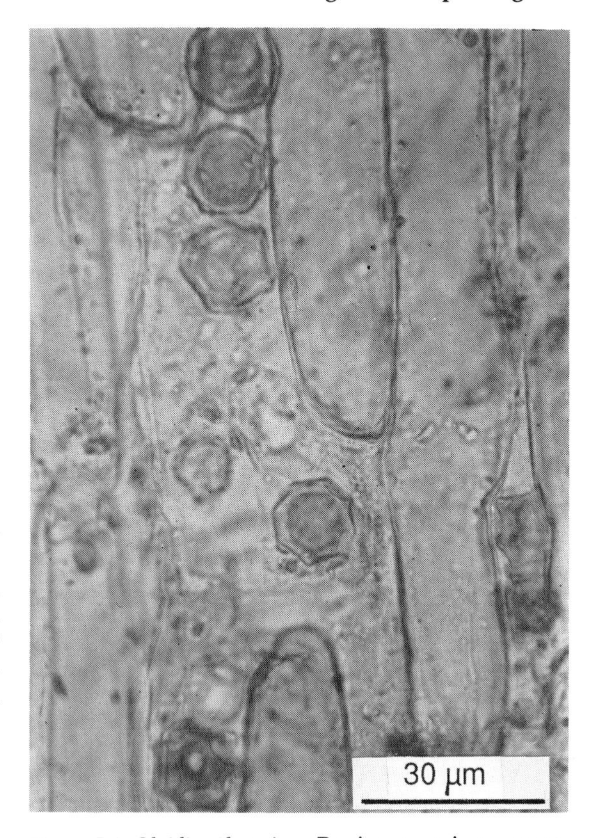

Figure 5.1 *Olpidium brassicae.* Resting spores in grass root.

5.6 MINOR SEEDLING OR ROOT PATHOGENS AND PARASITIC ROOT FUNGI

Isolations made from diseased and apparently healthy grass roots yield species of unspecialized facultative parasitic fungi such as *Colletorichum*, *Curvularia*, *Cylindrocarpon*, *Fusarium*, *Idrella*, *Microdochium*, *Phoma*, *Pythium* and *Rhizoctonia* spp. and many species of saprobes. Some of the facultative parasites are able to persist indefinitely on roots as mycelium or in soil as resting mycelium, spores or sclerotia, while some are short-lived saprobes. The primitive or morphologically simple, obligate, parasitic root-fungi, *Lagena, Ligniera, Olpidium* (Fig. 5.1), *Polymyxa* and *Rhizophidium* spp. may be found in root hairs and cortical cells of the hosts. They survive as spores in the soil and can be detected in the plants only by microscopic examination. Vesicular-arbuscular (VA) mycorrhiza are often also present (see Chapter 12). Like the primitive root fungi they persist mainly by spores, require a host plant for growth and development and are detectable only by microscopy. Some of the primitive root fungi are known vectors of plant viruses, while some of the facultative parasites produce phytotoxins. Members of both groups occur in complexes with bacteria and nematodes, some of which can cause disease or predispose roots to further invasion by the facultative parasites, now termed weak pathogens. Frequently, it is difficult, and sometimes it has proved impossible to determine the role these microorganisms play in pathogenesis even when disease symptoms are present because they are not distinctive.

Salt (1979) has combined the characteristics of these two groups of 'unspecialized facultative parasitic fungi' and 'primitive obligate root fungi' under the heading of 'minor pathogens' defined as 'saprophytes or parasites damaging only meristematic and cortical cells and surviving in the soil as saprophytes, as resting spores or sclerotia'. Pathogens of minor importance are excluded since they have a restricted host or environmental range, but includes those fungi, that under conditions unfavourable to the host or if concerned in complexes with other organisms, may cause disease.

Table 5.2 Minor seedling or root pathogens and parasitic root fungi

Species	Hosts/occurrence	Diagnosis/ Life history	Distinguishing characters	References
Cladochytrium (*Physoderma*) spp.	*Cladochytrium* (*Physoderma*) spp. have been reported causing seedling/root diseases and/or leaf streaks on grasses in Europe and N. America (Lundin, 1975; Sampson and Western, 1954; Smith, 1965; Sprague, 1950).			
including: *C. graminis* Büsgen (1887)	*C. graminis* was described from Germany causing infection and galling of grass roots (probably *Anthoxanthum* sp.). Resting spores with rough, granulated exospores c. 26–36 μm, rhizoids and septate, turbinoid cells were present (Cook, 1934; Sparrow, 1960). Massee (1913) probably described the same species from European and British material with infected leaves and/or roots and seed of several grass species including *Festuca ovina* and *Poa annua*. Resting spores were smooth, 29 × 34 μm diam. av.			
Cladochytrium (*Physoderma*) *caespitis* Griffon and Maublanc (1910)	*C. caespitis* occurred on yellowing and withering seedlings of *Lolium perenne* in France. Infection proceeded from the sheaths into adjacent tissues and into the roots. Resting spores were smooth 12–30 × 15–45 μm. Cook (1934, 1935) studied this fungus from *Agrostis* sp. at the Sports Turf Research Institute in England. It had resting spores similar to those from France, 15 × 31 μm diameter av.			
Physoderma agrostidis Lag. (Minden, 1915) See also *Physoderma* Leaf spot–p. 329	*P. agrostidis* from *Agrostis alba* in Germany had resting sporangia c 19 × 34 (av. 26) μm. The taxonomy of these fungi is confused. Sparrow (1960) retains *Cladochytrium* as a separate genus from *Physoderma* but Karling (1968) does not.			
Cylindrocarpon destructans (Zinssm.) Scholten Syn. *C. radicicola* Wollenw. Teleomorph *Nectria radicicola* Gerlach and L. Nilsson	A widely distributed, unspecialized facultative parasite found on root surfaces and in cortical cells of many plant species including turf grasses, weakly pathogenic, but involved in damping-off complexes and seed rot, some isolates low-temperature tolerant.	*Conidia* free, in false heads in pionnotes, cream in the mass, aseptate to 3-septate, cylindric, straight or slightly curved 7–38 × 2–7.5 μm. *Conidiophores* slightly penicillately or verticillately branched. *Chlamydospores* intercalary, numerous, catenulate, or in knots, globose, brown 10–16 μm. diam.	Conidia *Fusarium*-like but stouter, straight and without foot cell. Aerial mycelium in culture yellowish-white, floccose.	Gilman (1959) Thornton (1960) Salt (1979)
Idrella bolleyi (Sprague) von Arx Syns. *Microdochium bolleyi* (Sprague) de Hoog & Hermanides-Nijkof *Aureobasidium bolleyi* (Sprague) von Arx *Gloeosporium bolleyi* (Sprague)	Soil-or litter-borne with some isolates causing seed rot and root necrosis on a wide range of grasses and cereals.	*Conidia* aseptate, falcate to bean-shaped, 6–10 × 1.5–2.5 μm borne apically on ampulliform conidiogenous cells or on lateral pegs of hyaline hyphal cells without denticles. *Chlamydospores* dark brown in strands or aggregates (microsclerotial in host roots). *Colonies* in culture-slimy then pink or carrot-coloured, turning black.	Small falcate conidia mostly formed on hyphal pegs. Strands or aggregates of chlamydospores.	Von Arx (1981 a, b) Domsch and Gams (1972) de Hoog and Hermanides-Nijkof (1977) Murray and Gadd (1981) Salt (1977, 1979) Sprague (1950)

Table 5.2 Continued

Species	Hosts/occurrence	Diagnosis/ Life history	Distinguishing characters	References
*Lagena radicicola** Vanterpool & Ledingham Syn. *Lagenocystis radicicola* (Vanterpool and Ledingham) Copeland	Wide host range on wild grasses, cereals and dicotyledonous plants in N. Australia, Britain and Italy. Associated with stunting of lateral roots in Gramineae, but uncertain as cause of much root injury since other root fungi (e.g. *Pythium* spp.) may also be present. Thallus occupies cortical epidermal and root hair cells.	*Thallus* sac- or tube-shaped, one per cell 14–35 μm, sometimes coiled in host cell. Conjugation of thalli may occur. Neck of thallus forms exit tube and vesicle in which zoospores differentiate. *Zoospores* granular, bean-shaped, biflagellate 7–11 μm or 0.6 μm when encysted on host surface. Grows into host cell by a tube which enlarges to become the thallus.	Unlike *Polymyxa* and *Ligniera*, *Lagena* has bean-shaped zoospores similar to *Pythium* spp. Unlike *Pythium* spp. has thallus (zoosporangium) restricted to one cell. Not culturable axenically.	Barr and Slykhuis (1969) Macfarlane (1970) Salt (1977, 1979) Truscott (1933) Vanterpool and Ledingham (1930)

*The original description of *Lagena* mistakenly included the oogonial state of a *Pythium* and therefore the morphological and taxonomic knowledge of this fungus is deficient (Dr D.J.S. Barr, personal communication).

Species	Hosts/occurrence	Diagnosis/ Life history	Distinguishing characters	References
Ligniera pilorum. Fron and Gaillat (Plasmodiophorales) (Syn. *L. junci* (Schwartz) Maire and Tison Other *Ligniera* spp. are reported on Poaceae	On roots of many species of grasses including *Lolium perenne* and *Poa annua*. Reported to kill or severely stunt *L. perenne* seedlings in New Zealand (Latch, 1966). May or may not cause clubbing or hypertrophy of root hairs.	*Resting spores* thick-walled, spherical, 3.5–5 μm aggregated into loose cytosori, mainly in root hairs. *Zoosporangia* thin-walled, segmented, remaining so until maturity 11– × 7–18 μm. *Zoospores* few per sporangium, pyriform 3.5–7 × 3.5 μm biflagellate, liberated by sporangial rupture.	*Resting spores* in globose, to elliptical or irregular cytosori mainly in root hairs. Zoosporangia remaining segmented until maturity. No exit protuberances. Not culturable axenically.	Barr (1979) Cook (1926) Fron and Gaillat (1925) Karling (1968) Latch (1966) Maire and Tison (1911) Michail and Carr (1966a) Nicholson (1959) Salt (1977, 1979)
Olpidium agrostidis Sampson	On *Agrostis stolonifera*, Yorkshire, England, causing apical swellings of root hairs.	*Resting spores* elliptical to ovoid, double-walled, exospore smooth, 20–30 × 9–17 μm. *Zoosporangia* oval to elongate, solitary or aggregated in root hairs 30–68 × 10–23 μm, one or more exit tubes, short. *Zoospores* spherical, 2.2 μm diam., long flagellum.	*Resting spores* smooth, not stellate as for *O. brassicae*.	Sampson (1932)
Olpidium brassicae (Wor.) Dang. Syns. *Asterocystis radicis* de Wild. and *Olpidium radicicolum*	Wide host range on dicotyledonous and monocotyledonous hosts including careals and turf grass species in *Agrostis*, *Festuca*, *Lolium* and *Poa*, may	*Resting spores* spherical or oval, exospores with coarse wrinkles, endospore smooth 8–30μm diam., in cortex and root hairs. *Zoosporangia* one or	Sporangia with exit tubes and stellate resting spores. Cannot be grown in axenic culture.	Barr and Slykhuis (1969) Britton and Rodgers (1963) Endo (1961) Fron and Gaillat (1925) Guyot (1927) Salt (1977, 1979)

Table 5.2 Continued

Species	Hosts/occurrence	Diagnosis/ Life history	Distinguishing characters	References
de Wild.	cause hypertrophy of root hairs, stunting and stoppage of root growth. Associated with but not certainly the cause of chlorosis and necrosis of seedlings and mature grasses. Usually causes little damage. Vector of some viruses in dicotyledons.	more per cell in cortex and root hairs, thin-walled, usually spherical 12–20 μm to elongate 25–220 × 20–45 μm with one to several discharge tubes. *Zoospores* uniflagellate, spherical, 3 μm diam. invade root hairs and cortical cells.		Sampson (1932) Schwartz and Cook (1928) Sparrow (1960) Sprague (1950) Vanterpool (1930)
Olpidium radicale Schwartz and Cook	On *Poa annua*, other unspecified grasses, and dicotyledonous plants in wet soil, Kent, England. Other grasses susceptible.	*Resting sporangia* variable in shape, one or more per cell, 100–160 μm diam. *Zoosporangia* in cortical cells, globose to elongate 60–200 μm, thin-walled with one or several exit tubes. *Zoospores* discharged in host cell, may fuse and form resting sporangium or form zoosporangium.		Lange and Insunza (1977) Schwartz and Cook (1928)
Phialophora graminicola (Deacon) Walker Syns. *P. radicicola* Cain var. *Graminicola* Deacon *radicicola* sensu Scott (1970) (slightly 'obed hyphopodia) (fig. 159)	A common, avirulent parasite which colonizes grass stem bases, rhizomes and roots including turf grass species, in Britain and Australia. Potentially important in controlling take-all patch caused by *Gaeumannomyces graminis*.	*Cultures* slow growing on PDA (4–6 mm/day) *Phialides* produced singly, laterally on hyphae or in loose clusters, hyaline to pale brown, 5–20 × 2–4 μm. *Phialospores* hyaline to faint yellow, rounded at apex, more pointed at base, straight, slightly, rarely strongly, curved, 5–11 (14) × 1.5–2.5 pm, germinable. *Mycelium* pale brown 3–5 μm wide, in culture darkening to grey or dark grey, marginal hyphae sometimes recurved. *Hyphopodia* (a) abundant, brown, sub-globose to oval 6–15 μm in diam. or 9–18 × 6–15 μm with slightly lobed margins. (b) hyaline to pale	Slightly lobed, hyphopodia in culture on wheat coleoptiles. Phialospores not often strongly curved. Slow growth in culture at optimum temperature. Root cells may be filled with brown fungal cells resembling microsclerotia of *Idrella bolleyi*.	Deacon (1973, 1974) Scott (1970) Walker (1980, 1981)

Table 5.2 Continued

Species	Hosts/occurrence	Diagnosis/ Life history	Distinguishing characters	References
		brown, intercalary or terminal sub-globose 6–10 μm in diam.		
Phialophora sp. with lobed hyphopodia Syn. *P. radicicola* var. *radicicola* Sensu Deacon (1970).	Occurs commonly on grasses in Europe and Australia. Low pathogenicity on grass roots. May give control of take-all patch.	Not distinguishable morphologically or culturally from *G. graminis* var. *graminis*.	Lobed hyphopodia	Deacon (1974) Walker (1980, 1981) Wong and Siviour (1979)
Physoderma	see *Cladochytrium*			
Polymyxa graminis Ledingham	Roots of cereals and grasses, including *Agrostis* and *Cynodon* spp. Some specialization–Canadian isolates did not infect *Agrostis*, *Festuca*, *Lolium*, *Phleum* and *Poa* spp. Causes little hypertrophy and only slight root damage, but is a vector of wheat mosaic and other cereal viruses.	*Resting spores* hexagonal to polyhedral, 4–7 μm, smooth, light to dark brown in groups (cytosori) of a few to 200 + in root cells *Zoosporangia* large, up to 200 × 40 μm, segmented irregular with persistent smooth walls and discharge processes or tubes. *Zoospores* numerous, spherical, biflagellate 2.5–5.6 μm diam., discharged without vesicle formation.	Cytosori mainly in epidermal root cells. Differs from *Ligniera* in having larger, elongate zoosporangia with segments partially disappearing at maturity and having exit tubes of protuberances. Not culturable axenically.	Barr (1979) Britton and Rogers (1963) Dale and Murdoch (1969) Karling (1968) Ledingham (1939) Salt (1979) Slykhuis and Barr (1978)
Rhizophidium graminis Led.	Wide host range on roots of cereals, cultivated and wild grasses and dicotyledonous plants including turf grasses *Festuca rubra*, *Holcus lanatus* and *Lolium perenne* in N. America, Europe. Host-specific strains possibly occur. Causes little root damage, but possibly vector of viruses.	*Resting spores* globose, with hyaline to light-brown, two-layered wall, at first smooth, becoming rough when mature, some external to host root hair or root epidermal cells. *Zoosporangia* globose to ovate, external to host cells 6–75 μm with rhizoids within host cell. Dehises explosively, leaving cupulate remains of sporangium wall. *Zoospores* pyriform, 2 × 3 – 1.5 – 4 μm. Encyst on epidermal cells and root hairs which are then penetrate by a germ tube and branched rhizoidal system.	Spherical to ovate external sporangium dehiscing to leave cup-shaped walls. Rhizoids with in host. Cannot be grown in axenic culture.	Barr and Slykhuis (1969) Ledingham (1936) Macfarlane (1970) Salt (1979) Sparrow (1960) Sprague (1950)

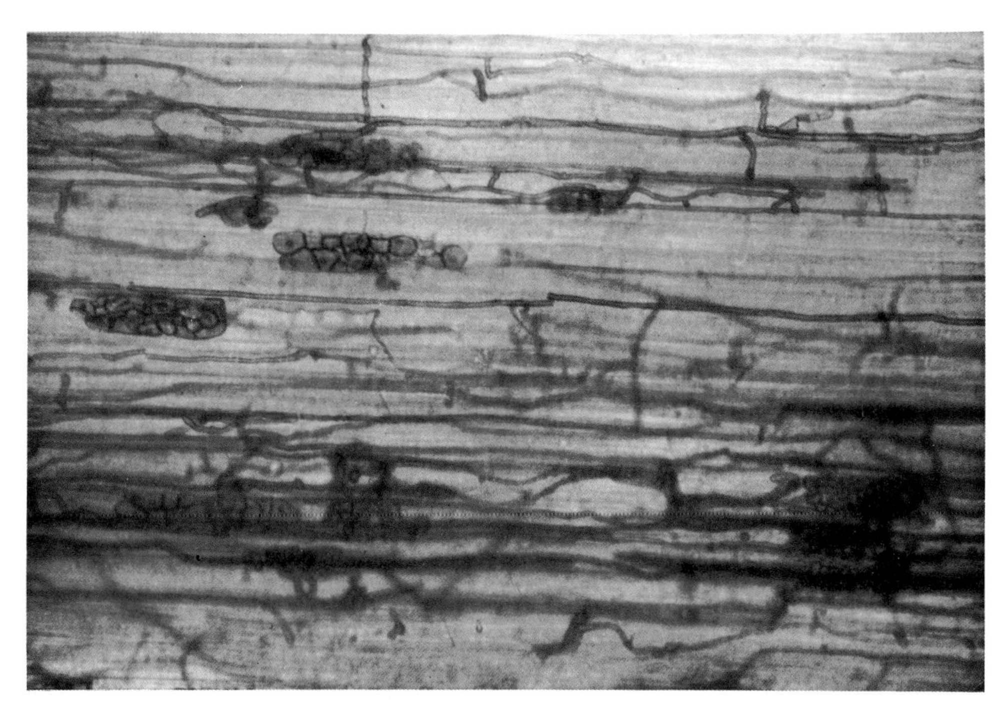

Figure 5.2 *Phialophora graminicola – mycelium and 'chlamydospores.'*

Årsvoll (1975) noted that at low temperatures, some low-temperature-tolerant fungi regarded as saprobes were capable of damaging grass seedlings (see p. 92). Rawlinson and Colhoun (1970) found that a mercury seed dressing fungicide protected oat seedlings from lesioning by soil-borne fungi, normally regarded as saprobes, but only under natural or simulated winter conditions. Browning root rot of cereals and grasses caused by *Pythium arrhenomanes*, in association with *Lagena radicicola*, now regarded as of minor importance (Vanterpool, 1952), is favoured by inadequate phosphate and excessive nitrogen, summer fallowing, cold, wet springs, followed by hot, dry weather.

The adaptability of a cultivar to latitude, climatic regime, soil type and fertility may influence its susceptibility to a minor pathogen as Bruehl (1955) showed in the case of barley cultivars and *Pythium* root rot. Early maturing cultivars from Egypt, Afghanistan and India, usually autumn sown, suffered more severely from a root rot caused by a complex with *P. arrhenomanes* as the main species than did cultivars adapted to the long photoperiod in late spring in South Dakota.

Cold, wet soils, nitrogen deficiency and crowding of plants predisposed grasses to root decay caused by *Idrella* (*Aureobasidium*) *bolleyi* (Sprague, 1948).

Although *Cochliobolus sativus* may act as a major pathogen of cereals and grasses, causing seed rot, seedling blight, crown and root rots it has frequently been isolated from apparently healthy and diseased wheat roots (Broadfoot, 1934; Simmonds and Ledingham, 1937).

Rhizoctonia solani Kuhn, *Fusarium culmorum* (W.G. Smith) Sacc. and *F. avenaceum* (Fr.) Sacc. which may, in certain circumstances behave as major pathogens of the Gramineae are also known as weak parasites or saprobes. However, combinations of some *Fusarium* spp. which are saprobic or weakly pathogenic alone (Smiley, 1980; Zogg, 1950) may act synergistically to cause more severe disease (Zogg, 1950).

The primitive root fungi are not considered very damaging to roots and often cause no macroscopic symptoms, but by damaging root hair cells they may interfere with the uptake of water which may cause stress in drought periods. Heavy infection of cabbage root cortical cells with *Olpidium brassicae* was shown to diminish the translocation of iron to the shoots, caused chlorosis and reduced yield (MacFarlane, 1974).

Several factors influence the colonization of seed, once sown in the ground, by minor pathogens. However, some of these which become established on the seed at sowing persist until the mature plant senesces (Salt, 1979). Delayed germination permits invasion by many species of fungi and bacteria which would not normally cause seed rot and damping-off. Exudates from seeds are the major source of nutrients for these microorganisms and the amount of exudation is increased by seed coat damage, high soil moisture in the early stages of germination and cold soil in the seeds of various plant species (Lynch, 1978).

The characteristics of some minor seedling or root pathogens and parasitic root fungi are shown in Table 5.2 and other fungi which may enter into complexes with these, e.g. *Cochliobolus sativus*, *Gaeumannomyces graminis, Microdochium nivale, Fusarium culmorum, Rhizoctonia solani, Pythium* spp. (Table 5.1) and elsewhere under appropriate headings.

5.7 CONTROL OF SEED ROTS AND SEEDLING DISEASES

1. Use seed of high germination capacity and energy without mechanical injuries. Grass seed of poor germination capacity and energy with mechanical injuries is at greater risk from seed rot and damping-off diseases than fresh, vigorous, intact seed because it is less able to resist and recover from attacks by fungi and bacteria. If the protective tissues of pericarp and testa are undamaged diffusion of organic nutrients is slowed and these are less available for the growth of rot organisms in the spermosphere and access to food reserves in the endosperm is denied. Slow germination maintains the seedling in a susceptible state for a longer period of exposure to pathogens.

2. Sow the seed at the correct seed rate and seeding depth. If sown too thickly disease spread is more rapid, especially after emergence. When sown too deeply, nutrient reserves in the seed become exhausted before the seedling is independent of them resulting in a weak, spindly, disease-susceptible plant.

3. Ensure a well-prepared, adequately fertilized seed bed to assist rapid, even germination and establishment. A moisture-retentive, firm, yet freely draining seed bed without low spots provides the optimum physical basis for even, rapid germination.

Adequate, balanced supplies of phosphorus and nitrogen, while not significant in early stages of germination are needed for rapid seedling and turf establishment and are an insurance against severe attacks of root disease due to *Pythium* spp. Excessive, unbalanced nitrogen may lead to blight of seedling turf resulting from attacks by complexes of *Microdochium nivale* and *Fusarium culmorum* in Britain (Smith, 1965) and by *Cochliobolus sativus* and *Curvularia* spp. under warmer temperatures in N. America (Madsen and Hodges, 1980).

4. Sow at times when conditions are unfavourable to the pathogens and most favourable to rapid seedling establishment, if possible. Temperature influences the incidence of seed rots and pre-emergence damping-off. In Britain, western Canada, mid-western USA (Smith, 1954; Smith, unpublished; Sprague, 1946) and New Zealand (Falloon, 1980b) grasses often establish poorly when sown in cold, wet soil in spring. Greater success usually attends autumn seeding when soil temperatures are highest. In seedling disease caused by *F. culmorum* and *M. nivale* it is likely that if there is abundant inoculum present in the soil this will favour infection even if temperature and moisture are suboptimal for disease development. Wet soils do not favour the incidence of seedling blight by *F. culmorum* even around its optimum temperature of c. 25 °C, probably because of the antagonistic fungi and bacteria which also find that a favourable temperature for development (Slykhuis, 1947). It is difficult to find conditions unfavourable for the activities of *Pythium* spp. since there are cool and warm-temperature species. Similar problems occur with rhizoctonia seed rot and damping-off caused by *R. solani* and *R. cerealis*. However, warm-season *R. solani* is not likely to cause problems in Britain, western Europe and areas with similar climates if sowing is done in warmed-up soil, while in hotter regions *R. solani* is less likely to cause seedling injury when sowing takes place under cool conditions. In situations where severe damping-off occurs when conditions for germination and establishment are otherwise satisfactory it is necessary to resort to fungicidal seed dressings or soil fumigation.

5. Control soil moisture, as far as possible, at a level which encourages rapid seed germination and growth yet does not favour pathogen development and spread. Seedling disease caused by *Pythium* spp. and other zoosporic fungi, is favoured by a saturated or over-saturated soil. Water logging is a greater problem on heavy than on light soils. To encourage germination, moisture is required at the seeding depth and at this stage a surface mulch may be advantageous. After brairding, surface moisture should be kept at a minimum and deeper, less frequent watering is indicated. Usually soils with a moisture content at field capacity and below favour the development of seedling disease caused by *R. solani* and *F. culmorum* (Colhoun and Park, 1964; Couch, 1973).

6. Effective use of fungicidal seed dressings (Table 5.3). Some workers have found marginally beneficial or no improvement in seedling stands of forage or turf grass species from seed treated with fungicidal seed dressings in controlled environments and/or field

Table 5.3 Seed dressing fungicides effective against damping-off in turf grasses

Pathogen	Grass species	Fungicide	Reference
Fusarium culmorum	*Festuca rubra* spp. *commutata*	thiram tetrachloroparabenzoquinone organo-mercurial	Smith (1957b)
F. culmorum	*F. rubra* spp. *commutata* *Cynosurus cristatus* *Phleum bertolonii*	thiram	Smith (1957b)
F. culmorum	*F. rubra* ssp. *commutata* *Cynosurus cristatus* *Phleum bertolonii*	thiram captan	Jackson (1958)
F. culmorum	*Lolium perenne*	benomyl/captan	Holmes (1977)
F. culmorum	*Lolium perenne*	carbendazim drazoxolon calcium peroxide	Gussin and Lynch (1983b)
Not specified	*Phleum pratense* *F. rubra* ssp. *rubra* and *commutata* *Agrostis tenuis* *Lolium perenne*	captan thiram	Falloon (1980a)
Not specified	*Lolium perenne*	captan thiram ethazole	Falloon (1980b)
Pythium aphanidermatum	*Lolium multiflorum*	ETMT	Freeman (1972)
Pythium spp. *Cochliobolus sativus*	*Agropyron cristatum*	organomercurial	Sprague (1946)

studies (Lewis and Clements, 1982); Mead, 1955; Michail and Carr, 1966a; Tribe and Herriott, 1969; Tyler, Murphy and MacDonald 1956). Fulkerson (1953) obtained erratic results with grasses. Other studies indicated considerable advantage in using fungicidal seed dressings on grasses (Andrews, 1953; Ehrenreich, 1958; Falloon, 1980a,b; Freeman, 1972; Holmes, 1977; Jackson, 1958; Kreitlow, Garber and Robinson, 1950; Smith, 1957a,b; Sprague, 1946).

Seed dressings based on contact fungicides applied as liquids, slurries or dry powders effectively control many of the seed-rotting fungi and pathogens carried on the surface of the seed or in the hulls, but not those which are within the seed, such as some of the loose smuts. They will also kill or inhibit these organisms in the soil in a limited zone round the seed. Systemic fungicides will also give this protection and control some of the pathogens within the seed. Usually seed-dressing fungicides show little control of post-emergence damping-off or seedling blight (Freeman, 1972; Mead, 1955; Smith, 1965; Tyler, Murphy and MacDonald, 1956). In some instances, however, increases in seedling and young plant vigour have been noted (Falloon, 1983; Rawlinson and Colhoun, 1970) following the use of captan or organo-mercury seed dressings.

Post-emergence damping-off caused by *Pythium* spp., *Cladochytrium* sp. and *F. culmorum* has been associated with nitrogen and phosphate deficiency in the soil (Smith, 1954, 1965). In many cases some recovery from infection followed the application of mono-ammonium phosphate at 85–170 g/100 m^2 in a suitable bulking carrier. Howard (1956) reported that malachite green will kill the mycelium of *Pythium* spp. when the fungus is active. Cheshunt compound at 37.5 g in 5 litres of water applied as a soil drench to saturate the surface soil has also proved effective against post-emergence damping-off caused by *F. culmorum* (Smith, 1965). Cases of seedling blight should be treated with the fungicide spray known to be effective (and of low phytotoxicity) to the dominant pathogen in mature turf.

5.8 REFERENCES

Andersen, N.A. (1982) The genetics and pathology of *Rhizoctonia solani*. *Ann. Rev. Phytopathol.*, **20**, 329–47.

Andrews, E.A. (1943) Seedling blight and root rot of grasses in Minnesota. *Phytopathology*, **33,** 234–9.
Andrews, E.A. (1953) Seedling blights and root rot of forage grasses. *Diss. Abs.*, **13,** 962.
Anon. (1960) Rules for testing seeds. *Proc. Off. Seed Anal.*, **49**(2), 1–71.
Årsvoll, K. (1975) Fungi causing winter damage in Norway. *Meld. Norg. LandbrHøgsk..*, **54**(9), 49pp.
Arx, J. von. (1981a) Notes on *Microdochium* and *Idrella. Sydowia*, **34,** 30–8.
Arx, J. von. (1981b) *The Genera of Fungi Sporulating in Pure Culture*, 3rd ed, J. Cramer, Vaduz, 424pp.
Baker, R. and Martinson, C.A. (1970) Epidemiology of diseases caused by *Rhizoctonia solani.* In *Rhizoctonia solani, Biology and Pathology* (ed. J.R. Parmeter, Jr) Univ. Calif. Press, Berkeley, pp. 172–88.
Barr, D.J.S. (1979) Morphology and host range of *Polymyxa graminis, Polymyxa betae* and *Ligniera pilorum* from Ontario and some other areas. *Can. J. Plant Pathol.*, **1,** 85–94.
Barr, D.J.S. and Slykhuis, J.T. (1969) Zoosporic fungi associated with wheat spindle streak mosaic in Ontario. *Can. Plant Dis. Surv.*, **49**(4), 112–13.
Barton, R. (1957) Germination of oospores of *Pythium mamillatum* in response to exudates from living seedlings. *Nature*, **180,** 613–14.
Bateman, D.F. (1970) Pathogenesis and disease. In *Rhizoctonia solani, Biology and Pathology* (ed. J.R. Parmeter, Jr) Univ. Calif. Press, Berkeley, pp. 161–71.
Bean, G.A. (1966) Observations and studies on fusarium blight disease of turf grass. *Phytopathology*, **56,** 583 (Abs.)
Bean, G.A. (1969) The role of moisture and crop debris in the development of fusarium blight of Kentucky bluegrass. *Phytopathology*, **59,** 479–81.
Beard, J.B. (1973) *Turfgrass; Science and Culture.* Prentice-Hall, New Jersey, 658pp.
Bell, A.A. (1967) Fungi associated with root and crown rot of *Zoysia japonica. Plant Dis. Rept.*, **51**(1), 11–14.
Bennett, F.T. (1928) On two species of *Fusarium, F. culmorum* (W.G. Sm.) Sacc. and *F. avenaceum* (Fries) Sacc. as parasites of cereals. *Ann. Appl. Biol.*, **15,** 213–44.
Bennett, F.T. (1933) *Fusarium* species on British cereals. *Fusarium nivale* (Fr.) Ces. = *Calonectria graminicola* (Berk. and Br.). *Wr. Ann. Appl. Biol.*, **20,** 272–90.
Bennett, F.T. (1935) *Fusarium* species on British cereals. Ann. Appl. Biol., **22,** 479–507.
Berner, K.E. and Chapman E.S. (1977) The cellulolytic activity of six Oomycetes. *Mycologia*, **69,** 1232–6.
Blair, I.D. (1942) Studies on the growth in soil and the parasitic action of certain *Rhizoctonia solani* isolates from wheat. *Can. J. Res. C.*, **20,** 174–85.
Blair, I.D. (1943) Behaviour of the fungus *Rhizoctonia solani* Kühn in the soil. *Ann. Appl. Biol.*, **30,** 118–27.
Bloom, J.R. and Couch, H.B. (1960) Influence of environment on diseases of turf grasses. 1. Effect of nutrition, pH and soil moisture. *Phytopathology*, **50,** 523–5.
Boerema, G.H. and Verhoeven, A.A. (1977) Checklist for scientific names of common parasitic fungi. Series 2b. Fungi on field crops, cereals and grasses. *Neth. J. Plant Pathol.*, **83,** 165–204.
Booth, C. (1971) *The genus Fusarium.* Commonw. Mycol. Inst., Kew, 237pp.
Bowen, G.D. and Rovira, A.D. (1976) Microbial colonization of plant roots. *Ann. Rev. Phytopathol.*, **14,** 121–44.
Brien, R.M. (1935) Three fungi causing 'brown patch' of lawns in New Zealand. *NZ J. Agric.*, **51,** 157–9.
Britton, M.P. (1969) Turf grass diseases. In *Turfgrass Science* (eds A.A. Hanson and F.V. Juska), Am. Soc. Agron. 14, pp. 288–335.
Britton, M.P. and Rodgers, D.P. (1963) *Olpidium brassicae* and *Polymyxa graminis* in roots of creeping bent in golf putting greens. *Mycologia*, **55,** 758–63.
Broadfoot, W.C. (1934) Studies on foot and root rot of wheat. 4. Effect of crop rotation and cultural practice on the relative prevalence of *Helminthosporium sativum* and *Fusarium* species indicated by isolations from wheat plants. *Can. J. Res.*, **10,** 115–24.
Bruehl, G.W. (1955) Barley adaptations to *Pythium* root rot. *Phytopathology*, **45,** 97–103.
Buchholz, W.F. (1942) Gross pathogenic effect of *Pythium graminicola, P. debaryanum* and *Helminthosporium* on seedlings of crested wheatgrass. *Phytopathology*, **32,** 2 (Abs.).
Buchholz, W.F. (1949) A comparison of gross pathogenic effect of *Pythium graminicola, Phythium debaryanum* and *Helminthosporium sativum* on seedlings of crested wheatgrass. *Phytopathology*, **39,** 102–16.
Burpee, L.L. (1980a) Identification of *Rhizoctonia* species associated with turf grass. In *Advances in Turfgrass Pathology* (eds B.G. Joyner and P.O. Larsen), Harcourt, Brace, Jovanovich, Duluth, Minn., pp. 25–8.
Burpee, L.L. (1980b) *Rhizoctonia cerealis* causes yellow patch of turf grass. *Plant Dis.*, **64,** 114–16.
Burr, T. and Stanghellini, M.E. (1973) Propagule nature and density of *Pythium aphanidermatum* in field soil. *Phytopathology*, **63,** 1499–501.
Campbell, W.A. (1949) A method of isolating *Phytophthora cinnamoni* from soils. *Plant Dis. Rep.*, **33,** 134–5.
Campbell, R., Newman, E.I., Lawley, R.A. and Christie, P. (1982) Relationship between a *Rhizoctonia* species and grassland plants. *Trans. Br. Mycol. Soc.*, **79**(1), 123–7.
Chang-Ho, Y. (1970) The effect of pea root exudate on the germination of *Pythium aphanidermatum* zoospore cysts. *Can. J. Bot.*, **48,** 1501–14.
Chi, C.C. and Childers, W.R. (1964) Penetration and infection in alfalfa and red clover by *Pellicularia filamentosa. Phytopathology*, **54,** 750–4.
Christensen, M.J. (1979) *Rhizoctonia* species associated with diseased turf grasses in New Zealand. *NZ J. Agric. Res.*, **22,** 627–9.
Christou, T. (1962) Penetration and host – parasite relations of *Rhizoctonia solani* in the bean plant. *Phytopathology*, **52,** 381–9.
Coker, W.C. and Patterson, P.M. (1927) A new species of *Pythium. J. Eliz. Mitch. Sci. Soc.*, **42,** 247–50.
Cole, H., Braverman, S.W. and Duich, D. (1968) Fusaria and other fungi from seeds and seedlings of Merion and other turf-type bluegrass. *Phytopathology*, **58,** 1415–19.
Coley-Smith, J.R. and Cooke, R.C. (1971) Survival and

germination of fungal sclerotia. *Ann. Rev. Phytopathol.*, **9,** 65–92.
Colhoun, J. and Park, D. (1964) Fusarium diseases of cereals. 1. Infection of wheat plants with particular reference to the effects of soil moisture and temperature on seedling infection. *Trans. Br. Mycol. Soc.*, **47,** 559–72.
Colhoun, J., Taylor, G.S. and Tomlinson, R. (1968) Fusarium diseases of cereals. 2. infection of seedlings by *F. culmorum* and *F. avenaceum* in relation to environmental factors. *Trans. Br. Mycol. Soc.*, **51** (3&4), 379–404.
Conners, I.L. (1967) *An Annotated Index of Plant Diseases in Canada*. Queen's Printer, Ottawa, 381pp.
Cook, R.J. and Bruehl, G.W. (1968) Relative significance of parasitism versus saprophytism in colonization of wheat straw by *Fusarium roseum* culmorum in the field. *Phytopathology*, **58,** 306–401.
Cook, W.R.I. (1926) The genus *Ligniera* Maire and Tison. *Trans. Br. Mycol. Soc.*, **11,** 196–213.
Cook, W.R.I. (1934) Some observations on the genus *Cladochytrium* with special references to *C. caespitis* Griffon and Maublanc. *Ann. Bot. Lond.*, **48,** 177–85.
Cook, W.R.I. (1935) An account of diseases of lawns caused by a parasitic fungus. *J. Bd. Greenkpg. Res.*, **5,** (16), 18–21.
Couch, H.B. (1973) *Diseases of Turfgrasses*. Krieger, Huntington, NY, 348pp.
Dahl, A.S. (1933) Effect of temperature on brown patch of turf. *Phytopathology*, **23,** 8 (Abs.).
Dale, J.L. and Murdoch, C.L. (1969) *Polymyxa* infection of Bermudagrass. *Plant Dis. Rept.*, **53**(2), 130–1.
Daniels, J. (1963) Saprophytic and parasitic activities of some isolates of *Corticium solani*. *Trans. Br. Mycol. Soc.*, **46,** 482–502.
Das, A.C. and Western, J.H. (1959) The effect of inorganic manures, moisture and inoculum on the incidence of root diseases caused by *Rhizoctonia solani* Kühn in cultivated soil. *Ann. Appl. Biol.*, **47,** 37–48.
Davey, C.B. and Papavisas, G.C. (1960) Effect of dry mature plant materials and nitrogen on *Rhizoctonia solani* in soil. *Phytopathology*, **50,** 522–5.
Deacon, J.W. (1973) Factors affecting occurrence of the Ophiobolus patch disease of turf and its control by *Phialophora radicicola*. *Plant Pathol.*, **22,** 149–55.
Deacon, J.W. (1974) Further studies on *Phialophora radicicola* and *Gaeumannomyces graminis* on roots and stem bases of grasses and cereals. *Trans. Br. Mycol. Soc.*, **63**(2), 307–27.
Dodman, R.L. and Flentje, N.T. (1970) The mechanism and physiology of plant penetration by *Rhizoctonia solani*. In *Rhizoctonia solani, Biology and Pathology*. (ed. J.R. Parmeter, Jr), Univ. Calif. Press, Berkely, pp. 149–60.
Domsch, K.H. and Gams, W. (1972) *Fungi in Agricultural Soils*. Longmans, London.
Drechsler, C. (1928) *Pythium arrhenomanes* n.sp. a parasite causing maize root rot. *Phytopathology*, **18,** 873–5.
Drechsler, C. (1936) *Pythium graminicolum* and *P. arrhenomanes*. *Phytopathology*, **26,** 676–84.
Drechsler, C. (1940) Three species of *Pythium* associated with root rots. *Phytopathology*, **30,** 189–213.
Drechsler, C. (1943) Two species of *Pythium* occurring in southern states. *Phytopathology*, **33,** 261–99.
Durbin, R.D. (1959) Factors affecting the vertical distribution of *Rhizoctonia solani* with reference to CO_2 concentration. *Am. J. Bot.*, **46**, 22–5.
Eherenreich, J.H. (1958) Effects of certain fungicides on seed germination and seed establishment of range and forage plants. *J. Range Mgmt.*, **11,** 22–7.
Endo, R.M. (1961) Turf diseases in southern California. *Plant Dis. Rept.*, **45,** 869–73.
Endo, R.M. and Colt, W.M. (1974) Anatomy, cytology and physiology of infection by *Pythium*. In *Proc. Am. Phytopathol. Soc.*, Minnesota, Vol. 1, pp. 215–23.
Falloon, R.E. (1980a) Effects of fungicide seed treatments on seedling emergence of pasture and turf grasses. *NZ J. Exp. Agric.*, **8,** 287–9.
Falloon, R.E. (1980b) Seedling emergence response in ryegrasses (*Lolium* spp.) to fungicide seed treatments. *NZ J. Agric. Res.*, **23,** 385–91.
Falloon, R.E. (1983) Increased herbage production from perennial ryegrass following fungicide seed treatment. *NZ J. Agric. Res.*, **26,** 1–5.
Flentje, N.T. (1957) Studies on *Pellicularia filamentosa* (Pat.) Rogers. 3. Host penetration and resistance and strain specialization. *Trans. Br. Mycol. Soc.*, **40,** 322–36.
Flentje, N.T., Dodman, R.L. and Kerr, A. (1963) Mechanism of host penetration by *Thanatephorus cucumeris*. *Aust. J. Biol. Sci.*, **16,** 784–99.
Flowers, R.A. and Hendrix, J.W. (1969) Gallic acid in a procedure for isolation of *Phytophthora parasitica* var. nicotianae and *Pythium* spp. from soil. *Phytopathology*, **59,** 725–31.
Freeman, T.E. (1972) Seed treatment for control of pythium blight of ryegrass. *Plant Dis. Rept.*, **56**(12), 1043–5.
Freeman, T.E. (1980) Seedling diseases of turf grasses incited by *Pythium*. In *Advances in Turfgrasses Pathology* (eds P.O. Larsen and B.G. Joyner), Harcourt Brace Jovanovich, Duluth, Minn., pp.41–4.
Freeman, T.E. and Horn, G.C. (1963) Reactions of turf grasses to attack by *Pythium aphandermatum* (Edson) Fitzpatrick. *Plant Dis. Rept.*, **47,** 425–7.
Frezzi, M.J. (1956) Species of pathogenic *Pythium* identified in the Argentine republic. *Rev. Invest. Agr. T.X.* (2), 241pp. (Spanish).
Fron, G. and Gaillat (Mlle) (1925) On the root scorch disease of Gramineae. *CR Acad. Agric. France*, **11,** 119–22. (R.A.M. 4, 353 (1925)).
Fulkerson, R.S. (1953) A preliminary study on the effect of some fungicides on the establishment of forage seedlings. *Can. J. Agric. Sci.*, **33,** 30–40.
Fulton, D.E., Cole, H.Jr and Nelson, P.E. (1974) Fusarium blight symptoms on seedling and mature Merion Kentucky bluegrass plants inoculated with *Fusarium roseum* and *Fusarium tricintum*. *Phytopathology*, **64,** 354–7.
Garrett, S.D. (1970) *Pathogenic Root-infecting Fungi*. Cambridge University Press, Cambridge, 294pp.
Gilman, J.C. (1959) *A Manual of Soil Fungi* 2nd edn, Constable, London, 450pp.
Gordon, W.L. (1959) Occurrence of *Fusarium* species in Canada. 4. Taxonomy and geographic distribution of *Fusarium* species on plants, insects and fungi. *Can. J. Bot.*, **37,** 257–90.
Gordon, W.L. and Sprague, R. (1941) Species of *Fusarium*

associated with root rots of Gramineae in the Northern Great Plains. *Plant Dis. Rept.*, **25**, 168–80.

Griffiths, E. and Siddiqui, M.A. (1958) Microbial antagonism of *Fusarium culmorum. Nature, Lond.*, **182**, 4640, 956.

Griffin, D.M. (1958) Influence of pH on the incidence of damping-off fungi. *Trans. Br. Mycol. Soc.*, **41**, 483–90.

Griffin, D.M. (1963) Soil moisture and the ecology of soil fungi. *Biol. Rev.*, **38**, 141–66.

Griffon, E. and Maublanc, A. (1910) A new chytrid parasite on ryegrass turf. *Bull. Soc. Mycol. France*, 317–20.

Gussin, E.J. and Lynch, J.M. (1983a) Root residues. Substrates used by *Fusarium culmorum* to infect wheat, barley and ryegrass. *J. Gen. Microbiol.*, **129**, 271–5.

Gussin, E.J. and Lynch, J.M. (1983b) Chemical control of *Fusarium culmorum* on ryegrass, *Lolium perenne. Trans. Br. Mycol. Soc.*, **81**(2), 426–9.

Guyot, A.L. (1927) Contribution to systematic and biologic studies of *Asterocystis radicis. Ann. des Epiphyties*, **13**, 79–93. (French). (*Rev. Appl. Mycol.*, 202–203, 1927).

Hadar, Y., Chet, I. and Henis, Y. (1979) Biological control of *Rhizoctonia* damping-off with wheat bran culture of *Trichoderma harzianum. Phytopathology*, **69**(1), 64–8.

Harrington, G.T. (1923) Use of alternating temperatures in the germination of seeds. *J. Agric. Res.*, **23**(5), 295–332.

Hendrix, F.F. Jr, and Campbell, W.A. (1970) Distribution of *Phytophthora* and *Pythium* species in the continental United States. *Can. J. Bot.*, **48**, 377–84.

Hendrix, F.F. Jr, and Campbell, W.A. (1973) Pythiums as plant pathogens. *Rev. Plant Pathol.*, **11**, 77–98.

Hendrix, F.F. Jr, Campbell, W.A., and Moncrief, J.B. (1970) *Pythium* species associated with golf turf grasses in the south and south-east. *Plant Dis. Rept.*, **54**(5), 419–21.

Hendrix, F.F. Jr and Papa, K.E. (1974) Taxonomy and genetics of *Pythium*. In *Proc. Am. Phytopathol. Soc.*, Minnesota, Vol. 1, pp. 200–7.

Henis, Y. and Ben Yephet, Y. (1970) Effect of propagule size of *Rhizoctonia solani* on saprophytic growth, infectivity and virulence on bean seedlings. *Phytopathology*, **60**, 1351–6.

Hickman, C.J. and Ho, H.H. (1966) Behaviour of zoospores in plant-pathogenic Phycomycetes. *Ann. Rev. Phytopathol.*, **4**, 195–220.

Hirane, S. (1960) Studies in pythium snow blight of wheat and barley, with special reference to the taxonomy of the pathogens. *Trans. Mycol. Soc. Japan.*, **2**(5), 82–7. (Japanese, En.).

Hodges, C.F. and Coleman, L.W. (1985) Pythium-induced root dysfunction of secondary roots of *Agrostis palustris. Plant Dis.*, **69**(4), 336–40.

Holmes, S.J.I. (1976) A comparative study of the infection of perennial ryegrass by *Fusarium nivale* and *F. culmorum* in sterilized soil. *Ann. Appl. Biol.*, **84**, 13–19.

Holmes, S.J.I. (1977) Grass establishment problems and their control by seed treatments. *Proc. Symp. Problems of Pest and Disease Control in a Northern Environment.* London. p. 7–8.

Holmes, S.J.I. (1979) Effect of *Fusarium nivale* and *F. culmorum* on the establishment of four species of pasture grass. *Ann. Appl. Biol.*, **91**, 243–50.

Holmes, S.J.I. and Channon, A.G. (1975) Glasshouse studies on the effect of low temperature on infection of perennial ryegrass seedlings by *Fusarium nivale. Ann. Appl. Biol.*, **79**, 43–8.

Hoog, G.S. de and Hermanides-Nijhof, E.J. (1977) Survey of the black yeasts and allied fungi. *Stud. Mycol.*, **15**, 178–222.

Howard, F.L. (1956) Quoted in South-western Turfletter. *US Golf Assoc. Green Sect.*, **1**, p3.

Ito, S. and Tokunaga, Y. (1935) Notae mycologicae asiae orientalis. *Trans. Mycol. Soc. Japan*, **14**, 11–33.

Jackson, N. (1958) Seed dressing trial, 1958. *J. Sports Turf Res. Inst.*, **9**(34), 455–8.

Jackson, N. (1964) Fusarium blight. Turf disease notes, 1964. *J. Sports Turf Res. Inst.*, **40**, 79.

Joyner, B.G. and Partyka R.E. (1980) Rhizoctonia brown patch; Symptoms diagnosis and distribution. In *Advances in Turfgrass Pathology* (eds P.O. Larsen and B.G. Joyner), Harcourt Brace Jovanovich, Duluth, Minn., pp. 21–3.

Karling, J.S. (1968) *The Plasmodiophorales.* 2nd rev. ed. Hafner, New York, 256 pp.

Kernekamp, M.F., de Zeeuw, D.V., Chen, S.M., Ortega, B.C., Tsiang, C.T. and Khan, A.M. (1952) Investigations on physiological specialization and parasitism of *Rhizoctonia solani. Tech. Bull. Minn. Agric. Expt. Sta.* 200.

Klisiewicz, J.M. (1968) Relation of *Pythium* spp. to root rot and damping off of sunflower. *Phytopathology*, **58**, 1384–6.

Kommedahl, T. and Young, H.C. (1956) Effect of host and soil substrate on the persistence of *Fusarium* and *Rhizoctonia* in soil. *Plant Dis. Rept.*, **40**, 28–9.

Kommedahl, T. and Siggeirsson, E.I. (1973) Prevalence of *Fusarium* species in roots and soils of grassland in Iceland. *Res. Inst. Nedri As, Hveragerdi, Bull.*, 14.

Kouyeas, H. (1963) *Ann. Inst. Phytopathol. Benaki*, NS, **5**(3), 207–37.

Kouyeas, H. (1964) Notes on species of *Pythium*, 2. *Ann. Inst. Phytopathol., Benaki*, NS **6**(1), 117–23.

Kreitlow, K.W., Garber, R.J. and Robinson, R.R. (1950) Investigations on seed treatment of alfalfa, red clover and Sudan grass for control of damping-off. *Phytopathology*, **40**, 883–98.

Kreutzer, W.A. (1972) *Fusarium* spp. as colonists and potential pathogens in root zones of grassland plants. *Phytopathology*, **62**, 1066–70.

Krywienczyk, J. and Dorworth, C.E. (1980) Serological relationships of some fungi in the genus *Pythium. Can. J. Bot.*, **58**, 1412–17.

Lange, L. and Insunza, V. (1977) Root inhabiting *Olpidium* species: The *O. radicale* complex. *Trans. Br. Mycol. Soc.*, **69**, 377–84.

Latch, G.C.M. (1966) Fungus diseases of ryegrasses in New Zealand. 2. Foliage and root and seed diseases. *NZ J. Agric. Res.*, **9**, 808–19.

Leach, C.M. and Pierpoint, M. (1958) *Rhizoctonia solani* may be transmitted with seed of *Agrostis tenuis. Plant Dis. Rept.*, **42**, 240.

Leach, L.D. (1947) Growth rates of host and pathogen as factors determining the severity of pre-emergence damping-off. *J. Agric. Res.*, **75**(576), 161–79.

Ledingham, G.A. (1936) *Rhizophidium graminis* n.sp. a

parasite of wheat roots. *Can. J. Res. C.* **14,** 117–121.
Ledingham, G.A. (1939) Studies on *Polymyxa graminis* n.gen., n.sp. a plasmodiophoraceous root parasite of wheat. *Can. J. Res. C*, **17,** 38–51.
Ledingham, R.J. and Vanterpool, T.C. (1967) Pythium root rot of wheat in Queensland soil. *Queensland J. Agric. Anim. Sci.*, **24,** 241–3.
Lewis, G.C. and Clements, R.O. (1982) The effect of fungicide seed dressings on the establishment of Italian ryegrass. Tests of Agrochemicals and Cultivars, 3. *Suppl. Ann. Appl. Biol.*, **100,** 86–7.
Lipps, P.E. and Bruehl, G.W. (1978) Snow rot of winter wheat in Washington. *Phytopathology*, **68,** 1120–27.
Lundin, P. (1975) Seedling diseases of turf grasses. *Weibulls Grastips.*, **18,** 27.
Lynch, J.M. (1978) Microbial interactions around imbibed seeds. *Ann. Appl. Biol.*, **89,** 165–7.
McCarter, S.M. and Littrell, R.H. (1968) Pathogenicity of *Pythium myriotylum* to several grass and vegetable crops. *Plant Dis. Rept.*, **53**(3), 179–83.
Macfarlane, I. (1970) *Lagena radicicola* and *Rhizophydium graminis*, two common and neglected fungi. *Trans. Br. Mycol. Soc.*, **55,** 113–16.
Macfarlane, I. (1974) *Olpidium* infection and iron deficiency in cabbage. *Rept. Rothamsted Expt. Sta., 1973*, **1,** 124.
Maciejowska-Pokacka, Z. and Paramentier, G. (1978) First record of a *Pythium* pathogenic to cereals and grasses in Belgium and its identification as *Pythium vanterpoolii. Parasitica*, **34**(1), 3–10.
Mäkelä, K. (1972) Seed-borne fungi on cultivated grasses in Finland. *Acta Agr. Fenn.*, **124**(2), 1–44.
McKenzie, F. and Taylor, G.S. (1983) *Fusarium* populations in British soils relating to different cropping practices. *Trans. Br. Mycol. Soc.*, **80**(3), 409–13.
Madsen, J.P. and Hodges, C.F. (1980) Nitrogen effects on the pathogenicity of *Drechslera sorokiniana* and *Curvularia geniculata* on germinating seed of Festuca rubra. *Plant Dis.* **70**(11) 1033–16.
Maire, R. and Tison, A. (1911) On some non-hypertrophying Plasmodiophoraceae. *C.R. Acad. Sci. Paris*, 206–8 (French).
Martinson, C.A. (1963) Inoculum potential relationships of *Rhizoctonia solani* measured with soil microbiological sampling tubes. *Phytopathology*, **53,** 634–8.
Martinson, C.A. (1965) Formation of infection cushions by *Rhizoctonia solani* on synthetic films in soil. *Phytopathology*, **55**(2), 129 (Abs).
Massee, G. (1913) A new grass parasite. *Cladochytrium graminis* (Büsgen). *Bull. Misc. Inf., Kew*, 205–7.
Matthews, V.D. (1931) *Studies on the Genus Pythium.* University of North Carolina Press, Chapel Hill, 136pp.
Mead, H.W. (1955) The effect of fungicides on seedling diseases of legumes and grasses in Saskatchewan. *Can. J. Agric. Sci.*, **35,** 329–36.
Michail, S.H. and Carr, A.J.H. (1966a) Italian ryegrass, a new host for *Ligniera junci. Trans. Br. Mycol. Soc.*, **49,** 411–18.
Michail, S.H. and Carr, A.J.H. (1966b) Effect of seed treatment on establishment of grass seedlings. *Plant Pathol.*, **15**(2), 60–4.
Middleton, J.T. (1943) Taxonomy, host range and geographic distribution of the genus *Pythium. Mem. Torrey Bot. Club*, **20**(1), 1–171.
Mitchell, D.J. (1978) Relationships of inoculum levels of several species of *Phytophthora* and *Pythium* to the infection of several hosts. *Phytopathology*, **68,** 1754–19.
Mordue, J.E.M. (1974) *Thanatephorus cucumeris.* C.M.I. Descript. Path. Fungi Bact. 406.
Murray, D.I.L. and Gadd, G.M. (1981) Preliminary studies on *Microdochium bolleyi* with special reference to colonization of barley. *Trans. Br. Mycol. Soc.*, **76**(3) 397–403.
Muse, R.R., Schmitthenner, A.F. and Partkya, R.E. (1974) *Pythium* associated with foliar blighting of creeping bentgrass. *Phytopathology*, **64,** 252–3.
Nicolson, T.H. (1959) Mycorrhiza in the Gramineae. 1. Vesicular-arbuscular endophytes with special reference to the external phase. *Trans. Br. Mycol. Soc.*, **42**(4), 421–38.
Nirenberg, H.I. (1981) A simplified method for identifying *Fusarium* spp. occurring on wheat. *Can. J. Bot.*, **59,** 1599–609.
Papavisas, G.C. and Davey, C.B. (1961) Saprophytic behaviour of *Rhizoctonia* in soil. *Phytopathology*, **51,** 693–9.
Papavisas, G.C. and Davey, C.B. (1962a) Isolation and pathogenicity of *Rhizoctonia* saprophytically existing in soil. *Phytopathology*, **52,** 834–40.
Papavisas, G.C. and Davey, C.B. (1962b) Activity of *Rhizoctonia* in soil as affected by carbon dioxide. *Phytopathology*, **52,** 759–66.
Park, D. (1977) *Pythium fluminum* sp. nov. with one variety and *P. uladhum* sp. nov. from cellulose in fresh-water habitats. *Trans. Br. Mycol. Soc.*, **69,** 225–31.
Parmeter, J.R. Jr (ed.) (1970) *Rhizoctonia solani, Biology and Pathology.* University of California Press, Berkeley, 255 pp.
Parmeter, J.R. and Whitney, H.S. (1970) Taxonomy and nomenclature of the imperfect state. In *Rhizoctonia solani, Biology and Pathology* (ed. J.R. Parmeter, Jr), University of California Press, Berkeley, pp. 7–19.
Perry, D.A. (1972) Seed vigour and field establishment. *Hortic. Abs.*, **42,** 334–42.
Pitt, D. (1964a) Studies on sharp eyespot disease of cereals. 1. Disease symptoms and pathogenicity of isolates of *Rhizoctonia solani* Kühn and the influence of soil factors and temperature on disease development. *Ann. Appl. Biol.*, **54,** 77–89.
Pitt, D. (1964b) Studies on sharp eyespot disease of cereals. 2. Viability of sclerotia: persistence of the causal fungus, *Rhizoctonia solani* Kühn. *Ann. Appl. Biol.*, **54,** 231–40.
Plaats-Niterink, A.J. van der. (1979) *Monograph of the Genus Pythium. Studies in Mycology*, 21. Centraalbureau. Schimmelcultures, Baarn, 242pp.
Rawlinson, C.J. and Colhoun, J. (1970) Chemical treatment of cereal seed in relation to plant vigour and control of soil fungi. *Ann. Appl. Biol.*, **65,** 459–72.
Richardson, M.J. (1979) *An Annotated List of Seed-borne Diseases.* Commonwealth Mycological Institute, Kew, 320 pp.
Richter, H. and Schneider, R. (1953) Researches on morphological and biological differentation of *Rhizocto-*

nia solani K. Phytopathol. Z., **20,** 167–226. (German).
Robertson, G. (1980) The genus *Pythium* in New Zealand. *NZ J. Bot.*, **18,** 73–102.
Roth, L.F. and Riker, A.J. (1943) Influence of temperature, moisture and soil reaction on the damping-off of red pine seedlings by *Pythium* and *Rhizoctonia. J. Agric. Res.*, **67,** 273–93.
Saladini, J. (1976) A study of *Pythium* species associated with turf grasses in Ohio: their prevalence and pathogenicity. PhD. Thesis, Ohio State University, 94pp.
Saladini, J.L. (1980) Cool versus warm season pythium blight and on other related *Pythium* problems. In *Advances in Turfgrass Pathology*. (eds P.O. Larsen and B.G. Joyner), Harcourt Brace Jovanovich, Duluth, Minn., pp. 37–9.
Salt, G.A. (1977) A survey of fungi in cereal roots at Rothamsted, Woburn and Saxmundham 1970–1975. *Rothamsted Exp. Sta. Rept., 1976*, **2,** 153–8.
Salt, G.A. (1979) The increasing interest in 'minor pathogens'. In *Soil-borne Plant Pathogens* (eds B. Schippers and W. Gams), Academic Press, New York, pp. 289–312.
Sampson, K. (1932) Observations on a new species of *Olpidium* occurring in the root hairs of *Agrostis. Trans. Br. Mycol. Soc.*, **17**(3), 182–94.
Sampson, K. and Western, J.H. (1954) *Diseases of British Grasses and Herbage Legumes*. 2nd Rev. eds., Cambridge University Press, 111pp.
Samuel G. and Garrett, S.D. (1932) *Rhizoctonia solani* on cereals in South Australia. *Phytopathology,* **22,** 827–36.
Sanford, G.B. (1938) Studies on *Rhizoctonia solani* Kuhn. 4. Effect of soil temperature and moisture on virulence. *Can. J. Res. C.*, **16,** 203–13.
Sanford, G.B. (1952) Persistence of *Rhizoctonia* in soil. *Can. J. Bot.*, **30,** 652–64.
Schmitthenner, A.F. (1980) *Pythium* species. Isolation, biology and identification. In *Advances in Turfgrass Pathology* (eds P.O. Larsen and B.G. Joyner), Harcourt, Brace Jovanovich, Duluth, Minn., pp. 33–36.
Schwartz, E.J. and Cook, W.R.I. (1928) The life-history and cytology of a new species of *Olpidium; Olpidium radicale* sp. nov. *Trans. Br. Mycol. Soc.*, **13,** 205–21.
Scott, P.R. (1970) *Phialophora radicicola*, an avirulent parasite of wheat and grass roots. *Trans. Br. Mycol. Soc.*, **55,** 163–7.
Sharma, B.B. and Wahab, S. (1976) Comparative study of certain hydrolases in *Luffa cylindrica* fruit infected with *Pythium aphanidermatum. Indian Phytopathol.*, **29,** 81–3.
Sherwood, R.T. (1970) Physiology of *Rhizoctonia solani. In Rhizoctonia solani, Biology and Pathology* (ed. J.R. Parmeter, Jr), University California Press, Berkeley, pp. 69–92.
Simmonds, P.M. and Ledingham, R.J. (1937) A study of the fungus flora of wheat roots. *Sci. Agric.*, **18,** 48–58.
Slykhuis, J.T. (1947) Studies on *Fusarium culmorum* blight of crested wheat and brome-grass seedlings. *Can. J. Res. C*, **25,** 15–80.
Slykhuis, J.T. and Barr, D.J.S. (1978) Confirmation of *Polymyxa graminis* as a vector of wheat spindle streak mosaic virus. *Phytopathology*, **68,** 639–43.
Smiley, R.W. (1980) Fusarium blight of Kentucky bluegrass: New perspectives, In *Advances in Turfgrass Pathology* (eds P.O. Larsen and B.G. Joyner), Harcourt Brace Jovanovich, Duluth, Minn., pp. 155–75.
Smiley, R.W. (1982) *Compendium of Turfgrass Diseases*. Am. Phytopathol. Soc., 102 pp.
Smith, J.D. (1954) Diseases of grass seedlings and young grass plants. In Turf disease notes, 1954, *J. Sports Turf Res. Inst.*, **8**(30), 448–50.
Smith, J.D. (1955) Control of pre-emergence damping-off. In Turf disease notes, 1955. *J. Sports Turf Res. Inst.*, **9**(31), p. 68–71.
Smith, J.D. (1956) Seed dressing trials, 1956. *J. Sports Turf Res. Inst.*, **9**(32), 244–250.
Smith, J.D. (1957a) The effect of fungicide seed dressings on the seedling establishment of turf grass species in soil inoculated with *Fusarium culmorum.* In The control of certain diseases of sports turf grasses in the British Isles. M.Sc. Thesis, University of Durham, Vol. 2, pp. 222–60.
Smith, J.D. (1957b) Seed dressing trial, 1957. *J. Sports Turf Res. Inst.*, **9**(33), 369–72.
Smith, J.D. (1965) *Fungal Diseases of Turf Grasses*, 2nd edn (Revised Jackson, N. and Smith, J.D), The Sports Turf Res. Inst. 97pp.
Sparrow, F.K. (1960) *Aquatic Phycomycetes*, 2nd edn, University of Michigan Press, 1187 pp.
Sprague, R. (1946) Controlling root rots of cereals and grasses. *Bimonthly Bull. N. Dak. Agr. Exp. Sta.*, **9**(2), 40–5.
Sprague, R. (1948) Gloeosporium decay in Gramineae. *Phytopathology*, **38,** 131–6.
Sprague, R. (1950) *Diseases of Cereals and Grasses in North America*. Ronald Press, New York, 538pp.
Stanghellini, M.E. (1974) Spore germination, growth and survival of *Pythium* in the soil. *Proc. Am. Phytopathol. Soc. Minn.,* **1,** 211–14.
Stanghellini, M.E. and Burr, T.J. (1973a) Germination *in vivo* of *Pythium aphanidermatum* oospores and sporangia. *Phytopathology*, **63,** 1493–6.
Stangehellini, M.E. and Burr, T.J. (1973b) Effect of soil water potential on disease incidence and oospore germination of *Pythium aphanidermatum. Phytopathology*, **63,** 1496–8.
Stanghellini, M.E. and Hancock, J.G. (1971) The sporangium of *Pythium ultimum* as a survival structure in soil. *Phytopathology*, **61,** 157–64.
Subramanian, L.S. (1928) Root rot and sclerotial disease of wheat. *Pusa Imp. Inst. Ag. Res. Bull.*, 177pp.
Thornton, R.H. (1956) *Rhizoctonia* in natural grassland soils. *Nature*, **177**(4501), 230–231.
Thornton, R.H. (1960) Growth of fungi in forest and grassland soils. In *The Ecology of Soil Fungi* (eds D. Parkinson and J.S. Waid), Liverpool University, Press, pp. 84–91.
Thornton, R.H. (1965) Studies of fungi in pasture soils. 1. Fungi associated with live roots. *NS J. Agric. Res.*, **8,** 417–19.
Traquair, J.A. and Smith, J.D. (1981) Spring and summer brown patch of turf grass caused by *Rhizoctonia solani* in western Canada. *Can. J. Plant Pathol.*, **3,** 207–10.
Tribe, A.J. and Herriott, J.D.B. (1969) Fungicide treatment

for herbage seed. Experimental Work, Edinburgh School Agric. p63.

Trow, A.H. (1901) Observations on the biology and cytology of *Pythium ultimum*. n. sp. *Ann. Bot.*, **15,** 269–312.

Truscott, J.H.L. (1933) Observations on *Lagena radicicola. Mycologia*, **25,** 263–5.

Tsao, P.H. (1970) Selective media for isolation of pathogenic fungi. *Ann. Rev. Phytopathol.*, **8,** 157–86.

Tu, C.C. and Kimbrough, J.W. (1975) Morphology, development and cytochemistry of the hyphae and sclerotia in the *Rhizoctonia* complex. *Can. J. Bot.*, **53,** 2282–96.

Tyler, L.J., Murphy, R.P. and MacDonald, H.A. (1956) Effect of seed treatment on seedling stands and hay yields of forage legumes and grasses. *Phytopathology*, **46,** 37–44.

Van Luijk, A. (1934) Researches on grass diseases. *Med. Phytopath. Lab. Willie Commelin Scholten, Baarn.*, **13,** 1–22.

Van Luijk, A. (1938) Antagonism between various microorganisms and different species of the genus *Pythium* parasitizing upon grasses and lucerne. *Med. Phytopath. Lab. Willie Commelin Scholten*, **14,** 43–82.

Vanterpool, T.C. (1930) *Asterocystis radicis* in the roots of cereals in Saskatchewan. *Phytopathology*, **20,** 677–80.

Vanterpool, T.C. (1938) Some species of *Pythium* parasitic on wheat in Canada and England. *Ann. Appl. Biol.*, **25,** 528–43.

Vanterpool, T.C. (1942) *Pythium* root rot of grasses. *Sci. Agric.*, **22,** 674–87.

Vanterpool, T.C. (1952) The phenomenal decline of browning root rot (*Pythium* spp.) on the Canadian prairies. *Sci. Agric.*, **32,** 443–52.

Vanterpool, T.C. and Ledingham, G.A. (1930) Studies on 'browning' root rot of cereals, the association of *Lagena radicicola* n. gen, n. sp. with root injury of wheat. *Can. J. Res.*, **2,** 171–94.

Vanterpool, T..C and Sprague, R. (1942) *Pythium arrhenomanes* on cereals and grasses in the Northern Great Plains. *Phytopathology*, **32,** 327–8.

Vanterpool, T.C. and Truscott, J.H.L. (1932) Studies on browing root rot of cereals. 2. Some parasitic species of *Pythium* and their relation to the disease. *Can. J. Res.*, **6,** 68–93.

Waid, J.S. (1957) Distribution of fungi within decomposing tissues of ryegrass roots. *Trans. Br. Mycol. Soc.*, **40,** 391–406.

Waid, J.S. (1974) Decomposition of roots. In *Biology of Plant Litter Decomposition, Vol. 1.* (eds C.H. Dickinson and G.J.F. Pugh), Academic Press, New York.

Walker, J. (1980) *Gaeumannomyces, Linocarpon, Ophiobolus* and several other genera of scolecospored Ascomycetes and *Phialophora* conidial states, with a note on hyphopodia. *Mycotaxon*, **11,** 1–129.

Walker, J. (1981) Taxonomy of take-all and related fungi. In *Biology and Control of Take-all* (eds M.J.C. Asher and P.J. Shipton), Academic Press, New York, pp. 15–74.

Waller, J.M. (1968) *Pythium* root rot. *Rept Rothamsted Exp. Sta. (1967)*, **1,** 39.

Waller, J.M. (1979) Observations on *Pythium* root rot of wheat and barley. *Plant Pathol.*, **28,** 17–24.

Warcup, J.H. (1951) The ecology of soil fungi. *Trans. Br. Mycol. Soc.*, **34,** 376–99.

Warcup, J.H. and Talbot, P.H.B. (1962) Ecology and identity of mycelium isolated from soil. *Trans. Br. Mycol. Soc.*, **45,** 495–518.

Waterhouse, G.M. (1967) *Key to Pythium Pringsheim.* Mycol. Pap. 109. Commonwealth Mycology Institute, Kew.

Waterhouse, G.M. (1968) *The genus Pythium Pringsheim.* Mycol. Pap. 110. Commonwealth Mycological Institute, Kew.

Waterhouse, G.M. and Waterston, J.M. (1964a) *Pythium arrhenomanes*. C.M.I. *Description Path. Fungi Bact.*, 39.

Waterhouse, G.M. and Waterston, J.M. (1964b) *Pythium intermedium.* C.M.I. *Descriptions Path. Fungi Bact.*, 40.

Weindling, R. and Fawcett, H.S. (1936) Experiments in the control of *Rhizoctonia* damping-off of citrus seedlings. *Hilgardia*, **10,** 1–16.

Weste, G. (1975) Comparative pathogenicity of root parasites to wheat seedlings. *Trans. Br. Mycol. Soc.*, **64**(1), 43–53.

Willetts, H.J. (1972) The morphogenesis and possible evolutionary origins of fungal sclerotia. *Biol. Rev.*, **47,** 515–36.

Wong, P.T.W. and Siviour, T.R. (1979) Control of ophiobolus patch in *Agrostis* turf using avirulent fungi and take-all supressive soils in pot experiments. *Ann. Appl. Biol.*, **92,** 191–97.

Zjawin, S., Halisky, P.M. and Funk, C.R. (1975) Seedling blight and melting-out disease of *Lolium perenne* caused by *Rhizoctonia solani. Proc. Am. Phytopathol. Soc.*, **2,** 110.

Zogg, H. (1950) On mixed infections in cereal foot rots. *Schweiz. Z. Pathol. Bakt.*, **13,** 574–9. (German).

Part three

Low-temperature fungal diseases of amenity turf grasses

6 *Low-temperature fungal diseases*

6.1 MICRODOCHIUM PATCH AND PINK SNOW MOULD

These are different symptoms caused by the same fungus, *Microdochium nivale*, (Fries) Samuels and Hallett, synonym *Fusarium nivale* (Fr.) Sorauer, which has a very wide distribution as a pathogen on the Gramineae.

Microdochium patch describes the disease as it develops in turf in the absence of a permanent snow cover and pink snow mould as it appears at the end of the winter after snow melt.

Microdochium patch disease is probably the commonest, most disfiguring and damaging disease of golf and bowling green turf in the colder seasons of the year in many parts of Western Europe and Scandinavia (Bennett, 1933 a, b; Björklund, 1971; Bourgoin *et al.*, 1974; Courtillot, 1976; De Leeuw and Vos, 1970; Gram, 1929; Gray, 1963; O'Rourke, 1976; Sampson, 1931; Schoevers, 1937; Skirde, 1970; Smith, 1953, 1959, 1965, 1974 a, b, 1975 a, b, 1978; Sprague, 1959; Welling and Jensen, 1970; Ylimäki, 1972) and Japan (Maki, 1976; Tahama, 1973). It is also commonly found on the coarser types of turf of football fields and cricket outfields (Smith, 1965).

Attacks weaken and may kill turf grass and permit invasion by weeds. In cool-temperate oceanic climates such as in the British Isles, coastal regions of the Eastern and Northwestern States of the USA, in British Columbia, New Zealand and Australia, fusarium patch disease is important and in some regions it may develop at any time of the year during cool, wet weather (Conners, 1938; Foster, 1949; Gould, 1957; Latch, 1973; Mebalds and Kellock, 1983; Smith, 1953). However, when the summers are drier, fusarium patch is most prevalent in autumn, winter and spring when the seasonal decline of grass growth occurs (Smith, 1965) as in New South Wales (Anon, 1967; Siviour, 1975) and Victoria (Jones, 1961) in Australia. It may develop in places where there is never any snow (USGA, 1957). Microdochium patch is probably more common than has been reported in autumn and spring in regions where it is usual to regard the pathogen as causing a snow mould disease. Meiners (1955) in the Pacific Northwest of the USA and Smith (1974b) in the Canadian prairies have noted autumn outbreaks subsequent to cold, rainy weather, sleet showers or temporary light snow cover developing on unfrozen ground. Since there is little grass growth at this time, recovery from these attacks is slow and under the permanent, winter snow cover snow mould develops.

Pink snow mould describes the symptoms of *M. nivale* on the leaves of grass after the disappearance of the permanent snow cover. The mycelium and the spores of the fungus assume a pink colour after exposure to light. The general term 'snow mould' may refer to attacks of over-wintering disease caused by either *M. nivale* or complexes of other low-temperature pathogens (q.v.) with or without *M. nivale*. 'Snaeskimmel' (Danish), 'snømugg' (Norwegian), 'snømøgel' (Swedish), 'sneeschimmel' (German), 'snaesveppur' (Icelandic), 'lumihome' (Finnish) and 'pourriture des neiges' or 'moissure des neiges' (French) generally refer to attacks by *M. nivale* appearing after snow melt.

In cool-temperate continental climates and at high elevations in warm-temperate ones across North America, pink snow mould is often an important over-wintering disease of turf grasses (Afanasiev, 1962; Broadfoot, 1936; Couch, 1962, 1973; Dahl, 1934; Fushtey, 1975; Gould *et al.*, 1977; Howard, Rowell and Keil, 1951; Lebeau, 1968; Madison, Petersen and Hodges, 1960; Meiners, 1955; Monteith and Dahl, 1932; Platford, Bernier and Ferguson, 1972; Smith, 1974b, 1978, 1980b; Stienstra, 1974; Vaartnou and Elliott, 1969; Weihing and Shurtleff, 1966; Wernham, 1941). The disease may also be important on grasses generally in regions with cool or cold continental climates in Europe (Årsvoll, 1973, 1975; Jamalainen, 1959, 1974; Skirde, 1970; Ylimäki, 1972) and northern Japan (Tomiyama, 1961; Yoshikawa, 1969).

6.1.1 Symptoms

(a) Microdochium patch

The disease first appears in turf as small patches, roughly circular and 2.5 to 5 cm in diameter, at first water-soaked and then yellow to orange–brown. Around the margins of these patches of dying and dead grass shoots and leaves there may be seen, in moist conditions, a faint fringe of white or pale pink mycelium which tends to mat together the aerial plant parts. Patches may increase in diameter up to 25 cm or so with irregular margins and adjacent patches may coalesce. Symptoms are illustrated in Fig. 4 (colour plate section). The fungus may penetrate as far as the crowns of the plants, but complete killing in North America may result from subsequent winter injury or from the activity of secondary invaders.

Patches may fill in gradually from surviving plants when the disease has passed over. This recovery from damage may be slow in winter or spring when leaf production is low. In some cases a patch may show concentric-ring or 'frog-eye' symptoms with fungal attack proceeding actively at the periphery and a green recovering centre with a ring of dead leaves between. If attacks occur before a moderate snowfall, and/or short duration snow cover, on snow melt, patches may have a bleached appearance with little increase in diameter having taken place under the snow. Under prolonged, thick snow cover existing patches increase in size, numerous latent infections develop and abundant aerial mycelium may be produced. This is not as common in the British Isles as in North America or continental Europe.

(b) Pink snow mould

As the winter snow cover melts, exposed patches appear bleached and are often covered with abundant, white mycelium which may mat together the leaf blades. The patches gradually turn pink because of colour change of the mycelium and the development of pink sporodochia of the fungus. The margins of the patches, especially on *Poa annua* turf may take on an orange–brown tone under moist conditions because of the continued activity of the pathogen (Fig. 4, colour plate section). Recovery may be slow if the weather is dry in spring.

6.1.2 The causal fungus

The most recent taxonomic classification of the fungus which has been commonly called *Fusarium nivale* is:

Teleomorph: *Monographella nivalis* (Schaffnit) E. Müller (1977).

Anamorph: *Microdochium nivale* (Fries) Samuels and Hallett (1983).

The name *Microdochium nivale* will be used in this text. Other names which have been used for the teleomorph are: *Calonectria nivalis*, *Griphosphaeria nivalis*, *Micronectriella nivalis*, *Calonectria graminicola*, *Nectria graminicola*, *Sphaerulina divergens*, *Monographella divergens*. The oldest name for the conidial state is *Lanosa nivalis* Fr. (Fries, 1849).

For further discussions on the morphology and classification of this fungus see Boerema and Verhoeven (1977), Gams and Müller (1980) Gerlach (1981) and Samuels and Hallett (1983). The name *Gerlachia nivalis* (Ces. ex Sacc) W. Gams and Müller (1980) is a misnomer.

M. nivale isolates vary considerably in morphology in culture, but grass isolates are pathogenic towards cereals and *vice versa* (Smith, 1957a, 1959, 1983). Although strains differ considerably in minimum temperatures for growth there is contradictory evidence regarding physiological specialization (Bennett, 1933a; Booth, 1971a). The perfect state has not been found on turf grasses (Smith, 1965, 1983). (But see Litschko and Burpee, 1987.)

On grass showing microdochium patch symptoms, aerial mycelium is white, cobwebby, occasionally knotted and often quite sparse, stretching from leaf to leaf, especially at the patch borders. In the case of pink snow mould, on emergence from the snow the mycelium is white, but often more felted, matting together the leaf blades, gradually changing to a pink colour. Conidia are borne in sporodochia (secondary conidiophores only). The conidiogenous cells are annellophores, but the annellations, formed at the narrowed tip of the conidiophore vary in sharpness in different isolates and are not visible with the light microscope (Gams and Müller, 1980). Sporodochia may be found on leaves and suspended in mycelium under moist conditions. Conidia are heelless, without any special foot-cell (Fig. 6.1).

On turf grasses, from British sources, one-septate conidia were most common, and aseptate, 2-septate and 3-septate quite common. Four- and 5-septate forms were less common. One-septate spores measured 8.0–18.0 μm × 1.8–3.0 μm (most 12.5–15.0 μm × 2.0–2.75 μm) (Smith, 1953). Three-septate conidia measured 19–30 μm × 3.5–5.0 μm (Booth, 1971a). Sprague (1950a) gives the measurements for aseptate as 8–12 × 2.0–2.8 μm, 1-septate as 13–18 × 2.4–3.0 μm, 3-septate as 19–27 × 2.8–3.8 μm and 4–7 septate as 19–30 × 2.5–4.0 μm. Nirenberg (1981) gives the maximum dimensions of *F. nivale* vars. *nivale* and *majus* as 30 μm long. Bennett (1933a) noted that conidia of a Continental isolate grown on the same media as a British strain were considerably longer and wider. Isolates from turf grass in Saskatchewan in the Canadian prairies had mainly 0- or 1-

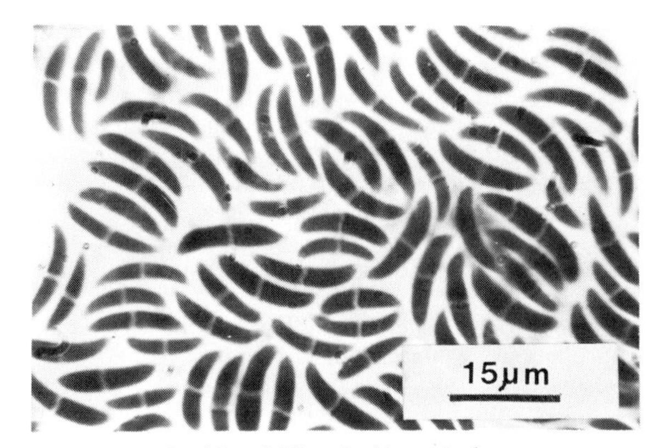

Figure 6.1 Conidia of *Microdochium nivale.*

septate conidia (Smith, 1983). There are no microconidia or chlamydospores. Conidia (Fig. 6.1) are curved (falcate), slightly broader in the lower half, narrowing upwards to a slightly curved, sharp apex and almost always heelless.

Only a brief exposure to light is necessary to initiate sporulation in culture (Sanderson, 1970; Smith, 1953). Turf grass isolates usually produce abundant loose or dense white or faintly salmon-coloured mycelium but some produce no spores. In other cases pink or salmon-pink, orange and finally rufous spore masses are produced on the surface of potato dextrose, potato sucrose or a semi-synthetic medium with yeast extract in good light (Smith, 1957a). Spore production may be encouraged by growing isolates at 13–17°C in 12h nuv light (Smith, 1981b). Nirenberg (1981) recommends the use of an agar medium with a low nutrient content and growth under continuous black light (nuv) at 17°C to encourage formation of conidia in sporodochia.

Perithecia of the teleomorph, *Monographella nivalis*, may develop in the leaf sheaths of cereals and in culture but could not be induced in turf grass isolates in culture on wheat straw at 20°C (Booth, 1973; Smith, 1983). Perithecia are oval or flattened oval, papillate, appearing as black dots, sometimes abundant, 150–260 μm across and up to 300 μm high with plectenchymatous walls 10–30 μm thick, gold to dark brown. Asci are parallel-walled or spindle-shaped, straight or curved with a thin wall and an amyloid, apical spore-discharge ring. Ascospores are hyaline, 2–4-celled, usually 8 ascus. They measure 50–70 × 7–9 μm (Müller, 1977).

6.1.3 Isolation of the fungus and pathogenicity testing

Mycelial isolates may be obtained by incubating infected plant shoots in cool (10°C) moist chambers and isolations made from spores on 'pink' leaves on the media suggested above. Turf may be inoculated with macerated cultures grown in potato–dextrose broth applied with an atomizer. The inoculated turf is then placed in a moist chamber or covered with well-moistened cellulose wadding or newspaper and polyethylene sheeting and incubated in the dark at 1–2 °C for 6–8 weeks. For individual plants the method described by Smith (1981a) may be used. For field inoculation, cultures of the fungus grown on a wheatmeal/ground dried grass/sand mixture (Smith, 1957a) or on sterile rye grain (Smith, 1976) have proved satisfactory.

6.1.4 Host range and susceptibility

In mature turf in Britain, *Agrostis canina* L.spp. *canina* Hwd., *A. canina* L. spp. *montana* (Hartm., A. stolonifera L., A. *tenuis* Sibth., *Cynosurus cristatus* L., *Festuca rubra* L. ssp. *commutata* Gaud., *F. rubra* L. ssp. *litoralis* Meyer, *F. rubra* L. ssp. *rubra*, *F. ovina* L., *F. longifolia* Thuill, *F. tenuifolia* Sibth., *Holcus lanatus* L., *Lolium perenne* L., *L. multiflorum* Lam., *Poa annua* L., *P. pratensis* L., *P. trivialis* L. and *Puccinellia maritima* (Huds.) Parl. have been found attacked (Smith, 1957a). Bennett (1933a) noted the disease on many species of weed grasses. Sprague (1950) lists 90 grasses as hosts for the fungus in North America while Howard, Rowell and Keil (1951) give 30 grass hosts of which half are used in turf. Couch (1973) has 85 grass hosts, many of which are forage grasses. Braverman (1986) has also reviewed information on the resistance of grasses to this pathogen for the period 1967–1983.

Annual meadow-grass or annual bluegrass (*Poa annua* L.) is the species in mown turf most commonly reported to be attacked by *M. nivale* in most countries. It is often killed out completely by pink snow mould in late winter. Although it is also severely damaged by microdochium patch, turf will heal up more rapidly than that of many other species during periods of mild weather which occur in autumn, winter and spring. In close mown turf *P. annua* still manages to set seed which ensures its persistence, but under these conditions it may also behave as a short-lived perennial. Contrasting views about the place of this species in turf have been presented by Haes (1956), Gibeault (1966), Engel (1977) and many other workers. One of the reasons for the rapid recovery of turf of *P. annua* from attacks of pink snow mould (and other snow mould diseases) and for the invasion by *P. annua* of turf of other species weakened by disease is because turf may be regenerated from seed of *P. annua* lying in the base of the turf. One of the main reasons for its susceptibility is probably its rapid response to applications of nitrogenous fertilizer with the concomitant liability to produce forced and sappy growth during mild periods in winter and spring. During this period, when humidity is high and the soil may be wet, forced herbage is very susceptible to attack. Smith (1953, 1954, 1957a) used this species for fungicide trials because it could be readily forced and infected. Turf in which *P. annua* is the dominant grass species must be managed most carefully if the disease is to be avoided.

Although it is possible to make some generalizations about the susceptibility of turf grass species and cultivars to microdochium patch and pink snow mould there are at least three difficulties. *M. nivale* may occur in complexes with other low-temperature fungi in various degrees of dominance (Mebalds and Killock, 1983; Smith, 1965); it may occur either as fusarium patch or pink snow mould or as both forms of the disease; the response of species and their cultivars is not the same under different climatic, soil and management conditions, the latter particularly relating to fertility and irrigation. The range of this variation is indicated by the response of different species and cultivars.

(a) *Agrostis* spp.–bentgrasses

Bentgrasses, extensively used in fine turf from subboreal to temperate regions show a range of susceptibility from very susceptible to moderately resistant. Considerable variations have been reported in the same species and cultivars depending on climate and management and therefore reference should be made to reports from different regions such as in North America–Wisconsin (Dahl, 1934), Michigan (Tyson, 1936), California (Madison, Petersen and Hodges, 1960), Rhode Island (Howard, Rowell and Keil, 1951), Western Washington (Goss, Gould and Brauen, 1974), Western and Eastern Washington (Gould *et al.*, 1978), British Columbia (Taylor, 1971–78), Saskatchewan (Smith, 1980a 1981b) and in Europe–British Isles (Jackson, 1962b; Jackson, 1964; Smith, 1958a, 1965), British Isles/Western Europe (RIVRO, 1978; STRI, 1978; Thuesen, 1975), North of Scotland (Gray and Copeman, 1975), West Germany (Skirde, 1970).

Gould *et al.* (1978) tested 160 lines of bentgrasses for resistance to snow mould injury and for turf quality in Western Washington and 138 of the most promising of these in Eastern Washington. They found, in general, that varieties of selections from northern climates had the greatest resistance to *M. nivale* and that the stolonized bentgrasses were more resistant to this pathogen than the seeded sorts. *A. tenuis* lines were less resistant than those of *A. canina* or *A. stolonifera*. Smith (1980a) found that *A. canina* and *A. stolonifera* entries were considerably more resistant to pink snow mould than those of *A. tenuis* in Saskatchewan. There is general agreement that of the commonly used cultivars of *A. stolonifera*, Penncross is one of the most resistant cultivars. Taylor (1971–1978) reported microdochium patch disease scores from many cultivars of bentgrass established vegetatively and from seed in coastal British Columbia. The variation in ratings obtained for the same cultivar from year to year even in one location illustrates the difficulties in evaluating their resistance to fusarium patch disease when general disease incidence is low. With moderate infection in 1975 and 1976 of up to 25%, several of the *A. tenuis* and *A. stolonifera* cultivars were little affected and few of the *A. canina* cultivars showed more than slight disease. With increased disease severity in late autumn and winter of 1976/77, the resistance of all *A. tenuis* and *A. canina* cultivars broke down. However, seeded Penncross, Emerald and Prominent and stolonized Nimisila, Huffine, MCC-3 and Smith 732 showed only slight damage. Data from Saskatchewan (Smith, 1980a) tended to confirm these conclusions. European data (RIVRO, 1978; STRI, 1978) indicated that *A. tenuis* cultivars, except Highland showed generally good resistance, *A. stolonifera* cultivars had fairly good resistance and *A. canina* cultivars except Aca 61 were of good resistance.

(b) *Festuca* spp. – fine-leaved fescues

Fine-leaved fescues are not used as extensively in monostands in fine turf as the bentgrasses which probably reduces their liability to attack by microdochium patch. Jamalainen (1951, 1954, 1955) considered *F. rubra* as one of the more resistant species to *M. nivale* in Finland. Gray and Copeman (1975) found that *F. rubra* was not infected in sheep-grazed, semi-natural grassland in the North of Scotland, but Skirde (1970) reported that *F. rubra* was moderately susceptible and other fine-leaved fescues were severely atacked in W. Germany. On the other hand, De Leeuw and Vos (1970) in the Netherlands, placed *F. rubra* in the least susceptible category. In Sweden, Jonsson and Nilsson (1973) found no correlation between artificial infection in the laboratory and natural infection in the field in the resistance to 12 *F. rubra* cultivars. Smith (1965) considered S.59 and Dawson resistant, New Zealand Chewing's moderately resistant and Cumberland Marsh susceptible to microdochium patch disease.

(c) *Lolium perenne* – perennial ryegrass

Most cultivars are regarded resistant or moderately resistant to microdochium patch disease in Britain (Smith, 1965) and reaction to this particular snow mould is usually not recorded, although general snow mould damage occurs (RIVRO, 1980; STRI, 1978; Taylor, 1971–1978). Resistance in this species is likely to break down under heavy snow covers (Gray and Copeman, 1975; Jackson, 1963; Jamalainen, 1974). Pasture types, which may be used in amenity turf may become susceptible if 'forced' by late applications of high dosages of N fertilizer (Gray and Copeman, 1975). The cvs Valinge (Jamalainen, 1951) and Ejer, Pleno, Barlenna, S.23, Lamora, Barenza and Vigor (Thuesen, 1975) were considered resistant to *M. nivale*.

(d) *Poa pratensis* – smooth-stalked meadow-grass or Kentucky bluegrass

Most cultivars are resistant or moderately resistant to microdochium patch disease (Jamalainen, 1955, 1974; smith, 1965). Under persistent heavy snow covers leaves may be attacked, but characteristically, crowns are not heavily damaged. Vargas, Beard and Payne (1972) reported some differential susceptibility in *P. pratensis* cultivars. Thuesen (1975) reported that Fylking had some resistance.

(e) Other turf grass species

Cynosurus cristatus, crested dogstail and *Poa trivialis*, rough-stalked meadow-grass are susceptible, but *Phleum pratense* is not usually damaged by *M. nivale* (Jamalainen, 1954; Smith 1965).

6.1.5 Epidemiology

(a) Source of inoculum for disease outbreaks

Although *M. nivale* is rarely isolated directly from soil (Gordon, 1952; McElroy, Jones and Rinehart 1954; Nicholls, 1956; Rawlinson and Colhoun, 1969) the fungus is regarded as soil-borne on infected straw (Bruehl *et al.*, 1966; Bruehl and Lai, 1966; Holmes, 1979; Jamalainen, 1962; Rawlinson and Colhoun, 1969; Sanderson, 1967; Sprague, 1950). It can be isolated from plants growing in soil. It is also seed-borne in cereals and grasses (Booth and Taylor, 1976; Matthews, 1971; Noble and Richardson, 1966; Rawlinson and Colhoun, 1969). *M. nivale* has been shown to be capable of surviving on naturally or artificially infected cereal straw for up to a year (Bruehl and Lai, 1966; Sanderson, 1967), and of spreading through non-sterile soil from infected agar discs and artificially infected straws (Booth & Taylor, 1976).

Dahl (1934) used infected grass clippings for plant inoculation; Smith (1957b) used an isolate from turf grass grown on a sterile wheatmeal/dried ground grass/sand mixture or on sterile rye grain (1976) to infect large turf grass plots. In the earlier British tests the inoculated turf was covered with moistened sackcloth to maintain humidity and increase susceptibility by reducing carbohydrate reserves. This was not necessary in the later Canadian tests where rye inoculum was applied in autumn without shading. Spore suspensions of *M. nivale* applied to the soil round *Lolium perenne* seedlings were not effective in causing disease, but mycelial macerates were (Holmes and Channon, 1975). Mycelial macerates of isolates from turf grasses, rye and wheat were infective towards *Poa annua* (Smith, 1957a) as were spore suspensions of turf grass, rye and wheat isolates on rye leaves (Smith, 1980a, 1983). Scattered, discrete leaf spots which appear to have developed under the snow cover typical of conidial infection may be found on winter cereal leaves (Smith, 1975a; Sprague, 1950; Walther and Foeke, 1981), but most infection of grass seedlings and mature turf appears to be from soil- or debris-borne mycelium. Splash dispersal of conidia by raindrops or in water films, mechanical operations such as brushing, scarifying, aerating or mowing or pedestrian traffic could carry the sticky conidia or the mycelium in leaf or soil fragments. These are all possible methods of spread of the disease where a snow cover is lacking. Dahl (1934) described the infection process by mycelium. In turf grass, wind dispersal by ascospores is not likely, since perithecia have not yet been shown to occur naturally. Bennett (1933a) suggested that in turf the fungus persists saprophytically, since there are no chlamydospores or sclerotia, as dark-brown aggregates of mycelium in plant residues. Petrini, Müller and Luginbühl (1979) and Perry (1986) considered that *M. nivale* is usually a harmless endophyte (on cereals?). Petrini, Müller and Luginbühl (1979) suggested that it only becomes pathogenic under a snow cover lasting for several weeks. However, as outbreaks may occur at any time of the year in climates like those of the British Isles (Smith, 1965) and where there is no snow (Årsvoll, 1973), it is likely that sub-clinical infections of crowns and roots may always be present. In some cases the fungus will develop in apparently healthy turf if it is shaded and kept moist (Smith, 1965).

(b) The effect of temperature

The wide range of temperature over which isolates of the fungus are active is probably the main factor responsible for its wide geographical range as a pathogen on the Gramineae (i.e. in all continents except Antarctica) (CMI Map 432: 1967). *M. nivale* is a mesophilic fungus with the ability to cause disease at the low end of its growth temperature range. The optimum and maximum temperature ranges for most isolates are 18–20 and 30–32.5 °C respectively (Bennett, 1933a; Broadfoot, 1938; Dahl, 1934; Ekstrand, 1955a; Endo, 1963; Smith, 1953, 1980b; Tasugi, 1935). Isolates differ in their minimum temperature for growth and the rate at which they will grow at low temperatures (Bennett, 1933b; Booth, 1971b; Smith, 1953). Mycelium can withstand temperatures down to −20 °C and remain viable (Bennett, 1933b). Some isolates will grow at −6 °C (Smith, 1980b).

In controlled environments, Dahl (1934) obtained most rapid infection of turf grass and cereals at 0–5 °C with slight, slow attacks at 15–20°. However, Endo (1963) found infection on Seaside *A. stolonifera*, grown on sterile quartz sand with no nutrient, slight after 2 weeks at 21 °C and 10, 45 and 10% at 15.5, 21 and 27 °C respectively, with no infection at 32 °C. However, the turf grass was grown under abnormal conditions. The most favourable meteorological conditions are those of moist (British) summer with shade temperatures 12–13 °C average minimum and 18–19 °C average maximum (Bennett, 1933a). In the USA, Couch (1962) has suggested a temperature range of 0–7.2 °C for optimum disease development with economically important outbreaks to 18.3 °C. No disease occurred on winter wheat with *M. nivale* at −1.5 °C (Bruehl and Cunfer, 1971). At 3 °C *M. nivale* became the dominant pathogen on turf when soil-heating cables were used to regulate temperature by Lebeau (1964). Several, short, intermittent periods at 0 °C arrested the recovery of *Lolium perenne* seedlings inoculated with *M. nivale* (Holmes and Channon, 1975). Generally, intermittent, short periods of lower temperatures within cool or cold, moist weather are more likely to predispose turf to attacks by microdochium patch than are continuous cold spells. In the absence of a snow cover, in autumn, winter and early spring attacks on aerial parts of the turf grasses frequently follow sharp frosts which check grass growth or damage aerial tissues (Smith, 1965).

(c) Effect of moisture and snow cover

Generally, wet weather, persistent drizzle or fog, especially if temperatures are low, temporary snow covers in autumn or early winter, sleet showers or hail and heavy dews during clear weather, especially in spring and autumn predispose turf to attacks by *M. nivale*. In Britain, Bennett (1933a) considered that *M. nivale* is favoured by the moist weather prevailing in summer and that the most numerous attacks occurred from May onwards. However, from late spring through summer, infections are masked to a greater or lesser extent by the vigorous leaf and shoot production in most cool-season grasses, while autumn and winter infections are not, as grass growth is then slower. Patches of the disease are more noticeable during autumn, winter and early spring (Smith, 1965).

There are few experimental data on the effect of irrigation on the incidence and severity of the disease. Madison, Petersen and Hodges (1960) found that morning irrigation increased the number of patches of disease more than afternoon watering in California on turf of Highland. *A. castellana.* Irrigation practice had little effect on turf of Seaside *A. stolonifera.* Under British conditions it is usual practice to complete watering of the turf in summer in time for the surface to dry off, so that cool, moist conditions, favouring infection will be presented for the least possible time.

A deep snow cover developing on unfrozen turf early in winter provides favourable environmental conditions for the growth of cold-tolerant turf grass pathogens, including *M. nivale.* Although the snow cover protects the grass from frost injury it also maintains a stable, humid climate favourable for mycelial growth of the fungus. The longer the cover persists the longer are the conditions likely to be favourable for substrate colonization and plant invasion. During the time the grass is covered with snow its carbohydrate reserves are being depleted. Under a deep snow cover there is insufficient light for photosynthesis, but respiration continues at measurable rates as low as −30°C (Meyer and Anderson, 1952; Scholander *et al.*, 1953). This drains carbohydrate reserves leading to lowering of both snow mould resistance and winter hardiness (Kneen and Blish, 1941; Tomiyama, 1961). The depth of the snow cover, whether the ground was covered before it froze and the subsequent air temperature determines the turf surface temperature (Bruehl *et al.*, 1966; Ylimäki, 1962). It is possible to determine only a broad, general relationship between the duration of snow cover and the incidence of different snow mould diseases, as Årsvoll has done for Norwegian grasslands, including amenity turf (1973). Between 1968 and 1971, out of 2401 sites 1410 showed damage by *M. nivale.* slight attacks occurred frequently after 30 days snow cover, but some sites showed injury when there was no snow cover. After 90 days snow cover, severity did not increase (Årsvoll, 1973). In the Canadian prairies pink snow mould assumes epidemic proportions only after several months of snow cover which melts to produce very wet conditions in spring (Lebeau, 1968). The disease is often common on turf against buildings with heated basements and over buried pipes or other buried heating systems. There is often a pre-hibernal history of microdochium patch disease.

(d) Effect of soil pH, lime applications and soil texture

The severity of microdochium patch disease was related to the amount of lime applied and the pH of the top

2.5 cm of turf in a field experiment over three seasons (Smith, 1958b). The higher the pH (over the test range of 4.3 to 7.2) the more severe was the infection. Slight disease occurred where no lime was applied. Apparently, strain differences occur, since Schaffnit and Meyer-Hermann (1930) found the optimum pH for the growth of their strain was 7.9, but Bennett (1933a) noted an optimum between 6.6 and 6.9 with a pH range of 2.5 to 13.0. The latter (1933a) found that the fungus would attack cereals and persist in all normal field soils of acid to alkaline reaction while Schaffnit and Meyer-Hermann (1930) found that their isolate would not grow in acid soils. Bennett (1933a) suggested that antagonistic organisms may have suppressed *M. nivale* in the case of acid soils. In Britain, the disease occurs on acid to alkaline turf, but its incidence is higher where alkaline amendments such as bone meal, steamed bone flour, sodium nitrate, nitrochalk, basic slag and urea-formaldehyde are used (Escritt and Legg, 1969; Escritt and Lidgate, 1962) or following the application of lime (Smith, 1958a).

Microdochium patch disease occurs on all soils from sands to clays, the heavier, less well-drained and usually more-fertile ones being more favourable to attacks. Recovery from the disease on sandy soils may take longer because of the lack of nutrients and inadequate water supply.

(e) Effects of nutrition

The most severe attacks of microdochium patch disease usually follow intensive turf management. In particular, Bennett (1933b) reported that fertile bowling greens were more severely affected than those of low fertility. Succulent, forced turf grass growth has little resistance to the disease and the most important factor in producing this type of growth is the application of fertilizers high in soluble nitrogen and 'unbalanced' with respect to potassium and phosphorus (Handoll, 1966). However, even 'balanced' fertilizers applied at a sufficiently high rate at the wrong time of the year will encourage this susceptible grass (Dawson and Gregg, 1936). Dahl (1934) found that ammonium sulphate applied with activated sewage sludge in late autumn predisposed putting-green turf to pink snow mould attack. Smith (1957a) showed that severity of microdochium patch disease in late winter was positively correlated with increasing dosage of ammonium sulphate applied in late autumn. Gould (1965) concurred with this finding in North America. The form of the nitrogen as well as the dosage influences the severity of attacks of *M. nivale*. Applications of organic nitrogen fertilizers such as cottonseed meal, dried blood or Milorganite at equivalent nitrogen dosage resulted in more severe disease than did ammonium sulphate, urea, sodium nitrate or calcium nitrate (Tyson, 1936). Applications of slow-release nitrogenous fertilizers have given equivocal results (Escritt and Lidgate, 1962; Madison, Petersen and Hodges 1960; Powell, Blaser and Schmidt, 1967) sometimes increasing disease severity or having little or no effect. Increasing rates of potassium from 0 to 3.3 kg/100 m^2 resulted in reduced disease when nitrogen levels were not greater than 6 kg/100 m^2. When nitrogen was increased to 10 kg/m^2 per season the disease severity increased sharply and there was little effect from potassium applications on Colonial (*A. tenuis*) turf (Goss, 1968). Potassium applications had a greater effect on the amount of disease than those of phosphorus (Dawson and Gregg, 1936; Goss and Gould, 1968). Moderate to high rates of nitrogen (2.9–9.8 kg/100 m^2) reduced disease severity in tests in the USA, when compared with none, but the highest rate, considered impractical for British conditions, reduced turf vigour and density (Brauen *et al.*, 1975). No microdochium patch was seen on turf plots treated with sulphur a 2.2 kg/100 m^2. In Finland, winter attacks by *M. nivale* on *L. perenne* were increased by nitrogen applications and reduced by those of potassium. Plant survival when calcium, but no nitrogen potassium or phosphorus was applied to soil with pH 4.7 was poor, but good at pH 6.5. Disease resistance appeared to be increased by manganese, copper and sulphur applications (Nissinen, 1970).

6.1.6 Recommendations for the control of the disease

(a) Cultural

(i) MOISTURE CONTROL. Remove surplus water as rapidly as possible by drainage, pricking, spiking, elimination of surface thatch, scarifying, vertical mowing and disperse dew by switching or hose dragging to assist in disease prevention.

Improve 'air drainage' by removing or opening up fences, herbaceous borders or rows of trees and reduce shade to encourage more rapid drying of the turf surface.

Do not smother the turf with heavy top dressings or leave turf uncut in late autumn or winter. Rake off mulches of fallen leaves.

(ii) MAINTENANCE OF TURF VIGOUR AND FERTILITY CONTROL. Maintain turf vigour suited to the season by adequate drainage and aeration, fertilization and top dressing appropriate to the turf species, climate and intensity of usage.

Judicious balanced fertilization is indicated especially on fine turf; avoid applications of quick-acting nitrogenous fertilizers in autumn or heavy applications of slower-acting nitrogenous fertilizer at any time. Frequent, light applications of fertilizer enable more precise fertility control to be achieved. Where turf irrigation is practised, a higher nutritional plane may be indicated. If it is necessary, for special purposes, to make an unseasonal application of nitrogenous fertilizer late in the year appropriate fungicide treatment should be given concurrently. There is less risk of microdochium patch disease causing severe damage on the coarser turf of football pitches than on fine turf since the grasses used on the former, often *Lolium perenne* and *Poa pratensis*, are less susceptible.

Applications of lime are likely to encourage the disease, but sulphur, ammonium sulphate or other acidifying materials tend to discourage it. Following liming a suitable fungicide should be used.

(iii) USE OF RESISTANT SPECIES AND CULTIVARS. The resistance of grass species and cultivars varies according to region, level of fertility and management practice. Local experience must govern their selection (Smith, 1980a; see Section 6.1.4). Since *Poa annua* the most susceptible species a major effort should be made to exclude it from turf by seed-bed treatment, by suitable management and fungicidal treatment of the turf species selected to prevent *P. annua* taking over diseased patches.

Reduce liability to severe disease in susceptible species, such as *Agrostis tenuis* by using them in polystands with less susceptible species such as *Festuca rubra*. There are stolonized types of *A. canina* and *A. stolonifera* which are less susceptible than seeded forms although they are not as commonly used. Other grass species are usually less susceptible than the bentgrasses and these should be used except where the finest turf is needed. There are now several new cultivars of *L. perenne* and *P. pratensis* which will stand close mowing and are wear tolerant (RIVRO, 1980; STRI, 1978, 1980) which may be used successfully in medium fine turf.

(b) Chemical control

Preventive applications of fungicides should be made on turf known or considered likely to be prone to attack.

The timing and number of fungicide applications will vary from region to region, but usually the first application should be in late summer or early winter before development of a permanent snow cover. In mild, snow-free climates winter and spring applications may be needed.

Include surrounds to playing areas, such as the aprons and collars of golf greens in the treatments, since they are often sources of infection.

(c) Suitable fungicides

Mixtures of mercurous and mercuric chlorides, the use of which is still permitted in some countries, are generally the most reliable and persistent fungicides for fusarium patch disease control. As preventatives using dry powders, the equivalent of approximately 45g a.i./100 m^2 application should be given. This dosage can be reduced considerably if a wettable powder formulation, based on microfine mercurous chloride (Smith, 1957c) is used as a spray. In order to check an attack immediately it is seen higher dosages are usually needed, applied to the whole of the susceptible area and not just to the disease patches.

Organo-mercury fungicides are also still permitted in some countries. They may give effective control if applied in a preventative role before the development of a snow cover. Of these, phenyl mercuric acetate may sometimes show phytotoxicity at effective dosages. There are effective substitutes for mercury fungicides for the control of *M. nivale*, notably quintozene, chlorothalonil, anilazine, mancozeb, maneb, benomyl, thiophanate-methyl, benzimidazoles, cadmium compounds, dichlorophen, iprodione and triadimefon (Fushtey, 1983; Smith and Mortensen, 1981), which may be registered for use in some countries. It is good practice to 'ring the changes' with these fungicides, especially if systemic materials are included in the programme, to reduce the risk of resistance to a particular fungicide developing and to take care of disease complexes (Davidson, 1986; Gould, Goss and Byther, 1977; Smith and Mortensen, 1981). Since *M. nivale* often occurs in complexes with other snow moulds alternative materials should also be used against the other pathogens.

Gould, Goss and Byther (1977) in the Pacific Northwest of the USA and Davidson (1986) in the Peace River region of Northern Alberta have emphasized the need for an early application of fungicide to control *M. nivale* and found that the substitution of a late, heavy application did not compensate for an early light treatment. In the prairies of western Canada, where the complex of snow moulds is different from the Pacific Northwest of the USA, effective control of pink snow mould is dependent on fungicidal prevention of microdochium patch disease before a permanent snow cover develops (Smith, 1974b).

Systemic materials such as benomyl, methyl thiophanate, thiabendazole and iprodione are effective against *M. nivale* (Davidson, 1986; Smith and Reiter, 1976) only if repeated applications are made during winter, autumn and spring. However, fungus tolerance to iprodione has been shown (Chastagner and Vassey, 1982) and is suspected in the Canadian prairies in the case of benomyl (Smith and Mortensen, 1981).

In Scandinavia, quintozene and organo-mercurials

were found effective against *M. nivale* (Jamalainen, 1960; Ylimäki, 1972) and in the Austrian Tirol, the order of fungicide effectiveness against fusarium patch disease was anilazine, benomyl, chlorothalonil, quintozene and mancozeb (Köck, 1976; Dr. L. Köck, personal communication). In Saskatchewan, in the Canadian prairies, a heavy autumn attack of microdochium patch was prevented by benomyl, dichlorophen, methyl thiophanate and quintozene, but the most effective control was obtained with mercurous/mercuric chloride, phenyl mercuric acetate, or quintozene (Davidson, 1986; Smith, 1976).

The use of benomyl or thiabendazole, which are effective against *M. nivale* alone, when the latter is associated with a *Typhula* sp. or *Coprinus psychromorbidus* snow mould against which little effectiveness is shown may lead to the control of *M. nivale*. However, with the release from this competition the other pathogen may become dominant and cause severe injury under a snow cover (Smith, 1976). On the other hand, quintozene in high dosage applied in autumn is effective against both *M. nivale* and basidiomycete snow moulds. When its activity has declined by next summer it acts selectively in suppressing antagonists to *M. nivale* allowing the latter to cause outbreaks of microdochium patch disease. Residual effects of some other fungicides have been noted (Smith and Reiter, 1976). Granular formulations of fungicides are sometimes as effective as spray forms in controlling *M. nivale*, but higher dosages of the active ingredients may be required.

6.2 COTTONY SNOW MOULD-LOW-TEMPERATURE BASIDIOMYCETE (LTB) AND SCLEROTIAL LOW-TEMPERATURE BASIDIOMYCETE (SLTB) PHASES OF *COPRINUS PSYCHROMORBIDUS*

In 1931 an unidentified basidiomycete was found associated with a *Fusarium* sp. and a *Rhizoctonia* sp. causing snow mould of golf greens at Edmonton, Alberta (Broadfoot, 1936, 1938; Broadfoot and Cormack, 1941). The same Basidiomycete was found on lawn turf in Saskatoon, Saskatchewan in 1932 (Vanterpool, 1944), and in Alaska (Lebeau and Logsdon, 1958) and in southern Manitoba (Allen, Bernier and Ferguson, 1976; Platford, Bernier and Ferguson, 1972). It is the common cause of snow mould on all classes of sports turf and lawns, particularly in the lower snowfall regions of western Canada (Lebeau, 1960, Smith, 1969a, 1972, 1973, 1974ba,b, 1975c, 1978, 1980a–d), but it also causes damage in the higher snowfall regions in the Canadian prairies. Although the pathogen was considered to have a geographical range restricted to western Canada and Alaska, it is probably more widespread than this (Spotts, Traquair and Peters, 1981; Traquair *et al.*, 1983). It is an unspecialized pathogen with a wide host range on winter cereals, grasses, forage legumes and weeds (Cormack, 1948; Lebeau, 1969; Vanterpool, 1944) and causes pear decay in cold storage (Spotts, Traquair and Peters, 1981). The fungus has two vegetative phases which are conspecific with *Coprinus psychromorbidus* (Redhead and Traquair, 1981; Traquair, 1980; Traquair and Smith 1982) but which differ morphologically and pathogenically and are designated as the low temperature basidiomycete (LTB) or the sclerotial low temperature basidiomycete (SLTB).

6.2.1 Cottony snow mould – Low-temperature basidiomycete (LTB) phase

(a) Symptoms

The disease is first seen at snow melt in spring as bleached patches of turf, often almost circular in outline and up to one metre in diameter (Fig. 5, colour plate section). Abundant mycelium is often present at the patch edge, but it may be quite sparse and greyish-white (Fig. 6.2). There are no sclerotia as with *Typhula* spp. and *Sclerotinia borealis*. Crowded patches may show a green strip of undamaged grass between them giving a 'crazy paving' effect (Fig 6.3). The disease may go unnoticed at snow melt, becoming apparent only when spring growth recovery of the turf takes place. Under humid conditions, saprophytic fungi, notably *Alternaria alternata* Kiessler and *Cladosporium herbarum* Link, may colonize leaf tissues killed by the LTB giving a grey appearance to the patches.

Patches of *Agrostis, Poa annua*, and *Festuca rubra* turf attacked by the LTB fungus usually do not recover and the space left often fills in with *Poa annua* and broad-leaved weeds from seed lying dormant in the turf surface. In all except the most susceptible cultivars of *Poa pratensis* (Smith 1975c, 1980b) recovery usually takes place, but it is sometimes as late as August before patches have healed up completely.

(b) The causal fungus *Coprinus psychromorbidus* (Redhead and Traquair, 1981)

Many attempts have been made by workers in western Canada to find a fruiting state of the fungus (Smith, 1977a). Traquair (1980) found a small agaric of the *Coprinus urticicola* complex of the section *Herbicolae* on the necrotic crowns of alfalfa (*Medicago sativa* L.), dug from the field, subjected to a

Figure 6.2 Patches of cottony snow mould (non-selerotial *C. psychromorbidus* playing field turf). (H.W. Mead).

Figure 6.3 Cottony snow mould. 'Crazy paving' effect of adjacent patches.

freezing test and allowed to recover in a greenhouse. By means of a 'di-mon' mating technique (Bruehl, Machtmes and Kiyomoto, 1975) he showed that the LTB and the *Coprinus* sp. were conspecific. Fruits of the fungus were later found on alfalfa plants in the field and on sheaths of a potted plant of *Agrostis stolonifera* L. in a greenhouse at Saskatoon in 1981 (Fig. 6, colour plate section). This fungus had been found in Manitoba in the Canadian prairies on leaf sheaths of Marquis wheat (*Triticum aestivum* L.) and decaying stems of nettle (*Urtica dioica* L.) and named *Coprinus urticicola* (Berk. and Broome) Buller (Hanna, 1939). However, the name was incorrectly applied (Redhead and Traquair, 1981) and the fungus from alfalfa (Traquair, 1980) and *Agrostis stolonifera* is at present named *C. psychromorbidus*.

The caps (basidiocarps) of *C. psychromorbidus* are 7–12 mm wide, conical to flat when mature with a narrow central boss. The edges of the cap are finely furrowed and when older they split and curl up. The cap surface is felted with curled back, orange–yellow to yellow–brown scales, the remains of a universal veil. There is a characteristic fungal smell. The *gills* are narrow, attached to a ring at the top of the stalk, grey at first, then blackening and deliquescing. The *stalk* (stipe) is 40–70 mm long and 2–3 mm wide when mature, usually slightly enlarged at the base with a hollow centre and watery, white, fibrous flesh. The surface of the stalk has a fine, silky to faintly frosted surface. *Buttons* are at first pure white with brown scales developing later. When mature and deliquescing, HCN is evolved by the basidiocarps. The *spore print* is chocolate–brown to brownish–black. *Basidiospores* (6.4–) 7.2–8.8 (–9.6) × 4.8–5.6 (–6.4) μm are pale brown or yellow–brown to blackish–brown when mounted in KOH or H_2O, smooth, elliptical to broadly elliptical in profile and generally elliptical or ovate in face view. They are thick-walled with an apical pore. Basidia are shortly club-shaped to almost sac-like, thin-walled, hyaline, 12.0–25.0 × 8.8–10.0 μm and four-spored. There are short *sterile basidial elements* (brachybasidioles) surrounding the basidia, *pleurocystidia* bridging the gills and *cheilocystidia* at their edges. Clamp connections are present throughout the ground hyphae of the basidiocarp tissue (from Traquair, 1980, modified).

In culture, isolates from spores or hyphae from basidiocarps are mesophilic, with optimum temperatures c. 22 °C. Growth is rapid at this temperature. The edge of the colony consists of radiating mycelium which may be sparse, adpressed to the surface or cottony. The aerial mycelium is white, woolly or cottony, becoming resupinate or felted as it ages with white hyphal knots. The mycelium has a strong fungal smell. No fruiting has been seen in culture. HCN is produced on soybean agar after 3 weeks (Traquair, 1980). Hyphae are hyaline, thin-walled with clamp connections, mostly 1.6–4.0 μm in width. The hyphal knots consist of compactly interwoven, swollen hyphae 4.0–8.0 μm wide with oil droplets. The outer hyphae of these knots are thick-walled and narrower, 1.2–4.8 μm wide.

The fungus may be isolated from mycelium on leaves, crowns and shoots of turf grasses by plating unwashed infected fragments on to potato dextrose or potato malt yeast extract agar (BASM) at 1 °C*. Alternatively, isolates may be obtained from mixed inocula on turf plugs by the process of eliminating other low-temperature tolerant fungi. These are sclerotia, in the case of *Typhula* spp. and *Myriosclerotinia borealis* and the characteristic spores and pink mycelium of *Microdochium nivale* when exposed to light (q.v.). The turf plug is incubated in a glass or plastic moist chamber at c. 4 °C under fluorescent or nuv light and then it is examined for a non-sporing, white basidiomycete with abundant clamp connections and cottony growth. Hyphae are similar in dimensions to those of mesophilic isolates from basidiocarps given above, i.e. finer than in most *Typhula* spp., *M. borealis* and *M. nivale*, but the density of aerial growth is very variable. There are no sclerotia or spores and rarely rudimentary basidia.

Ward, Lebeau and Cormack (1961) grouped isolates of the LTB into types A, B and C on their cultural appearance, supported by their reaction to temperature, pH, tolerance of antibiotics and HCN, their ability to produce HCN in culture and hosts and their pathogenicity. Type A isolates were slow growing, produced HCN in large quantities in plants and none in culture; they were moderately pathogenic on grasses and produced stroma-like bodies in culture (Ward, 1964b). Type B were more rapid growing than type A and produced abundant, fluffy, aerial mycelium and released much HCN in culture and were of equivalent pathogenicity to type A towards grasses. Type C was a heterogenous group of very fast-growing isolates which did not produce HCN and were not pathogens. The LTB isolates used by Smith in field fungicide and turf grass-resistance studies were highly pathogenic on grasses, winter cereals, and alfalfa (Smith, 1975c, 1976, 1980a). They were of Type B.

Broadfoot and Cormack (1941) found that the cardinal temperatures for the LTB to be −4, 15 and 26 °C, and Smith (unpublished) noted that the optimum temperature for most turf grass isolates was c. 15 °C. Ward, Lebeau and Cormack (1961) found that the optima for type A and B isolates was 12.5 °C and for type C 12.5 and 17.5 °C but response to temperature was variable for types B and C. All isolates grew at 0 and 25 °C.

(c) HCN and pathogenesis

Lebeau and Dickson (1953) found that an isolate of the LTB produced sufficient HCN on natural and artificial culture media to kill buds and crown tissue of alfalfa. The production of HCN over a wide range of temperature suggested that it was produced during active mycelial growth rather than by autolysis of its own hyphae, as with other basidomycetes (Robbins, Rolnick and Kavanagh, 1950). However, Ward and Lebeau (1962) later showed that whereas a type A isolate produced HCN during active growth, a type B isolate appeared to produce HCN by autolysis. Lebeau and Cormack (1956) showed that turf grasses infected with the LTB gave positive tests for HCN

* Why the fungus isolated from fruiting bodies should be mesophilic and that from overwintering disease patches should be low-temperature tolerant requires further investigation.

whereas uninfected controls and turf samples taken in spring contained no toxin. Lebeau, Cormack and Moffat (1959) found that alfalfa plants damaged by the LTB contained HCN in concentrations proportional to the severity of damage in the host. Ward (1964a) showed that the LTB accumulates a cyanogenic compound in growing mycelium before free HCN could be detected and this compound breaks down chiefly during autolysis. Ward and Thorn (1966) found that HCN could be produced throughout the mycelial growth of the fungus from L-glycine. Apparently HCN is a normal metabolic product of many basidiomycetes (Locquin, 1944, 1947; Bach, 1956).

HCN will damage grass seedlings if present in sufficient concentration (Filer, 1964; Traquair and McKeen, 1986). However, it is uncertain whether it initiates pathogenesis in the LTB. Although Lebeau, Cormack and Moffat (1959) stated that 'mycelium was not present in crown bud tissue until positive tests were obtained for HCN' they also admitted that it was difficult to prove that plant tissues had absorbed HCN before invasion by the fungus. Their data also showed that HCN production was highest at the end of a winter's growth when autolysis would also be high (Lebeau and Cormack, 1956; Lebeau, Cormack and Moffat, 1959).

Ward, Lebeau and Cormack (1961) noted that Type C isolates were highly compatible with other isolates in culture, but that in Type A and B isolates 'zones', were formed between adjoining colonies. These 'zones', where ribbons of green grass show as boundaries between adjacent diseased patches, are often noticed in attacks of the LTB in turf and are a useful diagnostic character (Fig. 6.3). They appear to be due to mutual or intraspecific antagonism (Lebeau, 1975; Smith and Årsvoll, 1975). It seems highly unlikely that they are directly due to HCN. In addition, to its toxic effects on plants, this chemical had been shown to inhibit the growth of several fungal species (Robbins, Rolnick and Kavanagh, 1950). If the HCN produced by the fungus during growth was responsible for the initiation of pathogenesis and the death of the patches, the occurrence of the green ribbons would be unlikely since around patch margins the fungus is very active metabolically. It seems possible that the fungal attack on the green ribbons is prevented by a water-diffusible mutual inhibitor to LTB growth (Smith and Årsvoll, 1975). Grass in patch centres may have been killed by HCN resulting from autolysis of LTB mycelium.

(d) Host range and cultivar resistance

Most species of *Bromus, Elymus* and *Agropyron*, including *A. cristatum* (L.) Gaertn., often used in low maintenance lawns in the Canadian prairies, are very resistant to resistant. Of the other turf grasses, *Poa pratensis* cultivars vary from high to moderately resistant, but *Poa trivialis* L. is susceptible (Fig. 6.4). *Festuca rubra* L. ssp. *rubra* is often more severely damaged than *F. elatior* L.; *F. rubra* L. var. *commutata* Gaud. and *F. ovina* L. are generally susceptible. *Phleum pratense* is moderately susceptible (Cormack, 1952b). *Agrostis* spp. and *Poa annua* are often used in fungicide tests because of their susceptibility (Smith, 1976). However, some cultivars of creeping bent (*A. stolonifera* L.) e.g. Northland (Lebeau, 1967) are less susceptible than other *Agrostis* spp. Smith (1975c, 1980d) found considerable differences in susceptibil-

Figure 6.4 Disease resistance to cottony snow mould (non-sclerotial *Coprinus psychromorbidus*). Differential resistance shown by cultivars of *Festuca rubra* (rear) and *Poa pratensis* (front).

ity to LTB in cultivars of *P. pratensis, F. rubra, F. ovina* and *F. longifolia* in inoculated turf grass plots. A type B isolate (Ward, Labeau and Cormack, 1961), grown on sterile rye grain was used as inoculum. *Poa pratensis* cultivars were generally more resistant than those of *F. rubra* (Fig. 6.4) and *F. ovina*. Susceptible lines often came from mild climates, where the range of pathogens was different from the Canadian prairies. Many of the resistant lines came from the USSR or the eastern Baltic and Canada (Smith, 1975c, 1980a,b). No strains completely resistant to the LTB were found, but selection for early winter dormancy promised to be of possible value in increasing field resistance to LTB snow mould in *Poa pratensis*. The cultivar Dormie (Smith and Cooke, 1978) appeared to derive its field resistance from this characteristic.

(e) Epidemiology

Turf established from seed is not usually damaged by the LTB in its first or second year (Cormack, 1952b; Smith, 1969b, 1975c). The form in which the pathogen oversummers is uncertain since there is no known sclerotial stage for the LTB, although there is for the SLTB which is conspecific. Stromatal cushions are produced in culture by Type A isolates and these may be more resistant to unfavourable conditions than mycelium, but Type B isolates which are found attacking turf grasses are mycelial only. Probably the fungus persists as mycelium in plant debris (Lebeau and Cormack, 1961) and that the disease develops from mycelial infection which, as in alfalfa, takes place before winter (Lebeau, Cormack and Moffat, 1959). Once established, the disease appears to be very persistent, returning in the same turf location. Traquair (1980) and Traquair and Smith (1982) found that the cultural characters of isolates of *C. psychromorbidus*, which is conspecific with the LTB and the SLTB, ranged from the completely mycelial LTB, through forms with mycelial knots and cushions to the SLTB sclerotial form. Some SLTB isolates lose the ability to produce sclerotia on repeated subculturing, but others have retained the sclerotial character through 10 years of repeated subculture on 'lean' medium, cornmeal, malt, yeast extract (Traquair and Smith, 1982). Possibly, some LTB isolates have pleomorphic SLTB oversummering stages. Deep snow covers are not so important in disease development as with snow moulds such as *Myriosclerotinia borealis* and *Typhula ishikariensis* (q.v.) and it is not necessary for the soil to be unfrozen for severe disease to develop (Cormack, 1948; Lebeau, 1964). These two factors may be of importance in pathogenesis in relation to the significance of HCN. Shallow snow covers would permit freer ventilation and escape of the toxicant more rapidly than deep ones, but the lower temperatures at the soil surface would probably favour absorption of the HCN by plant tissues (Lebeau and Dickson, 1953). Slow thawing at temperatures near freezing appears to favour disease development (Cormack, 1948). Nevertheless, LTB snow mould on turf grasses is usually more severe in years when a persistent snow cover develops before the soil is fully frozen (Smith, 1980d). In such years the likelihood of complexes of two or more snow moulds developing is also increased (see Section 6.7). It seem probable that the LTB is a 'low-grade' pathogen which needs to develop considerable amounts of mycelium by colonizing the dead host tissues before it can overwhelm living crown buds (Cormack, 1948; Lebeau and Dickson, 1953; Smith, 1980b).

Heavy shading by trees and the liberal, late-season use of nitrogenous fertilizer and irrigation water increases severity (Smith, 1969b), by delaying the acclimation of the turf grass.

(f) Control

(i) CULTURAL. The use of suitable resistant species and cultivars forms the main basis for cultural control. On minimum care turf, in farm yards and on rough lawns, crested wheatgrass, *Agropyron cristatum* (L.) Gaertn. is probably the best choice since it is very resistant to the LTB. It goes almost completely dormant in late fall but starts growth very early in spring. It may be used in a polystand with *Poa pratensis* if a denser turf is needed. On irrigated turf of domestic lawns, golf tees, fairways, and green collars, where LTB is prevalent resistant cultivars of *Poa pratensis* are suitable. Many of the newer cultivars which are stated to have resistance to snow mould have not been tested for resistance to LTB snow mould. Until such time as they have been proven, resistant cultivars such as Dormie (Smith and Cooke, 1978) Delta and Park should be sown. Merion and other very susceptible cultivars (Lebeau, 1976; Smith, 1975c, 1980a) should be avoided or used in low proportions in blends. There are no *Festuca rubra* or *F. ovina* cultivars comparable in resistance to that of the best *Poa pratensis* available and only Durar *F. longifolia* has high resistance (Smith, 1975c). Both Penncross and Seaside bentgrasses are severely attacked in golf greens, but the vegetatively propagated Northland *A. stolonifera* is often less severely damaged (Lebeau, 1967). Bentgrass greens in the Canadian prairies should always be protected by fungicides against the LTB if invasion of *P. annua* into patches of bentgrass killed by the snow mould is to be prevented. *Poa annua* itself is very susceptible to LTB snow mould.

Management practices should be suited to the type of turf involved. Any major or minor nutrient lack is likely to be reflected in poor recovery from the disease. Particular care should be taken to adjust the amount of nitrogenous fertilizer applied after July to the state of grass growth. If inorganic nitrogen such as ammonium sulphate (23-0-0) or ammonium nitrate (34-0-0) is applied after the end of July, use light applications (0.5 kg/100 m^2) and generally the total seasonal application should not exceed 2.5 kg/100 m^2. Inorganic nitrogen applied in September will delay dormancy and acclimation. Lebeau (1976 and personal communication 20 Jan. 1980) has found that the use of the slow-release, organic nitrogen fertilizer, crotonyl-di-urea (19-11-11) applied in combination with mercuric chloride or chloroneb fungicides resulted in good spring colour and good control of LTB snow mould. By making applications of nitrogenous fertilizer when lawn turf has gone 'dormant' in early winter it is possible to elicit more rapid spring green-up. Ammonium sulphate (23-0-0) alone at 0.125 or 0.25 kg N/100 m^2 or in combination with superphosphate (20% soluble phosphoric acid) at 0.15 or 0.30 kg P/100 m^2 did not significantly increase severity of LTB snow mould. However, phosphate alone should not be used since it doubled the severity of the disease (Smith, 1977b).

Field obsevations suggest that, as in alfalfa (Cormack, 1948), the disease is less severe on turf watered deeply (to 15 cm) in autumn than on shallowly watered or on unwatered turf. Since LTB occurs in regions where desiccation injury is also a major problem in spring, adequate soil moisture, but not waterlogging, resulting in a deep root system is essential if the risk of poor recovery from snow moulds and abiotic winter injury is to be reduced. Although other LTB will cause severe disease under shallow (10 to 15 cm) snow covers, long-lying snow drifts increase injury. Snow fences may sometimes be used to control the formation, position and depth of snow drifts (Darby, 1971). It is sometimes expedient to remove snow mechanically from prone locations in late winter or spring or to dust black or dark-coloured soot or fine fly ash to increase the rate of snow melt at that time (Dewey and Nielson, 1971). Soil heating with electrical resistance cables has been used effectively to control LTB (and *M. nivale*) attacks on very susceptible turf of *Agrostis tenuis* (Lebeau, 1964, 1967). Plots with minimum temperatures controlled at −3 °C or unheated were damaged by the LTB, those controlled at 3 or 6 °C were infected with *M. nivale*, but those maintained at a minimum of 0 °C were unaffected by snow mould. The cost of installation and operation limits the use of the method.

Polyethylene and other synthetic plastic sheeting and netting and brush, compost, straw or soil have been used to give protection to turf from cold injury and desiccation (Cooper, 1982; Evans, 1975; Lebeau, 1964; Ledeboer and Skogley, 1967; Watson and Wicklund, 1962) in the northern USA and Canada. In southern Alberta, polyethylene covers increased the effectiveness of organic mercurial fungicides, allowing a much lower dosage to be used to obtain effective control of LTB snow mould. Top dressing appears to have little effect on disease severity. The mycelium of the fungus may be seen growing through a light top dressing at snow melt.

Damage from LTB snow mould may be so severe on susceptible cultivars that it is necessary to returf or reseed the killed areas.

(ii) FUNGICIDAL. Broadfoot (1936) obtained effective control of snow mould, considered to be caused by a complex of the LTB, a *Fusarium* sp. (probably *M. nivale*) and a *Rhizoctonia* sp. (perhaps a secondary organism) with mercuric or mercurous chlorides, singly or in combination. A single application of 400 g a.i./100 m^2 or greater gave effective control. Lebeau, Cormack and Ward (1961) also found mercury chlorides the most effective materials. No difference in efficiency was noted between mercuric and mercurous chlorides.

Where a very susceptible cultivar, such as Merion *Poa pratensis* is used, combinations of the inorganic mercury chlorides are still the most effective materials (Allen, Bernier and Ferguson, 1975, 1976; Smith, 1976). Wettable powder formulations of these are available in addition to dry powders, usually formulated in the proportion of 2:1 mercurous to mercuric chloride. The wettable powders usually give as effective control at lower dosages (80–110 g a.i./100 m^2) as the dry powders, but more than one application is advisable in autumn and early winter (Smith, 1976, 1980b).

Quintozene (80–390g a.i./100m^2), chloroneb (140–280g a.i./100m^2), and thiram + oxycarboxin + carboxin (180–215 g a.i./100m^2) as wettable powders are very effective substitutes for inorganic mercury, except on very susceptible cultivars. At least two applications are required, in autumn and early winter (Smith, 1976, 1980c; Smith and Mortensen, 1981). Selection of a fungicide for LTB control should take into account complexes with other snow moulds. Smith and Mortensen (1981) obtained effective control of LTB snow mould complexed with microdochium patch disease on bentgrass turf with three applications of triadimefon (62g a.i./100m^2), quintozene (200g a.i./100m^2), mercuric/mercurous chloride (47g a.i./100m^2), fenarimol (62 g a.i./100 m^2), thiram + carboxin + oxycarboxin (183 g a.i./100 m^2), and borax (375 g a.i./100 m^2). Combinations of three fungicides, even those regarded as effective against the LTB (see above) gave conflicting results. In another test, where the dominant snow mould was *Microdochium nivale*, with the LTB also present, combinations of three fungicides were generally more effective than single materials applied thrice.

6.2.2 Cottony snow mould – Sclerotial low-temperature basidiomycete (SLTB) phase

An unidentified, sclerotial, low-temperature-tolerant basidiomycete, designated as the SLTB was found associated with snow mould damage to turf grass from southeastern Saskatchewan to the Peace River region of British Columbia (Smith, 1972, 1974a). It has also been found in spring, after snow melt, on diverse dying and dead hosts and on leaf and twig debris lying on the ground (Smith and Piening, 1980; Traquair and Smith, 1982).

(a) Symptoms

Patches of turf damaged by this snow mould are similar in shape to those caused by the LTB, but they are often smaller and injury more superficial. The mycelium is white, cobwebby, stretching from leaf to leaf in longer lawn turf, but particularly in that of close-mown swards of golf greens or aprons, much of the mycelium occurs as white, cream or tan knots on leaves and litter. These knots (Fig. 7, colour plate section) develop into thick-walled brown or black sclerotia, irregularly shaped, mostly 0.25 to 1.5 mm in size (Fig. 6.5 and 6.6e) which are found mainly on the side of host tissues or plant debris nearest to the ground. Honey-coloured exudates may ooze from the sclerotia.

(b) The causal fungus *Coprinus psychromorbidus* (Redhead and Traquair, 1982)

(i) METHODS OF ISOLATION. Fully imbibed sclerotia are surface sterilized with 70% ethyl alcohol for 1–2 minutes, rinsed in several changes of sterile water, sliced with a sterile scalpel and plated on CMMY agar (Smith, 1981a) and incubated at 1 °C in taped Petri dishes until mycelium is produced. Excised hyphae are then cultured on CMMY agar at 15 °C. Cultural success is often low and is reduced further by more active surface sterilizants such as hypochlorite. If bacteria are troublesome contaminants, add 5% glycerol to BASM agar (Smith, 1981a) and incubate sclerotial slices at −3 to −4 °C.

(ii) RATE OF GROWTH IN CULTURE. Growth on BASM agar in growth tubes is more rapid (1.3–3.3 mm/d) than on PDA or malt agar (radial growth 0.67–1.7 mm/d). For most isolates the optimum growth temperature is from 10 to 15°C, but some isolates have a very low optimum of 0–5°C with a depression of growth rates between 3 and 5°C (Traquair and Smith, 1982).

(iii) CULTURAL CHARACTERS. On cornmeal, malt, yeast extract agar, CMMY (Smith, 1981a) the colony margin is even, narrow, low-growing or uneven and with lanose mycelium. The aerial mycelium may be either sparse or lanose and felted. Growth is slower than that of the LTB. Compact hyphal knots, produced mainly on the surface, appear from c. 4

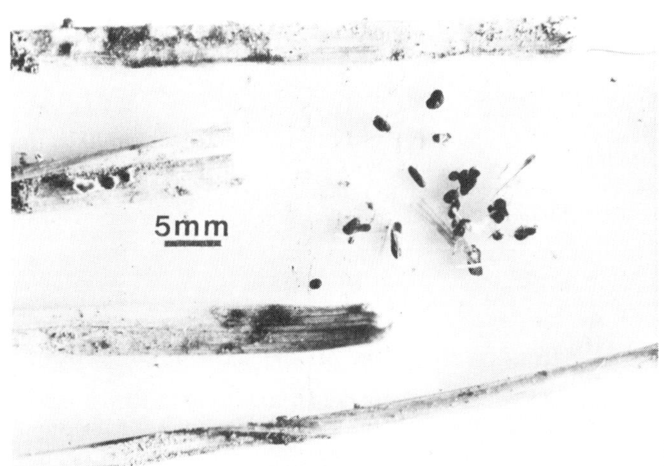

Figure 6.5 Sclerotia of *Coprinus psychromorbidus* from dead leaves of *Poa pratensis*.

weeks at 6°C. At first, these are cream of buff, changing as they develop to brown or black, hard, rounded or eliptical sclerotia up to c. 3 mm, sometimes with pale-brown exudates. At temperatures above 15°C a brown diffusate appears in the agar becoming more intense as the temperature increases to its maximum for growth of c. 25°C. On this medium some isolates have continued to produce sclerotia for 10 years of successive transfers, but on other media isolates may lose this ability.

(iv) ANATOMICAL CHARACTERS OF MYCELIUM AND SCLEROTIA In culture, hyphae are hyaline and thin-walled with clamp connections, frequently branching at the clamps from 1.8 to 5.0 μm wide becoming tightly interwoven when lanose, submerged hyphae may be contorted, up to 5.5 μm wide. Crystalline hyphal incrustations and intercalary swellings may be associated.

Sclerotial initials have hyaline mycelium with thin or only slightly thickened walls. Clamps occur at cross-walls. Hyphae are frequently branched, tightly interwoven, fused or coalesced. Internal hyphae inflate and become irregularly shaped up to 8.0 μm wide. As the sclerotial cortex darkens the hyphae of the medulla inflate further up to 15.0 μm, rarely more, to take on an elliptical or irregular shape. Cell contents are oily. Cortical hyphae are 5.6–8.0 μm wide, hyaline to yellowish, thick-walled, and angular in the layer nearest the medulla. The outer cells of the rind are angular and thick-walled, more compressed and compact than those of the rest of the cortex, but similar in form to those of the latter region, 2.5–6.0 μm wide with thick, dark-brown walls, but devoid of cell contents. When turgid and pricked under water the oily contents of the medullary cells are discharged into the water as a milky cloud. This is a good diagnostic feature for the SLTB as none of the other common sclerotial snow moulds show the character.

(v) CONSPECICITY WITH *C. PSYCHROMORBIDUS* Single-spore, monokaryon isolates of *C. psychromorbidus* which were dikaryotized by dikaryon LTB isolates were also dikaryotized by SLTB isolates (Traquair and Smith, 1982).

(c) Pathogenicity

Foliar symptoms of snow mould damage caused by the SLTB are similar to those caused by LTB *Coprinus*, but there is less mycelium and the fungus is not as aggressive as the latter. The disease may be produced on turf by inoculating field plots with isolates grown on sterilized rye grain during late summer. It also developed on acclimated winter wheat grown in sterile soil in controlled environment cabinets inoculated with cornmeal – soil – sand cultures incubated at 1°C for 3–4 days and then at – 3°C for 30 days.

Although *Agrostis* spp. turf grasses appear to be most severely damaged in the field by the SLTB *Coprinus* there is no detailed information on relative susceptibilities of other species and cultivars.

(d) Control

Usually the damage resulting from the SLTB is minor compared with that caused by the LTB phase and scarcely justifies the application of fungicides. However, it may occur in complexes (see p. 95) with the LTB and *Typhula* spp. Fungicides effective against the LTB phase will control the SLTB (Smith, 1974a).

6.3 FROST SCORCH, LEAF ROT, TIP BLIGHT, OR STRING OF PEARLS DISEASE

This disease has been reported on forage grasses, turf grasses and winter cereals from northern and eastern Europe, North America and northern Japan (Anon, 1931; Appel, 1933; Baudys, 1928, 1930; Bokura, 1926; Cash, 1953; Davis, 1933; Ekstrand, 1938; Flachs, 1935; Freckmann, 1919; Haskell, 1928; Heald, 1924; Howard, Rowell and Keil, 1951; Hungerford, 1923; Kreitlow, 1942; Matsumoto, 1928; Mikolajska, 1974; Mühle, 1953; Oettingen, 1934; Pape, 1926; Remsberg, 1940a; Remsberg and Hungerford, 1933; Sampson and Western, 1942; Smith, 1980f; Sprague, 1955; Stirrup, 1932; Stout, 1911; USDA Index of Plant Diseases, 1960; Van Luijk, 1934; Vang, 1945). It may be locally severe.

6.3.1 Symptoms

Infected plants may be stunted. Attacked leaves wilt, curl or roll, according to species and bleach almost white. Leaf tips may be narrowed to tendril-like form in species which normally have expanded leaves, and leaf bases may remain green. Sclerotia form in bead-like rows on the withered leaves (Kreitlow, 1942; Stout, 1911).

6.3.2 The causal fungus–*Sclerotium rhizoides* Auersw. (Botanizhe Zeitung 7:294–1849)

The sclerotia, borne superficially on the leaves may be oval, spherical, or oblong (Ekstrand, 1938; Stout, 1911) from 1 to 5 mm varying in size according to host. At first they are white or grey, darkening to almost black, becoming rough. In severe cases they are so abundant as to give a string-of-pearls appearance, in others they may be solitary on the leaves or in leaf axils (Kreitlow, 1942). The mycelium is white or grey, septate, coarse branched, without clamp connections, No fruits have ever been found (Baudys, 1930), naturally or in culture. The fungus grows well at 16°C (Stout, 1911), but three isolates studied by Smith* (1980f) were mesophilic. Sclerotia similar to those on grasses form in culture on lima bean agar or on the surface of sterilized bean pods (Stout, 1911). The fungus may be isolated by placing fragments of infected leaves on lima bean agar or potato-dextrose agar (Stout, 1911). Smith (umpublished) isolated a *Sclerotium* sp. (see below) by plating slices of surface-sterilized sclerotia on to BASM (potato malt) agar (see *Typhula*, p. 79). Attempts to obtain infection of leaves by mycelium were unsuccessful (Stout, 1911). Smith (unpublished) failed to obtain infection of seedlings of *Poa pratensis* L. and *Phleum pratense* L. grown in sterile culture inoculated with mycelium of *Sclerotium* sp. although sclerotia formed on root and shoot systems (see below).

Some of the reports of the disease from Norway and the USA (Davis, 1933; Ekstrand, 1938) appear to have been due to attacks by *Typhula* spp. (mainly *T. incarnata* Lasch ex Fr. – synonyms *T. graminum* auct. *non* Karst. and *T. itoana* Imai) according to Ekstrand, 1938, Remsberg, 1940), and others are doubtfully *S. rhizoides* (Anon, 1931: USDA Index of Plant Diseases, 1960). *T. graminum* and *Sclerotium rhizoides* were regarded by some workers as identical (Haskell, 1928), but this was disproved by Baudys (1930) and Jørstad (cf Ekstrand, 1938). Stout (1911) found that *S. rhizoides* from *Calamagrostis canadensis* in Wisconsin was identical with the fungus from European sources on a wide range of grass hosts including *Poa pratensis*. The first description of the fungus which appeared in 1849 was reviewed in the Botanische Zeitung 7: 294 and the Latin description is quoted by Stout (1911).

6.3.3 Host range and susceptibility

In turf grass *Agrostis* spp. (Ekstrand, 1938; Flachs, 1935; Kreitlow, 1942; Stirrup, 1932; Van Luijk, 1934) *Poa pratensis* (Kreitlow, 1942; Smith, 1980f; Stout, 1911) and fine-leaved *Festuca* spp. (Appel, 1933; Smith, 1980f; Van Luijk, 1934) are attacked by *S. rhizoides*. There is no published information on specific or varietal susceptibility in turf grasses.

* From *Festuca longifolia*, Agassiz, British Columbia, from *Poa pratensis* on the Saskatoon Campus of the University of Saskatchewan, and from an unidentified moorland grass at Alston in the Pennines, England.

6.3.4 Epidemiology

The mycelium is systemic and perennial. In the USA the disease becomes most conspicuous during April and May. The mycelium develops is the growing points and leaves are successively infected before they expand. The unaffected basal part of the blade becomes flattened out and the area is covered with mycelium just below the point where the next leaf in succession expands. The sclerotia are produced along the infected portion of the leaves or from the mycelium at the leaf base. Sclerotia mature in May and when ripe drop from the leaves with attached short strings of dry mycelium. They may germinate to produce mycelium, but their fate is uncertain. Infected leaves that appear at the same time as sclerotia from the previous season are present on the ground. The mycelium is perennial in the underground parts of the plant and may be a soil inhabitant. Infection of the aerial parts of the plant comes from these underground tissues (Stout, 1911).

The disease is said to be favoured by a long snow cover (Haskell, 1928), but there are conflicting opinions about the relationship of the disease to soil moisture. Oettingen (1934) observed that the disease does not occur in areas subject to regular spring and winter floods. Stirrup (1932) reported that it occurred on *Agrostis* spp. on an upland meadow where rainfall was unusually high. Mikolajaska (1974) found it was an important disease on river meadows needing drainage. Stout (1911) studied the disease in marsh meadows in Wisconsin.

6.3.5 Control

Improve fertility and raise soil pH for it seems probable that the disease is more severe where fertility (Oettingen, 1934; Stirrup, 1932) and soil pH is low (Stirrup, 1932). Howard, Rowell and Keil (1951) suggest collection of clippings when mowing diseased areas to remove inoculum from upper leaves and the use of a mercury fungicide if the disease is severe.

6.4 GREY OR SPECKLED SNOW MOULD OR TYPHULA BLIGHT

Several *Typhula* species or subspecies cause disease at low temperatures on overwintering plants. Four of these, *T. incarnata*, *T. ishikariensis* and its varieties *ishikariensis*, *idahoensis* and *canadensis* (Årsvoll and Smith, 1978; and see below) are important snow moulds of turf grasses (and also of forage grasses and winter cereals) from the cool temperate to the boreal regions of the northern hemisphere (Andersen, 1960; Årsvoll, 1975a; Årsvoll and Smith, 1978, 1979a,b; Dahlsson, 1973; Ekstrand, 1955a,c; Fushtey, 1980; Gould, Goss and Byther, 1977; Gulaev, 1948; Howard, Rowell and Keil, 1951; Imai, 1929, 1930, 1936; Jackson and Fenstermacher, 1969; Jackson, 1962a; Jamalainen, 1951, 1956, 1957, 1970, 1978; Jensen, 1970; Köck, 1976, and Dr L. Köck, pers. comm. 27 Feb. 1981; Kristinsson and Gudleifsson, 1976; Lebeau and Cormack, 1961; Lehmann, 1965; Dr N. Matsumoto, pers. comm. 7 Jan. and 12 Feb. 1981, 26 Nov. 1982; McDonald, 1961; Meiners, 1955; Ormrod, 1973; Potatosova, 1960; Røed, 1969; Remsberg, 1940a,b; Schmidt, 1976; Skipsna, 1958; Skirde, 1970; Smith, 1974a,b, 1975b, 1976, 1978, 1981; Sprague, 1950b, 1955; Stienstra, 1980; Tasugi, 1935; Taylor, 1972, 1974, 1975; Taylor and Fushtey, 1977; Tomiyama, 1961; Vaartnou and Elliott, 1969; Vang, 1945; Vanterpool, 1944; Volk, 1937; Wernham, 1941; Ylimäki, 1967, 1972).

The snow cover requirements, temperature optima and the lower osmotic pressure tolerance of their mycelium (Bruehl and Cunfer, 1971; Tomiyama, 1961) fit the *Typhula* snow moulds into different 'epidemiological niches' in relation to *Myriosclerotinia borealis* which is favoured by colder winters with a longer snow cover and *Microdochium nivale* which has a higher optimum temperature and which may cause severe damage in the absence of a snow cover (see section 6.1.5 (b)). The distribution of these pathogens in North America is not a simple latitudinal one as suggested by Ekstrand (1955) for Scandinavia and Lebeau and Cormack (1961) for Western Canada (Smith, 1974a).

Regional climates may be considerably modified by topography, aforestation and windbreaks which change the precipitation amounts and/or snow depth and duration. This may determine the incidence or dominance of a particular *Typhula* sp. In northeastern Saskatchewan (Lat. 54°N) for instance, *T. ishikariensis* var. *canadensis* is a common turf grass pathogen and it is important in the Peace River Region of northern Alberta and British Columbia (Davidson, 1986). On the other hand, *T. incarnata*, which can cause snow mould in the absence of a snow cover, is very uncommon in the Canadian prairies. However, *T. incarnata*, *T. ishikariensis* vars. *canadensis* and/or *ishikariensis* are important turf grass pathogens in southern Ontario (Fushtey, 1980), Minnesota (Stienstra, 1980; Sweets and Stienstra, 1976) and northern Wisconsin (Dr G. Worf, personal communication April, 1980; Smith, unpublished). In northern Idaho and northeastern Washington *T. incarnata* is the dominant species although *T. ishikariensis* is reported to be an occasional problem in the region (Gould, Goss and Byther, 1977; Smith, unpublished). In western Washington only *T. incarnata* is common (Gould *et al.*, 1978). Bruehl and Cunfer (1975) reported *T. idahoensis* Remsberg (*T. ishikariensis* (Imai) var. *idahoensis* Årsvoll and Smith) dominant on winter wheat on former grasslands or grass–sagebrush lands and *T. ishikariensis* (Imai) *T. ishikariensis* var. *ishikariensis* (Imai) Årsvoll and Smith, 1978) dominant on former forest lands in the Pacific Northwest of the USA.

T. ishikariensis Imai sensu Årsvoll and Smith (see below) has a circumpolar distribution in Asia, Europe and North America (Årsvoll and Smith, 1978; Imai, 1936; Dr N. Matsumoto, personal communication Jan. and 12 Feb. 1981). *T. idahoensis* Remsberg was described from Idaho and Montana (Remsberg, 1940a,b. It has a wider distribution that this in the Pacific Northwest of North America (Bruehl *et al.*, 1978; Bruehl and Machtmes, 1980), but its distribution limits are uncertain because of earlier taxonomic confusion (Årsvoll and Smith, 1978; Jamalainen, 1957; McDonald, 1961). Årsvoll and Smith (1978) have found that a morphologically distinguishable variety, *T. ishikariensis* Imai var. *canadensis* Smith and Årsvoll was widely distributed on grasses in Canada from British Columbia to Ontario and parts of the northern USA. To accommodate this variant Årsvoll and Smith proposed that the prior specific name *T. ishikariensis* should be used with *ishikariensis*, *idahoensis* and *canadensis* as varieties. Dr N. Matsumoto (pers. comm. 26 November, 1982) has found three biotypes of *T. ishikariensis* in northern Japan. One of these resembles var. *canadensis* in morphology, but is different genetically. The three varietal names in *T. ishikariensis* (sensu Årsvoll and Smith) will be used here.

Some isolates of *Typhula phacorrhiza* Fries (Smith, 1981a) have been found pathogenic on winter wheat (Seaman and Schneider, 1984) but others are non-pathogenic on Poaceae (Burpee *el al.*, 1985, 1987). An avirulent isolate from turf grass in Ontario has been shown antagonistic to *Typhula ishikariensis* in field tests (Burpee *et al.*, 1985, 1987).

6.4.1 Grey snow mould or typhula blight caused by *Typhula incarnata* Lasch ex Fr.

Snow mould of turf grasses caused by *T. incarnata* has been reported from Scandinavia generally (Jensen, 1970; Ylimäki, 1972), Norway (Årsvoll and Smith, 1979a; Smith, 1975b), Denmark (Jensen, 1974), Southern Sweden (Dahlsson, 1973; P. Lundin and P. Weibull, personal communication), The Netherlands (de Leeuw and Vos, 1970), Germany (Skirde, 1970),

Austria (Köck, 1976, and Dr L. Köck, personal communication), Switzerland (Smith, unpublished), the British Isles (Jackson, 1962a, 1963), Japan (Hosotsuji, 1977), western, central and eastern Canada (Fushtey, 1980; Platford, Bernier and Ferguson, 1972; Smith 1973, 1978 and unpublished; Vaartnou and Elliott, 1969), north western United States and northern Great Plains (Evans, 1973; Gould, Goss and Byther, 1977; Meiners, 1955; Sprague, 1950a), the mid-west and eastern United States (Jackson and Fenstermacher, 1969; Remsberg, 1940a; Wernham, 1941), New England (Dr J.M. Fenstermacher, personal communication).

The disease may develop in a mild form in winter in the absence of a snow cover, appearing first as patches of yellow–brown grass, 2–5 cm or so in diameter. Under a snow cover the patches increase in size and may become quite large, up to half a metre or so. On emergence from the snow the bleached grass leaves may be covered with a fairly sparse to dense white or greyish-white mycelium. The grey colour is probably due to atmospheric pollutants in the snow (Jackson and Fenstermacher, 1969). Faintly pink sclerotia, up to 5 mm in diameter may be found on or within the infected tissues at this time, often in the crowns of the plants (Fig. 8, colour plate section). They are sometimes firmly attached to the leaves and crowns of the plant. Where snow cover has been light and/or of short duration plant injury is often superficial, but this fungus may cause severe damage to plant crowns and even roots under a long, deep snow cover. Under such conditions other snow mould fungi are often present in adjacent patches.

(a) The fungus–*Typhula incarnata* Lasch ex Fr.

Corner (1950) considered that *T. itoana* from Japan (Imai, 1929) and *T. incarnata* from Europe, as redescribed by Donk (1933 from Rabenh. Fung. Europ. No. 1313) were identical. *T. incarnata* Lasch ex Fr. (Fries, 1836) has priority over. *T. itoana* Imai (Corner, 1950, 1970; Lehmann, 1964).

There is considerable confusion in the literature of *T. incarnata* with *T. graminum* (Lehmann, 1964; McDonald, 1961; Remsberg, 1940a, b; Tasugi, 1935; Vang, 1945, Volk, 1937) and other sclerotial fungi such as *Sclerotium rhizoides* Auersw. (Hungerford, 1923) and *S. fulvum* Fr. (Young, 1937). Røed (1969) showed that isolates of *T. graminum* Karst. from CBS Baarn, *T. itoana* from Canada (originally from Remsberg) *T. incarnata* Lasch ex Fr. from Japan (from Tomiyama) and three *T. itoana* isolates from winter rye and *Phleum pratense* were interfertile. He regarded these fungi as strains of *T. incarnata* Lasch ex Fr. Berthier (1974, 1977) places *T. incarnata* Lasch ex Fr.? in a new sub-genus *Microtyphula* in the genus *Typhula* (Fries emend. auct. Syst. Myc. 1821 1: 494).

Sclerotia of *T. incarnata* from plants are faintly pink and smooth when young, darkening to reddish-brown or dark-brown and wrinkling as they dry. They are variable in shape from subglobose to elongated, flattended or irregular, 1 – 5 × 0.5–3 mm (Fig. 8, colour plate section and 6.6c). There is no pedicel as in *T. phacorrhiza* Fries. In autumn, 1 to 3 sporophores may develop from each sclerotium. These are often pubescent with hairs clasping the sclerotium. They may be straight, slightly curved, simple or less commonly with branched stipes and clubs, 5–30 mm tall. The white stipes are 0.5–1 mm and the rose-coloured clubs 0.2–3 mm in diameter (Fig. 9, colour plate section). The clubs vary in shape from elongate-fusiform to flattened and branched. Basidia forming the hymenium on the clubs are tetrapolar, and broadly clavate, 5–8 × 27–35 μm with ovoid

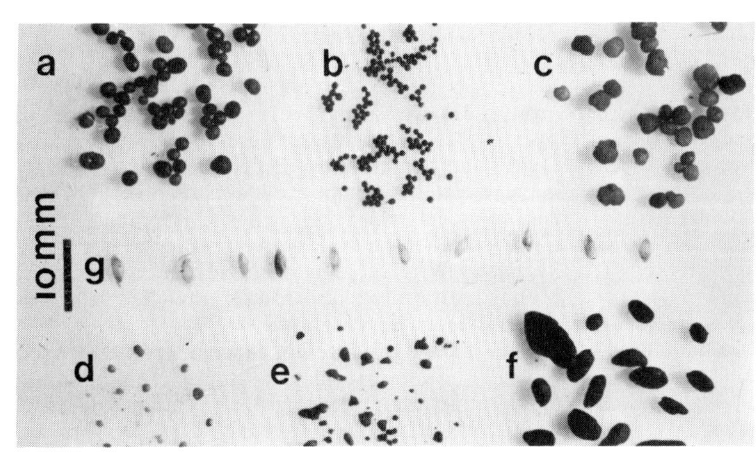

Figure 6.6 Sclerotia of snow mould fungi. (a) *Typhula ishikariensis* var. *ishikariensis* (b) *T. ishikariensis* var. *canadensis* (c) *T. incarnata* (d) *Acremonium boreale (Nectria tuberculariformis).* (e) *Coprinus psychromorbidus* (f) *Sclerotinia (Myriosclerotinia) borealis* (g) Seeds of *Poa annua* for size comparison.

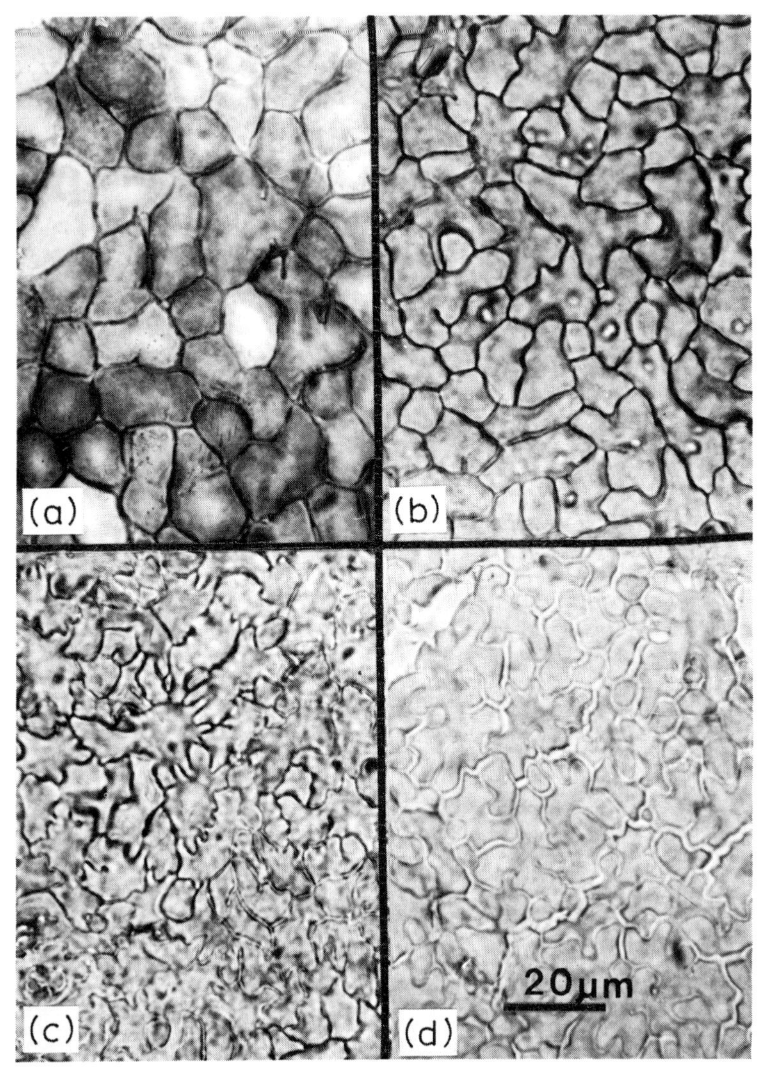

Figure 6.7 Rind patterns of sclerotia of *Typhula* spp.
(a) *Typhula ishikariensis* var. *ishikariensis*
(b) *T. ishikariensis* var. *canadensis*
(c) *T. ishikariensis* var. *idahoensis*
(d) *T. incarnata*

basidiospores flattened or incurved on one side above an apiculus (pip-shaped) 3–5 × 5–10 µm (Årsvoll and Smith, 1978, 1979a; Imai, 1936; Remsberg, 1940a; Smith, 1981a; Sprague, 1950).

The fungus grows well on malt, potato–malt or potato–dextrose agars. Sweets and Stienstra (1976) have examined the cultural requirements of *T. incarnata* (and *T. ishikariensis*) in detail. The optimum pH range for mycelial growth is 5–8 °C (Sweets and Stienstra, 1976; Tomiyama, 1961) with limits at 3 and 10 °C. Frozen agar will not support the growth of *T. incarnata*, probably because it requires a higher water potential than *M. borealis* which will continue growing on media that is frozen. The differences in response to temperature observed by Tomiyama (1961) and confirmed by Bruehl and Cunfer (1971) may be related to osmotic pressure tolerance. On agar culture media mycelial growth is white, radial, sometimes stranded and if derived from sclerotia the hyphae are dikaryotic with clamp connections. Monokaryotic mycelium from germinated basidiospores is much less vigorous.

Sclerotia are produced in concentric rings or piled centrally, often irregular in shape. The optimum temperature for mycelial growth is usually 9–15 °C and the range from below −6 to 21 °C (Årsvoll, 1975a;

Table 6.1 Temperature ranges for *Typhula* snow moulds in artificial culture

Species/variety	Temperature °C Minimum	Optimum	Maximum	Reference
Typhula incarnata Lasch ex Fr.	<0	9 to 12	18	Remsberg (1940a)
	<−5	10 to 15	22 to 23	Ekstrand (1955a)
	<−5	8 to 15	22 to 23	Tasugi (1935)
	0	10	25	Remsberg and Hungerford (1933)
	0	8	25 to 30	Volk (1937)
		10		Vang (1945)
	−6	9 to 12	21	Årsvoll (1975a)
	<−6	12 to 15	21	Årsvoll and Smith (1978)
	−7	7 to 15		Tomiyama (1961)
	−10	7	22	Lehmann (1965a, b)
T. ishikariensis Imai var. *ishikariensis* Årsvoll and Smith	<−6	9 to 12	20	Årsvoll and Smith (1978)
= '*T. borealis*' Ekstrand	−5	10	21	Ekstrand (1955a)
T. ishikariensis Imai var. *idahoensis* Årsvoll and Smith	6	9	18	Årsvoll (1975a)
	<−6	9 to 12	18 to 20	Årsvoll and Smith (1978)
	< 0	9 to 12	18	Remsberg (1940a)
	<−5	5	20	Dejardin and Ward (1971)
T. ishikariensis Imai var. *canadensis* Smith and Årsvoll	<−6	6 to 9	21	Årsvoll and Smith (1978)
	−5	5 to 10	15 to 20	Ekstrand (1955a)
= '*T. hyperborea*'	< 0	10	16	Potatosova (1960)

From Smith (1969a)—revised.

Årsvoll and Smith, 1978; Smith, 1980b) (Table 6.1). It is not always easy to distinguish single dark sclerotia of *T. incarnata* from those of *T. ishikariensis* vars., but those of *T. incarnata* are resilient when turgid and those of *T. ishikariensis* are brittle when pressure is applied with a needle point. The cortical cells of *T. incarnata* sclerotia are lobate, interlocking like a jigsaw puzzle. (Compare with *T. ishikariensis* vars. in Figs 6.7 a,b,c,d.)

(b) Isolation of the fungus and pathogenicity tests

Allow the sclerotia to imbibe water and then surface sterilize them with 70% ethyl alcohol or hypochlorite solution (1% chlorine) for 2 min and then wash 5–7 times in sterile water. Slice the sclerotia with a sterile scalpel and plate out on malt or potato – malt agar (Smith, 1981a). Incubate at 10–15°C.

For pathogenicity tests in controlled environments culture the fungus in PDA broth at 10–15°C, macerate the mycelium and apply to run-off on test grasses with an atomizer. Only one isolate should be used because mixing different ones may reduce pathogenicity (Årsvoll, 1976b; Smith and Årsvoll, 1975). Incubate plants in the dark in moist chambers of the kind suggested by Blomqvist and Jamalainen (1968), Årsvoll (1975a) or Smith (1981a) for up to 10 weeks at 1–2°C. Place plants on a greenhouse bench to recover before rating. Macerated mycelium gives more reliable infection than sclerotia as inoculum.

(c) Host range and susceptibility

T. incarnata is a weak, unspecialized pathogen (perthophyte) on the Gramineae (Bruehl, 1967a, b; Cormack and Lebeau, 1959; Hirai, 1956; Imai, 1929; Lehmann, 1965c; Tomiyama, 1961; Wernham and Chilton, 1943). Its sclerotia may form on dead or moribund parts of other plants. Host resistance is assumed to be non-specific (Lehmann, 1965c) and quantitative governed by several genes each possessing an additive effect (Årsvoll and Smith, 1979a). There are considerable differences in resistance between species and cultivars of grasses to *T. incarnata* (Årsvoll, 1977); Gould, Goss and Byther, 1978; Schmidt, 1976; Tezuka and Komeichi, 1980; Vargas, Beard and Payne, 1972; Wernham, 1941). However, fertility conditions of the turf when attacks develop and the duration of the snow cover have a considerable effect on damage severity, in some cases as important as inherent resistance of grass (Jackson and Fenstermacher, 1969). In diploid ryegrasses (*L. perenne*) those tested in Northern Japan from high latitudes showed best winter survival when *T. incarnata* was the most frequent pathogen. This rating did not change with increased autumn fertilization or with more frequent cutting. Good survival was correlated with poor autumn growth because of early dormancy (Tezuka and Komeichi, 1980). The effect of plant age on the winter hardiness of grasses has been reviewed by Årsvoll (1977). Usually first year grasslands are more severely damaged by snow moulds than older ones. Young plants were found to be more susceptible to snow moulds in pot experiments than larger, well-established ones (Årsvoll, 1977). Lehmann (1965c) noted that there was an interaction between winter hardiness and susceptibility. Winter hardiness and snow mould resistance are both increased by reserves of carbohydrates and prolonged deep snow cover subjects the plants to nutrient exhaustion and reduces disease resistance (Bruehl and Cunfer, 1971; Tomiyama, 1961). Pre-hibernal fertilization and management greatly influences carbohydrate reserves, dormancy, hardening and subsequently freezing tolerance and snow mould resistance of grasses. Hirai (1956) found that *T. incarnata* produced a cytotoxin which provoked a hypersensitive reaction in resistant plants. This was a simple acidic substance.

(d) Epidemiology

Although *T. incarnata* will make mycelial growth from −7 to 21°C (Årsvoll, 1975a; Lehmann, 1965b; Smith, 1981a; Table 6.1), the optimum, 9–15°C, for most isolates is a little higher than for other graminicolous *Typhula* spp., but lower than *M. nivale* (Smith, 1981a; Table 6.1). Bruehl and Cunfer (1971) found that *T. ishikariensis* var. *idahoensis* was virulent as low as −1.5°C on wheat, but that neither *M. nivale* nor *T. incarnata* caused injury at this temperature. Although grass plants may be damaged from late autumn to early spring in wet, cold climates when there is little or no snow cover, more often the fungus is active under deep, prolonged snow (Sprague and Rainey, 1950) especially when the turf surface remains unfrozen (Jackson and Fenstermacher, 1969; Köck, 1976).

Sclerotia form on and in diseased tissues as mycelial spread declines. This change probably starts before snow melt since immature sclerotia can be seen on emerging plants. These oversummering structures are resistant to high temperatures and desiccation and remain dormant until autumn.

Potatosova (1960a) found that sclerotia would germinate in 4–6 weeks if kept moist at low temperatures (1.4–13.3°C). In nature, sclerotial germination takes place in cool, moist weather in autumn either by production of vegetative mycelium or of sporophores. In laboratory tests in sand, the optimum temperature for mycelial germination was 2–10°C, and high substrate moisture and atmospheric humidity above 70% RH was necessary. Light is not necessary for mycelial germination (Lehmann, 1965b), but for basidiocarp production short wave ultra-violet light is needed (Lehmann, 1965b; Remsberg, 1940a; Tasugi, 1935).

Basidiospores are discharged during rainy or foggy weather (Sprague and Rainey, 1950) and dispersed by wind and rain. The optimum temperature for their germination is 9–20°C according to Tomiyama (1961), but 12–17°C for Lehmann (1965b) and they remained viable after 4 months storage at − 10 − 15°C. Basidiospores germinate to produce a haploid mycelium which is less pathogenic than when it is dikaryotized (Lehmann, 1965c).

Weather conditions largely determine the incidence and time of development of sporophores. In Rhode Island, this varies from late September to early November (Jackson and Fenstermacher, 1969). In England in 1963, in the heaviest recorded outbreak of the disease, sporophores developed in late September (Jackson and Fenstermacher, 1969). In the USSR, sclerotia germinated from September to October (Potatosova, 1960a). Sprague and Rainey (1950) reported that sclerotia germinated in October and spores were trapped until 21 December in Washington. In Central Norway and Switzerland, sporophores are freely produced in October, although sclerotia commence to germinate in Norway in September (Smith, unpublished). The extreme climate of the Canadian Prairies is very unfavourable for sporophore production of graminicolous *Typhula* spp. (Smith, unpublished). Basidiocarps are not often found on turf in western Washington (Gould, *et al.*, 1977). While some outbreaks of the disease probably result from basidiospore infection (Sprague and Rainey, 1950), mycelium produced by direct germination of sclerotia is more likely to start the disease (Hindorf, 1980; Lehmann, 1965c; Sprague and Rainey, 1950; Tasugi, 1935; Tomiyama, 1961).

(e) Control

Early studies in Michigan by Tyson (1936) showed that bentgrasses fertilized with inorganic nitrogen, e.g. ammonium sulphate, urea or sodium nitrate suffered less from snow mould than those where the same amount of nitrogen supplied as cottonseed meal, dried blood or dried digested sewage was given. Although the causal agent of the snow mould was not specified, symptoms of *T. incarnata*, common in Michigan (Vargas and Beard, 1971) are illustrated by

Tyson (1936). Practical experience suggests that, whatever the cause, snow mould damage to turf grass is aggravated by excessive or unbalanced nitrogenous fertilizer, especially if applied late in the growing season when it produces lush, unseasonable herbage growth (see above and Årsvoll and Larsen, 1977).

The general turf management methods used for the cultural control of other snow moulds should be used. See in particular, control of *Microdochium nivale* and *T. ishikariensis* (pp. 65 and 87).

Published data on the control of grey snow mould often do not specify the causal organism. This makes the evaluation of the effectiveness of the disease control given by resistant cultivars difficult. Regional evaluation of these is necessary as is the case with other snow moulds (Smith, 1980a). Unhardy cultivars, poorly adapted to local environment should be avoided (Årsvoll, 1977; Årsvoll and Smith, 1979a). Most of the common cultivars of *Poa pratensis* were susceptible in Michigan, but Adorno, Monopoly and some selections from New Jersey showed fairly good resistance (Vargas, Beard and Payne, 1972).

The bentgrasses (*Agrostis* spp.) are often severely damaged by *T. incarnata* (Smith, 1978; P. Weibull, personal communication; Wernham, 1941). Bentgrass lines, more resistant than the older cultivars such as Highland and Astoria (*A. tenuis* Sibth.), were identified by Gould, *et al.* (1978). The same strains were tested for resistance to *M. nivale* in western Washington (Table 6.2).

Table 6.2. Disease ratings for snow moulds caused by *T. incarnata* in western Washington and *M. nivale* in eastern Washington on the same lines of *Agrostis* spp. in golf-green turf. (Data calculated from Gould *et al.*, 1978).

Species	No. of entries	Average rating*	
		T. incarnata	*M. nivale*
Agrostis tenuis Sibth.	33	3.97	1.43
A. stolonifera L.	73	1.74	3.34
A. canina L. ssp. canina Hwd.	9	2.67	2.44
A. gigantea Roth.	3	5.00	1.33

* The rating was on a scale of 1 to 5, where 5 was least disease.

A. gigantea (redtop) lines were most resistant to *T. incarnata*; few of the creeping bentgrasses (*A. stolonifera*) showed much resistance and on average were the least resistant to *T. incarnata*, while several of the browntop bentgrasses (*A. tenuis*) showed considerable resistance. The velvet bentgrasses were intermediate. Towards *M. nivale*, redtop and browntop were most resistant but the correlation between resistance to *T. incarnata* and *M. nivale* was very weak except in the case of the velvet bentgrasses.

(f) Chemical control of snow mould caused by *T. incarnata*

Where T. incarnata causes yearly problems, fungicides are needed for effective control, especially on fine turf of golf and bowling greens composed of *Agrostis* spp. and *Poa annua*.

Mercurous and mercuric chlorides, singly and in combination have been used to control snow moulds of turf grasses, before *Typhula* spp. were known causes, from the late 1920s (Dahl, 1934; Evans, 1973; Monteith, 1927; Noer, 1944; Smith, 1965; Tyson, 1936; Wernham and Kirby, 1943). Dosages of up to 125 g a.i./100 m^2 (of the salts) were used, often in the proportion of 2:1 mercurous to mercuric chloride. However, in eastern Washington cadmium succinate, phenyl mercuric acetate and thiram fungicides were more effective than the inorganic mercurials against *T. incarnata* snow mould (Meiners, 1955). In Rhode Island, Jackson and Fenstermacher (1969) confirmed the effectiveness of phenyl mercuric acetate, mercurous and mercuric chloride mixtures and cadmium succinate fungicides in preventing this snow mould. Vargas and Beard (1970) found that a granular formulation of chloroneb gave better control than a wettable powder formulation of the same material. It was better than mercury chloride mixtures. In Idaho, where *T. incarnata* was the dominant snow mould, Gould *et al.* (1977) obtained best control of the disease with early autumn, late autumn and spring treatments with (1) benomyl and chlorothalonil, (2) mancozeb, mancozeb and chloroneb, (3) mancozeb, benomyl and chloroneb and (4) mancozeb, thiophanate methyl and chloroneb formulations. In Michigan, Vargas and Beard (1971) found that single applications of chloroneb or mercurous + mercuric chloride fungicides made one month before the development of a permanent snow cover were almost as effective as those made just before the snow cover developed. Taylor (1974) obtained as effective control of *T. incarnata* snow mould in the interior of British Columbia with one application of chloroneb as with mercurous + mercuric chloride. A wettable powder formulation of quintozene gave good control when applied at high dosage. Quintozene was also effective applied dry with a fertilizer.

Dickens and Butler (1977) obtained good to fair control of *Typhula* sp. (unspecified) on *Agrostis stolonifera* and *Poa pratensis* turf in Colorado with chloroneb, quintozene, iprodione, inorganic mercury chloride mixtures and benomyl (at abnormally high dosage) fungicides. Worf (1977) in Wisconsin found chloroneb alone unreliable against grey snow mould

(*Typhula* sp.) on turf of *A. stolonifera* and *P. annua*. Quintozene was more consistent, especially at high dosage and mercurous + mercuric chloride, combined with quintozene most dependable. In the following season, a granular mercury chloride fungicide, especially when combined with quintozene or chloroneb gave consistent control of typhula snow mould (Worf, 1977). In New York State, *T. incarnata* snow mould, on turf of *A. stolonifera* and *P. annua*, was controlled adequately only by inorganic mercury chloride and cadmium fungicides in 1977 (Smiley and Craven, 1978). However, in 1978, effective control of snow mould caused by the same fungus was obtained with triadimefon, and inorganic mercury chlorides with chloroneb and a hydantoin fungicide at high dosage (Thompson, Craven and Smiley, 1979). In the same region, Marion, Landscoot and Terry (1979) obtained best control of *T. incarnata* snow mould with thiram at high dosage, chlorothalonil + cadmium followed by combinations of chlorothalonil with thiram and chloroneb. In 1980, Marion, Vangellow and Yount (1981) obtained most effective control of the same snow mould with chlorothalonil, chlorothalonil + a sticker and combinations of thiram and cadmium with other fungicides. Iprodione was very efficient at high dosage, but completely ineffective in controlling the disease at low dosage. In New Hampshire, iprodione, quintozene, cadmium succinate, chlorothalonil + chloroneb and triadimefon were very effective against *T. incarnata* on bentgrasses (Nutter *et al.*, 1979). In New York State, Smiley and Craven (1981) controlled grey snow mould (*Typhula* spp.) with chlorothalonil, iprodione, triadimefon, mercurous + mercuric chloride and cadmium fungicides. In Ontario, Burpee and Goultry (1984) obtained excellent control of grey snow mould (*Typhula* spp?) with a granular product containing quintozene on creeping bentgrass turf.

6.4.2 Speckled snow mould caused by *Typhula ishikariensis* Imai and varieties

Many published records of typhula snow mould on turf grasses do not specify whether *T. incarnata* or *T. ishikariensis* was involved. *T. ishikariensis* snow mould develops where winters are longer and more severe than is the case with *T. incarnata* (Årsvoll, 1975a; Ekstrand, 1955a; Jamalainen, 1974; Kristinsson and Gudleifsson, 1975; Smith, 1981b). Since the snow cover develops earlier and remains longer in these regions, turf grasses probably suffer more because of the greater depletion of energy reserves (Tomiyama, 1955, 1961) than under the shorter duration snow cover necessary for *T. incarnata*. However, there is also evidence from infection experiments that *T. ishikariensis* is more aggressive towards grasses than *T. incarnata* (Årsvoll, 1976b, 1977; Cormack and Lebeau, 1959; Wernham and Chilton, 1943).

Imai (1930) described *T. ishikariensis* on wheat, grasses and red clover in Japan and Remsberg (1940a) described the similar graminicolous *T. idahoensis* in North America. Røed (1956) and Jamalainen (1964) in Scandinavia and McDonald (1961) in Canada considered them synonymous. Ekstrand (1955a) described two 'new species', *T. borealis* on winter cereals, and some non-grass plant species and *T. hyperborea* on cereals and grasses only from northern Sweden, Norway and Finland (Årsvoll, 1975a; Ekstrand, 1955a). McDonald (1961) suggested that these were also *T. ishikariensis*. No type material of these 'species' could be found (Årsvoll and Smith, 1978). Smith (1973, 1974a) found a 'new' graminicolous *Typhula* sp., temporarily called 'FW', in Western Canada. Årsvoll and Smith (1978) described this as *T. ishikariensis* Imai var. *canadensis* and at the same time reduced *T. idahoensis* Remsberg to a variety of *T. ishikariensis*. In 1973, *T. ishikariensis* Imai was identified from winter cereals in Idaho and Washington (Bruehl, Machtmes and Kiyomoto, 1975), but Bruehl and Machtmes (1980) consider this and var. *idahoensis* separate species, and do not recognize var. *canadensis*. *T. borealis* of Ekstrand (1955a) and *T. ishikariensis* Imai var. *ishikariensis* Årsvoll and Smith (1979a) may be the same fungus although genetical evidence is lacking. They are morphologically very similar. *T. borealis* and var. *ishikariensis* have a host range wider than the Poaceae (Årsvoll, 1975a; Ekstrand, 1955a). *T. hyperborea* (Ekstrand, 1955a) and *T. ishikariensis* var. *canadensis* appear to be similar morphologically and in host range (Årsvoll and Smith, 1978), being confined to grasses and cereals. An isolate of *T. borealis* from tulips received from Moscow Botanic Gardens in 1980 (Smith, unpublished) was similar morphologically to var. *ishikariensis*.

T. ishikariensis Imai sensu Årsvoll and Smith (1978) has been reported causing snow mould of turf grasses in northern Japan (Hosotsuji, 1977), northern Scandinavia (Årsvoll and Smith, 1979a; Jensen, 1970; Smith, 1975b; Ylimäki, 1972), Canada (Allen, Bernies and Ferguson, 1976; Årsvoll and Smith, 1978; Ekstrand, 1955c; Fushtey, 1975, 1980; Lebeau and Cormack, 1961; Smith, 1976; Vaartnou and Elliott, 1969; Vanterpool, 1944), northern United States (Gould *et al.*, 1977; Remsberg, 1940b; Smith, unpublished; Sweets and Stienstra, 1976; Wernham, 1941; Dr G.J. Worf, personal communication, 1980), north eastern USA (J.M. Fenstermacher, personal communication).

Since it has been found on grasses, cereals or other

plants in other northern or mountainous regions (Kristinsson and Gudleifsson, 1975; Lebeau and Logsdon, 1958; Potatosova, 1960; Procenko, 1967; Schmidt, 1976; Skipsna, 1958; Smith, unpublished) its range on turf grasses is likely to be wider than reported.

(a) The fungi

(I) *TYPHULA ISHIKARIENSIS* IMAI VAR. *ISHIKARIENSIS* ÅRSVOLL AND SMITH (1978) = *T. ISHIKARIENSIS* IMAI (1930) = *T. BOREALIS* EKSTRAND (1955A)? Sclerotia are erumpent, readily detached from the host, globose to subglobose or slightly flattened, light brown to almost black (0.3–) 0.5–1.5 (–2) mm diam., surface smooth to rough (Fig. 8, colour plate section). Rind cells are fairly regular in outline, moderately lobate, rarely digitate (Fig. 6.7a). Sporophores are 3–20 mm in length. In autumn, 1–3 sporophores are produced from each sclerotium; they are erect, straight or curved, rarely branched, fusiform, occasionally flattened and ramose (0.3–)0.5–1 (–3) mm broad, greyish-white to light brown, stipe filiform, slender, darker than the fertile portion. Basidiospores are similar in shape to those of *T. incarnata*, ovoid, ellipsoidal, flattened to incurved on one side above a pointed apiculus (5–) 6–8 (–11) μm. On agar media aerial mycelium is often sparse, sclerotia single, scattered or in concentric rings, on the surface or submerged in the agar (Årsvoll and Smith, 1978).

Ekstrand's description of *T. borealis* (1955) is very similar to the above, except that the sclerotia are slightly smaller, up to 1.5 mm and the basidiospores are slightly longer and narrower, 5.5–13.25 × 2.0–4.5 μm. The host range of *T. borealis* on winter cereals and grasses, clover, winter rape and beets is similar to that of *T. ishikariensis* var. *ishikariensis* (Imai 1930; Ekstrand, 1955a; Tomiyama, 1961; Ylimäki, 1969). The cardinal temperatures for *T. borealis* were : minimum, – 5; optimum, a little above 10; maximum, slightly above 20°C (Ekstrand, 1955a). The optimum pH for mycelial growth in culture for *T. ishikariensis* is 5–8 (Tomiyama, 1961) and 5–7 (Sweets and Stienstra, 1976).

(II) *TYPHULA ISHIKARIENSIS* IMAI VAR. *IDAHOENSIS* (REMSBERG) ÅRSVOLL AND SMITH (1978) = *T. IDAHOENSIS* REMSBERG (1940A). Sclerotia are erumpent, less easily detached from the host than those of var. *ishikariensis*, globose to subglobose or slightly flattened, brown to almost black, 0.5–2 mm diam., surface often rough and ridged; rind cells are irregular in outline (Fig. 6.7c). On sclerotia germinated outdoors in Norway, lobate, often digitate sporophores were 3–14 mm tall, clavulae elongate, fusiform, rarely ramose, sometimes inflated, 0.5–1.5 mm broad, greyish-white to light brown, stipe filiform, slender, darker than the fertile portion. Basidiospores are ovoid to ellipsoidal, flattened or incurved on one side above a pointed apiculus (6–) 7–9 (–13) × (2.5–) 3–5 (–8) μm (Årsvoll and Smith, 1978). Bruehl and Cunfer (1975) give measurements of collections of both vars. *idahoensis* and *ishikariensis*. Dimensions given by Remsberg (1940a) for *T. idahoensis* are similar to those above.

On agar culture media, mycelium is usually sparse, but occasionally abundant and floccose in some isolates. Sclerotia are single, clustered or tending to form in concentric rings, on the surface and submerged in the agar var. *idahoensis* is restricted to hosts in the Gramineae (Bruehl and Cunfer, 1975; Remsberg, 1940a).

On the basis of morphology Potatosova (1960) in the USSR considered that *T. idahoensis* Remsberg was synonymous with *T. graminearum* Gulaev (1948), *T. humulina* A. Kuznetsova (1953) and *T. borealis* Ekstrand (1940a). While the general description given by Potatosova (1960) agrees with that of *T. idahoensis* Remsberg (1940a), dimensions are different: sclerotia, 0.3–2 mm; clavulae, 0.5–5.0 × 0.3–1 mm; stipae, 3–7 × 0.1–0.5 mm; basidia, 27–31.5 × 5.8–7.7 μm; basidiospores, 9.7–10.6 × 4.6–5.2 μm.

(III) *TYPHULA ISHIKARIENSIS* IMAI VAR. *CANADENSIS* SMITH AND ÅRSVOLL (1978) = *T. HYPERBOREA* EKSTRAND (1955A)? Sclerotia are readily detached from the host or from abundant wefts of mycelium spanning the leaves (Fig. 12, colour plate section); they are globose to subglobose or elongate-oval, light brown to almost black (0.2–) 0.3–0.8 (–1.6) mm diam., surface smooth, often with attached hyaline hyphae when fresh, rind cells fairly regular in outline to very irregular, lobate, sometimes digitate (Fig. 6.7b). Sporophores are (1–) 3–6 (–11) mm tall, clavulae elongate-fusiform, sometimes inflated, (0.2–) 0.4–0.8 (–1.4) mm broad, greyish-white to light brown, stipe erect, slender, darker than the fertile portion (Fig. 10, colour plate section). Basidiospores are ovoid to ellipsoidal, flattened or incurved on one side above a pointed apiculus, (5–) 6–8 (–11) × (2–) 3–4 (–4.5) μm. Fig. 6.8. On agar media, aerial mycelium is abundant floccose; sclerotia are formed in the aerial mycelium, usually on the surface, but occasionally submerged in the agar, scattered or in radial rows rather than in concentric rings. The hosts are grasses, winter cereals and rarely forage legumes in northern North America (Årsvoll and Smith, 1978). In Japan (Dr N. Matsumoto, personal communication, 7 Jan. and 12 Feb. 1981 and 26 Nov. 1982) a fungus similar morphologically to var. *canadensis* has a similar host range.

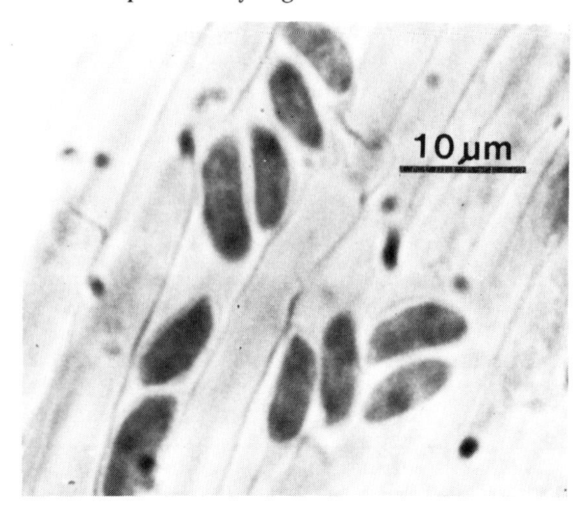

Figure 6.8 Basidiospores of *Typhula ishikariensis var. canadensis*.

T. hyperborea described by Ekstrand (1955a) from northern Scandinavia has similar sclerotia to the above, up to 1.5 mm in diameter with sporophores darker in colour than *T. borealis*. Basidiospores were 5.5–11.0 µm long × 2.75–5.75 µm wide. The aerial mycelium was often fluffy in culture and the hosts were grass species only. The description fits var. *canadensis*. The cardinal temperatures for *T. hyperborea* were: minimum, −5; optimum, 5–10; maximum, 15–20°C (Ekstrand 1955a). This is similar to those for var. *canadensis* (Årsvoll and Smith, 1978). The *Typhula* sp. reported by Vanterpool in 1932 from turf in Saskatoon (1943) with sclerotia 0.5–1.0 mm in diameter was probably the var. *canadensis*.

(b) Symptoms and diagnosis of *T. ishikariensis* vars.

In all varieties of *T. ishikariensis*, field symptoms are similar to those caused by *T. incarnata* (q.v.) except that sclerotia and sporophores of all vars. of *T. ishikariensis* are never red or pink as in *T. incarnata* and sclerotial characters are quite different. Compared with the orange–brown or pinkish sclerotia of *T. incarnata* those of *T. ishikariensis* range from light amber to very dark brown or almost black when dry. The sclerotia of the latter are never gelatinous or 'rubbery' like those of *T. incarnata* when swollen with water. The rinds of *T. incarnata* sclerotia are much more difficult to remove than those of *T. ishikariensis* vars. In *T. ishikariensis* vars. sporophores vary in colour in different collections of the same variety from greyish-white to light brown. The stipe base is always darker than the clavula.

It is very difficult to determine the variety of *T. ishikariensis* if only a few sclerotia are available. While there are differences in basidiospore morphology the fruiting stage may not be available. Ekstrand (1955a) used basidiospore size and shape successfully to separate the 'species' *T. borealis* and *hyperborea*, but basidiospore characters do not provide a convenient diagnostic tool. The slightly lower optimum temperature of var. *canadensis* and more abundant scattered sclerotia in culture distinguishes the variety from the other two. Probably the best single diagnostic feature is rind character (see text, Table 6.3 and Fig. 6.7). The rind may be separated from the medulla of a sclerotium with a needle when fully swollen in water and softened for 15–30 min in a nearly boiling 10% aqueous solution of potassium hydroxide. As a final resort use may be made of the dimon mating technique to separate vars. *idahoensis* and *canadensis* from var. *ishikariensis* (Smith and Årsvoll, 1978).

Bruehl and Machtmes (1980) who do not recognize var. *canadensis*, consider that there is a continuum of variation in cultural characters between *T. idahoensis* Remsberg and *T. ishikariensis* Imai. They suggest the use of *T. ishikariensis* sensu Årsvoll and Smith (1978) where there was no need to take the classification further. This is agreed. However, it is pointed out that the vars. *canadensis* and *idahoensis* which showed the greatest morphological differences were closest genetically (Årsvoll and Smith, 1978.)

(c) Host range and susceptibility

T. ishikariensis is considered to be a more aggressive snow mould than *T. incarnata*. Varieties *ishikariensis* and *canadensis* are reported to have wider host ranges than var. *idahoensis* (Årsvoll and Smith, 1978; Bruehl and Cunfer, 1975; Ekstrand, 1955a; Imai, 1930; Ylimäki, 1969). Monokaryons are less pathogenic than dikaryons in var. *idahoensis* (Kiyomoto and Bruehl, 1976). There was no differential virulence to host cultivars although dikaryons showed great differences in virulence on a particular wheat cultivar. This is also the case with var. *canadensis* on winter rye cultivars (Smith, unpublished). Most grasses are susceptible to *T. ishikariensis* (Årsvoll, 1975a; Kristinsson and Gudleifsson, 1975; Lebeau and Logsdon, 1958; Remsberg, 1940b; Schmidt, 1976; Smith and Årsvoll, 1975; Wernham and Chilton, 1943), including turf grasses in *Agrostis* spp., *Poa pratensis* L., *Festuca rubra* L., *Lolium perenne* L., *Festuca pratensis* Huds. and *Phleum pratense* L. Grass cultivars and strains from northern regions are usually more resistant than those from the south (Adachi, Miyashita and Araki, 1976; Andersen, 1966; Ekstrand, 1955a; Jamalainen, 1974). Vaartnou and Elliott (1969) found considerable differences in susceptibility to *T. ishikariensis* (probably var.

Table 6.3 Some morphological and cultural characters useful in differentiating between *Typhula ishikariensis* varieties

Character	var. *ishikariensis*	var. *idahoensis*	var. *canadensis*
Sclerotial attachment to plant tissue	Usually firmly attached	Tend to be superficial	Quite superficial, easily detached, from abundant aerial mycelium.
Sclerotial diameter (mm)	0.3–2	0.5–2	0.2–1.6
Sporophore height (mm)	4–20	3–14	1–11
Aerial mycelium on BASM* agar at 6 °C	Usually little	Usually little	Abundant
Sclerotial arrangement in culture on BASM* agar at 6 °C	Sclerotia in concentric rings	Sclerotia in concentric rings or in a pile	Sclerotia scattered abundant and suspended in aerial mycelium
Superficial appearance of rind of imbibed sclerotia	Often rough, but rarely ridged or wrinkled	Often wrinkled, ridged or with splits in surface.	Ridges rare, but with superficial mycelium attached
Rind cell characters	Digitate cells rare, cell outlines smoother than var. *idahoensis*; cells spherical in some isolates	Cells more irregular in outline than var. *ishikariensis*, digitate and lobate cells sometimes present; cell disjunctions and ridges	Rind cells resemble var. *ishikariensis* rather than *idahoensis*, but less rounded than the former and not so lobate as the latter. Rind ridges not common.

* Potato-malt agar.

canadensis) in cultivars of *F. rubra* and *P. pratensis* in forage and seed tests in northwest Canada. Older stands were more severely damaged suggesting inoculum build up. Although none of the grasses were killed, some were severely damaged, such as Golfrood, Duraturf and the Olds cultivars of *F. rubra* and Merion and Nugget in *P. pratensis*. Park *P. pratensis* tolerated the fungus while there were several tolerant clones in the Common sort. Highly resistant clones were found in Reptans and Boreal *F. rubra*. Methods for large-scale screening of grasses for resistance to *T. ishikariensis* varieties in controlled environments have been developed (Årsvoll 1976b, 1977; Årsvoll and Larsen, 1977; Cormack and Lebeau, 1959; Smith, 1981a; Smith and Årsvoll, 1975; Wernham and Chilton, 1943).

(d) Epidemiology

Sprague (1952, 1959) and Sprague and Rainey (1950) suggested that infection by basidiospores was possible. Ekstrand (1955a) considered that infection by *T. borealis* and *T. hyperborea* took place mainly by basidiospores through the leaf and shoot system. From there the fungi penetrated to the whole plant. This claim appears to have been based on observations rather than from experimental evidence. It seems more likely that mycelium from sclerotia is the main source of inoculum in *T. ishikariensis* as well as in *T. incarnata* (Cunfer and Bruehl, 1973). In var. *idahoensis*, after sporulation is complete, hyphae may emerge from the spent sporophores and grow like hyphae from sclerotia (Cunfer and Bruehl, 1973). Sporophores of *T. ishikariensis* var. *canadensis* have not yet been found in the field in western Canada on sclerotial inoculum of different isolates of the fungus applied to turf grasses or winter cereals which subsequently developed severe snow mould (Smith, 1975a; Smith and Reiter, 1976). Sporophores did not develop on sclerotia of different isolates applied to marked plots of turf in summer or on golf greens with a history of heavy attacks in north-eastern Saskatchewan. Many attempts to induce sporophore development on sclerotia of *T. incarnata* and *T. ishikariensis* vars. sown in pots in late summer and placed outside have always failed at Saskatoon in Canadian prairies (Smith, unpublished). This method of induction is reliable in eastern Washington (Christen, 1979; Christen and Bruehl, 1979; Cunfer and Bruehl, 1973) and in eastern Norway (Smith and Årsvoll, 1978). Sporophore production by both the above species is common in pastures and amenity turf in autumn in eastern and central Norway (Årsvoll, 1975a; Smith, unpublished). Basidiospores probably have a greater role to play in infection in much more temperate climates than in areas with more extreme climates, such as the Canadian Prairies. Christen (1979) noted that secondary sclerotia would form in the sporophores in crosses of var. *idahoensis* × var. *ishikariensis* which had remained infertile. The secondary sporophores, if produced in the field would provide a further survival mechanism; they remained viable for 11 months at 10 °C.

According to Potatosova (1960a), sclerotia of *T. ishikariensis* germinate in October at 1.4–4.6 °C and do not need a dormancy period, just low temperature and moisture. Those of var. *canadensis* will occasionally germinate in moist chambers in a cool greenhouse but generally there is a requirement for exposure to short-wave uv light from mercury vapour tubes or to daylight. The var. *idahoensis* is most virulent in the temperature range −1.5–1.5 °C (Bruehl and Cunfer, 1975).

Sclerotia are smaller and are produced in greater numbers in *T. ishikariensis* varieties, particularly in var. *canadensis* than in *T. incarnata* and are more easily detached from the host. The smaller size of the propagule is compensated for by greater numbers, perhaps an evolutionary modification favouring sclerotia as inoculum. The sporophores of *T. ishikariensis* are much shorter than those of *T. incarnata* and would be more sheltered in the turf. This would be advantageous when the microclimate was extreme, but disadvantageous for wind dispersal. It seems likely that the basidiospores main function is likely to be in the exchange of genetic material.

The small size and ease of detachment of the sclerotia from leaves in var. *canadensis*, in particular, may be a further modification in response to very extreme climates, allowing them to escape from very exposed positions on leaves to comparatively sheltered places in the sole of the turf. The smaller size and ease of detachment would also allow some short-range wind dispersal. Sclerotia of *Typhula* spp. are long-lived structures, retaining viability for many years in cold storage. Bruehl *et al.* (1966) observed that 8 years was not long enough to eradicate var. *idahoensis* or *T. incarnata* in wheat crops. McKay and Raeder (1953) and Bruehl *et al.* (1966) noted that snow mould increased with each successive wheat crop. Vaartnou and Elliott (1969) found field evidence for inoculum buildup in *T. ishikariensis* as the age of grass stands increased. Survival of var. *idahoensis* sclerotia buried in soil was little affected by storage at 1–2 °C except in the presence of clover, but considerably reduced at 24 °C when in the rhizospheres of pea and clover. Depth of burial had little effect on sclerotial survival but germination was more rapid in deeply buried sclerotia than those in the soil surface. There are psychrophilic soil-borne bacteria on sclerotial surfaces which can inhibit sclerotial germination, but do not antagonize growing mycelium. They may provide a mechanism of biological control operating within a specific crop rotation (Huber and McKay, 1968).

Smith and Årsvoll (1975) and Årsvoll (1975b) found that when macerated mycelium of dikaryotic isolates of *T. ishikariensis* var. *ishikariensis* from different geographic regions in Norway were mixed and used as inoculum against *Phleum pratense* L. in pathogenicity tests, significantly less severe disease was produced than when isolates were used separately. A similar reduction in aggressiveness was found in isolates of var. *canadensis* from different locations in Canada when these were mixed together. Mutual antagonism was demonstrated in culture in these two varieties, but not between different dikaryotic isolates of *T. incarnata*. In a more critical study, Årsvoll (1976b) showed that isolates of *T. ishikariensis* and *T. incarnata* from different regions of Norway, aggressive when used singly, lost much pathogenicity when isolates of each species were mixed. Considerable intraspecific antagonism was shown between isolates obtained from the same field and even between isolates from the same square metre in *T. ishikariensis*, although this mutual antagonism was not shown in all cases. In one field, all isolates from the centre of a one metre quadrat, and in another, two isolates from a 10 metre quadrat showed no antagonism suggesting that the isolates in the mixtures were of the same origin. It seems likely that intraspecific antagonism between *Typhula* isolates is due to the production of metabolites or staling substances which are mutually inhibitory to the different isolates. Variability in pathogenicity of the isolates does not seem to explain the effect of mixing isolates on reduction in disease severity. The mixing of isolates in these *Typhula* spp. and varieties is contra-indicated unless critical studies show that mutual antagonism is not involved (Årsvoll, 1976b; Smith and Årsvoll, 1975). The genetical and epidemiological implications of mutual antagonism are unclear.

(e) Effect of plant nutrients

Ekstrand (1955a) considered that adequate phosphorus, and to a lesser degree, potassium increased resistance to *T. borealis*. Lime applications had the effect of increasing resistance mainly by making the phosphorus available. Bruehl *et al.* (1966) found that moderate applications of ammonium nitrate and ammonium phosphate in autumn had no appreciable effect on the disease (in winter wheat), but adequate nutrition aided recovery. Årsvoll and Larsen (1977) in experiments in controlled environments found that attacks by *T. ishikariensis* on *Phleum pratense* increased with increasing nitrogen and were severe at low phosphorus concentrations, decreasing in severity with increasing phosphorus. Equivocal results were obtained with potassium. Acclimation improved plant resistance except with high nitrogen and low phosphorus. The effect of excessive nitrogen applications is to delay the onset of dormancy and winter acclimation (hardening-off), which in cool temperate, continental and boreal regions, where snow moulds such as *T. ishikariensis, T. incarnata, Myriosclerotinia borealis* and *Coprinus psychromorbidus* cause problems, is often signalled by a 'browning-off' of the turf (Smith, 1975c, 1980b, 1981b).

(f) Control

(i) GENERAL SNOW MOULD CONTROL METHODS

1. Avoid unbalanced or excessive nitrogenous fertilizer, particularly towards the end of the growing season, to allow the grasses to 'harden off'. Moderate applications of balanced fertilizer may be applied when the turf has reached near-dormancy, especially if fertility has declined, as indicated by soil analysis, to encourage rapid growth and recovery in spring.
2. Improve soil and air drainage, and in snowy regions, regulate snow drifting with ventilated fences placed to drop the snow clear of the turf, or if that is not possible, to promote an even snow cover. Remove snow drifts with a snow blower in spring, if feasible, but take care not to cause mechanical injury by wheeled or tracked vehicles. More rapid snow melting in spring may be promoted by the application of dark material such as soot, dark sand, or topdressing when the sun becomes stronger.
3. Use known resistant cultivars, or if not available, those which are well adapted to the regional climate. These are often of northern origin.
4. Apply suitable fungicides. These are indicated below. *T. ishikariensis* is less amenable to control by fungicides which are effective against *T. incarnata* and several applications may be needed in fall and early winter. Since *T. ishikariensis* may occur in complexes with other snow moulds, a fungicide programme should be chosen which will control all pathogens (see below and p. 92).

(ii) CHEMICAL CONTROL OF SNOW MOULD CAUSED BY *T. ISHIKARIENSIS* VARIETIES. *T. ishikariensis* varieties generally cause more severe turf grass injury and are more difficult to control with fungicides than *T. incarnata*. While *T. ishikariensis* may be the dominant snow mould it is often in complex with the others (Fushtey, 1975, 1980; Smith, 1973, 1974a, 1976; Stienstra, 1980; Taylor, 1974, 1975, 1976). Predominance of a particular snow mould may change with the season. Most snow mould fungicides show selective activity against particular groups of fungi suppressing some more than others (Hossfeld, 1974) or suppressing micro-organisms in the turf which are antagonistic to the snow mould(s) (Smith, 1976; Smith and Reiter, 1976; Fig. 6.9). Where snow mould complexes occur, prediction of dominance is difficult, even if it is possible to estimate the abundance of sporophores of a *Typhula* snow mould before a snow cover develops. Spots of disease caused by *Microdochium nivale* can be seen before a snow cover develops, but the latent infection from other snow moulds generally can not. Often there are no sporophores of *T. ishikariensis* to be found and, in any case, it is probable that mycelium derived from sclerotia is the main source of inoculum in both *T. incarnata* and *T. ishikariensis* (Cunfer and Bruehl, 1973; Hindorf, 1980; Lehmann, 1965c). Andersen (1966) found methoxymethyl mercuric chloride more effective than captan, quintozene or methoxymethyl mercuric salicylate in control of *T. ishikariensis* on a susceptible *Phelum pratense* cultivar. However, these materials were less effective when applied to a less-susceptible cultivar. Ylimäki (1972) reported effective control of snow moulds of turf grasses in Finland, including *T. ishikariensis*, with quintozene applied in autumn. In the interior of British Columbia, against snow mould caused by complexes of *T. canadensis*, *T. incarnata* and *M. nivale*, Taylor (1974, 1976) found that the non-mercurials, chloroneb (179 g a.i.) and quintozene (112.5 g a.i./100 m^2) would effectively control mild outbreaks of snow mould with one application. To control severe cases, up to 214 g a.i./100 m^2 of quintozene was used without turf damage. Applications of chloroneb and quintozene, combined with fertilizer in granular form, at the same dosage gave similar results to the spray application (Taylor, personal communication). In southern Ontario, Fushtey (1975) found that on turf of *Agrostis stolonifera* L. fungicides containing benomyl or re-

Figure 6.9 Increased development of *Typhula ishikariensis* var. *canadensis* mycelium on turf plot sprayed with thiobendazole.

lated compounds failed to give satisfactory control where *Typhula* spp. (*T. incarnata*, *T. ishikariensis* and *T. canadensis*) (Fushtey, 1980; Smith, unpublished) were involved. Fungicides containing chloroneb or chlorothalonil were effective. Late autumn application, probably not before 1 November, was suggested. In later studies in southern Ontario (Fushtey, 1980), where the three *Typhula* spp. or varieties were complexed with *M. nivale*, fungicides containing inorganic mercury chlorides, and phenyl mercuric acetate gave the best overall control. Where there was little disease (controls with less than 7%) all fungicides gave acceptable control. Where *T. canadensis* was the dominant snow mould and the disease was very severe (controls averaged 85%) quintozene (131.25 or 178.75 g a.i.), mercurous + mercuric chlorides (72 g Hg salts), an experimental wettable powder (Baymeb 6447, 60 g a.i./100 m^2), a quintozene granular material at twice normal dosage (16.9% quintozene) and a granular fungicide containing 0.68% phenyl mercuric acetate + 4.65% thiram at twice normal dosage gave excellent control. Conflicting results were obtained at two other sites. On one, where *T. ishikariensis* var. *canadensis* was also severe (controls averaged 95%) only chlorothalonil (194.4 g a.i.), mercurous + mercuric chlorides (75g a.i.) and the granular phenyl mercuric acetate + thiram gave practical control. On the other, where *T. ishikariensis* var. *ishikariensis* was moderately severe (controls averaged 65%) only benomyl and iprodione (at the highest dosage, 60 g a.i./100 m^2) failed to give effective control.

The most effective fungicides in Saskatchewan on putting-green and lawn turfs, where *T. ishikariensis* and another basidiomycete, *Coprinus psychromorbidus* (LTB phase – see Section 6.2) were complexed, were quintozene (200 g a.i./100 m^2), mercury chlorides (100 g a.i./100 m^2) and chloroneb (200 g a.i./100 m^2). Phenyl mercuric acetate was erratic in effect (Smith, 1976).

In Minnesota, where complexes of *M. borealis*, *T. incarnata* and *T. ishikariensis* occurred, the most effective fungicides were chloroneb with quintozene (79 + 91 g a.i./100 m^2) and chloroneb (178 g a.i./100 m^2) on turf of *Agrostis stolonifera* with *Poa annua* (Stienstra, 1980). Combinations of chloroneb with mercury chlorides and quintozene with mercury chlorides were also very effective.

(iii) BIOLOGICAL CONTROL OF *T. ISHIKARIENSIS*. Burpee *et al.* (1985, 1987) have shown that an avirulent isolate of *T. phacorrhiza* Fries applied as a culture on grain reduced the incidence of snow mould caused by *T. ishikarienesis* var. *ishikariensis* in *Agrostis stolonifera* turf of a golf green. This gives promise of providing a method for the biological control of this pathogen.

6.5 SCLEROTINIA SNOW MOULD

Sclerotinia snow mould causes damage to winter cereals and perennial grasses in Scandinavia, Finland, northern Europe, Asiatic parts of the USSR and the Ukraine, northern Japan, Alaska, some northern states of the USA and Canada (Adachi, Miyashita and Araki, 1976; Andersen, 1960; Årsvoll, 1973, 1975a, 1976a; Cormack and Lebeau, 1959; Ekstrand 1955a; Eleneff, 1926; Groves and Bowerman, 1955; Jamalainen, 1949, 1978; Khokhyrakoff, 1935; Lebeau and Logsdon, 1958; Pukhalski, 1937; Røed, 1960; Sakuma and Narita, 1963; Shavrova, 1972; Smith,

1975a; Solkina, 1939; Sprague, Fischer and Figaro, 1961; Tomiyama, 1955; Tupenevich, 1965; Tupenevich and Shusko, 1939; Uländer, 1910; Vleugel, 1917; Yakovlev, 1939).

Reports of the disease on turf grasses are less numerous than those on winter cereals and forage grasses. However, in regions where forage grasses are damaged by the fungus, turf attacks may be expected. The disease has been reported on turf grass in Norway, Sweden, Finland, Alaska and northwest Canada, the Prairie Provinces, southern Ontario and Minnesota (Allen, Bernier and Ferguson, 1976; Cormack, 1952a; Groves and Bowerman, 1955; Hansen, 1969; Jensen, 1970; Kallio, 1966; Smith, 1972, 1973, 1974a, 1978 and unpublished; Stienstra, 1974, 1980; Vaartnou and Elliott, 1969; Ylimäki, 1972).

6.5.1 Symptoms

At snow melt in spring, patches of grass up to 15 cm or so across, with water-soaked leaves and sparse grey mycelium appear. Patches may coalesce (Fig. 11 colour plate section). Infected leaves often bleach and wrinkle, becoming thread-like when exposed to light, but later darken with the growth of saprophytic fungi. Sclerotia of the fungus develop on and in the tissues. Plants of infected *Agrostis* spp. bearing sclerotia at snow melt are often dead.

6.5.2 The causal fungus

Teleomorph: *Myriosclerotinia borealis* (Bubak and Vleugel) (Kohn, 1979): synonym *Sclerotinia borealis* Bubak and Vleugel (Vleugel, 1917; Saccardo, Syll. Fung. Sect. 2, p. 1179, 1928). *S. graminearum* Eleneff ex Solkina is probably the same fungus (McDonald, 1961; Shavrova, 1972; Solkina, 1939). Uländer (1910) was the first to report the fungus causing damage to *Dactylis glomerata* and other grasses in northern Sweden.

There are no macroconidia, but globose, hyaline phialospores (spermatia) 2.5–3.5 μm have been found in culture only (Groves and Bowerman, 1955). They are produced endogenously in flask-shaped phialides borne in clusters on the mycelium. Microconidia have been reported in *S. graminearum* (Solkina, 1939).

Sclerotia of the fungus are found in sheaths, crowns, leaf axils, on the surface or within leaves. Their size varies according to that of the host and its nutrition and the organs on which they were formed. They are at first cream, putty-coloured or faintly pink, globular, elongate, flake-like or arched, darkening to black with a distinct rind, often up to 7 or 8 mm long and up to 3 to 4 mm wide when fresh. When the host tissues dry up they become wrinkled and readily detached. Sclerotia germinate in autumn to produce one or more, stalked, cup-shaped apothecia (Fig. 6.10) varying in colour from pale yellow, fawn, buff or even faint pink to pale brown. Apothecial discs vary from approximately 1–6 mm in diameter and their stalks from 1 to about 20 mm in height. The upper surface of the disc of the mature apothecium bears asci and filiform paraphyses, the latter slightly swollen at the free end and about 2 μm in diameter. The asci are cylindrical, tapering to a slender stalk, inoperculate with an apical pore. Asci vary in width from 8–17 μm and in length measure 150–300 μm. Ascospores are in one row, one-celled, hyaline, often unequal in size, elliptical to oval, sometimes slightly pointed. They are forcibly discharged.

Figure 6.10 Germinated sclerotia of *Myriosclerotinia borealis* with apothecia.

Naturally produced sclerotia appear to require a ripening period before germinating to produce fertile apothecia (Pukhalski, 1937; Solkina, 1939; Sprague, Fischer and Figaro, 1961; Tupenevich and Shusko, 1939; Yakovlev, 1939). Yakovlev (1939) considered exposure to sunlight was also needed. Exposure to light, either daylight or ultra-violet and maintenance of a very moist, cool environment have proved to be necessary procedures for the induction of apothecia in naturally produced and cultured sclerotia (Årsvoll, 1976a; Groves and Bowerman, 1955; Smith, 1981a; Sprague, Fischer and Figaro, 1961).

The temperature optima for ascospore germination in *S. graminearum* varied with different isolates from 3 to 16°C; spores would not germinate at 30°C and freezing at – 3°C did not kill spores (Solkina, 1939). In *M. borealis* the cardinal temperatures for spore germination are, minimum – 6, optimum <9 and maximum 21°C, on culture media. Ascospore germination is not reduced by storage for 3 years at – 23°C. Ascospore germination takes place from one or both ends and they will germinate when still contained in the ascus (Årsvoll, 1976a).

The fungus may be isolated from fully imbibed sclerotia after surface sterilization with 70% ethyl alcohol and/or 1% chlorine water for 2 min, followed by washing in several changes of sterile water. The sclerotia are then sliced with a sterile scalpel and the pieces planted on potato dextrose or cornmeal, malt-extract, yeast extract agar.

On culture media, *M. borealis* will grow at temperatures lower than – 7°C (Tomiyama, 1955; Ward, 1966) with temperature optima given by different workers between 3 and 15°C (Årsvoll, 1976a; Ekstrand, 1955a; Tomiyama, 1955). Ward (1966, 1968) in critical studies on the relationship between temperature, growth and respiration, found the optimum temperature for vegetative growth was 0°C, although the initial growth rate was higher at 5, 10 and 15°C. Ice crystals retarded, but did not prevent, mycelial growth on agar media. Tomiyama (1955) found that *S. graminearum* (*M. borealis*) was able to develop more quickly on frozen media than on unfrozen ones at the same temperature. The optimum temperature for sclerotial formation was 10°C with a range from 0°C to 15°C (Ward, 1966). The maximum temperature for growth is 15–20°C (Årsvoll, 1976a; Ekstrand, 1955a; Sakuma and Narita, 1963; Ward, 1966). The pH ranges for vegetative growth in culture show an optimum between 3 and 6 with little growth at 8 or 9 (Tomiyama, 1955; Ward, 1966). The ranges of pH for apothecial production were determined by Årsvoll (1976a) to be 3.5–7, with the optimum lying between 3 and 6. Yakovlev (1939) found that the fungus developed most profusely on acid soils and Demidova (1960) noted that no sclerotia germinated at pH 8.

6.5.3 Pathogenicity tests

Ascospore suspensions are usually more effective as inoculum than mycelium (Årsvoll, 1976a, 1977; Årsvoll and Larsen, 1977; Cormack and Lebeau, 1959; Sakuma and Narita, 1963; Smith, 1981a and unpublished; Sprague, 1961; Tomiyama, 1955). Sakuma and Narita (1963) found almost no difference in pathogenicity when using ascospores from sclerotia on different grass species in reciprocal infection studies. Årsvoll (1976a, 1977) and Årsvoll and Larsen (1977) found that more severe infection occurred at 0°C than at 2°C. Poor infection was obtained when mycelial macerates were used as inoculum at – 2°C (Smith, unpublished).

6.5.4 Host range and susceptibility

Most of the commonly used cool-season turf grasses are susceptible to *M. borealis* (Cormack, 1952a; Ekstrand, 1955a; Groves and Bowerman, 1955; Hoshiro and Hirashima, 1978; Jamalainen, 1949, 1970, 1974; Kallio, 1966; Lebeau and Logsdon, 1958; Nissinen and Salonen, 1972; Røed, 1960; Sakuma and Narita, 1963; Smith, 1972, 1974a; Tupenevich and Shusko, 1939; Vleugel, 1917). In Fenno-Scandia, *Poa trivialis, Phleum pratense, Festuca rubra* and *Lolium perenne* are most severely attacked, but great strain differences are shown. Generally, strains of grasses from the north of this region are more resistant than those of southern origin (Ekstrand, 1955a; Jamalainen, 1949, 1974). *Lolium perenne* is generally so susceptible to all snow moulds, including *M. borealis* and lacking in frost resistance that it does not overwinter well in Finland (Jamalainen, 1974a; Nissinen and Salonen, 1972) and because of susceptibility to *M. borealis* is not grown much in eastern Hokkaido in Japan (Sakuma and Narita, 1963). *Festuca rubra* is similarly affected in these two locations. Those grass cultivars which are not frost hardy, due to depletion of nutrients, were nearly all susceptible to *M. borealis* (Hoshiro and Hirashima, 1978). Differences in susceptibility were noted in *Poa pratensis* cultivars in Alaska: Merion was least resistant, Common most resistant with Park and Newport intermediate (Kallio, 1966). In a severe outbreak in southern Saskatchewan in 1971, Seaside *Agrostis stolonifera* was less severely damaged than Penncross (Smith, 1972). First-year stands of agricultural grasses have been found more susceptible than older ones to *M. borealis* (Andersen, 1960; Årsvoll, 1975a, 1976a, 1977; Blomqvist and Jamalainen, 1968; Jamalainen, 1949, 1970; Pohjakallio, Salonen and Antilla, 1963; Thörn, 1967).

6.5.5 Epidemiology

Most workers agree that the most severe damage from *M. borealis* occurs in years with deep, prolonged snow covers, particularly if that cover develops on unfrozen or lightly frozen soil (Adachi, Miyashita and Araki, 1976; Årsvoll, 1973; Ekstrand, 1955a; Elenev, 1926; Hoshiro and Hirashima, 1978; Jamalainen, 1949, 1974; Kallio, 1966; Khokhyrakoff, 1935; Lebeau and Cormack, 1961; Sakuma and Narita, 1963; Smith, 1974a; Sprague, Fischer and Figaro, 1961). However, Tomiyama (1955) found that *M. borealis* was common in Hokkaido, in northern Japan when there was comparatively little snow and that it was the prevalent pathogen under snow when the soil was frozen for a long time. This can be related to his experimental conclusion that the mycelium of *M. borealis* could grow more quickly on frozen agar media than on unfrozen (supercooled) media at the same temperature and more quickly than *Typhula incarnata* on frozen media (Tomiyama, 1955). Røed (1960) in Norway found more severe cases of *S. borealis* damage occurred after winters with rather thin snow cover on deeply frozen soil which may have caused greater frost injury because the grass was less well insulated. The effect of mycelial and ascospore infection have been discussed in Section 6.5.3.

Ascospore infection in autumn appears to be the most likely means of primary infection in the field. Spread from host plant under the snow seems to be through contact of infected leaves with healthy ones (Sakuma and Narita, 1963). Entry of mycelium into leaves may be via wounds (Tomiyama, 1955), through stomata or between cells (Årsvoll, 1976a).

Saturated soils and temperatures between approximately 6 and 12°C favour sporulation, but higher temperatures (greater than 15°C) inhibit it. Apothecia are not hard to find in all grassland types in regions with long periods of cool or cold, humid autumn weather which gradually shade off into the colder temperatures of winter (Årsvoll, 1973, 1975a; Røed, 1960; Smith, unpublished). However, where there are abrupt temperature changes and autumns are dry, as in the central prairie region of Canada (Saskatchewan), sufficiently long periods of favourable weather for apothecial production are probably rare. In the more humid autumn weather and in the Peace River region and in southern Alberta conditions for apothecial production are more favourable (Drs J.G.N. Davidson and L. Piening, personal communications). Probably, the criteria which govern ascospore production and infection are more critical in determining disease occurrence and severity than subsequent subniveal temperatures and snow depth (Årsvoll, 1976a; Ekstrand, 1955a; Khokhyrako, 1935). However, the fungus is a weak pathogen and an extended period of snow cover may lead to further weakening of host tissue resistance to invasion by depletion of nutrient reserves (Bruehl and Cunfer, 1971; Hoshiro and Hirashima, 1978; Tomiyama, 1955).

Significant injury from *M. borealis* in grassland in Norway occurred only when there were more than 170–180 days of snow cover which included 90 'ice days'. Where there were more than 180 days of snow cover, including 110–120 ice days during winter, grassland damage from *M. borealis* was more severe than that from *Typhula ishikariensis*. Where *M. borealis* injury occurred its incidence increased with increasing elevation, i.e., the opposite of abiotic winter injury (Årsvoll, 1973). In Japan, damage to grasses, caused by *M. borealis* is reported to be most severe in the eastern and northern part of Hokkaido, the principal northern island (Sakuma and Narita, 1963).

Although Lebeau and Cormack (1961) considered that *M. borealis* was confined to extreme northern regions of North America such as Alaska (latitude 60°N) and the Prince George region of British Columbia (latitude 54°N), the fungus has subsequently been found causing severe turf damage much further south, e.g. in southern and south-eastern Saskatchewan (Smith, 1972 and unpublished), southern Manitoba (Allen, Bernier and Ferguson, 1976) in southern Ontario (latitude 44°N, Smith, unpublished) and northern Minnesota (latitude 47°N, Stienstra, 1974, 1980). In western Canada epidemics of the disease occurred in 1971/1972 and 1973/1974, but more localized outbreaks may be expected at higher elevations in non-epidemic years. Although the fungus has a boreal and sub-boreal distribution the climatic effects of continental land masses on temperature and snowfall may extend its range southwards, as for example to northern Washington (latitude 40°N), where it was found at 1200 m elevation (Sprague, Fischer and Figaro, 1961).

Increased frequency of the disease has been noted on peat compared with mineral soils (Årsvoll, 1973; Demidova, 1960; Ekstrand, 1955a; Jamalainen, 1949; Yakovlev, 1939). This may be due to the effects of pH on mycelial growth or sporophore development, but lime applications have been reported to increase plant resistance perhaps by improving availability of phosphorus (Ekstrand, 1955a; Tupenevich and Shusko, 1939). Tomiyama (1955) found P deficiency increased disease severity and Ekstrand (1955a) and Årsvoll and Larsen (1977) noted that additional P decreased disease severity. While adequate nitrogen is needed to maintain plant vigour and encourage recovery from damage in spring (Jamalainen, 1957; Nissinen and Salonen, 1972; Sakuma and Narita, 1963; Shalavin,

1960), it has been shown that resitance to *M. borealis* may be expected to decline as the dosage of N is increased (Årsvoll and Larsen, 1977). The effect of potassium on resistance to the pathogen is uncertain (Årsvoll and Larsen, 1977; Ekstrand, 1955a; Tomiyama, 1955).

6.5.6 Control

The disease is likely to be a major problem in turf grasses only in continental North America north of the Canadian/USA border, perhaps in the colder parts of the Mid-West USA, and in northern parts of Norway, Sweden and Finland except in high snowfall regions or years. Its distribution in the USSR is likely to be similar to that in continental North America, while in Japan only the high snowfall regions of Hokkaido appear to be at great risk.

Turf grass cultivars from northern latitudes should be used where the disease is expected (Andersen, 1960; Jamalainen, 1974). *Poa pratensis* is the preferred species for lawn turfs in nothern regions and except in the most severe outbreaks, patch recovery usually takes place.

Festuca rubra is too susceptible for use where the disease is common. *Lolium perenne* lacks sufficient cold hardiness, apart from its susceptibility and should not be used. No resistant *Agrostis* species or cultivars appear to be available and although Colonial bentgrass (*A. tenuis*) is generally less susceptible than Seaside or Penncross (*A. stolonifera*) (Smith 1977a), where the disease is common, bentgrasses require protection with fungicides.

Lack of nitrogen and consequent low vigour and slow recovery from the disease should be avoided but excessive N fertilization in late summer and fall encourages fusarium patch. *Agrostis* turf which is very prone to attacks by *M. borealis* tends to become too acid but if injudiciously limed this will also encourage attack by *M. nivale*. Soil P should be maintained at an adequate level.

Snow fences should not be used around golf greens to trap snow where the disease is common. Attempt to speed the rate of drift melting by snow removal or by using dark materials, such as peat, soot or dark fine soil spread lightly on the drifts.

After a severe attack on bentgrass turf the carry-over of sclerotial inoculum can be reduced by allowing the turf to dry, brushing it and then picking up the dislodged sclerotia with a vacuum sweeper. Repeat the suction collection until few sclerotia are found.

Effective control of this snow mould in turf grasses is possible with fungicides. Quintozene (pentachloronitrobenzene), effective in the prevention of the disease in forages (Hansen, 1969; Jamalainen, 1970; Sakuma and Narita, 1963; Ylimäki, 1955) in northern Europe and Japan, was also found very effective on *Poa pratensis* turf in Alaska (Kallio, 1966) and on different types of amenity turf in western Canada (Smith, 1976; Smith and Reiter, 1976). Other effective fungicides are benomyl, methyl thiophanate, phenyl mercuric acetate with thiram or alone, thiram, thiram with carboxin and oxycarboxin and chlorothalonil (Kallio, 1966; Smith, 1976; Smith and Reiter, 1976). In Minnesota a combination treatment of quintozene and thiophanate methyl proved effective on golf-green turf (Stienstra, 1974) but neither mercurous/mercuric chloride mixtures nor chloroneb effectively controlled the snow mould (Stienstra, 1980). Tank mixes of chloroneb and quintozene, chloroneb and mercurous/mercuric chloride mixtures and quintozene with mercurous/mercuric chloride gave effective snow mould control when complexes of *Typhula* spp. (q.v.) and *M. borealis* were concerned (Stienstra, 1980).

Biological control of *M. borealis* with *Acrostalagmus roseus* has been demonstrated (Pohjakallio *et al.*, 1956) experimentally, but practical measures have not been developed.

6.6 OTHER LOW-TEMPERATURE-TOLERANT TURF GRASS PATHOGENS AND ANTAGONISTS

Many of the 55 species of fungi isolated by Dr R. Sprague from winter cereals during the cold months of December 1959 to April 1960 and December 1960 to April 1961 (Bruehl *et al.*, 1966) also occur on grasses. Of the low-temperature isolates of these tested for pathogenicity in cold chambers at 1–3 °C only *Microdochium nivale, T. incarnata* and *T. ishikariensis* var. *idahoensis* and a *Rhizoctonia* sp. were pathogenic. A *Rhizoctonia* sp. was isolated from damaged turf in winter in central and southern Alberta (Broadfoot, 1936) and a pathogenic *Rhizoctonia* sp. was isolated from patches of turf with snow mould injury in spring in Saskatchewan and northern Alberta (Traquair and Smith, 1981). Some of the isolates showed mycelial growth optima c. 20 °C, e.g. lower than summer isolates.

Årsvoll (1975a) determined the temperature/growth responses on agar media and the pathogenicity of fungi isolated from grasses suffering from winter injury on seedlings of *Phleum pratense* L. grown on Hoagland's agar in test tubes incubated at 0 and 3 °C. Those fungi which proved pathogenic were then tested on greenhouse-grown grass plants of seven species at the 4–6 leaf stage in pots, using the technique of Blomqvist and Jamalainen (1968) at temperatures of 0–5 °C (average 3 °C) and the damage assessed. In the temperature/growth response study, on culture media, of 33 species, 19 grew at −6 °C, 10 at −3 °C and 4 at 0 °C. At 0 °C none of the fungi had a mycelial growth capacity, expressed at percentage of growth at their optimum growth temperature, as high as *Myriosclerotinia borealis*, 80%, *Typhula ishikariensis* var. *ishikariensis*, 61% or *T. incarnata*, 27% *M. nivale* had only a 6% growth capacity of optimum, i.e. less than that of 13 other fungi which are not recognized as snow moulds. At 3 °C, 11 of the 'non-snow moulds' showed over 20% of optimum growth capacity. In the test-tube pathogenicity studies, where the disease assessment scale was 0 = no attack to 4 = very severe attack, not even the recognised snow moulds caused severe damage after 4 weeks on acclimated plants at 0 °C, but *Corticium fuciforme* (Berk.) Wakef (see Section 24.2.1 for taxonomy), 2.6, *Fusarium avenaceum* (Corda ex Fr.) Sacc., 2.5, *M. nivale*, 2.9, *T. ishikariensis* var. *ishikariensis*, 2.8, and a sterile hyphomycete, 2.6 caused moderately severe injury on non-acclimated seedlings. After 8 weeks, 14 'non-snow moulds' caused moderate to moderately severe damage on acclimated seedlings. These species included: *Ascochyta phleina* Sprague, 2.6, *Ascochyta* sp., 3.5, *Cercosporella herpotrichoides* Fron., 2.3, *Coniothyrium cerealis*, E. Müller, 2.0, *Corticium fuciforme*, 2.4, *Epicoccum purpurascens* Ehrenb. ex Schlecht., 2.4, *Fusarium avenaceum*, 3.0, *Fusarium culmorum* (W.G. Smith) Sacc., 2.4, *Fusarium equisiti* (Corda) Sacc., 2.5, *Hendersonia culmicola* Sacc., 2.4, *Mortierella hyalina* (Hartz) W. Gams, 2.3, *Mycocentrospora acerina* (Hartig) Dayton, 2.5, *Phoma eupyrena* Sacc. Wollenw., 2.0 and a sterile hyphomycete, 2.7. *T. ishikariensis* and *M. nivale* caused very severe injury of 4 and 3.9 respectively, but *T. incarnata* caused only moderately severe, 2.7 and *M. borealis*, although ascospores were used as inoculum, only slight, 1.4, injury on acclimated seedlings. In the pot test on eight grass species incubated at 0–5 °C, *Agrostis tenuis, Phleum pratense, Festuca rubra*, and *Poa pratensis* suffered most severely, *Festuca pratensis* was intermediate, while *Lolium perenne, Dactylis glomerata*, and *Bromus inermis* were least affected by the 'non-snow moulds'. Of the latter species, *C. fuciforme* (see p. 259 for taxonomy), *Dactylaria graminicola* Årsvoll, *Fusarium avenaceum, Fusarium equisiti*, the sterile basidiomycete and hyphomycete were almost as pathogenic as the recognized snow moulds.

Snow rot or snow blight caused by various *Pythium* spp. has been recognized for many years as a disease of winter wheat in Northern Japan (Ito and Tokimaga, 1935; Iwayama, 1933; Tomiyama, 1961) and more recently in Washington State (Bruehl, *et al.*, 1966; Lipps, 1980; Lipps and Bruehl, 1978, 1980) and in North Dakota (Stack, Jons and Lamey, 1979). It has not yet been found on winter wheat or on grasses in Canada. For its development it requires a higher humidity under a snow cover than the other snow moulds (Hirane, 1955; Iwaykiri, 1946; Lipps and Bruehl, 1978; Tomiyama, 1961). The occurrence of pythium snow blight as a turf disease appears likely.

There is need for further studies on low-temperature-tolerant fungi which are not now regarded as snow moulds. Although they do not usually cause obvious symptoms like the recognized snow moulds under a snow cover, they may cause considerable damage at the low temperatures prevailing in winter before a permanent snow cover develops and in spring at and after snow melt. Since many of them are mesophils they may be able to establish themselves before snow mould attacks develop and become important competitors for space and substrate with the regular snow moulds.

A brown root rot is caused by *Phoma sclerotiodes* (Preuss) ex Sacc. (*Plenodomus meliloti* Dearness and Sanford) which has a wide host range on plants from regions with severe winters, following the winter dormancy period, in western Canada, Finland and the USSR (Boerema and Van Kesteren 1981; Conners, 1967; Cormack, 1934; McDonald, 1955, Mead, 1962; Newton and Brown, 1924; Salonen, 1962; Sanford, 1933). It is pathogenic on grasses and cereals which have been exposed to low temperature (Henry and Berkenkamp, 1965; Knowles and Smith, 1981; Lebeau and Logsdon, 1958; Robertson, 1931; Smith, 1980e, 1981; Smith and Piening, 1980). Lebeau and Logsdon (1958) noted it on *Festuca rubra* L. and *Poa pratensis* L. in Alaska although they doubted whether it was a major pathogen.

Protopycnidia of the fungus have been collected from snow mould-damaged turf grass in western Canada since 1972 (Anon., 1979; Smith, 1981a; Smith and Piening, 1980). It was involved in severe brown basal rot of overwintering *Dactylis glomerata* (Knowles and Smith, 1981; Smith, 1980e).

On dead and dying grasses and winter cereals there is a brown basal rot and root necrosis. Greenish-black, globular pycnidial initials (protopycnidia) up to 0.5 mm diam. are found firmly attached to roots to 10 cm at least below the soil surface. Occasionally these are also found on the stem of dead hosts, mainly at or near ground level and on crown tissues. Necrosis is very noticeable where protopycnidia are attached.

The sclerotia-like protopycnidia (Smith and Piening, 1980) or pycnosclerotia (Boerema and van Kesteren, 1964) are greenish-black, subglobose to conoid, either single or clustered. On roots of grasses they may be slightly elongate and are dished on the adaxial surface where there is a very short attachment process. Maximum diameter of these protopycnidia is approximately 0.8 mm. At this stage there are no locules (Smith, 1981a). Mature pycnidia are similar in colour and shape to protopycnidia, but tend to be more hemispherical and they are solitary or confluent, the latter especially in culture, up to 2 mm in groups. The pycnidial wall, which in protopycnidia is thin and delicate, becomes much thickened as the pycnidia mature. The wall cells are usually 5- to 6-sided, 5–10 μm in diameter, their cavities becoming filled as they age. Locules develop in the ground tissue of the pycnidia at scattered points, the spores being discharged through beaked ostioles up to 200 μm long as a cream or yellow cirrus. The spores are hyaline, elliptical, biguttulate (4–) 4.5–6.5 (–8) × 2–3 (–3.5) μm. Conidiophores are cone-shaped and insignificant, formed round the periphery of the locule. In culture the mycelium is greyish-white, darkening to brownish-white with the pycnidia mostly superficial (Colotelo and Netolitzsky, 1964; Dearness and Sanford, 1930; Smith, 1981a).

To isolate the fungus detach protopycnidia from the roots with needles or use a blender run at slow speed to remove them from plant base or root tissues. Surface sterilize with 70% alcohol for 30 s. Wash in several changes of water and plate on to cornmeal + malt extract + yeast extract agar and incubate at 1 °C. There is no need to split the protopycnidia with a sterile scalpel as the mycelium will grow out from the superficial cells. If bacterial contaminants are troublesome plate the 'sterilized' protopycnidia on to potato – malt extract – yeast extract agar containing 5% glycerol and incubate at −3 to −4 °C. If protopycnidia on pieces of plant tissue are incubated in a moist chamber for several weeks at 15 °C, pycnidia and spores will develop. From the latter single cell isolations can be made. The optimum temperature for mycelial growth of most isolates from grasses and cereals is c 15°C. Sanford (1933) found the temperature range for mycelial growth to be 0–27°C with a optimum of 15–17°C. However, most isolates make appreciable growth at–7°C.

Pathogenicity on grasses and cereals has not been proven experimentally, although it is often the predominant fungus (Lebeau and Logsdon, 1958). In winter cereals it is often found in complex with the SLTB phase of *Coprinus psychromorbidus* (Smith and Piening, 1980 and unpublished). Sanford (1933) found that leguminous hosts were susceptible only during winter and spring dormancy stages, although dead *Avena sativa* L. roots were colonized saprophytically. However, not all grasses and winter cereals bearing protopycnidia of *P. sclerotiodes* are dead, so susceptibility in these plants may also be related to dormancy state.

The sclerotia of the hyphomycete *Acremonium boreale* Smith and Davidson are common on grasses and many other plants in Canada from British Columbia to Ontario (Smith, 1972, 1973, 1974a,b, 1981a; Smith and Davidson, 1979), on dead stems of herbaceous plants. They have also been found on forage legume debris in Norway (Smith and Davidson, 1979). The sclerotia of the fungus are frequently collected with those of other sclerotial snow moulds such as the SLTB phase of *Coprinus psychromorbidus*, *M. borealis* and *Typhula* sp. from patches of turf grass damaged by snow moulds in western Canada. It is weakly pathogenic towards unhardened grasses at 0 and 3°C (Smith and Davidson, 1979). Its main ecological significance seems to be as an invasive primary saprophyte on a wide range of plant substrates. Its antagonism to snow moulds and other fungi suggests that it may play a significant role in determining the nature and intensity of damage in snow mould complexes (Smith and Davidson, 1979).

The sclerotia-like stromata (Fig. 6.6d and Fig. 13, colour plate section) scattered on stems, leaves, bark and overwintered seeds, 0.1–2.0 mm long × 0.1–1.0 mm wide × 0.5–1.0 mm high in the middle, lenticular in surface view, pulvinate in longitudinal section; white at first, becoming bright orange, finally light brown and hard, occasionally green with superficial algal growth; pale orange internally; subepidermal and remaining so or becoming erumpent by longitudinal splitting of the host epidermis; easily removed from the substrate of some hosts, but often firmly attached to subcortical tissues of gramineous hosts; remaining sterile for long periods, composed of compacted *textura epidermoidea*, cells containing drops of orange pigment, walls 1.0–1.5 μm thick. The mycelium in the host or substrate is systemic.

Under cool, moist conditions, a continous layer of conidiophores develop from portions of the stromal surface; they are erect (7–) 20–30 μm long × 1.0 μm wide at the tip and 1.5 μm wide at the base, hyaline, smooth and branching. Conidiogenous cells are cylindrical to subulate, orthophialidic, monoblastic, terminal and arising as lateral branches of the main axis, often immediately subtending a terminal phialide and not delimited from the main axis by a septum, 7–16 (–30) μm long, hyaline with a smooth, slightly thickened, but not flared, collarette.

Cream, pink, orange, or orange–red masses of gloeoconidia may develop on the stromata. Conidia are unicellular with densely staining regions at each end, heteropolar, straight, polysymmetric, oblong to elliptic, sometimes waisted centrally, with or without a protuberant, flattened, basal abcission scar, hyaline, produced in basipetal succession and held in cream to orange slime. They measure (4.5–) 5.6–6.3 (–7.5) × 1.5–2.0 μm from natural stromata but with a larger size range from culture (Smith and Davidson, 1979).

Acremonium boreale has the teleomorph *Nectria tuberculariformis* collected in alpine boreal habitats in the Swiss Alps, Austria, Colorado and North Dakota. Several other similar species of low-temperature tolerant alpine-boreal fungal antagonists have also been described (Samuels *et al.*, 1984).

The fungus is very slow growing in culture even at its optimum temperature which, for five isolates on cornmeal, malt extract, yeast extract agar, CMMY ranged from 9 to 18°C. Appreciable growth in colony diameter was made at –6°C and good growth took place at 0°C. On 'lean' media such as CMMY the peripheral mycelium is usually submerged in the agar and the colony centre may be waxy or individual sclerotia-like stromata may develop, especially in tube cultures, at temperatures below 10°C. Aerial mycelium is sparse, but occasionally conidia develop on the stromata on conidiophores which are similar to those produced on 'natural' stromata. Conidia are produced in basipetal succession and congregate in heads at the tips of free conidiophores or 'slime-down' on the stroma.

Sporulation occurs from –3 to 20°C and spores germinate from –2.5 to 20°C, but at the upper end of the range a high proportion of spores become pear-shaped or spherical and fail to produce germ tubes or swell and burst.

To isolate *A. boreale* use fully imbibed sclerotia as for *Myriosclerotinia borealis*. Plate disinfected slices on CMMY medium and incubate at 0°C in a sealed container until mycelium is produced. Transfer to CMMY and incubate at 6–15°C. Stromata developing on this medium may produce conidiophores bearing conidia from which further cultures can be made.

For details of pathogenicity tests with this and other low-temperature fungi of low virulence see Årsvoll (1975a) and Smith and Davidson (1979).

6.7 SNOW MOULD COMPLEXES AND COMPETITION

Diverse species of snow moulds, such as *Microdochium nivale*, and/or the LTB and SLTB phases of *Coprinus psychromorbidus*, *Myriosclerotinia borealis*, *Acremonium boreale*, *Phoma sclerotiodes* and *Typhula* spp. and vars. may occur in complexes in single or overlapping patches of turf grass snow mould (Fushtey, 1975, 1980; Smith, 1974a, 1976, 1978; Smith and Mortensen, 1981). Other 'non-snow moulds' such as *Rhizoctonia* spp., *Cladosporium herbarum*, *Fusarium* spp. and miscellaneous leaf and crown pathogens and saprophytes may also be present (Broadfoot, 1936; Smith, unpublished; Traquair and Smith, 1981). The dominant snow mould can often be determined if sclerotia are present, but associated mycelial fungi may modify the incidence and severity of the disease. The 'non-snow moulds' may have been important in the fungal succession on or in the plants leading up to the snow mould attack. Where two or more low-temperature pathogens responsible for the injury are present they must have been able to compete successfully for the supply of nutrients, space, moisture, or susceptible plant tissues. This is direct competition. In a broader sense they compete in having survival structures such as sclerotia (e.g. in *Typhula* spp.), resting mycelium (*M. nivale*), or protopycnidia (*P. sclerotiodes*) with differing longevity. They may have different tolerances to drought, desiccation, moisture, heat and chemicals which affect their competitive saprophytic ability (Garrett, 1956). But commensal and symbiotic relationships may also occur between snow moulds and their associates, such as occur between other soil organisms (Brian, 1960; Matsumoto and Araki, 1982; Matsumoto and Sato, 1982) although little is known about these. One of the mechanisms by which successful competition may be achieved is antibiosis (Clark, 1965; Park, 1960).

Micro-organisms compete for possession of plant residues which can be regarded as energy or raw material in different states of vitality (Bruehl, 1975). The more vital, the less likely are the saprophytic organisms to be able to colonize it and take possession. This energy can be passively possessed, stored within resting structures, such as the sclerotia of many snow moulds. Energy can also be passively possessed when the pathogen produces no true dormant structures and dies out when its available substrate is gone. There seem to be no snow moulds of this latter type *M. nivale*, seems to be of the 'combination possession' kind (Booth and Taylor, 1976). The mycelium of the fungus attacks seedling tissues, invades aerial parts of grasses and sporulates on dead tissues. The short-lived spores can only be regarded as infective agents (Smith, 1953). However, some recent observations suggest that the fungus may also behave as a persistent endophyte in green tissues (Petrini, Müller and Luginbuhl, 1979). Inoculum may also be carried over unfavourable periods as plectenchymatic aggregates in plant residues (Bennett, 1933a) for considerable periods of time (Booth and Taylor, 1976). Resting structures are probably not very important in the persistence of this mesophilic snow mould, since suitable conditions for its development are presented at several seasons in moist temperate climates (Smith,

1953). *M. nivale* has an advantage over its competitors, e.g. the sclerotial snow moulds, in cool, moist weather because it can cause disease at higher temperatures than they can and so gains possession of the substrate. When temperatures are lower (e.g. 0–3°C), it is at a competitive disadvantage with, e.g., *T. incarnata*, since the mycelial growth capacity of *M. nivale* at 0°C and 3°C is only 8 and 18% of that at its growth temperature optimum compared with 27 and 50% of that of *T. incarnata* (Årsvoll, 1975a). Although *M. nivale* is a mesophilic fungus, it is still capable of slow growth at −6°C. Once the disease is established as microdochium patch in autumn it has established possession of tissues and, using them as a base, it persists and develops slowly under the snow. Matsumoto and Sato (1982) showed that *T. incarnata* had excellent competitive saprophytic ability, but that of *T. ishikariensis* was poor. The low competitive saprophytic ability of the latter was compensated for by greater virulence.

M. borealis, which can cause disease when the tissues of grasses are frozen, has an advantage over *Typhula* spp. which are slower growing at lower temperatures. At 0°C the mycelial growth capacity of *M. borealis* is 80% of that at its optimum, that of *T. ishikariensis* var. *ishikariensis* and of *T. incarnata* is 27% (Årsvoll, 1975a). The effect of freezing of the plant tissues is to increase the osmotic pressure of the cell sap which *M. borealis* tolerates better than *Typhula* spp. (Bruehl and Cunfer, 1971; Tomiyama, 1961; Volk, 1937), *T. ishikariensis* var. *canadensis* probably has an advantage over *T. incarnata*, with which it occurs in heavy snowfall areas of British Columbia (Smith, 1974b) in having a much lower optimum temperature for mycelial growth (Årsvoll and Smith, 1978). This may also account for the infrequent occurrence of *T. incarnata* in Saskatchewan (Smith, 1974a, 1978) which has lower temperatures in winter and a lower snowfall than the regions in British Columbia where *T. incarnata* occurs (Potter, 1965). The lower snow cover in Saskatchewan results in less moderation of the severe winter air temperatures under the snow cover (Smith, 1981b).

Sclerotia of snow moulds such as *M. borealis*, *Typhula* spp., the SLTB phase of *C. psychromorbidus* and the low-temperature antagonist, *Acremonium boreale* remain dormant during spring, summer and early autumn. If they did not do so, the energy reserves which give them a competitive advantage over mycelial, low-temperature fungi would be wasted since they are unable to invade plant tissues at temperatures higher than a few degrees above freezing. These fungi require moist conditions and prolonged low temperatures before their sclerotia will germinate. *M. borealis* and *Typhula* spp. also require light, and in the case of the latter, light of a specific wavelength (Remsberg, 1940a; Tasugi, 1935) before they will produce sporophores. None of these fungi have a pronounced dormant period once their sclerotia have parted company from their hosts or matured in culture (Årsvoll, 1976b; Årsvoll and Smith, 1978; Smith and Davidson, 1979; Traquair and Smith, 1982). *Typhula* spp., the SLTB phase of *C. psychromorbidus* and *A. boreale* will germinate mycelially in the dark if given adequate moisture and prolonged cool temperatures, but *M. borealis* usually will not. The effect of low temperatures and chilling on oversummering sclerotia has not been examined to the same extent as on overwintering sclerotia (Coley-Smith and Cooke, 1971). The role of associated micro-organisms in the 'sclerotiosphere' in suppressing sclerotial germination in *T. ishikariensis* var. *idahoensis* was shown by Huber and McKay (1968). The associated bacteria did not antagonize growing mycelium.

Different snow moulds may be found colonizing the same plant (Årsvoll, 1975a; Ekstrand, 1955a; Smith, unpublished). Ekstrand noted antagonism between some snow mould species inoculated on the same plate of culture medium and incubated at temperatures between −5 and 10°C. Little antagonism was noted between *M. nivale* and the others since its abundant aerial mycelium overgrew the boundaries between colonies at all temperatures between −2 and 5°C. Antagonism was apparent between *M. borealis* and *T. borealis* (*T. ishikariensis*) from −5 to 10°C, between *M. borealis* and *T. incarnata* from −2 to 5°C and between *T. borealis* and *T. incarnata* between −5 and 10°C. This was apparently an antibiotic effect. Smith and Davidson (1979) noted similar effects between the snow mould antagonist *Acremonium boreale* and *M. nivale*, *T. incarnata*, *T. ishikariensis* vars. *ishikariensis* and *canadensis* at temperatures of −3 to 10°C. The SLTB phase of *C. psychromorbidus* was little affected. Isolates of *Acremonium boreale* varied in antibiotic activity (Smith and Davidson, 1979), but their mycelial growth was not apparently affected by the snow mould pathogens. Other species of low-temperature tolerant alpine-boreal fungi produced diffusible materials in culture which inhibited the growth of mesophilic plant pathogens (Samuels *et al.*, 1984). Tomiyama (1961) reported antagonism between *M. borealis* and *T. incarnata* in culture and suggested that this might play an important role in determining their distribution.

Primary infection by one snow mould may prevent secondary invasion by another because of antibiotic effects (Ekstrand, 1955a), but not apparently in all turf grass diseases. For example, *M. borealis* will overgrow patches of disease caused by *T. ishikariensis*

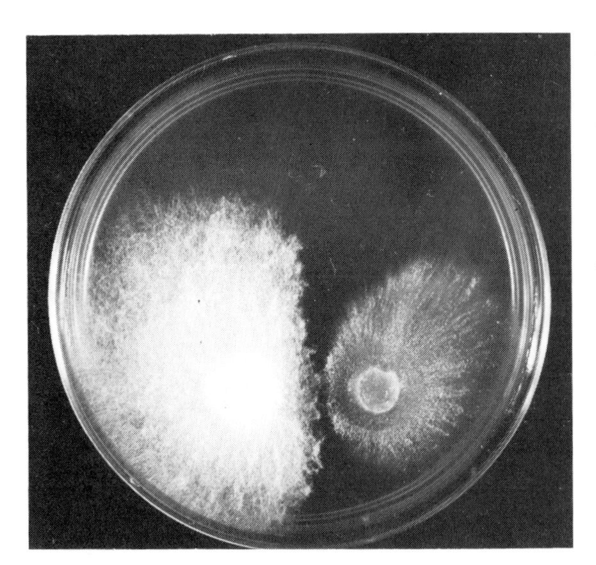

Figure 6.11 Barage effect between monokaryotic colonies of *Typhula incarnata* derived from the same dikaryotic isolate.

var. *canadensis* and the latter fungus will overgrow turf damaged earlier by *M. nivale*. Intraspecific competition (mutual antagonism) between colonies of isolates of the ascomycete snow moulds *M. borealis* and *M. nivale* seems slight, but that between isolates of the basidiomycetes *C. psychromorbidus* (LTB phase), *Typhula incarnata* (Fig. 6.11), *T. ishikariensis* vars. *ishikariensis* and *canadensis* has been found (Aårsvoll, 1976b; Lebeau, 1975; Smith and Aårsvoll, 1975). When macerated dikaryotic isolates of either *T. ishikariensis* var. *ishikariensis* or var. *canadensis* were mixed and used as inoculum there was significantly less severe disease than when the isolates were used separately (Smith and Aårsvoll, 1975). Mixtures of highly aggressive isolates of either *T. ishikariensis* var. *ishikariensis* or *T. incarnata* from different geographic regions of Norway were of much lower pathogenicity than when used separately. *T. ishikariensis* var. *ishikariensis* showed considerable mutual antagonism between isolates from four grassland fields and even from within one-square-metre plots in the fields. This showed considerable intraspecific and intravarietal variability. On the other hand no reduction in pathogenicity occurred when four isolates of *M. nivale* or *M. borealis* were mixed (Aårsvoll, 1976a).

When adjacent patches of snow mould caused by the LTB phase of *C. psychromorbidus* grow towards each other they often do not merge but remain like 'crazy paving' with ribbons of green grass between them (Fig. 6.3). This occurs on both fine putting-green turf and on lawns and it appears to be due to mutual antagonism (Smith and Årsvoll, 1975). Lebeau (1975) found that mixing pathogenic isolates of the same fungus, which also causes winter crown rot of alfalfa, resulted in lower virulence and mycelial growth of the mixture. It has been suggested (Colotelo and Ward, 1961) that the production of HCN in alfalfa plants infected by the LTB phase of *C. psychromorbidus* was due to the β-glucosidase activity of the fungus acting on the cyanogenetic substrates in the host. Pathogenesis was described by Lebeau and Cormack (1961) as being associated with the accumulation of toxic amounts of HCN in host tissues. Strains of the fungus differ in their cultural characters, pathogenicity (Ward, Lebeau and Cormack, 1961), and ability to liberate HCN in culture and host plants:

Type A Highly virulent, slow growers which produce no HCN in culture, but release large amounts of HCN in interaction with plants.

Type B Less virulent than A, grow rapidly, produce large quantities of HCN in culture, but smaller amounts in the host.

Type C Non-pathogenic, rapid growers, produce no HCN in culture or in conjuction with the host.

Although mixing of pathogenic isolates reduced fungal virulence it had no significant effect on the production of β-glucosidase, so no apparent relationship seemed to exist between pathogenicity and the secretion of the enzyme. However, virulence was restored and production of β-glucosidase resumed when an avirulent isolate was mixed with a virulent strain (Lebeau, 1975).

The biochemistry of the mutual antagonism shown by *Typhula* spp. and *C. psychromorbidus* is not understood. However, antagonism is usually ascribed to staling substances or antibiotics excreted by fungi into their substrate or volatile products which they elaborate. These metabolites accumulate to the extent that the growth of the fungus producing them stops or is restricted. HCN is an inhibitor of respiratory enzymes (Hutchinson, 1973) which has been shown capable of killing plant seedlings. While it is produced by many fungi (Bach, 1956; Filer, 1964; Lebeau and Cormack, 1961), Lebeau and Cormack (1961) were unable to show its production by *T. ishikariensis* var. *idahoensis*. The frequently seen mutual inhibition of LTB colonies in turf is unlikely to have resulted from the action of HCN produced by the adjacent colonies under the snow unless high concentrations of the toxicant were being produced by actively growing mycelia at the patch margins. Under such conditions the death of the intermediate

6.8 SUMMARY OF SNOW MOULDS OF TURF GRASSES

Table 6.4 Identification of snow moulds of turf grasses

Disease	Pathogen	Diagnosis
Microdochium patch and pink snow mould	Anamorph-*Microdochium nivale* (Fries) Samuels and Hallett.	Microdochium patch symptoms appear in cool autumn or spring weather particularly, as patches 2.5–5 cm across of water-soaked, yellow, orange–brown or brown grass which may coalesce.
	synonym-*Fusarium nivale* (Fr.) Sorauer.	In pink snow mould, after snow melt, patch centres bleach and have an orange–brown or brown margin with white mycelium. A pink colour may develop on the infected leaves (Fig. 4, colour plate section)
	Teleomorph-*Monegraphella nivalis* (Schaffnit) E. Müller	To identify the fungus incubate diseased plugs or isolate in full light or nuv at 14–17 °C and examine for spores produced in salmon-pink sporodochia. Grow on PDA or PSA medium. Spores of var. *nivalis* are typically 0- or 1-septate and heelless or nearly so, and this is the var. most common on turf. Spores var. *major* are similar in morphology, but multiseptate.
Sclerotinia snow mould	*Myriosclerotinia borealis* (Bub. and Vleug.) Kohn Synonym-*Sclerotinia borealis* (Bub. and Vleug.)	After snow melt in spring, patches of grass with water-soaked leaves and sparse grey mycelium appear. These are up to about 15 cm across. Patches may coalesce. Infected leaves bleach almost white, wrinkle, and may turn thread-like, darkening with saprophytic fungi. Sclerotia at first cream to putty-coloured or even faintly pink, globular, elongate or flake-like and arched with plant vascular remains attached, are found in sheaths, crowns and leaf axils, on or within leaves. They vary in size according to host, but are commonly up to 7–8 × 3–4 mm. When mature they turn black, wrinkle when dry and readily detach from the host. Infected plants are usually dead (Fig. 11, colour plate section; Figs 6.6 and 6.11).
Grey or speckled snow mould	*Typhula incarnata* Lasch ex Fr.	At or after snow melt, or sometimes after a cold, wet period in winter with little or no snow, discrete patches may be only 2–5 cm across, but may increase under a snow cover up to 0.5 m. At snow melt sparse to dense white to greyish-white mycelium may mat together the patches. Globular, to flattened-spherical, faintly pink sclerotia, up to 5 mm in diam. are in or on infected tissues of leaves, and plant bases. Sclerotia darken from pink to brown to reddish–brown or dark brown and wrinkle on drying and may be firmly attached. Sporophores to about 20 mm in height with pale pink or white stipes and pink to rose-coloured clubs may develop in moist autumn weather from sclerotia at the base of previously attacked turf. Cortical cells of sclerotia are lobate, interlocking like pieces of a jigsaw puzzle. In culture young sclerotia are white, then pink, but may turn chestnut brown and become irregularly-shaped and are formed in rings without much aerial mycelium. When fully swollen, they are resilient and gelatinous (Figs 8 and 9, colour plate section; Figs 6.6 and 6.7).
Speckled snow mould	*Typhula ishikariensis* Imai. vars.	Dark-coloured sclerotia give patches of disease a speckled appearance although field symptoms of *T. ishikariensis* vars. are similar to those caused by *T. incarnata*. Sclerotia of *T. ishikariensis* are never pink or red, but dark amber to dark chestnut when fresh and dark brown to almost black when dry. They are not gelatinous. Sporophores have greyish white clavulae shading into smoky-brown stipe bases.
	var. *ishikariensis* Årsvoll and Smith	Sclerotia are usually firmly attached to plant tissue, 0.3–2 mm diam. and sporophores are 4–20 mm tall. Little aerial mycelium is produced on potato-malt extract (BASM) agar and sclerotia are formed in concentric rings in Petri dish cultures. Sclerotial rinds often rough, but rarely ridged or wrinkled. Rind cells rarely digitate and spherical cells with thickened walls present in some isolates. Cell outlines smoother than in var. *idahoensis*.

	var. *idahoensis Årsvoll and Smith*	Sclerotia not usually firmly attached to plant, 0.5–2 mm diam. and sporophores 3–14 mm tall. Usually there is little aerial mycelium on BASM agar and sclerotia are formed either in concentric rings or in a central pile in Petri dish cultures. Sclerotial rinds often wrinkled, ridged, or with splits in the surface. Rind cells less regular in outline than var. *ishikariensis*, sometimes with digitate and lobate cells and with disjunctions and ridges.
	var. *canadensis* Smith and Årsvoll	Sclerotia superficial, easily detached, suspended in mycelium between leaves, 0.2–1.6 mm in diam. Sporophores 1–11 mm tall. In culture on BASM at 6 °C aerial mycelium is abundant with sclerotia suspended in it rather than in concentric rings on agar. Sclerotial rind sometimes rough, but without ridges and with superficial mycelium attached. Rind cells resemble those of var. *ishikariensis* rather than *idahoensis*, but less rounded than the former and not as lobate as the latter (Fig. 3 and 6, colour plate section; Figs 6.6 and 6.7).
Cottony snow mould (LTB snow mould)	*Coprinus psychromorbidus* Redhead and Traquair Non-sclerotial low-temperature tolerant basidiomycete	LTB phase: After snow melt in spring patches 15 cm or more emerge showing white, abundant to sparse mycelium, particularly at patch margins. Patches often do not coalesce and show green ribbons of undamaged grass between them, but symptoms vary on *Poa pratensis*, *P. annua* and *Agrostis* spp., the latter two species are usually killed and *P. pratensis* is very slow to recover. There are NO sclerotia and NO spores, but the mycelium which is fluffy in culture on BASM agar has abundant clamp connections at 6 °C.
Cottony Snow mould (SLTB snow mould)	*Coprinus psychromorbidus* Sclerotial low-temperature basidiomycete	SLTB phase: Disease patches are usually smaller than those of the LTB and the aerial mycelium less abundant, cobwebby stretching from leaf to leaf. Hyphae aggregate to form sclerotia on grass leaves and on the underside of leaves and twigs lying on the turf. Sclerotia are at first grey, turning grey–brown and finally charcoal-black, often irregularly shaped 0.25–1.5 mm, but usually not more than 1 mm in any dimension, sometimes in the form of flakes. Sclerotia may produce a pale-brown exudate. The sclerotial rind is made up of several layers of rounded, pigmented cells. When fully swollen in water the sclerotia exude milky contents when pierced with a needle. The fungus and the LTB are conspecific with *C. psychromorbidus*, but the SLTB is less pathogenic than the LTB and is also psychrophilic; optimum temp. 15°C (Figs 5 and 7, colour plate section; 6.2, 6.3
	Coprinus psychromorbidus	Fruiting stage: Mesophilic, occurs on oversummering plants of cereals, forage legumes and grasses as a saprophyte. May also fruit on wooden pegs or canes in greenhouse plant containers used for infested sod. The caps are typically of *Coprinus* type, delicate, 7–12 mm wide, conical to flat when mature with a narrow central boss. The stalk is 40–70 mm long and 2–3 μm wide with a slightly swollen base. The cap surface is felted with recurved, orange – yellow – brown scales (Fig. 6, colour plate section).
Brown root rot	*Phoma sclerotiodes* (Preuss ex Sacc.)	Grass plants show a basal brown root rot at snow melt with sclerotia-like protopycnidia up to c. 0.8 mm in diam. on lower crowns and roots down to c. 10 cm deep in soil. They are usually solitary on roots, greenish-black, globose to conoid or slightly elongated on grass roots, dished on the abaxial side with a short attachment process and without locules at this stage. The protopycnidial wall is thin and easily detached. Mature pycnidia are hemispherical, solitary, or confluent in culture with thickened walls with several locules which lead to beaked ostioles up to 200 μm long. Spores are discharged through the ostioles as a cream or yellow cirrus. Conidia are one-celled, hyaline, elliptical, biguttulate, $4–8 \times 2–3.5$ μm.

Table 6.4 Continued

Disease	Pathogen	Diagnosis
Frost scorch or 'string-of-pearls' disease	*Sclerotium rhizoides* Auersw.	Plants are stunted with wilted, curled, or rolled leaves, bleached white, tips tending to become tendril-like, bases remaining green. Sclerotia formed in bead-like rows, superficially on withered leaves, oval, spherical or oblong, 1–5mm diam., white or grey darkening to almost black and rough. No spores.
	Acremonium boreale Smith and Davidson (anamorph)	Sclerotia-like stromata of the anamorph scattered on stems, leaves, bark and overwintered seeds c. 0.1–2.0 × 0.1 mm, plano-convex to globose or spindle-shaped, white a first, becoming bright orange, then light brown and hard, sub-epidermal at first, then erumpent. Conidia develop on stromatic surface in cream, pink, orange, or orange–red masses, heteropolar, straight, polysymmetric, oblong to elliptic, sometimes waisted, c. 5.0–6.3 × 1.5–2.0 μm (Fig. 13, colour plate section).
	Nectria tuberculariformis (Rehn ex Sacc.) Winter (teleomorph)	Teleomorph develops on the same stroma as the anamorph. For diagnosis see Samuels *et al.*, 1984 and Smith, 1987.

Table 6.5 Distribution, predisposition and control of snow moulds of turf grasses

Disease and distribution	Predisposing factors	Cultural control	Fungicidal treatments
Microdochium patch and pink snow mould (*Microdochium nivale*) Boreal, sub-boreal, and temperate regions and at higher elevations in warmer regions of Europe, North America, Asia, and Australia. The snow mould with the widest climatic (geographical) range because of its mesophilic and psychrophilic adaptability.	Unbalanced or excess nitrogen fertilizer, particularly if applied late in the growing season. Alkaline turf surface and/or high surface moisture. Cold, humid weather or late autumn/early winter, temporary snow covers. Slow spring snow melt. Turf warming in winter from house basement, pipes, ducts, or heating cables. Covers or mulches to protect turf from frost. Susceptible species (or cultivars of generally resistant ones) e.g. *P. annua*, *P. trivialis*, and *Agrostis* spp. *P. pratensis* may be damaged, but seldom killed.	Let turf 'harden off' with balanced, but not excessive or late-season nitrogen fertilizer, keep mowing while leaf growth continues. Remove clippings and fallen leaves. Open up hedges and other dense windbreaks. Break up dew by switching or poling. Replace susceptible cultivars or species with more resistant ones. Control establishment or ingress of *P. annua*	Control disease before growth slows down in autumn or patches may remain until spring. Depending on region, this may require several fungicide applications from late summer until a permanent snow cover develops. 'Ring the changes' in fungicides to control components of snow mould complexes and to reduce the risk of *M. nivale* developing resistance to a particular fungicide. Effective fungicides are: Benomyl, mercurous + mercuric chlorides, phenyl mercuric acetate, phenyl mercuric acetate + thiram, oxycarboxin + thiram + carboxin, mancozeb, maneb, chlorothalonil, quintozene, iprodione, methyl thiophanate and triadimefon are generally effective.

Sclerotinia snow mould (*Myriosclerotinia borealis*) Regions with high snowfall and long duration; parts of Scandinavia, the USSR and Japan, in Alaska, western and central Canada, and north-central USA.	Severe damage occurs only in higher snowfall areas and years when snow cover is prolonged. Epidemics occur even when snow cover develops on frozen ground. Most common grasses are susceptible, but *Agrostis* cultivars are often severely damaged. Disease endemic on native grasses and sown species in ditches and other snow traps adjacent to mown turf which serve as infection reservoirs. Peat soils with low pH. Inadequate P and N.	Spread snow drifts to encourage rapid melting, especially on fine *Agrostis* turf. Control disease adjacent. Use cultivars of northern origin rather than southern ones which are less winter-hardy. Ensure that the phosphorus status of the soil is adequate, but that there is sufficient nitrogen for good spring recovery. Do not use *F. rubra* north of Lat. 54°N in western Canada. Reduce inoculum by turf scarification and pick up sclerotia with vacuum sweeper.	When autumn is moist and 'open' make at least two applications of fungicide before a permanent snow cover develops. Effective fungicides are benomyl, methyl thiophanate, chlorothalonil, thiobendazole, phenyl mercuric acetate, with or without thiram, and carboxin + thiram + oxycarboxin. Mercurous + mercuric chlorides and chloroneb are not reliable against the turf grass disease.
Grey snow mould (*Typhula incarnata*) Temperate to sub-boreal climates in Europe, Asia, and North America, but less common in more extreme climates in continental interiors, eg Prairies of Canada and higher snowfall regions of more northerly latitudes, probably due to competition from other fungi more tolerant of lower temperatures. Often in complex with *M. nivale* and/or *T. ishikariensis* vars.	Excessive or unbalanced nitrogenous fertilizer especially in organic form if applied late in the growing season. Cultivars which are not winter-hardy. Most *Poa pratensis* and *Agrostis stolonifera cultivars* are susceptible. *Poa annua* is very susceptible.	Allow turf to harden off in autumn for winter as for *M. nivale*. Ensure adequate, balanced, but not excessive or late nitrogenous fertilization. Use winter-hardy cultivars of northern origin. Control ingress of *P. annua*.	Fungicides should be applied in fall or early winter before the development of a permanent winter snow cover. In some less snowy regions, e.g. coastal areas, where *T. incarnata* is the main snow mould fungus only one fungicide application is needed. Timing of the application is not critical in places where *T. incarnata* is the principal component of complexes. Where *T. ishikariensis* is present in significant amounts one fungicide application is inadequate. Effective fungicides include: cadmium succinate, phenyl mercuric acetate + thiram; mercurous + mercuric chlorides; chloroneb, chlorothalonil and mancozeb in combination; quintozene, iprodione and triadimefon.
Speckled snow mould (*Typhula ishikariensis* vars.) var. *ishikariensis* – northern Scandinavia, northern Japan, northern North America. May be expected on turf grasses in	Longer and colder winters than favour *T. incarnata*, but similar fertility conditions to the latter. Most species and cultivars are susceptible, but especially those of	Similar to *T. incarnata*, but snow drift control is especially important under long snow covers. Remove snow from fine turf or speed melting with dark top-dressings.	*T. ishikariensis* vars. are less responsive to control by some fungicides effective against *T. incarnata* and several autumn and early winter applications may be

Table 6.5 Continued

Disease and distribution	Predisposing factors	Cultural control	Fungicidal treatments
Iceland, higher evaluations in northern Europe, Switzerland, the USSR, and northern Asia since it has been reported on grasses or other species from these regions. var. *idahoensis*–arid grassland areas in Washington, Idaho, Utah, and Montana in the USA, northern Japan. Reports on var. from other regions not confirmed may be vars. *ishikariensis* of *canadensis*. Lower snowfall areas than var. *ishikariensis* or *canadensis*. var. *canadensis*–moderate to heavy snowfall regions in western Canada, particularly with long snow covers, eastwards to Ontario and Minnesota, northern Japan.	southern origin, and when excessive nitrogen and irrigation delays the onset of winter dormancy.	Use cultivars with high winter hardiness especially those derived from local ecotypes. Ensure adequate soil phosphorus status.	needed. Effective fungicides are: chloroneb, quintozene, phenyl mercuric acetate, with and without thiram; carboxin + thiram + oxycarboxin; mercurous + mercuric chlorides: iprodione, chlorothalonil triadimefon. Results with benomyl and thiobendazole are erratic.
Cottony snow mould (*Coprinus psychromorbidus*) LTB and SLTB (sclerotial) phases of a highly variable fungal species. So far, the fungus has been found on turf grasses only in the Canadian Prairies, the Yukon, and Alaska.	Shallow to deep snow covers, slow melting of drifts in spring. Unbalanced, late nitrogenous fertilization and unbalanced phosphate fertilization when turf grasses are quiescent in winter Previous attacks. Most common turf grasses are susceptible.	Control snow deposition with ventilated fences and spread snow drifts or speed snow melting with dark-coloured top dressing. Use cultivars which show early onset of dormancy and encourage this by greatly reducing nitrogen use in fall and early winter. There are some field resistant *Poa pratensis* cvs., and *Festuca* spp., but not in other turf grasses. Dormant applications of complete fertilizers recommended to assist spring recovery.	Unless resistant *P. pratensis* cultivars are in use it may be necessary to use mercurous + mercuric chlorides to control severe outbreaks. Two applications of fungicide at least are needed, the last one as late as possible before the development of a permanent snow cover. Effective materials are: mercurous + mercuric chlorides, quintozene, chloroneb, oxycarboxin + thiram + carboxin, phenyl mercuric acetate, triadimefon, fenarimol. Borax is also effective in combination with other materials, but its use is risky because of possible phytotoxicity.

Some of the fungicides listed may not be registered for control of these diseases in different countries. Other materials may also be effective.

ribbon of green grass would also be expected. In the reduction in pathogenicity in the studies with the *Typhula* spp. (Årsvoll, 1976b; Smith and Årsvoll, 1975) it seems most likely that the mutual inhibition of the pathogenic isolates was due to a water-diffused inhibitor. Since macerated mycelium was used as inoculum, competition between cells or groups of cells of the mixed inoculum occurred reducing viability or aggressiveness. This may have been due to the production of metabolites or staling substances, which were mutually inhibitive to the different isolates. It is also possible that one or more of the isolates with lower pathogenicity than the other and with faster growth was able to compete more effectively for living space on the host tissues. Considerable variability in pathogenicity, in rate of growth and antibiotic activity occurs in both dikaryotic and monokaryotic isolates of graminicolous *Typhula* spp.

At present it is uncertain how the mutual antagonism between isolates of basidiomycete snow moulds might be used to control the diseases they cause as has been possible in the case of fairy rings caused by *Marasmius oreades* (Smith, 1980g)*. However, it is possible to start snow moulds of turf grasses on a field scale by inoculation with cultures of the fungi grown on sterile grain (Smith, 1975c; Smith and Reiter, 1976). It may be possible to use cultures of isolates of low virulence which will antagonize or outgrow wild strains in regions where the diseases are endemic. It is unwise to use mixed isolates of basidiomycete snow moulds in pathogenicity tests in an attempt to simulate natural conditions (Lebeau, 1975; Smith and Årsvoll, 1975).

The onset of near-dormancy in turf grasses in autumn or winter is shown by a decrease in leaf and shoot production and the senescence of tissues. In their moribund state they may be colonized by mesophilic primary saprophytes such as *Cladosporium herbarum* (Pers.) Link, *Alternaria alternata* (Fr.) Keissler, *Epicoccum nigrum* Link, *Leptosphaeria* and *Pleospora* spp. Leaf death and chlorophyll degeneration follows. Mesophilic colonists may modify leaf substrates needed by the cold-tolerant snow moulds which follow them (Blakeman, 1971) and also antagonize them (Fokkema, 1976). The selection of a turf grass which becomes quiescent or the withholding of nitrogen which may hasten onset of quiescence may improve field resistance through alternation of the ecological succession of organisms on leaves and shoots (Hudson, 1968). A study of these aspects appears to hold promise of the development of biological methods for the control of snow moulds of turf grasses (Smith, 1980b).

*Recently Burpee *et al.* (1987) have shown that an isolate of a low-temperature-tolerant saprotroph *Typhula phacorrhiza* will colonize senescent and non-senescent foliage of creeping bentgrass and suppress the development of *T. ishikariensis* snow mould. The mechanism of antagonism is uncertain.

6.9 REFERENCES

Adachi, A., Miyashita, Y. and Araki, H. (1976) On the varietal differences in the winter survival of perennial ryegrass, *Lolium perenne* L. *Hokkaido Nat. Agric. Exp. St. Bull.*, **114**, 173–93.

Afanasiev, M.M. (1962) Lawn diseases. *Coop. Ext. Serv., Mont. Sta. Coll., Bozeman.* Leaflet 59 (Revised), dupl. 4 pp.

Allen, L.R. (1975) Snow mould diseases of turfgrasses in Manitoba, the causal organisms and their control by applications of borax and other non-mercurial compounds. MSc. thesis, University of Manitoba, 140 pp.

Allen, L.R., Bernier C.C. and Ferguson, A.C. (1975) Chemical control of low temperature organisms. *Golf Superintendent*, (Aug.), 11–14.

Allen, L.R., Bernier, C.C. and Ferguson, A.C. (1976) Chemical control of low temperature organisms attacking turf grass. *Proc. Can. Phytopathol. Soc.*, **42**, 29.

Andersen, I.L. (1960) Investigation on the wintering of meadow plants in Northern Norway. I. *State Exp. Sta. Holt., Tromsø, Rept.* **27**, 635–60. (Norwegian, Eng.).

Andersen, I.L. (1966) Investigation on the wintering of meadow plants in Northern Norway. III. Some investigations on injuries to meadow plants caused by low-temperature fungi. *State Exp. Sta. Holt, Tromsø. Rept.* **33**, 20 pp. (Norwegian, En.).

Anon. (1931) Plant pathology at the Idaho Station. *Id. Agr. Exp. Sta. Bull*, **179**, 12 pp.

Anon. (1967) Thirty-seventh annual disease survey for the twelve months ending 30 June 1967. *NSW Dept. Agric. Sci. Serv., Biol. Br.* 52 pp.

Anon. (1979) *Plenodomus meliloti* Dearness and Sanford. *Agric. Can. Res. Br., Res. Sta.* Saskatoon. Rept. 289.

Appel, O. (1933) Grass diseases. *Deutsche Landw. Presse* **60**, 641. (RAM **13**, 382, 1934. German).

Årsvoll, K. (1973) Winter damage in Norwegian grasslands, 1968–1971. *Meld. Norg. LandbrHøgsk.*, **52,** 21 pp.

Årsvoll, K. (1975a) Fungi causing winter damage on cultivated grasses in Norway. *Meld. Norg. LandbrHøgsk.*, **54**, (9), 49 pp.

Årsvoll, K. (1975b) Biological resistance investigations in grass. *Nord. Jordbr. Forsk.*, **57**, 516–21. (Norwegian).

Årsvoll, K. (1976a) *Sclerotinia borealis*, sporulation, spore germination and pathogenesis. *Meld Norg. LandbrHøgsk.*, **55**, 11 pp.

Årsvoll, K. (1976b) Mutual antagonism between isolates of *Typhula ishikariensis* and *Typhula incarnata*. *Meld. Norg. LandbrHøgsk.*, **55**, (19), 6 pp.

Årsvoll, K. (1977) Effects of hardening, plant age and development in *Phleum pratense* and *Festuca pratensis* on resistance to snow mould fungi. *Meld. Norg. LandbrHøgsk.*, **56**, 14 pp.

Årsvoll, K. and Larsen, A. (1977) Effects of nitrogen, phosphorus and potassium on resistance to snow mould

fungi and in freezing tolerance in *Phleum pratense*. *Meld. Norg. LandbrHøgsk.*, **56** (29), 14 pp.
Årsvoll, K. and Smith, J.D. (1978) *Typhula ishikariensis* and its varieties, var. *idahoensis* comb. nov. and var. *canadensis* var. nov. *Can. J. Bot.*, **56**, 348–64.
Årsvoll, K. and Smith, J.D. (1979a) Descriptions of grass diseases. No. 5 Typhula blight, grey or speckled snow mould. *Typhula incarnata* Lasch ex Fr. *Weibulls Gräs-tips*, **22**, 2 pp. (Swedish, En.).
Årsvoll, K. and Smith, J.D. (1979b) Descriptions of grass diseases. No. 6. Typhula blight, Grey or speckled snow mould. *Typhula ishikariensis*.
Bach, E. (1956) The agaric *Pholiota aurea*: physiology and ecology. *Dansk. Bot. Arkiv.*, **16**, 1–220.
Baudys, E. (1928), Phytopathological notes, 4. *Ochrana Rostlin*, **8**, 151–62. (*R.A.M.* **8**, 289–290, 1929. Czech).
Baudys, E. (1930) *Sclerotium rhizoides* on grasses. *Mykologia, Prague*, **8**, 50–5. (*R.A.M.* **10,** 191–2, 1931, Czech).
Bennett, F.T. (1933a) *Fusarium* species on British cereals. *Ann. Appl. Biol.* **20**, 272–90.
Bennett, F.T. (1933b) Fusarium patch diseases of bowling and golf greens. *J. Bd. Greenkpg. Res.*, **3**, 79–86.
Berthier, J. (1974) The genus *Typhula* (Clavariaceae) and related genera. Classification – New species. *Extra Bull. Mem. Soc. Linn. Lyon*, **43**, 182–8. (French).
Berthier, J. (1977) Monograph of *Typhula* Fr., *Pistillaria* Fr. and neighbouring genera. *Soc. Linn. de Lyon*, 214 pp. (French).
Björklund, B. (1971) Snow mould in lawns (Sw. Eng.) *Weibulls Gräs-tips*, **12–14**, 22–23.
Blakeman, J.P. (1971) Alteration of resistance of plants to fungal disease by inducing changes in the chemical environment at the shoot surfaces. In *Altering the resistance of Plants to Pests and Diseases*. (ed. R.H.S. Woods) *Pest Articles News Summaries*, **17**, 240–57.
Blomqvist, H. and Jamalainen, E.A. (1968) Preliminary tests on winter cereal varieties of resistance to low-temperature parasitic fungi in controlled conditions. *J. Sci. Agric. Soc. Finland*, **40**, 88–95.
Boerema, G.H. and Van Kesteren, H.A. (1981) Nomenclature notes on some species of *Phoma* sect *Plenodomus*. *Persoonia*, **3**, 17–28.
Boerema, G.H. and Verhoeven, A.A. (1977) Check-list for scientific names of common parastic fungi. Series 2b. Fungi on field crops: cereals and grasses. *Neth. J. Plant Pathol.*, **83**, 165–204.
Bokura, V. (1926) Investigation on *Typhula graminum* Karst. *J. Plant Protection, Japan*, **13**, 476–89. (Cited by Remsberg, R. and Hungerford, C.W. 1933. In Japanese).
Booth, C. (1971a) *The Genus Fusarium*. Commonweath Mycology Institute Kew, 237 pp.
Booth, C. (1971b) *Micronectriella nivalis* (conidial state: *Fusarium nivale*). *C.M.I. Descriptions of Pathogenic Fungi and Bacteria*, **309**, 2 pp.
Booth, C. (1973) *Fusarium*. Laboratory guide to the identification of the major species. Commonw. Mycol. Inst. Kew, 57 pp.
Booth, R.H. and Taylor, G.S. (1976) Fusarium diseases of cereals. XI. Growth and saprophytic activity of *Fusarium nivale* in soil. *Trans. Br. Mycol. Soc.*, **66**, 77–83.
Bourgoin, B., Billot, C., Kerguelen, M., Hentgen, A. and Mansat, P. (1974) Behaviour of turf grass species in France. In *Proc. 2nd Int. Turfgrass Res. Conf.*, Blackburg, Va (ed C. Roberts), pp. 35–40.
Boyce, J.H. (1952). *Winter Injury and Snow Mold Problems in Canada*. Central Exp. Fm. Ottawa, 12 pp. (duplicated).
Brauen, S.E., Goss, R.L. Gould C.J. and Orton, S.P. (1975) The effects of sulphur in combinations with nitrogen, phosphorus and potassium on colour and fusarium patch disease of *Agrostis* putting turf. *J. Sports Turf Res. Inst.*, **51,** 83–91.
Braverman, S.W. (1986) Disease resistance in cool-season forage, range and turf grasses II *Bot. Rev.*, **52** (1) 1–112.
Brian, P.W. (1960) Antagonistic and competitive mechanisms limiting survival and activity of fungi in soil. In *The Ecology of Soil Fungi* (eds D. Parkinson and J.S. Waid) Liverpool University Press, Liverpool, pp. 115–29.
Broadfoot, W.C. (1936) Experiments on the chemical control of snow mould in Alberta. *Sci. Agric.*, **16**, 615–18.
Broadfoot, W.C. (1938) Snow mould of turf in Alberta. *J. Bd. Greenkpg. Res.*, **5**, 182–3.
Broadfoot, W.C. and Cormack, M.W. (1941) A low-temperature basidiomycete causing early spring killing of grasses and legumes in Alberta. *Phytopathology*, **31**, 1058–9.
Bruehl, G.W. (1967a) Lack of significant pathogenic specialization within *Fusarium nivale*, *Typhula idahoensis* and *incarnata* and correlation of resistance in winter wheat to these fungi. *Plant Dis. Rept.*, **51**, 810–19.
Bruehl, G.W. (1967b) Correlation of resistance to *Typhula idahoensis*, *T. incarnata*, and *Fusarium nivale* in certain varieties of winter wheat. *Phytopathology*, **57**, 308–10.
Bruehl, G.W. (1975) Systems and mechanisms of residue possession by pioneer fungal colonists. In *Biology and Control of Soil-borne Plant Pathogens* (ed. G.W. Bruehl) American Phytopathological Society, pp. 77–83.
Bruehl, G.W. and Cunfer, B. (1971) Physiologic and environmental factors that affect the severity of snow mold of wheat. *Phytopathology*, **61**, 792–9.
Bruehl, G.W. and Cunfer, B. (1975) *Typhula* species pathogenic to wheat in the Pacific North West. *Phytopathology*, **65**, 755–76.
Bruehl, G.W. and Lai, P. (1966) Prior colonization as a factor in the saprophytic survival of several fungi in wheat straw. *Phytopathology*, **56**, 766–8.
Bruehl, G.W. and Machtmes, R. (1980) Cultural variation within *Typhula idahoensis* and *T. ishikariensis* and the species concept. *Phytopathology*, **70**, 867–71.
Bruehl G.W., Machtmes, R., Kiyomoto, R. (1975) Taxonomic relationships among *Typhula* species as revealed by mating experiments. *Phytopathology*, **65,** 1108–14.
Bruehl, G.W., Machtmes, R., Kiyomoto, R. and Christen, A.A. (1978) Incompatibility alleles and fertility of *Typhula idahoensis*. *Phytopathology*, **68**, 1307–10.
Bruehl, G.W., Sprague, R., Fischer, W.R., Nagamitsu, M. Nelson, W.L. and Vogel, O.A. (1966) Snow molds of winter wheat in Washington. *Wash. Agric Exp. Sta. Bull.*, **677**, 50 pp.
Burpee, L.L. and Goultry, L.G. (1984) Evaluation of fungicides for control of pink and gray snow mold on creeping bentgrass. *Turfgrass Res.*, *Ann Rept. Ont. Agric.* Coll., Univ. Guelph, **3**, 6–7.

Burpee, L.L., Kaye, L.M., Goultry, L.G. and Lawton, M.B. (1985) Biological control of gray snow mold on creeping bentgrass. In *Proc. NE Div. Am. Phytopathol. Soc. 45th Ann. Mtg.* 6–8 Nov. Newport, R.I. p. 16 (Abstract).

Burpee, L.L., Kaye, L.M., Goultry, L.G. and Lawton, M.B. (1987). Suppression of gray snow mold on creeping bentgrass by an isolate of *Typhula phacorrhiza. Plant Dis.*, 71 (1), 97–100.

Cash, E.K. (1953) Checklist of Alaskan fungi. *Plant Dis. Rept.* Suppl. 219.

Chastagner, G.A. and Vassey, W.E. (1982) Occurrence of iprodione-tolerant *Fusarium nivale* under field conditions. *Plant Dis.*, **66**, 112–14.

Christen, A.A. (1979) Formation of secondary sclerotia in sporophores of species of *Typhula. Mycologia*, **71**, 1267–9.

Christen, A.A. and Bruehl, G.W. (1979) Hybridization of *Typhula ishikariensis* and *T. idahoensis. Phytopathology*, **69**, 263–6.

Clark, F.E. (1965) The concept of competition in microbial *ecology. Ecology of Soil-borne Plant Pathogens.* (eds K.F. Baker and W.C. Snyder), University of California Press, Berkeley.

Coley-Smith, J.R. and Cooke, R.C. (1971) Survival and germination of fungal sclerotia. *Ann. Rev. Phytopathol.*, **9**, 65–92.

Colotelo, N. and Netolitzsky, H. (1964) Pycnidial development and spore discharge of *Plenodomus meliloti. Can. J. Bot.*, **42**, 1467–9.

Colotelo, N. and Ward, E.W.B. (1964) β-glyoxidase activity and cyanogenesis in susceptibility of alfalfa to winter crown rot. *Nature*, **189**, 242–3.

Commonwealth Mycological Institute (1967) *Distribution Maps of Plant Pathogens*, 432.

Conners, I.L. (compiler) (1938) Snow mold. In *Cultivated grasses. 18th Ann. Rept. Can. Plant Dis. Surv.*, p. 26.

Conners, I.L. (1967) *An Annotated Index of Plant Diseases in Canada*. Queen's Printer, Ottawa. 381 pp.

Cooper, R.J. (1982) Protecting turf from winter injury. *Golf Course Mmt.*, **50**, 30–2.

Cormack, M.W. (1934) On the invasion of roots of *Medicago* and *Melilotus* by *Sclerotinia* sp. and *Plenodomus meliloti* D. & S. *Can. J. Res.*, **11**, 474–80.

Cormack, M.W. (1948) Winter crown rot or snow mold of alfalfa, clovers and grasses in Alberta. I. Occurrence, parasitism and spread of the pathogen. *Can. J. Res. Sect. C*, **26**, 71–85.

Cormack, M.W. (1952a) Snow mould (low-temperature *Sclerotinia*) on cultivated and other grasses at Prince George, B.C. *Can. Plant Dis. Survey (1951)*, **31**, 39.

Cormack, M.W. (1952b) Winter crown rot or snow mold of alfalfa, clovers or grasses in Alberta. II. Field studies on host and varietal resistance and other factors related to control. *Can. J. Bot.*, **30**, 537–48.

Cormack, M.W. and Lebeau, J.B. (1959) Snow mold infection of alfalfa, grasses and winter wheat by several fungi under artificial conditions. *Can. J. Bot.*, **37**, 685–93.

Corner, E.J.H. (1950) '*Clavaria* and allied genera'. *Ann. Bot. Mem.* 1. Oxford University Press, 740 pp.

Corner, E.J.H. (1970) Supplement to 'A monograph of *Clavaria* and allied genera' *Beih. Nova Hedwigia*, **33**, 299 pp.

Couch, H.B. (1962) *Diseases of Turfgrasses*. Reinhold, New York, 289 pp.

Couch, H.B. (1973) *Diseases of Turfgrasses*, 2nd edn. Krieger, New York, 348 pp.

Courtillot, M. (1976) Elements of recognition of turf diseases (Fr.) *Jardins de France* (12), 19–29.

Cunfer, B.M. and Bruehl, G.W. (1973) Role of basidiospores as propagules and observations on sporophores of *Typhula idahoensis. Phytopathology*, **63**, 115–20.

Dahl, A.S. (1934) Snow mold of turfgrasses as caused by *Fusarium nivale. Phytopathology*, **24**, 197–214.

Dahlsson, S-O. (1973) *Typhula* in turfgrasses in southern Sweden. *Svensk. Frotydning*, **46**, 68–69, 73. (Swedish).

Dahlsson, S-O (1975) Frightful tracks. *Weibulls Gräs-tips*, **18**, 11–14. (Swedish, En.).

Darby, D.E. (1971) Snow and wind control for farmstead and feedlot. *Can. Dep. Agric. Bull.*, **1461**, 21 pp.

Davis, W.H. (1933) Snow mold and brown patch caused by *Sclerotium rhizoides. Phytopathology*, **22**, 8. (Abs.).

Davidson, J.G.N. (1986) Chemical control and differentiation of the snow mold complex on a new bentgrass green. (Abs. in *Can. Soc. Hortic. Sci. 66th Ann. Conf. Agr. Inst. Can. Univ. Sask.* 6–10 July 1986. 1 p.)

Dawson, R.B. and Gregg, R. (1936) Fusarium patch disease of turf caused by *Fusarium nivale*. An interim report on the causes, prevention and cure. *Bd. Greenkpg. Res.*, 16 pp.

De Leeuw, W.P. and Vos, H. (1970) Diseases and injuries on turfgrasses in the Netherlands (German, En.). *Rasenturf-Gazon* (1), 65–67, 84.

Dearness, J. and Sanford, G.B. (1930) A new species of *Plenodomus. Ann. Mycologici*, **28**, 324–5.

Dejardin, R.A. and Ward, E.W.B. (1971) Growth and respiration of the psychrophilic species of the genus *Typhula. Can. J. Bot.*, **49**, 339–47.

Demidova, Z.A. (1960) Concerning the distribution and biology of *S. borealis* on winter rye in the Sverdlovsk region. *Trud. Inst. Biol. Ural. Fil. Akad. Nauk SSSR*, **15**, 17–45. (R.A.M. **41**, 518. 1962).

Dewey, W.G. and Nielsen, R.F. (1971) Control snow mold by speeding snow melt. *Crops Soils*, **23**, 8–9.

Dexter, S.T. (1933) Effect of several environmental factors on the hardening of plants. *Plant Physiol.*, **8**, 123–39.

Dexter, S.T. (1956) the evaluation of crop plants for winter hardiness. *Adv. Agron.*, **8**, 203–39.

Dickens, L.E. and Butler, J.D. (1977) Evaluation of fungicides for control of Typhula snow mold. *Am. Phytopathol. Soc. Fungicide and Nematicide Tests.*, **32**, 151.

Dobrozrakova, T.L. (1929) *Fusarium nivale* (*Calonectria graminicola*) on autumn-sown cereals near Leningrad. *Rev. Appl. Mycol.*, **8**, 771.

Donk, M.A. (1933) Revision of the Homobasidiomycetae-Aphyllophoraceae II. *Med. Bot. Mus. Herb. Roy. Univ. Utrecht*, **9**, 1–278.

Duben, J. (1978) Researches on the foot disease complex of winter wheat with particular reference to species of the genus *Fusarium* Link. Diss. Univ. Göttingen, 149 pp. (German).

Ekstrand, H. (1938) *Sclerotium rhizoides*, an imperfectly known sclerotial fungus occurring on grasses. *Vaxtskyddsnotiser, Vaxtskyddanst, Stockholm*, **3**, 39–41. (Swedish).
Ekstrand, H. (1955a) The overwintering of autumn-sown cereals and forage grasses. Summary of investigations carried out and programme for the future. *Statens Vaxtskyddsanstalt Med.*, **67**, 125 pp. (Swedish, En.).
Ekstrand, H. (1955b) The overwintering of autumn-sown cereals and forage grasses. III. Non-parasitic injuries and combinations of parasitic and winter injury. *Statens Vaxtskyddsanstalt. Med.*, **67**, 96–103.
Ekstrand, H. (1955c) Occurrence of winter-killing fungi in extra-Scandinavian countries. *Vaxtskyddsnotiser, Stockholm*, **3**, 55–6. (*R.A.M.* **35**(3), 192, 1956). (Swedish, En.).
Eleneff, P.F. (1926) Agricultural methods to combat snow mould of cereals. La defense des plantes, Leningrad, **3**, 39–42. (French).
Endo, R.M. (1963). Influence of temperature on rate of growth of five fungus pathogens of turfgrass and on rate of disease spread. *Phytopathology*, **53**, 857–61.
Engel, R.E. (1977) Annual bluegrass—Friend or foe. *Golf Superintendent*, **45**, 22–4.
Esau, K. (1957) Phloem degeneration in Gramineae affected by the barley yellow dwarf virus. *Am. J. Bot.*, **44**, 245–51.
Escritt, J.R. and Legg, D.C. (1969) Fertilizer trials at Bingley. *In Proc. 1st Int. Turfgrass Res. Conf. Sports Turf Res. Inst.* Bingley, England, pp. 185–90.
Escritt, J.R. and Lidgate, H.J. (1962) An investigation of the suitability of a urea formaldehyde fertilizer product for use on turf. *J. Sports Turf Res. Inst.*, **10**, 385–93.
Evans, G.E. (1973) Combat snow molds with mulches. *Golf Superintendent*, **41**, 13–17.
Evans, G.E. (1975) Winter mulch covers, spring vigor, and subsequent growth of *Agrostis. Agron. J.*, **67**, 449–54.
Filer, H.J. (1964) Parasitic and pathogenic aspects of *Marasmius oreades*, a fairy ring fungus. PhD. Thesis, Wash. State Univ., 75 pp.
Flachs, K. (1935) Some lesser known grass disease. *Nach. Schädl-bekampf., Leverkusen*, **10**, 57–62, 101–3. (German).
Fokkema, N.J. (1976) Antagonism between fungal saprophytes and pathogens on aerial plant surfaces. In *Microbiology of Aerial Plant Surfaces.* (eds C.H. Dickinson and F.T. Preece), Academic Press, London and New York.
Foster, W.R. (1949) Snow mold. In *Cultivated Grasses* (Compilers I.L. Conners and D.B.O. Savile), 29th Ann. Rep. Can. Plant Dis. Surv., 37.
Freckmann, W. (1919) A very little noticed disease of reed canary grass. *Mitt. d. Ver. z. Ford. d. Moork.*, **37**, 317. (Cited by E. Muhle, 1953).
Fries, E. (1836) *Epicrisis Systematis Upsaliae*, 1936/38, 610 pp.
Fries, E. (1859) *Summa Veg. Scand.*, (2), 495.
Fushtey, S.G. (1975) The nature and control of snow mold of fine turfgrass in southern Ontario. *Can. Plant Dis. Surv.*, **55**, 87–90.
Fushtey, S.G. (1980) Chemical control of snow mold in bentgrass turf in southern Ontario. *Can. Plant Dis. Surv.*, **60**, 25–31.
Fushtey, S.G. (1983) Fusarium patch control 1982–83. *Proc. 20th Ann. Conf. Western Canada Turfgrass Assn.* 27 Feb–2 Mar. 1983, pp. 69–71.
Gabran, O. (1939) The air exchange through a snow cover and its influence on the overwintering of the plants. *Meteor. Z.*, **56**, 354–6. (German).
Gams, W. and Müller, E. (1980) Conidiogensis of *Fusarium nivale* and *Rhynchosporium oryzae* and its taxonomic implications. *Neth. J. Plant Pathol.*, **86**, 45–53.
Garrett, S.D. (1956) *Biology of Root Infecting Fungi.* Cambridge University Press, 292 pp.
Gerlach, W. (1981). Present concept of *Fusarium* classification. In *Fusarium: Biology and taxonomy* (eds P.E. Nelson, T.A. Toussoun and R.J. Cook), Pennsylvania State University Press, University Park, Pa., 457 pp.
Gibeault, V.A. (1966) Investigations on the control of annual meadow-grass. *J. Sports Turf Res. Inst.*, **42**, 17–40.
Gordon, W.L. (1952) The occurrence of *Fusarium* species in Canada. II. Prevalence and taxonomy of *Fusarium* species in cereal seed. *Can. J. Bot.*, **30**, 209–351.
Goss, R.L. (1968) The effects of potassium on disease resistance. In *The role of Potassium in agriculture* (eds W.J. Kilner, S.E. Younts and C. Brady) American Society of Agronomy, Crop Science Society of America, Madison, pp. 221–4.
Goss, R.L. and Gould, C.J. (1968) Some interrelationships between fertility levels and fusarium patch disease of turfgrasses. *J. Sports Turf Res. Inst.*, **44**, 19–26.
Goss, R.L., Gould, C.J. and Brauen, S.E. (1974) Performance of bentgrasses under putting conditions. *Proc. 20th Ann. Rocky Mountain Regional Turfgrass Conf.*, Colo. State Univ., pp. 72–77 (Duplicated).
Gould, C.J. (1957) Turf diseases in Western Washington in 1955 and 1956. *Plant Dis. Rept.*, **41**, 344–7.
Gould, C.J. (1965) Fungicides used for turf grass disease control in the U.S.A. *J. Sports Turf Res. Inst.*, **41**, 32–9.
Gould, C.J., Goss, R.L. and Byther, R.S. (1977) Disease control in putting turf. *Coll. Agric. Wash. State Univ. Bull.*, EM **2050**, 2 pp.
Gould, C.J., Goss, R.L., Ensign, R.D. and Law, A.G. (1977) Snow mold control on turf grass with fungicides in the Pacific Northwest. Wash. State Univ., Coll. Agric. Res. Center Bull., **849**, 7 pp.
Gould, C.J., Goss, R.L., Law, A.G. and Ashworth, B. (1978) Disease resistance and quality of bentgrasses in Washington State. *U.S. Golf Assoc. Green Sect. Record*, Sept/Oct., 5–11.
Gram, E. (1929) Study of damping-off. III. Cereal and grass seed. *Tidsskr. Planteavl.*, **35**, 141–268. (Danish)
Gray, E.G. (1963) New and uncommon plant diseases and pests. Snow moulds in North Scotland. *Plant Pathol.* **12**, 184.
Gray, E.G. and Copeman, G.J.F. (1975) The role of snow moulds in winter damage to grassland in northern Scotland. *Ann. Appl. Biol.*, **81**, 247–51.
Groves, J.W. and Bowerman (1955) *Sclerotinia borealis* in Canada. *Can. J. Bot.*, **33**, 591–4.
Gulaev, (1948) *T. graminearum* Gulaev. Damping-off of pine seedlings in forest nurseries. *Trans. For. Husbandry*, **9**, 44–9. (Cited by Potatosova, E.G. 1960).
Haes, E.C.M. (1956) Annual meadow-grass in turf—An

appraisal. *J. Sports Turf Res. Inst.*, **32**, 216–18.
Handoll, C. (1966) Turf disease notes, 1966. *J. Sports Turf Res. Inst.*, **42**, 65–8.
Hanna, W.F. (1939) *Coprinus urticaecola* on stems of Marquis wheat. *Mycologia,* **31**, 250–7.
Hansen, I.R. (1969) Control of low-temperature fungi on grasses with fungicides. *Jord. Avling.*, **3**, 8 pp. (Norwegian).
Haskell, R.J. (1928) Diseases of cereal and forage crops in the United States in 1927. *Plant Dis. Rept.* Suppl. **62**, 302–53.
Heald, F.D. (1924) *Sclerotium rhizoides* Auersw. *Plant Dis. Rept.* Suppl. **35**, 270.
Henry, A.W. and Berkenkamp, B. (1965) Diseases of cereal crops. Root rot (*Plenodomus* sp.). *Can. Plant Dis. Surv.*, **45**, 42.
Hindorf, H. (1980) The significance of the sporophores of *Typhula incarnata* Lasch ex Fr. for the spread of winter barley rot. *Med. Fac. Landbouww. Rijksuniv. Gent.*, **45**, 121–7. (German).
Hirai, T. (1956) Studies on the nature of disease resistance in plants. A contribution to the knowledge concerning the mechanism of infection of plants by the weak pathogenic fungi under snow cover and the nature of disease resistance against fungi or viruses on the basis of histo- and cytochemical observations. *Forsch. Pflkr., Kyoto.*, **5**, 139–57. (*R.A.M.* **37**, 33–34 1985).
Hirane, S. (1955) Studies on the control of pythium snow blight of wheat and barley. *Bull. Agric. Imp. Bur. Min. Agric. Japan* (Cited by Tomiyama, K. 1961. Snow blight of winter cereals in Japan. *Recent Adv. Bot.*, Univ. Toronto Press **1**, 549–52.)
Holmes, S.J.I. (1979) The effects of *Fusarium nivale* and *F. culmorum* on the establishment of four species of pasture grass. *Ann. Appl. Biol.*, **91**, 243–50.
Holmes, S.J.I. and Channon, A.G. (1975) Glasshouse studies on the effect of low temperature on infection of perennial ryegrass seedlings by *Fusarium nivale. Ann. Appl. Biol.*, **79**, 43–8.
Hoshiro, M. and Hirashima, T. (1978) Studies on freezing resistance of pasture species. 1. Effect of freezing injury and sclerotinia snow-blight disease on wintering of several grasses in Nemuro-Kushiro district. *J. Jpn Grassland Sci.*, Sci., **23**, 289–94.
Hosotsuji, T. (1977) Control of diseases, insect pests and weeds of turf. *Jpn Pesticide Inf.*, **33**, 5–8.
Hossfeld, R. (1974) Promotion of typhula disease on winter barley through application of fungicides for control of straw break. *Nachrichtenblatt Deutsch. Pflanzen Shutzdienst.*, **26**, 19.
Howard, F.L., Rowell, J.B. and Keil, H.L. (1951) Fungus diseases of turfgrasses. *Rhode Island Agric. Exp. Sta. Bull.*, **308**, 56 pp.
Huber, D.M. and McKay, H.C. (1968) Effect of temperature, crop and depth of burial on the survival of *Typhula idahoensis* sclerotia. *Phytopathology*, **58**, 961–2.
Hudson, H.J. (1968) The ecology of fungi on plant remains above the soil. *New Phytol.*, **67**, 837–74.
Hungerford, C.W. (1923) A serious disease of wheat caused by *Sclerotium rhizoides* in *Idaho. Phytopathology,* **13**, 463–4.
Hutchinson, S.A. (1973) Biological activities of volatile fungal metabolites. *Ann. Rev. Phytopathol.*, **11,** 223-46.
Imai, S. (1929) On the Clavariaceae of Japan. *Trans. Sapporo Natl. Hist. Soc.*, **11**, 38–44.
Imai, S. (1930) On the Clavariaceae of Japan. II. *Trans. Sapporo Natl. Hist. Soc.*, **11**, 70–7.
Imai, S. (1936) On the causal fungus of the typhula blight of graminaceus plants. *Jap. J. Bot.*, **8**, 5–18. (*R.A.M.* **15**, 347–8, 1936).
Ito, S. and Tokimaga, Y. (1935) Notae Mycologicae Asiae orientalis 1. *Sapporo Natl. Hist. Soc. Trans.*, **14**, 11–13. (*R.A.M.* **15**, 57–58, 1936)
Iwayama, S. (1933) On a new snow rot disease of cereal plants caused by a *Pythium* sp. *Agric. Exp. Sta. Ioyama-Ken, Japan.* 20 pp. (*R.A.M.* **13**(1), 17, 1934).
Iwaykiri, R. (1946) Agriculture and horticulture, **21**, 567–8. (Cited by Tomiyama, K. 1961—wrong date).
Jackson, N. (1962a) Turf disease notes, 1962. *J. Sports Turf Res. Inst.*, **38**, 410–16.
Jackson, N. (1962b) Further notes on the evaluation of some grass varieties. *J. Sports Turf Res.* Inst., **10**, **38**, 394–400.
Jackson, N. (1963) Turf disease notes, 1963. *J. Sports Turf. Res. Inst.*, **11**, **39**, 26–28.
Jackson, N. (1964) Further note on the evaluation of some grass varieties. *J. Sports Turf Res. Inst.*, **40**, 67–75.
Jackson, N. and Fenstermacher, J.M. (1969) Typhula blight: Its cause, epidemiology and control. *J. Sports Turf Res. Inst.*, **45**, 67–73.
Jamalainen, E.A. (1949) Overwintering of Gramineae plants and parasitic fungi. 1. *Sclerotinia borealis*, Bubak and Vleugel. *J. Sci. Agric. Soc. Finland*, **21**, 125–142.
Jamalainen, E.A. (1951) The occurrence of overwintering fungi on pasture grasses in Finland. *Nord. Jordbr. Forskn.*, **1–2**, 529–34. (Swedish, En.).
Jamalainen, E.A. (1954) Overwintering of cultivated plants under snow. *F.A.O. Plant Prot. Bull.*, **2**, 102–5.
Jamalainen, E.A. (1955) *Fusarium* species causing plant diseases in Finland. *Acta Agric. Fenn.*, **83**, 159–72.
Jamalainen, E.A. (1956) Overwintering of plants in Finland with respect to damage caused by low-temperature pathogens. *Finn. State Agric. Res. Bd. Publ.*, **148**, 1–30.
Jamalainen, E.A. (1957) Overwintering of Gramineae plants and parasitic fungi. II. On the *Typhula* sp. fungi in Finland. *J. Sci. Agric. Soc. Finland*, **29**, 75–81.
Jamalainen, E.A. (1958) Experiments on the use of some chloro-nitrobenzene and organic mercury compounds for the control of low-temperature parasitic fungi on winter wheat. *J. Sci. Agric. Soc. Finland*, **30**, 251–63.
Jamalainen, E.A. (1959) Overwintering of Gramineae plants and parasitic fungi. 3. Isolations of *Fusarium nivale* from gramineous plants in Finland. *J.Sci. Agric. Soc., Finland*, **31**, 28–4.
Jamalainen, E.A. (1960) Low-temperature parasitic fungi of grassland and their chemical control in Finland. *Proc. 8th Int. Grassland Conf.*, 194–6.
Jamalainen, E.A. (1962) Trials on seed treatment of winter cereals in Finland. *Acta Agric. Fenn.*, **3**, 1–54.
Jamalainen, E.A. (1964) Control of low-temperature parasitic fungi in winter cereals by fungicidal treatment of stands. *Ann. Agric. Fenn.*, **3**, 1–54. (Finnish, En.).
Jamalainen, E.A. (1970) Overwintering of grasslands in

North Finland. *J. Sci. Agric. Soc. Finland*, **42,** 45–58. (Swedish, En.).

Jamalainen, E.A. (1974) Resistance in winter cereals and grasses to low-temperature parasitic fungi. *Ann. Rev. Phytopathol.*, **12,** 281–302.

Jamalainen, E.A. (1978) Wintering of field crops in Finland. *J. Sci. Agric. Soc. Finland*, **50,** 468–519. (Swedish, En.).

Jensen, A. (1970) Turf grass diseases and their significance in Scandinavia. *Rasen-Turf-Gazon*, **1**, 69–70. (German, En.).

Jensen, A. (1974) Grass diseases in Denmark. *Symp. Nord. Jordbruks Forskning, Ås, Norway*, 399–341. (Danish).

Johnston, W.J. and Dickens, R. (1976) Centipedegrass cold tolerance as affected by environmental factors. *Agron. J.*, **68,** 83–5.

Jones, L.C. (1961) Diseases of turf. In *Lawns and Playing Fields*. Reprint from *Victoria J. Agric.*, 32pp.

Jonsson, H.A. and Nilsson, C. (1973) Testing for *Fusarium* resistance. *Weibulls Gräs-tips*, **16,** 15–18.

Kallio, A. (1966) Chemical control of snow mold (*Sclerotinia borealis*) on four varieties of bluegrass (*Poa pratensis*) in Alaska. *Plant Dis. Rept.*, **50,** 69–72.

Khokhyrakoff, M. (1935) A little known disease of winter-sown cereals *(Sclerotinia). Plant Prot. Leningrad*, **4,** 94–7 (Russian) (*R.A.M.* 1936, 15, 566).

Kiyomoto, R.K. and Bruehl, G.W. (1976) Sexual incompatibility and virulence in *Typhula idahoensis. Phytopathology*, **66,** 1001–6.

Kneen, E. and Blish, M.J. (1941) Carbohydrate metabolism and winter hardiness of wheat. *J. Agric. Res.*, **62,** 1–26.

Knowles, R.P. (1961) comparison of grasses for dryland turf. *Can. J. Plant Sci.*, **41,** 602–6.

Knowles, R.P. and Smith, J.D. (1981) Orchard grass winter survival and stubble removal. *Agric. Can., Res. Br., Forage notes* **25,** 14.

Köck, L. (1976) Result of fungicide research in a disease-prone area. *Rasen-Turf-Gazon*, **7,** 77–79. (German, En., Fr.).

Kohn, L.M. (1979) Delimitation of the economically important plant pathogenic *Sclerotinia* species. *Phytopathology*, **69,** 881–6.

Kokkonen, P. (1942) *Maanmuokkans.* 180 pp. Helsinki. Cited by Ylimäki, A. (1962).

Kreitlow, K.W. (1942) *Sclerotium rhizoides* on grasses in Pennsylvania. *Plant Dis. Rept.* **26,** 360–1.

Kristinsson, H. and Gudleifsson, B.E. (1975) The activity of low-temperature fungi under the snow cover in Iceland. *Acta Bot. Isl.*, **4,** 44–57.

Kuznetzova, A. (1953) A new species of fungus, *Typhula humulina* A. Kusn. on underground stems of Hop. *Bot. Mat. Sect. Sporing Plants*, **9,** 142–5.

Latch, G.C.M. (1973) Diseases of turf and disease control. In *Turf Culture* (ed. C. Walker), NZ Institute for Turf Culture, pp. 167–176.

Lebeau, J.B. (1960) Resistance of legumes and grasses to low-temperature organisms. *Proc. 8th Int. Grassld Cong.*, 197–200.

Lebeau, J.B. (1964) Control of snow mold by regulating winter soil temperature. *Phytopathology*, **54,** 693–6.

Lebeau, J.B. (1966) Pathology of winter injured grasses and legumes in western Canada. *Crop Sci.*, **6,** 23–5.

Lebeau, J.B. (1967) Soil warming and winter survival of turf grass. *J. Sports Turf Res. Inst.*, **43,** 5–11.

Lebeau, J.B. (1968) Pink snow mold in Southern Alberta. *Can. Plant Dis. Surv.*, **48,** 130–1.

Lebeau, J.B (1969) Diseases affecting forage production in Western Canada. In *Proc. Can Forage Crops Symp.* (ed. K.F. Nielsen), Western Coop. Fertilizers Ltd. Calgary, pp. 123–8.

Lebeau, J.B. (1975) Antagonism between isolates of a snow mold pathogen. *Phytopathology*, **66,** 877–80.

Lebeau, J.B. (1976) Fall management of fine turf grass. *Res. Highlights, Agric. Can. Res. Sta.* Lethbridge, Alta. pp. 6–8.

Lebeau, J.B and Cormack, M.W. (1956) A simple method for identifying snow mold damage on turf grasses. *Phytopathology,* **46,** 298.

Lebeau, J.B. and Cormack, M.W. (1961) Development and nature of snow mold damage in western Canada. *Int. Bot. Congr.*, Montreal, 1959. In *Recent Advances in Botany*. University Toronto Press. Vol. 1 Sect. 5, pp. 544–9.

Lebeau, J.B., Cormack, M.W. and Moffat, J.E. (1959) Measuring pathogensis by the amount of toxic substance produced in alfalfa by a snow mold fungus. *Phytopathology*, **49,** 303–5.

Lebeau, J.B., Cormack, M.W. and Ward, E.W.B. (1961) Chemical control of snow mold of grasses and alfalfa in Alberta. *Can. J. Plant Sci.*, **41,** 744–50.

Lebeau, J.B. and Dickson, J.G. (1953) Preliminary report on production of hydrogen cyanide by a snow mold pathogen. *Phytopathology*, **43,** 581–2.

Lebeau, J.B. and Logsdon, C.E. (1958) Snow mold of forage crops in Alaska and Yukon. *Phytopathology*, **48,** 148–50.

Ledeboer, F.B. and Skogley, C.R. (1967) Plastic screens for winter protection. *Golf Course Rept.*, **35,** 22–33.

Lehmann, H. (1964) Systematic position and nomenclature of the cause of typhula rot (*Typhula incarnata* Lasch ex Fr.). *Monatsberichte Deutschen Akad. Wissenschaften Berlin*, **6,** 926–30. (German).

Lehmann, H. (1965a) The typhula rot as an overwintering disease of cereals. *Nachrichtenblatt Deutschen Pflanzenshutz Dienst. NS*, **19,** 141–5. (German).

Lehmann, H. (1965b) Researches on the typhula rot of cereals. I. On the physiology of *Typhula incarnata* Lasch ex Fr. *Phytopathol. Zeit.*, **53,** 255–88. (German).

Lehmann, H. (1965c) Researches on typhula rot of cereals. II. On the pathology of *Typhula incarnata* Lasch ex Fr. in diseased host plants. *Phytopathol. Zeitschr.*, **54,** 209–39. (German).

Lipps. P.E. (1980) A new species of *Pythium* isolated from wheat beneath the snow in Washington. *Mycologia*, **72,** 1127–33.

Lipps, P.E. and Bruehl, G.W. (1978) Snow rot of winter wheat in Washington. *Phytopathology*, **68,** 1120–7.

Lipps, P.E. and Bruehl, G.W. (1980) Reaction of winter wheat to pythium snow rot. *Plant Dis.* **64,** 555–8.

Litschko, L. and Burpee, L.L. (1987) Variation among isolates of *Microdochium nivale* collected from wheat and turf grasses. *Trans. Brit. Mycol. Soc.*, **89,** 252–6.

Locquin, M. (1944) The evolution and localization of hydrocyanic acid by basidiomycetes and ascomycetes. *Bull. Soc. Linn. Lyon*, **13,** 151–7. (French).

Locquin, M. (1947) The antibiotic activity of clytocybin, is it due to hydrocyanic acid? *Compte Rendu*, **225**, 893–4 (French).

Lorenzetti, F., Tyler, B.F., Cooper, J.P. and Breese, E.L. (1971). Cold tolerance and winter hardiness in *Lolium perenne*. Development of screening techniques for cold tolerance and survey of geographical varieties. *J. Agric. Sci. Camb.*, **76,** 199–209.

Van Luijk, A. ((1934) Investigations on grass diseases. *Med. Lab.* 'Willi *Commelin Scholten*', *Baarn, Netherlands*, **8,** 1–22.

Madison, J.H., Petersen, L.J. and Hodges, T.K. (1960) Pink snow mold on bentgrass as affected by irrigation and fertilizers. *Agron. J.*, **52,** 591–2.

Maki, Y. (1976) Utilization, research activities and problems of turf grass in Japan. *Rasen-Turf-Gazon.*, **7,** 24–8.

Marion, D.F., Landscoot, P.L. and Terry, K.J. (1979) New fungicides and fungicide combinations for gray snow mold control, 1978. *Am. Phytopathol. Soc. Fungicide and Nematicide Tests*, **34,** 17.

Marion, D.F., Vangellow, E. and Yount, E. (1980) New fungicides and fungicide combinations for gray snow mold control on putting greens, 1979. *Am. Phytopath. Soc. Fungicide and Nematicide Tests*, **36,** 140.

Matsumoto, T. (1928) On the sclerotial disease accompanying the so-called winter injury of barley. *J. Plant Protection, Japan*, **15,** 1–6. (Quoted by Remsberg, R. and Hungerford, C.W. 1933).

Matsumoto, N. and Araki, T. (1982) Field observation of snow mold pathogens of grasses under a snow cover in Sapporo. *Res. Bull. Hokkaido Natl. Agric. Exp. Sta.*, **135,** 1–10.

Matsumoto, N. and Sato, T. (1982) The competitive saprophytic abilities of *Typhula incarnata* and *T. ishikariensis*. *Ann. Phytopathol. Soc. Japan*, **48,** 419–24.

Matthews, D. (1971) A survey of certified ryegrass seed for the presence of *Drechslera* species and *Fusarium nivale* (Fr.) Ces. *NZ J. Agric. Res.*, **14,** 219–26.

McDonald, W.C. (1955) The distribution and pathogenicity of the fungi associated with crown and root rotting of alfalfa in Manitoba. *Can. J. Agric. Sci.*, **35,** 309–21.

McDonald, W.C. (1961) A review of the taxonomy and nomenclature of some low-temperature forage pathogens. *Can. Plant Dis. Surv.*, **41,** 256–60.

McElroy, C., Jones H.W. and Rinehart, E.A. (1954) An investigation of the soil microflora of two grassland plots. *Proc. Okla. Acad. Sci.*, **33,** 163–8.

McKay, H.C. and Raeder, J.M. (1953) Snow mold damage in Idaho's winter wheat. *Idaho Agric. Exp. Sta. Bull.* **200,** 5 pp.

Mead, H.W. (1962) Sweet clover. In Summary of the prevalence of plant diseases in Canada in 1961. (ed. D.W. Creelman), *Can. Plant. Dis. Surv.*, **42.**

Mebalds, M.I. and Kellock, A.W. (1983) Five fungal pathogens of *Agrostis* spp. turf in Victoria, Australia. *J. Sports Turf Res. Inst.*, **59,** 103–6.

Meiners, J.P. (1955) Etiology and control of snow mold of turf in the Pacific Northwest. *Phytopathology*, **45,** 59–62.

Meyer, B.S. and Anderson, D.B. (1952) *Plant Physiology*. Van Nostrand, New York, 784 pp.

Mikolajska, J. (1974) Studies on the appearance of parasitic fungi of grasses in relation to ecological changes in meadows of the Lyna River Valley. *Agric. Tech. Acad. Olszyln, Poland*, **6,** 50 pp. (*R.A.P.P.* **56**(6), 2300, 1975, Polish).

Monteith, J. (1927) Preventing snow mold injury on greens. *US Golf Assoc. Green Sect. Bull.*, **7**, 193–4.

Monteith, J. and Dahl, A.S. (1932) Turf diseases and their control. *Bull.* US *Golf Assn. Green Sect.*, **12**, 85–187.

Muhle, E. (1953) *The Diseases and Injuries of Fodder Grasses*. S. Hirzel Verlag, Leipzig, 167pp.

Müller, E. (1977) The taxonomic position of the snow mould of cereals. (German, Fr.). *Rev. Mycologie*, **41,** 129–34.

Newton, R. and Brown, W.R. (1924) Is the apparent winter-killing of sweetclover a result of disease injury. *Sci. Agric*, **5,** 93–6.

Nicholls, V. (1956) Fungi of chalk soils. *Trans. Br. Mycol. Soc.*, **39,** 233–8.

Nirenberg, H.I. (1981) A simplified method for identifying *Fusarium* spp. occurring on wheat. *Can. J. Bot.*, **59,** 1599–609.

Nissinen, O. (1970) Effects of different minerals on the resistance of English ryegrass to *Fusarium nivale* (Fr.) Ces. Preliminary results of laboratory experiments. *Peat Plant News*, **3,** 3–11.

Nissinen, O. and Salonen, A. (1972) Effect of *Sclerotinia borealis* on the wintering of grasses at the Muddusniemi Experimental Farm of the University of Helsinki at Inari in 1950–1965. I. The effect of weather conditions on the incidence of *S. borealis* and of species and the variety of the grass on the wintering of ley. *J. Sci. Agric. Soc. Finland*, **44,** 98–114. (Finnish, En.)

Nissinen, O. and Salonen, A. (1972a) Effect of *Sclerotinia borealis* on the wintering of grasses at the Muddusniemi Experimental Farm of the University of Helsinki at Inari in 1950–65. II. The effect of cultivation techniques on the wintering of leys. *J. Sci. Agric. Soc. Finland*, **44,** 115–125. (Finnish. En.)

Noble, M. and Richardson, M.J. (1966) *An Annotated List of Seed-borne Diseases*. 2nd edn. Commonwealth Mycology Institute Phytopath. Pap. **8**, 199 pp.

Noer, O.J. (1944) Fungicides for snow mold control. *Greenkpg. Rept.*, **12,** 13–14, 24–5.

Nordisk Jordbrugs Forskeres Forening (Northern Agricultural Research Workers Association) (1968) *Nord. Jordrugsforskning*, 360–4.

Nutter, F.W., Warren, C.W., Gottlieb, A.R. and Justus, S. (1979) Chemical control of fusarium and typhula snow molds. *Fungicide and nematicide tests*, **34,** 138.

O'Rourke, C.J. (1976) Diseases of grasses and forage legumes in Ireland. *An Foras Taluntais, Dublin*, 115pp.

Oettingen, H. von (1934) The 'string of pearls' sclerotial disease of grasses. *Nachricht. u. Schadlbekampf.*, **9,** 86–9 (German).

Olien, C.R. (1967) Freezing stresses and survival. *Ann. Rev. Plant Physiol.*, **18,** 387–408.

Ormrod, D.J. (1973) Diseases of lawns. *Publ. Br. Dep. Agric. BC*, 7 pp.

Pape, H. (1926) The sclerotium disease of meadowgrasses, especially of Reed Grass. *Illus. Landw. Zeit.*, **46,** 295–6. (*R.A.M.* **5,** 743.) (German).

Park, D. (1960) Antagonism – the background to soil fungi. In *The Ecology of Soil Fungi* (eds D. Parkinson and J.S.

Waid), Liverpool University Press, pp. 148–59.
Perry, D.A. (1986) Pathogenicity of *Monographella nivalis* to spring barley. *Trans. Br. Mycol. Soc.*, **86,** 287–93.
Petrini, O., Müller, E. and Luginbühl, M. (1979) Fungus as endophyte of green plants. *Naturwissenschaften*, **66,** 262. (German).
Platford, R.G., Bernier, C.C. and Ferguson, A.C. (1972) Lawn and turf diseases in the vicinity of Winnipeg, Manitoba. *Can. Plant Dis. Surv.*, **52,** 108–9.
Pohjakallio, O., Salonen, A. and Antila, S. (1963) The wintering of cultivated grasses at the experimental farms Viik (60° 10'N) and Muddusniemi (69° 5'N). *Acta Agric. Scand.*, **13**, 109–30. (*Herb. Abs.*, **33**(1987) 1963).
Pohjakallio, O., Salonen, A., Ruokola, A-L. and Ikaheimo, K. (1956) On the mucous mould fungus, *Acrostalagmus roseus* Bainier, as an antagonist to some plant pathogens. *Acta. Agric. Scand.*, **6,** 178–94. (*R.A.M.* **36,** 46–47. 1957.).
Potatosova, E.G. (1960) Fungi of the genus *Typhula* in the U.S.S.R. *Bot. J.*, **454,** 564–72. (Russian).
Potatosova, E.G. (1960a) Conditions of germination of *Typhula* sclerotia. *Zaschita rastenij ot vreditelej i bolezhej*, **7,** 40. (Russian).
Potatosova, E.G. (1960b) Typhulosis on winter crops. *Trud. Vses. Inst. Zasch. Rast.*, **14,** 135–42.
Potter, J.G. (1965) *Snow Cover. Climatological Studies No. 3.* Can. Dep. Transpt. Met. Br. Queen's Printer, Ottawa, 69 pp.
Powell, A.J., Blaser, R.E. and Schmidt, R.E. (1967) Physiological and color aspects of turf grasses with fall and winter nitrogen. *Agron. J.*, **59,** 303–7.
Procenko, E.P. (1967) *Typhula borealis* Ekstrand infecting tulips in the USSR. *Mycol. Pathol.* **1,** 107–9.
Pukhalshi, A.V. (1937) Injury to winter wheat and rye caused by the fungus *Sclerotinia*. *Bull. Appl. Bot. Select. Ser. A 21*, 53–61. (*R.A.M.* **16,** 526. 1937).
Rawlinson, C.J. and Colhoun, J. (1969) The occurrence of *Fusarium nivale* in soil. *Plant Pathol.* **18,** 41–5.
Redhead, S.A. and Traquair, J.A. (1981) *Coprinus* sect. *Herbicolae* from Canada, notes on extra-limital taxa, and the taxonomic position of a low-temperature forage crop pathogen from Western Canada. *Mycotaxon*, **13,** 373–404.
Remsberg, R.E. (1940a) Studies in the genus *Typhula*. Mycologia 30, 52–96.
Remsberg, R.E. (1940b) The snow molds of grains and grasses caused by *Typhula itoana* and *Typhula idahoensis*. *Phytopathology*, 30, 178–180.
Remsberg, R.E. and Hungerford, C.W. (1933) Certain sclerotium diseases of grains and grasses. *Phytopathology*, **23,** 863–74.
Ritchie, I.M. (1973) Selection of grasses and legumes for high altitude revegetation. In *Revegetation in the Rehabilitation of Mountainlands. Symp.* 16. J1-J3. N.Z. For. Serv.
RIVRO (1980) *55th Descriptive List of Varieties of Field Crops – 1980.* (55e Beschrijvende Rassenlijst voor Landbougewassen – 1980). RIVRO, Wageningen, 336pp.
Robbins, W.J., Rolnick, A. and Kavanagh, F. (1950) Production of hydrocyanic acid by cultures of a Basidiomycete. *Mycologia*, **42,** 161–6.
Robertson, H.T. (1931) The fungus *Plenodomus meliloti* causing a root rot of hollyhocks. *Rept. Dep. Agric. Can. Dominion Botanist*, (1930), **23.**
Røed, H. (1956) Parasitic winter injury to pasture growth and seed crops in Norway. *Nord. Jordbrugsforsk.*, **38,** 428–32. (Norwegian).
Røed, H. (1960) *Sclerotinia borealis* Bub. and Vleug., a cause of winter injuries to winter cereals and grasses in Norway. *Acta Agric. Scand.*, **10,** 74–82.
Røed, H. (1969) A contribution to the clarification of the relationship between *Typhula graminum* Karst. and *Typhula incarnata* Lasch ex Fr. *Friesia*, **9,** 219–25. (Norwegian).
Sakuma, T. and Narita, T. (1963) Studies on the snow blight or orchardgrass and other grasses caused by *Sclerotinia borealis*. *Hokkaido Agric. Exp. Sta. Rept.*, **9,** 68–84.
Salonen, A. (1962) *Plenodomus meliloti* Dearness and Sanford, found in Finnish Lappland. *J. Sci. Agric. Soc. Finland*, **34,** 169–72. (Finnish).
Sampson, K. (1931) The occurrence of snow mould on golfgreens in Britain. *J. Bd Greenkpg Res.*, **2**(5), 116–18.
Sampson, K. and Western, J.H. (1942) *Diseases of British Grasses and Herbage Legumes.* Cambridge University Press, 85 pp.
Samuels, G.J. and Hallett, I.C. (1983) *Microdochium stoveri* and *Monographella stoveri*, new combinations for *Fusarium stoveri* and *Micronectriella stoveri*. *Trans. Br. Mycol. Soc.*, **81,** 473–83.
Samuels, G.J., Rogerson, C.T., Rossman, A.Y. and Smith, J.D. (1984) *Nectria tuberculariformis, Nectriella muelleri, Nectriella* sp. and *Hyponectria sceptri*: low-temperature tolerant, alpine-boreal fungal antagonists. *Can. J. Bot.*, **62,** 1896–1903.
Sanderson, F.R. (1970) Fusarium diseases of cereals. 7. The effect of light on sporulation of *F. nivale* in culture. *Trans. Br. Mycol. Soc.*, **55**(1), 131–5.
Sanford, G.B. (1933) A rootrot of sweetclover and related crops caused by *Plenodomus meliloti* Dearness and Sanford. *Can. J. Res. C*, **8,** 337–48.
Schaffnit, E. and Meyer-Hermann, K. (1930) Contributions to the knowledge of the interrelations between cultivated plants, their parasites and their environment. On the influence of soil reaction on the mode of life of fungal parasites and the reaction of crop plants. *Phytopathol. Z.*, **2,** 99–166 (German).
Schmidt, D. (1976) Observations on snow mould affecting grasses. *Rev. Suisse d'Agric.*, **8,** 8–15.
Schoevers, T.A.C. (1937) Some observations on turf diseases in Holland. *J. Bd Greenkpg Res.*, **5**(16), 23–6.
Scholander, P.T., Flagg, W., Hock, R.J. and Irving, L. (1953) Studies on the physiology of frozen plants and animals in the Arctic. *J. Cell Comp. Physiol.*, **42,** 1–56.
Seaman, W.L. and Schneider, E.F. (1984) *Typhula phacorrhiza* on winter wheat in Ontario. Can. J. Plant Pathol., **6**(3) 267. (Abstract)
Shalavin. A.I. (1960) Protection of winter cereals from *Sclerotinia*. *Zasch. Rast. Moskva*, **5,** 23–4. (*R.A.M.* **40,** 218–219, 1961) (Russian).
Shavrova, L.A. (1972) Infection of introduced plants by fungi of the genus *Sclerotinia* in the Arctic north. *Byulleten' Glavogno Botanicheskogo Sada*, **84,** 106–9. (*R.A.P.P.* **52,** 44. 1973).

Siviour, T.R. (1975) The differential reaction of *Agrostis* spp. and *Cynodon dactylon* to applications of thiobendazole formulations and the efficiency of these and other selected fungicides for the control of dollar spot and fusarium patch of bent turf. *J. Sports Turf Res. Inst.*, **51**, 52–61.

Skipsna, J. (1958) Investigation on the snow moulds causing typhulosis (*Typhula itoana* Imai, *Typhula idahoensis* Remsberg) and their control in wheatfields in the west regions of the Latvian SSR. *Augsne Raza, Riga* **7**, 221–239. (*R.A.M.* **38**, 590, 1959).

Skirde, W. (1970) Turf grass diseases and their importance in the European inland transition area (German, En., Fr.). *Rasen-Turf-Gazon*, 1970 (1), 70–72.

Smiley, R.W. and Craven, M.M. (1978) Comparisons of fungicide formulations for controlling gray snow mold on golf course putting greens, 1977. *Am. Phytopathol. Soc. Fungicide and Nematicide Tests*, **33**, 138.

Smiley, R.W. and Craven, M.M. (1980) Comparisons of fungicides for controlling snow molds, 1979. *Am. Phytopathol. Soc. Fungicide and Nematicide Tests*, **35**, 155–6.

Smith, J.D. (1953) Fungi and turf diseases. 3. Fusarium patch disease. *J. Sports Turf Res. Inst.*, **8**, 29, 230–52.

Smith, J.D. (1954) Fusarium patch disease – Fungicide trials, 1954. *J. Sports Turf Res. Inst.* 8, **30**, 445–7.

Smith, J.D. (1957a) The control of fusarium patch disease with fungicides. In The control of certain diseases of sports turf grasses in the British Isles. M.Sc. Thesis, Univ. Durham 1, 1–96.

Smith, J.D. (1957b) Fusarium patch disease – Fungicide trials, 1957. *J. Sports Turf Res. Inst.* 9, **33**, 360–3.

Smith, J.D. (1957c) A new fungicide formulation for fusarium patch disease control. *J. Sports Turf Res. Inst.* 9, **33**, 364–6.

Smith, J.D. (1958a) The effect of species and varieties of grasses on turf disease. *J. Sports Turf Res. Inst.* 9, **34**, 462–6.

Smith, J.D. (1958b) The effect of lime applications on the occurrence of fusarium patch disease on a forced *Poa annua* turf. *J. Sports Turf Res. Inst.* 9, **34**, 467–70.

Smith, J.D. (1959) Turf diseases in the North of Scotland. *J. Sports Turf Res. Inst.* 10, **35**, 42–7.

Smith, J.D. (1965) *Fungal Diseases of Turf Grasses*. 2nd edn (Revised N. Jackson and J.D. Smith), The Sports Turf Res. Inst., Bingley, W. Yorks., England, 97pp.

Smith, J.D. (1969a) Overwintering disease of turf grasses. *Proc. 23rd N.W. Turfgrass Conf. Hayden Lake, Idaho*, pp. 65–67.

Smith, J.D. (1969b) Snow mold on lawns in Saskatoon. *Can. Plant Dis. Surv.*, **49**, 141.

Smith, J.D. (1972) Snow mold of turf grass in Saskatchewan in 1971. *Can. Plant Dis. Surv.*, **52**, 25–9.

Smith, J.D. (1973) Overwintering diseases of turf grasses in Western Canada. In *Proc. 25th N.W. Turfgrass Assoc. Conf.* Harrison Hot Springs, BC, pp. 96–103.

Smith, J.D. (1974a) Snow molds of turf grasses in Saskatchewan. In *Proc. 2nd Int. Turfgrass Res. Conf.* Blacksburg, Va. (ed. E.C. Roberts), 18–21 June, 1973. Am. Soc. Agron. and Am. Crop Sci Soc, pp. 313–24.

Smith, J.D. (1974b) Winter diseases of turf grasses. In *Summary 25th Ann. Nat. Turfgrass Conf.* Roy. Can. Golf Assn. Winnipeg, Man., pp. 20–5.

Smith, J.D. (1975a) Snow molds on winter cereals in Northern Saskatchewan in 1974. *Can. Plant Dis. Surv.*, **55**, 91–6.

Smith, J.D. (1975b) Scandinavian turf, pathologically speaking. *Weibulls Gräs-tips*, **18**, 15–18. (Swedish, En.)

Smith, J.D. (1975b) Scandinavian turf–pathologically speaking. *Greenmaster*, **11**, 5–6.

Smith, J.D. (1975c) Resistance of turf grasses to low-temperature basidiomycete snow mold and recovery from damage. *Can. Plant. Dis. Surv.*, **55**, 147–54.

Smith, J.D. (1976) Snow mold control in turf grasses with fungicides in Saskatchewan, 1971–74. *Can. Plant Dis. Surv.*, **56**, 1–8.

Smith, J.D. (1977a) LTB snow mold is probably not a graminicolous *Typhula* species. *Can. Plant Dis. Surv.*, **57**, 18.

Smith, J.D. (1977b) The effect of nitrogen and phosphorus fertilization on the incidence of LTB snow mold on lawn turf. *Rasen-Turf-Gazon*, **8**, 74–5.

Smith, J.D. (1978) Snow molds of grasses and winter cereals. *Can. Agric.*, **24**, 8–11.

Smith, J.D. (1979a) The conditions which favour snow mold injury to crops in Saskatchewan. *Agr. Can. Res. Br., Saskatoon Res. Sta.*, Highlights, 36–8.

Smith, J.D. (1979b) Meadow voles, turf damage and turf fungicides in Saskatchewan. *Greenmaster*, **15**, 8–9.

Smith, J.D. (1980a) Snow mold resistance in turf grasses and the need for regional testing. In *Proc. 3rd Int. Turfgrass Res. Conf.*, (ed. J.B. Beard), Am. Soc. Agron., Crop Sci. Soc. Am., Soil Sci. Soc. Am. and Int. Turfgrass Soc., pp. 275–82.

Smith, J.D. (1980b) Snow molds of turf grasses: Identification, Biology and Control. In *Advances in Turf grass Pathology. Proc. Symp. Turfgrass Diseases* 15–17 May Columbus, Ohio, (eds. B.G. Joyner and P.O. Larsen), Harcourt Brace Jovanovich, Duluth, Minn., pp.75–80.

Smith, J.D. (1980c) *Major Diseases of Turf grasses in Western Canada.* Ext. Div. Univ. Sask. Publ. 409, 13pp.

Smith, J.D. (1980d) The conditions which favor snow mold injury to crops in Saskatchewan. In *Research Highlights, 1979*, Research Br. Agr. Can. Res. Sta. Saskatoon, pp. 36–8.

Smith, J.D. (1980e) A low-temperature pathogen associated with winter injury of forages and other crops. *Research Highlights 1980*, Agric. Can. Res. Sta., Saskatoon. pp. 12–13.

Smith, J.D. (1980f) *A Sclerotial Fungus on Hard Fescue.* Research Summ. Western Reg, Res. Turfgrass Coord. Comm. (WRCC-11). Oregon. 23–25 June, 2 pp.

Smith, J.D. (1980g) Fairy rings: Biology, antagonism and possible new control methods. In *Advances in Turfgrass Pathology* (B.G. Joyner and P.O. Larsen), Harcourt Brace Jovanovich, Duluth, Minn., pp. 81–5.

Smith,. J.D. (1981a) Snow molds of winter cereals: Guide for diagnosis, culture and pathogenicity. *Can. J. Plant Pathol.*, **3**, 15–25.

Smith, J.D. (1981b) Some turf grass disease problems in Saskatchewan. *Greenmaster*, **17**, 5–7.

Smith, J.D. (1983) *Fusarium nivale* (*Gerlachia nivalis*) from cereals and grasses: is it the same fungus. *Can. Plant Dis. Surv.*, **63**, 25–6.

Smith, J.D. (1987) Winter hardiness and overwintering diseases of amenity turf grasses with special reference to the Canadian prairies. *Res. Br. Agric.*, Ottawa, Canada, *Tech. Bull.*, **12E**, 193 pp.

Smith, J.D. and Årsvoll, K (1975) Competition between basidiomycetes attacking turf grasses. *J. Sports Turf Res. Inst.*, **51**, 46–51.

Smith, J.D. and Cooke, D.A. (1978) Dormie Kentucky bluegrass. *Can. J. Plant Sci.*, **58**, 291–2.

Smith, J.D. and Davidson, J.G.N. (1979) *Acremonium boreale* n.sp., a sclerotial, low-temperature-tolerant snow mold antagonist. *Can. J. Bot.*, **57**, 2122–39.

Smith, J.D. and Mortensen, K. (1981) Fungicides for snow mold control in turf grass. Saskatchewan tests, 1979/1980. *Greenmaster*, **17**, 6–8.

Smith, J.D. and Piening, L. (1980) *Plenodomus meliloti*, SLTB and a *Pythium* associated with winter damage to winter cereals in Alberta and Saskatchewan. Rept. to Subcommitee on wheat, rye and triticale, Expert Committee on Grain Breeding, Winnipeg, 1980. 1 pp. (Mimeo.).

Smith, J.D. and Reiter, W.W. (1976) Snow mold control in bentgrass turf with fungicides, 1975. *Can. Plant Dis. Surv.*, **56**, 104–8.

Solkina, A.F. (1939) A study of the cycle of development of the fungus *Sclerotinia graminearum* Elen. *Plant Prot. Leningrad*, **18**, 100–108. (*R.A.M.* **18**, 582. 1939). (Russian).

Sports Turf Res. Inst. (1978) *Choosing Turfgrass Seed in 1978.* Sports Turf Res. Inst. Bingley, W. Yorks., 59 pp.

Sports Turf Res. Inst. (1980) *Turfgrass Seed, 1980.* Sports Turf Res. Inst. Bingley, W. Yorks., England, 105 pp.

Spotts, R.A., Traquair, J.A. and Peters, B.B. (1981) d'Anjou pear decay caused by a low-temperature basidiomycete. *Plant Dis.*, **65**, 151–3.

Sprague, R. (1950a) *Diseases of Cereals and Grasses in North America.* Ronald Press, New York, 538 pp.

Sprague, R. (1950b), Studies on the life-history of *Typhula* spp. on winter wheat in Washington. *Phytopathology*, **40**, 969. (Abs.).

Sprague, R. (1952) Contribution to the life history and control of snow molds of winter wheat in Washington. *Phytopathology*, **42**, 475. (Abs.).

Sprague, R. (1955) Check list of the diseases of grasses and cereals in Alaska. *Plant Dis. Rept. Suppl.*, **232**, 94–102.

Sprague, R. (1959) Epidemiology and control of snow mold of winter wheat and grasses in the Pacific Northwest of the USA. In *Proc. 9th Int. Bot. Cong., Montreal*. pp. 540–4. Univ. Toronto Rept. pp. 540–4.

Sprague, R., Fischer, W.R. and Figaro, P. (1961) Another sclerotial disease in Washington. *Phytopathology*, **51**, 334–6.

Sprague, R. and Rainey, C.J. (1950) Studies in the life-history of *Typhula* spp. on winter wheat in Washington. *Phytopathology*, **40**, 969 (Abstract).

Stack, R.W., Jons. V.L. and Lamey, H.A. (1979) Snow rot of winter wheat in North Dakota. *Phytopathology*, **66**, 5.

Stienstra, W.C. (1974) Snow molds in Minnesota golf turf. *Proc. Am. Phytopathol. Soc.*, **1**, 27 (Abs. 35).

Stienstra, W.C. (1980) Snow molds on Minnesota golf greens. In *Proc. 3rd Int. Turfgrass Res. Conf.* (ed, J.M. Beard), Am. Soc. Agron., Crop Sci. Soc. Am., Soil Sci. Soc. Am. and Int. Turfgrass Soc., pp. 271–4.

Stirrup, H.H. (1932) *Sclerotium rhizoides* in England. *Trans. Br. Mycol. Soc.*, **14**, 308.

Stout, A.B (1911) A sclerotium disease of blue joint and other grasses. *Univ. Wis. Agr. Exp. Sta. Res. Bull.*, **18**, 206–61.

Sweets, L.E. and Stienstra, W.C. (1976) The effect of the environmental factors on the growth of *Typhula* species and *Sclerotinia borealis*. *Proc. Am. Phytopathol. Soc.*, **3**, 228–9.

Sweets, L.E. and Stienstra, W.C. (1981) Factors affecting growth of *Typhula incarnata* and *T. ishikariensis* in culture. In *Proc. 4th Int. Turfgrass Res. Conf.* (ed. R.W. Sheard), Ont. Agr. Coll. and Intr. Turfgrass Soc., pp. 449–58.

Tahama, Y. (1973) On the main diseases of turfgrass. *J. Jpn Turfgrass Res. Assoc.*, **2**, 27–32.

Tàsugi, H. (1935) On the physiology of *Typhula graminum* Karst. *J. Imp. Agr. Exp. Sta. Nisigahara, Tokyo*, **2**, 443–60.

Taylor, D.K. (1971–1978) *Turfgrass Variety Trials* Res. Sta., Res. Br. Agr. Can. Agassiz. B.C. (Duplicated).

Taylor, D.K. (1972) Results of trials with fungicides for snow mold control on golf greens in the B.C. Interior, 1971–72. Agassiz Res. Sta Agr. Can. (Mimeo.). 1 pp.

Taylor, D.K. (1974) British Columbia Turfgrass Research Report. *Proc. NW Turfgrass Assoc. Sun River, Ore*, 2 pp.

Taylor, D.K. (1975) Snow mold control with fungicides. *Proc. 29th Northwest Turfgrass Assoc. Conf. Yakima, Wash.*, pp. 105–6.

Taylor, D.K. (1976) Report on snow mold trials in the B.C. Interior. *Turf Line News* (May), p. 26.

Taylor, D.K. and Fushtey, S.G. (1977) Chemical control of snow mold in fine turf. *Pesticide Res. Rept.*, 388.

Tervet, I.W. (1941) Turf injuries in Minnesota. *Minn. Agr. Exp. Sta. Rept. 1941* (Abs. in *Exp. Sta. Record*, **86**, 792–3).

Tezuka, M and Komeichi, M. (1980) Varietal differences in winter survival and productivity in late autumn of perennial ryegrass in Tenpoku District. *Bull. Hokkaido Prefect. Agric. Exp. Sta.*, **44**, 52–61. (Japanese, En.).

Thompson, D.C., Craven, M.M. and Smiley, R.W. (1979) Comparisons of fungicide formulations for controlling typhula blight on golf course fairways. *Am. Phytopathol. Soc. Fungicide Nematicide Tests*, **34**, 139.

Thörn, K,-G. (1967) Norrlandsk vallodling. II. Resultat av en vallinventering. 1959–61. *K. Skogs-o. LantbrAkad. Tidsk., Suppl.* **7**, 57 pp.

Thuesen, A. (1975) Research with grass species and varieties for turf, 1969–73. *Stat. Forsogsvirksomhed i Plantekultur. Med.*, **1180**, 4 pp. (Danish).

Tomiyama, K. (1955) Studies on the snow blight disease of winter cereals. *Hokkaido Natl. Agric. Exp. Sta. Rept.*, **47**, 1–234.

Tomiyama, K. (1961) Snow blight of winter cereals in Japan. *9th Int. Bot. Congr., Montreal, 1959.* In *Recent Advances in Botany*, University of Toronto Press, Vol. 1, Sect. 5, 549–52.

Traquair, J.A. (1980) Conspecificity of an unidentified snow mold basidiomycete and a *Coprinus* species in the Section *Herbicolae. Can. J. Plant Pathol.*, **2**, 105–15.

Traquair, J.A., Lebeau, J.B., Moffat, J.R. and Kokko, M. (1983) Northern distribution of LTB snow mold in Canada. *Can. Plant Dis. Surv.*, **63**, 1–2.

Traquair, J.A. and McKeen, W. (1986) Fine structure of root-tip cells of winter wheat exposed to toxic culture filtrates of *coprinus psychromorbidus* and *Marasmius oreades*. *Can. J. Plant Pathol.*, **8,** 59-64.

Traquair, J.A. and Smith, J.D. (1981) Brown patch of turf grass caused by *Rhizoctonia solani* in western Canada. *Can. J. Plant Pathol.*, **3**, 207–10.

Traquair, J.A. and Smith, J.D. (1982) Sclerotial strains of *Coprinus psychromorbidus*, a snow mold basidiomycete. *Can. J. Plant Pathol.*, **4**, 27–36.

Tupenevich, S.M. (1965) Damping-off of winter cereals. In *Distribution of Pests and Diseases of Crops in the USSR in 1964.* (eds I. Ya. Polyakov and A.E. Chumakov) Trudy Vses Inst. Zasch. Rast. 25, pp. 118–21

Tupenevich, S.M. and Shusko, V.N. (1939) Measures for preventing losses in winter cereals from *Sclerotinia graminearum. Plant Prot. Leningrad*, **18,** 85–99. (*R.A.M.* **18,** 581. 1939). (Russian).

Tyson, J. (1936) Snow mold injury to bentgrasses. *Quart. Bull. Mich. Agr. Exp. Sta.*, **19,** 87–92.

US Department of Agriculture (1960) Index of plant diseases in the United States. *USDA Agric Handbk.*, **165,** 531 pp.

Uländer, A. (1910) Redogorelse for verksamheten vid Sveriges Utsadforenings Filial i Lulea ar 1906–1909. *Sv. Utsadesforenings Tidskr.*, **20,** 33–53. (Swedish).

United States Golf Association (1957) Mid-continent Turf letter 4 (Oct.) (duplicated).

Vaartnou, H. and Elliott, C.R. (1969) Snow molds on lawns and lawn grasses in northwest Canada. *Plant Dis. Rept.*, **53,** 891–4.

Vang. J. (1945) *Typhula* species on agricultural plants in Denmark. *Aarsskr. Vet. Højsk.*, 46 pp. (*R.A.M.* **25,** 216–218, 1946, Dutch).

Vanterpool, T.C. (1944) Snow mold. in *23rd Ann. Rept. Can. Plant Dis. Surv. for 1943* (eds I.L. Conners and D.B.O. Savile), p. 40.

Vargas, J.M. and Beard, J.B. (1970) Chloroneb, a new fungicide for the control of typhula blight. *Plant Dis. Rept.*, **54,** 1075–7.

Vargas, J.M. and Beard, J.B (1971) Comparison of application dates for the control of typhula blight. *Plant Dis. Rept.*, **55,** 1118–19.

Vargas, J.M. Jr, Beard, J.B. and Payne, K.T. (1972) Comparative incidence of typhula blight and fusarium patch on 56 Kentucky bluegrass cultivars. *Plant Dis. Rept.*, **56,** 32–4.

Vleugel, J. (1917) On knowledge of the fungal flora in the vicinity of Umea and Lulea. III. *Sv. Bot. Tidsk.* **B11,** 304–24. (German).

Volk, A. (1937) Research into *Typhula graminum* Karst. *Z. Pfl. Krank.*, **47,** 338–65. (German).

Walther, H. and Foeke, I. (1981) *Fusarium nivale* Ces. ex Sacc. as the cause of leaf spots and withering on winter wheat, 1980. *Nachrichtenbl. Pflanzenshutz in der DDR*, **35**, 127–8. (German).

Ward, E.W.B. (1964a) On the source of hydrogen cyanide in cultures of a snow mold fungus. *Can. J. Bot.*, **42,** 319–27.

Ward, E.W.B. (1964b) The formation of stroma-like structures in cultures of a sterile low-temperature basidiomycete. *Can. J. Bot.*, **42,** 1025–30.

Ward, E.W.B. (1966) Preliminary studies of the physiology of *Sclerotinia borealis*, a highly psychrophilic fungus. *Can. J. Bot.*, **44,** 237–46.

Ward, E.W.B. (1967) Investigation of the low maximum temperature for growth of *Sclerotinia borealis. Proc. Can. Phytopathol. Soc.*, **34,** 26.

Ward, E.W.B. (1968) The low maximum temperature for growth of the psychrophile *Sclerotinia borealis*; evidence for the uncoupling of growth from respiration. *Can. J. Bot.*, **46,** 385–90.

Ward, E.W.B. and Lebeau, J.B. (1962) Autolytic production of hydrogen cyanide by certain snow mold fungi. *Can. J. Bot.*, **40,** 85–8.

Ward, E.W.B., Lebeau, J.B. and Cormack, M.W. (1961) Grouping of isolates of a low-temperature basidiomycete on the basis of cultural behaviour and pathogenicity. *Can. J. Bot.*, **39,** 297–306.

Ward, E.W.B. and Thorn, G.D. (1966) Evidence for the formation of HCN from glycine by a snow mold fungus. *Can. J. Bot.*, **44,** 95–104.

Watson, J.R. and Wicklund, L. (1962) Plastic covers protect greens from winter damage. *Golf Course Rept.*, **30,** 30–8.

Weibel, R.O. and Quisenberry, K.S. (1941) Field versus controlled freezing as a measure of cold resistance in winter wheat varieties. *J. Am. Soc. Agron.*, **33,** 336–43.

Weihing, J.L. and Shurtleff, M.C. (1966) Lawn diseases in the Midwest. Ext. Serv. Univ. Neb. Coll. Agr. Leaflet EC 66–1833, 19 pp.

Welling, B. and Jensen, A. (1970) Snow mould in lawns 1969/70. (Danish) *Stat. Plantepatolog. Forsg. Manedsoversigt.*, **431,** 45–50.

Wernham, C.C. (1941) New facts about eastern snow mold. *Phytopathology*, **31,** 940–3.

Wernham, C.C. and Chilton, St. J.P. (1943) Typhula snow mold of pasture grasses. *Phytopathology*, **33,** 1157–65.

Wernham, C.C. and Kirby, R.S. (1943) Prevention of turf diseases under war conditions. *Greenkp. Rept.*, **11,** 14–15, 26–27.

Worf, G.L. (1977) Chemical control of gray snow mold, 1976. *Am. Phytopathol. Soc. Fungicide and Nematicide Tests*, **32,** 150.

Worf, G.L. and Jalowitz, J. (1978) Chemical control of gray snow mold, 1978. *Fungicide and Nematicide Tests*, **34,** 140.

Yakovlev, A.G. (1939) A study of the biology of *Sclerotinia graminearum* on winter cereals. *Plant Prot. Leningrad*, **18,** 109–12 (*R.A.M.* **18**, 583, 1939) (Russian).

Ylimäki, A. (1955) On the effectiveness of penta- and tetra chloronitrobenzenes on clover rot. (*Sclerotinia trifoliorum*). *Acta. Agric. Fenn.*, **83,** 147–58. (Finnish, En.).

Ylimäki, A. (1962) The effect of snow cover on temperature conditions in the soil and overwintering of field crops. *Ann. Agric. Fenn.*, **1,** 192–216.

Ylimäki, A. (1967) Winter-killing fungi in Scandinavia. *Eucarpia, Fodder Crops Sect. Rept. of Meetg.*, Köln-Vogelsang, pp. 57–62.

Ylimäki, A. (1969) Typhula blight of clovers. *Ann. Agric.. Fenn.*, **8,** 30–7.

Ylimäki, A. (1972) Disease and overwintering problems of turf in Finland. (German, Fr., En.). *Rasen-Turf-Gazon*, **3**, 70–2.

Yoshikawa, I. (1969) The golf courses of Japan. *Proc. 1st Int. Turfgrass Res. Conf. Harrogate, England,* Sports Turf Res. Inst., Bingley, England, p. 7.

Young, P.A. (1929) Sclerotium blight (*Typhula graminum*). *Plant Dis. Rept.*, Suppl. **13**, 70.

Young, P.A. (1937) Sclerotium blight of wheat. *Phytopathology*, **27**, 1113–18.

Part four
Root diseases

7 *Fusarium blight (fusarium foliar blight, crown and root rot)*

Over the past 25 years fusarium blight has been documented in several regions of the United States as an increasingly troublesome but highly controversial disease that is particularly damaging to established *Poa pratensis* turf. First noted in the late 1950s in Pennsylvania (Couch and Bedford, 1966), this summer disease complex represents a major problem of golf fairways, amenity turf and home lawns in the mid-eastern and mid-western USA (Bean, 1969; Couch, 1976; Partyka, 1976) and also in California (Endo and Colbaugh, 1974; Endo *et al.*, 1973). Typical symptoms have been observed in Alberta and Saskatchewan but the incidence of this disease in the Canadian prairies has warranted little concern to date (Smith, unpublished).

Fusarium blight has been the subject of widely divergent but stoutly championed interpretations (Couch, 1973; Couch and Bedford 1966, Sanders and Cole 1981; Smiley 1980, 1983; Vargas 1981). Research by Smiley and Craven Fowler (1984), proving that identical symptoms may result from the pathogenic activities of fungi other than *Fusarium* spp., paves the way for a better understanding of what apparently is a complex of sequential or concurrent diseases. The name fusarium blight is retained to describe foliar blight, crown rot and root rot symptoms in situations where *Fusarium* spp. are the primary pathogens. Two additional diseases are now recognized, summer patch and necrotic ring spot, where *Fusarium* spp. are not involved or occur only as secondary invaders. A brief review tracing the vicissitudes of 'fusarium blight' over the years and the rationale for the current interpretation follows.

As familiarity with fusarium blight developed it was variously perceived to involve a foliar blight, a crown rot and a root rot but there was no consensus amongst investigators as to which of these symptoms, individually or in combination, predominated. In reporting the disease, Couch and Bedford (1966) assigned the name fusarium blight because of the preponderant leaf symptom pattern. No mention was made of crown or root involvement. Bean (1966) stated that the disease had both a leaf blight and a crown rot phase. This observation was endorsed by Couch (1973) but crown rot, together with root rot, was regarded as part of the second phase of disease development. Later, Couch, Garber and Fox (1979) included crown rot as part of the initial phase of the disease in conjunction with leaf blight but again no specific information was provided as to when crown tissues became involved. Fusarium blight occurred in California primarily as a severe crown rot (Endo and Colbaugh, 1974; Endo *et al.*, 1973). In the eastern USA the foliage blight symptom was of more common occurrence (Bean, 1966; Couch and Bedford, 1966) but crown rotting was deemed to play a major role in the demise of the grass host (Bean, 1969; Cole, 1976).

The nature of the fungal involvement in this disease has been a subject of regional difference, varied emphasis and heated controversy. Fusarium blight of turf grasses was first characterized by Couch and Bedford (1966) who implicated *Fusarium culmorum* (W.G. Smith) Sacc. and *F. poae* (Peck) Wollenw. as the incitants. To confirm the pathogenicity, immature grass plants in dew chambers were inoculated with isolates of the two fungi obtained from diseased turf. Some of the isolates were found to be highly aggressive leaf blighting pathogens and *Agrostis* spp. rated as the most susceptible hosts. Difficulties arose in relating this information to field conditions. Fusarium blight occurred primarily as a problem on *Poa pratensis* and predominantly as a disease of mature turf; symptoms seldom appeared until a turf stand was two or more years old (Bean, 1966; Cole, 1976; Cole *et al.*, 1973; Meyer and Berns, 1976; Smiley, Craven and Bruhn, 1980). Of greater significance, even though Fulton, Cole and Nelson (1974) were successful in producing patch symptoms on mature *Poa pratensis* cv 'Merion' sod in the greenhouse, reproduction of typical fusarium blight symptoms on mature turf in the field by artificial means was never accomplished.

In California, *F. culmorum* was recognized as the single species causing fusarium blight in that region (Endo and Colbaugh, 1974). In Michigan, both *Fusarium* species were implicated but extensive surveys suggested that the disease also involved an interaction with nematodes, in particular *Tylenchorhynchus dubius* (Butschli) Filipjev and *Criconemoides* spp. In greenhouse trials various nematicides significantly reduced the expression of symptoms associated with fusarium blight on *Poa pratensis*, prompting the conclusion that nematodes were the dominant pathogen in the relationship (Vargas, 1976; Vargas and Laughlin, 1972). Workers elsewhere were unable to confirm these findings (Cole, 1976; Couch and Bedford, 1966; Couch, Garber and Fox, 1979).

This situation was confounded further as a result of studies with the fungicides triadimefon and iprodione to control supposed fusarium blight. Both fungicides proved to be only slightly toxic to the alleged fusarial incitants *in vitro* but effectively prevented symptoms of the disease in the field. This paradox led to speculation that the fusaria may be involved in the disease but only following predisposition of the turf by the activities of other micro-organisms sensitive to these fungicides (Pennypacker, Sanders and Cole, 1982; Sanders *et al.*, 1978; Sanders, Burpee and Cole, 1978). Iprodione actually amplified the number of *Fusarium* propagules while triadimefon had variable but insignificant effects on propagule numbers (Smiley and Craven, 1979a, b). These authors concluded that whereas the materials may act indirectly by altering host metabolism (thereby alleviating stress or stimulating host defence mechanisms), there was the possibility that the primary incitant of fusarium blight was not among the fusaria.

Further studies lent support to this thesis. In an investigation of Fusarium blight and its relation to the various physical, chemical and microbial properties of *Poa pratensis* turf, Smiley, Craven and Bruhn (1980) found no correlation between the disease and any microbial group, including the fusaria. The disease was least severe when the percentage of *Fusarium*–infected crowns was highest. The sod pH and the numbers of *Fusarium* spp. propagules were significantly correlated with thatch decomposition, and the rate of thatch decomposition was the variable most closely associated with fusarium blight. It was proposed that phytotoxins released from the decomposing thatch, by fusaria or other micro-organisms, could be involved. Since it became impossible to show any direct association with the fusaria, this widespread and common disease was adjuged to be of complex aetiology, distinct from fusarium foliar blight, crown and root rot disease, and designated 'fusarium blight syndrome' (Smiley, 1980, 1983). The persistence of the Cornell researchers was rewarded recently with their discovery of at least two nonfusarial pathogens in New York State associated with symptoms typical of fusarium blight/fusarium blight syndrome (Smiley and Craven Fowler 1984, 1985). A *Phialophora* sp. (incorrectly identified as *P. graminicola* (Deacon) Walker) and *Leptosphaeria korrae* Walker and Smith were isolated from diseased *Poa pratensis* turf. Inoculation studies confirmed the two fungi as aggressive root and crown invaders of Kentucky bluegrass (and other turf grasses) leading to patch symptoms in mature turf identical to fusarium blight. The *Phialophora* fungus (now named *Magnaporthe poae*, Landschoot and Jackson) is most actively pathogenic at warmer temperatures and is the incitant of 'summer patch disease' (Smiley and Craven-Fowler, 1984). *L. korrae* has now been confirmed in several other locations in the United States as the cause of a cool temperature patch disease of 2–3-year-old sodded bluegrass lawns and designated 'necrotic ring spot' (Worf, Brown and Kachadoorian, 1983). Although differing in their temperature optima for inciting disease directly, both fungi apparently may also predispose turf grasses to invasion by *Fusarium* spp. during summer stress periods.

Thus at least three genera of fungi are associated with the summer turf ailment known commonly as 'fusarium blight'. The incidence, range and relative importance of fusarium leaf blight, crown and root rot (incited by *Fusarium* spp.), summer patch (incitant *M. poae*) and necrotic ring spot (incitant *L. korrae*), plus any interactions, awaits further clarification. Obviously much of the research reported over the past 25 years did not accommodate a multipathogen involvement, leading to misinterpretation of data and faulty conclusions. It should now become possible to explain various anomalies in the literature relating to the aetiology and chemical control of 'fusarium blight'.

7.1 SYMPTOMS

Early symptoms of fusarium blight (fusarium foliar blight, crown and root rot) first show as scattered lighter green or wilting patches 2–6 inches (5–15 cm) in diameter. Over a period of 36–48 hours and under environmental conditions favourable for disease development these patches become a dull reddish brown, fading to a tan, and finally to a light-straw colour. Elongate streaks, crescents and roughly circular patches up to 2–3 feet (0.6–1.0 m) in diameter may occur as the disease progresses. Individual leaves often exhibit characteristic lesions originating from the cut tip or distributed randomly over the entire leaf. One to many lesions can occur on a single leaf. Initially, these lesions appear as irregularly shaped dark-green blotches. They rapidly fade through light green to a reddish-brown to a dull-tan colour. Centres of lesions may bleach white, somewhat similar to leaf lesions caused by the dollar spot fungus. In size, individual lesions may extend up to 1 cm in length and involve the entire leaf width. Isolation of fungi from such lesions during the summer months yields predominantly species of *Fusarium*.

Patches are either uniformly blighted or may have a 'frog-eye or doughnut' appearance in which centre tufts of green, apparently healthy turf are surrounded by a ring of blighted turf. The frog-eye pattern was considered a useful diagnostic feature but it is not always present (Fig. 7.1).* When conditions are optimum for disease development, individual infection centres may coalesce and large areas of turf can be destroyed within 7–10 days of the initial appearance of symptoms. The pinkish mycelium of the fusaria can be seen on root and crown surfaces during periods of high soil moisture and these fungi may be isolated from the crown tissues of plants within or at the borders of the affected area.

In California, typical outbreaks of the disease occur in the absence of the foliage blight. Patches occur when foliage dies as a result of massive destruction of crowns and attached roots by *F. culmorum*. Stem bases and crowns develop a dark brown to black dry, firm rot, which may also affect the roots. Outbreaks of the disease in the eastern USA that follow this pattern, are not uncommon and, regardless of whether or not the initial foliage symptoms occur the end result is the same.

Impairment of function and demise of the crown region tissues appears to be the critical factor in the final expression of this disease. It is still uncertain whether this is due to the direct or indirect attacks by the fusaria, alone or in complex with other microorganisms.

*It has not been established with certainly in the Eastern United States that *Fusarium* spp. can produce the 'frog-eye or doughnut' symptom pattern without the presence of some other primary pathogen.

7.2 THE CAUSAL FUNGI

7.2.1 *Fusarium culmorum* (W.G. Smith) Sacc.; synonym: *Fusarium roseum* (LK) emend Snyd. and Hans. f. sp. *cerealis* (Cke). 'Culmorum'

Mycelium loosely cottony, carmine red with buff or off-white mycelium scattered through it; sporodochial masses golden to ochre, later darker; conidia thick, curved, bluntly pointed apex, slightly more tapering at base with prominent foot; 3–9-(usually 5-) septate, 4.5 – 6.5 × 30 – 60 μm (Sprague, 1950).

The following description of *F. culmorum* (W.G. Smith) Sacc. is adapted from Booth (1971). Single condia show rapid growth to produce floccose aerial hyphae; mycelium becomes yellow; red pigmentation on the surface of the medium diffuses throughout and the medium then resumes a deep reddish brown. This pigmentation changes from golden yellow on acid media through brilliant carmine becoming red to purple on alkaline media.

Microconidia are absent, macroconidia develop abundantly initially from phialides borne on loosely branched conidiophores, later in sporodochia. Phialides are 15–20 μm long by 5 μm at the base. The sporodochia often form as pionnotes with suppression of aerial mycelium. Macroconidia are 3–5-septate when mature, slightly curved and strongly dorsiventral with a pointed apex and a well-marked foot cell. The 3-septate macroconidia are 26–36 × 4–6 μm and the 5-septate types are 34–50 × 5–7 μm. Chlamydospores are oval to globose, generally intercalary, smooth to rough walled, 10–14 × 9–12 μm, formed singly, in chains or clumps.

7.2.2 *Fusarium poae* (Peck) Wollenw. apud Lewis; Synonym: *Fusarium tricinctum* (Cda) emend Synd. and Hans. f. sp. *poae*.

Mycelium cobwebby, carmine rose or buff, powdery with citron-shaped 0–1-septate microspores, 6–14 × 4–7 μm; macrospores narrowly sickle shaped, pointed at both ends, foot cell not prominent, 3–5-septate, 18–36 × 3.0–4.2 μm (Sprague, 1950).

The following description of *F. poae* (Peck) Wollenw. is adapted from Booth (1971).

Aerial mycelium appears hairy to felted and assumes a powdery appearance with the formation of microconidia. Later the aerial mycelium turns reddish brown. From below the cultures are white, yellow salmon to vivid red or vinaceous.

Formation of microconidia begins after about three days in single spore isolations. Microconidia may be formed initially from lateral pores or from small globose to doliiform lateral phialides 9–15 × 3–6 μm with an apical collar. Later they form on complex but compact conidiophores which under dry microscopic examination resemble bunches of grapes. These gradually become covered with slime. The microconidia are ampulliform 8–12 × 7–10 μm to globose 7–10 μm diameter.

Macroconidia do not form readily in all cultures. They are more frequently formed in single spore isolations. In strains forming macroconidia they develop from sporodochia formed on the surface of the agar. Individual macroconidiophores are rather slender with a much branched apex bearing 1–4 cylindrical to subulate phialides at the termination of each branch. The phialides measure 15–20 × 2.5 μm.

Macroconidia are curved falcate and slightly wider above the median septum. They are 3-septate when mature, 20–40 × 3–4.5 μm.

Chlamydospores are globose formed sparsely, singly or in chains, in older cultures.

7.2.3 Other *Fusarium* species

Although Couch and Bedford (1966) designated the two species described above as the incitants of *Fusarium* blight, other species of *Fusarium* have been associated with symptoms identical or similar to the disease. Sanders, Pennypacker and Cole (1980) isolated *F. acuminatum* Ell. and Keller. (syn. *F. roseum* Link ex Gray emend. Snyd and Hans. 'Acuminatum'), *F. crookwellense* Burgess, Nelson, and Toussoun (syn. *F. roseum* Link ex. Gray emend. Snyd. and Hans. 'Crookwell'), *F. equisiti* (Corda) Sacc. sensu Gordon var. *equiseti* (*F. roseum* Link ex. Gray emend. Synd. and Hans. 'Equiseti') from diseased *Poa pratensis* in Pennsylvania. Smiley and Howard (1976) found *F. acuminatum* Ell. and Keller. prevalent in Long Island and the same species was reported from Nebraska (Watkins, 1979). *F. graminearum* Schwabe and *F. heterosporum* Nees ex Fries are additional pathogenic species listed by Smiley (1983) as being associated with fusarium blight.

No information is available defining the role of these *Fusarium* species as disease incitants but it seems unlikely that they are primary pathogens on crowns and roots of *P. pratensis*.

7.3 HOST RANGE

The *Fusarium* species reported by Couch and Bedford (1966) as the incitants of fusarium blight and the associated fusaria listed above, are common soil and seed-borne pathogens with an extensive host range that includes many grasses and cereals (Booth, 1971; Couch, 1973; Sprague, 1950). The same species cause damping off and seedling blight in new plantings of turf grasses (see p. 30).

Couch and Bedford (1966) cited fusarium blight epidemics on turf of *Poa pratensis* L. cv. Merion and *Agrostis* spp. and in less severe proportions on *Festuca rubra* L. turf. The same authors demonstrated in greenhouse trials that *Agrostis tenuis* L. cv. Highland was most susceptible to isolates of the fusaria, with *P. pratensis* cv. Merion and *F. rubra* cv. Pennlawn ranking next in order of susceptibility. However, the marked susceptibility of *Agrostis* spp. has not been borne out in subsequent field observations. In practice, fusarium blight has achieved notoriety primarily as a disease of established intensively managed *P. pratensis* turf and presents only a minor problem on mature turf of *Lolium perenne* L. and *F. rubra* (Cole, 1976). Two warm-season grasses, *Eremochloa ophiuroides* (Munro.) Hack (Centipede grass) and *Zoysia japonica* Stead., may be subject to the disease in southern and central regions of the United States (Bell, 1967; Dale, 1967; Subirats and Self, 1972) and fusarium blight has been noted in one- to two-year-old stands of *Festuca arundinaceae* Schreb. in Texas (Colbaugh, personal communication).

In the light of recent knowledge the incidence, host range and distribution of fusarium blight are all in need of review since past records probably include many situations where the fusaria were wrongly diagnosed as the primary pathogens.

7.4 DISEASE DEVELOPMENT*

Species of *Fusarium* are common and widely distributed soil inhabitants capable of surviving as saprophytes in fragments of plant tissue or as chlamydospores (see Section 5.3.3). Fusaria are present in most soils before seeding any turf grass, but seed transmission as an additional source of primary inoculum may be important in the dissemination of highly virulent biotypes (Cole, Braverman and Duich, 1968). Since the disease usually affects mature stands of turf, dormant mycelium in previously infected plants or saprophytic mycelium present in the thatch constitute the main reservoirs of inoculum (Bean, 1966, 1969; Couch, 1973).

Bean (1969) postulated that infections of the crown region follow extensive saprophytic colonization of the crop debris. Invasive mycelium originating from these saprophytic colonies is cited by Couch (1973) as a source of leaf lesions, possibly exceeding spore infection in importance. Direct penetration and infec-

*Past observations on the development of fusarium blight must include many instances were the primary pathogen was misidentified.

tion through the ends of cut leaves have both been observed, but the latter is considered the most common portal of entry for the pathogens.

The incidence of fusarium blight on mature turf is very dependent on the prevailing environment conditions and is influenced markedly by the cultural practices imposed (Bean, 1966; Cole, 1976; Cutright and Harrison, 1970b; Partyka, 1976; Smiley, 1980; Turgeon, 1976). Couch and Bedford (1966) concluded, from their studies using artificial inoculation of greenhouse-grown seedling grass, that under periods of prolonged high humidity, severe foliar blighting will occur if the proper combination of *Fusarium* sp. genotype, plant genotype and air temperature is met. Certain isolates of these fusaria were shown to differ in their temperature requirements for optimum pathogenicity but generally daytime air temperature of 26.5–35 °C and night air temperatures of 21 °C and above were considered most conducive for disease development. Other investigators confirmed the requirement for temperatures in the 27–33 °C range when conducting similar inoculation studies (Cutright and Harrison, 1970b; Endo and Colbaugh, 1974; Fulton, Cole and Nelson, 1974; Subirats and Slef, 1972). These results from greenhouse studies were supported by field observations that sustained high temperatures (27–34 °C) and high humidity were prior conditions for the occurrence of fusarium blight outbreaks (Bean, 1966; Dale, 1967; Fulton, Cole and Nelson, 1974).

Daytime air temperatures in this range accompany temperatures in the turf microclimate which approach or exceed tolerable limits for growth of cool-season turf grasses. The effect is more pronounced on sites with reduced air circulation and under conditions of high humidity when transpirational cooling is restricted. Supra-optimal temperatures and concomitant heat stress, compounded frequently by drought stress, result in an increased rate of root maturation followed by death of the root system (Beard, 1973). The relationship between stress, impairment of root function and the onset of pathogenic activity by the causal fungi is open to conjecture, but observations on the geographical distribution and the local occurrence of fusarium blight indicate a close association between severe outbreaks of the disease and those locations where heat and/or drought stress conditions commonly obtain (Bean, 1966, 1969; Cole, 1976; Cutright and Harrison, 1970b; Partyka, 1976; Vargas, 1976). Endo and Colbaugh (1972), Cole (1976) and Sanders and Cole (1981) have drawn attention to the work of Cook and Papendick (1970, 1972) and Papendick and Cook (1974) on fusarium rot of dryland wheat. *Fusarium culmorum* was demonstrated to be an aggressive pathogen on wheat growing at low soil water potentials when plants were under internal moisture stress. A parallel situation is suggested for severe fusarium blight to occur.

Cultural practices which may influence the severity of the disease include: the provision of high levels of available soil nitrogen (Couch, 1976; Cole, 1976; Cutright and Harrison, 1970; Endo *et al.*, 1973; Funk, 1976; Turgeon and Meyer, 1974); inadequate calcium nutrition (Couch and Bedford, 1966); mowing height and return of clippings (Funk, 1976; Turgeon and Meyer, 1974; Turgeon, 1976). These factors may act by exacerbating stress conditions in the turf or contribute towards thatch build-up. Thatch is widely accepted as playing an important role in severe fusarium blight development (Cole, 1976; Couch, 1976; Funk, 1976; Partyka, 1976), yet no clear correlation is apparent between relative susceptibility to the disease and the thatch forming propensity of different *Poa pratensis* cultivars (Turgeon, 1976).

7.5 CONTROL

7.5.1 Cultural methods

Management practices which are advocated to reduce fusarium blight include the judicious use of nitrogen particularly during the spring; prevention and alleviation of drought stress by frequent watering and occasional syringing at midday in hot dry periods, improvement of free air flow over the turf, raising the mowing height and removal of clippings. Emphasis has been placed on the prevention of thatch build-up. Adherence to these practices may afford some improvement but additional chemical control measures are required in high risk areas.

7.5.2 Resistant varieties

Grass species and cultivars vary in their susceptibility to fusarium blight, and within *Poa pratensis*, currently the most important host, broad but inconsistent differences in cultivar reaction are possible. Expression of resistance is associated closely with the capacity of the particular genotype to tolerate stresses and the variable reactions are probably the result of attempts to grow a cultivar outside its area of adaption or to manage the variety in a manner that places it under extreme stress. By selecting for genotypes with wide stress tolerance, improved levels of race non-specific resistance of fusarium blight is deemed attainable (Funk, 1976).

The use of blends of cultivars or species mixtures to lessen the risk of disease is not always a reliable means of combatting fusarium blight but Gibeault *et al.* (1980) in plot trials in California over a three-year period, confirmed the benefits of including perennial ryegrass in mixtures with Kentucky bluegrass as a practical means of reducing fusarium blight incidence.

7.5.3 Chemical control

Couch (1964) recommended a preventive spray programme with mancozeb as a means of combatting fusarium blight, but Bean, Cook and Rabbitt (1967), Vargas and Laughlin (1972) and others reported no control from the use of this fungicide. Maneb proved equally ineffective in trials reported by Cutright and Harrison (1970a). Their trials did confirm the efficacy of an organo-mercury formulation in reducing the disease, previously reported by Bean, Cook and Rabbitt (1967), and they also were the first of many to demonstrate excellent control by the systemic fungicide benomyl. With control from contact fungicides proving at best erratic and unreliable, benomyl and the other related benzimidazole-derivative fungicides, methyl and ethyl thiophanate, were established in the early 1970s as the recommended fungicides against fusarium blight.

In the mid-1970s instances of fusarium blight were reported where the *Fusarium* species involved were tolerant of the benzimidazole fungicides (Burpee, Cole and Sanders 1976; Cole, 1976; Smiley and Craven, 1977; Smiley and Howard, 1976). It was noted that tolerance build-up to benzimidazole fungitoxicants by the fusaria were induced after only three years selection pressure of such fungicide use on golf fairway turf (Smiley and Craven, 1979b). Several additional chemicals including propiconazole, fenarimol, iprodione, triadimefon, triarimol, and DPX 4424, were evaluated extensively for turf diseases purported to be fusarium blight during the late 1970s and early 1980s and numerous trials on the control of fusarium blight by these chemicals have been reported in the APS Fungicide and Nematicide Tests 1976–82. Although effective, commercial development of triarimol and DPX 4424 has since been discontinued in the United States. Smiley (1980) ranked the commercially available fungicides in order of their efficacy against 'fusarium blight' as follows:

1. triadimefon, propiconazole, fenarimol
2. benomyl
3. methyl thiophanate
4. ethyl thiophanate, iprodione

Iprodione is the least effective of the new materials, especially at low application rates but triadimefon, propiconazole and fenarimol at low dosages provided long-term protection (Smiley, 1980). Couch, Garber and Fox (1979) have concluded that the best control is afforded by most of these fungicides when they are used at precise rates and correct timing in a preventive spray programme, commencing a few weeks before symptoms are apparent. The nature of the control afforded by iprodione, propiconazole and triadimefon was the subject of much conjecture since none of these fungicides is highly suppressive of growth or sporulation of *Fusarium* species (Sanders and Cole, 1981; Smiley, 1980). The varied response of *Fusarium* spp., *Magnaporthe poae* and *Leptosphaeria korrae* to these fungicides now offers an explanation for the many anomalies reported in the chemical control of 'fusarium blight'.

7.6 REFERENCES

Bean, G.A. (1966) Observations on Fusarium blight of turf grass. *Plant Dis. Rept.*, **50**, 942–5.

Bean, G.A. (1969) The role of moisture and crop debris in the development of Fusarium blight of Kentucky bluegrass. *Phytopathology,* **59**, 479–81.

Bean, G.A., Cook, R.N. and Rabbitt, A.E. (1967) Chemical control of Fusarium blight of turf grass. *Plant Dis. Rept.*, **51**, 839–41.

Beard, J.B. (1973) *Turfgrass Science and Culture*. Prentice-Hall, Englewood Cliffs, NJ, 658pp.

Bell, A.A. (1967) Fungi associated with root and crown rots of *Zoysia japonica. Plant Dis. Rept.*, **51**, 11–14.

Booth, C. (1971) *Genus Fusarium*. Commonwealth Mycol. Inst., Kew, Surrey, England, 237pp.

Burpee, L.L., Cole, H. Jr and Sanders, P.L. (1976) Control of benzimidazole tolerant *Fusarium* spp. on annual and Kentucky bluegrass. *Proc. Am. Phytopathol. Soc.*, **3**, 302. (Abs.)

Burpee, L.L., Sanders, P.L. Cole, H. Jr and Duich, J.M. (1978) Control of Fusarium blight with fungicides under fairway conditions in mixed Kentucky bluegrass-annual bluegrass, 1977. *Fungicide Nematicide Tests*, **33**, 140.

Cole, H., Jr (1976) Factors affecting Fusarium blight development. *Weeds Tress Turf,* **15**(7), 35–7.

Cole, H., Braverman, S.W., Duich, J.M. (1968) Fusarium and other fungi from seeds and seedlings of Merion and other turf-type bluegrasses. *Phytopathology*, **58**, 1415–19.

Cole, H. Jr, Forer, L.B., Nelson, P.E., Bloom, J.R. and Jodon, M.H. (1973) Stylet nematode genera and *Fusarium* species isolated from Pennsylvania turf grass sod production fields. *Plant Dis. Rept*, **57**, 891–5.

Cook, R.J. and Papendick, R.I. (1970) Soil water potential as a factor in the ecology of *Fusarium roseum* f. sp. *cerealis* 'Culmorum'. *Plants Soil*, **32**, 131–45.

Cook, R.J. and Papendick, R.I. (1972) Influence of water potential of soils and plants on root disease. *Ann. Rev. Phytopathol.*, **10**, 349–74.

Couch, H.B. (1964) Fusarium blight of turf grass. *Bull, NY State Turfgrass Assoc.*, **77**, 297–8.

Couch, H.B. (1973) *Diseases of Turfgrasses*, 2nd edn, R.E. Kreiger, Huntington, NY, 348pp.

Couch, H.B. (1976) Fusarium blight of turf grasses – an overview. *Weeds Tress Turf,* **15**(7), 9, 34–35.

Couch, H.B. and Bedford, E.R. (1966) Fusarium blight of turf grasses. *Phytopathology*, **56**, 781–6.
Couch, H.B., Garber, J.M. and Fox J.A. (1979) Relative effectiveness of fungicides and nematicides in Fusarium blight control. *Golf Business*, July, 12–16.
Cutright, N.J. and Harrison, M.B. (1970a) Chemical control of Fusarium blight of Merion Kentucky bluegrass turf. *Plant Dis. Rept.*, **54**, 771–3.
Cutright, N.J. and Harrison, M.B. (1970b) Some environmental factors affecting Fusarium blight of Merion Kentucky bluegrass. *Plant Dis. Rept.*, **54**, 1018–20.
Dale, J.L. (1967) A blight disease of *Zoysia*. *Plant Dis. Rept.*, **51**, 376.
Endo, R.M. and Colbaugh, P.E. (1972) Drought stress as a factor triggering fungal diseases of turf grass. *Calif. Turf Culture*, **22**, 21–3.
Endo, R.M. and Colbaugh, P.F. (1974) Fusarium blight of Kentucky Bluegrass in California. In *Proc. 2nd. Int. Turf. Res. Conf.* Am. Soc. Agron., Madison, Wis. 602 pp, pp. 325–7.
Endo, R.M., Baldwin, R. Cockerham, S. Colbaugh, P.E. McCain, A.H. and Gibeault, V.A. (1973) Fusarium blight, a destructive disease of Kentucky Bluegrass and its control. *Calif. Turf. Culture*, **23**, 1–2.
Fulton, D.E., Cole, H. Jr and Nelson, P.E. (1974) Fusarium blight symptoms on seedling and mature Merion Kentucky bluegrass plants inoculated with *Fusarium roseum* and *Fusarium tricinctum*. *Phytopathology*, **64**, 354–7.
Funk, C.R. (1976) Developing genetic resistance to Fusarium blight. *Weeds Tress Turf*, **15**(7), 41–4.
Gibeault, V.A., Autio, R. Spaulding, S. and Younger, V.B. and (1980) Mixing turf grasses controls Fusarium blight. *Calif. Turf. Culture*, **30**, 9–11.
Meyer, W.A. and Berns, F.H. (1976) Techniques for determination of Fusarium blight susceptibility in Kentucky bluegrass. *Weeds Trees Turf*, **15**(7), 44, 46–7.
Papendick, R.I. and Cook, R.J. (1974) Plant water stress and development of Fusarium root rot in wheat subjected to different cultural practices. *Phytopathology*, **64**, 358–63.
Partyka, R.E. (1976) Factors affecting Fusarium blight in Kentucky bluegrass. *Weeds Trees Turf*, **15**(7), 37–8.
Pennypacker, B.W., Sanders, P.L. and Cole, H. Jr (1982) Basidiomycetes associated with a patterned mid-summer wilt of bluegrass. *Plant Dis.* **66**, 419–20.
Sanders, P.L., Burpee, L.L. and Cole, H. Jr (1978b). Uptake translocation and efficacy of triadimefon in control of turf grass pathogens. *Phytopathology*, **68**, 1482–7.
Sanders, P.L., Burpee, L.L., Cole, H. Jr and Duich, J.M. (1978a) Control of fungal pathogens of turf grass with the experimental iprodione fungicide, RP 26019. *Plant Dis. Rept.*, **62**, 549–53.
Sanders, P.L. and Cole, H. Jr (1981) The fusarium diseases of turf grass. In *Fusarium: Diseases, Biology and Taxonomy* (eds P.E. Nelson, T.A. Toussoun and R.J. Cook), Pennsylvania State University Press, University Park and London, 457pp., pp. 198–205.
Sanders, P.L., Pennypacker, B.W. and Cole, H. Jr (1980) Histological study of 'Merion' Kentucky bluegrass showing symptoms of Fusarium blight. *Phytopathology*, **70**, 468. (Abstr.)
Smiley, R.W. (1980) Fusarium blight of Kentucky bluegrass: New Perspectives. In *Advances in Turfgrass Pathology* (eds B.G. Joyner and P.O. Larson), Harcourt Brace Jovanovich, Duluth, MN, pp. 155–75.
Smiley, R.W. (1983) *Compendium of Turfgrass Diseases.* American Phytopathology Soc., St Paul, MN, 102pp.
Smiley, R.W. and Craven, M.M. (1977) Control of benzimidazole-tolerant *Fusarium roseum* on Kentucky bluegrass. *Plant Dis. Rept.*, **61**, 484–8.
Smiley, R.W. and Craven, M.M. (1979a) *Fusarium* species in soil, thatch, and crowns of *Poa pratensis* turf grass treated with fungicides. *Soil Biol. Biochem.*, **11**, 355–63.
Smiley, R.W. and Craven, M.M. (1979b) *In vitro* effects of Fusarium blight controlling fungicides on pathogens of *Poa pratensis. Soil. Biol. Biochem*, **11**, 365–70.
Smiley, R.W. and Craven Fowler, M. (1984) *Leptosphaeria korrae* and *Phialophora graminicola* associated with Fusarium blight syndrome of *Poa pratensis* in New York. *Plant Dis.*, **68**, 440–2.
Smiley, R.W. and Craven Fowler, M. (1985) Techniques for inducing summer patch symptoms on *Poa pratensis. Plant Dis.*, **69**, 482–4.
Smiley, R.W., Craven, M.M. and Bruhn, J.A. (1980) Fusarium blight and physical, chemical, and microbial properties of Kentucky bluegrass sod. *Plant Dis.*, **64**, 60–2.
Smiley, R.W. and Howard, R.J. (1976) Tolerance to benzimidazole-derivative fungicides by *Fusarium roseum* on Kentucky bluegrass turf. *Plant Dis. Rept.*, **60**, 91–4.
Sprague, R. (1950) *Diseases of Cereals and Grasses in North America.* Ronald Press, New York, 537 pp.
Subirats, F.J. and Self, R.L. (1972) Fusarium blight of centipede grass. *Plant Dis. Rept.*, **56**, 42–4.
Turgeon, A.J. (1976) Effects of cultural practices on Fusarium blight. *Weeds Trees Turf*, **15**(7), 38–40.
Turgeon, A.J. and Meyer, W.A. (1974) Effects on mowing height and fertilization level on disease incidence in five Kentucky bluegrasses. *Plant Dis. Rept.*, **58**, 514–6.
Vargas, J.M. Jr (1976) The role of nematodes in the development of Fusarium blight. *Weeds Trees Turf*, **15**(7), 40–1.
Vargas, J.M., Jr (1981) *Management of Turfgrass Diseases.* Burgess, Minneapolis, 204pp.
Vargas, J.M. and Laughlin, C.W. (1972) The role of *Tylenchorhynchus dubius* in the development of Fusarium blight of Merion Kentucky bluegrass. *Phytopathology*, **62**, 1311–14.
Watkins, J.E. (1979) Fungicide suppression of Fusarium blight. *Am. Phytopathol Soc. Fungicide Nematicide Tests*, **34**, 147.
Worf, G.L., Brown, K.J. and Kachadoorian, R.V. (1983) Survey of 'necrotic ring spot' disease in Wisconsin lawns. *Phytopathology*, **73**, 839. (Abstr.)

8 *Spring dead spot*

In 1960 a disease of bermuda grass (*Cynodon dactylon* L.) was reported in Oklahoma by Wadsworth and Young (1960) which they named spring dead spot. Observed sporadically as early as 1936, the disease now is recognized as a major turf disease problem particularly in the northern-most areas of bermuda grass adaptation in the USA where winters are cold enough to induce dormancy of the host (Frederiksen, 1964; Lucas, 1980a; Wadsworth, Houston and Petersen, 1968). During the early 1960s a similar disease of undertermined cause occurred in New South Wales, Australia, which damaged couch grass (*C. dactylon*) turf very severely. This disease was also designated as spring dead spot and later was shown to be incited by either of two loculoascomycetous fungi, members of the genus *Leptosphaeria* (Smith, 1965, 1971a). The fungi implicated in New South Wales have not been associated with the disease elsewhere in Australia but one of the two pathogens, *L. korrae*, was confirmed recently as an incitant of spring dead spot of bermuda grass in California (Endo, Ohr and Krausman, 1985). To date no causal agent has been established for spring dead spot in the southeast and south central United States, or in Japan where *Zoysia* turf is subject to a similar disease.

8.1 SYMPTOMS

At the onset of spring growth well-defined circular patches of dead grass varying in size from a few centimetres to a metre or more in diameter become apparent in affected turf (Fig. 14, colour plate section). The foliage of the dead bermuda grass plants assumes a bleached straw colour while the stolons and roots are black and rotted. Contrasting markedly with the surrounding healthy turf, the unsightly dead patches persist for varying periods of time. In the southeastern United States, recolonization by the bermuda grass may be complete within one growing season. Recovered patches on greens may still be visible at the end of the summer as depressed or thin areas in the turf with very little thatch. Similarly, on longer turf, affected areas support grass that is somewhat stunted, but remains green later into the autumn than adjacent healthy grass (Lucas, 1980a,b). In the south central US several years may elapse before Bermuda grass will reestablish on the diseased patches (Wadsworth and Young, 1960). Weed grasses and broad-leaved weeds colonize the affected areas and contribute to the disfigurement of the turf. Examination of bermuda grass stolons traversing a diseased area reveals nodal roots that frequently are discoloured and malformed. Attendant shoots are stunted suggesting the presence of a persistent phytotoxin in the soil. Spring dead spot symptoms persist or recur on the same spot in successive years, the patches enlarging annually. Where annual recovery growth occurs, some healthy bermuda grass eventually may survive in the centre of old recurring patches, forming doughnut or ring-shaped diseased areas. The enlargement and coalescence of the multiple infection sites, so typical of severe spring dead spot disease, creates spectacular havoc in a turf for a few years, then the disease subsides and often disappears completely.

In New South Wales, the typical well-defined, circular patches of bleached dead grass occur in the spring, but, contrary to United States experience, mild symptoms resulting from progressive root- and stolon-rotting may mark the diseased patches in the preceding fall during periods of cold, wet weather. Signs of the causal agent are apparent. A septate, dark-brown mycelium in the form of runner hyphae and hyphal aggregates is associated with the diseased roots and stolons. Affected tissues become discoloured and numerous dark-brown flattened sclerotia occur on the stem bases, stolons and sometimes in the infected roots. Similar fungal signs and symptoms are reported for spring dead spot in California. The dead spots heal slowly as healthy grass grows in from the edges. Small patches may heal completely, but larger patches may be resprigged with healthy plants to consolidate healing. By mid-summer, symptoms of the disease usually have disappeared. The disease tends to recur on the same spots and the spots may increase in size and prevalence from year to year. Throughout the area of distribution, spring dead spot is a problem of mature turf, stands of two or more season's growth becoming increasingly vulnerable.

8.2 CAUSES

8.2.1 Undetermined

In the south eastern and south central USA, winter injury, plant parasitic nematodes, insects (white grubs and a scale insect), and various fungi have been mentioned in connection with spring dead spot (Dale and Diaz, 1963, Dale and McCoy, 1964; Kozelnicky, 1974; Kozelnicky, Hendrix and Powell, 1967; Wadsworth 1966; Wadsworth and Young, 1960). Members of the genera *Pythium*, *Fusarium*, *Curvularia*, *Helminthosporium and Gaeumannomyces* are the fungal pathogens most frequently found in association with diseased *C. dactylon* plants, but attempts to establish the pathogenicity of these fungi and reproduce spring dead spot symptoms in healthy turf have not met with any success to date (Diaz, 1964; Kozelnicky, 1974; Lucas, 1980a, b; McCoy, 1965; Wadsworth, 1966; Wadsworth, Houston and Peteson, 1968). Symptom suppression in response to preventive applications of certain fungicide and the tardy re-establishment of unthrifty bermuda grass plants on affected patches suggests the involvement of a fungal root pathogen(s) (see section 8.2.2) and/or the presence of a stable residual phytotoxin.

8.2.2 Fungi

In Australia, Smith (1965) isolated a long-spored Pleosporaceous fungus capable of reproducing the disease but subsequent work showed that a second closely related fungus with shorter spores was a much more common cause of spring dead spot in New South Wales. The long-spored fungus was identified initially as *Ophiobolus herpotrichus* (Fr.) Sacc., but both fungi were assigned later to the genus *Leptosphaeria* (Smith, 1971a; Walker and Smith, 1972).

In 1984, one of the two *Leptosphaeria* species named in Australia was confirmed as a cause of spring dead spot of bermuda grass in California (Endo, Ohr and Krausman, 1985). The fungus is now known to be widely distributed across the United States causing necrotic ring spot of *Poa pratensis* turf (pp. 129–132). Reports of its occurrence on bermuda turf was limited to California but the fungus was recently found causing spring dead spot on *C. dactylon* turf in Maryland (Dernoeden 1987, personal communication).

8.2.3 The causal fungi

(a) *Leptosphaeria narmari* Walker and Smith (1972)

Pseudothecia are clustered or single, flask-shaped to wide pyriform, thick necked, black, thick-walled, erumpent, $<800\,\mu m \times 650\,\mu m$. Asci are clavate with a foot, bitunicate, 8-spored, $100–155 \times 11–13$ μm. Ascospores are biseriate, pale brown, elliptical to fusiform, often bent, constricted at the central septum, $35–72 \times 4–6$ μm, 3–7 septate. Pseudoparaphyses are numerous, hyaline. Hyphae on the host are brown, septate, stranded, forming flattened, dark sclerotia, 40–400 μm diam. (For a detailed description see Walker and Smith, 1972.)

(b) *Leptosphaeria korrae* Walker and Smith (1972)

Pseudothecia are often closely packed, flask-shaped, globose, thick-walled, $<600 \times 500\,\mu m$. Asci are cylindrical to clavate with a foot, $145–200 \times 10–15\,\mu m$.

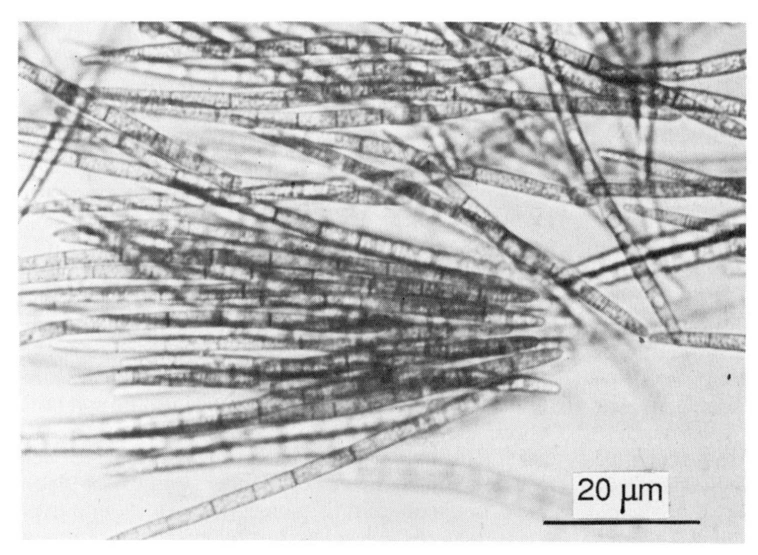

Figure 8.1 *Leptosphaeria korrae*-ascospores.

Ascospores are filiform, slightly twisted, lying parallel in bundles, pale brown, 120–180 (210) × 4–5.5 μm, 7–15 septate, widest at the middle, with most taper towards the base, rounded at the ends (Fig. 8.1). Pseudoparaphyses are numerous, hyaline. Hyphae on host stranded, forming flattened, dark sclerotia, 50–400 μm. (For a detailed description see Walker and Smith, 1972.)

Pseudothecia of both fungi are produced within the leaf sheaths of the host, but on older specimens, where host tissues have decayed, pseudothecia may appear superficial on the stolons. Dark-brown hyphae growing through the tissues blacken the stolons, and superficial mycelium in the form of runner hyphae and flattened sclerotic bodies, dark brown in colour, occur abundantly on stolon surfaces. The mycelial aggregations function as infection cushions and as resting sclerotia (Walker and Smith, 1972). Superficial mycelium may be grey in colour (Endo, Ohr and Krausman, 1985).

L. narmari and *L. korrae* grow readily in agar culture. On potato dextrose agar the optimum temperature for radial growth of both species is 25 °C. No asexual stage is known for either fungus (Walker and Smith, 1972).

8.3 HOST RANGE

The bermuda grasses, *Cynodon dactylon* (L) Pers., *C. magenisii* Hurcombe, *C. transvaalensis* Burtt-Davy and hybrids between these species, are affected by the disease (Wadsworth and Young, 1960). In addition to *C. dactylon*, *Axonopus compressus* Beauv., *Stenotaphrum secondatum* Kuntze and *Pennisetum clandestinum* Hochst are reported to be susceptible to spring dead spot disease in Australia (Smith, 1965). *Zoysia* spp. are damaged severely in Japan (Smiley, 1983).

8.4 DISEASE DEVELOPMENT

In the United States spring dead spot is confined to the transition zone where winter temperatures limit suitability of *C. dactylon* as a turf species (Madison, 1970). More specifically the geographic region falls between the 7°C and the 15°C November average daily temperature isotherms (Lucas, 1980a, b.) The former delineates roughly the northern range of *C. dactylon* in the USA while the latter marks an approximate line across the continent north of which freezing temperatures occur regularly and where *C. dactylon* is subject to winter dormancy.

Symptom suppression in response to preventive applications of certain fungicides (see Section 8.5) and the tardy re-establishment of unthrifty *C. dactylon* plants on affected patches suggest the involvement of fungal root pathogen(s) and/or the presence of stable residual phytotoxins. Efforts in the US to confirm a fungal pathogen or to characterize a toxin proved singularly unsuccessful (Kozelnicky, 1974; McCoy, 1965; Wadsworth, Houston and Peterson, 1968) until 1984 when *L. korrae* was established as an incitant of spring dead spot symtoms in California (Endo, Ohr and Krausman, 1985).

Spring dead spot in the south eastern and south central United States appears in the spring without evidence of any preliminary symptoms during the previous autumn. Evidently the damage occurs on the dormant turf during the mid-winter months as indicated by a 3-year study completed in North Carolina (Lucas, 1980b, c; Lucas and Gilbert, 1979). Timing of the damage was assessed by measuring the regrowth on plugs of turf sampled regularly through the winter from known diseased patches and adjacent healthy *C. dactyion* turf. Plugs from diseased and healthy areas regrew equally well when collected in December and placed in a warm greenhouse for one month. However, samples collected in late January or February and subjected to the conditions mentioned above, showed a marked reduction (64%) in shoot number on the plugs from the diseased areas as compared to those from healthy turf. Plugs from spring dead spot affected areas, treated in November with benomyl, compared equally with healthy turf in shoot recovery number when sampled in mid-winter. The positive effect of the fungicide was interpreted as benomyl possibly promoting winter hardiness of the host or working as a protectant against some fungus. Throughout the North Carolina investigations, lower root weights and delayed dormancy were common characteristics of the diseased *C. dactylon* truf areas. Lucas (1980b) concluded that *C. dactylon* is killed in mid-winter by cold weather and presented a theory to explain spring dead spot development. It involves the predisposition of *C. dactylon* to cold weather damage by some undetermined fungal species with properties akin to the basidiomycetous fungi that form small fairy rings. In the autumn, these saprophytic fungi, which degrade thatch and soil organic matter, could release sufficient nitrogen in the root zone to delay dormancy of the *C. dactylon* growing in the colonized area, and thus lead to its demise and accounting for the patterned symptoms.

Turf with ample thatch would favour the activity of saprophytic fungi and some basidiomycetes are suspected of releasing persistent toxins into the soil (See Chapter 30). Basidiomycetes associated with a patterned midsummer wilt of *Poa pratensis* turf have been documented (Pennypacker, Sanders and Cole, 1982).

The low-temperature damage theory is supported by observations that relate turf management practices to incidence of spring dead spot (Lucas, 1980a,b,c; Lucas and Gilbert, 1979). Intensive programmes employing large amounts of nitrogenous fertilizers and especially when applications are made in late summer, exacerbate the disease. Conversely, lowered nitrogen rates and concomitant increase in the amount of potassium reduces the disease incidence. Winter hardiness of the *C. dactylon* is seen as the principal factor being influenced by these practices. Other conditions noted by these authors which favour the disease include close mowing, thatch accumulation and soil compaction. Ineffective weed control and mistimed verticutting also were listed since both practices may impair bermuda grass recovery on affected areas.

While considering wide temperature variation in late winter and early spring as a possible predisposing factor Madison (1970) concluded that climate is not the major influence on spring dead spot incidence in the US. This opinion was based on observations that in the more southerly regions of the transition zone the disease still occurs but only on the most intensively managed greens turf. He postulated that once *C. dactylon* is debilitated by too-intensive management and/or by dense compacted soil, weak pathogens such as *Helminthosporium spiciferum* (Bain) Nicot can then bring about the death of the grass plants.

Spring root senescence of *C. dactylon* may play a part in the spring dead spot scenario. This is a recently discovered phenomenon where shortly after the first new leaf initiation (spring greenup) the entire root system may die, followed by a delay in new root initiation and replacement (DiPaola and Beard, 1980; DiPaola, Beard and Brawand, 1982). The limited root system supporting significant new shoot growth during this early period must place the grass host at risk to environmental stresses (e.g. low temperature injury) and directly or indirectly to the depredations of soil-borne pathogens, weak or otherwise, which could result in death of the plants.

In the regions where *Leptosphaeria* species have been confirmed as causal agents of spring dead spot, ascospores, sclerotia or infected plant debris are the presumed sources of primary inoculum. Once established in new swards of *C. dactylon* the fungi perennate on infected plants and generate symptoms when stands are two or more years old. Disease symptoms can appear in the autumn when the temperature ranges between 10 and 20°C and the grass host is making little or no active growth. At this time the growth of the pathogens is slow but the rate of root destruction exceeds the rate of root regeneration (Endo, Ohr and Krausman, 1985; Smith, 1971a,b).

8.5 CONTROL

8.5.1 Cultural methods

Madison (1970) concluded that spring dead spot is a management disease and implicated close mowing, high fertility, thatch accumulation and soil compaction as factors necessary for disease to occur. There is general consensus that procedures to correct any and all of the above are of benefit in reducing incidence of the disease. Limiting the overall use of nitrogen and omitting its application in late summer and utilizing higher amounts of potassium in the fertilizer programme have been strongly advocated by Lucas (1980a,b,c,), as a means to improving winter hardiness and so reducing the chances of a spring dead spot outbreak.

8.5.2 Resistant varieties

Differences in susceptibility between common and hybrid cultivars of *C. dactylon* have been reported (Wadsworth and Young, 1960), but efforts to select specifically for resistance to spring dead spot have not reached fruition to date.

8.5.3 Chemical control

Although a causal agent has not been determined in the south east and south central United States (see section 8.2.2), fungicides have been employed with some success in preventing spring dead spot. Nabam applied four times at monthly intervals commencing six weeks before the first autumn frost was reported to control the disease in Missouri (Kozelnicky, 1974). Benomyl at two to five times the normal rate consistently controlled spring dead spot when applied monthly in October, November and December. One application of benomyl at four times' normal rate applied in October or November was later found to be as efficient. High rates of quintozene were effective during the second year if five monthly applications were made commencing in July (Lucas, 1980a,b,c,d). Fungicides effective for necrotic ring spot control (see p. 131) should be evaluated for control of spring dead spot.

In Australia, Smith (1971b) reported control of spring dead spot (caused by species of *Leptosphaeria*) with nabam or thiram, applied every four weeks from the last month of summer until early spring. In more recent trials treatment of affected turf with preventive applications of mercury compounds, triadimefon, nabam, thiram, mancozeb and iprodione have given varying results. The last four compounds listed, especially iprodione, gave the more reliable disease control (Anon., 1983; Harris, 1983; Ford, 1983).

8.6 REFERENCES

Anon. (1983) Workshop on spring dead spot conducted by Bayer Aust. Ltd. *Bowling Greenkpr*, October, p. 4.

Dale, J.L. and Diaz, C. (1963). A new disease of bermuda grass lawns and turf. *Arkansas Farm Res.*, **12** (6), 6.

Dale, J.L. and McCoy, C.E. (1964). The relationship of a scale insect to death of bermudagrass in Arkansas. *Plant Dis. Rep.*, **48**, 228.

Diaz, C. (1964) The etiology of spring dead spot of bermudagrass. M.S. dissertation. Univ. of Arkansas.

Dipaola J.M. and Beard, J.B. (1980). Spring root dieback of warm-season turf grasses. *USGA Green Section Rec.*, **18** (4), 6–9.

DiPaola, J.M., Beard, J.B. and Brawand, H. (1982) Key events in the seasonal root growth of bermudagrass and St. Augustinegrass. *Hortscience*, **17**, 829–31.

Endo, R.M., Ohr, H.D. and Krausman, E.M. (1985) *Leptosphaeria korrae*, a cause of spring dead spot disease of bermudagrass in California. *Plant Dis.*, **69**, 235–7.

Ford, P. (1983) *Spring dead spot.* Australian Turf Research Inst. Mimeo, 5 pp.

Frederiksen, S. (1964). Spring dead spot, a progress report. *Golf Course Rep.*, **32**, 38–44.

Harris, J. (1983) Spring dead spot – another point of view. *Bowling Greenkpr*, Oct. pp. 3–4.

Kozelnicky, G.M. (1974) Updating twenty years of research; spring dead spot on bermudagrass. *USGA Green Section Rec.*, **12** (3), 12–15.

Kozelnicky, G.M., Hendrix, F.F. and Powell W.M. (1967) Soil organisms associated with spring dead spot of bermudagrass turf in Georgia (Abstr.) *Phytopathology*, **57**, 462.

Lucas, L.T. (1980a) Spring dead spot of bermudagrass. in *Advances in Turfgrass Pathology* (eds P.O. Larsen and B.G. Joyner), Harcourt Brace Jovanovich, Duluth, Mn. pp. 183–7.

Lucas, L.T. (1980b) Spring dead spot of bermudagrass. *USGA Green Section Rec.*, **18**, (3), 4–6.

Lucas, L.T. (ed.) (1980c) *Proc. Spring Dead Spot Bermudagrass* (*Cynodon* spp.) *Workshop*. North Carolina State Univ., Raleigh. 46 pp.

Lucas, L.T. (1980d) Control of spring dead spot of bermudagrass with fungicides in North Carolina. *Plant Dis.*, **64**, 868–70.

Lucas, L.T. and Gilbert, W.B. (1979) Managing dormant bermudagrass *Golf Course Mgmt.*, **47** (9), 14–23.

Madison, J.H. (1970) Do we have a new turf grass diseases? *J. Sports Turf Res. Inst.*, **46**, 17–21.

McCoy, R.W. (1965) A study of the etiology of spring dead spot. M.S. dissertation, Oklahoma State Univ. Stillwater. 56 pp.

Pennypacker, B.W., Sanders, P.L. and Cole, H. Jr (1982) Basidiomycetes associated with a patterned midsummer wilt of bluegrass. *Plant Dis.*, **66**, 419–20.

Smiley, R.W. (1983) *Compendium of Turfgrass Diseases.* Am. Phytopathol. Soc. 102 pp.

Smith, A.M. (1965) *Ophiobolus herpotrichus* a cause of spring dead spot in couch turf. *Agric. Gaz., NSW*, **76**, 753–8.

Smith, A.M. (1971a) Spring dead spot of couch grass turf in New South Wales. *J. Sports Turf Res. Inst.*, **47**, 54–9.

Smith, A.M. (1971b) Control of spring dead spot of couch grass turf in New South Wales. *J. Sports Turf Res. Inst.*, **47**, 60–5.

Wadsworth, D.F. (1966) Etiology of spring dead spot, a root rot complex of bermudagrass. PhD. dissertation, University of California, Davis 96 pp.

Wadsworth, D.F. and Young, H.C. Jr (1960) Spring dead spot of bermudagrass. *Plant Dis. Rept.*, **44**, 516–18.

Wadsworth, D.F., Houston, B.R. and Peterson, L.J. (1968) *Helminthosporium spiciferum*, a pathogen associated with spring dead spot of bermudagrass. *Phytopathology*, **58**, 1658–60.

Walker, J. and Smith, A.M. (1972) *Leptosphaeria narmari* and *L. korrae* spp. nov., two long spored pathogens of grasses in Australia. *Trans. Br. Mycol. Soc.*, **58**, 459–66.

9 *Necrotic ring spot*

Since the mid-1970s, a disease of *Poa pratensis* closely resembling take-all patch of bentgrasses (see page 134) has occurred with increasing frequency in the northeast, the midwest and the Pacific northwest United States (Chastagner *et al.*, 1984; Jackson 1984; Smiley, Fowler and Kane, 1984; Worf, Avenius and Stewart, 1982; Worf, Brown and Kachadoorian, 1983; Worf, Stewart and Avenius, 1986). Severe damage has been recorded most commonly on *P. pratensis* lawns in the early years following establishment from sod but other turf species are susceptible to infection and the disease may appear in younger seeded swards. Worf, Brown and Kachadoorian (1983) named the malady (then of undetermined cause) 'necrotic ring spot'. Necrotic ring spot is similar in symptom pattern to fusarium blight but is initiated at cooler temperatures, more akin to those favouring yellow patch (page 272), another disease it resembles and with which it has probably been confused.

Concurrent research by turf pathologists at several locations in the early 1980s established *Leptosphaeria korrae* Walker and Smith as the common incitant. The fungus is one of two *Leptosphaeria* species that cause spring dead spot of bermuda grass in Australia (Walker and Smith, 1972), and it was recorded on this host in California (Endo, Ohr and Krausman, 1984). In New York State the fungus is one of two (or more) root and crown pathogens that have been associated routinely with turf disease symptoms previously referred to as fusarium blight or 'fusarium blight syndrome' (Smiley, 1983, 1984; Smiley and Craven Fowler, 1984; Smiley, Fowler and Kane, 1984; Smiley, Kane and Fowler, 1985). *Leptosphaeria korrae*, *Magnaporthe poae* and perhaps other related fungi, figure commonly as the primary pathogens in this complex of diseases (page 117).

9.1 SYMPTOMS

Patches of diseased *P. pratensis* turf, 10 cm to a metre or more wide, become increasingly evident as the outer leaves of infected plants yellow and bleach. Younger, inner leaves may assume a bronze or purple to wine-red colour prior to turning brown as tillers or whole plants succumb to infection. Crater-like depressions are formed if all plants within the patch collapse and die but commonly, some centrally located plants survive to give a characteristic ring or frog-eye appearance to the patch (Fig. 15, colour plate section). Large areas of turf may be variously scarred or thinned as the infection centres coalesce. Symptoms can occur on turf in full sun or in shade throughout the growing season. Patch development is generally most prevalent during late spring and early autumn but periods of heat stress in July and August intensify the damage and the scarring becomes very obvious. Symptoms present in the antumn persist over the winter and, depending on the severity of the previous damage, affected turf recovers slowly in the following spring. Even where patch symptoms were mild and diffuse, plants on these sites are late to green up and growth is stunted as compared to surrounding healthy turf. Weed species invade the disease patches. Like take-all patch, symptoms of the disease on a particular site may increase in severity over a number of years and then subside, indicative of a decline phenomenon (Shipton, 1975).

Severely infected plants have little root hold as the fungus progressively invades roots, rhizomes and crowns. Dark-coloured stranded hyphae and laterally fused aggregates (plate mycelium) (Fig. 9.1) invest these organs which blacken as infection proceeds. Black pseudothecia, solitary or gregarious may be found on necrotic roots, rhizomes, leaf sheaths and crowns (Figs. 9.2, 9.3). Fruiting is sporadic and apparently restricted to tissues that have been dead for some time.

9.2 THE CAUSAL FUNGUS

Leptosphaeria korrae Walker and Smith (1972). For a description of the fungus see pp. 125–6.

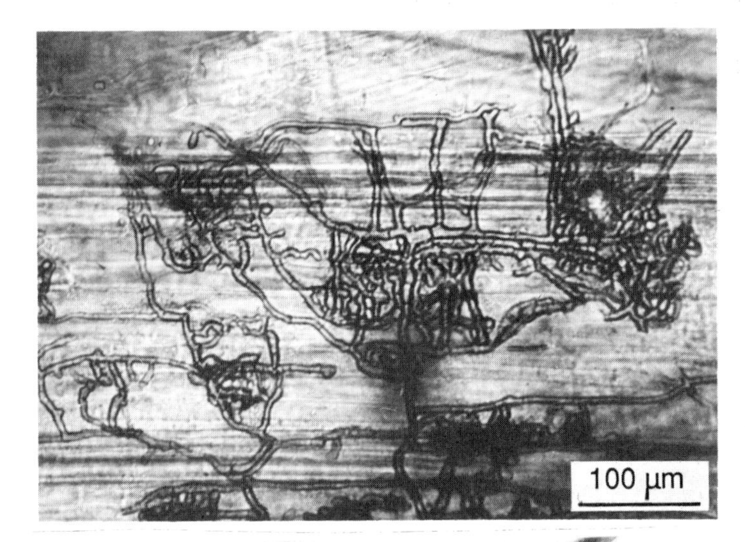

Figure 9.1 *Leptosphaeria korrae*–plate mycelium.

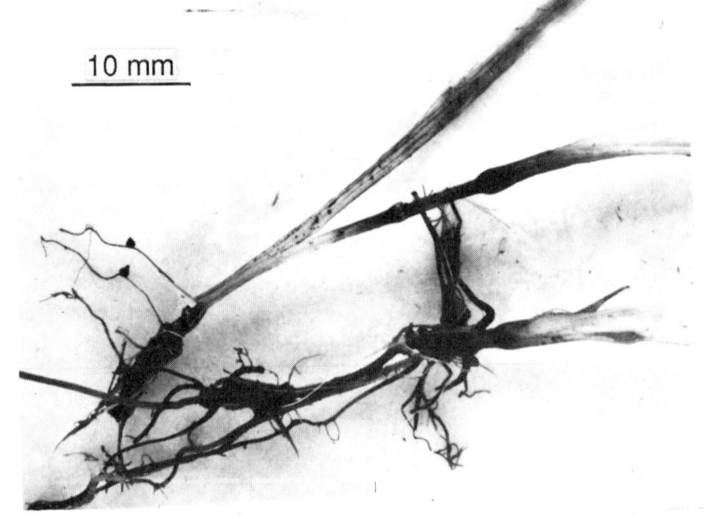

Figure 9.2 *Leptosphaeria korrae*–pseudothecia on grass roots.

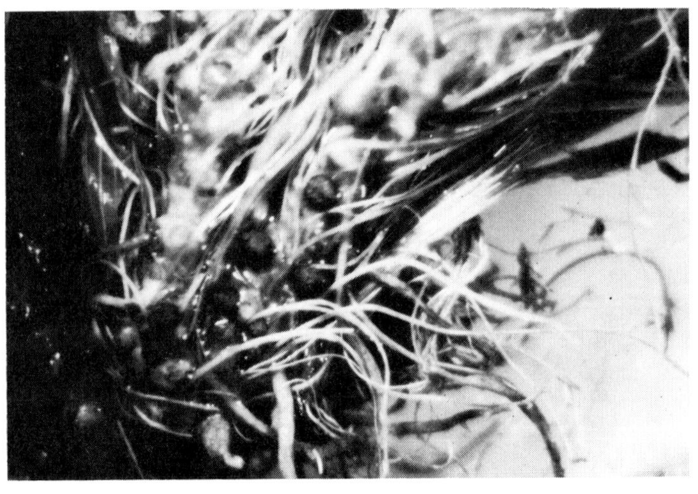

Figure 9.3 *Leptosphaeria korrae*. Pseudothecia in crowns of *Poa pratensis*.

American isolates of the fungus have been cultured on a variety of organic media. Growth rates at the optimum temperatures (20–28 °C) are slow (2.6–6.3 mm/day) with scant growth occurring outside the range 4–32 °C. Colonies are initially colourless, off-white to light grey, somewhat floccose, darkening with age through dark grey to almost black. On the selective Juhnke medium, a dark pigment diffuses from the very slow-growing colonies (Juhnke, Mathre and Sands, 1984; Smiley, Kane and Fowler, 1985; Worf, Stewart and Avenius, 1986). Fruiting has been induced on pieces of irradiated wheat leaf embedded in water agar and on sterilized oat grains (Jackson, 1984; Smiley and Craven—Fowler, 1984). Mature pseudothecia are produced in 2–3 months. No anamorphic state has been reported.

9.3 HOST RANGE

Grasses in the genera *Axonopus*, *Cynodon*, and *Eremochloa* were listed in Australia as hosts of *L. korrae* and isolates of the fungus from *Cynodon* proved pathogenic to the cereals oats, rice and wheat (Walker and Smith, 1972). In the United States, *L. korrae* has been isolated recently from *Cynodon* spp. in California showing symptoms of spring dead spot (Endo, Ohr and Krausman, 1984), and from several locations as a pathogen of *Poa pratensis* L. (Chastagner *et al.*, 1984; Jackson, 1984; Smiley and Craven Fowler, 1984; Worf, Avenius and Stewart, 1982; Worf, Brown and Kachadoorian, 1983; Worf and Stewart, 1985). Diseased plants of *Festuca rubra* L., *Agrostis* spp. (Jackson, 1984) and *Poa annua* L. (Jackson, 1984; Wolf, Stewart and Avenius, 1986) have also yielded *L. korrae*. These isolates, and additional isolates from *Poa pratensis* L., were pathogenic in varying degrees to cultivars of *Agrostis stolonifera, Festuca rubra, Festuca ovina* L., *Festuca arundinacea* Schreb, *Poa pratensis* L., *Lolium perenne* L. and *Cynodon dactylon* L., and to the cereals wheat and oats.

9.4 DISEASE DEVELOPMENT

In Australia the symptoms of spring dead spot of *C. dactylon* caused by *Leptosphaeria* spp. are initiated and become visible in the autumn when the temperature ranges between 10 and 20 °C and the grass host is making little or no active growth (Smith, 1971). Thus 'autumn dead spot' has been suggested as a more appropriate designation (Harris, 1983).

Nicrotic ring spot of *P. pratensis* turf evidently follows a similar pattern in the United States. Symptoms commonly appear under cool moist conditions in the autumn and spring, presumably when grass growth is arrested and root destruction can exceed root regeneration. Ascospores or infected plant debris are the probable primary inoculum sources. Inoculum may be present on harvested sod since symptoms have been confirmed on sod production sites. The disease is particularly evident 1–3 years after lawns are established from sod.

Conditions that favour necrotic ring spot development frequently approximate those favouring take-all patch. Cool temperatures, ample moisture and near-neutral soil pH at the turf soil interface often obtain, but the disease can occur over a soil pH range of 5.0–8.0 (Chastagner, 1985; Worf and Stewart, 1985).

9.5 CONTROL

Symptoms have been alleviated somewhat by applying ammonium sulphate and/or sulphur to affected turf at several locations in the northeastern US but Chastagner (1985) in Washington State and Worf and Stewart (1985) in Wisconsin, could not demonstrate any consistent improvement from application of sulphur or various nitrogen carriers. The Wisconsin trials did indicate the value of keeping affected turf well watered and adequately fertilized with nitrogen to promote recovery growth.

In Australia, Smith (1971) reported control of spring dead spot caused by *Leptosphaeria* spp. with late summer preventive applications of nabam or thiram. Inorganic and organic mercury fungicides, mancozeb and iprodione are additional fungicides used in preventive programmes with varying success (Anon., 1983; Harris, 1983). Nabam, quintozene and benomyl have been used effectively in preventive programmes against spring dead spot (incitant unknown) in the eastern United States (Lucas, 1980).

In vitro fungicide tests in New York State, Rhode Island and Washington State (Chastagner, 1984; Jackson, 1984; Smiley and Craven Fowler, 1984) have demonstrated that American isolates of *L. korrae* are sensitive to low concentrations of benomyl, propiconazole, fenarimol and phenyl mercury acetate. Iprodione and triadimefon were much less effective in suppressing colony growth. Field trials in Washington State have shown that single applications of fenarimol or propiconazole in late May provided effective control of the disease during the summer and early autumn. Triadimefon, benomyl and iprodione were generally ineffective.

Worf and Stewart (1985) reported disappointing results with triadimefon and other sterol-inhibiting compounds in field trials in Wisconsin. Promising control, obtained initially with iprodione and benomyl, was not borne out in more recent tests.

9.6 REFERENCES

Anon. (1983) Workshop on Spring dead spot conducted by Bayer Aust. Ltd. *Bowling Greenkpr.*, Oct. p.4.

Chastagner, G.A. (1984) Necrotic ring spot: Research on a new disease of bluegrass turf and its control in the Pacific Northwest. *Proc. 38th Northwest Turf Conf.*, pp. 94–5.

Chastagner, G.A. (1985) Symptoms of necrotic ringspot. *Grounds Maintenance*, **52,** 78.

Chastagner, G.A., Gross, R.L. Staley, J.M. and Hammer, W. (1984) A new disease of bluegrass turf and its control in the Pacific northwest. (Abstr.) *Phytopathology*, **74,** 811–12.

Endo, R.M., Ohr, H.D. and Krausman, E.M. (1984) The cause of the Spring dead spot disease (SDS) of *Cynodon dactylon* (L.) Pers. in California. *Phytopathology*, **74,** 812.

Harris, J. (1983) Spring dead spot – another point of view. *Bowling Greenkpr.*, Oct. p.3.

Jackson, N. (1984) A new cool season patch disease of Kentucky bluegrass turf in the northeastern United States. *Phytopathology*, **74,** 812. (Abstr.)

Juhnke, M.E., Mathre, D.E. and Sands, D.C. (1984) A selective medium for *Gaeumannomyces graminis* var *tritici*. *Plant Dis.*, **68,** 233–6.

Lucas, L.T. (1980) Control of Spring dead spot of Bermuda grass with fungicides in North Carolina. *Plant Dis.* **64,** 868–70.

Shipton, P.J. (1975) Take-all decline during cereal monoculture. In *Biology* and *Control of Soil Borne Plant Pathogens*. (ed. G.W. Bruehl), Am. Phytopathol. Soc., St Paul, MN, pp. 137–44.

Smiley, R.W. (1983) *Compendium of Turfgrass Diseases*. American Phytopathological Society, St Paul MN 136pp.

Smiley, R.W. (1984) 'Fusarium blight-syndrome' redescribed as a group of patch diseases caused by *Phialophora graminicola, Leptosphaeria korrae* or related species. *Phytopathology*, **74,** 811. (Abstr.)

Smiley, R.W. and Craven-Fowler, M. (1984) *Leptosphaeria korrae* and *Phialophora graminicola* associated with Fusarium blight syndrome of *Poa pratensis* New York. *Plant Dis.*, **68,** 440–2.

Smiley, R.W., Kane, R.T. and Craven-Fowler, M. (1985) Identification of *Gaeumannomyces*-like fungi associated with patch diseases of turf grasses in North America. *Proc. Vth Int. Turfgrass Res. Conf.*

Smiley, R.W., Craven-Fowler, M. and Kane, R.T. (1984) Characteristics of pathogens causing patch diseases of *Poa pratensis* in New York. *Phytopathology*, **74,** 811 (Abstr.)

Smith, A.M. (1971) Control of Spring dead spot of couch grass turf in New South Wales. *J. Sports Turf Res. Inst.*, **47,** 60–5.

Walker, J. and Smith, A.M. (1972) *Leptosphaeria narmari* sp. nov. and *L. korrae* sp. nov. two long spored pathogens of grasses in Australia. *Trans Br. Mycol Soc.*, **58,** 459–66.

Worf, G.L., Avenius, R.C. and Stewart, J.S. (1982) A *Gaeumannomyces*-like organism associated with diseased bluegrass in Wisconsin. *Phytopathology*, **72,** 975–6. (Abstr.)

Worf, G.L., Brown, K.J. and Kachadoorian, R.V. (1983) Survey of 'necrotic ring spot' disease in Wisconsin lawns. *Phytopathology*, **37,** 839. (Abstr.)

Worf, G.L. and Stewart, J.S. (1985) Bluegrass necrotic ring spot. *Am. Lawn Applicator*, 58–60, 62, 66.

Worf, G.L., Stewart, J.S. and Avenins, R.C. (1986) Necrotic ring spot disease of turf grass in Wisconsin. *Plant Dis.*, **70,** 453–8.

10 *Summer patch*

After several years of intensive investigation into the aetiology of fusarium blight, Smiley and Craven–Fowler (1984) established that fungi other than *Fusarium* species commonly function as primary pathogens of *Poa pratensis* in what transpires to be a complex of patch diseases similar or identical in symptom expression (see page 117). Two dematiaceous fungi have been implicated, *Leptosphaeria korrae* Walker and Smith and *Magnaporthe poae* (Landschoot and Jackson). The latter has a *Phialophora* anamorph, identified incorrectly as *P. graminicola* (Deacon) Walker (Landschoot and Jackson, 1987, 1988). The names necrotic ring spot (see page 129) and summer patch have been assigned to the diseases caused by the respective fungi. These soil-borne pathogens may occur singly or together in the same stand of turf to incite patch disease symptoms, and are presumed also to predispose susceptible turfs to infection or secondary colonization by *Fusarium* spp.

Members of the genus *Phialophora* Medlar including *P. graminicola* are encountered widely on the roots, creeping stems and crowns of grasses (Walker, 1980, 1981). These fungi have not been recognized as aggressive pathogens and in some instances their presence actually may confer advantages to the grass host by cross protecting against other, more virulent, root pathogens (Deacon, 1981). The confirmation of a *Phialophora* species as the cause of extensive and severe summer damage in Kentucky bluegrass turf directs attention to the need for careful screening of these fungi as pathogens over a range of environmental conditions before they are included in any biological control programmes.

10.1 SYMPTOMS

Affected grass plants in circular patches, irregular-shaped patches, rings or crescents, ten centimetres wide and up, contrast with the healthy plants as initially they assume a darker, grey–green colour and wilted appearance. White, banded heat-stress lesions may be present on leaf blades at this time. Deterioration of the turf within the patch proceeds rapidly as the infected plants wither, turn brown and die. The numerous infection centres that are usually present soon coalesce and severely disfigure the stand of turf (Fig. 16, colour plate section). Activity of the pathogen subsides with the onset of cool weather conditions but, in the absence of renovation procedures, badly scarred turf is slow to recover and the dead areas are prone to weed invasion.

Roots, creeping stems and crowns of diseased plants are variously invested by a network of brown, ectotrophic hyphae from which pale brown to hyaline penetration hyphae invade the underlying tissues. Aggregates of pigmented, swollen cells (plate mycelium) may be present on leaf sheaths and creeping stems, and hyphopodia may occur on root surfaces. Roots and rhizomes soon turn brown, necrotic and brittle as the vascular cylinder of these organs becomes plugged by a brown gum-like material. Growth cessation structures (Deacon, 1974) may completely fill pericyclic parenchyma cells. Small, inflated and occasionally poroid, modified hyphae are apparent on and in the cortical tissues.

10.2 THE CAUSAL FUNGUS

Magnaporthe poae, Landschoot and Jackson. Perithecia gregarious or single, immersed, sometimes superficial, black, body globose, 252–556 μm diameter, neck cylindrical, 357–756 μm long and 95–170 μm at widest point. Perithecial wall up to 47 μm thick, composed of several layers of brown, radially compressed isodiametric cells. External cells epidermoidal. Neck canal up to 40 μm diameter, lined with hyaline upwardly pointed periphyses in canal. Asci numerous, clavate, 63–108 μm long and 7–15 μm at widest point, cylindrical, short stalked, straight or slightly curved. 8-spored, unitunicate, apex tapering but rounded, apical pore with non-amyloid refractive ring. Ascospores fusoid, 24–39 μm long and 4–6 μm diameter, 3 septate at maturity, end cells hyaline,

Figure 10.1 Summer patch. Perithecia of *Magnaporthe poae* on wheat stems.

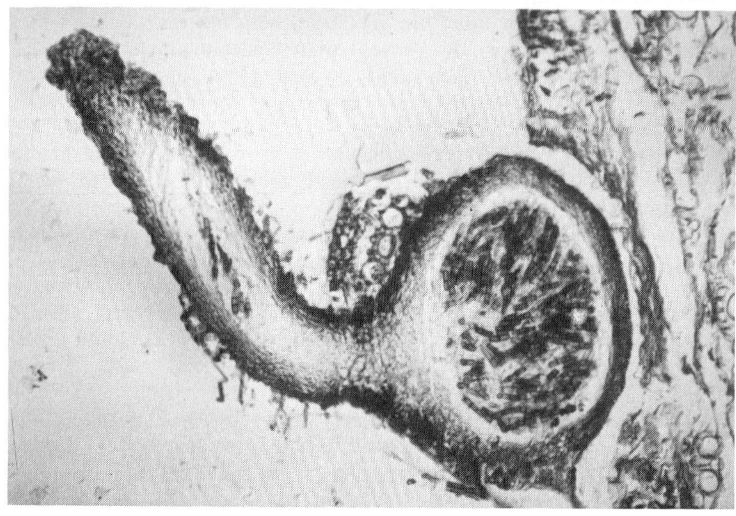

Figure 10.2 Summer patch. Longitudinal section of a perithecium of *Magnaporthe poae* showing asci and ascospores.

intermediate cells thick walled and germinating from one or both hyaline cells. Paraphyses arising from hymenium between asci, hyaline, septate sometimes branched, 64–112 μm long, 5–12 μm diameter near base tapering to 2–3 μm at tip. Superficial hyphae on the host roots sparce, brown, septate, 2–5 μm diameter, often with septa delimiting the lateral branches, single or in strands of 2–3. Infection hyphae hyaline, slender, permeating root tissue. Swollen hyphal cells often fill cortical cells. Hyphopodia globose, occurring singly or in aggregates on stem bases or on roots, 6–12 μm in diameter.

Cultures derived from single-ascospores maintained on half-strength PDA attained a diameter of 76 mm in 6 days at 28–30°C. Mycelium appressed, initially hyaline, then turning gray or olivaceous brown with dark, thick strands of mycelium radiating from the centre of the colony. Leading hyphae wavy, curling back towards centre of colony. Older colonies (7–8 weeks) appear olivaceous brown or black.

Phialides hyaline, 6–15 μm long, 2–5 μm diameter, straight or curved, tapering at tip, some with collarette. Collarette not visible in most, born on lateral branches or terminally, occurring singly or in clusters. Phialospores hyaline, 3–8 μm long and 1–3 μm wide, most slightly curved, some straight, commonly rounded at both ends.

Morphological features of *M. poae* are very similar to those of *M. rhizophila* Scott and Deacon (1983). However, *M. poae* differs from *M. rhizophila* in sexuality and ascospore size. The *Phialophora* conidial state differs from *P. graminicola* in phialospore and growth cessation structure conformation (Deacon, 1974). Cultures are faster growing with higher optimum and maximum growth temperatures and the fungus is an aggressive root pathogen at these higher temperatures.

A survey was conducted at the University of Rhode Island during 1985 and 1986 to identify the brown, ectotrophic fungi associated with roots and crowns of diseased grass plants in turf exhibiting summer patch symptoms. Fungi were isolated from turf samples collected from various locations in the eastern and central United States. Cultures of similar fungi isolated by other turf pathologists were solicited for the study.

The collection included cultures from New York State designated as *Phialophora graminicola* (Deacon) Walker (Fig. 5.2) and cited by Smiley and Craven–Fowler (1984) as the causal agent of summer patch disease. A comparison of these cultures (and similar biotypes from several other locations with cultures of *P. graminicola* derived from ascospores of *Gaeumannomyces cylindrosporus* Hornby, Slope, Gutteridge and Sivanesan (Jackson and Landschoot, 1986), revealed appreciable *in vitro* differences in morphology and physiology. The subsequent production of a teleomorphic state distinct from *G. cylindrosporus* provided confirmation that the reported incitant of summer patch (Smiley and Craven–Fowler, 1984, 1985) was named incorrectly (Landschoot and Jackson, 1987, 1988).

10.3 HOST RANGE

Smiley and Craven–Fowler (1984) trapped the fungus from diseased *Poa pratensis* L. turf using the cereals wheat and oats. Isolates so obtained were moderately to highly pathogenic to some cultivars of *Agrostis stolonifera, Festuca* spp. (fine-leaved species) *Lolium perenne* L. and *Poa pratensis* L. High temperatures were necessary for maximum disease expression (R.W. Smiley, 1985 private communication; Smiley and Fowler, 1985; Smiley, Craven–Fowler and Kane, 1985; Smiley, Craven–Fowler and Oknefski, 1985).

In the Rhode Island survey, the fungus has been isolated from *Poa pratensis*, and *P. annua*.

10.4 DISEASE DEVELOPMENT

Using controlled environment chambers, Smiley and Craven–Fowler (1985) were able to demonstrate that symptoms typical of summer patch disease could be induced experimentally. Patch symptoms developed if *P. pratensis* turf, inoculated with the pathogen, was kept continuously moist and incubated at 29 °C. Colonization and suppression of root growth by the fungus occurred at lower temperatures (21 °C) but the plants remained asymptomatic unless they were then subjected to the higher temperature regime. Abiotic lesioning of leaves, evident as white, necrotic bands, became apparent on root-infected plants shortly after they were stressed by high temperature and drought.

The environmental parameters detailed above match closely the field experience with summer patch disease. Symptoms are initiated during sustained hot rainy periods or when a spell of hot weather follows a period of heavy rainfall. Ascospores and conidia are potential inoculum sources but mycelium on infected turf grasses (and possibly on weed or agricultural grass species) probably serves as the primary survival mechanism and source for local spread. The disease has been noted in stands of *P. pratensis* at sod production sites and the fungus undoubtedly can be carried on harvested sod to new locations.

Conditions for development of summer patch, and the resulting symptoms, are similar to those described for 'fusarium blight' (see page 117). Arsenate herbicide stress has been shown to amplify the symptoms of summer patch (Smiley, Craven–Fowler and Oknefski, 1985) but drought stress is not necessarily a predisposing factor, a situation commonly listed as a prime requirement for fusarium blight to occur. The summer patch fungus is less tolerant of low moisture tensions than *L. korrae* and the fusariam turf pathogens but, like these fusaria, it is more tolerant of high temperatures (Smiley, Craven–Fowler and Kane, 1985a). Thus the possibility exists for *M. poae* to be the primary cause of damage (hot, wet conditions), for *M. poae* to interact with *L. korrae* (cool wet/hot wet conditions); for *L. korrae* and/or *M. poae* to interact with *Fusarium* spp. (cool wet/hot wet/hot dry conditions) and finally for *Fusarium* spp. possibly to act alone (hot dry conditions). All these situations would result in similar or identical symptoms that previously were considered to be fusarium blight.

10.5 CONTROL

Magnaporthe poae and *Leptosphaeria korrae* undoubtedly have been the inadvertent subjects of numerous field trials set up ostensibly to evaluate fungicides for the control of *Fusarium* spp. causing 'fusarium blight'. Since both these fungi, and the *Fusarium* spp., vary in their response to specific fungicides, the absence, presence or predominance of each pathogen in such situations could account for the puzzling inconsistencies in disease control reported for particular fungicides at different locations or in different years at the same location. The summer patch fungus is sensitive to more of the modern fungicides than is *L. korrae* (Smiley, and Craven Fowler, 1984). Laboratory tests have demonstrated that *in vitro* growth of *M. poae* is suppressed by triadimefon, benomyl, propiconazole and iprodione. Field trials tend to support these laboratory findings but more definitive work is required to determine the most effective chemical control agents, and where possible, cultural control measures.

10.6 REFERENCES

Deacon, J.W. (1981) Ecological relationships with other fungi: Competitors and hyperparasites. In *Biology and Control of Take-all* (eds M.J.C. Asher and P.J. Shipton), Academic Press. London, pp. 75–101.

Deacon, J.W. (1974) Further studies on *Phialophora radiciola* and *Gaeumannomyces graminis* on roots and stem bases of grasses and cereals. *Trans. Br. Mycol. Soc.*, **63,** 307–27.

Hornby, D., Slope, D.B., Gutteridge, R.J. and Sivanesan, A. (1977) *Gaeumannomyces cylindrosporus* a new ascoymcete from cereal roots. *Trans. Br. Mycol. Soc.*, **69,** 21–5.

Jackson, N. and Landschoot, P.J. (1986) *Gaeumannomyces cylindrosporus* associated with diseased turf grass in Rhode Island. *Phytopathology*, **76,** 654 (Abstr.).

Landschoot, P.J. and Jackson, N. (1987) *A Magnaporthe* sp. with a *Phialophora* conidial state causes Summer patch disease of *Poa pratensis* and *P. annua. Phytopathology,* **77,** 1734 (Abstr.).

Landschoot, P.J. and Jackson, N. (1988) *Magnaporthe poae* sp. nov. a hyphopodiate fungus with a *Phialophora* conidial state from grass roots in the United States. *Trans. Br. Mycol. Soc.* (in press).

Scott, D.B. and Deacon, J.W. (1983) *Magnaporthe rhizophila* sp. nov., a dark mycclial fungus with a *Phialophora* conidial state, from cereal roots in South Africa. *Trans. Br. Mycol. Soc.*, **81**(1), 77–81.

Smiley, R.W. and Craven–Fowler, M. (1984) *Leptosphaeria korrae* and *Phialophora graminicola* associated with Fusarium blight syndrome of *Poa pratensis* in New York. *Plant Dis.*, **68,** 440–2.

Smiley, R.W., Craven–Fowler, M. and Kane, R.T. (1985) Temperature and osmotic potential effects on *Phialophora graminicola* and other fungi associated with patch diseases of *Poa pratensis. Phytopathology*, **75,** 1160–7.

Smiley, R.W. and Craven–Fowler, M. (1985) Techniques for inducing Summer patch symptoms on *Poa pratensis. Plant Dis.*, **69,** 482–4.

Smiley, R.W., Craven–Fowler, M. and Oknefski, R.C. (1985) Arsenate herbicide stress and incidence of summer patch on Kentucky bluegrass turfs. *Plant Dis.*, **69,** 44–8.

Walker, J. (1980) *Gaeumannomyces, Linocarpon, Ophiobolus* and several other genera of scolecospored ascomycetes and *Phialophora* conidial states, with a note on hyphopodia. *Mycotaxon*, **11**(1), 1–129.

Walker, J. (1981) Taxonomy of take-all fungi and related genera and species. In *Biology and Control of Take-all.* (eds M.J.C. Asher and P.J. Shipton), Academic Press, London, pp. 15–74.

11 *Take-all patch (formerly Ophiobolus patch)*

Schoevers (1937) reported that in 1931 *Gaeumannomyces graminis* (Sacc.) V. Arx and Olivier, formerly *Ophiobolus graminis* (Sacc.) Sacc. was the cause of patching on golf greens at Eindhoven and severe patching on a lawn at Amersfoort in Holland. Monteith and Dahl (1932) briefly mentioned its occasional appearance in the USA. In the autumn and winter of 1951 plants of *A. tenuis* Sibth. and *A. stolonifera* L. from amenity turf in many parts of the British Isles were found to be heavily infected with *G. graminis* (Smith, 1952).

In Europe, *G. graminis* may cause severe damage to turf or golf fairways and greens, cricket outfields and tables, tennis courts, bowling greens and lawns in the United Kingdom and Eire, France, Spain, Belgium, the Netherlands, Switzerland, Norway, Denmark, Sweden and West Germany (Halcrow, 1965; Handoll, 1966; Hansen, 1964; Jackson, 1963, 1964; Nilsson and Smith, 1981; O'Rourke, 1976; Smith, 1952, 1956, 1957a, 1959, 1965). Although it was not recognized as an important cause of turf grass disease in North America until 1960 (Gould, Goss and Eglitis, 1961), in the Pacific Northwest it is considered second in importance on bentgrass turf of golf greens after microdochium patch (page 59, Gould, 1973). It affects turf from California to British Columbia (Gould, 1973; Ormrod, Hughes and Shoemaker, 1970; Wong, *et al.*, 1982) especially causing problems on new turf formed on forest or sandy soils (D.K. Taylor, personal communication, 1979; Smith, 1980). The disease is probably more widespread in North America than reported, since Jackson found it on bentgrass turf from Massachusetts laid on a golf green in Rhode Island and other locations in the eastern USA (1977, 1980, 1984) and Smith (1986) found the disease on *A. stolonifera* in a golf green in Saskatchewan. The disease is quite common in New South Wales, Australia on bentgrass golf greens (Anon, 1966, 1967; Smith, 1969; P.R. Wong, personal communication, 16 March 1979) and has also been found in Victoria, Tasmania and W. Australia (Wong and Siviour, 1979). The disease may have been confused with other patch diseases in the northeastern United States (Jackson, 1979).

11.1 SYMPTOMS

Patches of the disease are most noticeable during late summer, particularly if droughty, autumn and winter. If the winter is mild, patches may still be visible and the fungus active in the succeeding spring. The infected plants become bleached or bronzed. In fibrous turf, patches are first noticed when a few centimetres across as saucer-shaped depressions and these may increase to one metre or more over a number of years. The susceptible grasses in the centre are killed and their place is taken by less-suceptible or resistant grass species and weeds which are already present or which colonize the patches (Fig. 17, colour plate section). Round the patch margin, where the fungus is active, the bronzed ring of infected grasses has little root hold and the dead plants can be peeled off as a skin.

Infected roots show a dark-brown discoloration which in the early stages is located in the stele, the cortex appearing transparent, particularly when the roots are wet. Fine infection hyphae develop first in cortical tissues then penetrate to the stele. As the damage progresses roots turn dark-brown and under dry conditions become brittle and lack root hold. Runner hyphae, mycelial plates or mats and mycelial strands, like fine rhizomorphs occur on infected rhizomes, stolons, roots and under the basal leaf sheaths on the culm. In the latter location dark-brown or black, flask-shaped, beaked structures (perithecia), 0.25–0.5 mm in size develop and their necks protrude through the leaf sheaths (Fig. 11.1).

11.2 THE CAUSAL FUNGUS

Gaeumannomyces graminis (Sacc.) Arx and Olivier (Von Arx and Olivier, 1952; Walker, 1973, 1975) (formerly *Ophiobolus graminis* (Sacc.) Sacc.). The variety *avenae* (E.M. Turner) Dennis (Turner, 1940; Dennis, 1944; 1960) commonly causes take-all on oats in the field and attacks many grass hosts including turf grasses. The var. *tritici* Walker is often the cause of take-all in wheat and barley while the var. *graminis* is usually of low pathogenicity on temperate cereals. It attacks tropical grasses (Walker, 1975).

The mycelium of the fungus consists of plates or mats, formed of interwoven, dark-brown, septate hyphae which develop on plant crowns, stolons, rhizomes, upper parts of adventitious roots, culm bases and their sheathing leaves (Fig. 11.1). Runner-(macro-) hyphae, 3–6 mm in diameter, in flattened parallel strands, two or more cells wide, like narrow rhizomorphs, sometimes joining crusts of irregularly shaped hyphal cells, develop beneath the sheaths and on root surfaces. From these, hyphal branches develop which bear terminal, lateral or intercalary hyphopodia.* Infection may take place on all parts of the host plant below soil level by means of infection (micro-)hyphae which are paler in colour and finer than the runner hyphae (Walker, 1981). Infection hyphae may originate from single hyphopodia or from those which have aggregated into mycelial plates. Penetration of the host may be preceded by the formation of a hyphopodium and a narrow infection peg (Smith, 1952), but Russell (1934) reported that the formation of a hyphopodium (appressorium) was not necessary for host cell penetration.

Beneath the sheath and on the culm base, the dark-brown to black, smooth, flask-shaped, beaked, ostiolate perithecia, up to almost 0.5 mm diameter may be found with their curved necks protruding through the leaf sheaths. Asci are unitunicate, 8-spored, grouped in fascicles, elongate club-shaped, straight or curved (Fig. 11.1). In the case of var. *avenae* they measure 100–165 μm × 10–15 μm when mature. Asci are interspersed with hyaline septate paraphyses which are longer than the asci. Ascospores are also fascicled, hyaline, guttulate, aseptate and readily stainable with Direct blue 2B in lactophenol when young. When mature they are septate, faintly yellow in the mass and not so readily staining. In ten British collections, from *Agrostis* spp. and *Poa annua* in turf, ascospores ranged in length from 75 to 138 μm and asci were 103–165 μm. These lengths are characteristic of var. *avenae*. One isolate from *Poa annua* with typical disease symptoms and intermediate length asci of 78–122 μm was probably not var. *avenae* (Smith, 1956). European continental collections from *Agrostis* turf had ascospores of var. *avenae* type measuring 98–133 μm from Norway, 90–122 μm from Switzerland and 94–136 μm from the Netherlands. The Swiss and Netherlands isolates also had simple, unlobed hyphopodia (Nilsson and Smith, 1981). In the western USA, Gould, Goss and Eglitis (1961) reported that a collection from *A. tenuis* in Washington with ascospores 88–124 μm long should be considered as var. *avenae*. From the ascus and ascospore measurements given by Ormrod, Hughes and Shoemaker (1970) a collection from *A. tenuis* turf in British Columbia appear to be those of var. *avenae*. In the eastern USA Jackson (1977, 1980) found isolates from *Agrostis* turf in Rhode Island and Massachusetts with ascospores ranging from 80 to 151 μm of var. *avenae* with simple, unlobed hyphopodia. In Australia, Smith (1969) found the *avenae* variety with ascospores 107–130 μm on *A. tenuis* turf in New South Wales. P.T. Wong (personal communication, 16 March 1979) found two isolations of var. *avenae* on *Agrostis* sp. and *A. stolonifera* turf also in New South Wales with ascospore lengths of 78–128 μm and 75–132 μm respectively. However, another isolate from *A. stolonifera* on a golf green had 'intermediate' length spores. It was regarded as being of the var. *tritici*. In Saskatchewan on *A. stolonifera*, var *avenae* was involved with asci length 106–143 μm and ascospores 90–112 μm (Smith, 1986 and unpublished). Yeates (1986) found most of the isolates from *Agrostis* turf grass in Western Australia with ascospore lengths within the range of var. *avenae*.

Ascospores when fully mature germinate readily in water drops at room temperature from any cell along their length or from their ends. The fungus may be cultured from ascospores (Smith, 1956) or from mycelium on or in lesioned stem base, crown or root tissues after simple multiple washing or surface sterilization with materials such as mercuric chloride, silver nitrate or sodium hypochlorite. After treatment the sterilant is neutralized and tissue washed several times in sterile water and plated on agar medium. Cunningham (1981) has reviewed methods of isolation. A wide range of culture media may be used (Sivasithamparam and Parker, 1981), but both potato dextrose and glucose – salts – yeast extract serve will for routine purposes (Smith, 1956). For large scale pathogenicity studies maize – sand medium is satisfactory (Smith, 1956).

All three varieties of *G. graminis* have *Phialophora* states and isolates may produce curved, lunate to semicircular phialospores, in culture and from germinating ascospores (Walker, 1975, 1981). Smith (1965) found that they were produced freely in culture on sterile, chopped *Agrostis* leaves by an *avenae* isolate from *Agrostis* on abundant dark stromata in contact with the walls of a culture vessel. Usually phialospores do not germinate and their function is not known. There are also many *Phialophora*-like fungi on the roots of cereals and grasses with brown mycelium which could be mistaken for that of *G. graminis* (Nilsson, 1969; Deacon, 1973, 1974, 1976a, 1981). *Phialophora radicicola* Cain (Cain, 1972) was isolated from maize roots in Canada, but the same name has been applied to *P. radicicola* (Walker, 1975, 1980, 1981). Some of these, notably *Phialophora graminicola* (Deacon) Walker (Deacon, 1974; Walker, 1980) are of considerable interest since they may give protection against attacks of *G. graminis* vars. *tritici* and *avenae* (Wong, 1975).

11.3 HOST RANGE AND SUSCEPTIBILITY

There are extensive literature reviews of the host range and susceptibility of grass species to *G. graminis* by Nilsson (1969) and Nilsson and Smith (1981). Nilsson (1969) has also dealt with methods used to assess susceptibility in detail. Until 1940, when Turner (1940) published a description of the var. *avenae* any reports of susceptibility refer to the behaviour of grass species towards *G. graminis sensu lato*. Although there are now three described varieties of *G. graminis* it is often not possible to decide which variety is responsible for field cases of the disease, since to distinguish between var. *graminis* and var. *tritici* on one hand and var. *avenae* on the other, asci or ascospores must be measured. Perithecia may not be available or may be unripe. There are also isolates with intermediate length spores (Chambers and Flentje, 1967) and both vars. *tritici* and *avenae* may occur together in the same crop. The vars. *tritici* and *graminis* have been differentiated more recently (Walker, 1972), identification being based on hyphopodial characters (see previous footnote). Although var. *graminis* is common only on tropical grasses, Nilsson (1972) obtained isolates from southern Sweden which can produce lobed and/or unlobed hyphopodia on stem bases of grasses and cereals depending on temperature. In Britain and probably North America also the distribution of var. *avenae* and var. *tritici* is quite different. In Scotland and Wales they may be found together in the same cereal crop (Dennis, 1944; Willetts, 1961). The var. *avenae* has a wider distribution on grasses than as a cause of take-all of oats (Walker, 1975). Nevertheless in the cereals, var. *avenae* occurs mainly on oats. Although the variety was described from oats (Turner, 1940), it may have developed on the grasses and have its main reservoir there, perhaps on *Agrostis* spp. in the case of Britain, in the wetter north and west on the less base-rich soils where oats are more commonly grown. In these locations *Agrostis* grasslands are found which were never cropped to oats (Nilsson and Smith, 1981).

The var. *avenae* generally causes more severe damage and has a wider host range on grasses than vars. *tritici* and *graminis*, although there are exceptions (Hansen, 1964; Nilsson and Smith, 1981; Turner, 1940; E.M. Turner, personal communication 20 November 1951). There is not much published about cultivar resistance and information obtained before var. *avenae* was known to be implicated is not of much value in this respect (Nilsson and Smith, 1981). In pot tests against three var. *avenae* isolates, E.M. Turner (personal communication 20 Nov. 1951) found that only *Agrostis stolonifera* L., *Anthoxanthum odoratum* L., *Arrhenatherum elatius* (L.) J. and C. Presl., *Cynosurus cristatus* L., *Dactylis glomerata* L., *Holcus lanatus* L. and *Poa trivialis* L. had less than 50% of their seminal roots discoloured. Against three var. *tritici* isolates only *Alopecurus pratensis* L., *Festuca elatior* L., and *Lolium multiflorum* Lam. had more than 50% of their seminal roots discoloured. Towards an isolate of var. *avenae* from *A. stolonifera*, Smith (1957) found *A. tenuis, A. canina* L., *A. stolonifera*, *Lolium perenne* L. and *Poa trivialis* L. were quite susceptible, *C. cristatus*, *Festuca rubra* L. var. *commutata* Gaud, and *F. rubra* L. var. *trichophylla* Gaud. not so susceptible and *Poa pratensis* L. although showing root infection was the least susceptible. For Hansen (1964), *A. tenuis* was more severely damaged by var. *avenae* than var. *tritici*. *L. perenne*, *Phleum pratense* L. and *Poa pratensis* were highly resistant to both varieties. Deacon (1973a) found that 10 grasses tested against an *avenae* isolate ranged from the susceptible *A. tenuis*, *A. canina* and *A. stolonifera* to the resistant *Festuca rubra* and *Cynosurus cristatus*, a ranking similar to that found by Smith (1957a). Nilsson (private communication, 1979) found all grass species tested susceptible to isolates of var. '*graminis*' from southern Sweden, but they varied in susceptibility to var. *tritici*.

In practice it is *A. tenuis*, *A. stolonifera* and *A. canina* which are most frequently involved in take-all patch disease in Europe, North America and Australasia for these species are frequently used in fine turf and suffer most from attacks of the fungus (Gould, 1965; Smith, 1957a; Walker, 1975). *Poa trivialis* and *Poa annua* are usually less severely damaged. Although *Anthoxanthum odoratum*, *Deschampsia caespitosa* Beauv. *Festuca rubra*, (Smith, 1953), *Holcus lanatus* and *Poa pratensis* in take-all patches become infected, with *Poa annua* they are common species found surviving in the centres of patches when the *Agrostis* spp. have been killed out. Non-grass species in patches rarely show root infection (Smith, 1957).

11.4 EPIDEMIOLOGY

Although the disease in turf is not confined to the northern and western areas in Britain more cases have been reported from these places than from the south and east of the country (Smith, 1956). This may be in part due to the higher rainfall and humidity in the north and west and in part to the consequent more rapid leaching of soil bases necessitating more frequent liming (Smith, 1965). Effects of lime on soil pH and take-all patch are discussed below. In the Pacific Northwest of the USA, take-all patch occurs most seriously in the cool, moist areas west of the Cascade Mountains, although it has been known to cause trouble east of the mountains (Gould, 1973). The climate along the north-eastern seaboard of the USA, where the disease has been most recently reported in North America, is also fairly cool and moist. Although take-all is mainly a soil-borne disease, infection by wind-borne ascospores has been shown to occur on cereals on reclaimed polders in the Netherlands (Gerlagh, 1968). Seed transmission with grass seed was suspected in Scotland (Noble and Richardson, 1968; Smith, 1959).

11.4.1 Effect of liming and pH changes

There is a close relationship between the appearance of the disease and the application of lime, alkaline fertilizers such as basic slag, calcium cyanamide (Jackson, 1958; Smith 1957a), calcium ammonium nitrate (Anon, 1962) or even limey sand, particularly to *Agrostis* turf. Unlimed turf strips adjacent to limed areas sometimes escape the disease and take-all patch may often be induced experimentally by liming acid *Agrostis* turf (Smith, 1956, 1957a,b). O'Rourke (1976) reported a severe attack in *L. perenne* grown on calcareous marl exposed after peat had been removed from a peat bog in Co. Kildare, Eire. In the British Isles, Smith (1957a) found the disease on amenity turf in the pH range 4.3–7.5 with most severe cases in the range 5.1–7.5 (measured on samples to 100 mm depth). There are apparent contradictions to these findings since Gould, Goss and Eglitis (1963) reported that the disease occurred in Washington on unlimed experimental plots with a soil pH of 6.0 and Goss and Gould (1967) found that severity of take-all and pH of the upper 150 mm of soil was poorly correlated. Davidson and Goss (1972) found that lime applications to turf did not increase the severity of established take-all patch disease. Smiley (1972) reported that there was a poor correlation between bulk soil pH and take-all disease but a good correlation between rhizosphere pH and disease (Smiley and Cook, 1973) in wheat studies. Since most microbiological activity takes place in the top layers of the turf, Smith (1956, 1957a) sampled it to 50 mm depth and then partitioned the samples into 2 or 4 depths of 12.5 or 25 mm for pH determinations instead of using 100 or 150 mm deep samples, which is usual for chemical analyses. Deacon (1973a) also used 50 mm deep turf samples for pH and microbiological studies. The pH of the top 12.5 or 25 mm of the turf (less the live tops, but including the thatch) appeared to be critical in the incidence of the disease and the level below which it rarely occurred seemed to be pH 5.5 (Smith, 1957a). The usual form of lime to correct acidity in turf in Britain is ground limestone and the fineness of grist could be related to the pH of the top 25 mm, to the incidence of the disease and the migration down the turf profile of alkaline layers (Smith, 1957a). In higher rainfall areas in the north and west of Britain percolation of water and soil base leaching generally takes place more rapidly than in the south and east, especially on the thin soils of the uplands, requiring more frequent liming. The increased take-all incidence noted in wet years may be the result of increased rate of solution of lime and more rapid change in surface pH. The situation is probably aggravated on turf managed according to the 'acid theory' (Dawson, 1959) to discourage weeds and encourage fine-leaved small grass plants since this can lead to situations where it becomes so acid that fertilizer response can not be obtained. When such turf is limed, sudden changes in pH probably lead to rapid changes in the balance between the pathogen and its antagonists in the turf (Smith, 1965, Deacon, 1973a, and below). However, in the case of the disease in Saskatchewan (Smith, 1986), symptoms appeared on *A. stolonifera* several years after reconstruction of the golf green using a base of saline soil. The turf surface pH was 7.0 when the outbreak was seen.

11.4.2 Breaking of natural vegetation for turf formation

Take-all patch is common and damaging on newly established sown bent grass lawns in Coastal British Columbia (Ormrod, Hughes and Shoemaker, 1970; Smith, 1980) particularly on recently cleared woodland and on light soils (D.K. Taylor, personal communication, 1979). Although native and cultivated grasses vary in their susceptibility to *G.*

graminis many of them are excellent carriers of both the *tritici* and *avenae* varieties of the pathogen (Brooks, 1965; Chambers and Flentje, 1967, 1968; Kirby, 1922, 1925; Russell, 1934, 1939; Speakman and Lewis, 1978). Take-all is almost symptomless on many grasses (Brooks, 1965; Wehrle and Ogilvie, 1955) and on natural and long-established sown grassland an equilibrium state between pathogen and hosts develops. The breaking of long-undisturbed grassland mixes soil from different levels with their different populations of antagonists and pathogens (Deacon, 1973a; Sivasithamparam and Parker, 1981). Cultivation increases aeration and nitrification and this probably favours the pathogen (initially) (Garrett, 1937). Frequently these soils are deficient in lime and in preparation for seeding this material is incorporated into the seedbed giving a more intimate soil mixing than surface application and favouring severe outbreaks.

11.4.3 Fertilizers and fungicides

In pot studies with wheat, Garrett (1941) showed that wheat take-all was equally severe when deficiencies of either P or K or NPK occurred; adequate NPK reduced disease severity and N deficiency further reduced it. However, mineral nutrition affects both host and pathogen but in a different manner in an annual crop like wheat and a perennial turf grass cover. In the latter case the susceptible host is always present and acts as its own 'carrier' for both pathogen and antagonistic organisms. In wheat crops, kept substantially weed free, survival as a saprophyte or on a carrier host is necessary. Residual nutrients are evidently of more importance for the survival of *G. graminis* as a saprophyte where a grass carrier is not present between susceptible annual crops than in an infected perennial turf grass sward. There is no disease escape in perennial turf grasses resulting from a decline of potential inoculum as postulated by Garrett (1948) for cereals, although disease escape may result from the stimulation of new roots induced by nitrogen applications. As in cereals, the amount and form of nutrients, especially whether NH_4 or NO_3, in the rhizosphere (Smiley and Cook, 1973) is likely to influence the severity of root lesions.

Smith (1956) found, in a severe outbreak of the disease on *Agrostis* golf fairway turf, induced by liming, that applications of ammonium phosphate, ammonium sulphate or phenyl mercuric acetate fungicide checked the disease and the patches healed. Both fertilizers were more effective applied alone than with the fungicide. Similar results were obtained in experiments in New South Wales (Anon., 1962). By reducing the interval between spraying to 8–10 days in a later experiment Jackson (1958a) obtained good control with phenyl mercuric acetate. This is apparently related to the differential effect of phenyl mercuric acetate, a wide-spectrum biocide, on pathogen, antagonists and host plants. Gould, Goss and Eglitis (1961) suppressed the disease on *Agrostis* turf in Washington by applications of phenyl mercuric acetate. Goss and Gould (1967) found that the disease developed quickly on young 'Astoria' *Agrostis tenuis* turf after methyl bromide fumigation. This material probably eliminated antagonists as well as most of the inoculum of the pathogen (Dickson, 1947 and see below) and was followed by a regular spray programme with mercury fungicides which may also have affected the pathogen/antagonist balance. Under these conditions high N levels (9.8 kg/100 m^2) in the form of urea increased disease susceptibility at first, then severity declined suggesting disease escape by the production of abundant new roots. Both P and K had a suppressive effect on the disease. Chlordane, an insecticide, combined with ammonium sulphate was the best of several materials for the control of *G. graminis* for Gould, Goss and Miller (1966) in Washington but not in Britain. The most recent studies in Washington indicated that sulphur compounds were the most effective against take-all patch, but ammonium sulphate plus chlordane produced the quickest response. Ammonium sulphate alone was almost as good as the combination treatment (Davidson and Goss, 1972). Chlordane plus urea and lime with urea gave poor control. The mode of action of sulphur is uncertain. There may be a direct effect on the fungus or on its antagonists or through its effect on rhizosphere pH. Studies with wheat by Smiley and Cook (1973) suggested that control of the disease with ammonium forms of nitrogen is probably through the lowering of rhizosphere pH which suppressed *G. graminis*. Below pH 5 this is apparently direct inhibition while above pH 5 it is by soil microorganisms antagonistic to *G. graminis*. Fertilizers supplying nitrate nitrogen failed to suppress the fungus. Seed treatments with fungicides for control of take-all patch as for the wheat disease (Dolezal and Jones, 1980) appear impractical for a perennial crop such as turf grass.

11.4.4 Decline and antagonism

Following applications of lime to mature *Agrostis* turf on acid soils, three years may elapse before take-all patch may become serious under British conditions. Even if untreated, in succeeding years the disease declines in severity leaving behind patches of broad-leaved weeds and grasses other than *Agrostis* spp. (Smith, 1956). This is equivalent to 'take-all decline'

(TAD) in cereals (Slope and Cox, 1964).*

In most cases it has been impossible to find out with certainty whether overt symptoms of the disease were present before the lime or other alkaline materials were applied, since mild symptoms of the disease are difficult to recognize. Smith (1956) suggested that the pathogen is endemic on *Agrostis* spp. in turf, but active pathogenicity was suppressed by low pH. In mature turf equilibrium between the fungus and its antagonists has probably been reached. This may be disturbed by cultivating or liming.

From the results of surveys, Deacon (1973a,b) concluded that *G. graminis* is present only at low levels in sports turf, permanent pastures and leys in parts of Britain. He used wheat seedlings to detect *G. graminis* in soils with different cropping histories and in turves and cores from sports turf. *G. graminis* was found in five 25.4 mm diameter cores of 100 taken from 25 sites on four golf courses in West Yorkshire and in none of 68 taken from 19 sites on three golf courses in East Anglia. Much more extensive sampling of *Agrostis* turf is required to be certain of the distribution of the fungus. Scott (1970) isolated an avirulent *Phialophora*-like fungus from grass roots and Balis (1970) showed that inoculation of several grass species with a similar fungus reduced the extent of infection by *G. graminis* (*tritici*) of wheat following grasses. Deacon 1973a,b, 1974) found that there was an antagonistic *Phialophora*-like fungus similar to that of Balis (1970) in most of the grasslands surveyed, mainly in the southern and eastern counties and West Yorkshire in England. It was absent from turf with a pH of <c 4.5 and it was antagonistic towards *G. graminis* var. *avenae* on roots of *Lolium perenne* and *Avena sativa*. He suggested that the absence of this *Phialophora* sp. (*P. graminicola* (Deacon) Walker) could explain why such very acid turf often develops the disease after liming. There are several difficulties with this explanation. The growth of *G. graminis* and the *Phialophora* sp. is faster in alkaline or near alkaline than in acid conditions (Balis, 1970; Deacon, 1973a; Garrett, 1936, 1937; Kirby, 1925). At low pH the growth of both is restricted. The absence of the antagonist at <c 4.5 can not explain the occurrence of severe take-all patch in a soil pH range of 4.5–5.5, after liming which is common (Smith, 1956, 1957a, 1965). The results of studies on comparative competitive abilities of the two fungi at different pH levels must be evaluated before confidence can be placed in the explanation. The gradual decline in severity which takes place a few years after liming can be explained by (a) an increase in activity of the *Phialophora* reducing the infection of turf grass roots by *G. graminis* or (b) as a result of increasing acidity of the turf surface by solution and washing down of the lime (Smith, 1957) or its neutralization by atmospheric pollutants. This would slow down the activity of *G. graminis*. Smiley and Cook (1973) concluded from studies on take-all of wheat that reduction in disease resulted from direct inhibition of *G. graminis* at a rhizosphere pH less than 5 and indirect inhibition, possibly biological in nature above pH 5.0.

Fumigation of soil before sowing or planting turf grass is effective in many instances in killing weeds, insects, nematodes and seedling disease pathogens. Fumigants such as formalin, dazomet, chloropicrin or methyl bromide give only partial sterilization. In the Pacific Northwest of the USA the disease often appears on young bentgrass turf soon after initial methyl bromide fumigation of the soil (Gould, 1973) (and at the Sports Turf Research Institute – J.R. Escritt, personal communication). In Australia, take-all patch is only severe in the first or second year after fumigation (Corbett, 1962; A.M. Smith, cited in Wong and Siviour, 1979). The disease may disappear in the second, third or fourth year (after fumigation?) (Wong and Siviour, 1979).

Gould (1973) suggested that beneficial organisms competitive or antagonistic to *G. graminis* are either killed by the fumigation or are not normally present in the soil of forested areas. Corbett (1962) reported that in New South Wales, Australia, the only attacks by *G. graminis* have followed the use of methyl bromide fumigation of golf greens. If fumigation is followed by liming, take-all patch may develop and become severe, but the use of ammonium sulphate on re-established, sterilized greens prevents the establishment of the disease.

Micro-organisms other than *Phialophora*-like fungi are antagonistic to or compete with *G. graminis*. Sanford and Broadfoot (1931) showed that the pathogenicity of *G. graminis* var. *tritici* could be profoundly modified or controlled by many soil-inhabiting fungi and bacteria. Henry (1932) showed that unsterilized soil added to sterilized soil in pots inoculated with *G. graminis* var. *tritici* protected wheat plants against the pathogen at temperatures above 20°C and suggested that some of the 'antagonistic saprophytes' tested by Sanford and Broadfoot might be involved. More recently, Baker and Cook (1974) noted that bacteria and actinomycetes antagonistic to *G. graminis* were three times more abundant in TAD than in non-decline soil. Sivasithamparam and Parker (1978) showed that isolates of bacteria and actinomycetes reduced take-all disease in wheat

*Although it is difficult to relate decline in a perennial turf to any of the six categories of suppression suggested by Walker (1975).

growing in unsterile soil. Cook and Rovira (1976) suggested that fluorescent pseudomonads may be involved in specific antagonism against take-all in suppressive soils. Sivasithamparam, Parker and Edwards (1979) showed that there were changes in numbers of antagonistic bacteria on seminal and nodal roots of wheat which could be related to the development of resistance to *G. graminis*. Non-specific hyperparasites such as *Pythium oligandrum* Drechsler which show differences in pathogenicity towards *G. graminis* and antagonistic fungi (Deacon, 1976b) may influence the attack by the take-all fungus. Vampyrellid amoebae have been recorded perforating and killing pigmented hyphae of *G. graminis* in the Pacific northwest of the USA (Homma *et al.*, 1979; Old, 1979).

Wong and Siviour (1979) used avirulent fungi and take-all suppressive (TAS) soils to control *G. graminis* var. *avenae* in pot experiments with seedling turf of 'Pencross' *Agrostis stolonifera*. The addition of inoculum of *G. graminis*, an avirulent isolate of *G. graminis* var. *tritici* and two *Phialophora graminicola* varieties to sterilized pots of soil at the time of seeding completely controlled take-all patch disease. They artificially developed TAS soils by the repeated addition of live mycelium of *G. graminis* vars. *avenae*, *tritici* and *graminis* to soil. The addition of these partially controlled the disease. A top dressing of a 20 mm deep layer of a TAS soil developed from live mycelium of *G. graminis* var. *avenae* almost completely suppressed the disease.

11.5 CONTROL

11.5.1 Resistant grasses

In cool temperate regions only some species or cultivars of cool-season grasses such as *Agrostis* spp., *Festuca rubra* L. ssp. *commutata* Gaud., *F. rubra* L. ssp. *litoralis* and *Poa annua* L. are suited to the close mowing and other management practices demanded for high standard golf greens. In some parts of the world reliance is placed on monostands of *Agrostis* spp. for putting surfaces and sometimes golf tees and fairways. Most *Agrostis* spp. cultivars are very susceptible to take-all patch. Dilution of the *Agrostis* spp. with *F. rubra* cultivars, which are usually tolerant of the disease and which is a common practice in some regions, probably reduces severity of damage. *P. annua*, usually susceptible, sometimes escapes being killed, but is often found colonizing patch centres, establishing from seed in the base of the turf. As a first-aid measure, diseased fine bentgrass or bentgrass/fescue turf may be repaired with fescue sod or seed. For close mown golf tees a close-mowing tolerant cultivar of *Poa pratensis* (such as 'Nugget' or 'Dormie'), a species which is tolerant of the disease, may be used. Where the disease is a known hazard on grass hockey fields or cricket outfields mono- or polystands of *Poa pratensis* and *Festuca rubra* cultivars would provide disease-tolerant turf. *Agrostis* spp. and *Poa trivialis* should be avoided as they are susceptible. National lists for agronomically suited species and cultivars should be consulted for individual regions.

11.5.2 Use of lime or alkaline materials

The indiscriminate use of lime, shell sand or alkaline fertilizers, such as basic slag or alkaline top dressings should be avoided. Where lime is needed it should be applied as evenly as possibly and at carefully calculated dosage based on soil tests. Even low dosages of alkaline materials can raise the pH of the immediate turf surface considerably, but it does not need much of a lift for take-all patch to develop. For example, if the top 13 mm of an acid *A. tenuis* turf is raised above 5.5 there is a considerable risk that take-all patch will develop. No other treatment is needed to incite the disease. This is particularly the case in the west and north of the British Isles on poor fertility, thin, peaty soils. In other parts of the world, e.g. the Pacific northwest of the USA, the west coast region of Canada and in Australia other factors, such as fumigation with methyl bromide, aggravate the effect of liming or use of particular fertilizers.

The choice in some cases may be between using lime and obtaining a grass cover of some kind and risking the disease appearing, or withholding lime and having poor herbage without the disease. In this situation more slowly available lime such as shell sand or coarse ground limestone with particle sizes between 0.8 and 1.25 mm (passing 20 but not 30 meshes per inch) should be used. This may give the improvement required in acid turf without causing the disease to flare up. This should be done in conjunction with treatment to reacidify the upper soil layers (see below). Rather than having to take the step of liming very acid turf, a fertilizer and top dressing programme which maintains the surface pH at an acceptable level (say, pH 5.5–6.0) by alternating acid-tending and alkaline-tending materials should be adopted. This will reduce the risk of the disease occurring although its occurrence can not be ruled out.

11.5.3 Fertilization

In practice, lime is usually applied to established turf in the autumn or winter. Ammonium sulphate or sulphur may be conveniently applied, in spring, to reacidify the turf surface. These materials also lower the pH round the roots of younger turf, reducing the risk of take-all patch after soil amendment with lime or after soil fumigation. The use of a balanced fertilizer programme is indicated, but particular attention must be paid to phosphate level. Nutritional balance is of greater importance where turf is formed on soil which carried natural vegetation, such as coniferous woodland, or on light, sandy soils where nutrient deficiencies are likely to occur.

11.5.4 Fungicides

Smith (1956, 1956a,b) and Jackson (1958b) found that some measure of control of take-all patch resulted from the use of soil drenches or sprays with fungicides based on phenyl mercuric acetate. These fungicides are on longer manufactured or permitted in some countries. Triadimefon, triadimenol, carbendazim, fenarimol chlorothalonil and iprodione are suggested as possible alternatives (Anon., 1985; Dernoedon, O'Neill and Murray, 1981; Line, Sitton and Waldher 1981; Sanders *et al.*, 1978; Sanders, Burpee and Cole, 1978).

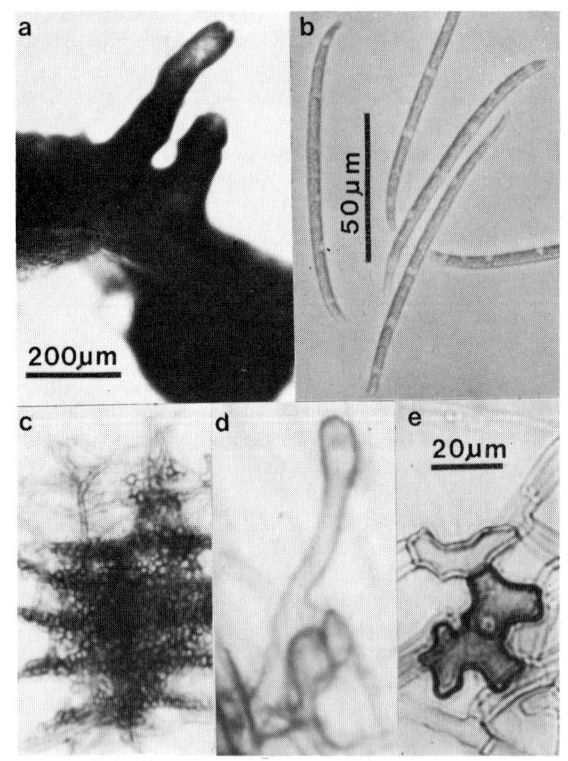

Figure 11.1 *Gaeumannomyces graminis* var. *avenae*
(a) Necked perithecia (b) Ascospores (c) Plate mycelium (d) Simple hyphopodia (e) Lobed hyphopodium of *G. graminis* var. *graminis* for comparison.

11.6 REFERENCES

Anon. (1962) Ophiobolus patch disease. *J. Sports Turf Res. Inst.*, **10** (38), 467–8.

Anon. (1966) Plant disease survey for the twelve months ending 30th June 1966. *36th Annual rept.* NSW Dept. Agric., Div. Sci. Serv., Biol. Br.

Anon. (1967) 37th annual plant disease survey for the twelve months ending 30th June 1967. NSW Dept. Agric., Div. Sci. Serv., Biol. Br.

Anon. (1983) Workshop on spring dead spot. Bayer Aust. Ltd. *Bowling Greenkpr.*, Oct. 1983, p.4.

Anon. (1985) *Directory of Amenity Chemicals.* 2nd edn Br. Agrochem. Assoc. and Sports Turf Res. Inst. Agr. Trg. Bd., Bourne Ho. Beckenham, Kent BR3 4PB. 35 p.

Arx, I.A. von and Olivier, D.L. (1952) The taxonomy of *Ophiobolus graminis* Sacc. *Trans. Br. Mycol. Soc.*, **35** (1) 29–33.

Baker, K.F. and Cook, R.J. (1974) *Biological Control of Plant Pathogens.* W.H. Freeman, San Francisco.

Balis, C. (1970) A comparative study of *Phialophora radicicola*, an avirulent fungal root parasite of grasses and cereals. *Ann. Appl. Biol.*, **66**, 59–73.

Brooks, D.H. (1965) Wild and cultivated grasses as carriers of the take-all fungus (*Ophiobolus graminis*). *Ann. Appl. Biol.*, **55**, 307–16.

Cain, R.F. (1952) Studies of Fungi Imperfecti. I. *Phialophora. Can. J. Bot.*, **30**, 338–43.

Chambers, S.C. and Flentje, N.T. (1967) Studies on oat-attacking and wheat-attacking isolates of *Ophiobolus graminis* in Australia. Studies on variation with *Ophiobolus graminis. Aust. J. Biol. Sci*, **20** (5), 927–40.

Chambers, S.C. and Flentje, N.T. (1968) Saprophytic survival of *Ophiobolus graminis* on various hosts. *Aust. J. Biol. Sci.*, **21** (6), 1153–61.

Cook, R.J. and Rovira, A.D. (1976) The role of bacteria in the biological control of *Gaeumannomyces graminis* by suppressive soils. *Soil Biol. Biochem.*, **8**, 269–73.

Corbett, D. (1962) Greenkeeping in New South Wales. *J. Sports Turf Res. Inst.*, **10** (38), 416–20.

Cunningham, P.C. (1981) Isolation and culture. In *Biology and Control of Take-all.* (eds M.J.C. Asher and P.J. Shipton), Academic Press., London., pp. 103–23.

Davidson, R.M. and Goss, R.L. (1972) Effects of P,S,N, lime, chlordane and fungicides on Ophiobolus patch disease of turf. *Plant Dis. Rept.*, **56**(7), 565–7.

Dawson, R.B. (1959) *Practical Lawncraft and Management of Lawn Turf.* 5th edn, Crosby Lockwood, London

Deacon, J.W. (1973a) Factors affecting occurrence of the Ophiobolus patch disease of turf and its control by *Phialophora radicicola. Plant Pathol.*, **22**(4), 149–55.

Deacon, J.W. (1973b) *Phialophora radicicola* and *Gaeumannomyces graminis* on roots of grasses and cereals. *Trans. Br. Mycol. Soc.*, **61**(3), 471–85.

Deacon, J.W. (1974) Further studies on *Phialophora radicicola* and *Gaeumannomyces graminis* on roots and stem bases of grasses and cereals. *Trans. Br. Mycol. Soc.*, **63**, 307–27.

Deacon, J.W. (1976a) Biology of the *Gaeumannomyces graminis–Phialophora radicicola* complex on roots of the Gramineae. *3rd EPPO Conference on Pathological*

Organisms in Cereal Monocultures, Gembloux, 17–19 June, 1975, 349–63.

Deacon, J.W. (1976b) Studies on *Pythium oligandrum*, an aggressive parasite of other fungi. Trans Brit. Mycol. Soc. 66, 383–391.

Deacon, J.W. (1981) Ecological relationships with other fungi: Competitors and hyperparasites. In *Biology and Control of Take-all* (eds M.J.C Asher and P.J. Shipton), Academic Press, London, pp. 75–101.

Dennis, R.W.G. (1944) Occurrence of *Ophiobolus graminis* var. *avenae* on wheat crops in the field. *Ann. Appl., Biol.*, **31**, 100–1.

Dennis, R.W.G. (1960) *British Cup Fungi and Their Allies. An Introduction to Ascomycetes.* Ray Society, London.

Dernoeden, P., O'Neill, N. and Murray, J. (1981) Evaluation of fungicides for control of ophiobolus patch and residual effects against dollar spot, 1980. *Am. Phytopathol. Soc. Fungicide Nematicide Tests.* **36**(278), 136.

Dickson, J.G. (1947) *Diseases of Field Crops.* McGraw-Hill, New York, 429 pp.

Dolezal, W.E. and Jones, J.P. (1980). A systemic seed treatment control of take-all disease of wheat. *Arkansas Farm Res.*, **29**(5), 10.

Garrett, S.D. (1936) Soil conditions and take-all of wheat. *Ann. Appl. Biol.*, **23**, 667–99.

Garrett, S.D. (1937) Soil conditions and the take-all disease of wheat. II. The relation between soil reaction and soil aeration. *Ann. Appl. Biol.*, **24**, 747–51.

Garrett, S.D. (1941) Soil conditions and take-all disease of wheat. Survival of *Ophiobolus graminis* on roots of different grasses. *Ann. Appl. Biol.*, **28**, 325–32.

Garrett, S.D. (1948) Soil conditions and take-all of wheat. 9. Interaction between host, plant nutrition, disease escape and disease resistance. *Ann. Appl. Biol.*, **35**, 14–17.

Gerlagh, M. (1968) Introduction of *Ophiobolus graminis* into new polders and its decline. *Neth. J. Plant Pathol.*, **74**, Suppl. 2, 97 pp.

Goss, R.L. and Gould, C.J. (1967) Some interrelationships between fertility levels and Ophiobolus patch disease in turf grasses. *Agron. J.*, **59**, 149–51.

Gould, C.J. (1965) Research on turf grass diseases in Washington State., U.S.A. *J. Sports Turf Res. Inst.*, **11**(40), 43–50.

Gould, C.J. (1973) Ophiobolus patch: Bane to bentgrass. *Golf Superintendent*, **41**(3), 3 pp. (reprint).

Gould, C.J., Goss, R.L. and Eglitis, M. (1961) Ophiobolus patch disease in Western Washington. *Plant Dis. Rept.*, **45**(4). 296–7.

Gould, C.J. Goss, R.L. and Eglitis, M. (1963) Ophiobolus patch on turf, *Golf Course Rept.*, (May) 3 pp.

Gould, C.J., Goss, R.L. and Miller, V.L. (1966) Effect of fungicides and other materials on the control of Ophiobolus patch disease on bentgrass. *J. Sports Turf Res. Inst.*, **42**, 41–8.

Halcrow, J.G. (1965) Turf disease notes, 1965. *J. Sports Turf Res. Inst.*, **41**, 53–8.

Handoll C. (1966) Turf disease notes, 1966. *J. Sports Turf Res. Inst.*, **42**, 65–8.

Hansen, L.R. (1964) A comparison between *Ophiobolus graminis* Sacc. and *Ophiobolus graminis* Sacc. var. *avenae* E.M. Turner. *Meld. Norg. LandbruksHøgskole.*, **43**(8), 11pp. (Norwegian).

Henry, A.W. (1932) Influence of soil temperature and soil sterilization on the reaction of wheat seedlings to *Ophiobolus graminis* Sacc. *Can. J. Res.*, **7**, 198–203.

Homma, Y., Sitton, J.W., Cook, R.J. and Old, K.M. (1979) Perforation and destruction of pigmented hyphae of *Gaeumannomyces graminis* by vampyrellid amoebae from Pacific Northwest field soils. *Phytopathology*, **69**, 1118–22.

Jackson, N. (1958a) Turf disease notes, 1958. *J. Sports Turf Res. Inst.*, **9**(34), 453–4.

Jackson, N. (1958b) Ophiobolus patch disease fungicide trial, 1958. *J. Sports Turf Res. Inst.*, **9**(34), 459–61.

Jackson, N. (1963) Turf disease notes, 1963. *J. Sports Turf Res. Inst.*, **11**(39), 26–8.

Jackson, N. (1964) Turf disease notes, 1964. *J. Sports Turf Res. Inst.* **11**(40), 76–80.

Jackson, N. (1977) Ophiobolus patch in Rhode Island. *Proc. Am. Phytopathol. Soc.*, **4**, 189 (Abs. NE 41).

Jackson, N. (1979) More turf diseases; old dogs and new tricks. *J. Sports Turf Res. Inst*, **55**, 163–167.

Jackson, N. (1980) Take-all patch (Ophiobolus patch) of turf grasses in the north eastern United States. In *Proc. III Int. Turf Grass Res. Conf.* (ed. J.B. Beard), American Soc. Agron., Crop Sci. Soc. of America and Int. Turf grass Soc., Madison, Wisconsin, pp. 421–424.

Jackson, N. (1984) A new cool season patch disease of Kentucky bluegrass turf in the northeastern United States. *Phytopathology*, **74**, 812. (Abstr.).

Kirby, R.S. (1922) The take-all disease of cereals and grasses. *Phytopathology*, **12**, 66–8.

Kirby, R.S. (1925). The take-all disease of cereals and grasses caused by *Ophiobolus cariceti* (Berkeley and Broome) Saccardo. *Cornell Agr. Exp. Sta. Mem.* **88**, 55 pp.

Line, R.F., Sitton, J.W. and Waldher, J.T. (1981) Control of take-all of wheat with Bayleton and Baytan, 1981. *Am. Phytopathol. Soc. Fungicide Nematicide Tests*, **37**(234), 120.

Monteith, J. Jr and Dahl, A.S. (1932) Turf diseases and their control, *Bull. US Golf Assoc. Green Sect.*, **12**(4), 156–7.

Nilsson, H.E. (1969) Studies of root and foot rot diseases of cereals and grasses. I. On resistance to *Ophiobolus graminis* Sacc. *Lantbrukshögskolans Annaler,* **35**(3), 275–807.

Nilsson, H.E. (1972) The occurrence of lobed hyphopodia on an isolate of the take-all fungus, *Ophiobolus graminis* Sacc. on wheat in Sweden. *Swedish J. Agric. Res.*, **2**, 105–18.

Nilsson, H.E. (1974) Attack of a different isolate of the root-killing fungus (*Gaeumannomyces graminis*). *Nordisk Jordbrugs Forskning*, **56**, 358–60. (In Swedish).

Nilsson, H.E. and Smith, J.D. (1981) Take-all of grasses. In *Biology and Control of Take all*, (eds M.J.C. Asher and P.J. Shipton), Academic Press, London, pp. 433–48.

Noble, M. and Richardson, M.J. (1968) An annotated list of seed-borne diseases. *Commonw. Mycol. Inst. Phytopathol. Pap.* 8, 191pp.

Old, K.M. and Patrick, Z.A. (1979) Giant soil amoebae, potential biocontrol agents. In *Soil-borne plant pathogens.*

(eds. B. Schippers and W. Gams), Academic Press, London, pp. 617–28.
Ormrod, D.J., Hughes, E.C. and Shoemaker, R.A. (1970) Newly recorded fungi from colonial bentgrass in coastal British Columbia. *Can. Plant. Dis. Surv.*, **50**(3), 111–12.
O'Rourke, C.J. (1976) *Diseases of Grasses and Forage Legumes in Ireland.* An Foras Taluntas (The Agricultural Institute of Ireland, Carlow), 115 pp.
Russell, R.C. (1934) Studies in cereal diseases 10. Studies of take-all and its causal organism *Ophiobolus graminis* Sacc. *Bull. Can. Dept. Agric.*, **170** NS. 64 pp.
Russell, R.C. (1939) Pathogenicity tests with cultures of *Ophiobolus graminis* Sacc., *Sci. Agric.*, **19**, 662–9.
Sanders, P.L., Burpee, L.L. and Cole, H. Jr (1978) Uptake translocation and efficacy of triadimefon in control of turf grass pathogens. *Phytopathology*, **68**, 1482–7.
Sanders, P.L., Burpee, L.L., Cole, H. Jr and Duich, J.M. (1978) Control of fungal pathogens of turf grass with the experimental iprodione fungicide, RP 26019. *Plant. Dis. Rep.*, **62**, 549–53.
Sandford, G.B. and Broadfoot, W.C. (1931) Studies of the effects of other soil-inhabiting micro-organisms on the virulence of *Ophiobolus graminis* Sacc. *Sci. Agric.*, **11**, 512–28.
Schoevers, T.A.C. (1937) Some observations on turf diseases in Holland. *J. Bd. Greenkpg. Res.*, **5**(16), 23–26.
Scott, P.R. (1970) *Phialophora radicicola*, an avirulent parasite of wheat and grass roots. *Trans. Br. Mycol. Soc.*, **55**, 163–7.
Sivasithamparam, K. and Parker, C.A. (1978) Effects of certain isolates of bacteria and actinomycetes on *Gaeumannomyces graminis* var. *tritici* and take-all of wheat. *Aust. J. Bot.*, **26**, 773–82.
Sivasithamparam, K., Parker, C.A. and Edwards, C.S. (1979) Bacterial antagonists to the take-all fungus and fluorescent pseudomonads in the rhizosphere of wheat. *Soil. Biol. Biochem.*, **11**, 161–5.
Sivasithamparam, K. and Parker, C.A. (1981) Physiology and nutrition. In *Biology and control of Take-all.* (eds M.J.C. Asher and P.J. Shipton), Academic Press. London, pp. 125–50.
Slope, D.B. and Cox, J. (1964) Continuous wheat growing and the decline of take-all. *Rep. Rothamsted Exp. Sta., 1963*, Pt. 1, 108.
Smiley, R.W. (1972) Relationship between rhizosphere pH changes induced by root absorption of ammonium versus nitrate-nitrogen and root disease with particular reference to take-all of wheat. Ph.D. Thesis. Washington State University Pullman, 68pp.
Smiley, R.W. and Cook, R.J. (1973) Relationship between take-all of wheat and rhizosphere pH in soils fertilized with ammonium versus nitrate nitrogen. *Phytopathology*, **63**, 882–90.
Smith, A.M. (1969) An oat-attacking strain of take-all in N.S.W. *J. Aust. Inst. Agric. Sci.*, **35**, 270–1.
Smith, J.D. (1952) A patch disease of sports turf caused by *Ophiobolus graminis* var. *avenae*. E.M. Turner. *J. Sports Turf Res. Inst.*, **8**(28), 140–3.
Smith, J.D. (1953) Turf disease notes. *J. Sports Turf Res. Inst.*, **8**(29), 259–60.
Smith, J.D. (1956) Fungi and turf diseases. 6. Ophiobolus patch disease. *J. Sports Turf Res. Inst.*, **9**, 180–202.
Smith, J.D. (1957a) A study of Ophiobolus patch disease of sports turf in relation to its occurrence and control. In The control of certain diseases of sports turf grasses in the British Isles. Vol 2. M.Sc. Thesis. University of Durham, King's College, Newcastle-upon-Tyne, pp. 184–221.
Smith J.D. (1957b) Turf disease notes, 1957. *J. Sports Turf Res. Inst.*, **9**(33), 373–4.
Smith, J.D (1959) Turf disease in the North of Scotland. *J. Sports Turf Res. Inst.*, **10**(35), 42–7.
Smith, J.D. (1965) *Fungal Diseases of Turf Grasses.* 2nd edn (Revised N. Jackson and J.D Smith), A Sports Turf Research Institute Publication. 97pp.
Smith, J.D. (1980) *Major Diseases of Turfgrasses in Western Canada.* Extension Div. Univ. Sask. Publn. 409, 14pp.
Smith, J.D. (1986) New or uncommon turf grass diseases in Saskatchewan. *Greenmaster*, **22**(4), 15, 17.
Speakman, J.B. and Lewis, B.G. (1978) Control of take-all disease of wheat by grasses and legumes. *Plant Pathol.*, **27**, 58–65.
Turner, E.M. (1940) *Ophiobolus graminis* Sacc. var. *avenae* var. n. as the cause of take-all or whiteheads of oats in Wales. *Trans. Br. Mycol. Soc.*, **24**(3–4), 269–81.
Walker, J. (1972) Type studies on *Gaeumannomyces graminis* and related fungi. *Trans. Br. Mycol. Soc.*, **58**, 427–57.
Walker, J. (1973) *Gaeumannomyces graminis* var. *graminis*, var. *avenae* and var. *tritici. CMI. Descriptions of pathogenic fungi and bacteria.* 381, 382, 383.
Walker, J. (1975) Take-all diseases of Gramineae: A review of recent work. *Rev. Appl. Plant Pathol.*, **54**(3), 113–44.
Walker, J. (1980) *Gaeumannomyces, Linocarpon, Ophiobolus* and several other scolecospored Ascomycetes and *Phialophora* conidial states with a note on hyphopodia. *Mycotaxon*, **11**, 1–29.
Walker, J. (1981) Taxonomy of take-all fungi and related genera and species. In *Biology and Control of Take-all* (eds M.J.C. Asher and P.J. Shipton), Academic Press, London, pp. 15–74.
Wehrle, V.M. and Ogilvie, L. (1955) Effect of ley grasses on the carry over of take-all. *Plant Pathol.*, **4**, 111–13.
Willets, H.J. (1961) A comparison between *Ophiobolus graminis* and *Ophiobolus graminis* var. *avenae. Trans Br. Mycol. Soc.*, **44**, 504–10.
Wong, P.T.W. (1975) Cross protection against wheat and oat take-all by *Gaeumannomyces graminis* var. *graminis. Soil Biol. Biochem.*, **7**, 189–94.
Wong, P.T.W., Rasmussen-Dykes, C., Perotti, L.E. and Brown, W.M. (1982) Occurrence of Ophiobolus patch of turf in Colorado. *Phytopathology*, **72**, 976.
Wong, P.T.W. and Siviour, T.R. (1979) Control of Ophiobolus patch in *Agrostis* turf using avirulent fungi and take-all suppressive soils in pot experiments. *Ann. Appl.*, Biol., **92**, 191–7.
Worf, G.L., Avenisu, R.C. and Stewart, J.S. (1982) A *Gaeumannomyces*-like organism associated with diseased bluegrass in Wisconsin. *Phytopathology*, **72**, 975–6. (Abstr.).
Yeates, J.S. (1986) Ascospore lengths of Australian isolates of *Gaeumannomyces graminis. Trans. Br. Mycol. Soc.*, **86**(1), 131–6.

12 *Mycorrhizae*

In mycorrhizal associations, highly specialized, soil-inhabiting fungi enter into long-term, symbiotic relationships with the roots of higher plants which for the most part are mutually beneficial to both partners. Although the association is truly parasitic on the part of the fungus, it is not pathogenic, except under special circumstances. The fungus obtains nutrients from the plant; it brings, under adverse growing conditions, some essential nutrients, such as phosphate and nitrogen to the roots. Although turf grasses in the genera *Agropyron, Agrostis, Anthoxanthum, Poa* and *Puccinellia* have been reported as mycorrhizal (Crush, 1973; Harley, 1959; Lawley, Newman and Campbell, 1982; Mosse, 1972; Nicolson, 1958, 1959), information on the relationship between fungus and host is scanty when compared with some other agricultural crops, e.g. citrus, cereals and vegetables. Their role in the establishment and maintenance of amenity turf has been little studied. Most of the grass mycorrhizae of interest are of the vesicular–arbuscular (VA) type (see below).

12.1 SYMPTOMS

Ectomycorrhizal infections (see below) cause macroscopically visible root distortion. With VA mycorrhizae it is necessary to examine roots microscopically after suitable clearing and staining treatment (Phillips and Hayman, 1970). Signs that the latter are infected are usually indefinite in unstained specimens even when most of their length is involved. Plant growth may be retarded when grasses are growing under adverse conditions and the host/fungus symbiotic balance is disturbed such as after soil fumigation (Menge, 1982). The fungus may then utilize nutrients needed for healthy plant growth resulting in chlorosis or mineral-deficiency symptoms. Under very wet soil conditions grasses are often without mycorrhizae needed to make available mineral nutrients in short supply (Gerdemann, 1968).

12.2 THE CAUSAL FUNGI

Three kinds of mycorrhizae may be recognized:

1. Ectomycorrhizae, where a sheath of fungus mycelium covers the root surface with some intercellular hyphae in the cortex.
2. Endomycorrhizae, in which a web of fungal hyphae pervading the soil round the root is associated with a denser mycelial growth in the root cortex.
3. Ectendomycorrhizae, a type intermediate in character between the other two (Gerdemann, 1968).

Most interest centres around endomycorrhizae which are common and important in host plant nutrition. Some endomycorrhizae have septate hyphae, but others are aseptate, zygomycetous members of the Endogonaceae with most of the species in the genera *Glomus, Acaulospora, Sclerocystis, Scutellospora* and *Gigaspora*. Fewer than 130 species has been described to the end of 1982 (Mosse, 1973; Scheuck and Peres, 1988; Trappe, 1982; Trappe and Schenck, 1982). These endophytic fungi are referred to as vesicular–arbuscular (VA) mycorrhizal fungi because of the characteristic fungal structures they produce within the host.

Considerable differences are apparent between species of VA fungi in the distribution of the fungus in the root, in the characteristics of the hyphae and in the possession and morphology of spores, sporocarps, vesicles and arbuscules. VA mycorrhizae have an endophytic hyphal system connected with an external system in the soil around the host.

In *Gigaspora*, Gerdemann and Trappe, and *Scutellaspora* (Walker and Sanders), spores are produced singly and terminally on a bulbous suspensor cell. They are usually globose or subglobose, $<200->600\,\mu m$ diam. Auxiliary cells on coiled hyphae are produced in the soil. Germination takes place either directly through the spore wall or from a germination shield. One or more germ tubes emerge from the spore wall near the suspensor-like cell. There are no sporocarps. Auxiliary cells are rarely produced in infected roots. Species have a world-wide distribution and are likely to be found in native grasslands where hosts occur (Gerdemann & Trappe, 1975; Trappe, 1982). In *Acaulospora* Gerdemann and Trappe, spores similar to those of *Scutellospora* are produced singly, but by the lateral budding of the stalk of an inflated end (terminus) of a hypha which collapses after its contents are transferred to the spore. The spores are $<100 - >400\,\mu m$ diam. and their walls have many layers. Several germ tubes are produced from peripheral compartments at the base of the spore. Vesicles and arbuscules are present in the host. There are no sporocarps. The 'honey-coloured sessile spores' of Mosse (1973) found in soil belong to *Acaulospora*. Single spores are likely to be found where ever suitable hosts are present (Gerdemann and Trappe, 1975; Trappe, 1982). In *Glomus* Tulasne and Tulasne, spores are produced, usually terminally on undifferentiated hyphae, or in sporocarps, in roots or free in the soil. Chlamydospores are $<15 - >300\,\mu m$ diam; sporocarps from $<1\,mm - <20\,mm$. Most species produce endomycorrhizae with vesicles and arbuscules. Species distribution is world-wide. The most common endomycorrhizal fungi are *Glomus* spp. (Gerdemann and Trappe, 1975; Trappe, 1982). *Sclerocystis* Berkeley and Broome is closely related to *Glomus*, but differs in having spores arranged in a sporocarp around the periphery of a central aggregation of hyphae. Chlamydospores are $<50 - >200\,\mu m$ long. Sporocarps are $<200 - <700\,\mu m$. One species, *S. rubiformis* Gerd. is widely distributed in temperate regions, other species are tropical (Gerdemann and Trappe, 1975; Trappe, 1982).

12.3 MYCORRHIZAL DEVELOPMENT

Infection of the host takes place by a hypha from a germinating spore, from the mycelium of living roots or from vesicles in infected root residues in the soil. The hypha, which may or may not produce an appressorium in contact with the hosts root epidermis, penetrates the epidermis through or between the cells and colonizes the cortex causing very little host cell reaction (Carling and Brown, 1982). In the colonizing process the intercellular or intracellular VA mycorrhizae may form complicated coils or loops and intercalary or terminal structures with fine branches called arbuscules. These structures may function as haustoria. Vesicles which may be inter- or intracellular, are usually terminal, oval or globular, thin- or thick-walled hyphal swellings, and are formed in the cortex of the host by the endophyte. They behave as nutrient storage organs or may become modified to thick-walled chlamydospores and serve as reproductive structures (Gerdemann, 1968; Carling and Brown, 1982). Endomycorrhizal fungi are obligate parasites and although their spores may remain viable in soil for long periods, they require live roots on which to complete their development.

12.4 EFFECTS OF MYCORRHIZAE ON GRASSES

In grasses, as in most other higher plants, the establishment of the VA mycorrhiza condition by the roots is routinely beneficial to the host. This is especially the case in marginal habitats where soils are deficient in nutrients. They are particularly important in phosphate nutrition (Mosse, 1973; Sparling and Tinker, 1975; Tinker, 1975), assist in the uptake of non-mobile ions (Menge, 1982) and increase water movement to the roots since the web of their hyphae outside the roots extend beyond the root-hair uptake zones into undepleted soil (Gerdemann, 1975; Ruehle and Marx, 1979; Sanders and Tinker, 1973).

Bouteloua gracilis (H.B.K.) Lag. and Steud., a warm-season range grass, used in specific low-fertility situations as a roadside grass in N. America (Beard, 1973) is native to rangeland where soil mositure is often a limiting factor to growth. Inoculation with its VA mycorrhiza *Glomus fasciculatum* (Thaxter sensu Gerdemann) Gerdemann and Trappe, increased photosynthetic rate, which was related to reduction in stomatal resistance and reduction in mesophyll resistance to CO_2 uptake. Water transport through the plant and transpiration rate was greatly increased. The physiological changes following mycorrhizal infection could influence the survival of the species in marginal habitats (Allen, Moore and Christensen, 1981).

In some situations, the activities of mycorrhizae may restrict grass growth. In New Zealand, Crush (1973) showed that *Lolium perenne, Dactylis glomerata* and *Anthoxanthum odoratum* increased in shoot weight when inoculated with the VA mycorrhiza *Glomus (Rhizophagus) tenuis* (Greenhall) Hall when soil available P was extremely deficient. When P was not so severely limiting, growth was depressed. It was suggested that the fungus was able to grow in the rhizosphere with little endotrophic mycelium, competing with the plant roots for a meagre supply of the nutrient. Bulwalda and Goh (1982) inoculated *Lolium perenne* plants with *Gigaspora margarita* Becker and Hall. Plant top yield was reduced by the endomycorrhiza. P concentration was not limiting and total oxidizable C, soluble sugar content and C/N ratios were lower in mycorrhizal plants. This suggested that competition for photosynthate C had resulted in growth depression, possibly due to the diversion of C from the roots to the external mycorrhiza. Under certain circumstances, VA mycorrhizae may become pathogenic and they also react with plant pathogens although this has not yet been demonstrated under field conditions (Dehne, 1982).

In grassland, neighbouring plants may influence each other's VA mycorrhizae and the fungi and bacteria in the rhizosphere. In British, neutral grassland species Christie, Newman and Campbell (1978) found that the abundance of root-surface bacteria and fungi on a particular host was often significantly changed by the presence of other plant species. VA mycorrhizae were altered less than bacteria and fungi when the grass species *Anthoxanthum odoratum, Lolium perenne, Plantago lanceolata* and *Trifolium repens* were grown together or with different partners. Lawley, Newman and Campbell (1982) grew three grasses of acid, infertile soils, *Agrostis tenuis, Deschampsia flexuosa* and *Festuca ovina* in pots, separately and in two-species mixtures. It was found that when the grasses were grown together they could influence each others' root micro-organisms, but there was no significant effect of a partner species on bacteria and less effect on root-surface fungi than had been noted in neutral grassland plants. Mycorrhizal infection was influenced by the partner species, but not in a regular manner. In *Agrostis* it was not affected by the partner species. In *Deschampsia* it was increased (when compared with monoculture) when *Festuca* was the partner, and in *Festuca* it was reduced when *Agrostis* was the partner. When grown in competition with *Holcus lanatus*, in fine sand which was a poor supplier of P, inoculation with *Glomus (Endogone) fasciculatus* reduced the growth of *Lolium perenne* to a greater extent than root competition alone. *L. perenne* does not normally respond well to inoculation with this mycorrhizal fungus yet the incidence of infection of its root system was increased by inoculation and by intermingling with *H. lanatus* roots (Fitter, 1977).

12.5 CONTROL

Generally, there is insufficient information on the deleterious effects of the mycorrhizae of amenity turf grasses to consider control measures. Where turf is to be established under marginal conditions, especially where soil fertility is very low, the use of locally adapted cultivars of symbiont populations is expedient. VA mycorrhizae are sensitive to fumigant biocides such as chloropicrin, formaldehyde, dazomet, methyl bromide, metam and vorlex, to contact fungicides such as quintozene, thiram and dicloran and to several systemic materials (Menge, 1982; Rhodes and Larsen, 1981). Some nematicides have little effect on the soil mycorrhizal population. There are some data linking the application of pesticides (fumigant biocides, fungicides, nematicides and herbicides) directly and indirectly to mycorrhizal infection. Some of these materials influence host root exudates. Pesticides that increase root exudation may increase mycorrhizal infection and those that decrease the exudates may decrease infection by mycorrhizae (Menge, 1982; Ratnayake, Leonard and Menge, 1978). Fungicides and fumigant biocides in particular, can influence the population of soil fungi, bacteria and nematodes, some of which are hyperparasites of mycorrhizae or are root pathogens, either increasing or decreasing particular species or groups. In turn, some of the latter are able to influence mycorrhizal infection (Daniels and Menge, 1980; Menge, 1982; Mosse, 1962). The differential effects of many of the pesticides are not yet predictable in amenity turf.

12.6 REFERENCES

Allen, M.F., Moore, T.S. Jr and Christensen, M. (1981) Comparative water relations and photosynthesis of mycorrhizal and non-mycorrhizal *Bouteloua gracilis* H.B.K. Lag. and Stoud. *New. Phytol.*, **88**, 683–93.

Beard, J.B. (1973) *Turfgrass: Science and Culture.* Prentice-Hall, Englewood Cliffs, NJ, 658pp.

Bulwalda, J.G. and Goh, K.M. (1982) Host–fungus competition for carbon as a cause of growth depressions in vesicular–arbuscular mycorrhizal ryegrass. *Soil Biol. Biochem.*, **14**, 103–6.

Carling, D.E. and Brown, M.F. (1982) Anatomy and physiology of vesicular–arbuscular and non-mycorrhizal roots. *Symp. Mycorrhizae Plant Dis. Res., Phytopathology*, 72(8), 1108–114.

Christie, P., Newman, E.I. and Campbell, R. (1978) The influence of neighbouring grassland plants on each others endomycorrhizas and root surface microorganisms. *Soil Biol. Biochem.*, **10**, 521–7.

Crush, J.R. (1973) Effect of *Rhizophagus tenuis* mycorrhizas on ryegrass, cocksfoot and sweet vernal. *New Phytol.*, **72,** 965–73.

Daniels, B.A. and Menge, J.A. (1980) Hyperparasitization of vesicular-arbuscular mycorrhizal fungi. *Phytopathology*, **70,** 584–8.

Dehne, H.W. (1982) Interaction between vesicular–arbuscular mycorrhizal fungi and plant pathogens. In *Symp. VA Mycorrhizae Plant Dis. Res., Phytopathology*, **72**(8), 1115–19.

Fitter, A.H. (1977) Influence of mycorrhizal infection on competition for phosphorus and potassium by two grasses. *New Phytol.*, **79,** 119–25.

Gerdemann, J.W. (1968) Vesicular–arbuscular mycorrhizae and plant growth. *Ann. Rev. Phytopathol.*, **6,** 397–418.

Gerdemann, J.W. (1975) In *The development and function*

of roots. (eds J.G. Torrey and D.T. Clarkson), Academic Press, New York, 575pp.

Gerdemann, J.W. and Trappe, J.M. (1975) Taxonomy of the Endogonaceae. In *Endomycorrhizas* (eds F.E. Sanders, B. Mosse, and P.B. Tinker), Academic Press, New York, pp. 35–51.

Harley, J.L. (1959) *The Biology of Mycorrhiza*. Plant Science Monographs. Leonard Hill, London, 233pp.

Lawley, R.A., Newman, E.I. and Campbell, R. (1982) Abundance of endomycorrhizas and root surface microorganisms on three grasses grown separately and in mixtures. *Soil Biol. Biochem*., **14**, 237–40.

Menge, J.A. (1982) Effect of soil fumigants and fungicides on vesicular-arbuscular fungi. *Phytopathology*, **72**(8), 1125–32.

Mosse, B. (1962) The establishment of vesicular–arbuscular mycorrhiza under aseptic conditions. *J. Gen. Microbiol*., **27**, 509–20.

Mosse, B. (1972) Effects of different *Endogone* strains on the growth of *Paspalum notatum*. *Nature, Lond*., **239**, 221–3.

Mosse, B. (1973) Advances in the study of vesicular–arbuscular mycorrhizas. *Ann. Rev. Phytopathol*., **11**, 171–96.

Nicolson, T.H. (1958) Vesicular–arbuscular mycorrhiza in the Gramineae. *Nature, Lond*., **181**, 178.

Nicolson, T.H. (1959) Mycorrhiza in the Gramineae. 1. Vesicular–arbuscular endophytes, with special reference to the external phase. *Trans. Br. Mycol. Soc*., **42**, 421–38.

Phillips, J.M. and Hayman, D.S. (1970) Improved procedures for clearing roots and staining parasitic and vesicular–arbuscular mycorrhizal fungi for rapid assessment of infection. *Trans. Br. Mycol. Soc*., **55**, 158–61.

Ratanayake, M., Leonard, R.T. and Menge, J.A. (1978) Root exudation in relation to supply of phosphorus and its possible relevance to mycorrhizal formation. *New Phytol*., **81**, 543–52.

Rhodes, L.H. and Larsen, P.O. (1981) Effect of fungicides on mycorrhizal development of creeping bentgrass. *Plant Dis*., **65**(2), 145–7.

Ruehle, J.W. and Marx, D.H. (1979) Fiber, food, fuel and fungal symbionts. *Science*, **206**, 419–22.

Sanders, F.E. and Tinker, P.B. (1973) Phosphate flow into mycorrhizal roots. *Pesticide Sci*., **4**, 385.

Schenck, N.C. and Perez, Y. (1988) *Manual for the identification of VA mycorrhizal fungi*. INVAM, Univ. of Florida, Gainesville, 241 pp.

Sparling, G.P. and Tinker, P.B. (1975) Mycorrhizas in Pennine grassland. In *Endomycorrhizas* (eds F.E. Sanders, B. Mosse and P.B. Tinker), Academic Press, New York, 545–60.

Tinker, P.B. (1975) Soil chemistry of phosphorus and mycorrhizal effects on plant growth. In *Endomycorrhizas* eds. F.E. Sanders, B. Mosse and P.B. Tinker, Academic Press, New York, pp. 353–71.

Trappe, J.M. (1982) Synoptic keys to the genera and species of zygomycetous mycorrhizal fungi. In *Symp. Va Mycorrhizae* and *Plant Dis. Phytopathology*, **72**(8), 1102–8.

Trappe, J.M. and Schenck, N.C. (1982) Taxonomy of the fungi forming endomycorrhizae. A. Vesicular–arbuscular mycorrhizal fungi (Endogonales) in *Methods and Principles of Mycorrhizal Research* (ed N.C. Schenck), American Phytopathology Society, pp. 1–9.

13 *Fungal endophytes*

Many grasses, including some of the common turf-forming species are infected by fungal endophytes. These parasites occur as non-haustoriate, intercellular mycelia, systemic to varying degree within the tissues of leaves, stems, roots and seeds. The presence of endophytes in grasses and sedges has been known for many years (Freeman, 1903; McLennan, 1920; Neill, 1941; Neubauer and Remer, 1902; Sampson, 1933, 1937, 1939; Vogl, 1898). Of the species currently identified, most are clavicipitaceous fungi in the genera *Epichloe*, *Balansia*, *Balansiopsis* and *Atkinsonella*, all within the tribe Balansiae. Host plants infected with the various parasites may remain asymptomatic indefinitely, or are asymptomatic for most of the growing season until the intercellular mycelium emerges from the plant in localized areas. Superficial stromata are formed that bear the characteristic fruiting structures, first conidia and then perithecia. *Epichloe typhina*, the cause of choke disease (page 203) is the most familiar member of this fungal group that, until recently, had received little attention, largely because their economic impact was considered of minor importance (Rykard, Bacon and Luttrell, 1985). The situation has changed dramatically of late with the realization that serious toxicoses in sheep and cattle result when the animals graze on forage and weed grasses that are heavily infected with fungal endophytes. Many genera of warm-season perennial weed grasses, and tall fescue (*Festuca arundinaceae* Shreb), are now known to harbour endophytic parasites that can produce toxic ergot alkaloids (Bacon *et al.*, 1986) or, in the case of perennial ryegrass (*Lolium perenne* L.) a series of tremorgenic neurotoxins, named lolitrems (Gallagher, White and Mortimer, 1981; Gallagher *et al.*, 1982). Tall fescue and perennial ryegrass are extensively utilized pasture grasses and the respective endophyte-induced toxicoses, fescue toxicity syndrome (Bacon *et al.*, 1977) and ryegrass staggers (Fletcher and Harvey, 1981), are of major economic significance in the United States and New Zealand (Siegel *et al.*, 1984a; Siegel, Latch and Johnson, 1985). The infected grasses, both of which remain consistently asymptomatic, were considered initially to carry the asexual or *Sphacelia* state of *E. typhina* (Bacon *et al.*, 1977; Neill, 1941; Sampson, 1933). The endophyte in tall fescue is now referred to as *Acremonium coenophialum* (Morgan-Jones and Gams, 1982), and the related fungus in perennial ryegrass is designated *A. lolii* (Latch, Christensen and Samuels, 1984). The two *Acremonium* species produce conidia in artificial culture but sporulation in or on the host plant has not been documented and transmission of both endophytes is through the seed (Bacon *et al.*, 1977; Latch and Christensen, 1982). Artificial transmission of the fungi to their respective hosts, *A. lolii* to tall fescue, *A. coenophialum* to perennial ryegrass and both to *Festuca rubra* L., has been accomplished by inoculating meristem tissue of juvenile seedlings (Latch and Christensen, 1985) or by the infection of callus tissue (Johnson, Bush and Seigel, 1986; Torrello, personal communication). Mycelium of the endophytes can be detected within the leaf sheaths, stems and seeds by staining (Bacon *et al.*, 1977; Clark, White and Paterson, 1983; Welty, Azvedo and Cook, 1986) or by enzyme-linked immunosorbent assay (ELISA) (Johnson *et al.*, 1982). The presence of other endophytic fungi in these two grasses has been recorded in New Zealand (Latch, Christensen and Samuels, 1984). A *Gliocladium*-like species was isolated from perennial ryegrass and a *Phialophora*-like species was found in tall fescue. The New Zealand researchers also isolated *E. typhina* from *Festuca rubra* and the same fungus as a 'non-choke-inducing endophyte' (NCI) was found in high frequencies in seed lots of *Festuca rubra* in the United States (Funk *et al.*, 1983, 1985). White and Cole (1985, 1986) in a survey of forage grasses that included several fine-leaved *Festuca* species found what appeared to be the tall fescue endophyte, *A. coenophialum*, in *F. arizonica* Vasey and a different, as yet unnamed fungus, in *F. versuta* Beal. *Festuca ovina* L. and *F. idahoensis* Elmer two of the important sward components on western rangelands were free of endophytic fungi when sampled in this study. Health problems with grazing animals on fine-fescue swards have not been reported but the potential for endophyte-related toxicoses cannot be ignored (White and Cole, 1985).

Whereas toxin-producing endophytes in forage grasses present a real problem in the agricultural sector, the same fungi offer a novel approach to reducing losses from insect herbivores in turf situations. New Zealand scientists (Mortimer and DiMenna 1983; Prestidge, Pottinger and Barker, 1982) were the first to report that the endophyte *A. lolii* was associated with the resistance of perennial ryegrass to the Argentine stem weevil, *Listronotus bonariensis* (Kuschel). Tall fescue infected with *A. coenophialum* showed similar resistance to the weevil (Barker, Pottinger and Addison, 1983) and to the oat aphid *Rhopalosiphum padi* L. (Latch, Christensen and Gaynor, 1985). Studies in the United States provided substantiating evidence that endophyte presence in grasses could enhance their resistance to leaf and stem-feeding insects. When infected by *A. lolii*, perennial ryegrass was documented as having good resistance to billbugs (*Spenophorus* sp.) (Ahmad and Funk 1983; Funk *et al.*, 1985), to sod webworms (*Crambus* spp.) (Funk *et al.*, 1983) and to fall armyworm (*Spodoptera frugiperda* J.E. Smith) (Hardy, Clay and Hammond, 1985). Endophyte-infected cultivars of Chewings fescue (*F. rubra* L. subsp. *commutata* Gaud) and hard fescue (*F. longifolia* Thuill.) demonstrated improved summer performance and probable resistance to hairy chinch bug (*Blissus leucopterus hirtus* Montandon) (Funk *et al.*, 1985). Clay, Hardy and Hammond (1985) showed that, in addition to the *Acremonium* species in tall fescue and perennial ryegrass, members of three other endophytic genera (*Atkinsonella*, *Balansia* and *Myriogenospora*), present respectively in Texas wintergrass (*Stipa leucotricha* Trin. and Ruper.), sandbur (*Cenchrus echinatus* L.) and dallisgrass (*Paspalum dilatatum* Poir.) all deterred feeding by the fall armyworm. For most of the grasses, survival and weights of fall armyworm larvae fed infected leaves were significantly lower, and larval duration was significantly longer compared to larvae fed on uninfected leaves. The same authors used fall armyworm larvae in preferred feeding tests as a simple bioassay to detect the presence of endophyte in perennial ryegrass and suggested that the method could be widely adopted for screening pasture grasses suspected of causing cattle toxicoses (Hardy, Clay and Hammond, 1985). Similar assay techniques have utilized aphids (*Rhopalosiphum padi* and *Schizaphis graminum Rondani*) to detect the tall fescue endophyte (Johnson *et al.*, 1985; Latch, Christensen and Gaynor, 1985a), and house crickets (*Acheta domestica* L.), as the test insect for the ryegrass endophyte (Ahmad *et al.*, 1985). In these assays the test insects were unable to survive when confined on the appropriate endophyte-infected plants.

Efforts are taking place to characterize the chemical constituents generated by the grass/endophyte associations and to relate them to interactions between infected grasses and the feeding animal (Bacon *et al.*, 1986; Bush *et al.*, 1982; Gallagher, White and Mortimer, 1981; Gallagher *et al.*, 1984; Johnson *et al.*, 1985; Siegel, Latch and Johnson, 1985). Of particular note is the variety of compounds, some related, some totally unrelated that endophyte-infected grass species may produce and there is evidence that the compounds affecting insect feeding behaviour may not necessarily be the same as those associated with toxicoses of grazing animals. In endophyte-infected perennial ryegrass, a basic indole derivative, peramine, was isolated as the feeding deterrent to Argentine stem weevil rather than the chemically different, staggers-producing neurotoxic lolitrems. Thus the possibility may exist for producing an endophyte-infected ryegrass, resistant to stem weevil, but in which the principal toxin to livestock is absent (Rowan and Gaynor, 1986).

Whether or not the dilemma of endophyte infection can be resolved satisfactorily for the agricultural sector, in turf situations endophyte presence can confer valuable protection against the depredations of various insects. Already endophyte-containing cultivars of tall fescue, fine-leaved fescues and perennial ryegrass are being developed and marketed (Funk *et al.*, 1985; Siegel, Latch and Johnson, 1985). In addition to insect resistance it is claimed that endophyte-infected turf grass plants may display enhanced heat and drought tolerance, improved vigour and recuperative ability, increased sward density, persistence and competitiveness against weed invasion, enhanced stress tolerance and disease resistance (Hurley and Pompei, 1984). Supporting evidence exists for most of the benefits listed (Funk *et al.*, 1983, 1985; Hurley *et al.*, 1984; Latch, Hunt and Musgrave, 1985; but the relationship between endophyte and disease resistance remains largely unresolved. White and Cole (1986) reported that the growth of the turf grass pathogen *Rhizoctonia cerealis* was inhibited *in vitro* by culture filtrates of *A. coenophialum*, *A. lolii* and an unnamed endophyte of *Festuca versuta.* However, the presence of an endophyte in perennial ryegrass (*A. lolii*?) did not restrict pathogenesis and subsequent sporulation by *Drechslera dictyoides* on this host (Cromey and Cole, 1984). Spaced plants of perennial ryegrass maintained for three years in a Rhode Island turf nursery developed severe choke disease symptoms each year and were similarly affected by *D. dictyoides* (Jackson, unpublished).

If endophyte presence proves to be of little consequence as a resistance mechanism to the common turf diseases then fungicides will continue as the routine control measure and this raises the question as to the impact of fungicides on the endophytes in turf grasses. Systemic fungicides, especially the benzimidazoles and the ergosterol biosynthesis inhibitors, control the *Acremonium* endophytes in laboratory agar bioassays (Harvey, Fletcher and Emms, 1982; Siegel *et al.*, 1984a,b, 1985). The same materials, applied either as foliar treatments or as soil drenches, were used to eradicate the endophytes from plants growing in pots in the greenhouse (Latch and Christensen, 1982; Siegel *et al.*, 1984b). Such control has not been obtained under field conditions (Siegel, Latch and Johnson, 1985). A temporary reduction in the endophyte levels in tall fescue, which lasted for about one year, was afforded by the use of propiconazole or triadimefon (Williams *et al.*, 1984). Thus it seems unlikely that disease control programmes in managed turf would affect endophyte content adversely. Where eradication of the endophyte is desired then chemical or heat treatments of the infected seed are the most reliable methods (Latch and Christensen, 1982; Siegel *et al.* 1984, 1985).

13.1 REFERENCES

Ahmad, S. and Funk, C.R. (1983) Bluegrass bill bug (Coleoptera: Curculionidae) tolerance of ryegrass cultivars and selections. *J. Econ. Entomol.*, **76,** 414–16.

Ahmad, S., Govindarajan, S., Funk, C.R. and Johnson-Cicalese, J.M. (1985) Fatality of house crickets on perennial ryegrass infected with a fungal endophyte. *Entomol. Exp. Appl.*, **39,** 183–90.

Bacon, C.W., Porter, J.K. Robbins, J.D. and Luttrell, E.S. (1977) *Epichloe typhina* from toxic tall fescue grasses. *Appl. Environ. Microbiol.*, **34,** 576–81.

Bacon, C.W., Lyons, P.C., Porter, J.K., Robbins, J.D. and Luttrell, F.S. (1986) Ergot toxicity from endophyte-infected grasses: a review. *Agron. J.*, **78,** 106–16.

Barker, G.M., Pottinger, R.P. and Addison, P.J. (1983). Effect of tall fescue and ryegrass endophytes on Argentine stem weevil. *Proc. N.Z. Weed Pest Cont. Conf.*, **36,** 216–19.

Bush, L.P., Cornelius, P.L., Buckner, R.C., Chapman, R.A., Burrus, P.B., Kennedy, C.W., Jones, T.A. and Saunders, M.J. (1982) Association of *N*-acetyl loline and *N*-formyl loline with *Epichloe typhina* in tall fescue. *Crop Sci.*, **22,** 941–3.

Clark, E.M., White, J.F. and Patterson, R.M. (1983) Improved histochemical techniques for the detection of *Acremonium coenophialum* in tall fescue and methods of *in vitro* culture of the fungus. *J. Microbiol. Methods*, **1,** 149–55.

Clay, K., Hardy, T.N. and Hammond, A.M. Jr (1985) Fungal endophytes of grasses and their effects on an insect herbivore. *Oecologia*, **66,** 1–5.

Cromey, M.G. and Cole, A.L.J. (1984) An association between a *Lolium* endophyte and *Drechslera dictyoides*. *Trans. Br. Mycol. Soc.*, **83,** 159–61.

Fletcher, L.R. and Harvey, I.C. (1981) An association of a *Lolium* endophyte with ryegrass staggers. *NZ Vet. J.*, **29,** 185–6.

Freeman, E.M. (1903) The seed-fungus of *Lolium temulentum* L., the darnel, *Phil. Trans. Roy. Soc. London B*, **196,** 1–27.

Funk, C.R., Halisky, P.M., Johnson, M.C., Seigel, M.R., Stewart, A.V., Amhad, S., Hurley, R.H. and Harvey, I.C. (1983) An endophytic fungus and resistance to sod webworms: association in *Lolium perenne L. Bio/ Technology*, **1,** 189–91.

Funk, C.R., Halisky, P.M., Amhad, S. and Hurley, R.H. (1985) How endophytes modify turf grass performance and response to insect pests in turf grass breeding and evaluation trials. In *Proc. Fifth Int. Turf Res. Conf.*, (ed. Lemaire F.), Avignon, France, p. 137–45.

Gallagher, R.T., White, E.P. and Mortimer, P.H. (1981) Ryegrass staggers: isolation of potent neurotoxins Lolitrem A and Lolitrem B from staggers producing pastures. *NZ Vet. J*, **29,** 189–90.

Gallagher, R.T., Hawkes, A.D., Holland, P.T., McGaveston, D.A., Pansier, E.A. and Harvey, I.C. (1982) Ryegrass staggers: the presence of lolitrem neurotoxins in ryegrass seed. *NZ Vet. J.*, **30,** 183–4.

Gallagher, R.T., Hawkes, A.D., Steyn, P.S. and Vleggaar, R. (1984) Tremorgenic neurotoxins from perennial ryegrass causing ryegrass staggers disorder of livestock: structure elucidation of lolitrem B. *J. Chem. Soc. Chem. Commun.*, **1984,** 614–16.

Hardy, T.N., Clay, K. and Hammond, A.M. Jr (1985) Fall army worm (Lepidoptera: Noctuidae): a laboratory bioassay and larval preference study for the fungal endophyte of perennial ryegrass. *J. Econ. Entomol.*, **78,** 571–5.

Harvey, I.C., Fletcher, L.R. and Emms, L.M. (1982) Effect of several fungicides on the *Lolium* endophyte in ryegrass plants, seed and in culture. *N.Z.J. Agric. Res.*, **25,** 601–6.

Hurley, R.H. and Pompei, M. (1984) The turf manager's friendly fungus: endophyte. *Grounds Maintenance*, **19**(18), 16.

Hurley, R.H., Funk, C.R., Halisky, P.M., Saha, D.C. and Johnson-Cicalese, J.M. (1984) The role of endophyte enhanced performance in grass breeding. *Proc. 28th Grass Breeders Work Conf.* College Station, Texas.

Johnson, M.C., Dahlman, D.L., Seigel, M.R., Bush, L.P., Latch G.C.M., Potter, D.A. and Varney, D.R. (1985) Insect feeding deterrents in endophyte-infected tall fescue. *Appl. Environ. Microbiol.*, **49,** 568–71.

Johnson, M.C., Bush, L.P. and Seigel, M.R. (1986) Infection of tall fescue with *Acremonium coenophialum* by means of callus culture. *Plant Dis.*, **70,** 380–2.

Johnson, M.C., Pirone, T.P., Seigel, M.R. and Varney, D.R. (1982) Detection of *Epichloe typhina* in tall fescue by means of enzyme-linked immunosorbent assay. *Phytopathology*, **72,** 647–50.

Latch, G.C.M. and Christensen, M.J. (1982) Ryegrass endophyte, incidence and control. *N.Z. J. Agric. Res.*, **25,** 443–8.

Latch, G.C.M. and Christensen, M.J. (1985) Artificial

infection of grasses with endophytes. *Ann. Appl. Biol.*, **107,** 17–24.

Latch, G.C.M., Christensen, M.J. and Samuels, G.J. (1984) Five endophytes of *Lolium* and *Festuca* in New Zealand. *Mycotaxon*, **20,** 535–50.

Latch, G.C.M., Christensen, M.J. and Gaynor, D.L. (1985) Aphid detection of endophyte infection in tall fescue. *N.Z. J. Agric. Res.*, **28,** 129–32.

Latch, G.C.M., Hunt, W.F. and Musgrave, D.R. (1985) Endophytic fungi affect growth of perennial ryegrass. *N.Z. J Agri. Res.*, **28,** 165–8.

McLennan, E. (1920) The endophytic fungus of *Lolium*. Part 1. *Proc. Roy. Soc. Victoria* (*N.S.*), **32,** 252–301.

Morgan-Jones, G. and Gams, W. (1982) Notes on Hyphomycetes. XLI. An endophyte of *Festuca arundinaceae* and the anamorph of *Epichloe typhina*, New taxa in one of two new sections of *Acremonium. Mycotaxon*, **15,** 311–18.

Mortimer, P.H. and DiMenna, M.E. (1983) Ryegrass staggers: Further substantiation of a *Lolium* endophyte aetiology and the discovery of weevil resistance of ryegrass pastures infected with *Lolium* endophyte. *Proc. N.Z. Grassl. Assoc.*, **44,** 240–3.

Neill, J.C. (1941) The endophytes of *Lolium* and *Festuca. N.Z. J. Sci. Technol.*, **23A,** 185–95.

Neubauer, H. and Remer, C. (1902) Uber die von H Vogl entdeckte Pilzschicht in *Lolium* fruchten. *Zentralbl. Bakteriol. Abt.*, **22,** 652–3.

Prestidge, R.A., Pottinger, R.P. and Barker, G.M. (1982) An association of *Lolium* endophyte with ryegrass resistance to Argentine stem weevil. *Proc. 35th N.Z. Weed Pest. Contr. Conf.*, **35,** 119–22.

Rowan, D.D. and Gaynor, D.L. (1986) Isolation of feeding deterrents against Argentine stem weevil from ryegrass infected with the endophyte *Acremonium lolii. J. Chem. Ecol.*, **12,** 647–58.

Rykard, D.M., Bacon, C.W. and Luttrell, E.S. (1985) Host relations of *Myriogenospora atramentosa* and *Balansia epichloe* (Clavicipitaceae). *Phytopathology*, **75,** 950–6.

Sampson, K. (1933) The systemic infection of grasses by *Epichloe typhina* (Press) Tul. *Trans. Br. Mycol. Soc.*, **18,** 30–47.

Sampson, K. (1937) Further observations on the systemic infection of *Lolium. Trans. Br. Mycol. Soc.*, **21,** 84–97.

Sampson, K. (1939) Additional notes on the systemic infection of *Lolium. Trans. Br. Mycol. Soc.*, **23,** 316–19.

Siegel, M.R., Latch, G.C.M. and Johnson, M.C. (1985) *Acremonium* fungal endophytes of tall fescue and perennial ryegrass: significance and control. *Plant Dis.*, **69,** 179–83.

Siegel, M.R., Johnson, M.C., Varney, D.R., Nesmith, W.C., Buckner, R.C., Bush, L.P., Burrus, P.B., II, Jones, T.A. and Boling, J.A. (1984a) A fungal endophyte in tall fescue: incidence and dissemination. *Phytopathology*, **74,** 932–7.

Siegel, M.R., Varney, D.R., Johnson, M.C., Nesmith, W.C. Buckner, R.C. Bush, L.P., Burrus P.B., II and Hardison, J.R. (1984b) A fungal endophyte of tall fescue: Evaluation of control methods. *Phytopathology*, **74,** 937–41.

Vogl, A. (1898) Mehl und die auderen Mehlprodukte der Cerealien und Leguminosen. *Z. Nahrungs-mittel-Untersuchung. Hyg. Warenkunde*, **12,** 25–9.

Welty, R.E., Azevedo, M.D. and Cook, K.L. (1986) Detecting viable *Acremonium* endophytes in leaf sheaths and meristems of tall fescue and perennial ryegrass. *Plant Dis.*, **70,** 431–5.

White, J.F. and Cole, G.T. (1985) Endophyte host associations in forage grasses. 1. Distribution of fungal endophytes in some species of *Lolium* and *Festuca. Mycologia*, **77,** 323–7.

White, J.F. and Cole, G.T. (1986) Endophyte–host associations in forage grasses. IV. The endophyte of *Festuca versuta. Mycologia*, **78,** 102–7.

Williams, M.J., Backman, P.A. Crawford, M.A., Schmidt, S.P. and King, C.C. (1984) Chemical control of the tall fescue endophyte and its relationship to cattle performance. *NZ J. Exp. Agric.*, **12,** 165–71.

Part five
Rusts

14 *Rusts*

14.1 INTRODUCTION

All turf grass species are attacked by rusts and some of these may cause severe damage, especially on very susceptible grass species and cultivars. However, in cool-temperate regions, regularly mown turf of cool-season grasses is often attacked too late in the season to suffer severe injury. This is in contrast to the situation in forage or seed crops, in hay or seed stands, or in lightly grazed pasture, where mature aerial tissues may be devastated by rust attacks.

14.1.1 General symptoms

The first signs of rust attacks on grasses are the development of small chlorotic flecks on leaves, sheaths, culms and/or inflorescences. The location of the lesions is characteristic of particular species (see Table 14.1). Infection may take place on adaxial, abaxial or on both leaf surfaces. The initial lesions may then extend, mainly in a longitudinal direction, in rows parallel with the leaf veins. In black rust, *Puccinia graminis*, and yellow or stripe rust, *P. striiformis*, the lesions extend into stripes, but in some of the leaf rusts e.g. *P. poae-nemoralis*, or in attacks of rusts on resistant hosts the lesions may remain discrete. Sori of spores develop on the lesions, in some species erupting early through the host epidermis which forms a frill around the powdery spore contents. In other rusts the spores remain covered for a longer period. On grass hosts most rusts have two types of spores, but in warmer regions, e.g. tropics or subtropics only the urediniospores may develop. Urediniospores are yellow, orange, brown or reddish in colour and in a severe attack the whole turf may assume this colour. Shoes and implements disturbing the herbage become coated with spores. Severe attacks by rusts may result in chlorosis and wilting of the grass. Later, less conspicuous brown or black teliospores of the fungi may develop in the uredinia or in separate telia.

14.1.2 Rust fungi

There are about 120 genera with nearly 5000 species of rusts, obligate parasites within the order Uredinales of the Basidiomycetes, throughout the world. However, of these only 10 or 12 species may cause significant turf grass problems. The rusts show great variations in their life cycles with up to five different spore forms. When all the spore forms are found on one plant the rust is described as *autecious*; when they are on different hosts they are called *heteroecious*. Usually the two host species are quite unrelated. After more than two centuries of study there is still disagreement on the names of the spores and numbering of the spore stages. This disagreement relates to conflicting views on the importance of morphology and spore origin in their nomenclature (Hiratsuka and Sato, 1982). The following spore states will be used here:

Symbol	Spore state	Spore
0	pycnium	pycniospore
I	aecium	aeciospore
II	uredinium	urediniospore
III	telium	teliospore
IV	basidium	basidiospores

When all spore states are produced the rust is described as *macrocyclic*, although the aecial host may not be known, e.g. as in *Puccinia striiformis*. On the other hand, a *microcyclic* rust produces only telia or telia and pycnia. All turf grass rusts are *heteroecious*, that is, the teliospores do not reinfect the grass host, although the non-grass, aecial host may not be known. In some rusts the number of host species is large e.g. black rust, *P. graminis*, stripe rust, *P. striiformis* and crown rust, *P. coronata*, while others occur naturally on one or a few grass species only, e.g. *P. festucae* on *Festuca* spp. and *P. cynodontis* on *Cynodon* spp. Some rusts, e.g. *P. graminis*, *P. coronata* and *P. striiformis*, have *formae speciales* (f. spp.) confined to one species or group of host species. A *forma specialis* is rarely genetically homogeneous, but consists of physiological races, each of which carries genes controlling virulence towards particular host genotypes which differ widely in pathogenicity and are of different geographical origin. They may

Table 14.1 Some rusts of turf grasses – Summary

Common name	Pathogen	Principal turf grass hosts	Alternate hosts	*Uredinia*	*Telia*	References – sources of disease resistance
Crown rust	*Puccinia coronata* Cda. *sensu lato*	*Agropyron*, *Agrostis*, *Festuca* spp. *Holcus lanatus*, *Lolium perenne*, *Poa pratensis*	*Rhamnus cathartica*, *R. frangula* *Berchemia*, *Eleagnus* spp. and *Shepherdia* spp.	*Uredinia* amphigenous, scattered or in stripes. *Urediniospores* spherical to ellipsoidal, 14–39 × 10–35 μm with pale-yellow walls, abundantly and finely echinulate. *Germ pores* few, scattered (Fig. 14.7) *Paraphyses* thin-walled, swollen-headed, around sorus periphery.	*Telia* dark brown, in stripes, hypophyllous, long covered with epidermis. *Teliospores* dark brown, two-celled with a crown of finger-like processes up to 12 μm long at apex of upper cell, sometimes slightly waisted at septum, 13–60 × 14 – 20 μm. (Figs 14.2 and 14.7). *Pedicel* dark brown, short. *Germ pores* none.	Bourgouin *et al.*, 1974; Braverman, 1967, 1977; Bundessortenamt, 1979; Cagas, 1978, 1982; Conners, 1967; Eshed and Dinoor, 1981; Kopec *et al.*, 1983; Latch, 1966; Lancashire and Latch, 1970; Murdoch *et al.*, 1973; O'Rourke, 1976; RIVRO, 1980; Schmidt, 1980
Rust of fescues	*Puccinia crandallii* Pam. and Hume	*Festuca rubra* L. ssp. *rubra*, *F. rubra* spp. *commutata* *F. idahoensis* *F. ovina*, and *Poa* spp.	*Symphoricarpos* spp.	*Uredinia* adaxial, light brown. *Urediniospores* ellipsoid to nearly globose, cinnamon-brown, echinulate, mostly 30–37 × 24–28 μm. *Germ pores* many, large, scattered. *Paraphyses* none.	*Telia* adaxial, blackish-brown, early-exposed. *Teliospores* ellipsoid, slightly constricted at septum, chestnut-brown mostly 40–50 × 20–26 μm, wall up to 12 μm thick at apex. *Germ pores* apical and septal.	Cummins, 1971; Hardison, 1963,
Bermuda grass rust	*Puccinia cynodontis* Lacroix ex Desm.	*Cyndon* spp.	None knownin some countries, but with a wide host range on species in Euphorbiaceae, Plantaginaceae, Ranunculaceae, Saxifragaceae, Scrophulariaceae, Valerianaceae, Violaceae	*Uredinia* cinnamon-brown, epiphyllous *Urediniospores* globoid, 20–26 × 19–23 μm, finely verrucose, *Germ pores* 2–3, equatorial *Paraphyses* none (Fig. 14.13).	*Telia* mainly hypophyllous, early exposed, brown-black, powdery. *Teliospores* chestnut brown ellipsoidal; 28–60 × 15–25 μm, upper cell pointed with wall 1.5–2.5 μm thick, 6–12 μm thick at top. (Fig. 14.13). *Germ pores* apical and septal. *Pedicel* pale yellow, longer than spore.	Baltensperger, 1962; Braverman and Oakes, 1972; Mulder and Holiday, 1971; Vargas *et al.*, 1967
Rust of fine-leaved fescues	*Puccinia festucae* Plowr.	*Festuca rubra*, *Festuca ovina*	*Lonicera* spp.	*Uredinia* light brown, on chlorotic streaks, epiphyllous. *Urediniospores* globoid to ellipsoidal, 22–32 × 18–	*Telia* epiphyllous, dark brown early exposed; no paraphyses. *Teliospores* ellipsoidal to globoid, slightly waisted at septum, upper cell shorter and darker	Conners, 1967; Cummins, 1971; Jørstad, 1961; Wilson and Henderson, 1966;

				28 μm. Cell walls yellow-brown, finely echinulate. *Germ pores* scattered, many (Fig. 14.1). *Paraphyses* none	brown than lower with top drawn out to 1–6, finger-like processes, usually erect and up to 17 μm long, 50–80 × 17–21 μm. Cell walls light brown. *Germ pores* none (Fig. 14.1). *Pedicel* short, similar in colour to spore, persistent.	Woodcock and Clarke, 1983.
Black or stem rust	*Puccinia graminis* Pers.	*Agropyron*, *Agrostis*, *Lolium* and *Poa* spp., *Anthoxanthum odoratum*, *Festuca arundinacea*, *Phleum pratense*	*Berberis* spp.	*Uredinia* brown, in long stripes, frilled, mostly on sheaths and culms. *Urediniospores* ellipsoidal or ovate, 16–34 × 12–22 μm, walls yellow–brown, finely echinulate, *Germ pores* 3–4, equatorial (Fig. 14.4). *Paraphyses* none.	*Telia* similar to uredinia, but black. *Teliospores* club-shaped to ellipsoidal with rounder or blunt top, slightly waisted near septum, lower cell longer than upper, 27–68 × 12–25 μm. Cell walls brown, thickened to 12μm on top. *Germ pore* in upper cell near apex, in lower cell inconspicuous, near septum (Fig. 14.1). *Pedicel* long, bleached or colourless below.	Badahur *et al.*, 1972. Bourgouin, 1974; Bourgouin *et al.*, 1974; Braverman, 1966; Braverman and Oakes, 1972; Britton and Bulter, 1965; Britton and Cummins, 1959; Cagas, 1975; Conners, 1967; Endo, 1961; Funk *et al.*, 1981; Goss and Law, 1977; Johnson-Cicalese *et al.*, 1983; Jørstad, 1961; Meyer, 1982; Misra *et al.*, 1965; NE-57, 1977; Pfleger, 1973; Watkins *et al.*, 1981; Weibull, 1977; Bakker and Vos, 1975;
Brown fleck rust	*Puccinia poae-nemoralis* Otth. syn *P. poae-sudeticae* Jørst.	*Agrostis*, *Festuca*, *Poa* spp., *Anthoxanthum odoratum*	Most races persist by wintering uredinia with aecial hosts unknown, but *P. arrhenatheri* has aecia and pycnia on *Berberis vulgaris*	*Uredinia* orange – brown to cinnamon – brown, small, on small chlorotic spots, epiphyllous and on sheaths. *Urediniospores* globoid to ellipsoidal, 19–27 × 16–23 μm, walls pale, finely and closely echinulate. *Germ pores* 7–11, scattered, indistinct. *Paraphyses* capitate or clavate constricted and bent below the head (Fig. 14.10).	*Telia* mostly hypophyllous, covered with epidermis, dark brown, Paraphyses scanty. *Teliospores* ellipsoidal or clavate with blunt or rounded top, sometimes irregular chestnut brown, paler brown, 25–48 × 15–26μm. Apex of top cell thickened to 8 μm. *Germ pores* none. *Pedicel* Yellow–brown, short and persistent. *Mesospores* may be present. Teliospores are often not produced (Fig. 14.10).	Bundessortenamt, 1979; Conners, 1967; Funk *et al.*, 1981; Gjaerum and Andersen, 1983 Goss and Law, 1977; NE-57 Tech. Res. Comm. 1977; RIVRO, 1980; Smith, 1980; STRI, 1980; Svensson, 1978; Tronsmo, 1982

Table 14.1 Continued

Common name	Pathogen	Principal turf grass hosts	Alternate hosts	*Uredinia*	*Telia*	References–sources of disease resistance
Orange stripe rust	*Puccinia poarum* Niels.	*Poa* spp.	*Petastites* spp., *Tussilago farfara*, *Senecio* spp.	*Uredinia* often scanty, oval to oblong, orange–yellow, mainly epiphyllous, sometimes on sheaths, soon without epidermal covering, on necrotic spots. *Urediniospores* globose to ellipsoidal, 18–37 × 14–26 μm, walls pale yellow, finely warted. *Germ pores* 5–8, inconspicuous, scattered. *Paraphyses* typically none.	*Telia* amphigenous, oval or oblong, long covered with epidermis. *Teliospores* club-shaped to cylindrical, rounded or flattened on top with little or no restriction at septum, 30–77 × 14–28 μm. Cell walls brown, lighter towards the stalk, apex thickened to 6 μm. *Germ pores* none. *Pedicel* yellow–brown, short and persistent.	Bakker and Vos, 1975; Bourgouin, 1974; Bundessortenamt, 1979; Conners, 1967; Gjaerum and Weibull, 1983; Goss and Law, 1977; Greene and Cummins, 1967; RIVRO, 1980; STRI, 1980; Weibull, 1977;
Brown or leaf rust	*Puccinia recondita* Rob. ex Desm. *sensu lato* syn. *P. rubigo-vera* Wint. *sensu lato*	*Festuca arundinacea, F. rubra, Holcus lanatus, Poa annua, P. pratensis*	Many species in Boraginaceae, Ranunculaceae, Crassulaceae.	*Uredinia* light-brown, scattered, small, usually epiphyllous. *Urediniospores* globose to ellipsoidal, 20–36 × 16–28 μm, walls light brown, slightly thickened round germ pores. *Germ pores* scattered, many (Figs 14.5 and 14.9). Euchinulation fine, many. *Paraphyses* none (but see telia).	*Telia* brown – black, often in long stripes, long-covered with epidermis. Dark-brown paraphyses divide sori into sections. *Teliospores* usually 2-celled, occasionally 1, 3 or more, club-shaped, cylindrical with upper cell obtusely or obliquely pointed, slightly waisted at septum, 32–75 × 12–25 μm, wall brown, thickened to 7 μm at top of upper cell. *Germ pores* none (Figs 14.3 and 14.9). *Pedicel* persistent, pale, short.	Braverman, 1967; Conners, 1967;
St Augustine grass rust	*Puccinia stenotaphri* Cummins	*Stenotaphrum secundatum*	None known	*Uredinia* amphigenous, yelow to cinnamon-brown. *Urediniospores* oval to ellipsoid, mostly 30–40 × 25–28 μm, echinulate. *Germ pores* 4–5, equatorial. *Paraphyses* peripheral, thick-walled, cylindrical.	*Telia* blackish, long-covered. *Teliospores* chestnut-brown, club to oblong, mostly 44–60 × 19–26 μm, wall up to 5.5 μm thick at apex. *Germ pores* apparently none. *Pedicel* brownish, persistent. *Paraphyses* none.	Cummins, 1967 Lee and Toler, 1972; Watson, 1957.

Disease	Pathogen	Hosts	Alternate host	Uredinia	Telia	References
Yellow stripe rust	*Puccinia striiformis* Westened. syn. *P. glumarum* Erikss. and Henm.	*Agrostis* spp., *Lolium perenne* *Poa pratensis*	None known. Uredinia winter on winter-green foliage	*Uredinia* yellow, oval, usually epiphyllous, occasionally on culms and inflorescences, usually in long, chlorotic stripes. *Urediniospores* ellipsoidal, 20–34 × 12–26 µm with pale yelllow walls, densely echinulate. *Germ pores* many, scattered. *Paraphyses* saccate, ephemeral.	*Telia* brown – black, in long stripes on leaf blades and sheaths, long covered with epidermis with dark brown paraphyses dividing the sori into compartments. *Teliospores* flattened, club-shaped, club-shaped or flattened on top, waisted near septum, 30–70 × 12–26 µm. Top of upper cell thickened to 5 µm. Occasionally there are 3–4, but usually 2 cells. *Germ pores* none. *Pedicel* pale to deep yellow.	Braverman, 1967; Britton and Cummins, 1956; Conners, 1967; Gaskin and Nunez, 1977; Goss and Law, 1977; Harivandi and Gibeault, 1980; Meyer, 1977, 1982; Murdoch *et al.*, 1973a; Tollenaar and Houston, 1967; Weibull, 1977.
Zoysia rust	*Puccinia zoysiae* Diet.	*Zoysia japonica*, *Z. matrella*, *Z. tenuifolia*	*Paederia* sp.	*Uredinia* epiphyllous, bright-yellow when fresh, paler when dry. *Urediniospore* obovate or ellipsoidal, 17–22 × 15–18 µm may have walls thickened to 8 µm apically. *Germ pores* many, indistinct, scattered. *Paraphyses* typically none.	*Telia* dark brown, early exposed, amphigenous. *Teliospores* dark brown, 28–42 × 15–14 µm with top wall of upper cell thickened to 7 µm. *Germ pores* apical and septal. *Pedicel* thick-walled, persistent, to ca 100 µm long.	Freeman, 1965; Gudauskas and McCarter, 1966; Haygood and Spenser, 1979; Juska and Kreitlow, 1972; Kazelnicky and Garrett, 1966; Kreitlow *et al.*, 1965.
Leaf rust	*Uromyces dactylidis* Otth syn. *U. poae* Rabh. ex Marcucci	Agrostis spp., *Cynosurus cristatus, Festuca rubra, Poa annua, P. pratensis, P. trivialis*	*Ranunculus* spp.	*Uredinia* amphigenous, in stripes, yellow – brown. *Urediniospores* globoid to ellipsoidal, 20–34 × 15–18 µm. Cell walls pale yellow–brown, finely echinulate (Fig. 14.14). *Germ pores* many, scattered. *Paraphyses* none.	Telia amphigenous in immersed stripes. *Teliospores* first formed in old uredinia, then in special telia, long covered with epidermis, brownish-black with brown paraphyses which divide telia into compartments. Teliospores are 1-celled, ellipsoidal, often irregular, rounded, flattened or lop-sided on top, 14–34 × 12–24 µm, thickened on top to 3 µm (Fig. 14.14). *Germ pores* none. *Pedicel* pale, as long as spore, collapsing.	Conners, 1967; Jørstad, 1961.

compete with each other, hybridize or mutate to give new races with different virulence from their parents on cultivars or strains within grass species.

Only the principal rusts of cool-season turf grasses will be dealt with here (Table 14.1). For amplification of the morphological and taxonomic information consult papers or texts by Arthur and Cummins (1962), Anikster and Wahl (1979), Blumer (1963), Cummins (1971), Cummins and Hiratsuka (1983), Gaumann (1959), Gjaerum (1974), Grove (1913), Henderson and Bennell (1979), Hiratsuka and Sato (1982), Ullrich (1977) and Wilson and Henderson (1966). For details of distribution of rusts on cool-season grasses used in turf consult regional lists. References are given in Table 14.1 to sources of disease resistance. Many of the grass rusts have several synonymous specific or subspecific names and this can lead to confusion when comparing North American and European information. Some of the more common synonyms are given.

14.1.3 Life cycles of grass rusts

As an example, in the heteroecious rust, *Puccinia graminis*, pycnia are produced on haploid mycelium, usually in the upper leaf-surface tissues of an alternate host, usually a *Berberis* sp., in late spring or early summer. In the pycnia small, single-celled, hyaline spores, haploid spermatia, are produced in basipetal succession. They function as gametes, attaching to compatible flexuous hyphae that function as receptive hyphae or trichogynes, of aecial primordia. Following dikaryotization of the mycelium, an aecium, bell-shaped in section, is formed, usually in the tissues of the lower leaf surface of the alternate host. Inside this structure, single-celled, dikaryotic aeciospores are produced in chains. Aeciospores are carried to grasses in the vicinity by the wind, germinate, gain entry to the host via stomata, and brown, linear uredinia containing single-celled, dikaryotic urediniospores are produced. The latter are the main means of local and long-distance dissemination of the rust. They also infect the primary host via stomata. Later in the season two-celled teliospores are produced in the urediniosori or in the dark brown or black, linear teliosori which, like the urediniosori, occur mainly on the grass culms in *P. graminis*. Teliospores are the overwintering spores of *P. graminis*. They are at first dikaryotic then diploid. In spring they germinate and produce haploid basidiospores of two mating types on 4-celled basidia and these are carried to the alternate host by the wind, germinate and penetrate through the cuticle of the upper leaf surface and initiate haploid pycnia and aecial primordia. There are many variations in life cycles in different rusts. For example, the life cycle of *P. striiformis* is completed on grass hosts and so is that of the *formae speciales* of *P. brachypodii* var. *poae-nemoralis*, except that of the form on *Arrhenatherum* spp. which has an alternate host on *Berberis vulgaris*.

14.2 CROWN RUST – *PUCCINIA CORONATA*

The name of this rust is derived from the 'crown' of processes on the top cell of the teliospores (Figs 14.2 and 14.7). However, it should be noted that *P. festucae* and a

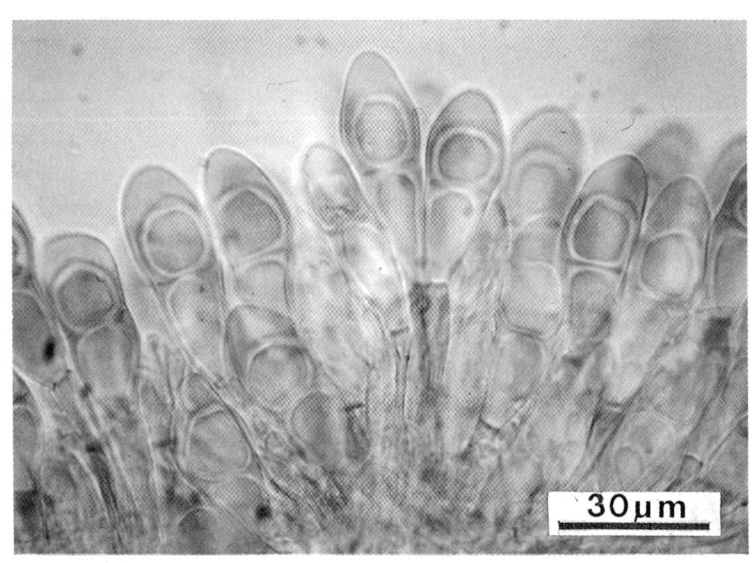

Figure 14.1 *Puccinia graminis* teliospores.

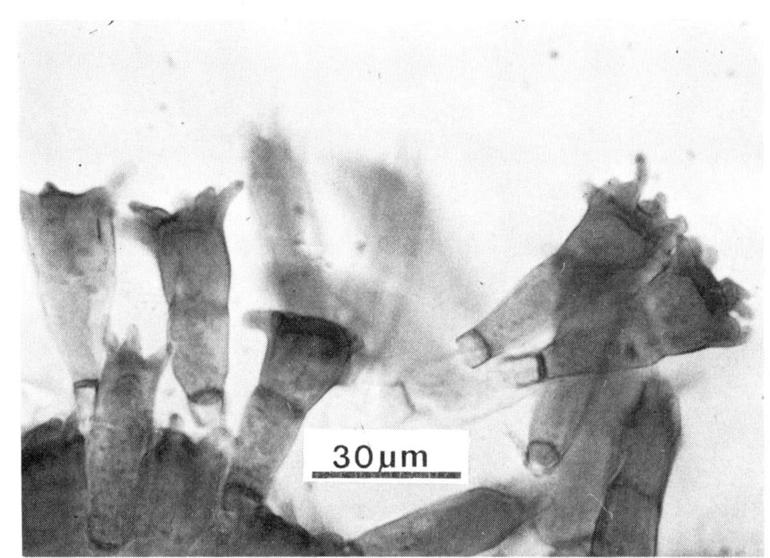

Figure 14.2 *Puccinia coronata* teliospores.

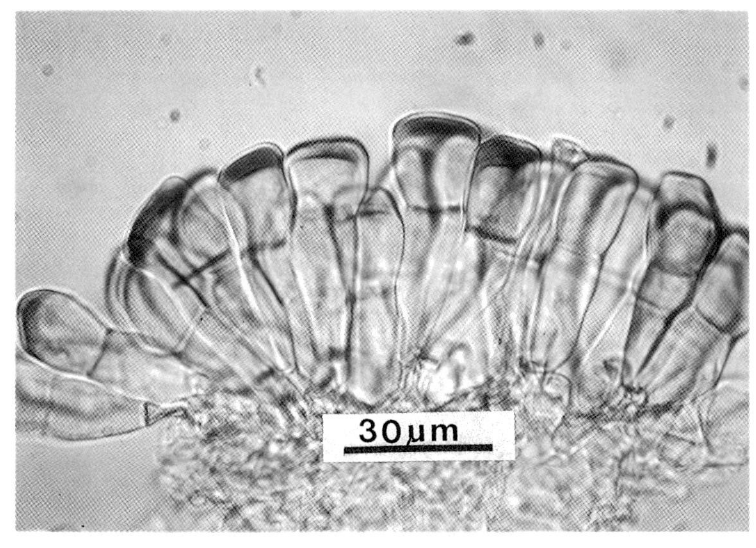

Figure 14.3 *Puccinia recondita* teliospores.

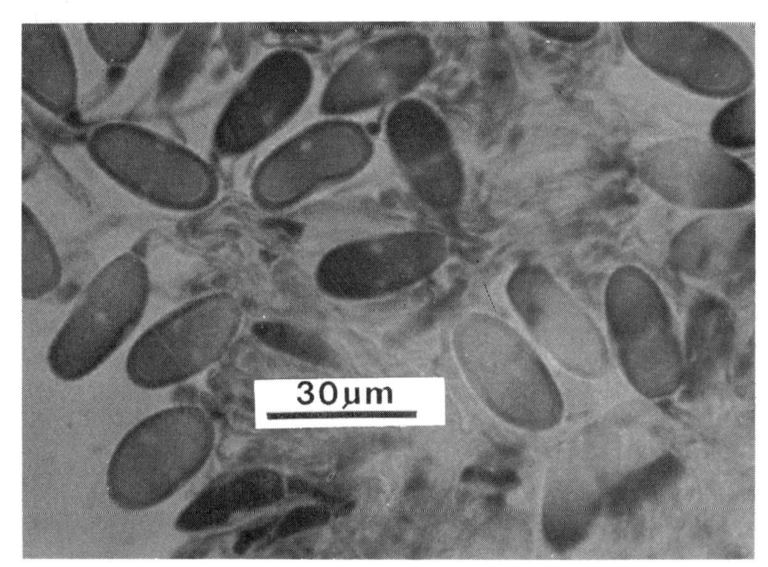

Figure 14.4 *Puccinia graminis* urediniospores stained to show equatorial germ pores.

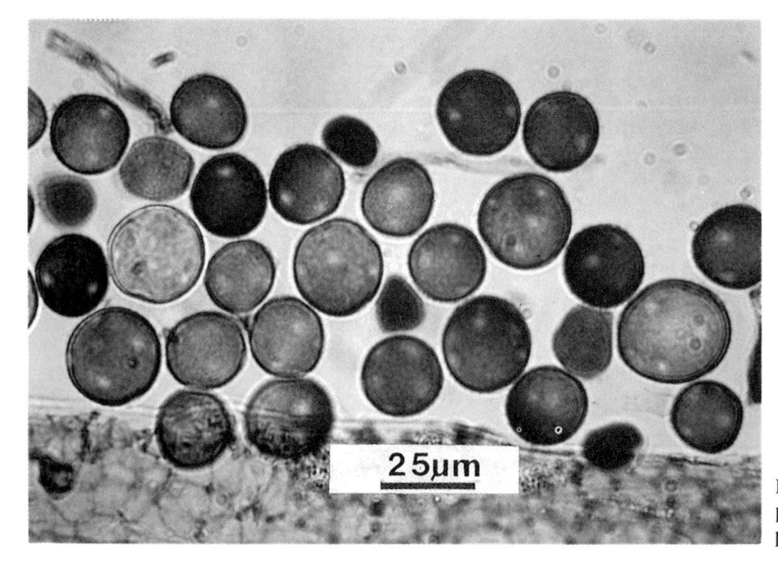

Figure 14.5 *Puccinia recondita* urediniospores stained to show scattered germ pores.

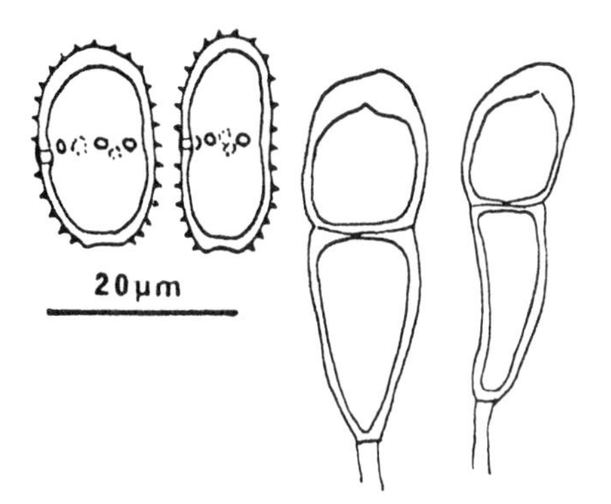

Figure 14.6 *Puccinia graminis*. (Urediniospores have equatorial germ pores.)

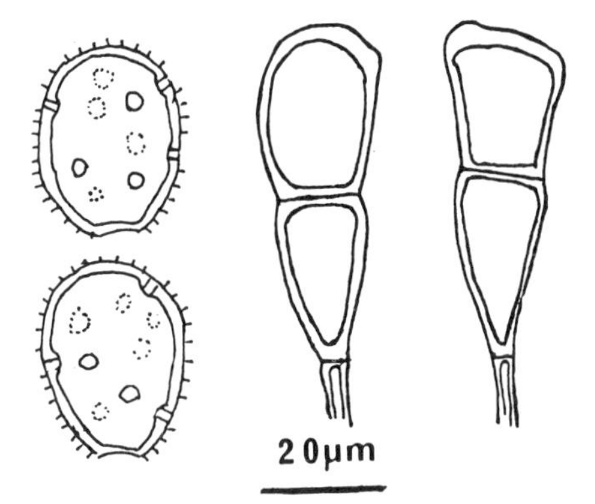

Figure 14.8 *Puccinia striiformis*.

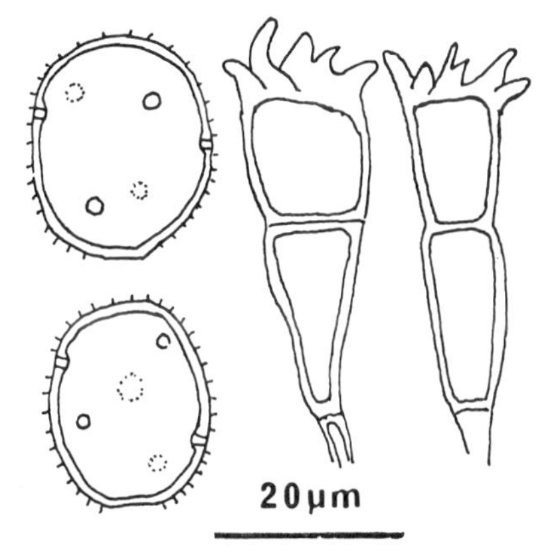

Figure 14.7 *Puccinia coronata*. (Teliospores with apical processes.)

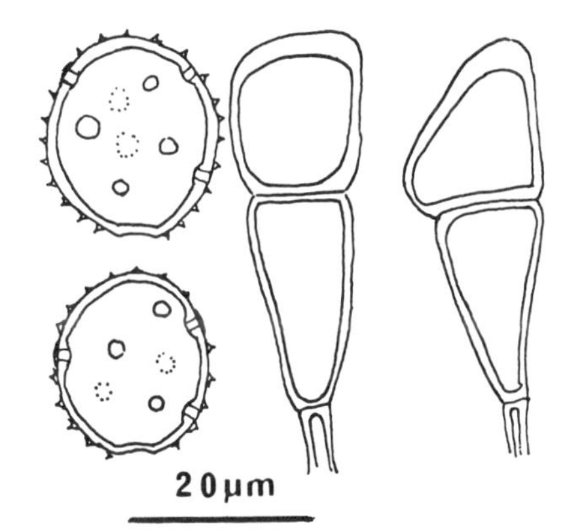

Figure 14.9 *Puccinnia recondita*.

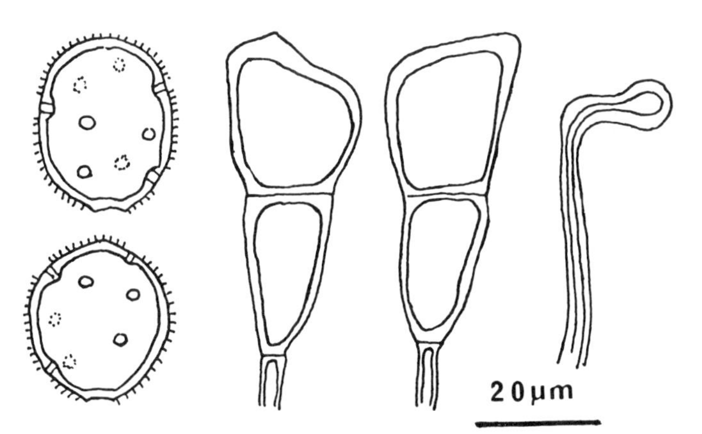

Figure 14.10 *Puccinia poae-nemoralis* (right, bent paraphysis).

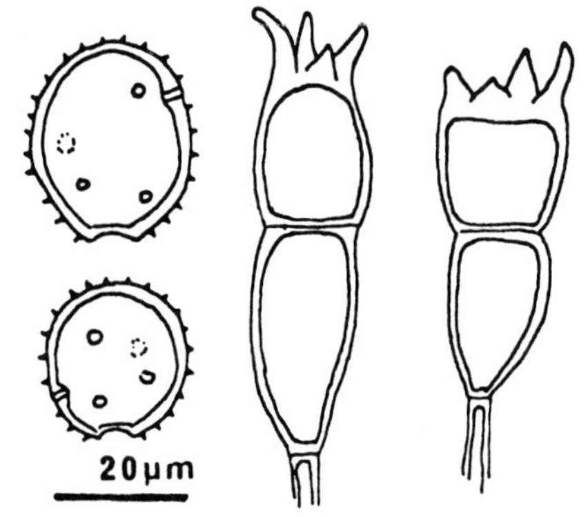

Figure 14.11 *Puccinia festucae* (Teliospores with apical processes).

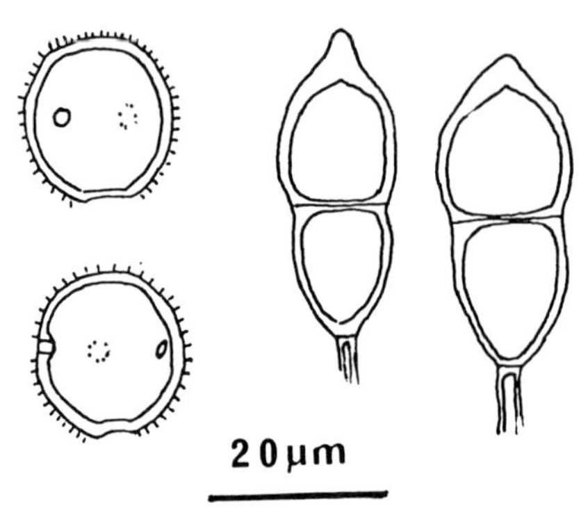

Figure 14.13 *Puccinia cynodontis*. (Teliospores with pointed apical cell).

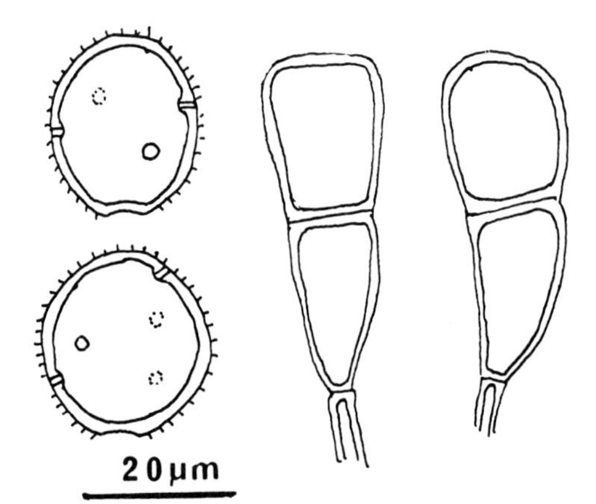

Figure 14.12 *Puccinia poarum*.

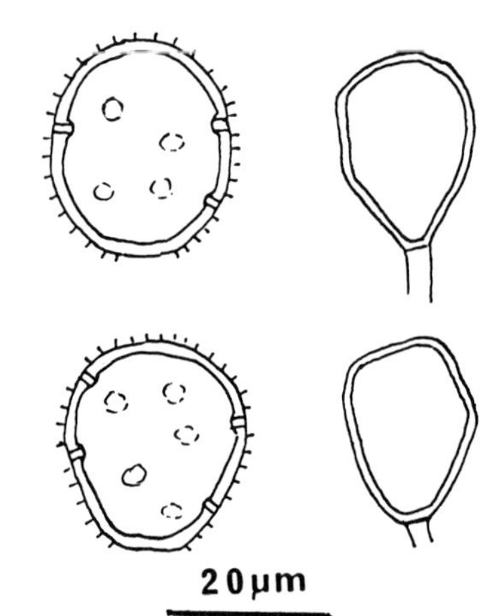

Figure 14.14 *Uromyces dactylidis*. (Teliospores are 1-celled).

few other *Puccinia* spp. (Cummins, 1971) also have a similar arrangement of processes (Fig. 14.11). Scattered, bright-orange uredinia occur on both adaxial and abaxial leaf surfaces of grasses and may be produced all winter in mild climates. Telia, aecia and pycnia may be absent as in New Zealand (Latch, 1966). When present, telia occur on abaxial leaf surfaces alongside uredinia, but are linear and black. Both uredinia and telia may be abundant on the major turf grass hosts, *Lolium perenne*, *L. multiflorum* and *Festuca arundinacea* and the rust is common and often devastating world-wide where these species are grown (Braverman, 1967; Kopec, Funk and Halisky, 1983; Kreitlow and Meyers, 1947; Latch, 1966; Nyquist, 1962; Plumb, Jenkyn and Bowen, 1976; Smith, 1965). The pycnial and aecial stages have been found on species in genera of the Rhamnaceae and Eleagnaceae (Simons, 1970). *P. coronata* has a very wide host range, most of the genera being in the Pooideae which includes all the important cool-season turf grass species (Braverman, 1967; Cagas, 1978; Cummins, 1971; Eshed and Dinoor, 1981; Fischer and Levine, 1941; Fraser and Ledingham, 1933; Grove, 1913; Mühle, 1971; Sampson and Western, 1954; Wilson and Henderson, 1966). There have been and still are many uncertainties and confusion in the taxonomy and specialization on hosts of *P. coronata* (Cummins, 1971; Simons, 1970). Neither the use of morphological characteristics nor host reactions have allowed satisfactory groupings of strains of the fungus (Brown, 1937; Eshed and Dinoor, 1981; Fraser and Ledingham, 1933). It seems to be a matter of opinion whether in the trinomial the category of variety or *forma specialis* is used. Below this level, there are many races or forms which react in very variable ways with grass hosts.

14.2.1 Disease development

P. coronata, like *P. graminis*, is a macrocylic, heteroecious rust. The importance of the alternate hosts, commonly *Rhamnus* spp., in the occurrence of the disease varies considerably in different regions, seasons and with the proximity of the alternate host to the grass host (Simons, 1970). *Rhamnus* is present in Britain, but is unimportant. The rust overwinters either as teliospores (Griffiths, 1953) or as urediniospores and mycelium (Sampson and Western, 1954). In Ireland, although urediniospores and mycelium are the main source of spring infection of grasses, *Rhamnus* species are considered important for genetic recombination within the fungus (O'Rourke, 1976). In eastern Canada a few *Rhamnus* bushes were found able to start epidemics of certain races over a large area, but the other races present were probably blown in from the United States where they were predominant (Conners and Savile, 1943, 1952). Craigie (1939) showed that urediniospores of *P. coronata* could be carried great distances by the wind to western Canada. The movement of urediniospores from the southern states of the USA to the north occurs in stages. In New Zealand, where crown rust is the major disease of perennial ryegrass (Latch, 1966) the aecial state is unknown (Cunningham, 1931) and the telia do not occur on ryegrass, so the fungus persists as urediniospores and mycelium.

Direct contact with water drops is necessary for urediniospore germination with cardinal temperatures of 0–2°C minimum, optima between 18 and 22°C and maximum 35°C (Durrell, 1918; Hemmi and Abe, 1933). Leaf penetration by urediniospores germ tubes (in oats) occurred after 3 hours at 20–23°C, but no infection was noted below 4°C or above 31°C (Marland, 1941). Epidemics are favoured by hot autumn or spring weather, favouring spore dispersal, with persistent dews suitable for spore germination and infection (see *P. graminis* and *P. striiformis*).

14.2.2 Host resistance

In *L. perenne*, which is a highly heterozygous, open-pollinated species, strains have been developed with 'field' resistance by selecting and combining resistant plants selected where several natural infections with *P. coronata* f. sp. *lolii* were regularly experienced (Gibbs, 1966). Within cultivars similarly, the proportion of resistant plants may be raised by one cycle of selection (Wilkins and Carr, 1971). Although major, individual genes giving a high degree of vertical resistance have been reported (K.J. McVeigh, 1975; see Hayward, 1977), polygenic (or horizontal) resistance which is inherited quantitatively, is considered to be more important because major gene resistance may be eroded (Wilkins, 1978). This is a well-known danger in inbred grasses where pathogens, such as *P. coronata*, may evolve new, more virulent races, overcoming host resistance (Kopec, Funk and Halisky, 1983; Wilkins, 1978b). Great variation has been found in cultivars and selections of forage and turf-type *L. perenne* in resistance to *P. coronata* by Kopec, Funk and Halisky (1983), providing the basis for germplasm selection for horizontal and vertical resistance. The cultivars Elka, Gator, Loretta, Prelude, Premier, Fiesta, Delray, Pennant and Palmer showed the best resistance in turf grass trials in New Jersey (Funk, 1983).

In the United States, most *Festuca pratensis* Huds. collections were found susceptible, while those of *F. arundinacea* Schreb., a species used as a turf grass in cool- to warm-humid regions, were resistant to *P. coronata* (Kreitlow and Myers, 1947). Two biotypes of *P. coronata* var. *festucae*, with some differences in the size and shape of the teliospores were recognized in Wales on *F. pratensis* and *F. arundinacea*, respectively by Wilkins and Carr (1971). In Czechoslovakia, Cagaš (1978) noted the high resistance of *F. arundinacea* to *P. coronata* var. *coronata* compared with *F. pratensis*. In Switzerland, Schmidt (1980) found that *P. coronata* populations on *F. pratensis* and *F. arundinacea* were very specialized on their host–plant species. From 37 lines of *F. arundinacea* tested, only one Swiss ecotype contained plants susceptible to rust from *F. pratensis*, but *F. pratensis* was even more resistant to the crown rust from *F. arundinacea*. Resistance in *F. arundinacea* to *P. coronata* is transferable to intergeneric hybrids with *L. multiflorum* (Wilkins, Carr and Lewis, 1974) and to *F. pratensis* (Cagaš, 1982).

14.2.3 Control

See pp. 174–175.

14.3 CYNODON RUST–*PUCCINIA CYNODONTIS*

The small, scattered or congregated, cinnamon-brown uredinia of this rust develop mainly on the abaxial leaf surfaces of the grass host, sometimes on purplish lesions. The telia are also mainly abaxial, early exposed, brownish-black and powdery. The disease rarely kills bermuda grass plants, but may cause severe leaf damage, especially in complexes with other leaf spot pathogens such as *Cochliobolus cynodontis* and *Drechslera gigantea* (Litzenberger and Stevenson, 1957; Vargas, Young and Saari, 1967). The rust has a world-wide distribution in warm-temperate to tropical regions (Cummins, 1971; McNabb, 1962; Mulder and Holliday, 1971).

14.3.1 Disease development

P. cynodontis is heteroecious with a plurivorous aecial state (Guyot and Viennot-Bourgin, 1943; Jørstad, 1958), but the latter may not be present or is not necessary for rust survival over the long dormant period common to its bermuda grass host. The thick-walled urediniospores are long-lived and quite able to remain viable for the duration of the host's dormancy (McNabb, 1962; Vargas, Young and Saari, 1967). Urediniospores germinate well over a wide range of temperatures (optimum 20°C), but high light intensities, particularly greater than 200 ft-c (lumen ft^{-2}), reduce germination. At least 10h of leaf wetness and a low light intensity are necessary for heavy infection, especially early in the leaf-wetting period. Light intensities above 200 ft-c reduced infection and increased the necessary duration of the leaf-wetness period (Vargas, Young and Sarri, 1967). Four races of the rust were found in collections made in Oklahoma, being identified on three clones of *Cynodon dactylon*.

14.3.2 Host resistance

C. dactylon strains in commercial use in the United States appeared less susceptible than *C. magenisii* or *C. transvaalensis*, but there is considerable variation in disease reaction (Baltensperger, 1962; Braverman and Oakes, 1972; Vargas, Young and Saari, 1967.)

14.4 RUST OF FINE-LEAVED FESCUES–*PUCCINIA FESTUCAE*

The uredinia of this rust are small, scattered, inconspicuous and yellow–brown on light-coloured leaf flecks on adaxial leaf surfaces. The telia are small, elongate, dark-brown, early exposed and powdery, also on adaxial leaf surfaces. Although not usually common it has been found on fine-leaved *Festuca* spp. in north-western and northern Europe, northern Asia, Alaska and Australia. The alternate hosts are *Lonicera* spp. (Conners, 1967; Cummins, 1971; Gjaerum, 1974; McAlpine, 1906; Wilson and Henderson, 1966; Woodcock and Clarke, 1983). There is little information on disease development and host resistance. It is confined to the fine-leaved fescues.

14.5 BLACK OR STEM RUST–*PUCCINIA GRAMINIS*

The prominent, powdery, black telia of this rust and also the brown uredinia occur mainly on the culms of grasses, particularly in seed crops, hence the two common names. In mown turf the uredinial state is most abundant on the upper leaves. In a severe attack it may be so abundant that the turf assumes a brown tinge. In dry, hot weather infected grasses may shrivel and collapse.

P. graminis in the inclusive sense attacks many grasses in the subfamily Pooideae. Urban (1967) revised the classification of *P. graminis*, mainly on the basis of urediniospore length into two subspecies and two varieties of one of the latter. They are *P. graminis* ssp. *graminis* var. *graminis*, *P. graminis* ssp. *graminis*-var. *stakmanii* and *P. graminis* ssp. *graminicola*. *P. graminis* ssp. *graminicola* is the cause of almost all infection on turf grasses (Savile, 1981; Savile and Urban, 1982; D.B.O. Savile, personal communication (Fig. 18, colour plate section).

In Britain black rust is of minor importance on cultivated grasses although often found on *Agrostis* spp. (Batts, 1951a,b; Sampson and Western, 1954; Wilson and Henderson, 1966). It has a higher temperature optimum than *P. striiformis* (Zadoks, 1965). In warmer climates, e.g. in N. America, it may cause severe damage to some *Poa pratensis* cultivars (Britton and Cummins, 1959; Britton and Butler, 1965; Conners, 1967; Pfleger, 1973). It is of minor importance in summer and early autumn in France, south of Paris on *Festuca rubra* and *Lolium perenne* (Bourgoin, 1974). It is rarely found on *Poa pratensis* in Sweden (Weibull, 1977). In Czechoslovakia, where *Phleum pratense* may be severely damaged by black rust, *Poa annua* was found to be a conspicuous permanent infection source (Cagaš, 1975). It has long been regarded as an important pathogen on *Phleum* spp. in the United States (Braverman, 1966, 1967).

14.5.1 Disease development

The alternate hosts for *P. graminis, Berberis* spp. may be absent where the rust occurs on grasses. The non-grass host is probably most important for the development of new physiological races by hybridization. These hybrids may be capable of overcoming the genetic resistance of previously resistant grass species and cultivars. However, there are other possible means for the production of new races in the absence of the alternate host, by mutation, heterokaryosis or parasexuality which are non-sexual mechanisms (Webster, 1974).

Teliospores are usually uncommon on mown grass, but when present, as in seed crops, the complete life cycle may be completed if the alternate host is in close proximity*. In warm climates, e.g. in California or as far north as Indiana in the United States, or some temperate climates, such as most of Norway (H.B. Gjaerum, personal communication), the fungus survives as urediniospores or as dormant mycelium in living leaves (Britton, 1958; Endo, 1961). Further north in North America the primary inoculum for grasses, as for cereals, is likely to be urediniospores, wind-borne from the south. In colder climates, epidemics may be delayed until late summer, when grass growth has declined. Even when the alternate host is present, its role in the development of epidemics may be unimportant. For example, it has been shown that *P. graminis* spp. *graminis* var. *graminis* (of wheat) does not overwinter in Britain. When epidemics of black rust occur there, which is rarely, the primary inoculum is of urediniospores carried in on winds from Continental Europe and the Western Mediterranean (Ogilvie and Thorpe, 1966; Zadoks, 1965). However, the sources of primary inoculum for races on grasses where the fungus is killed in winter, and an alternate host is absent, is uncertain.

Optimum temperatures for urediniospore germination (15–24°C), appressorium formation (16–26°C), penetration and vesicle formation c 29°C, in the substomatal cavity for *P. graminis* (on wheat) are higher than that for *P. striiformis* generally (Zadoks, 1965). The urediniospores of *P. graminis* are less sensitive to the lethal effects of sunlight than those of *P. striiformis* which has an important bearing on the viability of primary inoculum airborne for long distances (Maddison and Manners, 1972).

Climatic conditions, favourable for the initiation of epidemics of black rust, are found in bright, cloudless weather with high radiation at night, resulting in dew which persists for several hours on the grass after sunrise. Under these conditions, low light intensities (300 foot candles), favourable temperatures for germination and appressorium formation permit rapid penetration of the host via stomata. Gradual reduction in leaf surface wetness and higher light intensities (over 500 foot candles) allow the most rapid penetration of host cell walls (Givan and Bromfield, 1964; Rowell, Olien and Wilcoxon, 1958; Sharp *et al.*, 1958; Zadoks, 1965).

*At Ottawa, Canada *P. graminis* ssp. *graminicola* was found on *Agropyron repens* only near *Berberis* sp. (D.B.O. Savile, personal communication).

14.5.2 Host resistance

Although many cool- and warm-season turf grass species are susceptible in differing degrees to the *formae speciales* (or subspecies and varieties) and physiological races of *P. graminis*, in mown turf it is the susceptibility of *P. pratensis* cultivars which causes most concern to turf managers. This is particularly the case in North America where this species is extensively used in lawn-type turfs in the northern parts of the United States and in Canada. Particular attention was drawn to the problem of stem rust resistance in *P. pratensis* because of the extensive use of the Merion cultivar after its introduction in about 1952. Although it was superior in many ways to the land race sorts then available it proved extremely susceptible to *P. graminis* (Ray, 1953; Rogerson and King, 1954). Ranges of infection response from susceptible to immune have been noted in *P. pratensis* cultivars (Britton and Butler, 1965; Elliott, 1963; Pfleger, 1973).

Older cultivars, Newport, Delta, Belturf and Park showed high to moderate resistance while Merion was very susceptible (Pfleger, 1973). Highly resistant accessions in the 57 lines tested showed fewer and smaller uredinia and on resistant lines these were surrounded by necrotic leaf tissue. In the Northeast, Midwest and southern California regions of the United States, Merion, Birka and Touchdown cultivars of *P. pratensis* may be severely damaged by *P. graminis*, especially when soil fertility is low (Endo, 1961; Funk *et al.*, 1978; Meyer, 1982). In recent tests in New Jersey, HV62, Midnight, Bristol, Ba 72–500 and Talent were consistently highly resistant, while Merion, Mystic, Birka and Touchdown were among the most consistently susceptible (Johnson-Cicalese, Clarke and Funk, 1983). In the northwestern United States, Kenblue, Newport, Parade, Park and Rugby were found resistant in mown turf (Goss and Law, 1977). Differences in response of *P. pratensis* cultivars suggests that different races of *P. graminis* may be involved in the different regions and points the need for regional testing (Britton and Cummins, 1959; Britton and Butler, 1965; Johnson-Cicalese, Clarke and Funk, 1983; Meyer, 1982; Watkins *et al.*, 1981).

Commonly, polystands are established from blends of different cultivars of *P. pratensis* to improve their adaptability to local edaphic, climatic and management conditions. They are also used in attempts to take advantage of differential resistance of the component cultivars to diseases and insects. Watkins *et al.* (1981) found that stem rust resistance in blends usually approximated to the mean rust severity of the different cultivars when grown as monostands.

P. graminis may severely damage seed crops of *Lolium perenne* in western Oregon, reducing yields in some cases by more than 90%. Synthetics developed from crosses of resistant clones from old turf areas and turf-type cultivars have shown suitable agronomic characteristics and rust resistance, the latter controlled by multiple genes (Meyer, 1982). *P. graminis* may also cause damage on seed crops of *Phleum pratense*, occasionally used in amenity turf. Resistant lines were identified (Braverman, 1966; Nielsen and Dickson, 1958), but the appearance of new and aggressive races of the fungus has made breeding for disease resistance more problematical (Cagaš, 1975).

14.5.3 Control

See pp. 174–175.

14.6 BROWN FLECK RUST–*PUCCINIA POAE-NEMORALIS*

The yellow, later yellow–brown or reddish-brown uredinia of this rust occur mainly on adaxial leaf surfaces, but also develop on leaf sheaths. Initially they may be scattered, round or oval, on chlorotic and later, brown spots of leaf tissue, but in severe attacks they develop in dense groups and may be accompanied by leaf wilting. In the uredinia and surrounding them are the characteristic, thick-walled, capitate, usually bent paraphyses. Dark brown telia, long covered with epidermis are sometimes not produced (Jørstad, 1961; McNabb, 1962; Weibull, 1977; Wilson and Henderson, 1966). Both urediniospores and teliospores are difficult to distinguish from those of *P. poarum*, but separation of the two species may be made readily on the basis of paraphysis characters (see below). (Fig. 19, colour plate section).

14.6.1 Disease development

P. poae-nemoralis sensu *lato* is a polymorphic rust predominantly parasitic on *Poa* spp. often in cool-temperate to arctic regions. Heteroecism is not obligatory in most races and the fungus can overwinter as mycelium or urediniospores (Hardison, 1963; McNabb, 1962; Wilson and Henderson, 1966). This rust and *Puccinia poarum* Niels. are widespread and are the most important rusts on *Poa* spp. in western Europe, especially on *P. pratensis* (Bakker and Vos, 1975; Gjaerum and Andersen, 1983; Gjaerum and Weibull, 1983; O'Rourke, 1976) and further east (Frauenstein, 1970; Mühle, 1971; Teuteberg; 1974). In southern Scandinavia, heavy attacks may occur in July, but further north these may take place in August–September. Increasing temperatures inland favour heavier attacks compared with the coast where lower temperatures are the norm (Gjaerum and Andersen, 1983). Two peaks of urediniospore production occur in the Netherlands in mid-June and mid-September (Bakker and Vos, 1975). The rust appears in autumn on *P. pratensis* in France south of Paris (Bourgoin, 1974) and is a late-season rust in the Canadian prairies (Smith, 1980), and probably causes little damage to mown turf (Gjaerum and Andersen, 1983). Late, severe attacks may reduce winter hardening of *P. pratensis* (Tronsmo, 1982). Seed production of the latter species may be severely reduced in western Europe and in the northwestern United States by the rust (Gjaerum and Andersen, 1983; Hardison, 1943, 1963; Hardison and Andersen, 1965).

14.6.2 Host resistance

There is considerable variation in resistance to *P. poae-nemoralis* in cultivars of *Poa pratensis.* Resistance to this rust was not necessarily correlated with resistance to *P. poarum* (Bakker and Vos, 1975; Weibull, 1977) or to *P. poarum* and *P.*

striiformis (Weibull, 1977) which have different periods of attack (Bakker and Vos, 1975; Bourgoin, 1974; Weibull, 1977). There seems little agreement on the degree of resistance of cultivars of *P. pratensis* to the same rust in different locations in Europe.

14.7 ORANGE STRIPE RUST–*PUCCINIA POARUM*

The yellow – orange, short-lived uredinia of this rust are often small, scattered or in short rows, formed on necrotic spots, mainly on upper leaf surfaces and occasionally on the culms. The long-covered, oval to linear, black – brown telia are more abundant, on both leaf surfaces. Heavily infected leaves may wilt (Weibull, 1983b). Although the host range is wide, the rust causes most concern in amenity turf on *Poa annua*, *P. pratensis* and *P. trivialis* on which it may cause epidemics in western Europe and Scandinavia, especially on *P. pratensis* (Bakker and Vos, 1975; Gjaerum and Weibull, 1983; Jørstad, 1961; O'Rourke, 1976; Weibull, 1977). It is also found from western Europe to Japan but is absent from New Zealand (Cummins, 1971; McNabb, 1962; Mühle, 1971; Ullrich, 1977; Wilson and Henderson, 1966; Woodcock and Clarke, 1983). Bourgoin (1974) recorded *P. poarum* causing severe disfigurement of *P. pratensis* turf in France, attacks being longer and more severe the further south the species is grown. However, it generally does not do much damage to short-mown turf elsewhere. It is not a common rust on grasses in Canada or the United States (Conners, 1967), although *P. liatridis* (Arth. and Fromme) Bethel ex Arth. with aecia on *Liatris* (Eupatorieae) rather than *Petasites* and *Tussilago*, which Cummins put in synonomy with *P. poarum*, is found on *Agrostis* spp. (Arthur, 1934; Conners, 1967). Pycnia and aecia of *P. poarum* are commonly found on *Tussilago* in Europe on upper and lower leaf surfaces respectively. *P. poarum* may be confused with *P. poae-nemoralis* since their uredinia and telia are similar morphologically. However, *P. poae-nemoralis* has paraphyses in its freely produced uredinia and often produces no telia whereas *P. poarum* has no (or very few peripheral) paraphyses in usually scantily produced uredinia, but produces abundant telia.

14.7.1 Disease development

P. poarum is obligately heteroecious (Wilson and Henderson, 1966) and severe attacks on grasses may occur when they are adjacent to the alternate host (O'Rourke, 1976; Weibull, 1977). Two aecial generations develop each year, in spring and late summer, followed by the occurrence of uredinia and telia on the grass host (Bakker and Vos, 1975; O'Rourke, 1976; Wilson and Henderson, 1966). It overwinters as teliospores which germinate in early spring, producing basidiospores which infect the alternate host. In the Netherlands the periods of maximum production of urediniospores of *P. poarum* and *P. poae-nemoralis* are different, which affects disease incidence on resistant and susceptible cultivars of *P. pratensis* and hence disease ratings (Bakker and Vos, 1975).

14.7.2 Host resistance

Limited information is available on the resistance of cultivars of *P. pratensis* in mown turf in Europe (Table 14.1 and STRI, 1980).

14.7.3 Control

See pp. 174–175.

14.8 BROWN OR LEAF RUST – *PUCCINIA RECONDITA*

Urediniospore walls are dirty orange – brown, not hyaline, as for example, in *P. striiformis* and the uredinia occur mainly on (adaxial) leaf surfaces, hence the common name. Well-developed paraphyses are lacking in uredinia. The brownish-black telia occur mainly on abaxial leaf surfaces, but are also found on culms.

P. recondita is a species complex with a wide range in morphology and a very wide temperate grass host range. There are many synonyms (Cummins, 1971). Species and *formae speciales* within the complex have been recognized on the basis of grass and related alternate hosts (Cummins, 1971; Grove, 1913; McNabb, 1962; Wilson and Henderson, 1966). Aecial stages of members of the complex occur on several hosts in the Crassulaceae, Boraginaceae, Ranunculaceae and Hydrophyllaceae but their roles in the epidemiology of the disease is uncertain. Although the rust is heteroecious, the alternate host may be absent when grasses are infected (McNabb, 1962; Zadoks, 1965). Members of the complex are widespread on turf grass species, but little is known of their importance in mown turf. *P. recondita* f. sp. *agrostidis* Oud. produces uredinia and telia on *A. stolonifera* and *A. tenuis* in Great Britain and Ireland adjacent to aecia on *Aquilegia,* but it is not common (Grove, 1913; Wilson and Henderson, 1966).

14.8.1 Disease development

The uredinial state of *P. recondita* is more resistant to cold and less thermophilic than that of *P. graminis* (Stock, 1931) which allows of a wider range than the latter (Zadoks, 1965). As in other grass rusts, the urediniospores germinate best when free water is present (Hemmi and Abe, 1933). Knowledge of the source of inoculum for epidemics is scanty, but sporulating sori have been reported under snow covers, where the uredomycelium is protected by them from winter killing. Long-distance transport of urediniospores by wind is probable. Overwintering as dikaryotic mycelium has been reported from European countries (Zadoks, 1965). In New Zealand no alternate host has been found (McNabb, 1962).

14.8.2 Host resistance

There is no reliable information on grass host and cultivar resistance because of the occurrence of the species complex.

14.8.3 Control

See pp. 174–175.

14.9 YELLOW STRIPE RUST – *PUCCINIA STRIIFORMIS*

This disease is so-called because of the bright-yellow uredinia which often occur in long, chlorotic stripes on adaxial leaf surfaces, leaf sheaths, culms and inflorescences. It has a world-wide distribution, in temperate climates on cereals and wild and cultivated grasses (Mulder and Booth, 1971). There are a large number of physiological races and these have been categorized as varieties on the basis of slight morphological differences (Manners, 1960) or more commonly as *formae speciales* (Eriksson, 1894; Tollenaar, 1967). In north west Europe cereal races seldom occur on grasses (Hassebrauk, 1965; Manners, 1960; Zadoks, 1965), but in mountain valleys in Oregon and California *P. striiformis* f. sp. *tritici* is present on wild grasses which serve as oversummering hosts for outbreaks of the rust on winter wheat (Schaner and Powelson, 1973; Tollenaar and Houston, 1967). The stripe rust isolated from the very susceptible cultivar Merion of *Poa pratensis* in California is different from that on wheat and the grasses and was described as *P. striiformis* f. sp. *poae*. It forms uredinia only on *Poa* spp. (Tollenaar, 1967; Tollenaar and Houston, 1967). Courtillot (1978) reported the f. sp. *poae* on *P. pratensis*, from France. The form was first recorded in Europe in 1973 by Ullrich (1976). The identity of the special forms of *P. striiformis* in many countries is uncertain (Savile, 1983).

14.9.1 Disease development

P. striiformis is a heteroecious rust whose aecial host is unknown and perhaps no longer functional. Teliospores often do not mature, particularly in mown turf, and when they do their dormancy period is very short and the basidiospores produced seem incapable of infecting a host (Gäumann, 1959; Rapilly, 1979). Carry over of infection from one season to the next is by dikaryotic mycelium and urediniospores in cereals and grasses (Gassner and Piechel, 1934; Schaner and Powelson, 1973). Although long-distance spread of primary inoculum by wind is possible (Zadoks, 1965), it is more likely that the disease is endemic (Rapilly, 1979). Urediniospores are dispersed locally, often in clusters, by raindrop impact, rain and splash. Winds with blasts above 1 m/s play a major role in disease spread. The duration of the period between urediniospore germination, penetration and sporulation (latent period) depends on climate, host resistance and parasitic virulence and varies between about 3 and 12 weeks (Rapilly, 1979). *P. striiformis* has generally been regarded as a rust of cool climates or high elevations in warmer ones (Hassebrauk, 1965; Mulder and Booth, 1971; Zadoks, 1965). However, different *formae speciales* show optima for urediniospore germination between 6 and 24 °C (Manners, 1960; Schroeder and Hassebruk, 1964; Tollenaar and Houston, 1967). For *P. striiformis* f. sp. *poae* the optimum ranges from 15 to 18°C (Tollenaar and Houston, 1967). Although the fungus is not killed by moderate sub-zero temperatures, low temperatures extend the latent period and delay the appearance of the disease next spring. Hot, dry, sunny weather may rapidly kill the fungus, the urediniospores being vulnerable to the effects of exposure to ultra-violet linght (Maddison and Manners, 1972). Urediniospores germinate poorly in water, but do so freely in a nearly saturated atmosphere which results in condensation droplets on grass leaves (Tu and Hendrix, 1967).

14.9.2 Host resistance

While other *formae speciales* of *P. striiformis* infect many grass species, some severely (Humphrey, Hungerford and Johnson, 1924; Manners, 1960; Sanford and Broadfoot, 1932; Schaner and Powelson, 1973; Tollenaar and Houston, 1967), some are resistant. On the other hand, most *Poa* spp. were susceptible to f. sp. *poae* (Britton and Cummins, 1956). Gaskin and Nunez (1977) found that none of the 68 cultivars of *Poa pratensis* they tested, in the area of San Luis Obispo, California, where the rust is endemic, had any resistance to f. sp *poae*. In Hawaii, Newport and Kenblue were moderately susceptible and six other lines were only slightly affected by *P. striiformis* (f. sp. not specified; Murdoch, Laemmlen and Parvin, 1973a). In Sweden, only Geronimo of 14 cultivars of *P. pratensis* tested was consistently resistant to an unspecified f. sp. of the rust (Weibull,

1977) in a row test. In Sweden, since 1977, only slight damage to *Poa pratensis* from *P. striiformis* was noted, in hot, dry summer weather. Severe damage to the same grass was recorded in south-west England and at Christchurch, New Zealand in July 1978 and January 1983 respectively (P. Weibull, personal communication, 27 Oct. 1983). In mown turf at San Jose, California 42 of 56 cultivars of *P. pratensis* were rust free, including Geronimo (see above), while seven cultivars, including Merion, were severely damaged. In this location, 12 of 37 cultivars of *Lolium perenne* showed moderate rust symptoms caused by *P. striiformis* (f. sp. not specified, Harivandi and Gibeault, 1980). In the Pacific northwest of the United States, and northern California, *P. striiformis* is the most serious pathogen of *P. pratensis* in seed production fields and low maintenance turf. In the Willamette Valley of Oregon seed reductions of more than 50% and severe turf deterioration occurred in some varieties. Some experimental varieties and hybrids showed good resistance (Meyer, 1977, 1982).

14.9.3 Control

See pp. 174–175.

14.10 ZOYSIA RUST – *PUCCINIA ZOYSIAE*

The small, linear, yellow – orange uredinia of this rust occur mainly on adaxial leaf surfaces and may be so abundant on susceptible *Zoysia* spp. that the turf takes on an orange cast. Plants may be severely stunted and the turf thinned. The amphigenous, brownish-black, early-exposed telia are less common. *Zoysia* spp. in warmer climates in Manchuria, mainland China, Taiwan, Japan and the United States are reported to be infected (Cummins, 1971; Cummins and Lee Ling, 1950; Freeman, 1965; Gough and McDaniel, 1969; Gudauskas and McCarter, 1966; Juska and Kreitlow, 1972; Kreitlow, Juska and Haard, 1965; Wu, 1979).

14.10.1 Disease development

Although *P. zoysiae* is heteroecious, with a *Paederia* sp. as an alternative host, this is absent in North America. The alternative host is unnecessary as the rust is able to persist over the long winter dormant period of *Zoysia* in the mycelial or uredinial state which is capable of surviving at −18°C. The highest incidence of the disease is usually found on shaded lawn areas where the dew persists for the prolonged period needed for spore germination and infection. Light intensity does not appear to be a factor in the infection process (Cummins, 1971; Juska and Kreitlow, 1972; Kreitlow, Juska and Haard, 1965).

14.10.2 Host resistance

Juska and Kreitlow (1972) found that Emerald (*Zoysia japonica* × *Z. tenuifolia*) and Beltsville 101 were the least susceptible of 14 *Zoysia* lines to artificial inoculation. However, Freeman (1965) found Meyer and Emerald the most severely affected and *Z. matrella* was least affected, whereas Gudauskas and McCarter (1966) noted that lawns of *Zoysia japonica* Meyer, *Z. matrella* and *Z. japonica* × *Z. tenuifolia* Emerald equally affected, in Florida and Alabama respectively.

14.11 UROMYCES LEAF RUST – *UROMYCES DACTYLIDIS*

The uredinia of this rust are yellowish-brown, powdery and occur in rows on both leaf surfaces of *Poa* spp., but mainly on the adaxial leaf surfaces of *Festuca* spp. There are no paraphyses in the uredinia, but brown, subepidermal paraphyses divide the telia into locules. Telia are sunken, shiny dark brown, long covered with epidermis, scattered or arranged in lines, amphigenous on leaves of *Poa* spp., but adaxial in *Festuca* spp. Of the three varieties recognized by Cummins (1971), var. *poae* is probably less common in *Poa* spp. than *P. poae-nemoralis* and *P. poarum*. There are also vars. on *Festuca* and *Agrostis* spp. The one-celled teliospores which may be found in old uredinia as well as telia distinguish this species from the latter two. Distribution is world-wide (Cummins, 1971; Gjaerum, 1974; Kreitlow, Graham and Garber, 1953; McNabb, 1962; O'Rourke, 1976; Ullrich, 1977; Weibull, 1977; Wilson and Henderson, 1966).

14.11.1 Disease development

U. dactylidis is a polymorphous heteroecious rust with *Ranunculus* spp. as alternate hosts. In New Zealand alternation is probably facultative, although pycnial and aecial stages have been found (McNabb, 1962).

14.11.2 Host resistance

Little of significance has been reported on cultivar resistance in turf grass species.

14.11.3 Control

See pp. 174–175.

14.12 SOME OTHER RUSTS OF AMENITY TURF GRASSES

Puccinia crandalii Pam. and Hume is very abundant in western Canada and the western United States where the aecial hosts, *Symphoricarpos* spp. are plentiful. The aecia are often very conspicuous on

Symphoricarpos in grassland (Arthur and Cummins, 1962; D.B.O. Savile, personal communication). *P. crandalii* has been reported affecting seed production of *Festuca rubra* in the western United States (Hardison, 1963).

P. stenotaphri Cummins may cause slight to moderate damage on turf of *Stenotaphrum secundatum* in hot climates in Africa, India, the Caribbean and the southern United States (Freeman, 1967; Lee and Toler, 1972; Watson, 1957).

14.13 SOME USEFUL DIAGNOSTIC FEATURES OF COMMON, COOL-SEASON RUST SPECIES ON TURF GRASSES

Puccinia coronata has thin-walled paraphyses round the periphery of the uredinia compared with *P. poae-nemoralis*, which has many thick-walled or capitate paraphyses with narrow or bent necks.

P. coronata has orange uredinia whereas those of *P. graminis* and *P. recondita* are reddish-brown.

P. coronata and *P. festucae* have finger-like outgrowths from the top cell of the teliospore which distinguish them from other grass rusts. However, *P. festucae* and *P. coronata* can be distinguished most readily by alternate host species and grass host range.

P. festucae has many fewer paraphyses in the uredinia than *P. poae-nemoralis*.

P. striiformis differs from *P. coronata* in the long, striped uredinia on leaf blades of the former and the scattered uredinia of the latter.

P. graminis uredinia and telia commonly occur in long stripes on the straw except on mown lawn grasses whereas in other species the sori are usually on the leaf blades. Urediniospores of *P. graminis* have equatorial germ pores while in most other species they are scattered.

P. recondita may be distinguished from *P. coronata* because of the colour of the uredinia (see above). *P. recondita* has no paraphyses and the uredinia are scattered and light brown while those of *P. coronata* are scattered or in stripes and are orange in colour.

P. poarum in the uredinial state can be distinguished from *P. poae-nemoralis* by the sparse ephemeral uredinia of the former, developing on light-coloured leaf flecks.

Uromyces dactylidis differs from *P. poae-nemoralis* in lacking paraphyses in the uredinia and having one-celled teliospores.

14.14 CONTROL OF RUSTS ON AMENITY TURF GRASSES

There are very few experimental data on which to base specific recommendations for the control of rusts on amenity turf grasses by cultural methods. Generally, turf grass rusts do less damage when vigorous grass growth is promoted by adequate nitrogenous fertilization and irrigation. Conversely, reduction in vigour of leaf growth, such as occurs towards the end of the growing season, or which results from low fertility and drought stress, often shows in increased rust incidence (Beard, 1973; Britton, 1969; Cheeseman, Roberts and Tifany, 1965; Davis, Caldwell and Gist, 1960; Vargas, 1981).

Nitrogen is a key nutrient in disease management but adequate information on the effects of nitrogen applications in the control of turf grass rusts is lacking. In cereals, Gassner and Hassebrauk (1934) found that the susceptibility of wheat to *P. striiformis* was increased by high soil nitrogen, if phosphate and potassium were deficient. Raines (1922) concluded that the increased susceptibility of wheat to rusts, when excessive nitrogen was applied to the soil, was partly due to increased stand density and lengthened duration of the period of exposure to infection. General statements have been made that high soil nitrogen favours rust infections in plants (Goodman, Kiraly and Zaitlin, 1967) and that rusts develop most readily on vigorous plants (Butler and Jones, 1949; Dickson, 1959). On the other hand, Cheeseman, Roberts and Tifany (1965) found that a low nitrogen treatment on Merion *Poa pratensis* turf combined with high moisture stress reduced grass growth and favoured infection with *Puccinia graminis*. The effect of nitrogen varies with the susceptibility of the particular grass to rust (Gäumann, 1959). In New Zealand, Lancashire and Latch (1970) found that more severe *P. coronata* developed on the susceptible Grasslands Ruanui *Lolium perenne* at low soil N level (22 kg/ha of N) while it was more severe on the resistant *L. perenne* × *L. multiflorum* Grasslands Ariki at a higher N level (157 kg/ha of N). Broom *et al.* (1975) found no more severe *P. coronata* on S.24 *L. perenne* grassland at high N (100 kg/ha), but frequently mown, unfertilized pathways were heavily rusted in autumn. Lam and Lewis (1982) found that when *P. coronata* was moderately severe in 1978, there was much less damage to herbage of S.24 *L. perenne* in no N plots than in those receiving 200 kg/ha of nitrogen or greater whether measured as the percentage of infected tillers or percentage of infected leaves. However, regardless of N levels, older leaves were more severely infected than younger ones. In autumn 1979, when *P. coronata* incidence was very severe (90–100% tiller infection), no effects of N fertilization were discernible. Drought stress greatly reduced grass growth, precluded mowing and allowed *P. coronata* inoculum build-up and dispersal to all N levels of turf. High rust levels in autumn were usually associated with low levels of *Drechslera* leaf spots and vice versa. It cannot be assumed, in every instance, that effective control of turf grass rust will result from the maintenance of an adequate soil N level.

The prevention of drought stress appears as important as assuring an adequate state of soil fertility in the control of turf grass rusts. Cheesman, Roberts and Tifany (1965) reported that unirrigated turf of Merion *P. pratensis*, fertilized with either 125 or 600 kg/ha nitrogen as ureaform, was almost destroyed whereas frequently irrigated turf receiving the same fertilization did not become infected.

Frequent mowing of rapidly growing turf grass with collection and removal of the clippings also removes rust inoculum and may reduce disease incidence. Kazelnicky and Garrett (1966) and Parris (1971) found that zoysia rust caused by *P. zoysiae* could be effectively controlled by this means without the use of fungicides. The ability of turf to withstand periods of drought stress is influenced more by the height of mowing than by its frequency. The lowering of mowing height can be directly correlated with reduction of root growth, plant vigour and carbohydrate reserves (Troughton, 1957). The greater ability of longer mown turf to withstand drought stress can be related to its deeper root system. Lower mowing heights are sometimes associated with increased rust severity.

Rusts are often more severe on turf in shaded situations or near hedges and fences where air movement is poor and dew or other moisture is persistent. These environments provide the humid conditions suitable for urediniospore germination and the initiation of epidemics. Shaded areas may favour those rusts with low light requirements for urediniospore (e.g. *P. striiformis*) germination, or they may modify the response of the grass plant to infection. Where shading is the result of an overshadowing tree or hedge the grass may have been subjected to debilitating drought stress (Kreitlow, Juska and Haard, 1965). The remedies lie in improving 'air drainage', opening up the tree canopy or hedge and, if necessary, modifying irrigation.

Control of rusts of mown turf is possible by the use of disease resistance in cultivars of *Poa pratensis* to *P. graminis*, *P. striiformis*, *P. poae-nemoralis* and *P. poarum*; in *Lolium perenne* and *Festuca arundinacea* to *P. coronata*; in *Cynodon* spp. to *P. cynodontis* and in *Zoysia* spp. to *P. zoysiae*. For the most recent review of rust resistance in cool-season grasses see Braverman (1986). The occurrence of different rust races in different parts of the world and their ability for genetic recombination or mutation to produce new virulent forms necessitates the use of regional testing to determine resistance and genetic shift. The use of blends of cultivars of the same species should not be regarded as a means of rust control unless all components show high individual rust resistance. However, mixtures of host species such as *Festuca rubra* spp. *rubra* with *P. pratensis* or *Agrostis tenuis* with *F. rubra* ssp. *commutata* may be very effective in reducing overall rust damage where only one species is susceptible to the prevalent rust species.

Effective fungicides are available for the control of rusts in turf grass seed crops and in mown turf. The impetus for their development came from attempts to control cereal rusts by chemical means. The earlier history of these has been reviewed by Dickson (1959) and Peturson, Forsyth and Lyon (1958). Hardison (1963, 1976) reported that although field burning of crop debris gave partial control of *P. poae-nemoralis*, it gave inadequate control of *P. striiformis* in Oregon. Commercially acceptable control of *P. striiformis, P. poae-nemoralis* and *P. graminis* in *Poa pratensis; P. graminis* and *P. coronata* in *Lolium perenne* and *P. graminis* and *P. crandalii* in *Festuca rubra* ssp. *rubra* and *F. rubra* spp. *commutata* was obtained with sprays of nickel sulphate with maneb or zineb in seed crops (Hardison, 1963, 1975). At low dosages, one application of triadimefon or three applications of benodanil gave excellent control of *P. striiformis* (Hardison, 1975), superior to that given by oxathiin and thiazole compounds which, although effective in rust control were often too phytotoxic to seed crops (Hardison, 1971a,b). In Czechoslovakia Cagaš (1981) obtained good control of *P. poae-nemoralis* in seed crops of *P. pratensis* with triadimefon.

In mown turf, effective control of rusts has been reported with formulations of cycloheximide, cycloheximide with thiram, chlorothalonil, carboxin, mancozeb, maneb, oxycarboxin, triadimefon, and zineb (Couch and Cole, 1956; Courtillot, 1978; Gough and McDaniel, 1969; Murdoch, Laemmlen and Parvin, 1973b; Smith, 1980; Watkins, Houfek and Shearman, 1981; Worf, 1982; Worf and Huibregste, 1981).

14.15 REFERENCES

Anikster, Y. and Wahl, I (1979) Coevolution of the rust fungi on Gramineae and Liliaceae and their hosts. *Ann. Rev. Phytpathol.*, **17**, 367–403.

Arthur, J.C. (1934) Manual of the rusts in the United States and Canada. Science Press, Lancaster, PA, 438pp.

Arthur, J.C. and Cummins, G.B. (1962) *Manual of the Rusts in the United States and Canada*. Hafner, New York, 438 pp. (Suppl. 24 pp).

Badahur, P., Singh, S., Goel, L.B., Sharma, S.K., Sinha, V.C., Ahmad, R.E. and Singh, B.P. (1973) Impact of grass introduction on cereal rusts in India. *Indian J. Agric. Sci.*, **43**(3), 287–90.

Bakker, J.J. and Vos, H. (1975) Resistance of *Poa pratensis* L. varieties to *Puccinia poarum* Niels. and *Puccinia brachypodii* var. *poae-nemoralis* (Otth.) Cummins and Greene. *Rasen-Turf-Gazon,* **6,** 35–8.

Baltensperger, A.A. (1962) Evaluation of Bermuda grass varieties and strains. *Ariz. Agric. Exp. Sta. Rept*, **203,** 3–19.

Batts, C.C.V. (1951a) Physiological specialization of *Puccinia graminis* Pers. in south-east Scotland. *Trans. Br. Mycol. Soc.*, **34,** 533–8.

Batts, C.C.V. (1951b) The distribution, host range and seasonal development of *Puccinia graminis* in south-east Scotland. *Trans. Bot. Soc. Edinburgh*, **36,** 48–57.

Beard, J.B. (1973) *Turfgrass: Science and Culture*. Prentice-Hall, Englewood Cliffs, NJ, 638pp.

Blumer, S. (1963) *Rust and Smut Fungi on Cultivated Plants*. Gustav Fischer, Jena, 379pp. (German).

Bourgoin, S. (1974) The behaviour of the principal turf grasses under French climatic conditions. *J. Sports Turf Res. Inst.*, **50,** 65–8.

Bourgoin, B., Billot, C., Kerguelen, M., Hentgen, A. and Mansat, P. (1974) Behaviour of turf grasses in France. In *Proc. 2nd. Int. Turgrass Res. Conf.* (ed. E.C. Roberts), Am. Soc. Agron., Crop Sci. Soc. Am., pp. 35–47.

Braverman, S.W. (1966) Sources of resistance to stem and leaf rust among *Phleum* species. *Plant Dis. Rept.*, **50**(11), 849–851.

Braverman, S.W. (1967) Disease resistance in cool-season forage, range and turf grasses. *Bot. Rev.*, **33,** 329–78.

Braverman, S.W. (1977) Sources of resistance to crown rust in meadow fescue accessions. *Plant Dis. Rept.*, **61,** 463–5.

Braverman, S.W. (1986) Disease resistance in cool-season forage, range and turf grasses. *Bot. Rev.*, **52**(1), 1–112.

Braverman, S.W. and Oakes, A.J. (1972) Disease resistance in warm-season forage, range and turf grasses. *Bot. Rev.*, **38,** 491–544.

Britton, M.P. (1958) The identity, epiphytology and control of stem rust of Merion bluegrass. Ph. D. Thesis, Purdue University.

Britton, M.P. (1969) Turf grass diseases. In *Turfgrass Science* (eds, A.A. Hanson and F.V. Juska), Am. Soc. Agron. 14, pp. 288–335.

Britton, M.P. and Butler, J.D. (1965) Resistance of seven Kentucky bluegrass varieties to stem rust. *Plant Dis. Rept.*, **49,** 708–10.

Britton, M.P. and Cummins, G.B. (1956) The reaction of species of *Poa* and other grasses to *Puccinia striiformis. Plant Dis. Rept.*, **40,** 643–45.

Britton, M.P. and Cummins, G.B. (1959) Sub-specific identity of the stem rust fungus of Merion bluegrass. *Phytopathology*, **49,** 287–9.

Broom, E.W., Heard, A.J., Griffiths, E. and Valentine, M. (1975) Effect of fungicides and fertilizer on yield and diseases of grass swards. *Plant Pathol.*, **24,** 144–9.

Brown, M.R. (1937) A study of crown rust, *Puccinia coronata* Corda in Great Britain. 1. Physiologic specialization in the uredospore stage. *Ann. Appl. Biol.*, **24,** 504–27.

Butler, E.J. and Jones, S.G. (1949) *Plant Pathology*. Macmillan, London.

Bundessortenamt, (1979) *Descriptive Variety List for Turf-grasses.* Alfred Strothe, Verlag, Hanover, 198 pp. (German).

Cagaš, B. (1975) Host specialization of *Puccinia graminis* Pres. ssp. *graminicola* Urban. *Phytopathol. Z.*, **84,** 57–65.

Cagaš, B. (1978) The resistance of meadow fescue and tall fescue to *Puccinia coronata* var. *coronata. Ochrana Rostlin*, **14,** 106–12. (*R.A.P.P.* **58,** 2279, 1979). (Czech.)

Cagaš, B. (1981) The use of Bayleton in the protection of the seed cultures of smooth-stalked meadow grass. *Sbor Uvtiz-Ochrana Rostlin*, **17**(1), 61–6. (Czech, Russ., En., Ger.)

Cagaš, B. (1982) Tall fescue/*Festuca arundinaceae* Schreb./ from plant disease resistance viewpoint. *Proc. Conf. Grassland Res. Inst.* Banska Bystrica. pp. 182–8. (Czech).

Cheeseman, J.H., Roberts, E.C. and Tifany, L.H. (1965) Effects of nitrogen level and osmotic pressure of the nutrient solution on incidence of *Puccinia graminis* and *Helminthosporium sativum* infection in Merion bluegrass. *Agron. J.*, **57,** 599–602.

Conners, I.L. (1967) *An Annotated Index of Plant Diseases in Canada.* Can. Dept. Agric. Res. Br. Queen's Printer, Ottawa, 381pp.

Conners, I.L. and Savile, D.B.O. (1943) Twenty-second Ann. Rept. *Can. Plant Dis. Surv. (1942)*, **22,** 9–10.

Conners, I.L. and Savile, D.B.O. (1952) *Can. Plant Dis. Surv. (1951)* **32,** 22–3.

Couch, H.B. and Cole, H. Jr (1956) Five compounds tested for control of Merion bluegrass rust. *Plant Dis. Rept.*, **40,** 103–5.

Courtillot, M. (1978) Estimation of the effects of fungicides in lawns attacked by rust, powdery mildew and snow mould. *Phytiatrie-Phytopharmacie*, **27,** 191–7. (*R.A.P.P.* **58,** 5913, 1979). (French).

Craigie, J.H. (1939) Aerial dissemination of plant pathogens. *Proc. Pacific Sci. Congr.*, **6,** 753–67.

Cummins, G.B. (1945) Tropical plant rusts. *Bull. Torrey Bot. Club,* **72,** 123 pp.

Cummins, G.B. (1956) Host index and morphological characterization of the grass rusts of the world. *Plant Dis. Rept. Suppl.* 237, 52 pp.

Cummins, G.B. (1971) *The Rust Fungi of Cereals, Grasses and Bamboos.* Springer-Verlag, Berlin 570 pp.

Cummins, G.B. and Hiratsuka, Y. (1983) *Illustrated Genera of Rust Fungi.* Am. Phytopath. Soc. 152 pp.

Cummins, G.B. and Lee Ling (1950) An index of the plant rusts recorded for Continental China and Manchuria. *Plant Dis. Rept.*, Suppl. 196.

Cunningham, G.H., (1931) *The Rust Fungi of New Zealand.* J. McIndoe, Dunedin, 261 pp.

Davis, R.R., Caldwell, J.L. and Gist, G.R. (1960) Caring for your lawn. *Ohio Agric. Exp. Sta. Res. Bull.*, 271.

Dickson, J.G. (1959) Chemical control of cereal rusts. *Bot. Rev.*, **25,** 486–513.

Durrell, L.W. (1918) Factors influencing the uredospore germination of *Puccinia coronata. Phytopathology*, **8,** 81–2.

Elliott, E.S. (1963) Susceptibility of Kentucky bluegrass selections to some fungus diseases. *Proc. W. Va. Acad. Sci.*, **35,** 29–32.

Endo, R.M. (1961) Turf grass diseases in southern California. *Plant Dis. Rept.*, **45**, 869–73.
Eriksson, J. (1894) On the specialization of parasitism by cereal rust fungi. *Ber. Deut. Bot. Ges.*, **12**, 292–331.
Eshed, N. and Dinoor, A. (1981) Genetics of pathogenicity in *Puccinia coronata*; The host range among grasses. *Phytopathology*, **71**, 156–63.
Fischer, G.W. and Levine, M.N. (1941) Summary of the recorded data on the reaction of wild and cultivated grasses to stem rust (*Puccinia graminis*), leaf rust (*P. coronata*) in the United States and Canada. *Plant Dis. Rept. Suppl.* 130, 30pp.
Fraser, W.P. and Ledingham, G.A. (1933) Studies of the crown rust, *Puccinia coronata* Corda. *Sci. Agric.*, **13**, 313–23.
Frauenstein, K. (1970) The major pathogens of bluegrass (*Poa pratensis*) in the German Democratic Republic. *Nachrbl. Dt. PflanzShutzdienst, Berlin.* NF **24**(1), 5–9 (German, Russ., En.).
Freeman, T.E. (1965) Rust of *Zoysia* spp. in Florida. *Plant Dis. Rept.*, **49**, 382.
Freeman, F.E. (1967) Diseases of southern turf grasses. *Florida Agric. Exp. Sta. Tech. Bull.*,713.
Funk, C.R. (1983) Ryegrasses reviewed. *Golf Course Mangmt.*, **51**, (4), 81, 83, 85.
Funk, C.R., Engel, R.E., Duell, R.W. and Dickson, W.K. (1978) Kentucky bluegrasses for New Jersey turf. *Rutgers Turfgrass Proc.*, **9**, 120–42.
Funk, C.R., Engel, R.E. and Duell, R.W. (1981) Kentucky bluegrasses and their culture in New Jersey lawns. *Rutgers Turfgrass Proc.*, **12**, 117–37.
Gaskin, T.A. and Nunez, J. (1977) Reaction of *Poa* spp. to stripe rust (*Puccinia striiformis*). *Plant Dis. Rept.*, **61**, 588–9.
Gassner, G. and Hassebrauk, K. (1934) The influence of mineral salt nutrition on the susceptibility behaviour of standard variety collections used for the determination of cereal rust. *Phytopathol. Z.*, **7**(1), 63–72. (German).
Gassner, G. and Piechel, E. (1934) Investigations into the question of uredo overwintering of cereal rust in Germany. *Phytopathol. Z.*, **7**, 355–92 (German).
Gäumann, E. (1959) The rust fungi of Central Europe. *Beitr. Kryptogamenfl. Schweitz*, **12**, 407pp. (German).
Gibbs, J.G. (1965) A note on the pathogenicity of *Puccinia coronata* and *P. graminis* in New Zealand. *NZ J. Bot.*, **3**, 229.
Gibbs, J.G. (1966) Field resistance in *Lolium* sp. to leaf rust (*Puccinia coronata*). *Nature*, **209**, 420.
Givan, C.V. and Bromfield, K.R. (1964) Light inhibition of uredospore germination in *Puccinia graminis* var. *tritici*. *Phytopathology*, **54**, 382.
Gjaerum, H.B. (1974) *Nordic Rust Fungi*. Fungiflora, Olso, 321pp (Norwegian).
Gjaerum, H.B. and Andersen, I.L. (1983) Descriptions of grass diseases No. 14 *Puccinia poae-nemoralis* Otth. Weibulls Gräs-tips 22–25 (Dec.) 2pp.
Gjaerum, H.B. and Weibull, P. (1983) Description of grass diseases No. 15 *Puccinia poarum* Niels. Weibulls Gräs-tips 22–25 (Dec.) 2pp.
Goodman, R.N. Kiraly, Z. and Zaitlin, M. (1967) *The Biochemistry and Physiology of Infectious Plant Disease.* Van Nostrand, New Jersey, 354pp.
Goss, R.L. and Law A.G. (1977) Turf varieties differ in performance. *Golf Superintendent*, **45**(7), 27–9.
Gough, F.J. and McDaniel, M.E. (1969) *Zoysia* rust in Texas. *Plant Dis. Rept.*, **53**, 232.
Greene, H.C. and Cummins, G.B. (1967) *Puccinia holcina* and *P. poarum* redefined. *Mycologia*, **59**, 45–57.
Griffiths, D.J. (1953) Varietal resistance and susceptibility of oats to crown rust. *Plant Pathol.*, **2**, 73–7.
Grove, W.B. (1913) *The British Rust Fungi (Uredinales)*. Cambridge University Press, 412pp.
Gudauskas, R.T. and McCarter, S.M. (1966) Occurrence of rust on *Zoysia* species in Alabama. *Plant Dis. Rept.*, **50**, 885.
Guyot, A.L. and Viennot-Bourgin, G. (1943) Concerning the aecidial form on *Valerianella* corresponding with *Puccinia cynodontis* Desm. *An. Ec. Agric. Grignon. Ser.*, **3**(3), 100–5.
Hardison, J.R. (1943) The occurrence of amphispores in the leaf rust of bluegrasses. *Mycologia*, **35**, 79–82.
Hardison, J.R. (1963) Commercial control of *Puccinia striiformis* and other rusts in seed crops of *Poa pratensis* by nickel fungicides. *Phytopathology*, **53**, 209–16.
Hardison, J.R. (1971a) Relationships of molecular structure of 1,4, oxathiin fungicides to chemotherapeutic activity against rust and smut fungi in grasses. *Phytopathology*, **61**, 731–5.
Hardison, J.R. (1971b) Chemotherapy of smut and rust pathogens in *Poa pratensis* by thiazole compounds. *Phytopathology*, **61**, 1396–9.
Hardison, J.R. (1975) Control of *Puccinia striiformis* by two new fungicides BAY MEB 6447 and BAS 31702F. *Plant Dis. Rept.*, **59**, 652–5.
Hardison, J.R. (1976) Fire and flame for plant disease control. *Ann. Rev. Phytopathol.*, **14**, 335–79.
Hardison, J.R. and Anderson, W.S. (1965) Effects of symmetrical dichlorotetrafluoroacetone and nickel sprays on rust control, yield and germination of seed of *Poa pratensis*. *Phytopathology*, **55**, 1337–64.
Harivandi, M.A. and Gibeault, V.A. (1980) Rust on Kentucky bluegrass and perennial ryegrass cultivars. *Calif. Turfgrass Cult.*, **30**, 21.
Hassebrauk, K. (1965) Nomenclature, geographic distribution and host range of yellow rust, *Puccinia striiformis* West. Mitt. Biol. bund. Aust. *Land-u-Forstw.*, **116**, 75pp. (German, En.).
Haygood, R.A. and Spencer, J.A. (1979) Geographical and seasonal distribution of turf grass diseases in Mississippi *Plant Dis. Rept.*, **63**, 852–5.
Hayward, M.D. (1977) Genetic control of resistance to crown rust in *Lolium perenne* and its implications in breeding. *Theor. Appl. Genet.*, **51**, 49–53.
Henderson, D.M. and Bennell, D.P. (1979) *Notes from the Royal Botanic Garden, Edinburgh* (1979), **37**, 475–501.
Hemmi, T. and Abe, T. (1933) On the relation of air humidity to germination of urediniospores of some species of *Puccinia* parasitic on cereals. *Forsch. Geb. Pflanzenkrank. (Kyoto)*, **2**, 1–9. (Abs. in *Rev. Appl. Mycol.*, **13**, 83–4).
Hiratsuka, Y. and Sato, S. (1982) Morphology and tax-

onomy of rust fungi. In *The Rust Fungi* (eds K.J. Scott and A.K. Chakravorty), Academic Press, New York, London, 288 pp.

Humphrey, H.B., Hungerford, C.W. and Johnson, A.G. (1924) Stripe rust (*Puccinia glumarum*) of cereals and grasses in the United States. *J. Agric. Res.*, **29**(5), 209–27.

Johnson-Cicalese, J.M., Clarke, B.B. and Funk, C.R. (1983) Comparative reaction of Kentucky bluegrasses to stem rust in turf trials at Adelphi, New Jersey during 1981 and 1982. *Rutgers Turfgrass Proc.*, 1983, 87–95.

Jørstad, I. (1958) Uredinales of the Canary Islands. *Skr. Norske. Vidensk. Akad.*, **1**(2), 1–182.

Jørstad, I. (1961) Distribution of the Uredinales within Norway. *Nytt. Magasin Bot.*, **9**, 61–134.

Joshi, L.M. and Payak, M.M. (1963) A *Berberis* aecidium in Lahaul Valey, Western Himalayas. *Mycologia*, **55**, 247–50.

Juska, F.V. and Kreitlow, K.W. (1972) Conditions affecting the development and perpetuation of *Zoysia* rust. *Plant Dis. Rept.*, **56**(3), 227–9.

Kazelnicky, G.M. and Garrett, W.N. (1966) The occurrence of *Zoysia* rust in Georgia. *Plant Dis. Rept.*, **50**, 389.

Kopec, D.M., Funk, C.R. and Halisky, P.M. (1983) Sources of resistance of crown rust within perennial ryegrass. *Plant Dis.*, **67**, 98–100.

Kreitlow, K.W., Graham, J.H. and Garber, R.J. (1953) Diseases of forage grasses and legumes in the north eastern States. *Penn. State Agric. Exp. Sta. Bull.*, **573**, 42pp.

Kreitlow, K.W. and Meyers, W.M. (1947) Resistance to crown rust in *Festuca elatior* and *F. elatior* var. *arundinacea*. *Phytopathology*, **37**, 59–63.

Kreitlow, K.W., Juska, F.V. and Haard, R.T. (1965) A rust on *Zoysia japonica* new to North America. *Plant Dis. Rept.*, **49**, 185–6.

Lam, A. and Lewis, G.C. (1982) Effects of nitrogen and potassium fertilizer application on *Drechslera* spp. and *Puccinia coronata* on perennial ryegrass (*Lolium perenne*) foliage. *Plant Pathol.*, **31**, 123–31.

Lancashire, J.A. and Latch, G.C.M. (1970) The influence of nitrogenous fertilizer on the incidence of crown rust (*Puccinia coronata* Corda) in two ryegrass cultivars. *NZ J. Agric. Res.*, **13**, 287–93.

Latch, G.C.M. (1966) Fungus diseases of ryegrasses in New Zealand. *NZ J. Agric. Res.*, **9**, 394–409.

Laundon, G.F. (1967) Terminology in the rust fungi. *Trans. Br. Mycol. Soc.*, **50**, 189–94.

Lee, T.A. and Toler, R.W. (1972) St. Augustine rust in Texas. *Plant Dis. Rept.*, **56**(7), 630–2.

Litzenberger, S.C. and Stevenson, J.A. (1957) A preliminary list of Nicaraguan plant diseases. *Plant Dis. Rept.*, Suppl. 243, 10pp.

Maddison, A.C. and Manners, J.G. (1972) Sunlight and viability of cereal rust uredospores. *Trans. Br. Mycol. Soc.*, **59**, 429–33.

Manners, J.G. (1960) *Puccinia striiformis* Westend. var. *dactylidis* var. nov. *Trans. Br. Mycol. Soc.*, **43**, 65–8.

Marland, A.T. (1941) Time required for infection of oat plants by uredospores (*Puccinia coronifera*). *Bull. Plant Protection, Leningrad*, **17**, 134–7. (Russian) (Abstr. in *Rev. Appl. Mycol.*, **18**, 389).

McAlpine, D. (1906) *The Rusts of Australia*. Government Printer, Melbourne, 349 pp.

McNabb, R.F.R. (1962) The graminicolous rust fungi of New Zealand. *Trans. Roy. Soc. NZ* (*Bot.*), **1**, 235–57.

McVeigh, K.J. (1975) Breeding for resistance to crown rust (*Puccinia coronata* Corda var. *lolii* Brown) in turf type perennial ryegrass (*Lolium perenne*). Ph.D. Thesis, Rutgers University, New Brunswick.

Meyer, W.A. (1977) Studies on new stripe rust resistant Kentucky bluegrass cultivars. *Abs. Am. Soc. Agron.*, p.122.

Meyer, W.A. (1982) Breeding disease-resistant cool season turf grass cultivars for the United States. *Plant Dis.*, **66**(4), 341–4.

Misra, D.P., Ahmad, S.T. and Singh, S. (1965) A natural occurrence of specialized forms of *Puccinia graminis* and *P. striiformis on Lolium perenne. Indian Phytopathol.*, **18**, 214–15.

Mühle, E. (1971) *Diseases and Injuries of Fodder Grasses*. S. Hirzel Verlag, Leipzig, 422 pp.

Mulder, J.L. and Booth, C. (1971) *Puccinia striiformis. C.M.I. Descriptions of Pathogenic Fungi and Bacteria*, **291**, 2pp.

Mulder, J.L. and Holliday, P. (1971) *Puccinia cynodontis. C.M.I. Descriptions of Pathogenic Fungi and Bacteria*. Commonw. Mycol. Inst. Kew, **292.** 2 pp.

Murdoch, C.L., Laemmlen, F.F. and Pavin, P.E. (1973a) Response of Kentucky bluegrass and perennial ryegrass cultivars to rust (*Puccinia striiformis* and *P. coronata*) in Hawaii. *Plant Dis. Rept.*, **57**(3), 215–16.

Murdoch, C.L., Laemmlen, F.F. and Parvin, P.E. (1973b) Fungicidal control of *Puccinia striiformis* on three cultivars of *Poa pratensis* in Hawaii. *Plant Dis. Rept.*, **57**, 217–19.

NE-57 Technical Research Committee. (1977) Northeastern turf grass evaluation of Kentucky bluegrasses (*Poa pratensis* L.) 1968–73.

Nielsen, E.L. and Dickson, J.G. (1958) Evaluation of timothy clones for stem rust reaction. *Agron. J.*, **50**, 749–52.

Nyquist, W.E. (1962) Rust resistance in ryegrass. *Calif. Agric.*, **16**(5), 15.

Ogilvie, L. and Thorpe, I.G. (1966) Black stem rust of wheat in Great Britain. *Proc. 1st Cereal Rust Conf. Cambridge*, 1964. 172–6.

O'Rourke, C.J. (1976) *Diseases of Grasses and Forage Legumes in Ireland*. The Agricultural Institute (An Foras Taluntais) Carlow, Ireland. 115 pp.

Parris, G.K. (1971) Practical control of two lawn grass diseases without fungicides by removal of clippings. *Plant Dis. Rept.*, **55**(9), 778–9.

Peturson, B., Forsyth, F.R. and Lyon, C.B. (1958) Chemical control of cereal rusts. II. Control of leaf rust of wheat with experimental chemicals under field conditions. *Phytopathology*, **48**, 655–7.

Pfleger, F.L. (1973) Reaction of *Poa pratensis* introductions to *Puccinia graminis. Plant Dis. Rept.*, **57**, 595–8.

Plumb, R.T., Jenkyn, J.F. and Bowen, R. (1976) Pathogens occurring on ten varieties of ryegrass at Rothamsted in 1973–74 *J. Br. Grassld. Soc.*,, **31**(2), 65–7.

Raines, M.A. (1922) Vegetative vigour of the host as a

factor influencing susceptibility and resistance to certain rust diseases of higher plants. 1. Introduction. *Am. J. Bot.*, **9,** 183–203.

Rapilly, F. (1979) Yellow rust epidemiology. *Ann. Rev. Phytopathol.*, **17,** 59–73.

Ray, W.W. (1953) Leaf rust of Merion Bluegrass in Nebraska. *Plant Dis. Rept.*, **37,** 578.

RIVRO (1980) *55 Descriptive Variety List of Agricultural Crops.* RIVRO Wageningen, 336pp.

Rogerson, C.T. and King, C.L. (1954) Stem rust of Merion bluegrass in Nebraska. *Plant Dis. Rept.*, **38,** 57.

Rowell, J.B., Olien, C.R. and Wilcoxon, R.D. (1958) Effect of certain environmental conditions on infection of wheat by *Puccinia graminis. Phytopathology*, **48,** 371–7.

Sampson, K. and Western, J.H. (1954) *Diseases of British Grasses and Herbage Legumes.* 2nd edn, Cambridge University Press, 118pp.

Sanford, G.B. and Broadfoot, W.C. (1932) The relative susceptibility of cultivated and native hosts Alberta to stripe rust. *Sci. Agric.*, **13,** 714–21.

Savile, D.B.O. (1981) The supposed rust of *Echinochloa. Mycologia*, **23,** 1007–8.

Savile, D.B.O. (1983) *Puccinia striiformis. Fungi Canadensis*, **250,** 2 pp.

Savile, D.B.O. and Urban, Z. (1982) Evolution and ecology of *Puccinia graminis. Preslia*, **54,** 97–104.

Schaner, G. and Powelson, R.L. (1973) The oversummering and dispersal of inoculum of *Puccinia striiformis* in Oregon. *Phytopathology*, **63,** 13–17.

Schroeder, J. and Hassebrauk, K. (1964) Studies on the germination of the uredospores of yellow rust (*P. striiformis*). *Zbl. Bakt. Ser. 2*, **118** (6–7), 622–657. (Abs. in *Rev. Appl. Biol*, **44,** 3306). (German).

Schmidt, D. (1980) Specialization of crown rust (*Puccinia coronata* Cda.) on *Festuca pratensis* Huds. and *F. arundinacea* Schreb. *Ber. Schweiz. Bot. Ges.*, **90**(1–2), 55–60. (German).

Sharp, E.L., Schmidt, C.G., Staley, J.M. and Kingsolver, C.H. (1958) Some critical factors involved in establishment of *Puccinia graminis* var. *tritici. Phytopathology*, **48,** 469–74.

Simons, M.D. (1970) Crown rust of oats and grasses. *Am. Phytopathol. Soc. Monograph*, **5,** 47pp.

Smith, J.D. (1965) *Fungal Diseases of Turfgrasses.* 2nd edn. (revised N. Jackson and J.D. Smith). The Sports Turf Research Institute, 97pp.

Smith, J.D. (1980) *Major Diseases of Turfgrasses in Western Canada.* Univ. Sask. Extension Dept. Publ. 409, 14pp.

Solheim, W.G. (1958) *Puccinia poae-nemoralis* Otth on *Phleum pratense* L. and other grass hosts in Wyoming. *Plant Dis. Rept.*, **42**(4), 533.

Stock, F. (1931) Investigations on the germination and germ tube growth of the uredospores of some cereal rusts. *Phytopathol. Z.*, **3**(3), 231–79. (German).

STRI (1980) *Turfgrass seed, 1980.* The Sports Turf Research Institute, 105 pp.

Svensson, R. (1978) *Varietal Studies with Turfgrasses, 1968–1976.* Swedish University Agric. Sci. Information Centre, Alnarp. 111pp. (Swedish).

Tajimi, A. (1976) On the host plant of timothy rust, *Puccinia graminis* ssp. *phlei-pratensis. Bull nat. Grassld. Res. Inst.*, **9,** 25–40. (Abs. in *Rev. Appl. Plant Pathol.*, **56**(7), 3083. (Japan, En.)

Teuteberg, A. (1974) in *Ann. Rept. Biol. Inst. Agric. Forestry*, Berlin & Brunswick, pp.9 98–9. (Abs. in *Rev. Appl. Plant Pathol.*, **55,** (6), 2466.)

Tollenaar, H. (1967) A comparison of *Puccinia striiformis* f.sp. *poae* on bluegrass with f.sp. tritici and f.sp. *dactylidis. Phytopathology*, **57,** 418–20.

Tollenaar, H. and Houston, B.R. (1966) *In vitro* germination of uredospores of *Puccinia graminis* and *P. striiformis* at low spore densities. *Phytopathology*, **56,** 1036–9.

Tollenaar, H. and Houston, B.R. (1967) A study on the epidemiology of stripe rust, *Puccinia striiformis* West. in California. *Can. J. Bot.*, **45,** 291–307.

Tronsmo, A.M. (1982) Effects of low temperature hardening on resistance to biotic and abiotic stress factors in grasses. Dr. Scient. Thesis. Agric. Univ. Norway. 71pp.

Troughton, A. (1957) *The Underground Organs of Herbage Grasses.* Commonw. Bureau Pastures Field Crops, Hurley, Berks., Bull. 44, 163 pp.

Tu, J.C. and Hendrix, J.W. (1967) The summer biology of *Puccinia striiformis* in southern Washington. 1. Introduction of infection during the summer. *Plant Dis. Rept.*, **51,** 911–14.

Ullrich, J. (1976) First occurrence of yellow rust (*Puccinia striiformis* Westend) on smooth-stalked meadowgrass in Europe and its distinction from other rusts on *Poa* species. *Nachr. Deut. Pflshutzdienst.* (*Braunschweig.*), **28,** 177–80. (German).

Ullrich, J. (1977) The rust fungi on fodder grasses and turf grasses in mid-Europe. *Mitt. Biol. Bundesanst. Land. Forstwirtsch. Berlin*, **175,** 71pp. (German).

Urban, Z. (1967) The taxonomy of some European graminicolous rusts. *Ceska Mycologie*, **21,** 12–16.

Vargas, J.M. (1981) *Management of Turf Diseases.* Burgess, Minneapolis, 204pp.

Vargas, J.M., Young, H.C. and Saari, E.E. (1967) Effect of light and temperature on urediosopore germination, infection and disease development of *Puccinia cynodontis* and isolation of pathogenic races. *Phytopathology*, **57,** 405–9.

Watkins, J.E., Shearman, R.C., Houfek, J.A. and Riordan, T.P. (1981) Response of Kentucky bluegrass cultivars and blends to a natural stem rust population. *Plant Dis.*, **65**(4), 345–7.

Watkins, J.E., Houfek, J.A. and Shearman, R.C. (1981) Control of stem rust on Kentucky bluegrass, 1980. *Fungicide Nematicide Tests. Am. Phytopathol. Soc.*, **36,** 146–8.

Watson, A.J. (1957) A new rust for the West Indies. *Plant Dis. Rept.*, **41**(6), 547.

Webster, R.K. (1974) Recent advances in the genetics of plant pathogenic fungi. *Ann. Rev. Phytopathol.*, **12,** 331–53.

Weibull, P. (1977) Rust fungi attacking smooth-stalked meadow grass, *Poa pratenis. Weibulls Gräs-tips*, **20,** 3–7 (Swedish, En.).

Wilkins, P.W. (1978a) Specialization of crown rust on highly and moderately resistant plants of perennial ryegrass. *Ann. Appl. Biol*, **88,** 179–84.

Wilkins, P.W. (1978b) Specialization of crown rust (*Puccinia coronata* Cda.) on clones of Italian ryegrass (*Lolium multiflorum* Lam.). *Euphytica*, **27**, 837–41.

Wilkins, P.W. and Carr, A.J.H. (1971) Crown rust of grasses. *Rept,. Welsh Plant Breeding Sta. for 1970*, pp. 40–1.

Wilkins, P.W., Carr, A.J.H. and Lewis, E.J. (1974) Resistance of *Lolium multiflorum/Festuca arundinacea* hybrids to some diseases of their parent species. *Euphytica*, **23,** 315–20.

Wilson, M. and Henderson, D.M. (1966) *British Rust Fungi*. Cambridge University Press, 384 pp.

Woodcock, T. and Clarke, R.G. (1983) List of diseases recorded on field crops and pastures in Victoria before 30 June 1980. *Dept. Agric., Govt. Victoria, Tech. Rept.* Series 65, 31 pp.

Worf, G.L. (1982) Fungicide evaluation to control rust on ryegrass, 1981. *Fungicide Nematicide Tests, Am. Phytopathol. Soc.*, 37, 157.

Worf, G.L. and Huibregste, D.H. Jr (1981) Fungicide evaluation for control of rust on ryegrass, 1980. *Fungicide Nematicide Tests, Am. Phytopathol. Soc.* **36,** 151.

Wu, Wen-Shi, (1979) Diseases of lawn grass, *Zoysia tenuifolia.* Phytopathologist & Entomologist, Nat. Taiwan Univ. pp. 53–7.

Zadoks, J.C. (1965) Epidemiology of wheat rusts in Europe. *FAO Plant Prot. Bull.*, **13**(5), 97–108.

Part six
Smuts

15 *Smut diseases*

The smuts comprise a diverse and widely distributed group of fungi that are of major economic importance, particularly as pathogens of the Poaceae. A conspicuous feature of most smut fungi is their production of sori in host tissues that, on maturation, release an abundance of black or dark-coloured, dusty teliospores. Cereals and grasses are hosts to a large number of smut species which variously affect inflorescences, stems and leaves to bring about reductions in grain yield and dry matter production. Smut diseases confined to the inflorescence and culm are of obvious concern to the turf grass seed producer and several of common occurrence are mentioned briefly. Smut diseases of the vegetative organs are of greater significance in managed turf and are accorded a detailed treatment more befitting their importance.

15.1 LEAF SMUTS

Species of *Entyloma* cause leaf spot smut and leaf blister smut on the foliage of some cool season grasses but, although occasionally damaging, severe outbreaks have not been reported with any great frequency (Fischer, 1951, 1953b; Fushtey and Taylor, 1977; Gould, Brauen and Goss, 1977; Smith, 1980; McKenzie and Latch, 1981). In contrast, two other leaf smut fungi figure prominently as widely occurring, destructive pathogens on a range of cool-season grasses, particularly the turf-forming species of *Agrostis* and *Poa*. *Ustilago striiformis* and *Urocystis agropyri* establish systemic infections that adversely affect form and function of the grass host, in addition to physically disrupting leaf tissues when sporulation of the invading organism occurs. The ability of flag and stripe smut-infected plants to withstand unfavourable environmental conditions is impaired and secondary invasion by other pathogens may compound the damage in a smut-infected turf.

15.1.1 Blister or spot smut

Blister or spot smuts caused by *Entyloma* spp. occur on many of the grasses throughout the world. Of the turf grasses, *Agrostis, Poa, Holcus*, and *Phleum* spp. were reported as suscepts of *E. dactylidis* by Fischer (1953a). Whether attacks by other graminicolous species of *Entyloma* are referrable to *E. dactylidis* is uncertain, but the latter has a world distribution from sub-tropical to sub-boreal regions on members of the grass family (Boerema, Van Kesteren and Dorenbosch, 1962; Conners, 1967; Fushtey and Taylor, 1977; Gould, Brauen and Goss, 1977; Jørstad, 1963; Kucnierz, 1976; Mankin, 1969; McKenzie and Latch, 1981; Ramakrishnan and Srinivasan, 1950; Smith, 1980; Zundel, 1953). Dickson (1947) observed that spot smut (as *E. irregulare*) is of widespread distribution on *Poa* spp. in Europe and North America, especially in the north western region of the latter. Similarly, *E. crastophilum*, occurred on *Phleum* spp., *Agrostis* spp. and some other grasses over this same geographic region. Fischer (1953a) included both these *Entyloma* species under *E. dactylidis. Poa pratensis* L. is the turf grass species which is most often reported attacked (Boerema, Van Kesteren and Dorenbosch, 1962; Fischer, 1953a,b; Fushtey and Taylor, 1977; Gould, Brauen and Goss, 1977; Smith, 1980), but Jørstad (1963) recorded it on *P. annua* L. in Iceland, and Sprague (1955) and McKenzie and Latch (1981) found it on that species in Alaska and New Zealand respectively. Smith (unpublished) noted the fungus on *Festuca longifolia* Thuill. from turf in British Columbia.

(a) Symptoms

On *P. pratensis* the disease first appears as spots of water-soaked tissue on the lower leaf surfaces. In these areas, raised olive-green, brown, and later almost black blisters develop containing the sori of the smut spores. Individual spots may be surrounded by chlorotic leaf tissue which gives an overall yellowish cast to the infected turf. Later the turf may take on a greyish appearance as the sori mature and darken. The disease may grow out with the approach of spring, but where turf was severely infected some leaves containing sori may be found (Fig. 20, colour plate section) during the summer.

Blister smut is mainly a disease of leaf laminae

where the sori are most commonly found (Fischer, 1953a; Fushtey and Taylor, 1977; Gould, Brauen and Goss, 1977; Smith, 1980), but they may also occur in leaf sheaths and lower parts of the inflorescence (Fischer, 1953a; McKenzie and Latch, 1981). The sori vary in shape from circular or oval to almost rectangular; they are amphigenous, dark olive green to brown to almost black, pseudoparenchymatous, usually delimited by the leaf veins, 0.25–2.0 mm in length × 0.1–0.3 mm thick, lying between the two epidermal layers. Unlike stripe and flag smut (q.v.) the leaves are not usually ruptured by the sori.

(b) The causal fungus

Entyloma dactylidis (Pass.) Cif. (*Bull. Soc. Bot. Ital.* **2,** 55 (1924)).

Synonyms: *E. crastophilum* Sacc. (Michel. 1: 540. (1879)). *E. irregulare* Johans. (Ofvers. Vet.-Akad. Forhandle.: 159 (1884)). For full synonomy see Fischer (1953a).

The spores are globose, polygonal or polyhedral 7–20 × 8–17 μm, adherent in irregularly shaped groups (Fig. 15.1). The exospore is smooth and sheathless. Spore morphology varies considerably. Conidia, reported in other *Entyloma* spp. have not been reported in *E. dactylidis* (Fischer, 1953a). McKenzie and Latch (1981) listed teliospores at 10–14 × 8–13 μm in size with an exospore thickness of 0.5 μm. They described the basidiospores as hyaline and allantoid, 7–16 × 1.5–3 μm in size.

Epidemiologically, the disease seems to be favoured by mild winters (Fischer, 1951; Fushtey and Taylor, 1977; Gould, Brauen and Goss, 1977) and severe damage to *Poa pratensis* has been reported in the winter months especially in the Pacific northwest of North America. However, exceptionally it may cause moderate damage elsewhere, e.g. in eastern Washington (Fischer, 1951) and Rhode Island (Jackson, unpublished). Although recorded at Saskatoon in the Canadian prairies (Smith, 1980) after a mild winter, damage was slight. Fischer (1951) described a situation where *P. pratensis* only was infected in a mixed sward with *Festuca rubra* spp. *commutata*, *Lolium perenne*, and *Poa annua*, and suggested that there may be physiological races of the fungus since the latter three species are suscepts. However, cross inoculation trials with isolates of *E. dactylidis* from six species of grass in New Zealand, including *Poa pratensis, Poa annua, Agrostis stolonifera* and *Holcus lanatus*, indicated to McKenzie and Latch (1981) that the isolates were all one species. Gould, Brauen and Goss (1977) and Fushtey and Taylor (1977) noted differential susceptibility to the disease in field plots of cultivars of *P. pratensis.* Baron, Galaxy, Parade, and Victa were most susceptible, and Delft, Cougar, Adorno, Enporo, Fylking, Majestic, Merion, Nugget, Onar and Sydsport were the most resistant.

The disease is not likely to be of major importance except in very mild winters, since it occurs in the season when turf utilization is low. However, severe attacks may slow down spring recovery. No chemical control has been reported, but the use of resistant cultivars should be attempted where the disease is prevalent.

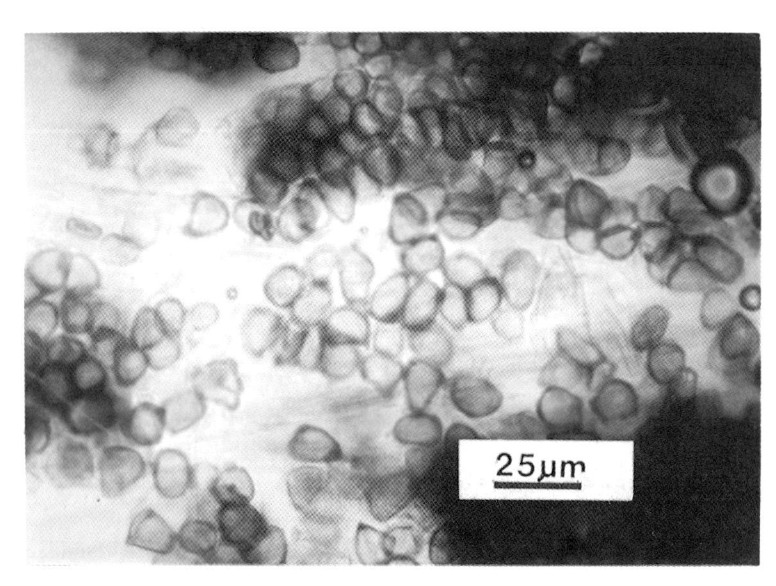

Figure 15.1 *Entyloma dactylidis*. Masses of irregularly-shaped teliospores.

15.1.2 Stripe smut

Stripe smut is of world-wide distribution on a large number of cool-season grasses (Fischer, 1953a; Zundel 1953). Many of the common turf grasses are hosts but damage by this fungus is most extensive and severe in turf formed by species of *Poa* and *Agrostis*. Stripe smut is a major turf disease of *Poa pratensis* L. turf in the US where widespread use of very susceptible cultivars, e.g. Merion, have contributed to the common occurrence of the pathogen.

(a) Symptoms

On unmown stands of grass, diseased plants may show various degrees of stunting, an erect, stiff growth of leaves and reduced inflorescence production. In mown turf these characteristics are less obvious but a conspicuous symptom common to both situations is the appearance of narrow, linear, interveinal, yellow–green streaks along the leaf blades and sheaths (Fig. 21, colour plate section). The streaks mark the initiation of the telial sori beneath the epidermis of the host. As the sori develop they assume a lead-grey appearance. When mature, rupture of the overlying epidermal layer exposes the black dusty masses of teliospores. Maturation of contiguous, interveinal sori along the length of the leaf results in the characteristic long, linear black stripes. Rupturing of the epidermis is soon followed by splitting or shredding of laminae as leaves necrose and curl from the tips downwards.

The description above is typical of stripe smut on *Poa pratensis* turf. However, on close-mown leaves of vigorously growing *Agrostis stolonifera* L., the disease may be difficult to recognize because of the paucity of sori and their small size (Hodges, 1969a). Even with the onset of drought or heat stress, often the only indication of a smut problem in bentgrass turf, is an unthriftiness and poor colour as sporadic leaf necrosis occurs on the infected plants. In contrast, smutted bluegrass turf under similar adverse growing conditions deteriorates markedly. Leaf necrosis intensifies, many infected plants die, and these swards become thin, ragged in appearance and of poor colour.

As in other grasses, sori primarily occur on the leaf laminae and sheaths of *Poa pratensis* and *Agrostis stolonifera*, but subterranean sori on rhizomes of the former have been reported (Duran, 1968; Ruben, 1968), and internodal sori may occur occasionally on the stolons of the latter, Hodges (1969a). Sori may also form on the inflorescences of many grasses including the above-mentioned species (Clinton, 1900; Davis, 1924; Fischer, 1940a; Hodges, 1970c,1972). In general, stripe smut infection inhibits the emergence, the number and size of inflorescences. Inflorescences formed may abort or be distorted and seed set is greatly reduced. On *Agrostis stolonifera*, sori have been observed in rachis, glumes and anthers but not in other organs of the inflorescence. In *Poa pratensis*, sori may occur in all parts of the inflorescence and in the developing caryopses. In both grasses, seed from smutted plants can produce seedlings that are smut free (Hodges, 1970c, 1971b, 1972).

Symptoms of stripe smut in a sward of *Poa pratensis* are indistinguishable macroscopically from those caused by flag smut. Morphological differences due to infection by these two leaf smuts have been demonstrated in field-collected plants grown under greenhouse conditions. Critical studies of the two diseases on the Merion cultivar of *Poa pratensis* confirmed that, although some effects on the plant are common to both smuts, i.e. reduced lateral branching, decreases in total dry matter and reduced inflorescence production, others are specific. Compared with healthy plants, stripe smut-infected plants have elongated, more upright, thicker leaves with broader blades; tillering is largely extravaginal, the root system is reduced somewhat and there is a reduction in inflorescence production. Infected plants have been shown to have a lower content of free amino acids. Although this may account for some of the morphological changes, the leaf characteristics are considered suggestive of hyperauxiny.

Morphological changes is plants infected by flag smut include dwarfing of the leaves and narrowing of the blades, predominantly intravaginal tillering, near absence of inflorescences and a severely reduced root system. Plants infected by *U. agropyri* have severely reduced levels of both soluble sugars and free amino acids which could well account for all the morphological changes brought about by this pathogen (Hodges, 1969c, 1970b, 1971b; Hodges and Robinson, 1977).

These morphological differences were only apparent on plants grown singly under greenhouse conditions and are not obvious in the field. Both fungi may occur together in a sward and mixed infections on the same plant have been reported (Clinton, 1904; Halisky, Funk and Bachelder, 1966, Niehaus, 1968; Thirumalachar and Dickson, 1949). Positive diagnosis requires microscopic examination of the teliospores.

(b) The causal fungus

Ustilago striiformis (Westend) Niessl (Hedwigia 15: 1, 1876). The extensive synonymy of this fungus has been outlined by Fischer (1953a).

Teliospores are chiefly globose to subglobose or ellipsoid, rather dark olive brown, more or less

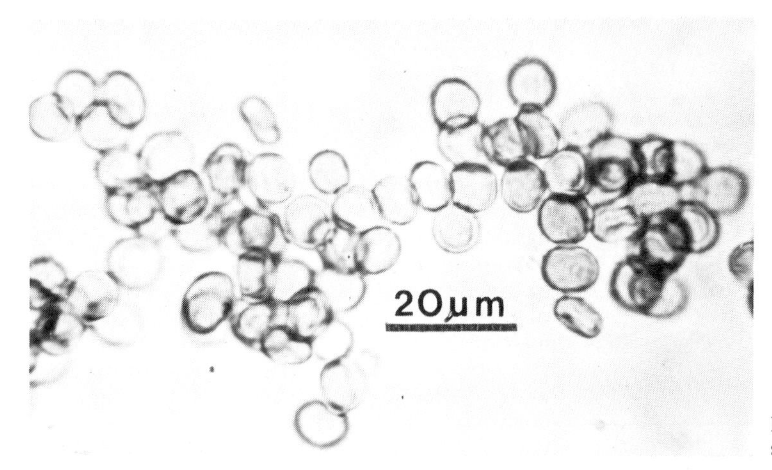

Figure 15.2 Teliospores of *Ustilago* striiformis.

prominently echinulate, 9–11 µm in diameter (Fischer, 1953a) (Fig. 15.2). Zundel (1953) gives teliospore size as 9–14 µm diameter. Mordue and Waller (1981b) report a mean spore diameter of 10.8 µm with walls 0.5 µm or less thick and spines 0.5–1.0 µm long.

In *Agrostis stolonifera* mycelium is intercellular with hyphae 1–2.4 µm in diam., moderately branched and septate. Septa are simple or rarely with clamp connections. Narrow hyphae 1.0 µm diameter occur in the stolon internodes connecting the nodes. The larger diameter hyphae, with short intercellular, knoblike protruberances, occur in leaves and also adjacent to sori in stolon internodes. Metamorphosis of the mycelium into teliospores is rapid and almost complete in leaves and stolon internodes leaving only traces of mycelium near sori. Mature teliospores measure 9–12 µm in diameter (Hodges, 1969a).

Hodges and Britton (1970) reported that the mycelium in the crown tissue of *Poa pratensis* was also intercellular. However, Thirumalachar and Dickson (1953) found both intra- and intercellular mycelium in parenchymal tissues present in the crowns of mature *Phleum pratense* L. plants. Both types of mycelial colonization were also demonstrated in the crowns of seedlings of several grasses including *Agrostis alba* L. and *Poa pratensis*. Cells of the phloem parenchyma were penetrated frequently and lobed to irregularly shaped hyphae filled the cell lumina.

The fungus may be grown on agar media to produce viable teliospores. Critical temperatures vary between isolates but are approximately 5 °C minimum, 10–18 °C optimum and 36 °C maximum (Fischer, 1940a; Kreitlow, 1943; Leach, Lowther and Ryan, 1946; Thirumalachar and Dickson, 1953).

(c) Disease development

Perennation of stripe smut is accomplished by dormant mycelium in infected plants or by teliospores in the thatch or soil. Teliospores in sori on components of turf grass inflorescences, including the caryopsis, are possible sources of primary inoculum into new seedings (Hodges, 1970c, 1972). Sori and 'mycelium reservoirs' occurring respectively in internodes and nodes of *Agrostis stolonifera* stolons may also serve to disseminate the fungus when turf of this species is established by vegetative means (Hodges, 1969a). Infected sod may be shipped considerable distance and the new turf areas then develop stripe smut. Contaminated turf machinery and teliospores rendered airborne by such machinery (rotary mowers, dethatchers) contribute to the local spread of the fungus.

Fresh teliospores of some races of *U. striiformis* can germinate directly but others may require an after-ripening period of several months (Fischer and Holton, 1957). The erratic germination of teliopores suggested to Thirumalachar and Dickson (1953) that an inhibitor was involved, but the vegetative condition of the host, the season of year when the teliospores were collected and the incubation temperature all influenced ability to germinate. Some germination occurred over a wide range of temperatures (4–32 °C) but the optimum varied considerably with the isolate. A minimum of 7 °C, an optimum of 22 °C and a maximum of 35°C was reported for *U. striiformis* by Stakman and Harrar (1957).

Environmental conditions which favour germination of grass seeds are also favourable to the germination of teliospores (Couch, 1973). On germination, teliospores produce one or more short, simple or branched germ tubes, the promycelia. Each comprises a few cells on which may be borne haploid sporidia. Fusion between compatible haploid elements, i.e. between sporidia, between sporidia and promycelial cells or between promycelial cells themselves, must occur before the dikaryotic infection hyphae can be produced (Fischer and Holton, 1957; Thirumalachar and Dickson, 1953). Infection hyphae may then penetrate directly into or between the epidermal cells of coleoptiles and the axillary buds of crowns and rhizomes. The mycelium develops systemically in the infected bud, moving intercellularly (Hodges, 1969a; Hodges and Britton, 1970), or both inter- and intracellularly (Thirumalachar and Dickson, 1953) to colonize all subsequent plant organs except the roots (Davis, 1924; Hodges, 1976b). Sporulation in the leaves may begin to appear 5–6 weeks after infection (Fischer, 1940a; Hodges and Britton, 1969).

Hodges and Britton (1970) described how the mycelial development within *Poa pratensis* cv. Merion is dependent on the infection site. Coleoptile infection results in the colonization of the primary crown. With the possible exception of a few escaping buds (Hodges, 1976b), in a highly susceptible host all new vegetative growth from this infected crown will be invaded by smut mycelium. Infection of axillary buds on rhizome nodes also results in mycelial colonization of the developing crown and all subsequent growth therefrom. In contrast, infection via an axillary bud issuing from a primary crown is restricted to that developing bud and mycelium does not grow back to the parent crown. Growth and development of the pathogen is in the direction of plant growth (Hodges and Britton, 1970).

Systemic infection of shoots and rhizomes from infected plants, together with further primary infections of the axillary buds of healthy plants, serves to perpetuate and spread the disease in mature stands of *Poa pratensis*. Cultivars of *Poa pratensis* may differ however, in their capacity of support systemic proliferation of the pathogen (Hodges, 1976b).

In *Poa pratensis* the fungus perennates in infected crowns and meristems, but in *Agrostis stolonifera* this function is accomplished by 'mycelial reservoirs' present in all the nodes of infected plants. Growth of the pathogen from an infected node into the adjoining axillary bud is influenced by the position of that bud on the stolon in relation to the apical meristem. Progressively more smutted stolons are produced from axillary buds on the nodes away from the apical meristem, confirming that the apical dominance mechanism which controls bud development and growth also controls development of the pathogen in *Agrostis stolonifera* (Hodges, 1969a, 1971a, 1973). The development of the pathogen into axillary buds is influenced also by temperature (Hodges, 1970a). Low night/day temperatures of 7.2–12.8°C allow the pathogen to grow in the developing stolons but intercalary formation of teliospores is slowed. Temperatures in the range 15.6–21.1°C are optimum for the growth of smutted stolons; cool periods in the spring and autumn provide these optimum temperatures for mutual development of host and pathogen. Higher temperatures of 23.9–29.4°C are optimum for sorus production. Temperatures in the range 32.2–37.8°C inhibit pathogen growth but the fungus is not eradicated from the stolons.

Cool periods in the spring and autumn also favour stripe smut in *Poa pratensis* turf and temperatures in the range 10–16°C have been reported as most conducive to symptom expression (Kozelnicky, 1969; Kreitlow, 1943). Prolonged high temperatures above 32°C bring about a remission of smut symptoms but these high temperatures also promote leaf necrosis and, combined with drought stress, the death of many infected plants (Davis, 1924; Fischer, 1940a; Kreitlow, 1943; Kreitlow and Juska, 1959; Kreitlow and Myers, 1944; Leach, Lowther and Ryan, 1946). Whereas leaf necrosis and death of infected *Poa pratensis* plants have been reported to be a direct function of temperature (Davis, 1924; Kreitlow, 1943), Hodges (1970a) proposed that a combination involving temperature, humidity and water stress probably results in the leaf necrosis and demise of *Agrostis stolonifera* plants infected with stripe smut. Evidently this stress complex is involved in the thinning of *Poa pratensis* swards since irrigation has been shown to increase dramatically the survival of stripe smutted Merion plants (Hodges, 1977). Irrigation will maintain sward density over the short term by prolonging the longevity of infected plants but there is a more lasting benefit. A number of shoots on stripe-smutted plants may escape systemic infection and irrigation stimulates their production and their establishment as healthy replacements in the sward (Hodges 1976b, 1977).

Paradoxically, management procedures which ensure high quality swards during the summer months promote stripe smut in *Poa pratensis* turf (Hodges, 1977; Hodges and Britton, 1969, 1970). Irrigation and fertilizer practices which alleviate heat stress and avert dormancy also ensure that infected plants survive to produce more infected shoots until the disease may reach epiphytotic proportions. Consequently the disease is often a problem in older intensively managed turf. The progressive increase in stripe smut as turf gets older has been documented for several cultivars of *Poa pratensis* (Halisky, Funk and Bachelder, 1966). For the cultivar Merion the number of smutted tillers per square foot increased from 8 to 112, a 14-fold increase between the third and fifth year under turf maintenance.

Although the relationship between host nutrition and stripe smut is not clearly established in published reports, nitrogen application to *Poa pratensis* turf affects disease expression and intensity. Lukens (1966) reported urea applied in late May to be an effective disease control treatment for stripe smut but Halisky, Funk and Bachelder (1966) demonstrated that such spring nitrogen applications afforded only temporary suppression of symptoms and resulted in greater stripe smut incidence the following autumn. Hull, Jackson and Skogley (1979) found that *Poa pratensis* cv Merion turf treated at high rates with incomplete fertilizers was the most heavily diseased. Turf receiving high N-P-K fertilizer was less diseased than turf receiving high N alone or N plus P or K, indicating that stripe smut severity in a susceptible *Poa pratensis* cultivar is a function of fertility level and balance.

(d) Control

(i) CULTURAL PRACTICES Stripe smut is seldom a cause for concern in turf newly established from seed, but it becomes an increasing problem as the turf ages. Application of a balanced (N-P-K) fertilizer at rates adequate to maintain good growth may offset the damage caused by stripe smut but too generous use of nitrogen alone will exacerbate the disease. Smutted plants are especially vulnerable to the lethal effects of heat and drought stress. An effective watering programme is needed during hot dry weather to ensure the survival of infected plants and to encourage production of healthy tillers. Although these cultural measures will help sustain sward density, a high incidence of diseased plants will still be apparent. In a susceptible turf, stand reduction of the stripe smut incidence to insignificant levels can only accomplished by chemical control means.

(ii) RESISTANT CULTIVARS *Ustilago striiformis* shows a high degree of pathogenic specialization at both species and cultivar level (Fischer, 1940a; Fischer and Shaw, 1953; Thirumalachar and Dickson, 1953). Six *formae speciales* have been recognized and designated *agrostidis*, *dactylidis*, *holci hordei*, *phlei* and *poae* according to the grass genera they infect (Halisky, Funk and Bachelder, 1966). Wide differences in susceptibility have been demonstrated among cultivars of *Poa pratensis* (Gaskin, 1966; Hodges, 1976a; Halisky, Funk and Bachelder, 1966; Kreitlow and Juska, 1959) and in *Agrostis stolonifera* (Gaskin 1965a; Hodges, 1969b). The degree of susceptibility of *Poa pratensis* cultivars may vary with location reflecting the distribution and predominance of strains of the pathogen. Merion, long recognized as a very susceptible cultivar may be equalled or surpassed in this characteristic by some of the more recent introductions (Hodges, 1976a).

Blending of three or four improved *Poa pratensis* cultivars is advocated to lessen the risk of severe sward deterioration possible when a single susceptible cultivar is used to form a turf.

(iii) FUNGICIDES Until the advent of systemic fungicides, effective chemotherapeutants were not available for controlling stripe smut. Carboxin and oxycarboxin showed initial promise in greenhouse trials for stripe smut control (Hardison, 1967), but under field conditions they were impractical (Hardison, 1971a). Benomyl, however, afforded excellent protection against stripe smut in turf of both *Poa pratensis* (Cole, Massee and Duich, 1970; Halisky, Funk and Babinski, 1968, 1969; Hardison, 1968; Jackson, 1970; Vargas, 1972; and others); and *Agrostis stolonifera* (Hardison, 1968; Robinson and Hodges, 1972). The related benzimidazole fungicide thiophanate-methyl has given moderate to good control in the field (Jackson, 1974; Vargas, 1972), but thiophanate-ethyl has proved to be of little value in controlling the disease. (Jackson and Fenstermacher, 1974; Hardison, 1971c; Vargas, 1972). The pyrimidine fungicides triarimol and, later, fenarimol gave excellent, prolonged protection (Cole *et al.*, 1978; Hardison, 1971a, 1972; Jackson and Fenstermacher, 1974; Vargas, 1972). Triarimol is no longer available in the US. Triadimefon showed similar effectiveness and persistence in control of the disease in extensive trials (Cole *et al.*, 1978; Hardison, 1974; Jackson and Dernoeden, 1979; Sanders *et al.*, 1978).

Most of the systemic fungicides giving good control of stripe smut have little or no action against leaf spotting fungi in the genus *Helminthosporium*. Smutted *Poa pratensis* plants become vulnerable to leaf spot disease and the symptoms may be exacerbated by applications of some systemic fungicides. Combination of the systemic fungicides with an effective leaf spot control material has been advocated (Jackson, 1970). Quintozene has proved of value for this function and excellent control of stripe smut and leaf spot has been demonstrated from a single tank-mix application of quintozene plus either benomyl, triarimol, fenarimol, or triadimefon when treatment is made in the late autumn (Jackson and Fenstermacher, 1974; Jackson and Dernoeden, 1979). Treatment with systemic fungicides during the growing season will also control stripe smut but repeated application may be needed.

15.1.3 Flag smut

Flag smut occurs throughout the wheat growing areas of the world as a common pathogen of that cereal and also as a pathogen on a broad range of cool-season grasses, including turf-forming species of *Agrostis*, *Festuca* and *Poa* (Fischer, 1953a; Zundel, 1953). The fungus has received most attention as a turf pathogen in the United States where it is the cause of a disease of *Poa pratensis*, the symptoms of which are identical to those of stripe smut. Both of these leaf smuts initiate primary infections in a similar manner but *Urocystis agropyri* is considered to occur less frequently in turf and generally is regarded as a minor pathogen of *Poa pratensis* (Allison and Chamberlain, 1946; Gaskin, 1965b). The possibility that newly introduced cultivars may have a better capacity to support flag smut and hence upgrade the importance of this fungus as a pathogen of *Poa pratensis* has been suggested (Hodges, 1976a, 1977).

(a) Symptoms

Grass plants in unmown stands may be dwarfed or contorted and produce few normal seedheads when infected with flag smut. Sori develop extensively on the upper leaves including characteristically, the flag leaf and occasionally on the culm. In managed turf the typical, linear interveinal sori occur in the leaves traversing both laminae and sheaths. Initially appearing as dull white to yellowish grey streaks, the sori mature to a lead grey becoming black as the epidermis ruptures to reveal the teliospore masses. Leaves shred and curl to provide symptoms and macroscopic signs identical with those of stripe smut. The exposed spore masses reportedly have a brownish appearance as opposed to the much darker-coloured stripe smut (Thirumalachar and Dickson, 1949), but this feature is not considered a reliable distinction and microscopic examination of the teliospores is advocated for positive identification (Kreitlow, 1948; Niehaus, 1968). Differing morphological and developmental characteristics of *Poa pratensis* in response to infection by the two smuts have been described for greenhouse-grown spaced plants (Hodges, 1969c, 1970b, 1971b; Hodges and Robinson, 1977) (see page 185), but these differences are not discernible in the field.

Sori of flag smut may occur on the inflorescences of grasses (Davis, 1922; Hodges, 1971b). Infection of *Poa pratensis* by flag smut has been shown to inhibit the rate of appearance, number and size of inflorescence. Unlike *Ustilago striiformis*, lengthening the period of floral induction did not decrease the inhibitory effect of flag smut infection on inflorescence production. Inflorescences which were formed could be distorted or abort, and seed set was much reduced. However, it was concluded that the disease is not disseminated by embryo-infected, seed (Hodges 1969c, 1970b, 1971b).

(b) The causal fungus

Urocystis agropyri (Preuss) Schroter. (Abh. Schles. Ges. Abth. Nat. Med. 1869–72, **7**. 1869). Synonyms *U. festacae* Ule., *U. poae* (Liro) Padw, *U. tritici* Korn. For an extended synonomy of the fungus, see Fischer (1953a).

Spore balls are globose to elongate, mostly 18–35 μm × 35–40 μm composed of 1–6, mostly 1–3 teliospores and a completely investing cortex of hyaline to brownish, smaller sterile cells. The teliospores are globose to subglobose, dark reddish brown to olivaceous brown, smooth, 12–18 μm diam (Fig. 15.3) (Fischer, 1953a). Mordue and Waller (1981a) list a range of 18–38 (mean 26.5) μm diam. for the spore balls, 3–12 μm diam. for the sterile cells and 11–20 (mean 15.5) μm diam. for the teliospores, the latter with walls 1.5 μm thick. The outer wall of the sterile cells may collapse with age giving a ridged effect to the covering layer (Zundel, 1953).

Noble (1924) reported on the mycelial characteristics of the fungus in wheat plants. Mycelium was both inter- and intracellular in the early stages of infection becoming predominantly intercellular, mostly 1.5–2 μm diam. Intracellular mycelium occurred in older tissues (e.g. parenchymatous cells at the nodes) functioning possibly as haustoria or ‘representing a characteristic form of dormant mycelium’.

Dikaryotic mycelium in the leaves is concentrated in the interveinal regions, replacing the mesophyll cells. Teliospore-forming hyphae coil and branch and the branches further interwine to form knots of hyphal cells which in turn form the spore balls. Central fertile cells with dense cytoplasm (the teliospores) are enveloped by vacuolate cells from the outer layers of the sporogenous knot in which the nuclei degenerate (the sterile cells) (Fischer, 1953a; Thirumlachar and Dickson, 1949).

The fungus can be cultured on agar media but typical teliospores have not been observed (Thirumalachar and Dickson, 1949).

(c) Disease development

The fungus overwinters as dormant mycelium in infected plants or by teliospores in the thatch and soil. Teliospores from sori borne on components of the inflorescence and contaminating the seed may also serve as primary inoculum (Hodges, 1971b). The

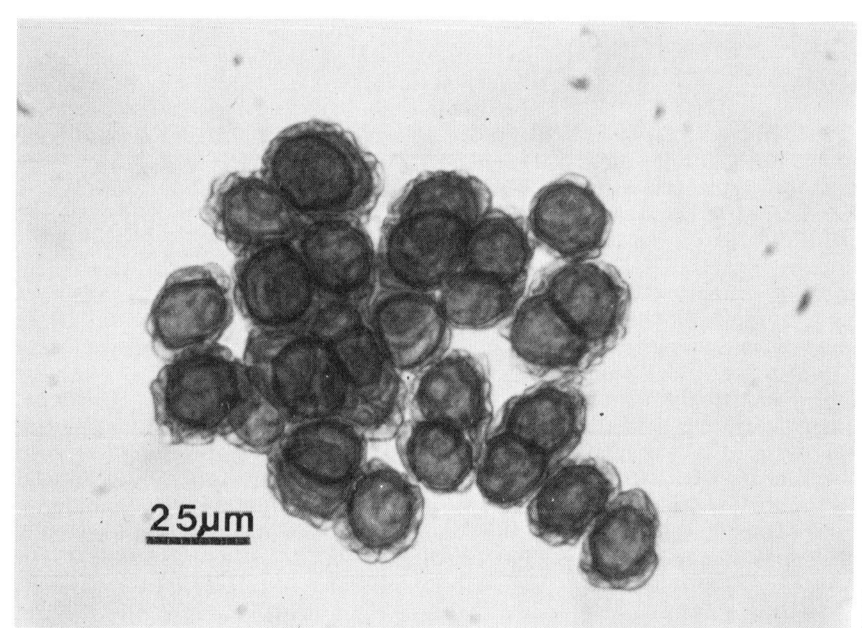

Figure 15.3 Teliospores of *Urocystis agropyri.*

fungus may be dispersed on infected sod. As with stripe smut, contaminated turf machinery and teliospores rendered air-borne by turf machinery contribute to the local spread of the fungus.

Teliospores of *Urocystis agropyri* are long lived and exhibit variable dormancy phenomena similar to those found in *Ustilago striiformis*. Various substances and materials (e.g. benzaldehyde, plant infusions) have been reported as stimulants to spore germination (Fischer, 1953a), but the teliospores can be germinated over water on slides in 5–45 days depending on the temperature regime imposed (Thirumalachar and Dickson, 1949). Soil temperatures of 18–24°C proved optimum for the germination of flag smut teliospores from wheat and 14–21°C was optimum for infection of wheat seedlings by this organism (Noble, 1924). After reviewing the results of several investigators Purdy (1965) listed the cardinal soil temperatures for infection of wheat by flag smut as 5° minimum, 20° optimum and 28°C maximum.

Germination of the teliospores of flag smut from *Agrostis alba* L. has been documented by Thiraumalachar and Dickson (1949). A single germ tube, the promycelium, issues from the teliospore bearing at the apex a whorl of 3–4 sporidia. Single haploid nuclei are usually present in each sporidium. Fusion between compatible sporidia occurs while they are still attached to the promycelium and the paired nuclei then migrate into the infection hyphae, arising usually from the conjugation tube. Dikaryotic infection hyphae also may arise directly on the promycelium from two, binucleate sporidium-like structures, indicating pairing of nuclei without the production of a conjugation tube.

Infection hyphae penetrate directly into or between the epidermal cells of the coleoptiles and the axillary buds of crowns and rhizomes. The mycelium develops systemically within the tissues of the infected bud; this is typically intercellular, but both inter-and intracellular mycelium has been observed in infected wheat plants. Development of the fungus keeps pace with the vegetative growth of the host and within 4–6 weeks sori begin to form within the leaves (Noble, 1924). Persistence and spread of flag smut in a *Poa pratensis* turf follows closely that described for stripe smut but Hodges (1976a) has demonstrated that *Urocystis agropyri* initiates primary infections of this host much more aggressively than *Ustilago striiformis*. Given this apparent advantage, it is difficult to reconcile the classification of flag smut as a minor pathogen of *Poa pratensis* (Allison and Chamberlain, 1946; Gaskin, 1965b). A higher mortality rate amongst flag smut-infected plants has been postulated by Hodges, (1976a) to explain the much lower incidence of this disease as opposed to that of stripe smut; in effect, flag smut is self-limiting in *Poa pratensis* turf. The same author points out that if inadvertent development of selections of this host are made that can better support *Urocystis agropyri*, this pathogen could be potentially more serious than *Ustilago striiformis* under conditions of intense management (Hodges, 1977).

No information is available regarding the influence of host nutrition on the severity of flag smut in turf. Application of lime has been shown to increase flag smut infection of wheat and the effect was magnified if superphosphate or farmyard manure was applied in addition. The effect was possibly the result of increased soil pH afforded by all the treatments, but calcium levels in the host plant (wheat) also apparently influenced the disease expression. Wheat plants deficient in calcium showed a marked reduction in the development of flag smut. Deficiencies of nitrogen or potassium showed a similar tendency, but a deficiency of phosphorus had the reverse effect (Fischer and Holton, 1957).

Figure 15.4 Sheath smut (*Ustilago hypodytes*) on *Agropyron repens*.

(d) Control of the disease

(i) CULTURAL PRACTICES Cultural practices which aid in the reduction of flag smut in turf have not been documented specifically but it is assumed that those listed for stripe smut (page 188) also obtain for flag smut.

(ii) RESISTANT CULTIVARS Physiological specialization occurs in *Urocystis agropyri* (Dickson, 1947; Fischer and Holton, 1957). Within *Poa pratensis* there is a considerable range in susceptibility to the pathogen (Gaskin, 1965b; Hodges, 1976a). Of 21 cultivars evaluated for their reaction to flag smut Merion proved the most susceptible. In these same trials only A. 34, Cougar and Nugget seemed to show resistance to both flag and stripe smut (Hodges, 1976a).

(iii) FUNGICIDES Since 1966, when Hardison first reported on the chemotherapy of *Urocystis agropyri* in *Poa pratensis* cv Merion, with two 1,4-oxathiin compounds, more effective control with several other systemic fungicides has been demonstrated, Hardison (1966, 1968, 1971a, 1971b, 1971c, 1972, 1974). These same fungicides also control *Ustilago striiformis* (see page 188).

15.2 SHEATH SMUT

Ustilago hypodytes (Schlecht.) (Fries Syst. 3: 518. 1832).

Synonomy: Angus (1956) disagreed with the decision of Fischer and Hirschhorn (1945) to divide the *U. hypodytes* complex into the species, *U. spegazzinii* Hirsch, *U. williamsii* (Griff.) Lavrov, *U. halophila* Speg. and *U. nummularia* Speg. Angus (1956) determined that *U. spegazzini* and *U. nummularia* were conspecific with *U. hypodytes* and suggested that the name *U. halophyta* should be temporarily retained for the smut on *Distichlis* spp. and found that *U. williamsii* was morphologically and culturally distinct. For further details of synonomy see Ainsworth and Sampson (1950) and Angus (1956) and Fischer (1953a).

U. hypodytes is a widely distributed sheath and culm smut of many species of grasses, mostly of low agricultural value in Europe (Ainsworth and Sampson, 1950; Mühle, 1971; Sampson and Western, 1942) on some important forage species in North America (Conners, 1967; Fischer, 1953a, USDA, 1960) and occurs in other parts of the world (Angus, 1956). In Western Canada the smut attacks the Fairway cultivar of *Agropyron cristatum* (L.) Gaertn. used in unirrigated pastures and coarse lawn turf (Smith, unpublished), considerably reducing plant vigour.

Sori occur in grass stems extending from node to successive node and sometimes involves spikelets, but inflorescences are usually absent in smutted plants (Fig. 15.4). There is no special sorus covering although the spores are at first covered by the leaf sheaths. The spores gradually weather away leaving a bare culm. The teliospores are spherical to oval sometimes with an inconspicuous smooth or crenellated cap at each pole, yellow to brown, smooth to faintly verrucolose, (3) 3.5–5 (6) μm in diameter.

Infection is systemic and a perennial mycelium occurs in the rhizomes (Bond, 1940). The disease is not seed-borne and the teliospores have a long incubation period before germination (Fischer and, Holton, 1957). No chemical control has been reported, but fungicides effective against stripe smuts (q.v. page 188) should bc tricd.

15.3 HEAD SMUTS

15.3.1 Covered smuts

Tilletia decipiens (Pers.) Korn. (Hedw. **16**, 30, 1877) (for synonomy see Ainsworth and Sampson, 1950, Fischer, 1953a).

Infected plants are stunted. Ovaries are replaced by spore masses, held together by the pericarp and partly concealed by the glumes, about 1 mm long. Spore masses have a foetid (rotten fish) smell and when mature at the end of the summer, rupture and the brown, powdery teliospores infest adjacent seed. Teliospores are globose to sub-globose, brown with deep, irregular reticulations (3–5 μm wide and 2–4 μm deep) 26–32 μm in diameter. It is widespread on *Agrostis canina, A. stolonifera*, and *A. tenuis* in Britain (Ainsworth and Sampson, 1950) and present on *A. tenuis* from Nova Scotia, Canada, and St Pierre and in a seed lot of the latter species received in Canada from the USA (Conners, 1967).

Tilletia pallida (G.W. Fischer. *Mycologia*, **30**, 393, 1938).

Spore masses and spores lighter in colour than those of *T. decipiens*. Teliospores are a little smaller, with irregular, conical spines c.3 μm in exospore, 25–28 μm. Found on *Agrostis canina* L. from Canada and on that species and *A. stolonifera* L. from the USA (Fischer, 1938; USDA, 1960; Conners, 1967).

Tilletia caries (DC) Tul. (*Ann. Sci. Nat. Bot.*, **3**, 113, 1815) (for synonomy see Ainsworth and Sampson, 1950; Mordue and Waller, 1981c). Spore masses are dark-brown and foetid. Teliospores are light brown, globose to sub-globose, reticulations wide and shallow (2–4 μm wide and 0.51 μm deep), 14–20 μm diameter. Found on *Poa* spp. and *Agropyron cristatum* in North America (USDA, 1960).

Tilletia controversa Kuhn in Rabenhorst, (*Fungi Europaei* No. 1896, 1974) (for synonomy see Duran and Fischer, 1961). Spore masses are powdery, pale to dark reddish to dark brown, consisting of spores and sterile cells. Teliospores are globose to subglobose yellow to reddish-brown 16–25 μm diam. With ploygonal reticulations of wall which are larger and with deeper reticulation than those of *T. caries* (Fig. 15.5). Occurs occasionally on turf grass species *Agropyron, Festuca, Holcus, Lolium* and *Poa* causing dwarfing and bunting.

Tilletia foetida (Wallr.) (Liro Maanviljelys Talondellinen Koelaitos Vuoikirja 1915–1916, **27**, 1920) (for synonomy see Fischer, 1953a; Mordue and Waller, 1981d). Spore masses brown, teliospores light to dark brown without reticulations, 15–26 μm diameter. *T. caries* and *T. foetida* arc rcportcd to bc ablc to overwinter on perennial grasses. Both can infect *Agropyron* spp. (Fischer, 1936, 1939) and *T. caries* is capable of infecting *A. cristatum* in the field (Ainsworth and Sampson, 1950). *T. foetida* will infect *Lolium perenne* L. (Bressman, 1932). Spore balls of *T. caries* were found in a lot of *Poa pratensis* L. seed received in Canada from Washington State, USA (Smith, unpublished).

Tilletia lolii Auers. (Klotzsch-Rabenhorst, *Herb. Viv. Myc.* 1899, 1954). Sori partly hidden by the

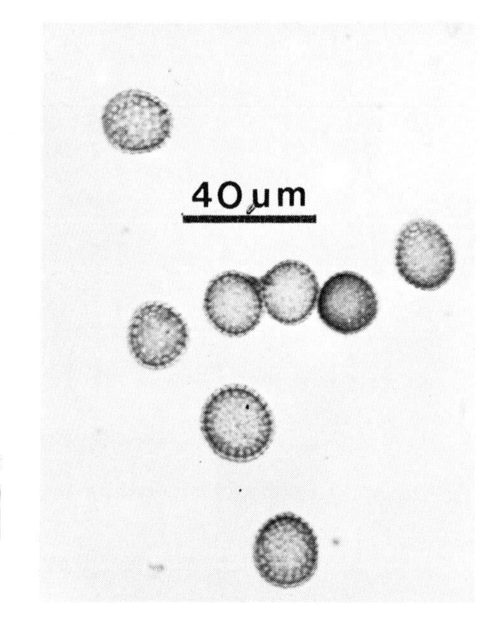

Figure 15.5 Teliospores of *Tilletia contraversa* from *Poa pratensis*.

glumes, 5–7 mm long. Spore mass powdery, brown, foetid when fresh, teliospores globose to sub-globose, light-brown, reticulate (shallow, small mesh, 2–4 μm wide, 2–3 μm deep) 18–22 μm diameter. *Lolium* spp., including *L. perenne* are susceptible (Sampson and Western, 1954; Ainsworth and Sampson, 1950).

Tilletia holci (Westend.) Schroet. (*Cohn Beitrag. Biol. Pflanz.*, **2**, 365, 1877) (for synonomy see Ainsworth and Sampson, 1950; Fischer, 1953a). Sori in ovaries partly hidden by glumes, 2–3 μm long. Spore masses brownish-black, slightly foetid. Teliospores globose to sub-globose, dark-brown, reticulate (ridges prominent and thick, depressions hexagonal, 4–6 μm wide, 2–3 μm deep, 21–26 μm diameter). On *Holcus lanatus* L. and *H. mollis* L. in Britain (Ainsworth and Sampson, 1950) and on *Anthoxanthum odoratum* L. and *H. lanatus* in North America, South America, Europe and New Zealand (Conners, 1967; USDA, 1960; Waller and Mordue, 1983).

(a) Disease development

The teliospores of these *Tilletia* spp. are dispersed and infest 'clean' seed and the soil following break up of the sori. This may take place before harvest, but usually during this and subsequent operations. Teliospore germination occurs when the seed sprouts. A haploid promycelium develops from which a crown of branches with sporidia arise. These may be abstricted or remain attached. Dikaryotic sporidia develop by fusions between sporidia or with the promycelium. Entry to the host by the diploid hyphae is via immature seedling tissues, normally the plumule. Growth within the host is at first intra-and later intercellular. Eventually the mycelium may reach the inflorescence primordium where it develops in step with the hosts rate of growth which may be affected by its presence. Eventually the fungus replaces ovarian tissues with sori of teliospores.

Covered smuts or bunts caused by *Tilletia* spp. do not usually cause significant damage to turf grass seed crops and may be prevented with fungicidal seed dressings similar to those used for cereals. Where soil-borne inoculum is involved fungicidal seed dressings are less effective.

15.3.2 Loose smuts

Ustilago bullata Berk. (Hooker, *Flora of New Zealand*, **2**, 196, 1855).

This is a complex of closely related smuts on grasses, components of which were formerly accorded specific status according to host. These were reduced to one species with eight physiological races by Fischer (1940b, 1953a).

Sori replacing complete floral parts, sometimes including glume bases, are at first covered with the host epidermis which ruptures to expose the powdery-black spore mass which is sometimes coherent for a time. (Fig. 15.6). Eventually only the bare rachis and shreds of the glume bases may be left. Teliospores are globose to sub-globose, yellow–brown to dark brown, minutely verrucose to granular or smooth, 6–12 μm in diameter (Fig. 15.7).

Found on a wide range of grass hosts in many parts of the world including N. and S. America, Europe, Asia Minor, and Australasia (Ainsworth and Sampson, 1950; Cunningham, 1924; Falloon, 1979; Fischer, 1940b; Fraser and Scott, 1926; Liro, 1938; McAlpine, 1910). Of the grasses commonly used in coarse,

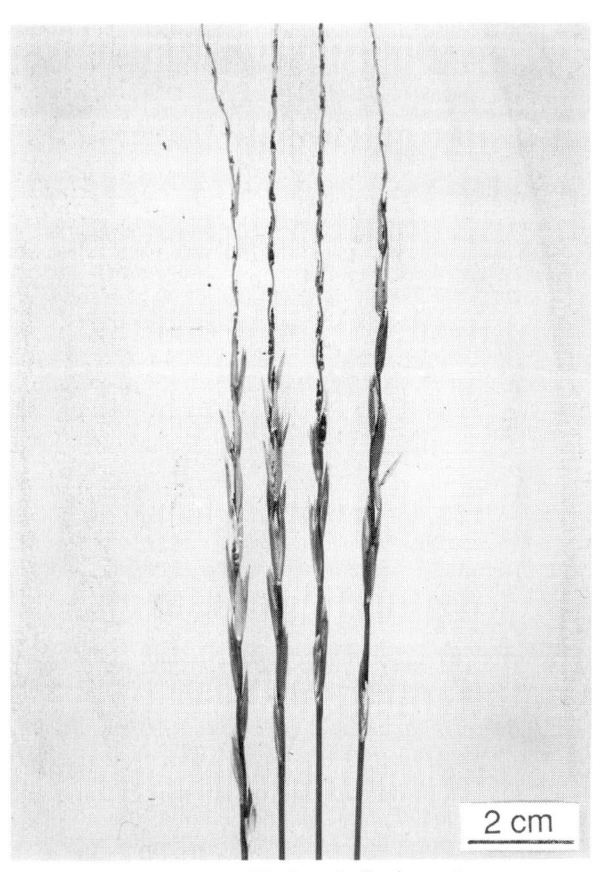

Figure 15.6 Loose smut (*Ustilago bullata*) on *Agropyron* sp.

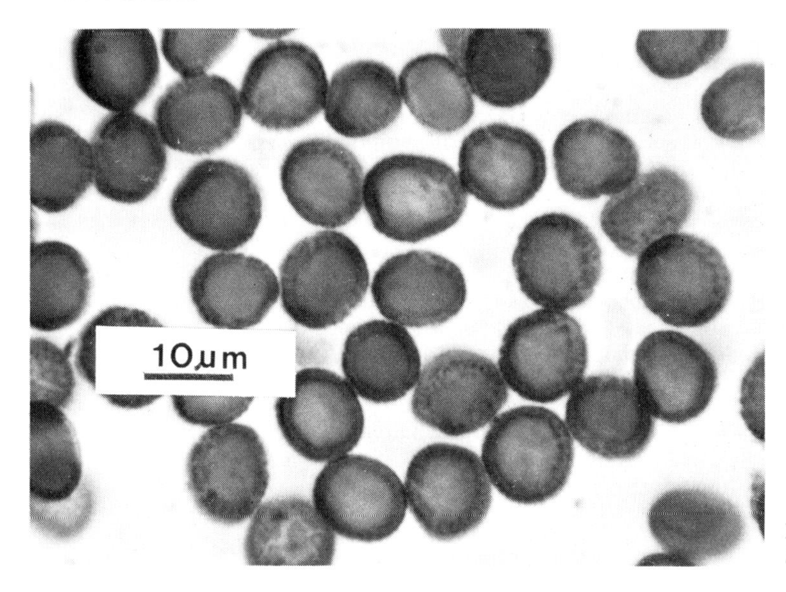

Figure 15.7 Teliospores of *Ustilago bullata.*

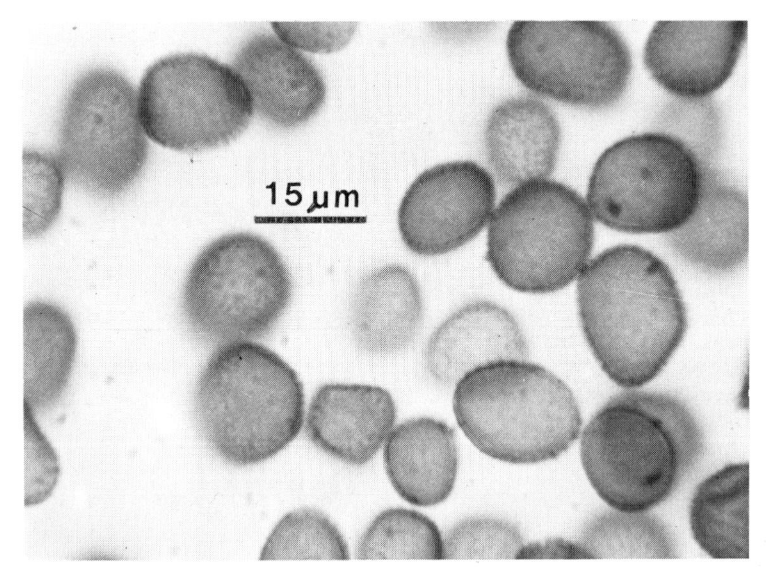

Figure 15.8 Teliospores of *Ustilago macrospora.*

unirrigated turf only *Agropyron cristatum* (L.) Gaertn and *Elymus junceus* Fisch. are likely seed crop hosts.

Ustilago trebouxii H.&P. Sydow (*Ann. Mycol.*, **10**, 214, 1912) (for synonomy see Fischer, 1953a).

Sori can be found as narrow stripes in sheaths and blades of upper leaves and the inflorescence is usually distorted with the seed replaced by brown, powdery spore masses which are eventually dispersed leaving a bare rachis. Teliospores are globose to sub-globose, faintly echinulate or smooth, light brown, 4–7 µm in diameter. Reported from Oregon on *Poa pratensis* L.

Ustilago macrospora Desm. *(Pl. Crypt. de France*, 2127, 1850).

Sori occur as brown or black striae, mainly in leaves and leaf sheaths which become shredded towards the tips at maturity. The inflorescence is either aborted or severely distorted and may also contain sori (Fig. 15.8). Teliospores are spherical to elongate or angular,

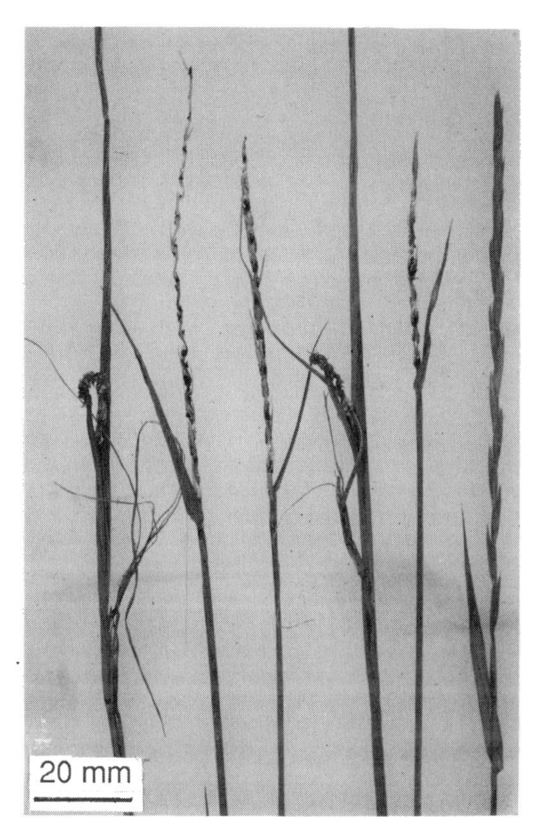

Figure 15.9 Loose smuts on *Agropyron trachycaulum*. Distorted inflorescences–*Ustilago macrospora*. Erect inflorescences–*U. bullata* (Two on right).

yellow–brown, coarsely echinulate to verrucose, 8–19 μm in diameter, mostly larger than those of *U. bullata* with which the inflorescence symptoms may be confused (Fig. 15.9). Of hosts in *Agropyron, Bromus, Elymus, Phalaris*, and *Calmagrostis* spp. only *Agropryon cristatum* (L.) Gaertn. in the turf grasses is reported infected (Ainsworth and Sampson, 1950; Fischer, 1953a).

Ustilago cynodontis (Pass.) (*Henn. Bot. Jahrb.* (*Engler*) **14,** 369, 1891) (for synonymy see Zambettakis, 1963).

Sori completely or partially destroy the inflorescence and occasionally destroy the upper leaves. Teliospores vary in colour from olive-yellow to brownish-black, and are very variable in size even from the same inflorescence, being, 5–11 μm, globose to sub-globose, smooth or with faint echinulations. On *Cynodon dactylon* (L.) Pers. in S. Europe, Asia, Africa, North and South America (Fischer, 1953; Mulder and Holliday, 1971; Rizvi and Kafi, 1969; Sharma and Agrawal, 1978; Zembettakis, 1963). According to Zambettakis (1963) *C. dactylon* is also smutted by the separate species *U. paraguariensis* Speg., *U. hitchcockiana* Zundel, and *Sorosporium cynodontis* Ling.

Ustilago affinis Ell. & Ev. (Cockerell, *Bull, Torr. Bot. Club*, **20,** 297, 1893) (for synonomy see Fischer, 1953a).

Sori affect only the inflorescence which is distorted. Teliospores are olive to yellow–brown, ovoid to irregular, 4–8 μm with a smooth, thin exospore. On *Stenotaphrum secundatum* (Walt.) Kuntze, mainly in Florida (Freeman, 1967)

(a) Disease development and control

The head smuts with powdery teliospores which are mostly dispersed before harvest are called 'loose smuts' to distinguish them from those where the spore masses remain largely intact until harvest, e.g. the *Tilletia* spp. (bunts) and the 'covered smuts'. However, although embryo or pericarp infection may result at flowering time, from germination of wind-borne loose smut spores on receptive floral parts or pericarp of other grass flowers, the teliospore inoculum may also be seed-borne between the glumes, paleae or on the surface of the caryopses. In the case of covered smuts, teliospores are produced in coherent sori enclosed in the glumes and they are not dispersed at flowering time. They are dependent on weathering, harvesting, or seed cleaning for liberation. Since they are not as powdery as loose smuts, they tend to be dispersed as aggregates. Such aggregates do not find as ready access to the spaces between glumes and caryopsis so the spores tend to be externally seed-borne and more accessible to the action of contact-type fungicides. Internal infections are amenable to control by systemic fungicides only.

Germination of the teliospores of *Ustilago* differs from that of *Tilletia* (see p. 193). In *Ustilago*, typically the germ tube or promycelium segments into four cells and the sporidia are borne on these both terminally and laterally. The sporidia are haploid and dikaryotization arises from the fusion of the cells of the promycelium, sporidia, or hyphae which develop from them. Teliospores may have a resting period before germination and some are long-lived, e.g. *T. caries* (18 years), *T. foetida* (25 years), *U. bullata* (10 years) (Fischer, 1936). Seedling infection by seed-borne spores and/or mycelium has been demonstrated in *U. bullata*, (Falloon, 1979), *U. striiformis*, and *U. cynodontis* (Fischer and Holton, 1957). The presence of systemic smut mycelium in perennial grass (e.g. in the rhizomes of *U. cynodontis* (Zambettakis, 1963) reduces the effectiveness of fungicidal treatments in control (Falloon, 1980).

15.4 REFERENCES

Ainsworth, G.C. and Sampson, K. (1950) *The British Smut Fungi (Ustilaginales)*. Commonwealth Mycological Institute, 137 pp.

Allison, J.L. and Chamberlain, D.W. (1946) Distinguishing characteristics of some forage grass diseases prevalent in the north central states. *US Dept. Agric. Circ. 747*; 16p.

Angus, A. (1956) A taxonomic investigation of *Ustilago hypodytes* (Schlecht) Fr. and its allies. *Trans. Br. Mycol. Soc.*, **39**, 115–24.

Boerema, G.H., Van Kesteren, H.A. and Dorenbosch, M.M.J. (1962) New fungus records for the country. In *Jaarboek 1961*, Versl. Pl-Ziekt. Dienst, Wagenigen 136, pp. 110–11.

Bond, T.E.T. (1940) Observations on the disease of sea lyme-grass. (*Elymus arenarius L.*) caused by *Ustilago hypodytes* (Schlecht) Fr. *Ann. Appl. Biol.*, **27**, 330–7.

Bressman, E.N. (1932) *Lolium* infected with bunt of wheat. *Phytopathology*, **22**, 865–6.

Clinton, G.P. (1900) Leaf smut of timothy, red top and bluegrass. The smuts of Illinois agricultural plants. *Illi. Agric., Expt. Sta. Bull.*, 57 pp.

Clinton, G.P. (1904) North American Ustilagineae. *Proc. Boston, Soc. Nat. Hist.*, **31**, 329–529.

Cole, H., Massee, L.B. and Duich, J.M. (1970) Control of stripe smut in 'Merion' Kentucky bluegrass turf with benomyl. *Plant Dis. Rept.*, **54**, 146–50.

Cole, H. Jr, Burpee, L.L., Sanders, P.L. and Duich, J.M. (1978) Stripe smut control with a single spring dormant season application of fungicide, 1977. *Fungicide Nematicide Tests*, **33** (265), 142.

Conners, I.L. (1967) *An Annotated Index of Plant Diseases in Canada and Fungi Recorded on Plants in Alaska, Canada and Greenland.* Publ. Res. Br. Canada Dept. Agric. 1251, Queen's Printer, Ottawa, Ont., 381pp.

Couch, H.B. (1973) *Diseases of Turfgrasses*, 2nd edn, R.E. Krieger, Huntington, NY, 348pp.

Cunningham, G.H. (1924) The Ustilagineae, or 'Smuts' of New Zealand. *Trans. NZ Inst.*, **55**, 397–433.

Davis, W.H. (1922) *Urocystis agropyri* on red top. *Mycologia*, **14**, 279–81.

Davis, W.H. (1924) Spore germination of *Ustilago striaeformis. Phytopathology*, **14**, 251–67.

Dickson, J.G. (1947) *Diseases of field crops.* McGraw-Hill, New York and London, 429pp.

Duran, R. (1968) Subterranean sporulation by two graminicolous smut fungi. *Phytopathology*, **58**(3), 390.

Duran, R. and Fischer, G.W. (1961) *The genus Tilletia.* Washington State University, 138pp.

Falloon, R.E. (1979) Seedling and shoot infection of *Bromus catharticus* by *Ustilago bullata. Trans. Br. Mycol. Soc.*, **73**, 49–56.

Falloon, R.E. (1980) Fungicidal control of *Ustilago bullata* seedling and shoot infection of *Bromus catharticus. NZ J. Exp. Agric.*, **8**(2), 173–7.

Fischer, G.W. (1936) The susceptibility of certain wild grasses to *Tilletia tritici* and *Tilletia levis. Phytopathology*, **26**, 867–86.

Fischer, G.W. (1938) Some new grass smut records from the Pacific Northwest. *Mycologia*, **30**, 385–95.

Fischer, G.W. (1939) Studies on the susceptibility of forage grasses to cereal smut-fungi. III. Further data concerning *Tilletia levis* and *T. tritici. Phytopathology*, **29**, 575–91.

Fischer, G.W. (1940a) Fundamental studies of stripe smut of grasses (*Ustilago striaeformis*) in the Pacific Northwest. *Phytopathology*, **30**, 93–118.

Fischer, G.W. (1940b) Host specialization in head smut of grasses, *Ustilago bullata. Phytopathology*, **30**, 991–1017.

Fischer, G.W. (1951) A local wintertime epidemic of blister smut *Entyloma crastophilum* in Kentucky bluegrass at Pullman, Washington. *Plant Dis. Rept.*, **35**, 88.

Fischer, G.W. (1953a) *Manual of North American Smut Fungi.* Ronald Press, New York, 343p.

Fischer, G.W. (1953b) Smuts that parasitize grasses. In *Plant Diseases.* US Dept. Agric. Yearbook of Agriculture 1953, pp. 280–4.

Fischer, G.W. and Hirschorn, E. (1945) A critical study of some species of *Ustilago* causing stem smut on various grasses. *Mycologia*, **37**, 236–66.

Fischer, G.W. and Holton, G.S. (1957) *Biology and Control of the Smut Fungi.* Ronald Press, New York, 622 p.

Fischer, G.W. and Shaw, C.G. (1953) A proposed species concept in the smut fungi, with application to North American species. *Phytopathology*, **43**, 181–8.

Fraser, W.P. and Scott, G.A. (1926) Smut of western rye grass. *Phytopathology*, **16**, 473–7.

Freeman, T.E. (1967) Diseases of southern turf grasses. *Fla. Agr. Expt. sta. Tech. Bull.*, **713**, 31pp.

Fushtey, S.G. and Taylor, D.K. (1977) Blister smut in Kentucky bluegrass at Agassiz, B.C. *Can. Plant Dis. Surv.*, **57**, 29–30.

Gaskin, T.A. (1965a) Varietal reaction of creeping bentgrass to stripe smut. *Plant Dis. Rep.*, **49**, 268.

Gaskin, T.A. (1965b) Varietal resistance to flag smut in Kentucky bluegrass. *Plant Dis. Rept.*, **49**, 1017.

Gaskin, T.A. (1966) Evidence for physiological races of stripe smut (*Ustilago striiformis*) attacking Kentucky bluegrass. *Plant Dis. Rept.*, **50**, 430–1.

Gould, C.J., Brauen, S.E. and Goss, R.L. (1977) Entyloma blister smut on *Poa pratensis* in the Pacific Northwest. *Proc. Am. Phytopathol. Soc.*, **4**, 205–6 (Abstr.)

Halisky, P.M., Funk, C.R. and Babinski, P.L. (1968) Control of stripe smut in Kentucky bluegrass turf with a systemic fungicide. *Plant Dis. Rept.*, **52**, 635–7.

Halisky, P.M., Funk, C.R. and Babinski, P.L. (1969) Chemical control of stripe smut in 'Merion' Kentucky bluegrass. *Plant Dis. Rept.*, **53**, 286–8.

Halisky, P.M., Funk, C.R. and Bachelder, S. (1966) Stripe smut of turf and forage grasses – its prevalence, pathogenicity, and response to management practices. *Plant Dis. Rept.*, **50**, 294–8.

Hardison, J.R.(1966) Chemotherapy of *Urocystis agropyri* in Merion Kentucky bluegrass (*Poa pratensis* L.) with two derivatives of 1, 4-oxathiin. *Crop. Sci.*, **6**, 384.

Hardison, J.R. (1967) Chemotherapeutic control of stripe smut (*Ustilago striiformis*) in grasses by two derivatives of 1, 4-oxathiin. *Phytopathology*, **57**, 242–5.

Hardison, J.R. (1968) Systemic activity of fungicide 1991, a derivative of benzimidazole, against diverse grass diseases. *Plant Dis. Rept.*, **52**, 205.

Hardison, J.R. (1971a) Chemotherapeutic eradication of *Ustilago striiformis* and *Urocystis agropyri* in *Poa pratensis* by root uptake of α-(2, 4-dichlorophenyl)-α-phenyl-5-pyrimidine methanol (EL-273). *Crop Sci.*, **11**, 345–7.

Hardison, J.R. (1971b) Chemotherapy of smut and rust pathogens in *Poa pratensis* by thiazole compounds. *Phytopathology*, **61**, 1396–9.

Hardison, J.R. (1971c) Control of *Ustilago striiformis* and *Urocytis agropyri* in *Poa pratensis* by thiophanate fungicides. *Phytopathology*, **61**, 1462–4.

Hardison, J.R. (1972) Control of *Ustilago striiformis* and *Urocystis agropyri* in *Poa pratensis* 'Merion' by minute doses of pyrimidine compounds. *Plant Dis. Rep.*, **56**, 55–7.

Hardison, J.R. (1974) Control of stripe smut and flag smut in Kentucky bluegrass by a new systemic fungicide, BAY MEB 6447. *Crop Sci.*, **14**, 769–70.

Hodges, C.F. (1969a) Symptomatology and histopathology of *Agrostis palustris* infected by *Ustilago striiformis* var. *agrostidis*. *Phytopathology*, **59**, 1455–7.

Hodges, C.F. (1969b) Additional varieties of *Agrostis palustris* infected by *Ustilago striiformis* var *agrostidis*. *Plant Dis. Rept.*, **53**, 298–9.

Hodges, C.F. (1969c), Morphological differences in Kentucky bluegrass infected by stripe smut and flag smut. *Plant Dis. Rept.*, **53**, 967–8.

Hodges, C.F. (1970a) Influence of temperature on growth of stripe smutted creeping bentgrass and on sorus development of *Ustilago striiformis*. *Phytopathology*, **60**, 665–8.

Hodges, C.F. (1970b) Comparative morphology and development of *Poa pratensis* infected by *Ustilago striiformis* and *Urocystis agropyri*. *Phytopathology*, **60**, 1794–7.

Hodges, C.F. (1970c) Stripe smutted inflorescences on Kentucky bluegrass. *Plant Dis. Rept.*, **54**, 206–7.

Hodges, C.F. (1971a) Sporadic development of *Ustilago striiformis* in axillary buds from stolon nodes of perenially infected *Agrostis palustris*. *Phytopathology*, **61**, 940–2.

Hodges, C.F. (1971b) Floral induction and development in *Poa pratensis* infected with *Ustilago striiformis* var. *poae* and *Urocystis agropyri*. *Phytopathology*, **61**, 1373–6.

Hodges, C.F. (1972) Effect of infection by *Ustilago striiformis* var *agrostidis* on inflorescence development on *Agrostis palustris*. *Phytopathology*, **62**, 583–4.

Hodges, C.F. (1973) Evidence for developmental control of *Ustilago striiformis* by apical dominance in perenially infected stolons of *Agrostis palustris*. *Phytopathology*, **63**, 146–8.

Hodges, C.F. (1976a) Comparative primary infection characteristics of *Ustilago striiformis* and *Urocystis agropyri* on cultivars of *Poa pratensis*. *Phytopathology*, **66**, 1111–15.

Hodges, C.F. (1976b) Development of healthy shoots from *Poa pratensis* systemically infected by *Ustilago striiformis* and *Urocystis agropyri*. *Plant Dis. Rept.*, **60**, 120–1.

Hodges, C.F. (1977) Influence of irrigation on survival of *Poa pratensis* infected by *Ustilago striiformis* and *Urocystis agropyri*. *Can. J. Bot.*, **55**, 216–18.

Hodges, C.F. and Britton, M.P. (1969) Infection of Merion bluegrass, *Poa pratensis* by stripe smut, *Ustilago striiformis*. *Phytopathology*, **59**, 301–4.

Hodges, C.F. and Britton, M.P. (1970) Directional growth and the perennial characteristics of *Ustilago striiformis* in *Poa pratensis*. *Phytopathology*, **60**, 849–51.

Hodges, C.F. and Robinson, P.W. (1977) Sugar and amino acid content of *Poa pratensis* infected with *Ustilago striiformis* and *Urocystis agropyri*. *Physiol. Plant.*, **41**, 25–8.

Howard, F.L., Rowell, J.B. and Keil, H.L. (1951) Fungus diseases of turfgrasses. *Bull. R.I. Agric. Expt. Sta.*, **308**, 56pp.

Hull, R.J., Jackson, N. and Skogley, C.R. (1979) Nutritional implications of stripe smut severity in Kentucky bluegrass turf. *Agron. J.*, **71**, 553–5.

Jackson, N. (1970) Evaluation of some chemicals for control of stripe smut in Kentucky bluegrass turf. *Plant Dis. Rept.*, **54**, 168–70.

Jackson, N. and Dernoeden, P.H. (1979) Fall fungicide application for control of stripe smut and leaf spot, 1977. *Fungicide Nematicide Tests*, **34** (307), 143.

Jackson, N. and Fenstermacher, J.M. (1974) Fungicidal control of stripe smut and melting out with consequent maintenance of sward density in 'Merion' Kentucky bluegrass turf. *Plant Dis. Rept.*, **58**, 573–6.

Jørstad, I. (1963) Icelandic parasitic fungi apart from Uredinales. *Skr. norske Vidensk Akad. Mat. Nat. Klasse, N.S.*, **10**, Oslo Univ. Press, 72pp.

Kozelnicky, G.M. (1969) First report of stripe smut on bluegrass in Georgia. *Plant Dis. Rept.*, **53**, 580.

Kreitlow, K.W. (1943) *Ustilago striiformis*. II. Temperature as a factor influencing development of smutted plants of *Poa pratensis* L. and germination of fresh chlamydospores. *Phytopathology*, **33**, 1055–63.

Kreitlow, K.W. (1948) *Urocystis agropyri* on *Phleum pratense*. *Phytopathology*, **38**, 158–9.

Kreitlow. K.W. and Juska, F.V. (1959) Susceptibility of Merion bluegrass varieties to stripe smut (*Ustilago striiformis*). *Agron. J.*, **51**, 596–7.

Kreitlow, K.W. and Myers, W.M. (1944) Prevalence and distribution of stripe smut of *Poa pratensis* in some pastures of Pennsylvania. *Phytopathology*, **34**, 411–15.

Kucnierz, J. (1976) (pub. 1977). Species of Uredinales and Ustilaginales rare and new for Poland collected in the area of the Pieniny Mountains. *Mycologica*, **12**(2), 257–60. Inst. Pl. Prot. Krakow, Poland.

Leach, J.G. Lowther, C.V. and Ryan, M.A. (1946) Stripe smut (*Ustilago striaeformis*) in relation to bluegrass improvement. *Phytopathology*, **36**, 57–72.

Liro, J.J. (1938) Die Ustilagineen Finnlands II. *Ann. Acad. Sci. Fenn. Ser. A*, **42**, 720 pp.

Lukens, R.J. (1966) Urea, an effective treatment for stripe smut on *Poa pratensis*. *Plant Dis. Rept.*, **49**, 361.

Mankin, C.J. (1969) Diseases of grasses and cereals in South Dakota (A check list). *S. Dakota Agric. Expt. Sta. Tech. Bull.*, **35**, 28pp.

McAlpine, D. (1910) *The Smuts of Australia: Their Structure Life History, Treatment, and Classification*. Department of Agriculture, Melborne, 288 pp.

McKenzie, E.H.C. and Latch, G.C.M. (1981) New plant disease record in New Zealand: *Entyloma dactylidis* and *E. brizae* on grasses. *N.Z.L. Agric. Res.*, **24**(3–4), 397–400.

Mordue, J.E.M. and Waller J.M. (1981a) *Urocystis agropy-*

ri. C.M.I. Descriptions of Path. fungi and bacteria. No. 716.
Mordue, J.E.M. and Waller, J.M. (1981b) *Ustilago striiformis*. C.M.I. Descriptions of Path. fungi and bacteria. No. 717.
Mordue, J.E.M. and Waller, J.M. (1981c) *Tilletia caries*. C.M.I. Descriptions of Path. fungi and bacteria. No. 719.
Mordue J.E.M. and Waller, J.M. (1981d) *Tilletia foetida*. C.M.I. Descriptions of Path. fungi and bacteria. No. 720.
Mühle, E. (1971) *Diseases and Pests of Forage Grasses*. S. Hirzel Velag, Leipzig, 422 pp.
Mulder, J.L. and Holliday, P. (1971) *Ustilago cynodontis*. C.M.I. Descriptions of Path. fungi and bacteria. No. 297.
Niehaus, M.H. (1968) Relative amount of stripe and flag smut on Merion Kentucky bluegrass. *Plant Dis. Rept.*, **52,** 633–4.
Noble, R.J. (1924) Studies on the parasitism of *Urocystis tritici* Koern., the organism causing flag smut of wheat. *J. Agric. Res.*, **27,** 451–90.
Purdy, L.H. (1965) Flag smut of wheat. *Bot. Rev.*, **31,** 565–606.
Ramakrishnan, T.S. and Srinivasan, K.V. (1950) Two grass smuts. *Curr. Sci.*, **19**(7), 216–17.
Rizvi, K. and Kafi, A. (1969) Diseases of lawns in Karachi and their control. *J. Agric. Res. Counc. Pakistan*, **20**(1), 33–9.
Robinson, P.W. and Hodges, C.F. (1972) Effect of benomyl on eradication of *Ustilago striiformis* from *Agrostis palustris* and on plant growth. *Phytopathology*, **62**(5), 533–5.
Ruben, D. (1968) Subterranean sporulation by two graminicolous smut fungi. *Phytopathology*, **58,** 390.
Sampson, K. and Western, J.H. (1942) *Diseases of British Grasses and Herbage Legumes*. Cambridge University Press, 85 pp.
Sampson, K. and Western, J.H. (1954) *Diseases of British Grasses and Herbage Legumes*. Cambridge University Press, 118 pp.
Sanders, P.L., Burpee, L.L., Cole, H. Jr and Duich, J.M. (1978) Uptake, translocation and efficacy of triadimefon in control of turf grass pathogens. *Phytopathology*, **68,** 1482–7.
Sharma, K.D. and Agrawal, D.K.(1978) Cultural studies on *Ustilago cynodontis* (Press.) Henn. *Indian J. Microbiol.*, **18**(1), 66.
Smith, J.D. (1980) *Major diseases of Turf grasses in Western Canada*. University of Saskatchewan Publ. 409, 14 pp.
Sprague, R. (1955) Check list of the diseases of grasses and cereals in Alaska. *Plant Dis. Rep. Suppl.*, **232,** 94–101.
Stakman, E.C. and Harrar, J.G. (1957) *Principles of Plant Pathology*. Ronald Press, New York, 581 pp.
Thirumalachar, M.J. and Dickson, J.G. (1949) Chlamydospore germination, nuclear cycle and artificial culture of *Urocystis agropyri* on red top. *Phytopathology*, **39,** 333–9.
Thirumalachar, M.J. and Dickson, J.G. (1953) Spore germination, cultural characters and cytology of varieties of *Ustilago striiformis* and the reaction of hosts. *Phytopathology*, **43,** 527–35.
USDA (1960) *Index of Plant Diseases in the United States*. Agric. Handbook 165. US Department of Agriculture, 531pp.
Vargas, J.M., Jr (1972) Evaluation of four systemic fungicides for control of stripe smut in 'Merion' Kentucky bluegrass turf. *Plant Dis. Rept.*, **56,** 334–6.
Waller, J.M. and Mordue, J.E.M. (1981) *Tilletia controversa*. C.M.I. Descriptions of Path. fungi and bacteria. 746. 2pp.
Waller, J.M. and Mordue, J.E.M. (1983) *Tilletia holci*. C.M.I. Descriptions of Path. fungi and bacteria. 747 1 pp.
Watson, L. (1972) Smuts on grasses: some general implications of the incidence of Ustilaginales on the genera of Gramineae. *Queensland Rev. Biol.*, **47**(1), 46–62.
Zambettakis. C. (1963) The smuts of bermudagrass. *Rev. Mycol.*, **28**(5), 312–48.
Zundel, G.L. (1953) *The Ustilaginales of the World*. Pennsylvania State University Contrib. 176, 410 pp.

Part seven
Inflorescence diseases (other than smuts)

16 *Some important inflorescence diseases (other than smuts)*

16.1 BLIND SEED

This inflorescence disease was first recorded on rye, *Secale cereale* L. (Prillieux and Delacroix, 1892; Rehm, 1900), but occurs on many wild, forage and turf grasses (Fischer, 1944; Hardison, 1962; Neill and Hyde, 1939, 1942; Rehm, 1900; Sampson and Western, 1954; Sprague, 1950) notably in continental Europe (De Tempe, 1966; van Dijk, 1967), the British Isles (Lafferty, 1948, Noble and Gray, 1945; Sampson and Western, 1954; Wright, 1967), Australasia (Latch, 1966; McGee, 1971; Neill and Hyde, 1939; Wright, 1967) and western United States (Fischer, 1944; Hardison, 1962; Sprague, 1950). Although many grass and some of the cereal species have been shown to be susceptible by inoculation (Calvert and Muskett, 1945; Hardison, 1962) it is the ryegrasses which suffer most severely. *Alopecurus pratensis* L., *Bromus* spp., *Dactylis glomerata* L. and the cereals *Avena sativa* L. and *Hordeum vulgare* L. are resistant. When weather conditions are cool and moist at flowering time widespread infection of grass flowers may take place, as happened in Northern Ireland in 1946. In that year nearly 50% of the seed of S.23 and S.24 cultivars of perennial ryegrass, *Lolium perenne* L. was not viable (Wright, 1956). In the wet summer of 1965, an epidemic of the disease developed in the Netherlands, notably on Italian and Westerwolds ryegrass, *L. multiflorum* Lam. seed crops; 19.2% of ryegrass crops were infected (De-Tempe, 1966). In New Zealand, in years of severe infection, although these are now the exception, seed germination in ryegrasses has been reduced by up to 50% (Latch, 1966). In Oregon, by 1944, nearly 90% of fields were infected (Hardison, 1963). The disease has occasionally been so severe as to make profitable ryegrass seed growing problematical in several countries. However, the disease is not systemic, the seed being the only part of the plant affected and its presence in sown seeds does not affect the value of the grass for pasture or turf. Infected seeds cannot be removed by seed cleaning.

16.1.1 Symptoms

In flowering grasses the disease is not usually easy to detect since infected inflorescences and seeds are often normal in size and not much different from healthy ones. However, if the lemma and palea are removed from a flower an infected caryopsis will be seen to be more or less shrunken and may be covered with a pale pink, slimy exudate which, as the seed matures, gradually dries to a waxy consistency, rusty brown in colour. Macro- and microconidia can be seen in water mounts of such seeds viewed microscopically. Microscopical examination of washings and spore counts from inflorescences provides a more reliable method for estimating infection frequency and severity (Fischer, 1944; Hyde, 1938) than the use of a diaphanoscope. On the diaphanoscope severely diseased seeds are opaque, but opacity is not a reliable guide in determining percentage infection since less severely infected seeds often appear normal while weathered uninfected seeds may be opaque.

After overwintering, infected seeds give rise to stalked apothecia which appear about the same time as the grasses are flowering (Fig. 16.1).

16.1.2 The fungus

Teleomorph – *Gloeotinia granigena* (Quel.) Schum. (Schumaker, *Mycotaxon*, **8**, 125, 1979).

Synonyms: *Gloeotinia temulenta* (Prillieux and Delacroix), (Wilson Noble and Gray *in Trans. Br. Mycol. Soc.*, **37**, 29–32, 1954);

Phialea mucosa Gray in (*Trans. Br. Mycol.*, **25**, 329–333, 1942);

Phialea temulenta Prillieux and Delacroix in (*Bull. Soc. Mycol. Fr.*, **8**, 22–3, 1982).

Anamorph – *Endoconidium temulentum* Prillieux and Delacroix in (*Bull. Soc. Mycol. Fr.*, **7**, 116–17, 1891).

The earlier synonymy of the pathogen has been considered in detail by Wilson, Noble and Gray (1954) who erected the new genus *Gloeotinia* of the

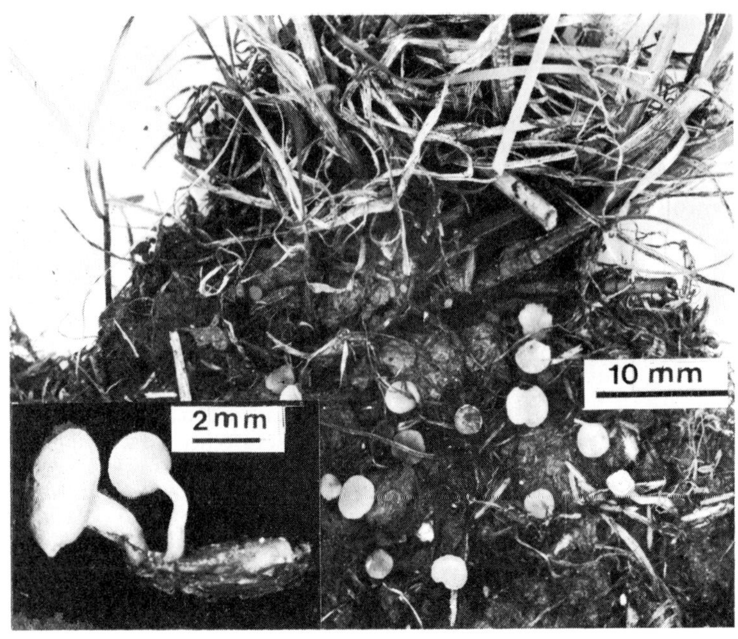

Figure 16.1 Blind seed disease of *Lolium perenne* caused by *Gloeotinia granigena*.
(a) Apothecia on a sod from a heavily infected crop.
(b) Seed of *L. perenne* with apothecia of *G. granigena*.
(Source: Northern Ireland Dept. Agric.)

Sclerotiniaceae to accommodate *G. (Phialea) temulenta*.

The small, fleshy apothecia develop from colourless, septate, intertwining hyphae, 3–4 μm wide in the pericarp, testa and endosperm of the seed. Usually there are one or two, but may be up to seven apothecia per seed. These structures at first have pale pink to cinnamon-coloured discs which darken with age to cinnamon, 1–3.5 mm in diameter, almost closed at first, but eventually becoming cupulate and later flattened or recurved with a smooth margin. The stalks are cylindrical, 1–8 mm long and 0.4 mm in diameter (Fig. 16.1).

Asci are 8-spored, cylindro-clavate, 66–166 × 3.3.–7 μm, slightly thickened and with a non-staining pore at the apices. The slightly pointed, elliptical, unicellular, biguttulate, smooth ascospores, 7.6–13 × 3–6 μm, lie in uniseriate rows towards the top of the asci. The asci are accompanied by simple, hyaline paraphyses.

Macroconidia are cylindrical to slightly crescentic, with rounded ends, containing several oil globules, 11–21 × 3.3–6 μm. They develop in succession from the apices of short hyphal outgrowths from the pericarp and form a pink slime. Microconidia, 3.4–4.8 × 2.7–3.2 μm are formed in pink, cushion-like sporodochia on the surface of the caryopses; they have not been seen to germinate (Gray, 1942). Slightly different measurements are given by Prillieux and Delacroix (1892), Neil and Hyde (1942) and by Wilson, Noble and Gray (1945) from culture.

16.1.3 Epidemiology

Apothecia develop successively over a 3- to 7-week period from the undifferentiated mycelial stroma, which permeates the shed or sown seed (Wright and Sproule, 1965). Ascospore liberation is also prolonged, since asci ripen over an extended period. This coincides with the general time of ryegrass flowering when air-borne ascospores (Neill and Armstrong, 1955) enter between the open flower scales and initiate primary infections of the ovary. These are visible 9–14 days later on the developing caryopsis as a slimy matrix containing abundant macroconidia. Secondary infections result from the dissemination of the sticky macroconidia and microconidia by contact or rain splash. Cold, wet conditions favour the development of the disease by extending the periods of apothecial production, pollination and ripening of ryegrass and increasing the frequency of secondary conidial infections (Wilson, Noble and Gray, 1945). Infections occurring early in floral development destroy the embryo and no seed develops. Later infections, occurring when the embryonic tissues and endosperm have developed, result in the typical non-viable 'blind' seeds. Infection still later than this may leave the embryo untouched, but conidia are produced on the surface of these viable seeds.

16.1.4 Control

Storing seed for 2 years will eliminate the fungus completely and planting seed deeper than usual, so that it is covered to a depth of at least 13 mm, effectively reduces primary infections (Hardison, 1963). A treatment with water at 45–46 °C for 2 h will eradicate the pathogen without seed damage (De Tempe, 1966) and a vapour-heat treatment has also been reported as effective (Miller and McWhorter, 1948) by Couch (1973). However, the control given by such measures and by the deep ploughing of seriously infected fields and removal of as much 'light' seed as possible at harvest time is largely nullified if sources of primary infection remain. Annual burning of straw and stubble of those fields which were severely infected or which had been cropped for 3 years was very successful in controlling the disease in *Lolium* spp. and *Festuca arundinacea* Shreb. seed production fields in Western Oregon, provided that an even burn was achieved (Hardison, 1963). Burning also controlled ergot (p. 207), and other seed crop diseases (see this part). However, crop debris burning is not an effective control measure where sources of infection remain in a district in the form of ryegrass pastures or in the edges of many small fields as in Ireland (O'Rourke, 1976) where seed often comes from maiden crops (Wright, 1967). Large-scale field burning is not environmentally acceptable if smoke pollution affects large urban areas. So-called 'mobile field sanitizers' have been designed to reduced smoke pollution in Oregon (Hardison, 1976) and yet still to give an effective burn. Effective burning of grass crop debris is not feasible in some wetter climates where seed is produced, such as in parts of the British Isles.

Traditional fungicidal seed treatments are largely ineffective, because the fungal stroma, or pseudosclerotium is very deep-seated in the seed. Alternative methods of chemical control have been tried. Several chemicals suppressed the apothecium development when applied as soil surface drenches in pot tests, but failed to give control in the field. Some foliar sprays or granular applications effective in pots gave poor control in the field, probably because of interference from the foliage. However, a 15% granular formulation of sodium azide at 22.4 kg/ha gave complete control of apothecial production, but date of application was critical (Hardison, 1978). McGee (1971) obtained 90% control of apothecial production with benomyl at 5.6 kg/ha in perennial ryegrass seed crops in Australia.

In New Zealand specific attempts to control the disease were not made, but from 1937 seed samples were collected before harvest, the extent of infection determined and the grower advised whether to take the ryegrass seed crop or to use it for hay (Hyde, 1945). After 1964, the disease ceased to be of major importance. Weather conditions in the previous 10 years probably had been unfavourable for infection during the flowering period. Increasing use of nitrogenous fertilizer to increase seed production probably contributed to lower infection by changing the physiological susceptibility of the plant rather than by suppressing apothecial production (Hampton and Scott, 1980a,b; Scott, 1974).

Breeding for disease escape as a means of disease control with early or late flowering strains avoiding the main ascospore shower during grass flowering or selecting for time of diurnal floret opening was suggested by Corkill (1952). Unfortunately the period of ascospore discharge shows annual and field-to-field variation and ascospore distribution and floret opening periodicity generally coincide. Wright (1967) considered this method had limited possibilities and by using artificial epidemics induced by spraying plants with cultures of macroconidia, selected four plants of ryegrass showing resistance, from a population of 50 000. Resistance was inherited polygenically, genes governing resistance being mostly dominant. It was hoped that by backcrossing, this resistance could be transferred into susceptible but agronomically desirable cultivars. The same clones were shown to have resistance to infection from isolates of *G. temulenta* from British and New Zealand sources (Sproule and Faulkner, 1974). A backcross programme to introduce resistance to disease in parents of a reconstituted S.24 *L. perenne* cultivar was successful (Wright and Faulkner, 1982).

16.2 CHOKE OR CAT'S TAIL DISEASE

In this disease the grass inflorescence may become girdled with a fungal stroma which 'chokes' or delays flower emergence and may also restrict seed production. The thickening caused by the fungus resembles an animal tail. Unlike blind seed, ergot and twist diseases, in choke the causal fungus is systemic and perennial in stem bases, stolons and rhizomes, but not roots (Sampson, 1933). The host range is very wide (Andersen, 1978; Conners, 1967; Kohlmeyer and Kohlmeyer, 1974; Koponen and Mäkela, 1976; Mühle, 1971; Oudemans, 1919; Pshedetskaya, 1971; Sampson, 1933; Sampson and Western, 1954; Sprague, 1950; Vladimirskaya, 1928) and includes most of the commonly used, cool-season turf grasses. It is probably as common in North America as in Europe (Kohlmeyer and Kohlmeyer, 1974). Although recorded mainly from the Northern Hemisphere, it also occurs in New Zealand (Hedley and Braithwaite, 1978). The disease is seed-borne in *Agrostis* spp., *Festuca rubra* L. and probably in *F. ovina* L., *Poa bulboa* L. and other *Poa* spp. (Harberd, 1961; Mühle and Frauenstein, 1970; Richardson, 1979; Sampson and Western, 1954; Wernham, 1942). Although the disease is most often seen on wild grasses, especially on those in hedgerows (Sampson and Western, 1954) and in shaded places (Kohlmeyer, 1956), it is occasionally seen in pastures and on infrequently mown amenity turf. Seed crops of forage grasses may become severely diseased and seed yield may be severely reduced in the cooler northerly regions of Europe and the USSR (Eriksson, 1904; Kirby, 1961; Large, 1954; Mühle, 1971; Sampson and Western, 1954).

In the turf grass species *Agrostis tenuis* Sibth. and *A. stolonifera* L. Bradshaw (1959) found that infection by the fungus usually resulted in complete panicle suppression, and an associated large increase in tiller density without any plant vigour reduction. This would be an advantage in pastures or amenity turf where vegetative growth is needed. Harberd (1961) found no apparent loss in vigour in infected wild plants of *Festuca rubra* grown in observation plots, in fact some appeared more vigorous than uninfected ones. In *F. rubra*, infected plants may produce considerable amounts of seed, but a high proportion of the seeds may be infected by the fungus. In one case, 99% of the seedling progeny from infected plants were themselves infected (Sampson and Western, 1954).

16.2.1 Symptoms

The fungal stroma is at first white or cream-coloured, around sheaths or culms, up to 5 cm in length, thickening and changing to orange in colour as the perithecia develop (Fig. 22, colour plate section).

16.2.2 The causal fungus

Teleomorph – *Epichloë typhina* (Press. ex Hook.) Tul. (For synonomy see Boerema and Verhoeven, 1977).

Anamorph – *Acremonium typhinum* (Morgan-Jones and Gams, 1982) not *Sphacelia typhina* Sacc.

The mycelium of the fungus may be found in spring in leaves and shoots of plants where it has perennated. In stained, longitudinal sections the hyphae appear little branched and intercellular. In preparation for fruiting, hyphae grow out from leaves in young tillers and cover them with a delicate weft of mycelium. This eventually develops into a stromatic sheath filling the interlaminar spaces and often in later-flowering grasses preventing the young inflorescence from emerging (Sampson, 1939; Sampson and Western, 1954). The conidial stromata produce conidia on undifferentiated hyphae. Conidia are ovate $4-5\times3\,\mu m$ ($3-9\times1-3\,\mu m$, Sprague, 1950). The maturing orange stromata contain numerous immersed perithecia with prominent, papillate ostioles containing asci each with 8, many-celled ascospores measuring $120-200\,\mu m$ (Sampson and Western, 1954). Sprague (1950) give dimensions of $300-600\,\mu m$ high $\times 250\,\mu m$ diam. for perithecia and cylindrical asci measuring $150-230\times6-9\,\mu m$. Ascospore discharge is forcible and diurnal with a maximum towards evening (Ingold, 1948). Ascospores may germinate on agar medium to give secondary conidia and the fungus grows well in culture, producing conidia, but ascocarps do not develop (Vladimirskaya, 1928; Sampson and Western, 1954).

16.2.3 Disease development

Kirby (1961) found that the formation of a stroma by the fungus requires active but not too rapid growth at a particular stage in the development of the flower apex. Extensive growth of the systemic mycelium occurred only when the flowering apex was between the 'double ridge' and 'spikelet-primordium' stages. No choking occurred if the apex grew rapidly between these stages and the fungal invasion was confined to panicle branches and spikelets. Slow apex growth when very small resulted in its death and no stroma was formed.

In *F. rubra*, hyphae growing up the flowering stem pass into all parts of the flower, penetrating to the tissues of the ovary. Sometimes the stamens may be so heavily infected that pollen does not form. The disease is transmitted by seed that is not too severely damaged. Latent infection may persist for several seasons with some plants or tillers producing little or no seed some with infected panicles showing no clinical signs (Sampson, 1933).

The disease is transmitted by infected clones propagated vegetatively and usually every plant derived from an infected clone subsequently develops the disease. Each of the bulbils of *Poa bulbosa* from an infected plant will carry mycelium. *F. rubra*, which does not usually flower in the seedling year may not show external evidence of infection until the second or third year (Sampson and Western, 1954).

In *Dactylis glomerata*, where the disease is not seed-borne, it has been shown that ascospores and conidia would germinate on the cut surfaces of the stubble in a saturated atmosphere and the mycelium would grow down the pith of the stems. Infection resulted in the development of stromata next spring (Western and Cavett, 1959).

The stromata of *E. typhina* are parasitized by the fly *Phorbia (Pegohylemia) phrenione* Séguy. Both the fungus host and the insect parasite are widely distributed in Europe and North America and occur also in Mexico and East Asia. The insect does not depend on any particular grass species but occurs wherever the fungus is found and feeds on the condidial stroma. Conidia which pass through its digestive tract are still capable of germination. It has been suggested that the fly might be able to spread the disease to healthy grasses (Kohlmeyer and Kohlmeyer, 1974).

16.2.4 Control

There is no satisfactory method of control for the disease. Roguing plants showing stromata is usually unsuccessful. In those species where the disease is not seed-borne, older, heavily infected crops should be ploughed up and reseeded. Early season application of giberellic acid to valuable breeding material of *D. glomerata* enabled some seed to be obtained probably because the treatment induced plants to elongate fast enough to keep ahead of the systemic mycelium (Emecz and Jones, 1970).

Latch and Christensen (1982) found that they could obtain seedlings free from *Lolium* endophyte by treating infected seeds with the fungicides propiconazole or prochloraz at 0.5 g/kg of seed and infection-free plants were procured by drenching the growing medium in pots with benomyl suspension at 0.1 g/l of soil. Fungicidal eradication of *E. typhina* from infected seeds or plants may possibly be achieved with systemic fungicides.

16.3 ERGOT

Ergot is a common disease of grass inflorescences. Ergot of rye, caused by *Claviceps purpurea* (see below) has been recognized and well documented from earliest times (Andersen, 1972; Anerud, 1939; Barger, 1931). In most grasses some of the seed is replaced by a dark sclerotium, the ergot (Luttrell, 1981). Seed yield is reduced, because ergoty inflorescences produce fewer seeds. Seed quality is reduced and market value is lowered or a parcel of shipment may be rejected if the stipulated tolerance of ergots is exceeded. The alkaloids in ergot sclerotia of *C. purpurea* in forage grass, feed or milled grain may cause chronic or acute poisoning of animals or humans (Barger, 1931; Riggs, Henson and Chapman, 1968; Seaman, 1981). Some of these alkaloids are used in medicine and the hallucinatory drug D-lysergic acid diethylamide (LSD) is derived from ergots.

There are many species of *Claviceps* causing ergots on Poaceae and Cyperaceae which can be distinguished (with difficulty) by their morphology and host range. The species are grass-tribe specific (Langdon, 1954; Loveless, 1971). Many of these occur on warm-season turf grass species, e.g. *C. cynodontis Langdon on Cynodon dactylon* (L.) Pers., *C. paspali* Stev. and Hall on *Paspalum* spp. and *C. yanagawensis* on *Zoysia japonica* Steudel. *C. purpurea* (Fr.) Tul. is the common species on cool-season turf grass seed crops, all species of which are susceptible.

16.3.1 Common ergot–the causal fungus

The teleomorph,-*Claviceps purpurea* (Fr.) Tul. (Tulasne in *Ann. Sci. Nat. Bot., III*, **20,** 45, 1853), with an anamorph *Sphacelia segetum* Lev. (Leveille in *Mem. Soc. Linn. Paris*, **5,** 578, 1823) has a world-wide distribution on many species of grasses and cereals from the sub-boreal to the warm-temperate regions (Andersen, 1972; Berkenkamp, 1976; Brady, 1962; Frauenstein, 1967; Grasso, 1962; Latch, 1966; Mühle, 1971; Seaman, 1981). Severe epidemics have been reported from northern Europe, USSR, southwestern USA and southwest England (Andersen, 1972; Brummer, 1937; Chester and Lefebre, 1942; Jenkinson, 1958; Mühle, 1971; O'Rourke, 1976; van Dijk, 1967).

16.3.2 Symptoms and epidemiology

The disease is first seen as sticky drops of 'honeydew' which ooze from inflorescences at flowering time. In early maturing grasses, e.g. some *Poa* spp. and *Festuca rubra* L., the disease may not progress beyond the honeydew stage (Sprague, 1950), but in most the honeydew stage is replaced by a pseudo-parenchymatous structure, the sclerotium or ergot (Fig. 23, colour plate section). The size and shape of the sclerotium is a host-dependent character (Loveless and Peach, 1974). Ergots range in size from 2 mm long on small-seeded *Agrostis* spp. to 18 or 20 mm in the large-seeded *Festuca arundinacea* Schreb. (Sampson and Western, 1954). They ripen at the same time as the seed or a little later and may be shed soon after maturity at which time they have a dark-purple rind and a greyish-white interior. They may vary in shape from nearly seed-shaped, almost hidden by the floral parts to prominent, protruding, blunt horn-shaped.

Sclerotia fall to the ground or are harvested with the seed and remain dormant until the next growing season. Germination is activated by a chilling period

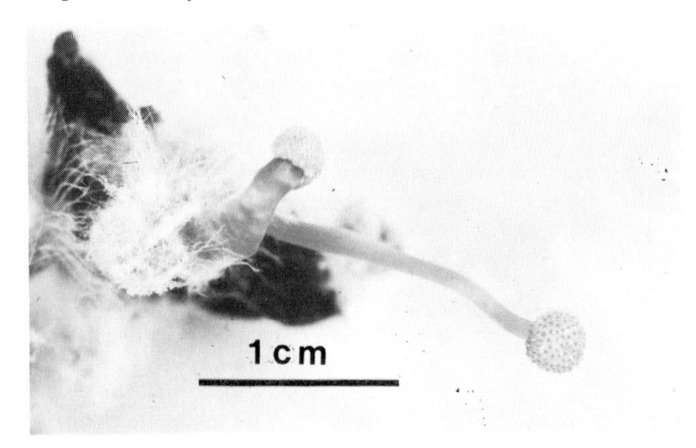

Figure 16.2 Germinated sclerotium (ergot) of *Claviceps purpurea*. Stalked stromata with perithecia in the heads.

and imbibition of water, but the sclerotia will not germinate until a suitable temperature of 10–25 °C obtains for a sufficient length of time for stroma formation (Mitchell and Cooke, 1968a,b). Germination of sclerotia takes place from stromatal cushions.

On germination, flesh-coloured, stalked stromata, 5–25 mm long are produced. These have numerous, flask-shaped perithecia embedded in their spherical heads (Fig. 16.2). The ostioles of these perithecia are marked by papillae. Many stromata may develop from each sclerotium. Large ergots have a higher percentage germination than smaller ones and produce a proportionately greater number of clavae (Cooke and Mitchell, 1966). Deep burial of sclerotia reduces their longevity and if it is at a depth greater than the length of the clavae it will prevent ascocarp formation (Bretag and Merriman, 1981; Brown, 1947; Pshedetskaya, 1979).

Perithecia contain hyaline, slightly curved, club-shaped asci 100–125 × 4 μm. Asci have eight filiform asci, which are septate when mature, and measure 50–76 × 0.6–0.7 μm (Sprague, 1950). The spores may be forcibly discharged several centimetres into the air where they are disseminated by wind, or they may be exuded into a mucilaginous matrix (Colotelo and Cook, 1977) which may aid dispersal by insects.

Ascospores which are deposited in open grass flowers germinate and their germ tubes penetrate between the cells of the stigma and intercellular hyphae grow down into the ovary wall eventually killing and replacing ovarian tissues with a fungal stroma (Butler, 1966; Luttrell, 1980, 1981). For successful primary infection, synchronization of sclerotium germination and ascospore discharge with the early stages of host flowering seems necessary (Butler and Jones, 1961). This does not always occur and neither does a suitable physiological state of the host coincide with heavy spore discharges which are necessary for heavy infection (Brown, 1947; Petch, 1937). Stromata mature about one week after first signs of germination, which occurs in the Northern Hemisphere from early to late June (Jenkinson, 1958; Petch, 1937). Active spore discharge may vary from 14 to over 50 days; new crops of stromata may be produced from the same sclerotium (Cooke and Mitchell, 1967). Favourable temperatures for mycelial growth, 20–30°C within the host are higher than those for stroma production, 10–25°C (Cooke and Mitchell, 1967).

A few days after infection the 'honeydew' or *Sphacelia* stage appears (Campbell, 1957). A sticky exudate collects on the infected florets and this contains large numbers of elliptical, aseptate conidia, 2–3 × 4–6 μm, produced from the whitish, slimy mass of mycelium permeating the ovary. The honeydew, which contains a high concentration of several different sugars (Mower and Hancock, 1975b) attracts certain insects, sticks spores to their bodies and acts as a preserving agent for the conidia during dry, warm weather. The sugar flow through the sphacelium moves conidia to the outside of the flower (Mower and Hancock, 1975a). Except during conditions of rain or heavy dew the low water potential or honeydew outside the grass glumes is probably a major factor preventing colonization by *Fusarium* spp. and other fungi while it is sufficiently high within them to allow mycelial growth and conidial germination (Cunfer, 1976; Mower, Snyder and Hancock, 1975). The conidia of the sphacelial stage are disseminated by insects and rain splash and start secondary infections. On germination they may produce secondary spores. Infection of the grass flower by conidia proceeds in a manner similar to that by ascospores. Luttrell (1980) has described the formation of the ergot in detail.

C. purpurea appears unique among *Claviceps* spp.

since it has been recorded on more than 200 grass species in 50 tribes and 17 genera (Langdon, 1954). Identification has been arbitrary in many cases because of difficulties in distinguishing between species and *C. purpurea* may have been recorded where other taxa were concerned (Loveless, 1971). Stäger (1908), who failed to obtain infection in many instances of cross inoculation from different grass hosts, postulated the existence of different physiological races of *C. purpurea*. Brown (1947) showed by inoculation with pure culture that a single strain of *C. purpurea* attacked *Secale cereale*, *Hordeum vulgare*, *Triticum aestivum*, *Avena sativa*, *Bromus inermis* Leyss., *Agropyron* spp., *Elymus* spp., *Poa pratensis* L. and *Phleum pratense* L. but not *Zizania aquatica* L. (wild rice) in Manitoba. Campbell (1957) found that an isolate from *S. cereale* infected 46 indigenous and forage grasses in the field and in the greenhouse. With the exception of one of three isolates from *Glyceria borealis* (Nash) Batchelder all isolates from 32 grass species collected from British Columbia to southern Ontario infected rye, wheat and barley. Although variability in conidial colonies in culture was noted no evidence was found indicating the occurrence of species specific races. Loveless (1971) obtained evidence, based on conidial morphology, of British isolates showing a close similarity between species composition of groups of grasses and the biological forms suggested by Stäger's cross-inoculation studies with European ergot collections (Stager, 1908). Although the form of the sclerotia of *C. purpurea* depends upon the host, the size of the conidia and ascospores is controlled genetically by the fungus (Loveless and Peach, 1974). Analysis of the alkaloid content of a large number of samples of naturally occurring ergot sclerotia from cereal and grass hosts showed the existence of host-restricted strains which were characterized by their particular spectra of alkaloids (Mantle, Shaw and Doling, 1977). In cross-inoculation experiments from grasses to wheat these authors found that only 3 of 40 grass isolates failed to infect, compatible with results obtained by Campbell (1957). However, they could divide the isolates into highly or weakly infective which Campbell (1957) did not. Isolates from different hosts have been accorded varietal rank in Japan (Tanda, 1979a, b). The biological race situation is still somewhat confused.

16.3.3 Control

The effectiveness of control of *C. purpurea* in grass-seed crops is restricted by lack of resistance to the disease in available cultivars and by the multiplicity of sources of infection in a wide range of other susceptible weed and cultivated grasses and cereals which may be adjacent. Cultural practices aimed at reducing the amount of available inoculum such as deep burial of shed sclerotia by ploughing, the sowing of ergot-free seed of high germination or the cutting and removal of flowers from primary infection centres (preventing conidial spread) will aid somewhat in reducing the loss from this disease. Where sclerotia remain in the field from a heavy ergot infection, burial by ploughing to a depth of 5–8 cm after a cereal crop, or shallower after a small-seeded grass, will reduce the emergence of stromata. However, ingress of ascospores carried by wind for considerable distances may still result in some infection. Sanitation by the burning of crop residues is an effective, but sometimes environmentally unacceptable means of eliminating many shed ergots and other disease and insect pests (Hardison, 1976). During wet seasons these measures are often ineffective. Seedling diseases and herbicide injury may result in poor stands with susceptible late tillers. Chemicals, notably sodium azide, are available which will suppress ascospore development, but the economics of their use are doubtful (Hardison, 1977) and biological control of ergot using a *Fusarium* sp. has been demonstrated experimentally (Mower, Snyder and Hancock, 1975). In mown turf the disease is not of any significance unless irregular mowing allows honeydew or ergot formation.

16.4 SILVERTOP, WHITEHEADS OR WHITE-EARS

These diseases are frequently of major importance in amenity grass seed production in Europe and North America, but particularly in *Agropyron cristatum*, *Festuca rubra*, *F. ovina*, *F. arundinacea* and *Poa pratensis* (Berkenkamp and Meeres, 1975; Hardison, 1980; Wetzel, 1971 and B. Cagas, personal communication, 25 March 1983). Many other grass species are affected (Gagné, Richard and Gagnon, 1984) and sometimes heavy seed losses may result.

16.4.1 Symptoms

The disease described as 'silvertop' in North America may or may not be the equivalent of 'whiteheads' or 'white-ears' in Europe (Wetzel, 1971).

In silvertop, bleached, silvery-white inflorescences and stalks, above the terminal node become noticeable at flowering time (Fig. 24, colour plate section). Spikelets have no seed. Below the flower head culm damage may be seen as stem restrictions or fractures or insect punctures low down in the culm, above the top or penultimate node. In most cases, damage to the inflorescence stalk is apparent only when it is pulled from the sheath. The portion of the stalk above the node is damaged and discoloured. The remainder of the plant may appear healthy. White-ear symptoms may be similar. In whiteheads, where individual culms, tillers or plants may be dead and bleached from the crown upwards, injury from fungal root, crown or stem base diseases or crown and root insects may be apparent.

16.4.2 Causes

1. The silvery or bleached inflorescence and lack of seed results from interruption or changes in supply of nutrients to the developing inflorescence. This may be due to physiological changes in the plant resulting from late spring frosts, severe temperature fluctuations, long drought periods, soil composition, acute deficiency or excessive supply of mineral nutrients (Wetzel, 1971).
2. Insects which feed on stems by puncturing sheaths and so damage stems at the base of culms, notably plant bugs (Miridae and Capsidae) have been implicated in western North America in the occurrence of silvertop (Arnott and Bergis, 1967; Hardison, 1959, 1980; Petersen and Vea, 1971).
3. Thrips and mites, notably *Aceria tulipae* Keifer and *Siteroptes graminum* Reuter (1900) which are able to penetrate between the sheath and flower stalk have been associated with silvertop. They appear to be late invaders after injury has taken place and plants without silvertop symptoms may show considerable populations of these organisms (Berkenkamp and Meeres, 1975; Hardison, 1959; Holmes, Swailes and Hobbs, 1961).
4. *Fusarium poae* (Pk.) Wr. in association with the mite *Siteroptes graminum* as a vector was suggested as the cause of silvertop by Leach (1940). The fungus alone (Keil, 1942) or in association with the mite (Keil, 1946; Petersen and Vea, 1971) have been shown capable of inciting silvertop symptoms. However, silvertop symptoms may develop in the absence of the fungus, which is a weak pathogen (Sprague, 1950). The demonstration that silvertop could be effectively controlled by insecticides, but not miticides, indicated that the fungus and the mite were not primary causes in some regions. Both fungus and mite may be later tissue invaders (Hardison, 1959, 1980).

16.4.3 Disease development and control

Grass seed production results in considerable crop debris and this and the perennial nature of the crops tend to lead to the build up of insect pests and plant diseases over several years of continuous grass culture.

Marked reductions in the incidence of silvertop, whiteheads and white-ears have been reported following the use of chlorinated hydrocarbon and organophosphorus insecticides (Arnott and Bergis, 1967; Wetzel, 1971). Some of the effective, persistent chlorinated hydrocarbons may not now be used in many countries. The use of fire sanitation in grass seed production is widely practised (Hardison, 1980; Wetzel, 1971). It was shown effective against silvertop in *Festuca rubra* in Pennsylvania by Keil (1942) and became widely practiced in Oregon in the late 1940s giving effective control of the disease in western North America in *Agrostis tenuis, F. rubra* spp. *rubra* and *commutata, F. arundinacea, Lolium perenne, Poa pratensis* and *P. trivialis* (Hardison, 1980). Field burning is the preferred method for control of grass seed crop pests and diseases, where practicable. Otherwise, insecticides recommended for use against the particular insects found associated with the disease should be applied before the inflorescences appear from the sheaths (Craig, 1973).

16.5 STEM EYESPOT OF FESCUES

This disease is found mainly on *Festuca* spp., but other grass species growing within a heavily infected fescue stand may show slight stem eyespot lesions and some species of other grass genera are susceptible on inoculation. The disease is a major cause of seed crop loss in creeping red fescue, *Festuca rubra* L. ssp. *rubra* in the Peace River region of northern Alberta and British Columbia (Smith and Elliot, 1970). It may severely damage native and sown fescues in western North America from Alaska to California and has an arctic/alpine distribution, mainly on fine-leaved *Festuca* spp. from N. America to Iceland, Norway, Sweden, Switzerland and Southern France (Smith, 1971, 1976a,b 1980; Smith and Elliott, 1970; Smith and Shoemaker, 1974)*. Although the disease was originally described as a leaf and pedicel blight (Sprague, 1950) on *F. idahoensis* Elmer from Idaho, it is now known as an eyespot, particularly on flowering stems and as an inflorescence blight. Since the leaf blades are infrequently lesioned it has little potential as a cause of disease of mown fescue turf. Through its effect in reducing seed production it may restrict the regeneration of the native *F. idahoensis* and *F. ovina* L. (Rydb.) Gleason which are important components

*Roadside plants of *F. rubra* collected by J.D.S. on 12 February 1984 in the Rakai forge near Ashburton, New Zealand showed typical eye spot lesions of *D. festucae*.

of some native grasslands of dry areas in western N. America (Smith, 1971a,b).

16.5.1 Symptoms

On flowering stems of the current season symptoms vary from vague brown or purple, linear spots to sharp, brown linear streaks or linear eyespots with dark margins and light-coloured centres (Fig. 16.3a). Lesions on leaf sheaths often correspond with those on the stem inside. The inflorescence branches, glumes and lemmae may be lesioned. Leaf blade lesions are rare. Bleached culms of the previous season may show dark, irregular staining or may be completely blackened (Smith, 1971b; Smith, Elliot and Shoemaker, 1968).

16.5.2 The causal fungus

Teleomorph – *Didymella festucae* (Weg.) (Holm in *Sven. Bot. Tidskr.*, **47,** 520–5, 1953).
Synonym – *Didymosphaeria* (*Massariopsis*) *festucae* Weg (in *Mitt. Thurgau. Naturforsch. Ges.* **12,** 170–83, 1896).
Anamorph – *Phleospora idahoensis* Sprague (in *Mycologia*, **40,** 177–93, 1948).
(For discussion of taxonomy see Smith and Shoemaker, 1974).

Pycnidia are inconspicuous, immersed, intra- or sub-epidermal, with thin walls, 100–150 μm diam. often without definite ostioles on current seasons stems, but more obvious, blister-like, in rows on overwintered, blackened stems. Conidia are dis-

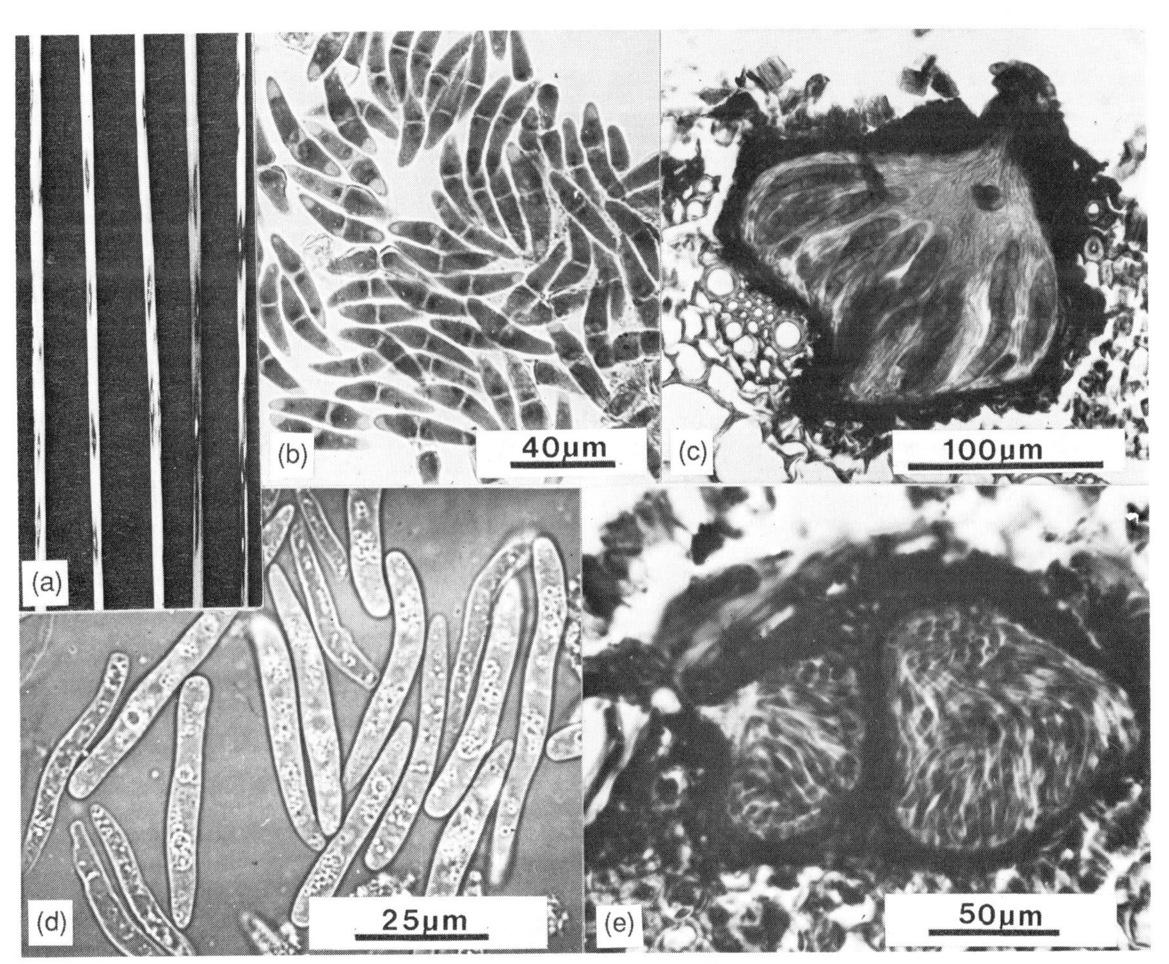

Figure 16.3 Stem eyespot of *Festuca rubra* caused by *Didymella festucae* (a) Stem eyespot symptoms; (b) Ascospores; (c) Perithecia; (d) Conidia; (e) Pycnidia.

charged in late spring or summer as a pink cirrus, hyaline, curved, basally truncate, variously guttulate, borne on short, truncate conidiophores. Conidia measure 18–105 × 3.0–9.0 μm from N. America, but smaller from Europe, 0–3 septate. Microspores are small, rare, of unknown function (Smith and Shoemaker, 1974) (Fig. 16.3 a to e).

Ascocarps occur with overwintered pycnidia between vascular strands of sheaths and culms, subepidermal, flattened-globose with a rudimentary clypeus, thin walls and low, conical ostiolar protuberance, approx. 180–430 μm diam. × 140–190 μm high in *F. rubra*, ripening in summer. Asci are cylindrical or club-shaped, 51–147 μm × 15–27 μm, with eight, biseriate ascospores which are two-celled, hyaline, spindle-shaped, basal cell larger than upper, 18–63 μm × 4.5–14.0 μm (Fig. 16.3b and c) (Smith and Shoemaker, 1974).

16.5.3 Disease development

Conidia are carried sparingly on the seed of *F. rubra* (Smith, Elliot and Shoemaker, 1968) and fungal mycelium in infected straw and lemmae of the seed. These are probably the means of long-distance transmission of the disease. Dispersal of conidia and ascospores by rain splash is likely within the crop, but frequently these spores germinate *in situ* on debris of the previous season and the mycelium of the fungus grows out from last season's lesions on stem debris when it is wetted (Smith, 1971b). The new season's flowering shoots growing through this infested debris become infected at an early stage, through contact with the fungus mycelium, mainly through the leaf sheaths. Lesions on the flowering stems correspond with initial infections which grew through the leaf sheaths. Leaf blades do not become infected so frequently as sheaths because they are protected by the latter. However, leaf blades are susceptible to infection as indicated by inoculation studies (Smith and Shoemaker, 1974). Severity of infection increases with the age of the crop. The disease spreads gradually along verges of highways sown with *F. rubra*, in western Canada, probably in infected stem fragments carried by mowing which is usually done after the inflorescences have emerged. The disease is common on the native fescue species in Oregon in the fescue seed growing areas. It has been found on 'wild' *F. rubra* near infected *F. idahoensis* in seed-growing areas yet *F. rubra* seed crops and *F. rubra* sown in roadside verges there were free from the disease (Smith, 1971b). The practice of annual straw and stubble burning in Oregon (Hardison, 1976), climatic differences and the lower susceptibility of Chewing's fescue, *F. rubra* L. ssp. *commutata* Gaud. (Smith and Elliott, 1972) commonly grown in Oregon, may explain the lower disease incidence on *F. rubra* seed crops there than in the Peace Region of Canada, where *F. rubra* L. ssp *rubra*, creeping red fescue, is grown for seed.

16.5.4 Control

No *Festuca* spp. appear immune, but some cultivars of *F. pratensis* Huds. *F. arundinacea* and *F. rubra* spp. *commutata* are less susceptible than most of those of *F. rubra* ssp. *rubra*. There is little resistance in creeping red fescue lines so far field tested. The use of high seeding rates (6–10 kg/ha) and high applications of nitrogen (50–70 kg/ha) applied in late autumn appear to minimize some of the effects of the disease (Elliott and Hennig, 1974). Crop sanitation by burning straw and stubble gives some control, but after harvest this may not be feasible or effective because of wet conditions. Burning to reduce debris is recommended as part of stand rejuvenation (Elliott, 1974). None of these operations gives much control of eyespot, but they tend to improve yields from the first crop after treatment. Field control of the disease has been obtained with applications of benomyl, biloxazol, fenarimol, polyoxins B and D and maneb. However, at fungicide dosages which give disease control, seed yield depression was noted (Davidson and Kleingebbinck, 1981).

16.5.5 Other grass diseases caused by *Phleospora* spp.

A leaf spot of *Agropyron repens* L. Beauv. and *Elymus canadensis* L. caused by the actively parasitic fungus *Phleospora graminearum* Sprague and Hardison (Hardison and Sprague, 1943; Sprague, 1950) from Michigan has also been found on *Festuca elatior* L. in Japan (Nishihara, 1972). Spots on leaves are elongate with brown margins and light centres. Pycnidia are immersed in the tissues, at first inconspicuous without ostioles, then erumpent and ostiolate, sub-globose, 90–160 μm. Spores are yellowish-hyaline, widest at the base, tapering to a blunt point, 30–55 μm × 3.3–5.6 μm, 1–6 septate (Hardison and Sprague, 1943).

Phleospora muhlenbergiae Sprague and Solheim in Solheim causes an indefinite leaf spotting on *Muhlenbergia arizonica* (Sprague, 1950).

16.6 TWIST

This disease distorts flowering tillers of a wide range of cultivated grasses and cereals and may also damage leaves of vegetative tillers (Atanasoff, 1925; Becker, 1955; Dennis and Foister, 1942; Grove, 1935; Mäkelä and Koponen, 1975; Mühle, 1953, 1971; Parberry, 1970; Protsenko, 1957; Rainio, 1936; Sampson and Western, 1954; Smith, 1965; Sprague, 1950; Stieltjes, 1933; Walker and Sutton, 1974). In northern countries it may cause considerable grass seed crop losses (Mühle, 1953). The disease is occasionally encountered in turf. The fungal stroma on leaves of *Agrostis* spp. and *Poa annua* L. may be associated with leaf galls caused by the nematode *Ditylenchus graminophilus* Goodey (Jackson, 1958). Twist has a worldwide distribution.

16.6.1 Symptoms

Infected flowering shoots become twisted and malformed and may become enveloped in an irregular black stroma which contains the spore bodies (pycnidia) of the fungus. Exsertion of the inflorescence is partly or fully arrested and seed production is impared (Fig. 25, colour plate section). Irregular stripe-like lesions may develop on adjoining leaves. These lesions darken and produce pycnidia at maturity. Eventually, infected leaves become discoloured, wither and die.

16.6.2 The causal fungus

Anamorph – *Dilophospora alopecuri* (Fr.) Fries (*Summ. Veg. Scand.*, 419 (Fig. 2). 1849). (For synonomy see Walker and Sutton, 1974.)

Pycnidia in the mesophyll of young lesions on expanded leaves are brown, globose and glabrous, usually unilocular, 160–220 μm diam. with walls several cells thick. Pycnidia may be aggregated in rows between the veins on older lesions on leaves and leaf sheaths. Conidia are formed in succession from phialidic conidiogenous cells which lack any definite collarette. Conidia are produced in succession in mucilage, cylindrical or tapered to the base, hyaline, 0–3 septate, smooth, with or without guttules, up to 13 μm long × 1.5 μm wide. The most characteristic feature of these conidia are the simple or branched appendages (Walker and Sutton, 1974). There is circumstantial evidence that *Lidophia graminis* (Sacc.). Walker and Sutton is the teleomorph of *D. alopecuri* (Mäkelä and Koponen, 1975; Walker and Sutton, 1974), but experimental proof of conspecificity is lacking. The two fungi have been found together on the same stroma or plant. Branches which form on broken ascospores of *L. graminis* at germination bear a resemblance to those on the pycnidiospores of *D. alopecuri*. The genetical connection between *D. alopecuri* and *Mastigosporium album* Riess claimed by Rainio (1936) has been disproven (Sampson and Western, 1938; Schaffnit and Wieben, 1928; Sprague, 1938).

16.6.3 Disease development

The disease is seed-borne on grass and cereal hosts (Richardson, 1979). Atanasoff (1925) and more recently, Protsenko (1957) found evidence that the appendages on the spores were a mechanism for nematode dispersal of the fungus, but these nematodes (*Anguillulina* spp.) are not necessary for disease transmission (Kotte, 1934; Schaffnit and Wieben, 1928; Stieltjes, 1933). Seed-borne infection is internal and external. The fungus is also soil-borne and carried in infected straw (Becker, 1955). Conidia are dispersed by rain splash.

16.6.4 Control

Little attention has been given to methods of control of the disease in the United States since it is not sufficiently important (Sprague, 1950). This is not the case in northern Europe where *Phleum pratense* may be frequently attacked (Mühle, 1953) and the disease may cause severe injury to cereals.

Cultural control measure such as high cutting, stubble burning and longer rotations have been suggested (Becker, 1955; Sprague, 1950). Dry seed disinfectants are ineffective (Becker, 1955).

16.7 REFERENCES

Andersen, I.L. (1972) Ergot (*Claviceps purpurea* (Fr. Tul.) on Gramineae plants in the county of Troms (Northern Norway). *Ny. Nord.*, **59**, 34–43.

Andersen, I.L. (1978) Choke disease, a rare parasite on grass species in northern Norway. *Norden*, **82**, 238–9. (Norwegian).

Anerud, K. (1939) Ergot and ergotism. *Landtmannen, Uppsala*, **23**, 49, 1185–8. (*R.A.M.* **16**, 272–3, 1940.) (Swedish).

Arnott, D.A. and Bergis, I. (1967). Causal agents of silvertop and other types of damage to grassland crops. *Can. Entomol.* **99**, 660–70.

Atanasoff, D. (1925) The Dilophospora disease of cereals. *Phytopathol.*, **15**, 11–40.

Bacon, C.W., Porter, J.K., Robbins, J.D. and Luttrell, E.S. (1977) *Epichloë typhina* from toxic tall fescue. *Appl. Environ. Microbiol.*, **34** (5), 576–87.

Barger, G. (1931) *Ergot and Ergotism.* Gurney and Jackson, London, 279 pp.

Becker, A. (1955) Observation on the occurrence of the plumed spore disease in the years 1951–53 in the Eifel region and the carrying out measures of control and prevention. *NachrBl. Dtsh. PflSchDienst.* (*Braunschweig*) *Stuttgart*, **7.6**, 100–4. (German).

Berkenkamp, B. (1976) Diseases of timothy in Alberta. Timothy Plus. *Phleum Newsletter*, **5**, 1 pp.

Berkenkamp, B. and Meeres, J. (1975) Observations in

silvertop of grasses in Alberta. *Can. Plant Dis. Surv.*, **55**, 83–4.
Boerema, G.H. and Verhoeven, A.A. (1977) Check-list for scientific names of common parasitic fungi. Series 2b Fungi on field crops, cereals and grasses. *Neth. J. Plant Pathol.*, **83**. 165–204.
Bradshaw, A.D. (1959) Population differentiation in *Agrostis tenuis* Sibth. II Incidence and significance of infection by *Epichloë typhina*. *New Phytol.*, **58**, 310–15.
Brady, L.R. (1962) Phylogenetic distribution of parasitism by Claviceps species. *Lloydia*, **25**, 1–6.
Bretag, T.W. and Merriman, P.R. (1981) Effect of burial on the survival of sclerotia and production of stromata by *Claviceps purpurea*. *Trans Br. Mycol. Soc.*, **77**, 658–60.
Brown, A.M. (1947) Ergot in cereals and grasses. In *Proc. Can. Phytopathol. Soc.*, **15**, 15. (Abstr.)
Brummer, V. (1937) Observations on outbreaks of fungal disease on timothy in Finland. *J. Sci. Agric. Soc. Finland,* **9**, 165–80.
Butler, E.J. and Jones, S.G. (1961) *Plant Pathology*. Macmillan, London.
Butler, G.M. (1966) Vegetative structures. In The Fungi, vol. 2. (eds. G.C. Ainsworth and S.A. Sussman), Academic Press, New York and London, pp. 83–112.
Calvert, E.L. and Muskett, A.E. (1945) Blind seed of ryegrass *Phialea temulenta* (Prill. and Delacr.) *Ann. Appl. Biol.*, **32**, 329–43.
Campbell, W.P. (1957) Studies on ergot infection in gramineous hosts. *Can. J. Bot.*, **35**, 315–20.
Chester, K.S. and Lefebre, C. (1942) Ergot epiphytotic in south western pastures. *Plant Dis. Rept.*, **26**, 408–10.
Colotelo, N. and Cook, W. (1977) Perithecia and spore liberation of *Claviceps purpurea*: Scanning electron microscopy. *Can. J. Bot.*, **55**, 1257–9.
Conners, I.L. (1967) *An Annotated Index of Plant Diseases in Canada.* Queen's Printer, Ottawa, 381 pp.
Cooke, R.C. and Mitchell, D.T. (1966) Sclerotium size and germination in *Claviceps purpurea. Trans Br. Mycol. Soc.*, **54**, 95–100.
Cooke, R.C. and Mitchell, D.T. (1967) Germination pattern and capacity for repeated stroma formation in *Claviceps purpurea. Trans Br. Mycol. Soc.*, **50**, 257–83.
Corkill, L. (1952) Breeding for resistance to blind seed disease (*Phialea temulenta* Prill. and Delacr.) in ryegrass (*Lolium* spp.). *Proc. 6th Int. Grassld. Congr.*, pp. 1578–84.
Couch, H.B. (1973) *Diseases of Turfgrasses*. Krieger, New York, 348 pp.
Craig, C.H. (1973) *Insect Pests of Legume and Grass Crops in Western Canada.* Info. Div. Agric. Can. Publ. 1437–47 pp.
Cunfer, B.M. (1976) Water potential of ergot honeydew and its influence upon colonization by microorganisms. *Phytopathology*, **66**, 449–52.
Cunningham, I.J. (1958) Non-toxicity to animals of ryegrass endophyte and other endophytic fungi in New Zealand grasses. *NZ J. Agric. Res.*, **1** (4), 489–97.
Davidson, J.G.N. and Kleingebbinck, H.W. (1981) Fungicidal control of stem eyespot (Didymella festucae) of creeping red fescue. *Proc. Joint Mtg. Can. Microbiol. Soc., Can. Phytopathol. Soc.* Abs. 36 p. 76.
Dennis, R.W.G and Foister, C.E. (1942) List of diseases of economic plants recorded in Scotland. *Trans. Br. Mycol. Soc.*, **25**, 266–306.
Elliott, C.R. (1974) Burning creeping red fescue seed fields. *N.R.G. News*, Agr. Can. Res. Sta., Beaverlodge, Alta. Sept. 1 pp.
Elliott, C.R. and Henning, A.M.F. (1974) Fertilizing grasses for seed production. *N.R.G. News* Agr. Can. Res. Sta., Beaverlodge, Alta. Sept. 1 pp.
Emecz, T.I. and Jones, G.D. (1970) Effect of gibberellic acid on inflorescence production in cocksfoot plants with choke (*Epichloë typhina*). *Trans. Br. Mycol. Soc.*, **55**, 77–82.
Eriksson, J. (1904) Choke in timothy. *Medd. Lantbr. Akad. Exp.*, **81**, 16. (Swedish).
Fischer, G.W. (1944) The blind seed disease of ryegrass in Oregon. *Phytopathology*, **34**, 934–5. (Abstr.)
Frauenstein, K. (1967) Studies on the incidence of ergot, *Claviceps purpurea* (Fr.) Tul. on *Poa pratensis. Z. PflKrank. Pfl. Path. Shutz.*, **74**, 443–59 (German).
Gagné, S., Richard, C., Gagnon, C. (1984). Silvertop of grasses: state of knowledge. *Phytoprotection*, **65**, 45–52 (French, En).
Grasso V. (1962) *Claviceps* on Italian Gramineae. *Ann. Speriment. Agric.*, **6**, 1–43.
Gray, E.G. (1942) *Phialea mucosa* sp. nov. The blind seed fungus. *Trans. Br. Mycol. Soc.*, **25**, 329–33.
Grove, W.B. (1935) *British Stem and Leaf Fungi. I Sphaeropsidales.* Cambridge University Press, 488 pp.
Hampton, J.G. and Scott, D.J. (1980a) Blind seed disease in New Zealand. I. Occurrence and evidence for use of nitrogen as a control measure. *NZ J. Agric. Res.*, **23**, 143–7.
Hampton, J.G. and Scott, D.J. (1980b) Blind seed disease of ryegrass in New Zealand. II. Nitrogen fertilizer; effect on incidence, and possible mode of action, *NZ J. Agric. Res.*, **23**, 149–53.
Harberd, D.J. (1961) Note on choke disease in *Festuca rubra. Rept. Scot. Plant Breed. Res. Sta.*, pp. 47–51.
Hardison, J.R. (1959) Evidence against *Fusarium poae* and *Siteroptes graminum* as causal agents of silver top of grasses. *Mycologia*, **51**, 712–28.
Hardison, J.R. (1962) Susceptibility of Gramineae to *Gloeotinia temulenta. Mycologia*, **54**, 201–16.
Hardison, J.R. (1963) Control of *Gloeotinia temulenta* in seed fields of *Lolium perenne* by cultural methods. *Phytopathology*, **53,** 460–4.
Hardison, J.R. (1976) Fire and flame for plant disease control. *Ann. Rev. Plant Pathol*, **14**, 335–79.
Hardison, J.R. (1977) Chemical control of ergot in field plots of *Lolium perenne. Plant Dis. Rept.*, **61,** 845–8.
Hardison, J.R. (1978) Chemical suppression of *Gloeotinia temulenta* apothecia in field plots of *Lolium perenne. Phytopathology*, **68**, 513–16.
Hardison, J.R. (1980) Role of fire for disease control in grass seed production. *Plant Dis.*, **6**, 641–5.
Hardison, J.R. and Sprague, R. (1943) A leaf spot of grasses caused by a new species of *Phleospora. Mycologia*, **35**, 185–8.
Hedley, J. and Braithwaite, M. (1978) Choke, a disease of grasses caused by *Epichloë typhina* (Pers.) Tul. *Horticulture in New Zealand,* **9**, 6–9.

Holmes, N.D., Swailes, G.E. and Hobbs, G.A. (1961) The Eriophyid mite, *Aceria tulipae* (K) (Acarina: Eriophydae) and Silvertop in grasses. *Can. Entomol.* **93**, 644–7.

Hyde, E.O.C. (1938) Detecting *Pullularia* infection in rye-grass seed crops. *NZ J. Agric.*, **52**, 301–2.

Hyde, E.O.C. (1945) Ryegrass seed. Preharvest examination for blind seed disease and estimation of germination capacity. *NZ J. Agric.*, **70**, 271–5.

Ingold, G.T. (1948) The water-relations of spore discharge in *Epichloë*. *Trans. Br. Mycol. Soc.*, **31**, 277–80.

Jackson, N. (1958) Root and leaf galls. In *Turf disease notes, 1958. J. Sports Turf Res. Inst.* **9** (34), 454.

Jenkinson, J.G. (1958) Ergot infection of grasses in the Southwest of England. *Plant Pathol.* **7**, 363–71.

Johnson, M.C., Pirone, T.P., Siegel, M.R. and Varney, D.R. (1982) Detection of *Epichloë typhina* in tall fescue by means of of enzyme-linked immunosorbent assay. *Phytopathology*, **72** (6), 647–50.

Keil, H.L. (1942) Control of silvertop of fescue by burning. *Plant Dis. Rept.*, **26**, 259.

Kirby, E.J.M. (1961) Host-parasite relations in the choke disease of grasses. *Trans. Br. Mycol. Soc.*, **44**, 493–503.

Kohlmeyer, J. (1956) Observations on the habit of *Epichloë typhina*. (Pers.) Tu. *Ber. Dtsch. Bot. Ges.*, **69**, 149–57. (German).

Kohlmeyer, J. and Kohlmeyer, E. (1974) Distribution of *Epichloë typhina* (Ascomycetes) and its parasitic fly. *Mycologia*, **66**, 77–86.

Koponen, H. and Mäkelä, K. (1976) *Phyllachora graminis, P. silvatica, Epichloë typhina* and *Acrospermum graminum* on grasses in Finland. *Karstenia*, **15**, 46–55.

Kotte, W. (1934) The plumed spore disease of cereals. *Nachr. Schadlbekampf. Leverkusen.*, **9**, 170–4.

Lafferty, H.A. (1948) Blind seed disease of rye-grass. *J. Dept. Agric. Repub. Ire.*, **45**, 192–201.

Langdon, R.F.N. (1954) The origin and differentiation of *Claviceps* species. *Univ. Queensland Pap. Dept. Bot.*, **3**, 61–8.

Large, E.C. (1954) Surveys for choke in cocksfoot seed crops 1951–53. *Plant Pathol.* **3**, 6–11.

Latch, G.C.M. (1966) Fungus diseases of ryegrasses in New Zealand. II Foliage, root and seed diseases. *NZ J. Agric. Res.*, **9**, 808–19.

Latch, G.C.M. and Christensen, M.J. (1982) Ryegrass endophyte, incidence, and control. *NZ J. Agric. Res.*, **25**, 443–8.

Leach, J.G. (1940) *Insect Transmission of Plant Diseases*. McGraw-Hill, New York, 615 pp.

Loveless, A.R (1971) Conidial evidence for host restriction in *Claviceps purpurea*. *Trans. Br. Mycol. Soc.*, **56**, 419–34.

Loveless, A.R. and Paech, J.M. (1974) Evidence for genotypic control of spore size in *Claviceps purpurea*. *Trans. Br. Mycol. Soc.*, **63**, 612–16.

Luttrell, E.S. (1980) Host-parasite relations and the development of the ergot sclerotium in *Claviceps purpurea*. *Can. J. Bot.*, **58**, 942–58.

Luttrell, E.S. (1981) Tissue replacement diseases caused by fungi. *Ann. Rev. Phytopathol.*, **19**, 373–89.

Mäkelä, K. and Koponen, H. (1975) *Lidophia graminis* (Sacc.) Walker and Sutton and *Dilophospora alopecuri* (Fr.) Fr. on grasses in Finland. *Acta Agric. Scand.*, **25** (2), 169–75.

Mantle, P.G., Shaw, S. and Doling, D.A. (1977) Role of weed grasses in the etiology of ergot disease in wheat. *Ann. Appl. Biol.*, **86**, 339–51.

McGee, D.C (1971) The effect of benomyl on *Gloeotinia temulenta* under laboratory and field conditions. *Aust. J. Exp. Agric. Anim. Husb.*, **11**, 693–5.

Miller, P.W. and McWhorter, F.P. (1948) The use of vapor-heat as a practical means of disinfecting seeds. *Phytopathology*, **38**, 89–101.

Mitchell, D.T. and Cooke, R.C. (1968a) Some effects of temperature on germination and longevity of sclerotia in *Claviceps purpurea*. *Trans. Br. Mycol. Soc.*, **51**, 721–9.

Mitchell, D.T. and Cooke, R.C. (1968b) Water uptake, respiration pattern and lipid utilization in sclerotia of *Claviceps purpurea* during dormancy and germination. *Trans. Br. Mycol. Soc.*, **51**, 731–6.

Mower, R.L. and Hancock, J.G. (1975a) Mechanism of honeydew formation by *Claviceps spp. Can. J. Bot.*, **53**, 2826–34.

Mower, R.L. and Hancock, J.G. (1975b) Sugar composition of ergot honeydews. *Can. J. Bot.*, **533**, 2813–25.

Mower, R.L., Snyder, W.C. and Hancock, J.G. (1975) Biological control of ergot by *Fusarium*. *Phytopathology*, **65**, 5–10.

Mühle, E. (1953) *Diseases and Pests of Cultivated Forage Grasses Planted for Seed Production.* S. Hirzel Verlag, Leipzig. 167 pp. (German).

Mühle, E. (1971) *Diseases and Injuries of Fodder Grasses.* S. Hirzel Verlag. Leipzig 422 pp. (German).

Mühle, E. and Frauenstein, K. (1970) Observations on the occurrence of choke disease. *Z. Pfl. Krankh. Pfl. Schutz.*, **77** (4–5), 177–185. (German).

Neill, J.C. and Armstrong, C.S. (1955) An aerial survey of ascospore distribution of blind seed disease of ryegrass *Gloetinia (Phialea temulenta) NZ J. Sci. Tech.*, **A37**, 106–9.

Neill, J.C. and Hyde, E.O.C. (1939) Blind seed disease of ryegrass. *NZ J. Sci. Tech.*, **A20**, 281–301.

Neill, J.C. and Hyde, E.O.C. (1942) Blind seed disease of ryegrass-II. *NZ J. Sci. Tech.*, **A24**, 65–71.

Nishihara, N. (1972) A new disease of *Festuca elatior* caused by *Phleospora graminearum. Bull. Nat. Grassld. Res. Inst. Japan.*

Noble, M. and Gray, E. (1945) Blind seed disease of ryegrass. *Scot. J. Agric.* **25**, 94–7.

O'Rourke, C.J. (1976) *Diseases of Grasses and Forage Legumes in Ireland.* An Foras Taluntais (The Agricultural Institute), Carlow, 115 pp.

Oudemans, C.A.J.A. (1919–1924) *Enumeration Systematica Fungorum.* 1–5 The Hague.

Parberry, D.G. (1970) *Dilophospora alopecuri* in Australia. *Search ANZAAS*, **1** (2), 83.

Petch, T. (1937) More about *Claviceps*. *Naturalist (London)*, pp. 20–25.

Petersen, A.G. and Vea. E.V. (1971) Silvertop of bluegrass in Minnesota. *J. Econ. Entomol.*, **64**, 247–52.

Peyronel, B. (1930) Fungal symbiosis of the *Lolium* type in certain Gramineae of the genus *Festuca. Nuovo Giorn. Bot. Ital. NS*, **37** (3), 643–8. (Italian).

Prillieux, E. and Delacroix, G. (1892) *Phialea temulenta*, the ascospore stage of *Endoconidium temulentum. Bull. Soc. Mycol. Fr.*, **8**, 22–3. (French).

Protsenko, E.P. (1957) A new case of joint infection of grasses by fungi and nematodes. *Bull. Bot. Gdn., Moscow*, **29**, 91–3. (Russian). (*R.A.M*, **37**, 414, 1958.)

Pshedetskaya, L.I. (1971) Physiological and ecological characteristics of *Epichloë typhina* strains. *Tnidyvses. Inst. rast.*, **29**, 36–41. (Russian).

Pshedetskaya, L.I. (1979) The overwintering conditions and features of sclerotial germination of the ergot *fungus. Referativnyi Zhurnal, Biologiya*, **21**, 63–58. (*R.A.P.P.* **58** (12), 5913, 1979.)

Rainio, A.J. (1936) On the Dilophospora disease of *Phleum pratense* L. and *Alopecurus pratensis L. Valt. Maatralousk. Julk.*, **87**, pp. (German, Finn.) (R.A.M. **16**, 184–5 1937.).

Rehm, H (1900) Ascomycetes exs. fasc 27. *Hedwigia* **39**, 192.

Richardson, M.J. (1979) *An Annotated List of Seed-borne Diseases*, 3rd edn. Commonw. Mycol. Inst., Kew. Int. Seed Test. Assoc., Zurich, 320 pp.

Riggs, R.K., Henson, L. and Chapman, R.A. (1968) Infectivity of and alkaloid production by some isolates of *Claviceps purpurea. Phytopathology*, **58**, 54–5.

Sampson, K. (1933) The systemic infection of grasses by *Epichloë typhina* (Pers.) *Tul. Trans. Br. Mycol. Soc.*, **18**, 30–47.

Sampson, K. (1939) Additional notes on the systemic infection of *Lolium. Trans. Br. Mycol. Soc.*, **23** (4), 316–19.

Sampson, K. and Western, J.H. (1938) Note on the supposed connection between *Mastigosporium album* Riess and *Dilophospora alopecuri* (Fr.) Fr. *Trans. Br. Mycol. Soc*, **22**, 168–73.

Sampson, K. and Western, J.H. (1954) *Diseases of British Grasses and Herbage Legumes*. 2nd edn British Mycològical Society. Cambridge University Press, 118 pp.

Schaffnit, E. and Wieben, M. (1928) Research on the cause of the plumed spore disease *Dilophospora alopecuri* (Fr.) Fr. *Forsch. Geb. PflKrank. Immunitat. PflR*, **5**, 3–8 (German).

Scott, D.J. (1974) Blind seed disease. Preharvest testing to cease. *NZ J. Agric.*, **129**, 19.

Seaman, W.L. (1981) *Ergot of Grains and Grasses*. Canada Agriculture Publ. 1438, 14 pp.

Smith, J.D. (1965) *Fungal Diseases of Turf Grasses*, 2nd edn (Revised N. Jackson, and J.D. Smith), The Sports Turf Research Inst., 97 pp.

Smith, J.D. (1971a). *Phleospora* stem eyespot of fescues in Oregon and the *Didymella* perfect stage of the pathogen. *Plant Dis. Reptz.*, **55**, 63–7.

Smith, J.D. (1971b) *Phleospora idahoensis* on native *Festuca* spp. in the northwestern Great Plains. *Can. J. Bot.*, **49**, 377–81.

Smith, J.D. (1976a) *Phleospora idahoensis* and *Didymella festucae* from Gotland, Sweden. *Bot. Notiser, Stockholm*, 129–200.

Smith, J.D. (1976b) *Didymella festucae* on *Festuca* spp. in Norway. *Blyttia*, **34**, 99–102. (Norwegian, En.).

Smith, J.D. (1980) *Didymella festucae* and *Phleospora idahoensis* in Southwest Iceland. *Res. Inst. Nedri As, Hveragerdi, Iceland, Bull.*, **32**, 12 pp.

Smith, J.D. and Elliott, C.R. (1970) Stem eyespot on introduced *Festuca* spp. in Alberta and British Columbia. *Can. Plant Dis. Surv.*, **50**, 84–7.

Smith, J.D. and Elliott, C.R. (1972) Didymella stem eyespot of *Festuca* spp. in northern Alberta and British Columbia in 1970 and 1971. Can. Plant Dis. Sun., **52** (2), 39–41.

Smith, J.D., Elliott, C.R. and Shoemaker, R.A. (1968) A stem eyespot of red fescue in Northern Alberta. *Can. Plant Dis. Surv.*, **48**, 115–19.

Smith, J.D. and Shoemaker, R.A. (1974) *Didymella festucae* and its imperfect state, *Phleospora idahoensis* on *Festuca* species in western North America. *Can. J. Bot.*, **52**, 2061–74.

Sprague, R. (1938) Two *Mastigosporium* leaf spots on Gramineae. *J. Agric. Res.*, **57**, 287–99.

Sprague, R. (1950) *Diseases of Cereals and Grasses in North America.* Ronald Press, New York, 538 pp.

Sproule, T.R.M and Faulkner, J.S. (1974) The reaction of eleven genotypes of *Lolium perenne* to British and New Zealand strains of *Gloeotinia temulenta. Plant Pathol.* **23**, 144–7.

Stäger, R. (1908) On the biology of ergots. *Cent. Bakt.* II, **20**, 272–99. (German).

Stieltjes, D. (1933) Dilophospora disease of grains and grasses. *Tijdschr. Plantenziekt.*, **39**, 8, 200–6. (Dutch).

Tanda, S. (1979a) Mycological studies on ergot in Japan (part 6). A physiological race of *Claviceps purpurea* Tul. var. *alopecuri* Tanda collected from *Trisetum bifidum* Ohwi. *J. Agric. Soc. Tokyo*, **23**, 207–13.

Tanda, S. (1979b) Mycological studies of ergot in Japan. Two varieties of *Claviceps purpurea* parasitic on *Agrostis* spp. *J. Agric. Soc. Tokyo*, **23**, 215–21.

Tempe, J. de (1966) Blind seed disease of ryegrass in the Netherlands. *Neth. J. Plant Pathol.*, **72**, 299–310.

Van Dijk. G.E. (1967) Common grass diseases in Europe and their importance. *Eucarpia Forage Crops Section Meetg, Køln-Vogelsang*, 20–21 Sept. 1967, pp. 1–19.

Vladimirskaja, N. (1928) Contribution to the biology of *Epichloë typhina* Tul. *Zasch. Rast. Vredit.*, **5** (3–4), 335–47. (*R.A.M.* **8**, 313–314, 1929.) (Russian).

Walker, J. and Sutton, B.C. (1974) *Dilophia* Sacc. and *Dilophospora* Desm. *Trans. Br. Mycol. Soc.*, **62** (2), 231–241.

Wernham, C.C (1942) *Epichloë typhina* on imported fescue seed. *Phytophathology*, **32**, 1093.

Western, J.H. and Cavett, J.J. (1959) The choke disease of cocksfoot (*Dactylis glomerata*) caused by *Epichloë typhina* (Fr.) Tul. *Trans. Br. Mycol. Soc.*, **42**, 298–307.

Wetzel, T. (1971) White-ears. In *Diseases and Pests of Forage grasses* (ed. E. Muhle), S. Hirzel Verlag, Leipzig, pp. 49–94. (German).

Wilson, M., Noble, M. and Gray, E.G. (1945). The blind seed disease of ryegrass and its causal fungus. *Trans. Roy. Soc. Edin.* **61**, 327–40.

Wilson, M., Noble, M. and Gray, E.G. (1954) *Gloeotinia*–a new genus of the Sclerotiniaceae. *Trans. Br. Mycol. Soc.*, **37**, 29–32.

Wright, C.E. (1956) Blind seed disease of ryegrass I–Summary of preliminary investigations into techniques of evaluating degrees of resistance to *Phialea temulenta* in *Lolium perenne. Res. Exp. Rec. Min. Agric. N. Ire.*, **6**, 1–14. (Cited by Wright, 1967).

Wright, C.E. (1967) Blind seed disease of ryegrass. *Euphytica*, **16,** 122–30.

Wright, C.E. and Faulkner, J.S. (1982) A backcross programme introducing resistance to blind seed disease (*Gloeotinia temulenta*) into the cultivar S 24 of the cross-pollinated species *Lolium perenne*. *Rec. Agric. Res. Dept. Agric. N. Ire.*, **30**, 45–52.

Wright, C.E. and Sproule, T.R.M. (1965) Blind seed disease of ryegrass. VI. Resistance of *Lolium perenne* varieties to blind seed disease (*Gloeotinia temulenta*) *Rec. Agric. Res. Min. Agric. N. Ire.*, **14**, 5–29. (Cited by Wright, 1967.)

Part eight

Foliar blights and patches

17 *Anthracnose and colletotrichum basal rot*

Disease caused by or associated with *Colletotrichum graminicola* (Ces.) Wils. has been reported on a wide range of grasses and cereals from many parts of the world. (Boning and Wallner, 1936; Bruehl and Dickson, 1950; Chowdhury, 1936; Conners, 1967; Couch, 1973; Dickson, 1956; Duke, 1928; Edgerton and Carvajal, 1944; Grove, 1937; Kreitlow, Graham and Garber, 1953; Mühle, 1953; Nishihara, 1972; Noble and Richardson, 1968; Sanford, 1935; Selby and Manns, 1909; Sprague, 1950; USDA, 1960; Wilson, 1914; Winter, 1940; Wolff, 1947). Disease symptoms may include lesioning or blighting of aerial parts (anthracnose) and/or a culm, crown and root rot, that generally becomes apparent as the host matures. The ubiquitous occurrence of this fungus on senescing and dead Poaceae is reflected in the listing of most turf grasses as hosts but the number of species subject to severe damage is limited. Anthracnose symptoms of sufficient intensity and extent on turf grasses to warrant attention have been reported primarily in North America with *Poa annua L. the* species most commonly involved. In Western Europe, coastal British Columbia and Washington, basal rot symptoms of *Poa annua*, rather than leaf and crown lesioning predominate.

Wilson (1914) found the fungus common on blue-grasses in New Jersey and Sprague and Evaul (1930) reported that it was the most abundant fungus in association with *Drechslera poae* Baudys and a *Fusarium* sp. in a disease epidemic on *Poa annua* in turf in the same state in 1928. They found *C. graminicola* pathogenic on *Poa annua* in greenhouse studies, although details of these were not reported. Wolff (1947) showed that it was mildly pathogenic on the aerial parts of seedlings of some species and cultivars of *Agrostis* spp. No studies on *Poa annua* infection were documented. Inoculum was applied to aerial parts of the grasses as a spore suspension and not as soil inoculum although the pathogenicity of the fungus to corn seedlings when applied to the soil was reported. In field observations Wolf was never able to confirm instances of turf grass disease where demise of the grass plants was attributable solely to *C. graminicola.* Invariably the fungus was in association with '*Helminthosporium*' spp. and other fungi recognized by Wolff as saprophytes on fine turf.

Smith (1954) followed the process of seminal root infection by *C. graminicola* of *Poa annua* seedlings grown from sterilized seed and recorded the infection process in roots and shoot bases of plants of the same species when the fungus was the dominant incitant of basal rot. Alexander (1969) reported that *C. graminicola* was damaging to *Poa annua* turf in South Carolina and Couch (1973) considered the fungus a serious problem on most of the turf grasses in common use. Like Wolff (1947), he commented on the association with '*Helminthosporium*' spp., but he described *C. graminicola* essentially as a pathogen of aerial plant parts. Suprisingly, *Poa annua* was not recorded by Couch (1973) or listed in the USDA Index of Plant Diseases (1960) as a host species. Couch (1979a,b) now questions, on the published experimental evidence from infection studies, whether damage to and death of *Poa annua* during warm summer months, previously ascribed to *C. graminicola*, is in fact caused by this fungus. In his opinon heat stress causing wilting is primarily responsible for the summer demise of *Poa annua.* Vargas (1976a,b, 1978, 1980, 1981) and Vargas and Detweiler (1976) dispute this hypothesis and now consider that *C. graminicola* attack is part of a disease syndrome on *P. annua* which is referred to by Vargas (1981) as HAS decline, in which '*Helminthosporium*' spp., *C. graminicola* and senescence of the grass are all involved. This position is endorsed by Bolton and Cordukes (1981) who reported on destructive outbreaks of anthracnose on *P. annua* in Eastern Canada. Damage to the turf resulted from a foliar blighting in which *Bipolaris sorokiniana* and *Curvularia lunata* were also involved. A survey of selections of *Poa annua* from different locations by these authors in 1981 demonstrated marked differences in susceptibility to *C. graminicola.*

17.1 SYMPTOMS

17.1.1 Foliage blight

Leaves and shoots of turf grasses, particularly the fine-leaved fescues, but also bentgrasses, bluegrasses and perennial ryegrass are blighted and killed back (Fig. 26, colour plate section) in irregularly shaped reddish to bleached patches up to several feet in diameter. The colour may lead to confusion with red thread disease except that there are no red 'needles' or pink mycelium. Instead, dark acervuli of *C. graminicola* develop on the dead and dying tissues. On several turf grass species artificially inoculated in growth chambers with spore suspensions of *C. graminicola*, Bolton and Cordukes (1981) described the uniform leaf symptoms produced as consisting of yellow to brown lesions on blades and sheaths that later support abundant setose acervuli.

17.1.2 Helminthosporium – anthracnose – senescence (HAS) syndrome

According to Vargas (1981), the HAS disease syndrome on *P. annua*, in which *C. graminicola* is concerned, starts as yellow – bronze patches of turf, 30 to 60 cm diameter, darkening to bronze at temperatures of 26°C and above, when humidity is high, with irregular brown or purplish-black lesions on the blades, some of which are caused by *Bipolaris sorokiniana*, but the acervuli of *C. graminicola* are also present. Little damage is done in cool weather but patches remain yellow, and the disease progresses rapidly if weather conditions are favourable. Root systems of diseased plants are sparse. Plants suffering from wilt may be distinguished from those suffering from HAS decline in that the former turn dark blue or purple.

17.1.3 Basal rot

Under cool conditions *P. annua* turf takes on a measly, piebald appearance when attacked. In a dense turf of this species individual plant or tillers may turn yellow or small patches may increase in size up to about 15 cm diameter with irregular outlines. In a thin, starved turf the yellowed leaves and shoots may be scattered within the patch.

Single, well-tillered individual plants may show infected and uninfected tillers. The disease develops slowly, the older leaves become discoloured first, starting with the tips and progressing down the blades to the leaf sheaths. The central leaf of this shoot usually is the last to show colour change, the tip may remain green for some time after the base has changed colour. Eventually the central leaf turns yellow, orange and finally red. In early stages of infection there are usually no obvious lesions on the leaf blades. The disease starts as a basal rot, and a dark-brown discolouration occurs at the bases of leaf sheaths of infected plants. The sheath bases turn black and the whole shoot can be separated easily from the crown. The black discolouration is due to the dark-coloured dendroid hyphae and dense mycelial aggregates of the fungus (Figs 17.1 and 17.2). Dark patches and lines occur mainly on sheaths and occasionally they continue on the leaves. Inflorescence formation is prevented in severely infected plants.

Dark patches and lines of the fungus occur on the stem below the crown and on adventitious roots. Root systems are often poor and necrotic in severely

Figure 17.1 Acervuli of *Colletotricum graminicola* on central shoot base of *Poa annua* (I.R. Evans).

Figure 17.2 Basal rot of *Poa annua* caused by *Colletotrichum graminicola*).

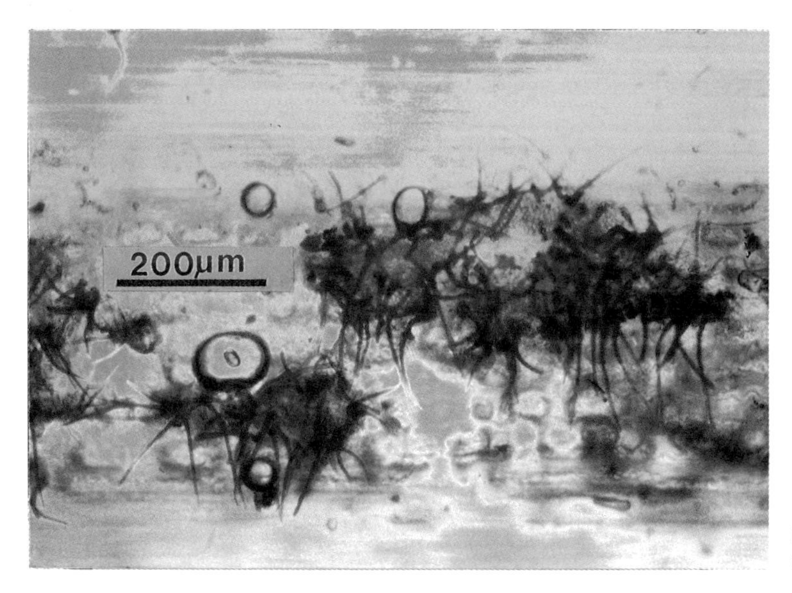

Figure 17.3 *Colletotrichum graminicola*. Acervuli with setae.

infected plants and at this stage leaves show narrow, dark lines and minute patches on abaxial and adaxial surfaces. Later, leaf blades darken and rot away. Acervuli with dark-brown setae (Fig 17.3.) visible with a hand lens, develop on necrotic tissues (Smith, 1954).

Similar symptoms have been recorded in cool weather on *Agrostis stolonifera*, *A. canina* and *Poa annua* from golf greens in the New England region of the United States (Jackson and Herting, 1985). The symptoms on bentgrass turf appear as diffuse patches, irregular in shape from 1–2 cm up to 50 cm or more, grey–green to tan, fading to a dull brown. Older leaves of individual plants are affected first and the discolouration progresses successively to the youngest emerging leaf. Infection of *A. canina* does not produce the orange or red colouration, typical of *C. graminicola* on *Poa annua* under similar conditions. A necrosis and blackening of basal sheath and crown tissues, extending into some of the adventitious roots becomes apparent. The affected plant parts support investing and invading mycelium, acervuli, dense black stromatal masses (sclerotia) and numerous appressoria. Roots are reduced in length on severely affected plants and both cortical and stelar tissues are blackened. Mature acervuli bearing conidia occur on deteriorating or dead leaf blades, sheaths, crown tissues and roots. Symptoms are apparent in late spring and early summer as temperatures range between 15°C and 25°C. Incidence of the disease lessen as temperatures increase, but symptoms on existing outbreaks persist into the summer.

17.2 THE CAUSAL FUNGUS

Teleomorph: *Glomerella graminicola* Politis (1975);
Anamorph: *Colletotrichum graminicola* (Ces.) (Wilson, 1914)–wrongly *C. graminicolum.*

Sutton (1966) and Politis (1975) discussed the confusion in the literature over the identity of *C. graminicola* and *C. falcatum* Went which are very similar morphologically. Von Arx and Muller (1954) regarded them as synonymous, but Sutton (1968) separated them on the basis of appressorial characters. Confirmation of the separate taxa based on their production of distinctive teleomorph states was presented by Politis (1975). Earlier problems of taxonomy have been considered by Wolff (1947) and Smith (1954).

The teleomorph *Glomerella graminicola* has been found only in culture on sterilized corn leaves, inoculated with monoconidial isolates of *C. graminicola* (Politis and Wheeler, 1972; Politis, 1975).

Perithecia are black, beaked, bristled, erumpent, globose to sub-globose, formed in black stromata, 195–575 μm high × 170–470 μm diam. Asci are 8-spored, pedicillate, clavate-cylindrical, inoperculate, unitunicate, thin-walled, 70–125 × 9–19 μm, in rosettes with paraphyses. Ascospores are hyaline, curved, aseptate, guttulate, 16–29 × 4–10 μm. (For a detailed description see Politis, 1975.)

Acervuli of *C. graminicola* from *P. annua* are dark brown to black, 20–200 μm in diam. when separate, but may become confluent in a dark, continuous stroma. They are at first immersed then erumpent on shoot bases, leaf sheaths and leaves. They are usually setose, the dark-brown almost straight setae developing either around or within the acervulus. Setae are 0 to 5-septate (frequently aseptate) and taper to an acute point from a swollen base 85–300 μm × 6–8 μm wide above the swollen base. Conidia from *Poa annua, Festuca rubra* and *F. ovina*[1] are fusoid, curved, hyaline, non-septate, with a central nucleus and granular contents, mostly 19–30 μm × 2.5–6.0 μm from plant material and 22–35 μm × 4–7 μm from culture. Spore measurements for isolates obtained in New England from *A. palustris* and *P. annua* were 13–32 × 3 to 6 μm and 21–32 × 3 to 6 μm respectively (Herting 1982). Appressoria from *P. annua* are irregularly shaped, aseptate, hyaline when young, turning dark-brown when mature, thick-walled, with one or more thin areas or pores, 6–22 × 6–13 μm. They develop at the ends of hyphae as swellings which are later cut off by cross walls. Appressoria are formed on the surface of most infected tissues often in regular lines above the junction of adjacent cells (Smith, 1954; Lapp and Skoropad, 1978). In morphology they resemble those figured by Sutton (1968) for *C. graminicola* isolates from sorghum and maize and by Herting (1982) for *A. palustris.*

Smith (1954) isolated the fungus by plating fragments of diseased tissues on to glucose – boric agar or

[1] From Yorkshire, England; Seattle, USA: Agassiz, Canada: Oslo, Norway (Smith, unpublished).

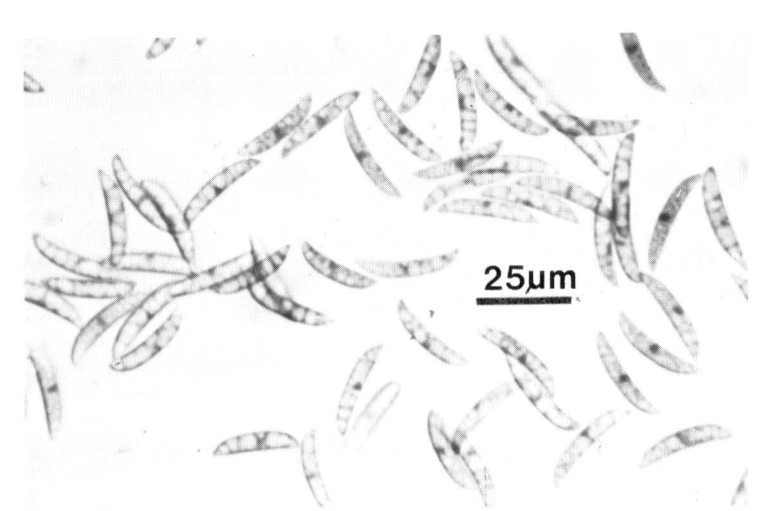

Figure 17.4 Conidia of *Colletotrichum graminicola* (nuclei stained).

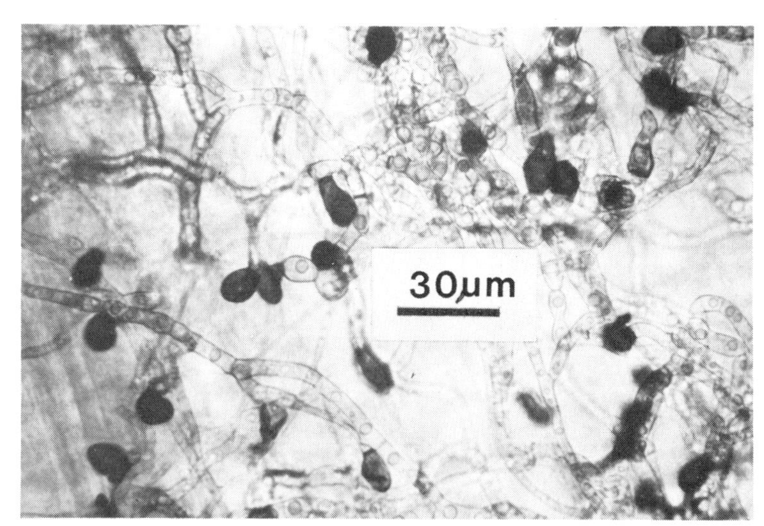

Figure 17.5 Appressoria of *Colletotrichum graminicola.*

by streaking out spores from acervuli on this medium. Satisfactory growth and sporulation occurs on minimal salts medium with glucose and yeast extract, potato dextrose and grass extract agar. Appressoria are produced freely on glucose–boric agar. Isolates obtained from *A. stolonifera* grew satisfactorily on potato–dextrose agar to produce setose acervuli with an abundance of spores (Herting, 1982). The optimum temperature for mycelial growth of a British isolate from *P. annua* was 22 °C. This value was lower than those found by Wolff (1947) and Bruehl and Dickson (1950) (27 °C and 28 °C respectively) or of an isolate from Holland of 28 °C. (Smith, 1954). The approximate optima for five New England isolates from *A. stolonifera* was 25 °C (Herting, 1982). In culture the mycelium is generally grey with grey, faint green, pink or brown colony centres depending on isolate. Diurnal zonation occurs, but light is not necessary for sporulation.

17.3 PATHOGENICITY TESTS

Wolff (1947) obtained infection of species of *Agrostis* in greenhouse experiments with isolates from *Agrostis* spp. and Sudan grass using spore suspensions applied to leaves as inoculum. Pathogenicity was considered mainly in terms of leaf infection. Symptoms were different from field cases. The disease progressed more rapidly when the temperature was between 15 and 30°C than at 10–25°C. No disease developed when the tests were conducted at 10–18°C.

Smith (1954) obtained infection of seminal roots of *P. annua* seedlings raised under sterile conditions from seed sterilized with mercuric chloride, by planting the seed in agar medium on which a culture of the fungus was growing. Since he was concerned only with the basal rot disease it was considered that pathogenicity was proven and that the infection process in *P. annua* was similar to that shown by Bruehl and Dickson (1950) for cereals and other grasses. Using isolates of *C. graminicola* obtained from *A. stolonifera* and *Poa annua*, Herting (1982) conducted a series of inoculation trials involving seedlings and more mature plants of both donor species. Mycelial and spore inocula were employed in the trials. Under the axenic conditions adopted none of the procedures, including the method reported by Smith (1954) were effective in reproducing the basal sheath and crown rot symptoms typical of the disease in the field. Temperature regimes chosen to approximate the conditions obtaining in field outbreaks (varying over the range 10–25 °C) resulted in very mild infection of superficial tissues of the test grasses. Infection was improved only slightly at the higher temperature and when inorganic nutrients were supplied to the plants. It was concluded that the conditions which predispose turf to severe basal sheath and crown rot remain obscure.

Field observations that higher temperatures figure prominently in leaf infections of some turf grasses by this fungus have been corroborated in greenhouse trials. Bolton and Cordukes (1981) reported the successful inoculation of *Poa annua*, *Poa pratensis*, *Festuca rubra* and *Agrostis tenuis* seedlings in moist chambers maintained at 30–33 °C when sprayed with conidial suspensions. Yellow to brown lesions developed on the laminae and sheaths. *Agrostis stolonifera* was found to be immune in these trials.

17.4 HOST RANGE

C. graminicola is a very common fungus on many species of turf grasses from temperate to sub-tropical regions of the world (Bolton and Cordukes, 1981; Bruehl and Dickson, 1950; Conners, 1967; Dahlsson, 1977, 1978; Gould, 1956; Howard, Rowell and Kiel, 1951; Noviello, 1963; Sprague, 1950; Smith, 1954, 1955, 1965, 1980; USDA 1960). Whereas it is reported as most damaging on *Poa annua* L. under conditions of high humidity and warm temperatures in the United States (Vargas, 1981), under much cooler temperatures it may also cause a basal rot on that species. Basal rot symptoms occur on *Poa annua* when growth is minimal or arrested in late autumn through to early spring, and have been recorded in the British Isles (Smith, 1954, 1955, 1965); Saskatchewan, coastal British Columbia and the Pacific Northwest (C.J. Gould, personal communication, Smith, 1980 and unpublished), Rhode Island (Jackson and Herting, 1985), Sweden (Dahlsson, 1977) and eastern Norway (Smith, unpublished). The first record of a *Colletotrichum* sp. associated with disease of basal leaves and roots of *Poa annua* is from a lawn in Huddersfield in November 1936, cited by Smith (1955). Under similar environmental conditions basal rot symptoms may occur on turf composed of *Agrostis* spp. in particular *A. stolonifera* cv. Penncross and *A. canina* L. cv. Kingstown in the New England region of the United States (Jackson and Herting, 1985). *Festuca rubra* L. cultivars are occasionally severely infected in coastal British Columbia (Smith, 1980 and D.K. Taylor, personal communication) and on that species in summer in Saskatchewan (Noviello, 1963). It has been reported to attack severely *Lolium perenne* L. in roadside verges in the US (Duell and Schmidt, 1974).

17.5 DISEASE DEVELOPMENT

The mode of infection and progress of the disease on *P. annua* was examined in detail by Smith (1954). From pores in the appressoria, infection pegs penetrate root tissues between adjacent cells of the epidermis and the hyphae pass into the cortex and the stele. Appressoria may be found on inner root tissues when the cortex has disintegrated, but generally not when the cortical cells are healthy or only slightly attacked. The initial attack on the plant bases appears to take place from soil-borne inoculum at or near soil level. Epidermal cells and cortex of the stem below the crown are invaded and overrun and the fungus passes into the vascular tissues, the cortex apparently offering little resistance to attack. The inner cortical tissues become heavily infected, bundle sheath, fibres, xylem vessels and phloem cells being plugged with mycelium. Few hyphae were found in the medulla. Infection of the main stems may proceed by way of one of the adventitious roots. Infection of a root may take place by way of the connection between the infected vascular system and parenchyma of the shoot and that root. Bruehl and Dickson (1950) considered that the nodal plate below the crown primordium in *Sorghum vulgare* Pers. functioned as a partial barrier to the progress of *C. graminicola* from the sub-coronary region into the crown. Since the central portions of the shoot cylinder show severe infection earlier than the outer leaf sheaths, it may be inferred that the younger tissues of the shoot primordium are more readily invaded from the sub-coronal region that are the older sheathing tissues.

Leaf and leaf-sheath infection is probably from two sources: (a) from spores produced in acervuli on tissues near soil level carried to sheaths and leaves and (b) by hyphae growing upwards from the lower infected tissues. Leaf-tip infections which occur when the intervening leaf blade is free from them are almost certainly due to spores. The infection process of *Poa annua* by spores was not examined but penetration of the grass host occurs by means of appressoria and the direct ingress of penetration hyphae (Skoropad, 1967). Politis and Wheeler (1973) and Politis (1976) have described the ultrastructure of the direct penetration process on maize and oat leaves. Bruehl and Dickson (1950) found that the germ tubes of *C. graminicola* on Sudan grass (Sorghum vulgare var. sudanense (Piper) Hitchc.) did not penetrate leaves by way of wounds or stomata.

C. graminicola persists saprophytically in crop debris that previously was parasitized (Vizvary and Warren, 1981). In mild climates it may also survive as hyphae or conidia in diseased plants (Wolff, 1947), but in the absence of plant residues lysis of these fungal structures occurs within a few days (Vizvary and Warren, 1981). Survival time of the appressoria of *C. graminicola* is temperature dependent (Skoropad, 1967). Below 15–20°C these structures remain dormant and may play an important role in the turf environment as resting structures or in latent infections.

The foliage blight symptom develops during hot humid weather. In western Canada, July, August and September are the most likely months for its development. The fine-leaved fescues have been recorded as severely affected in British Columbia, but *P. annua* may be severely blighted in hot summer weather in Saskatchewan and Alberta (Noviello, 1963; D.K.

Taylor, personal communication; Smith, unpublished).

The HAS syndrome or anthracnose of aerial parts of turf grasses is a disease occurring in warm or hot humid weather. According to Vargas (1980, 1981) the key environmental factor for the development of severe HAS decline appears to be a night-time temperature of 27°C or above for two to three consecutive nights. Although the problem may also develop when daytime temperatures are 30°C or above for several consecutive days it is usually not widespread unless the night-time temperature conditions are fulfilled.

The basal rot of *P. annua* is found as a cool season disease in Britain, Norway, Sweden, Washington, British Columbia and parts of New England. It was found in April and December in Britain (Smith, 1954), in October in Norway (Smith, unpublished), in December in Seattle, western Washington, and Vancouver, British Columbia (C.J. Gould, personal communication) and in October to December, and April to June in Rhode Island (Jackson, unpublished). The disease occurred on *Agrostis* spp. in Rhode Island and adjacent states over the same period (Jackson and Herting, 1985). In western Washington and Vancouver, British Columbia, parasitic nematodes were found associated with diseased grass. A similar connection between anthracnose and parasitic nematodes was proposed by Jackson and Herting (1985) after trials in Rhode Island with nematicide/fungicide combinations proved more effective in alleviating disease symptoms than fungicides alone. Root galls caused by *Ditylenchus radicicola* Greef. have been found on *P. annua* plants infected with *C. graminicola* (Smith, 1955). The disease is often associated with *P. annua* in overcompacted turf at the edges of pathways on aprons of golf greens, and on cricket squares where turf is deficient in nutrients, but it occasionally also develops on senescent turf well supplied with nitrogen. On experimental plots it is favoured by applications of vermiculite (Smith, 1956).

17.6 SPECIES AND CULTIVAR RESISTANCE

Jade, Diamond, Highlight, Belmonte, Bolero, Ensylva, and Pennlawn fine-leaved *Festuca* spp. were less susceptible than Dawson, Paramir, Engina and Bergond to anthracnose at Agassiz Research Station in British Columbia (Smith, 1980, D.K. Taylor, personal communication). Vargas (1981) reported that there were considerable differences in susceptibility between cultivars of commonly used cool-season turfgrasses and Bolton and Cordukes (1981) demonstrated marked differences in susceptibility by selections of *Poa annua* from different locations.

17.7 CONTROL

To control basal rot disease, reduce traffic, relieve compaction by aerifying or spiking and improve fertility if necessary. *P. annua* should be replaced with other turf grasses. Jackson and Herting (1985) found the combinations of triadimefon or benomyl with chlorothalonil were effective in reducing the severity of disease symptoms and further improvement occurred when the fungicides were combined with nematicide treatment.

Vargas (1980) reported that in HAS syndrome on *P. annua* effective control is possible by providing the growing conditions appropriate to accommodate to the agronomic character of the grass and by the use of suitable fungicides such as chlorothalonil, mancozeb, maneb and zinc sulphate or the systemics benomyl, thiophanate-ethyl and thiophanate methyl at 7–14 day intervals from July to September. The timely and moderate use of nitrogen (Vargas, Detweiler and Hyde, 1977) but particularly the preventive use of the fungicide triadimefon gave best control of the disease (Danneberger *et al.*, 1983).

17.8 REFERENCES

Alexander, P.M. (1969) Anthracnose, serious disease problems. *USDA Greens Sect. Rec.*, **7** (5), 8–9.

Arx, J.A. Von and Muller, E. (1954) The genera of amerosporous pyrenomycetes. *Beitr. Kryptogamenfl. Schwiez.*, **11** (1), 434 pp.

Bolton, A. T. and Cordukes, W.E. (1981) Resistance to *Colletotrichum graminicola* in strains of *Poa annua* and reaction of other turf grasses. *Can. J. Plant Pathol.*, **3**, 94–6.

Boning, K. and Wallner, F.W. (1936) Foot rot and other disorders on corn caused by *Colletotrichum graminicolum* (Ces.) Wils. *Phytopathol. Zeit.*, **9**, 99–110.

Bruehl, G.W. and Dickson, J.G. (1950) Anthracnose of cereals and grasses. *US Dept. Agric. Tech. Bull.*, **1005**, 37pp.

Chowdhury, S.C. (1936) A disease of *Zea mays* caused by *Colletotrichum graminicolum.* (Ces.) Wils. *Indian J. Agric. Sci*, **6**, 833–43.

Conners, J.L. (1967) *An Annotated Index of Plant Diseases* in Canada. Res. Br. Can. Dep. Agr. Publ. 1251, 381pp.

Couch, H.B. (1973) *Diseases of Turfgrasses*. Krieger, New York, 348pp.

Couch, H.B. (1979a) Heat stress, not anthracnose is scourge of *Poa annua. Weeds Trees Turf*, **18** (6), 47–56.

Couch, H.B. (1979b) Is it anthracnose or is it wilt? *The Greenmaster*, **15** (5), 3–6.

Dahlsson, S.-O. (1977) Anthracnose on a Swedish golf-green. *Weibulls Gras-tips*, **20** (Dec.) 15–16.

Dahlsson, S.-O. (1978) Anthracnose, *Colletotrichum graminicola* (Ces.) Wils., on turf. *Zeitschrift für Veget. technik.*, **1**, 30–1.

Danneberger, T.K., Vargas, J.M. Jr, Reike, P.E. and Street, J.R. (1983) Effect of carrier, rate and timing of

nitrogen fertilization and fungicide application on anthracnose development of *Poa annua* L. *Agron. J.*, **75**, 35–8.
Dickson, J.G. (1956) *Diseases of Field Crops*. McGraw-Hill, New York, 518 pp.
Duell, R.W. and Schmidt, R.M. (1974) Grass varieties for roadsides. *Proc.* 2nd *Int. Turfgrass Res. Conf.*, pp. 541–50.
Duke, M.M. (1928) The genera *Vermicularia* and *Colletotricum* Cda. *Trans. Br. Mycol. Soc.*, **13**, 156–84.
Edgerton, C.W. and Carvajal, F. (1944) Host–parasite relations in red rot of sugar cane. *Phytopathology*, **34**, 827–37.
Gould, C.J. (1956) Turf disease in Western Washington. *Proc. NW Turf. Conf.* (Reprint).
Grove, W.B. (1937) *British Stem and Leaf Fungi*, Vol II, Cambridge University Press., pp. 230–45.
Herting, V.J. (1982) The pathogenicity of some isolates of *Colletotrichum graminicola* (Ces.) Wilson on *Agrostis palustris* Huds. and *Poa annua* L. MS dissertation, Univ. of Rhode Island Kingston 120 pp.
Howard, F.L., Rowell, J.B. and Kiel, H.L. (1951) Fungus diseases of turf grasses. *Rhode Island Agric. Expt. Sta. Bull.*, **308**, 56 pp.
Kreitlow, K.W., Graham, J.H, and Garber, R.J. (1953) Diseases of forage grasses and legumes in the northeastern states. *Penn. State Agric. Expt. Sta. Bull.*, **573**, 42 pp.
Jackson, N. and Herting, V-J. (1985) *Colletotrichum graminicola* as an incitant of anthracnose/basal stem rotting of cool-season turf grasses. In *Proc. 5th Int. Turfgrass Res. Conf.* August 1985, Avignon, France, pp. 647–55.
Lapp, M.S. and Skoropad, W.P. (1978) Location of appressoria of *Colletotrichum graminicola* on natural and artificial barley leaf surfaces. *Trans. Br. Mycol. Soc.*, **70** (2), 225–8.
Morgan, O.D. (1956) Host-range studies on tobacco anthrancose caused by a species of *Colletotrichum. Plant Dis. Rept.*, **40**, 908–15.
Muhle, E. (1953) Die Krankheiten und Schadlingen der zur Samengewinnung angebauten Futtergraser, S. Hirzel Verlag, Leipzig, 167 pp. (Diseases and pests of cultivated forage grasses planted for seed production) (German).
Nishihara, N. (1972) Anthracnose of *Arrhenatherum elatius* caused by *Colletotrichum graminicolum. J. Japan Soc. Grassland Sci.*, **18** (3), 209–11.
Noble, M. and Richardson, M.J. (1968) *An Annotated List of* Seed-borne Diseases. 2nd edn Commonwealth Mycol. Inst. Phytopathological Pap. 8. 191 pp.
Noviello, C. (1963) Occurrence of lawn disease at Saskatoon, 1963. *Can. Plant Dis. Surv.*, **43** (3), 215.
Politis, D.J. (1975) The identity and perfect state of *Colletotrichum graminicola. Mycologia*, **67**, 56–62.
Politis, D.J. (1976) Ultrastructure of penetration by *Colletotrichum graminicola* of highly resistant oat leaves. *Physiol. Pl. Pathol.*, **8**, 117–22.
Politis, D.J. and Wheeler, H. (1972) The perfect stage of *Colletotrichum graminicola. Plant Dis. Rept.*, **56**, 1026–7.
Politis, D.J. and Wheeler, H. (1973) Ultrastructural study of penetration of maize leaves by *Colletotrichum graminicola. Physiol. Pl. Pathal.*, **3**, 465–71.
Sanford, G.B. (1935) *Colletotrichum graminicolum* (Ces.) Wils. as a parasite of the stem and root tissues of *Avena sativa. Sci. Agric.*, **15**, 370–6.
Selby, A.D. and Manns, T.F. (1909) Studies in diseases of cereals and grasses. *Ohio Agric. Expt. Sta. Bull.*, 203.
Skoropad, W.P. (1967) Effect of temperature on the ability of *Colletotrichum graminicola* to form appressoria and penetrate barley leaves. *Can. J. Plant Sci.*, **47**, 431–4.
Smith, J.D. (1954) A disease of *Poa annua. J. Sports Turf Res. Inst.*, **8**(30), 344–53 and 20 figs.
Smith, J.D. (1955) Turf disease notes. 1955. *J. Sports Turf Res. Inst.*, **9** (31), 60–75.
Smith, J.D. (1956) Turf disease notes. 1956. *J. Sports Turf Res. Inst.*, **9** (32), 233–4.
Smith, J.D. (1965) *Fungal Diseases of Turfgrasses*, 2nd edn (Revised N. Jackson and J.D. Smith). Sports Turf Res. Inst., 97 pp.
Smith, J.D. (1980) *Major Diseases of Turfgrasses in Western Canada.* Ext. Div. Univ. Sask. Publ. 409, 14 pp.
Sprague, R. (1950) *Diseases of Cereals and Grasses in North America.* Ronald Press, New York, 538 pp.
Sprague, H.B. and Evaul, E.E. (1930) Experiments with turfgrasses in New Jersey. *New Jersey Agric. Expt. Sta. Bull.*, **497**, 55 pp.
Sutton, B.C. (1966) Development of fructifications in *Colletotrichum graminicola* (Ces.) Wils and related species. *Can. J. Bot.*, **44**, 887–97.
Sutton, B.C. (1968) The appressoria of *Colletotrichum graminicola* and *C. falcatum. Can. J. Bot.*, **46**, 873–6.
USDA (1960) *Index of Plant Diseases in the United States.* US Dept. Agric. Handbook 125, 531 pp.
Vargas, J.M. Jr (1976a) Anthracnose. In *Proc. 16th Illinois Turfgrass Conf.* Urbana, I11., pp. 61–2.
Vargas, J.M. Jr (1976b) Disease poses threat to annual bluegrass. *Golf Superintendent*, **44**, 42–5.
Vargas, J.M. Jr (1978) Anthracnose: Key to the summer survival of annual bluegrass. *Grounds Maintenance*, **13** (11), 30, 32, 72.
Vargas, J.M. Jr (1980) Anthracnose – Rediscovering the wheel. Greenmaster, **16**, 22–4.
Vargas, J.M. Jr (1981) *Management of Turf Diseases.* Burgess, Minneapolis, 204 pp.
Vargas, J.M. Jr and Detweiler, R. (1976) Turfgrass disease research report. In *Proc. 46th Mich. Turfgrass Conf.*, E. Lansing, Mich., pp. 7–23.
Vargas, J.M. Jr and Detweiler, R. and Hyde, J. and (1977) Anthracnose fertility – fungicide interaction study. In *Proc. 47th Mich. Turfgrass Conf.*, E. Lansing, Mich., pp. 3–12.
Vizvary, M.A. and Warren, H.L. (1982) Survival of *Colletotrichum graminicola* in soil. *Phytopathology*, **72** (5), 522–5.
Wilson, G.W. (1914) The identity of anthracnose of grasses in the United States. *Phytopathology*, **4** (2); 106–13.
Winter, A.G. (1940) Causes of footrot on wheat, barley, rye and oats. (*Collectotrichum graminicolum* (Ces.) Wils.). *Phytopathol. Zeitschr.*, **13**, 282–92.
Wolff, E.T. (1947) An experimental study of *Colletotrichum graminicolum* on fine turf. PhD. dissertation. Penn State Univ., University Park 78 pp.

18 *Copper spot disease*

In 1943 a zonate leaf spot was reported on several varieties of sorghum (*Sorghum vulgare* Pers.) in the south-western United States (Bain and Edgerton, 1943). The disease, caused by the fungus *Gloeocercospora sorghi*, is now recognized as one of the most prevalent foliar maladies on *Sorghum* spp. in the Gulf states and is present in the northern US sorghum belt (Odvody, Dunkley and Edmunds, 1974). The fungus is of widespread distribution in tropical Africa and also occurs in Central and South America, India, Japan and northeast Australia (Mulder and Holliday, 1971). The same fungus causes a sporadic, but often severe summer disease of bentgrass (*Agrostis* spp.) known as copper spot (Wernham and Kirby, 1943; Keil, 1946; Howard, Rowell and Keil, 1951). This turf disease occurs in both eastern and southwestern coastal regions of the US, but outbreaks are most frequent in the northeastern region where velvet bentgrass (*Agrostis canina* L.) is the common host.

18.1 SYMPTOMS

The fungus causes a foliar blighting which starts as small reddish or brown, water-soaked lesions on individual leaves. Adjacent lesions enlarge, coalesce and soon involve the whole leaf which shrivels somewhat and turns a darker reddish-brown colour. Aggregations of blighted leaves form coppery and orange-coloured patches in the turf 2.5–7.5 cm in diameter (Fig. 27, colour plate section).

In wet weather the colour of the patches is intensified due to the presence of numerous salmon-pink, gelatinous spore masses on the infected leaves. When dry these sporodochia are bright orange or red and assume a characteristic crown shape.

Minute lenticular sclerotia are distributed within the necrotic tissues of leaves and sheaths, but often are hard to see without the aid of a hand lens. The overall reddish hue of the patches as opposed to a bleached straw colour serves to separate copper spot symptoms from those of dollar spot disease. The outlines of the patches are less well defined in copper spot. Also lines of spots or streaks formed by merging infection sites may follow the direction of mowing, indicating how the slimy *G. sorghi* spore masses served to disseminate the fungus on the rollers of the mower (Fig. 45).

18.2 THE CAUSAL FUNGUS

Gloeocercospora sorghi Bain and Edgerton ex Deighton (1971); Synonym: *G. sorghi* Bain and Edgerton (1943).

No teleomorph has been reported for this fungus.

There is no stroma, hyphae are septate, branched, 2–3.5 μm, wide, internal in the leaves. Sporodochia are amphigenous, pulvinate, <50 μm, developing over stomata, composed of branched, hyaline hyphae, short moniliform cells and ovoid to flask-shaped cells, 5–10 × 3–4 μm. The latter function as conidiophores, each producing single, smooth, hyaline, filiform, 1–7 septate conidia which may be slightly obclavate or widest in the middle, 20–195 × 1.4–3.2 μm, aggregated into slimy, salmon-coloured masses. Abundant, black, lenticular to spherical sclerotia, 0.1–0.2 mm form in necrotic host tissues. (For a more detailed description see Deighton, 1971.)

The fungus grows rapidly on many culture media and sporulates most profusely (Bain and Edgerton, 1943; Marion, 1974) on oatmeal agar (Riker and Riker, 1936). The optimum temperature for growth is 28–30°C. Dean (1966) noted that the cultural characteristics of the pathogen were better maintained by storing dry sclerotia. Marion (1974) reported that oatmeal agar cultures could be stored at 10°C for up to 60 days and still provide an abundance of viable spores on subculturing. The fungus may also be maintained on V-8 juice agar (Miller, 1955) under continuous fluorescent light at 22–25°C to produce spore inoculum of constant pathogenicity (Myers and Fry, 1978).

18.3 HOST RANGE

The reported turf grass hosts of *G. sorghi are Agrostis canina* L., *A. stolonifera* L., *A. tenuis* Sibth (Howard, Rowell and Keil, 1951; Sprague, 1950) and

Cynodon dactylon L. Pers. (Tarr, 1962).

18.4 DISEASE DEVELOPMENT

G. sorghi overwinters as sclerotia and modified thick-walled mycelium in the thatch of *Agrostis* spp. turf (Marion, 1974). Vegetative growth resumes with the advent of warm, wet weather in the spring when the soil temperatures at 2.5 cm depth reach a minimum of 17 °C for seven days (Howard, Rowell and Keil, 1951). Couch (1973) stated that air temperatures in the range 19–24 °C initiate active growth of the organism and that the pattern of development essentially parallels that of dollar spot disease. Field observations in Rhode Island suggest that although symptoms of both diseases can occur together in a turf stand, usually the dollar spot symptoms will occur first and overlap the later-developing copper spot, indicating a higher temperature requirement for optimum pathogenicity in *G. sorghi.* On sorghum, Dean (1966) reported an optimum temperature for growth and pathogenesis in the range 26.5–30 °C.

Vegetative hyphae from sclerotia are reported to enter the grass blades through stomata; disease symptoms develop witin 24 hours and the production of sporodochia occurs 24–48 hours later (Howard, Rowell and Keil, 1951). Dean (1966, 1968) showed that sclerotia of *G. sorghi* from sorghum overwintered successfully in crop debris and each germinated to produce a sporodochium with conidia. In inoculation studies using sorghum as the host, Bain and Edgerton (1943) and Dean (1966) concluded that germinating conidia enter via the stomata. No appressoria were found and there was no evidence of direct penetration of epidermal cells. In contrast, Myers and Fry (1978) gave a detailed account of direct penetration of sorghum leaves by this fungus. Ingress was initiated from a single appressorium or from appressorial aggregations followed by the development of subcuticular hyphae and subsequent penetration of the epidermal cell at about 30 hours after inoculation. Direct penetration of trichomes and entry through stomata was also observed.

According to Marion (1974) the disease is severe on *A. canina* turf under atmospheric conditions that ensure deposition of moisture on leaf surfaces and/or those conditions conducive to exudation of plant fluids. On cultivated turf, a combination of both moisture deposition and plant exudation was assumed to form the leaf surface fluids. He reported that the sporodochia, borne in rows between the leaf veins and on the tips of mown leaves, may be engulfed by leaf surface fluid droplets. Spore masses so engulfed disseminate readily by mechanical means (e.g., rain splash, mowers, shoes, etc.) to adjacent healthy plants. Mycelial growth was promoted in the fluid droplets and hyphae bridged from leaf to leaf via the droplets to spread infections laterally.

A. canina turf maintained under high nitrogen fertility regimes generated leaf surface fluids which enhanced spore germination by two- to three-fold compared to those produced by non-fertilized turf. This enhancement was related to increased total amino nitrogen (TAN) content of the leaf fluids in the high fertility regime. The amine component (most probably glutamine) was markedly stimulatory but individual amino acids varied from stimulatory to inhibitory in their effect on spore germination. Arginine was one of the most active amino acids in promoting germination. Marion (1974) concluded that the higher incidence of copper spot with high nitrogen fertility regimes noted in Rhode Island, was related to the increased TAN levels in the leaf surface fluid and concomitant improvement in inoculum potential of *G. sorghi.*

The capacity of the fungus to produce large quantities of secondary spore inoculum soon after the onset of primary infection provides the means whereby copper spot disease may reach epidemic proportions in a relatively short period of time if environmental conditions are suitable.

18.5 CONTROL

18.5.1 Cultural

Velvet bentgrass (*A. canina*) is especially prone to copper spot disease. Since turf of this species performs optimally under low to moderate nitrogen regimes, excessive use of nitrogen fertilizers will lead to trouble both agronomically and from a disease susceptibility standpoint. Interactions with other major turf nutrients and copper spot incidence have not been established but low soil pH is a factor. The fungus is most destructive to *A. canina* turf growing in soils between pH 4.5 and 5.5; liming to reduce the acidity of the soil will reduce the incidence of copper spot (Howard, Rowell and Keil, 1951).

18.5.2 Fungicides

Organic mercury and cadmium compounds are efficient fungicides for control of copper spot (Howard, 1947; Howard, Rowell and Keil, 1951; Keil, 1946). Trials conducted at the University of Rhode Island over the course of several years have proved the efficacy of chlorothalonil, anilazine, iprodione and cadmium-containing compounds as contact fungicides which afford excellent protection against copper spot when applied on a 7–10 day spray programme. The systemic materials benomyl, triadimefon, thiabendazole, methyl and ethylthiophanate all control the disease and provide longer protection (Dernoeden and Jackson, 1980; Jackson and Dernoeden, 1979). Two or three applications on a 14–28 day schedule commencing in June give good control of copper spot.

18.6 REFERENCES

Bain, D.C. and Edgerton, C.W. (1943) The zonate leaf spot, a new disease of sorghum. *Phytopathology*, **33,** 220–6.

Couch, H.B. (1973) *Diseases of Turfgrasses,* 2nd edn Kreiger, Huntington, New York, 348 pp.

Dean, J.L. (1966) Zonate leafspot of sorghum. Doctoral dissertation Louisiana State University, Baton Rouge, 89 pp.

Dean, J.L. (1968) Germination and overwintering of sclerotia of *Gloeocercospora sorghi. Phytopathology*, **58,** 113–14.

Deighton, F.C. (1971) Validation of the generic name *Gloeocercospora* and the specific names of *G. sorghi* and *G. inconspicua. Trans. Br. Mycol. Soc.*, **57,** 358–60.

Dernoeden, P.H. and Jackson, N. (1980) Evaluation of fungicides for control of copperspot and dollarspot on velvet bentgrass. 1979. *Fungicide Nematicide. Tests*, **35** (310), 148.

Howard, F.L. (1947) An organic cadmium fungicide for turf disease. *Greenskeeper. Rept.*, **15,** 10.

Howard, F.L., Rowell, J.B. and Keil, H.B. (1951) Fungus diseases of turf grasses. *Rhode Island Agric. Expt. Sta. Bull.*, **308,** 56 pp.

Jackson, N. and Dernoeden, P.H. (1979) Fungicides for copper spot control. 1977. *Fungicide Nematicide Tests*, **34** (287) 133.

Keil, H.L. (1946) Control of copper spot on five turf grasses. *Phytopathlogy*, **36,** 403.

Marion, D.F. (1974) Leaf surface fluid composition of velvet bentgrass as affected by nitrogen fertility and its relationship to inoculum viability of *Gloeocercospora sorghi* and severity of copper spot disease. Doctoral dissertation. University of Rhode Island, Kingston, 91 pp.

Miller, P.M. (1955) V-8 juice agar as a general purpose medium for fungi and bacteria. *Phytopathology*, **45,** 461–2.

Mulder, J.L. and Holliday, P. (1971) *Gloeocercospora sorghi.* CMI Descriptions of path. fungi and bacteria. No. 300.

Myers, D.F. and Fry, W.E. (1978) The development of *Gloeocercospora sorghi* in sorghum. *Phytopathology*, **68,** 1147–55.

Odvody, G.N., Dunkley, L.D. and Edmunds, L.K. (1974) Zonate leaf spot in the Northern Sorghum belt. *Plant Dis. Rept.*, **58,** 267–8.

Riker, A.J. and Riker, R.S. (1936) *Introduction to Research on Plant Disease.* John S. Swift, St Louis, 117 pp.

Sprague, R. (1950) *Diseases of Cereals and Grasses in North America.* Ronald Press, New York, 538 pp.

Tarr, S.A.J. (1962) *Diseases of Sorghum, Sudan Grass and Broom Corn.* Commonwealth Mycology Institute, Kew, Surrey, 380 pp.

Wernham, C.C. and Kirby, R.S. (1943) Prevention of turf diseases under war conditions. *Greenskeeper. Rept.*, **11** (4), 14–15, 26–27.

19 *Dollar spot disease*

Dollar spot caused by the fungus *Sclerotinia homoeocarpa* F.T. Bennett is a common and destructive disease of both cool and warm-season grasses in most of the locations wherever these grasses are utilized to form amenity turf.

In Britain the disease is confined almost entirely to slender *Festuca rubra* ssp *rubra* turf and other reports of the occurrence of dollar spot should be viewed with suspicion unless made by a competent observer familiar with the symptoms and preferably with access to facilities for diagnosis by cultural methods in the laboratory. Elsewhere, in Australasia, North and Central America and continental Europe a wide range of turf grasses is affected.

Dollar spot figures prominently in the northern regions of North America as a disease of *Agrostis* spp. putting greens and may also severely damage various other turf grasses comprising golf fairways, home lawns and amenity turf areas. It apparently is of infrequent occurrence in the coastal region of the Pacific Northwest (Chastagner, 1982, personal communication; Gould, 1966–67) but is well represented elsewhere in the United States (Gould, 1964) extending into the south where the warm-season turf grasses may be severely affected. If left unchecked, dollar spot disease may result eventually in death of infected plants, impairing the playing surface of turf and rendering it unsightly. Thinned areas so produced, thus become susceptible to invasion by undesirable weed species.

19.1 SYMPTOMS

On fine texture, close-mown turf the disease is first seen as very small spots, 1–2 cm in diameter, involving injury to the leaf blades and sheaths of a few adjacent tillers. On fine-leaved fescues, injured areas on the needle-like leaves are at first yellow–green then bleached. Lesions of this nature may affect distal and proximal ends of the blade with an intermediate portion remaining green or yellow–green for a time. In many instances a narrow dark band cuts off the shrunken bleached regions from the remaining green portion. On other fine turf species minute, tan-coloured or bleached flecks and lesions may be discernable. Leaf lesion symptoms on fine turf are usually short-lived due to further encroachment by the fungus which causes infected leaves to shrivel and discolour. The individual spots, first brown, later bleached or straw coloured, increase in size to encompass an area of turf about 5 cm in diameter. If unchecked, the spots become sunken and, where many occur together, they may coalesce to affect large areas of turf.

Dollar spot disease may also show on close-mown red fescue turf as a general diffuse discoloration rather than as discrete spots. The diffuse symptoms have been noted in the UK on experimental plots and under playing conditions where the turf had been severely infected the previously year.

The symptoms of dollar spot on some fine turf species (e.g. *F. rubra*, *Agrostis canina* L.) may be confused with the early symptoms of red thread disease. Patches of the latter have less well-defined margins than those of dollar spot. A careful examination of plants affected by the red thread fungus, even in the absence of the red stromata usually will reveal the pink-stranded mycelium. The copper spot fungus, *Gloeocercospora sorghi* (Bain and Edgerton) produces symptoms on *A. canina* which resemble dollar spot but the spots of the former have a distinctive reddish tinge due to the copious spore production of the causal agent (Fig. 27, colour plate section). The coloured sporodochia are visible with a hand lens.

On higher mown turf of coarser-textured species the individual spots are initiated in much the same way as described for fine turf but they become significantly larger. The spots or patches, more irregular in outline, may reach a diameter of 15 cm or more. Proliferation of these symptoms may result in the overall blighting of large areas of turf. Distinctive leaf lesions are readily discernible on the coarser textured turf species. These lesions are light tan coloured or bleached, often delineated by a reddish brown border (Fig. 28, colour plate section). Some extend downwards from the leaf tip, 2 cm or more in length, and involve all or part of the tissues. Intermediate lesions on leaves are usually shorter, often radiating from one leaf margin and frequently extending to the full width of the lamina.

Activity of the fungus appears to be confined to the aerial portion of the grass plant although a metabolite(s) toxic to roots has been demonstrated (Endo, Malca and Krausman, 1964; Kerr, 1956a). Signs of the fungus in the form of delicate, cobweb-like growths of mycelium suspended between the infected grass blades are a common feature of dollar spot while the early morning dew is present. The abundant production of aerial mycelium may be generated by incubating a diseased plug of turf in a moist chamber or by inverting a glass dish over an area of infected turf in the field.

19.2 HOST RANGE

Bennett (1937) stated that the disease occurred on *Festuca*, *Agrostis* and *Poa* spp. from Britain, America and Australia. North American workers (Bain 1962; Britton, 1969; Couch, 1973; Freeman, 1967; Haygood and Spencer, 1979; Hodges, Blaine and Robinson, 1975; Howard, Rowell and Keil, 1951; Sprague, 1950; Wilkinson, Martin and Larsen, 1975) record it on *Agrostis alba* L., *A. stolonifera* L., *A. canina* L., *A. tenuis* Sibth, *A. gigantea* Roth., *Cynodon dactylon* (L) Pers., *Digitaria ischaemum* (Schreb) Muhl., D. Sanguinalis (L.) Scop., *Eremochloa ophiuroides* (Munro) Hack, *Festuca ovina* L., *F. rubra* L., *F. arundinacea* Schreb, *Lolium perenne* L., *L. multiflorum* Lam., *Poa annua* L., *P. pratensis* L., *Paspalum notatum* Flugge., L. *Stenotaphrum secundatum* (Walt.) Kuntze., *Zoysia japonica Steud.*

In Australia, the first authentic record of the disease was made in 1934 in Queensland on *Digitaria didactyla*. It has been recorded since in all Australian States except Tasmania, and now constitutes a major turf disease problem on turf composed of *Agrostis* spp. (Reilly, 1969). Dollar spot is also an established disease in New Zealand on *Agrostis* spp. turf (Boesewinkel, 1977), and on turf in Japan (Hosotsuji, 1977).

Although dollar spot is established in continental Europe (France and Spain), occurring mainly on *Agrostis* spp., in the British Isles the disease is restricted to *Festuca rubra* L. spp *rubra* of sea-marsh origin. The fungus has been isolated from *Poa annua* in a sea-marsh turf bowling green where the *F. rubra* L. spp. *rubra* was heavily infected, and has been noted on other cultivars of this same species. Symptoms of the disease where produced on *Festuca rubra* L. ssp. *commutata* Gaud. (Chewings' fescue), *Agrostis tenuis* Sibth., *A. stolonifera* L. var. *compacta* Hartm., *Holcus lanatus* L., *Poa annua* L. and *P. pratensis* L. (cv. Cambridge) by inoculation (Smith, 1955). With the exception of Chewings' fescue the disease has not been seen on these species under turf conditions in the UK.

19.3 THE CAUSAL FUNGUS

Sclerotinia homoeocarpa F.T. Bennett (1937).

Bennett (1937) who examined in detail isolates from British, American and Australian sources showed that the fungus existed in several distinct strains which differed in their capacity for spore production in culture. The American and Australian isolates that he studied apparently were sterile but among the British isolates one strain yielded ascospores and conidia, one ascospores only and other rudimentary sporiferous structures.

The characteristics of teleomorph state generated on wheatmeal agar by an isolate from diseased turf in the English Midlands were described as follows (Bennett, 1937).

Apothecia cupulate 0.5–0.8 mm to disc- or funnel-shaped, 1–1.5 mm in diam., pale cinnamon to dark brown with prosenchymatous exciple, arising from microsclerotia or expansive sclerotial flakes or patches. Stalk cylindrical, slender flexuous from 5–10 mm or more long arising singly or in cluster, simple or branched in the upper part. Asci cylindroclavate, inoperculate 140–170 × 10.4–11.5 μm commonly 250–165 × 10.4 μm. Ascospores 8, uniseriate, hyaline, oblong elliptical, bi-guttulate, unicellular, a delicate median septum often during germination 15.6–16.9 × 5.2–6.5 μm commonly 16.0 × 5.5 μm. Paraphyases few, cylindro-clavate, sparsely septate, 80–120 × 2.0 – 2.2 μm (Fig. 29, colour plate section).

Conidial fructification cupulate, 0.4–0.6 mm in diam., light-brown, stalk 4–6 mm long, pale cinnamon. Conidia borne singly on hyphae within the cup, hyaline, one-septate, constricted at the septum, apex either rounded or narrowing to a blunt point; 15.6–20.8 × 4.5–5.2 μm commonly 20–21 × 4.6 μm, average 19.5 × 4.7 μm. Micronidia not observed.

Mycelium short, compact, white with cinnamon-coloured, floccose growth at the top of slant cultures. Sclerotial structures black, from small flakes to extensive patches, parchment-like, formed by conversion of superficial hyphae of the white mycelium into a mosaic of small thick-walled cells. Ascophores and conidiophores are typically erumpent from sclerotial structures, occasionally superficial on the edges or when sclerotial structures are excessively thin.

An ascigerous strain from bowling green turf in Northumberland, England, had the following characters.

Apothecia cupulate 0.4–0.7 mm in diameter, cinnamon to brown in colour on stalks 4–6 mm long, simple. Asci as for perfect strain but 180–220 × 10.4–

12.0 μm. Ascospores as for perfect strain but 18.2–26.0 × 7.8–9.0 μm commonly 19.5–20.8 × 7.8–8.0 μm. Microconidia spherical, hyaline, 1.5–2.0 μm in minute cream-coloured pustules, not known to germinate. Mycelium from sparse to abundant, white, faintly tinted bluish-green or chalcedony-yellow in different strains; cinnamon-coloured floccose hyphae at the tops of slant cultures and cinnamon-coloured hyphae among the white as the mycelium ages. Sclerotial structures as for perfect strain.

Fenstermacher (1970, 1980) and other turf grass pathologists working with American isolates of *S. homoeocarpa* have been singularly unsuccessful in their efforts to induce any fertile fruiting bodies of this fungus in culture. Microconidia, apothecial initials and immature apothecia have occurred occasionally on some isolates but viable conidia or ascospores have not been reported.

Working with English isolates in 1971, Jackson (1973) attempted to reproduce Bennett's findings and was successful in obtaining fertile apothecia in culture which resembled closely those described by Bennett as the 'ascigerous strain'. None of the isolates which fruited produced asexual spores. In the report of this work Jackson also discussed the taxonomic status of *S. homoeocarpa*. While concluding that the fungus was incorrectly placed in the genus *Sclerotinia* no name change was suggested until wider investigations concerning the generic nomenclature within the family Sclerotiniaceae have been resolved. Kohn (1979) concurs that the fungus is not a member of the genus *Sclerotinia* and suggests the epithet has been applied to species with apothecia identified as belonging to *Lanzia* and *Moellerodiscus*. The implication that dollar spot symptoms are caused by more than one fungus was also suggested by Jackson (1973).

Sporocarps or sclerotia have not been found under natural conditions. The 'sclerotial flakes' and 'microsclerotia' described by Bennett are better interpreted as stromal structures. The flakes, usually 1–4 cells thick, of darkly pigmented cells are the rinds of stromata which encompass and delineate part of the colonized substrate. In agar cultures these stromatal walls may assume an interrupted annular arrangement, cutting deeply into the medium. Stromata may be found occasionally encompassing portions of leaf and stem tissues of turf grasses killed by *S. homoeocarpa*. Walls similar to those found on culture media mark these structures (Fenstermacher, 1980).

For routine diagnosis of the fungus from British sources, Smith (1955) recommended plating incubated leaf fragments showing the mycelium of the fungus on to glucose-boric agar (Smith, 1954b) or glucose-salts yeast extract (Smith, 1953). In culture, British isolates produce at first a downy then denser and cottony, faintly cinnamon-white aerial growth. In inverted petri dishes the aerial mycelium forms bridging strands or wefts from the agar surface to the petri dish cover. The mycelium becomes more dense and mats on the surface of the medium into a scattered felted layer, cinnamon-brown in colour and interspersed with dark, olive-green or black, thin, plate-like structures in most isolates. These structures are the developing surface rind(s) of stromata. With some first isolations and with continuous subculturing of other isolates the dark cinnamon felted mycelium does not always develop prior to the stromatal rind initiation. Stromata generally assume an interrupted annular arrangement on the medium with blackened rind cutting deep into the medium perpendicular to the agar surface. In some cases, the agar surface is completely covered with rind but commonly the stromata in surface view are irregular in shape and unequal in size. Some inward turning at edges of the submerged rind marks the gradual envelopment of portions of the medium to complete the stromata. Satisfactory growth of aerial mycelium for routine diagnosis (showing the characteristic cinnamon colour) develops in 7 to 10 days. Stromatal initiation takes longer but may commence in 14–21 days from plating.

Bennett (1937) chose wheatmeal agar to grow British, Australian and American isolates of *S. homoeocarpa* and, disregarding a few minor differences, he stated that they all conformed in the main cultural features. Bennett's description is compatible with that presented above. In the US, routine isolation of the fungus from surface-sterilized leaf lesions onto various 'starchy' media (e.g. wheatmeal agar, oat agar, potato dextrose agar) has resulted in vegetative cultures of *S. homoeocarpa* which generally conform. The initial colour of the felt-like mycelium may vary to include shades of brown, yellow, grey and olive but as they age the characteristic cinnamon colour is usually represented followed by the typical stromatal development.

19.4 DISEASE DEVELOPMENT

The incidence of dollar spot disease shows a seasonal fluctuation in most years with most new infections occurring in the late spring and early summer, and again in the autumn. From the onset of primary symptoms in a turf area, build up of the disease may be rapid and once established it is most persistent. Commonly, the spots fade out in the winter to be almost indistinguishable from non-infected turf. However, when infections occur late in the growing season symptoms may be visible throughout the winter and persist until the spring flush of grass growth. Patches of thinned turf in the late spring may indicate where the disease occurred the previous autumn and often mark the sites for renewed disease activity.

Overwintering of the fungus is reported to occur as sclerotia or paper-thin sclerotial flakes (Britton, 1969; Couch, 1973), but according to Fenstermacher (1980) these reports probably represent an extrapolation of laboratory observations to the field. After a detailed study of the perennation of the fungus in Rhode Island he confirmed that stromata do occur in nature but he concluded that *S. homoeocarpa* survives the winter months as persistent mycelia in plant debris, with or commonly without the presence of well-defined visible stromata. Stromata have been collected in both the US and the UK during the late autumn and early spring mainly from fine-leaved fescue turf. Isolation of the fungi from the stromatized tissues yielded cultures with mycelial characteristics similar to *S. homoeocarpa*. Apothecia originating from stromata on fescue leaves have also been collected from the wild in both countries. Though similar, they were not conspecific with *S. homoeocarpa* and cultures generated from these ascocarps failed to produce disease symptoms when used as inoculum on the fescue host, Jackson (1973), Fenstermacher, (1980). Kerr (personal communication) suggested that fruiting may occur on sedges and grasses in the vicinity of the turf. In support, he quoted Whetzel (1946) who reported finding what appeared to be the same species (*S. heterocarpa* = *homoeocarpa*?) on *Carex* and several other plants. However, diligent searchings of diseased turf and adjacent vegetation by turf pathologists in several geographical locations have not produced confirmatory evidence to date. The possibility of ascospore or conidial infection cannot be ruled out but the occurrence must be extremely low or non-existent. Halisky, Myers and Wagner (1981) provided data relating dollar spot incidence and the presence of thatch, supporting the contention that the fungus persists vegetatively in plant debris and that seasonable outbreaks are the result of favourable conditions for mycelial infection. Transport of infected plant parts on implements and footwear evidently serves to spread the disease to newly established turf areas (Smith, 1955).

The various biotypes of the fungus from different geographical locations each grow over a wide range of temperatures but they vary in their growth optima (Bennett, 1937; Endo, 1963; Fenstermacher, 1970; Freeman, 1967). This variation may represent adaptation by the fungus for optimum disease development under the environmental conditions prevailing in the different locations. Bennett (1937) concluded that the British biotypes are better suited to cooler conditions than their American or Australian counterparts. High-temperature-tolerant biotypes exist in Florida (Bain, 1962; Freeman, 1959), but, in general, the disease is favoured by moderate temperatures (15–25 °C) and periods of high humidity (Couch, 1973; Endo, 1963; Howard, Rowell and Keil, 1951).

Monteith and Dahl (1932) reported that the fungus gains entry into the plant via cut leaf tips and through stomates. Appressorial formation has been observed by Endo (1966) and these structures may aid in stomatal penetration. Mycelium is freely produced on the necrotic tissues and extends radially from the initial lesion, bridging across to adjacent leaves. The ability of the pathogen to bridge air spaces is most noticeable in culture. Why the mycelial growth ceases to extend beyond a particular diameter (and hence form the 'dollar spot' symptom on fine turf) has not been explained satisfactorily. Possibly it may be due to accumulation of staling materials by the fungus that inhibit further growth. If the invasion of leaf tissues is checked in the early stages of disease development then a vigorously growing turf can recover quite quickly. In the absence of any disease control and especially if turf vigour is low, then infected plants are killed by fungus and the affected areas will be slow to heal.

Kerr (1956), in Australia, has shown that *S. homoeocarpa* mycelium, when introduced into soil, grew vigorously for a limited peiod and in the process produced a toxin that caused a root necrosis on the seedlings of several species (including wheat) without the fungus actually penetrating the plant. Similar observations were made in the United States by Endo and his co-workers (Endo, 1963; Endo, Malca and Krausman, 1964; Endo and Malca, 1965). A toxin produced by the dollar spot fungus destroyed the apical meristems and cortical tissues of creeping bentgrass roots growing in quartz sand culture. The roots showed a similar deterioration when exposed to D-galactose but this compound has not been confirmed as the toxic principle produced by the fungus (Malca and Endo, 1965). Root damage resembling that observed in the laboratory was observed in the field by Endo and Malca (1965) on turf with well-established foliar symptoms of dollar spot.

In describing the disease under UK conditions, Smith (1955) remarked on the capacity of *S. homoeocarpa* to continue its pathogenic activities during warm dry summer weather, with dews sufficing to support growth of the fungus. Howard Rowell and Keil (1951) had reported earlier that the fungus is active most noticeably when the turf is not growing rapidly due to low soil moisture. The implication from these field observations that drought stress may predispose turf to dollar spot disease was demonstrated experimentally by Couch and Bloom (1960) who showed that an increase in susceptibility can be detected when the soil moisture content reaches 75% of field capacity.

Like red thread disease, dollar spot incidence and severity is influenced by nitrogen fertility practices. On the sea-marshes of northwestern England, dollar spot is noted most frequently on the tops of ridges and on areas that some time previously have been stripped of turf and have subsequently recolonized with *Festuca rubra* and *Agrostis* spp. (Smith and Jackson, 1965). The turf has a much less vigorous growth in these areas. Musser (1950) reported that under American conditions turf low in vigour and in particular, short of nitrogen, is highly susceptible to the disease (Fig. 30, colour plate section). This pronouncement has been endorsed by many other investigators including (Altman, 1965; Endo, 1966; Freeman, 1969; Markland, Roberts and Frederick, 1969; Monteith, 1929; Roberts, 1963, 1967). Monteith and Dahl (1932) suggested that an adequate supply of readily available nitrogen enabled the grass to recover quickly from the depradations of dollar spot. This view was supported by Couch and Bloom (1960) despite their observations in greenhouse trials of an increase in susceptibility of *Poa pratensis* to dollar spot as nitrogen levels increased. They concluded that such an effect is probably masked in the field by the superior recuperative powers of the nitrogen-stimulated host.

Endo (1966) demonstrated experimentally that *S. homoeocarpa* requires a food base to support saprophytic growth prior to infection. He suggested that nitrogen-starved turf is more likely to provide such a food base in the form of senescent or dead foliage than turf fertilized adequately with nitrogen. Smith (1955, 1956, 1957a), working in the UK found that although applications of ammonium sulphate alone were not effective in controlling established infections they assisted in the suppression of symptoms secured by an efficient fungicide. Applications of ammonium sulphate made in late summer and early autumn may render turf more susceptible in the following spring and summer even when an efficient fungicide has been used. In one experiment (Smith, 1957a) it was found that several applications of ammonium sulphate at a low rate were more effective than less frequent higher rates in suppressing symptoms. In general, levels of nitrogen fertilizer use in the US (2–5 kg N per 100 m^2) are much higher than those adopted elsewhere. In the UK, in the 1950s, they seldom exceeded 0.5–1 kg N per 100 m^2 per annum.

There is some indication that the nitrogen source may also influence the development of dollar spot. Trials in New Jersey showed that dollar spot symptoms were reduced most by activated sewage sludge (Cook, Engel and Bachelder, 1964). These reductions in disease incidence were greater than those resulting from an equivalent amount of inorganic nitrogen. Similar trials in Iowa, endorsed these findings (Roberts, 1967; Markland, Roberts and Frederick, 1969), but again the extra benefit obtained from the complex organic fertilizers was not explained.

Other major turf nutrients, in particular phosphorus and potassium have little documented influence on dollar spot. Correlations have been made between high carbohydrate levels within the plant and increased susceptibility to dollar spot disease. It has been suggested that potassium, through its effect on carbohydrate mobilization with the plant, may alter host susceptibility and that turf should be amply supplied with this element to reduce dollar spot incidence (Markland, Roberts and Frederick, 1969). Smith (1955) and Freeman (1967) have shown small and inconsistent reductions of dollar spot symptoms in field trials with potassium fertilizers.

Soil pH apparently has little effect on disease development (Couch and Bloom, 1960). Bennett (1937) showed that there was no important difference in the rate of mycelial growth of the fungus within the range of pH 4–7. He concluded that any difference soil pH may have on the incidence of the disease would be an indirect one by influencing the growth and vigour of the host plants and so affecting their disease resistance.

In the UK dollar spot disease is confined almost entirely to *Festuca rubra* in particular those of sea-marsh origin. *Agrostis stolonifera* var. *compacta* Hartm., which is resistant to the disease, is a common constituent with the susceptible *Festuca rubra* ssp. *rubra* varieties in sea-marsh turf. Heavy attacks of dollar spot disease decimate the *F. rubra* in these mixed swards leaving the resistant bentgrass with volunteer grasses and weeds. As the amount of *F. rubra* in the herbage declines so does the amount of dollar spot. This process may continue until little *F. rubra* and dollar spot can be found in a turf which has been down for several years (Smith, 1957d). Elsewhere, the fungus has a much wider host range but wide differences between and within species for susceptibility to dollar spot exist (Cole, Perkins, and Duich, 1967; Cole, Massie and Duich, 1968a; Cole *et al.*, 1969; Monteith and Dahl, 1932) and there is considerable variation in the pathogenicity of isolates of the fungus to different hosts (Massie, Cole and Duich, 1968). Expression of resistance by cultivars of the different turf grasses may vary extensively between geographical locations and be further influenced by local cultural conditions (Hodges, Blaine and Robinson, 1975); fertility levels (Turgeon and Meyer, 1974; Wilkinson, Martin and Larsen, 1975), herbicide use (Karr, Gudauskas and Dickens, 1979) and mowing height (Turgeon and Meyer,

1974; Wilkinson, Martin and Larsen, 1975). The reader is encouraged to consult local listings of dollar spot-resistant grasses suited for their particular location.

19.5 CONTROL

19.5.1 Cultural control

Since drought-stressed turf is particularly vulnerable to infection careful attention should be given to irrigation practice. The soil should be maintained near field capacity for maximum benefit. Early morning removal of dew and guttated water by switching (poling) is recommended for close-mown, fine turf. Barriers to the free flow of air that delay drying of the turf surface should be avoided.

Nitrogenous fertilizers should be used judiciously during the spring and summer to maintain turf in vigorous growth. In the UK, where ammonium sulphate is commonly used to suppress symptoms on sea-marsh turf, dressings are best applied regularly in small amounts rather than in infrequent, larger amounts. Provided that the soil is maintained near neutrality by the use of sea- or shell sand, or occasional light liming in areas where atmospheric pollution is high, up to four or five applications of 15 g/m^2 of ammonium sulphate may be made during the spring and summer. In most cases these applications should not continue beyond the end of August because later applications may increase dollar spot incidence in the following year. There is also the risk of microdochium patch disease occurring during the autumn and winter months with late nitrogen applications.

19.5.2 Fungicides for the control of the disease

Monteith and Dahl (1932) reported that the disease could be readily controlled with mercury fungicides particularly those containing calomel. They found copper and sulphur combinations ineffective. Harrington (1941) reported thiram as an efficient fungicide but other investigators failed in most instances to demonstrate any benefit from this material in reducing dollar spot symptoms (Howard, 1955; Vaughn and Klomparens, 1952; Wilson and Grau, 1952). Smith (1954b, 1955) showed that cadmium, phenyl mercury or inorganic mercury were much superior to thiram in controlling the disease. Cadmium in inorganic and organic form has been used for several decades in the US with conspicuous success as a dollar spot fungicide (Couch, 1973; Davis, Engel and Snyder, 1955; Gould, 1966, 1967; Howard, 1947; Howard, Rowell and Keil, 1951). However, in the mid-1960s biotypes of *S. homoeocarpa* highly tolerant to both cadmium and mercury fungicides were isolated from turf in Rhode Island (Jackson, 1966). Reports from elsewhere in the eastern US soon confirmed that the cadmium-tolerant biotype of the fungus was widespread (Cole, Taylor and Duich, 1968a,b; Massie, Cole and Duich, 1968). Fortunately, alternative organic contact fungicides were available at that time. Chlorothalonil, cycloheximide and anilazine proved particularly effective although isolated instances of tolerance by the fungus to anilazine were reported (Nicholson *et al.*, 1971). Introduction of the benzimidazole-type systemic fungicides in the late 1960s was a significant contribution to the list of materials available for dollar spot control. Initially, excellent prolonged protection was afforded by low rates of benomyl, thiabendazole and the thiophanates (Goldberg, Cole and Duich, 1970; Massie, Cole and Duich, 1968; Reilly, 1969; Siviour, 1975). The resilient dollar spot fungus soon demonstrated marked tolerance to any and all of these related compounds (Cole, Warren and Sanders, 1974; Warren, Sanders and Cole, 1974). Repeated and exclusive use of cadmium or the benzimidazole fungicides was shown to exert a selection pressure on the natural populations of *S. homoeocarpa* such that biotypes tolerant to the fungicide are favoured and gradually become dominant (Warren, Sanders and Cole, 1977). Several fungicides have been evaluated since which show excellent dollar spot control. These include iprodione, triadimefon, triarimol, fenarimol and propiconazole. Numerous reports on the efficacy of these and other fungicides for dollar spot control appear in recent issues of the *Fungicide and Nematicide Reports* published by the APS. Cadmium fungicides are still available for use in the US but were withdrawn from the European market in the late 1960s. Mercurial fungicides are no longer recommended and generally are unavailable for dollar spot control.

The date of the appearance of symptoms varies with season and location but once the daytime air temperatures consistently reach 65–70°F then preventive fungicide applications should commence on turf of known susceptibility. A treatment programme involving repeated application of an effective fungicide(s) every 7–10 or 10–21 days is required depending on the fungicide used. It is a common practice, especially where budget constraints are a major consideration, to wait for the initial symptoms to occur and then commence the fungicide spray programme. In such cases, however, heavier initial rates and a shorter time interval between sprays is recommended. In order to avoid possible build-up of tolerance, repeated use of a single fungicide is to be avoided. A programme involving fungicides in either tank-mix combinations or on alternate spray schedules should be adopted.

Hall (1984) noted that during epidemics of dollar spot disease in southern Ontario epidemics developed in a stepwise fashion. He found that an increase in the infection rate occurred after two consecutive days of wet weather if the average temperature was at or above 22°C or after a period of three or more consecutive days if the average temperature was 15°C or greater. An application of benomyl 1 and 2 days after two infection periods gave an acceptable level of disease control as effective as that obtained from a regular preventive schedule of seven sprays.

19.6 REFERENCES

Altman J. (1965) Nitrogen in relation to turf diseases. *Golf Course Rept.*, **33**(5), 16–30.

Bain, D.C. (1962) Sclerotinia blight of Bahia and Coastal Bermuda Grasses. *Plant Dis. Rept.*, **46,** 55–6.

Bennett, F.T. (1937) Dollar spot disease of turf and its causal organism *Sclerotinia homoeocarpa* n. sp. *Ann. Appl. Biol.*, **24,** 236–57.

Boesewinkel, H.J. (1977) New plant disease records in New Zealand: records in the period 1969–76. *N.Z. J. Agric. Res.*, **20** 583–9.

Britton, M.P. (1969) Turf grass diseases. In *Turfgrass Science* (eds A.A. Hanson and F.V. Juska), American Society of Agronomy, Madison, Wisconsin, pp. 288–335.

Cole, H. Jr, Duich, J.M., Massie, L.B. and Barber, W.D. (1969) Influence of fungus isolate and grass variety on Sclerotinia dollar spot development. *Crop Sci.*, **9,** 567–70.

Cole, H. Jr, Massie, L.B. and Duich, J.M. (1968a) Bentgrass-varietal susceptibility to Sclerotinia dollar spot and control with l-(butyl carbomoyl)-2-benzimidazole carbamic acid, methyl ester, a new systemic fungicide. *Plant Dis. Rept.*, **52,** 410–14.

Cole, H. Jr, Perkins, A.T. and Duich, J.M. (1967) Sclerotinia dollar spot on bentgrass-varietal susceptibility to infection and influence of variety on fungicide effectiveness. *Plant Dis. Rept.*, **51,** 40–2.

Cole, H. Jr, Taylor, B. and Duich, J.M. (1968b) Evidence of differing tolerances to fungicides among isolates of *Sclerotinia homoeocarpa. Phytopathology*, **58,** 683–6.

Cole, H. Jr, Warren, C.G. and Sanders, P.L. (1974) Fungicide tolerance – a rapidly emerging problem in turf grass disease control. In *Proc 2nd Int. Turfgrass Res. Conf.*, 19 June 1973, Blacksburg, VA, pp. 344–9.

Cook, R.N., Engel, R.E. and Bachelder, S. (1964) A study of the effect of nitrogen carriers on turf grass disease. *Plant Dis. Rept.*, **48,** 254–5.

Couch, H.B. (1973) *Diseases of Turfgrasses.* Krieger, Huntington, NY, 348pp.

Couch, H.B. and Bloom, J.R. (1960) Influence of environment on diseases of turf grasses. II. Effect of nutrition, pH and soil moisture on Sclerotinia dollar spot. *Phytopathology*, **50,** 761–3.

Davis, W.H., Engel, R.E. and Snyder, H.D. (1955) 1954 Turf fungicide trials. *Golf Course Rept.*, **23**(1), 32–4.

Endo, R.M. (1963) Influence of temperature on rate of growth of five fungus pathogens of turf grasses and on rate of disease spread. *Phytopathology*, **53,** 857–61.

Endo, R.M. and Malca, I. (1965) Morphological and cytohistological responses of primary roots of bentgrass to *Sclerotinia homoeocarpa* and D-galactose. *Phytopathology*, **55,** 781–9.

Endo, R.M. (1966) Control of dollar spot of turf-grass by nitrogen and its probable basis. *Abstr. Phytopathology*, **56,** 877.

Endo, R.M., Malca, I. and Krausman, E.M. (1964) Degeneration of the apical meristem and apex of bentgrass roots by a fungal toxin. *Phytopathology*, **54,** 1175–6.

Fenstermacher, J.M. (1970) Variation within *Sclerotinia homoeocarpa.* F. T. Bennett. M.S. Dissertation, University of Rhode Island, Kingston. 68pp.

Fenstermacher, J.M. (1980) Certain features of dollar spot disease and its causal organism *Sclerotinia homoeocarpa.* In *Advances in Turf grass Pathology.* (eds P.O. Larsen and B.G. Joyner), Harcourt, Brace Jovanovich, Duluth, MN, pp. 49–53.

Freeman, T.E. (1959) Florida isolates of dollarspot fungus stand hot weather. *Floridia Agric. Exp. Sta. Res. Rept.*, **4,** 3.

Freeman, T.E. (1967) Diseases of southern turf grasses. *Florida Agric. Exp. Sta. Tech. Bull.*, **713,** 31 pp.

Freeman, T.E. (1969) Influence of nitrogen sources on growth of *Sclerotinia homoeocarpa.* (Abstr.) *Phytopathology*, **59,** 114.

Goldberg, C.W., Cole, H. Jr and Duich, J.M. (1970) Comparative effectiveness of thiabendazole and benomyl for control of Helminthosporium leaf spot and crown rot, red thread, Sclerotinia dollar spot and Rhizoctonia brown patch of turf grass. *Plant Dis. Rept.*, **54,** 1080–4.

Gould, C.J. (1964) Turf grass disease problems in North America. *Golf Course Rept.*, **32**(5), 36, 38, 40–43, 46, 48, 50, 52, 54.

Gould, C.J. (1966–67) Use of fungicides in controlling turf grass diseases. *Golf Superintendent*, **34**(9), (10), **35**(1).

Halisky, P.M., Myers, R.F. and Wagner, R.E. (1981) Relationship of thatch to nematodes, dollar spot and fungicides in Kentucky bluegrass turf. In *Proc. 4th Int. Turfgrass Res. Conf.* (ed. R.W. Sheard), University of Guelph and International Turfgrass Society, pp. 415–20.

Hall, R. (1984) Relationship between weather factors and dollar spot of creeping bentgrass. *Can. J. Plant Sci.*, **64,** 167–74.

Harrington, G.E. (1941) Thiuramdisulphide for turf diseases. *Science* (*N.S.*), **93,** 2413, 311.

Haygood, R.A. and Spencer, J.A. (1979) Geographical and seasonal distribution of turf grass diseases in Mississippi. *Plant Dis. Rept.*, **63,** 852–5.

Hodges, C.F., Blaine, W.M. and Robinson, P.W. (1975) Severity of *Sclerotinia homoeocarpa* blight on various cultivars of fine leaved fescues. *Plant Dis. Rept.*, **59,** 12–14.

Hosotsuji, T. (1977) Control of diseases, insect pest and weeds of turf. *Japan Pesticide Information Leaflet*, no. 33.

Howard, F.L. (1947) An organo-cadmium fungicide for turf disease. *Greenkpr. Rept.*, **15**(2), 10.

Howard, F.L. (1955) What's new in turf grass diseases and their control. *Golf Course Rept.*, **21**(3), 18–24.

Howard, F.L., Rowell, J.B. and Keil, H.L. (1951) Fungus diseases of turf grasses. *Rhode Is. Agric. Exp. Stn. Bull.*, **308,** 56 pp.

Jackson, N. (1966) Dollar spot disease and its control, with special reference to changes in the susceptibility of *Sclerotinia homoeocarpa* to cadmium and mercury fungicides. *Proc. 7th Illinois Turfgrass Conf.*, p. 21.

Jackson, N. (1973) Apothecial production in *Sclerotinia homoeocarpa* F.T. Bennett. *J. Sports Turf Res. Inst.*, **49,** 58–63.

Karr, G.W. Jr, Gudauskas, R.T. and Dickens, R. (1979) Effects of three herbicides on selected pathogens and diseases of turf grasses. *Phytopathology*, **69,** 279–82.

Kerr, A. (1956) Some interactions between plant roots and pathogenic soil fungi. *Aust. J. Biol. Sci.*, **9,** 45–52.

Kohn, L.M. (1979) Delimitation of the economically important plant pathogenic *Sclerotinia* species. *Phytopathology*, **69,** 881–6.

Malca, I. and Endo, R.M. (1965) Identification of galactose in cultures of *Sclerotinia homoeocarpa* as the factor toxic to bentgrass roots. *Phytopathology*, **55,** 775–80.

Markland, F.E., Roberts, E.C. and Frederick, L.R. (1969) Influence of nitrogen fertilizers on Washington creeping bentgrass, *Agrostis palustris* Huds. II Incidence of dollar spot *Sclerotinia homoeocarpa* infection. *Agron. J.*, **61,** 701–5.

Massie, L.B., Cole, H. Jr and Duich, J.M. (1968) Pathogen variation in relation to disease severity and control of dollar spot of turf grass by fungicides. *Phytopathology*, **58,** 1616–19.

Monteith, J. Jr (1929) Some effects of lime and fertilizer on turf diseases. *USGA Green Section Bull.*, **9,** 82–99.

Monteith, J. Jr and Dahl, A.S. (1932) Turf diseases and their control. *USGA Green Section Bull.*, **12,** 85–186.

Musser, H.B. (1950) *Turf Management.* McGraw-Hill, New York, 354 pp.

Nicholson, J.F., Meyer, W.A., Sinclair, J.B. and Butler, J.D. (1971) Turf isolates of *Sclerotinia homoeocarpa* tolerant to dyrene. *Phytopathol. Z.*, **72,** 169–72.

Reilly, D. (1969) Control of dollar spot disease on turf in New South Wales with the systemic fungicide benomyl. *J. Sports Turf Res. Inst.*, **45,** 63–6.

Roberts, E.C. (1963) Relationships between mineral nutrition of turf grass and disease susceptibility. *Golf Course Rept.*, **31**(5), 52–7.

Roberts, E.C. (1967) Turf quality and disease resistance with nitrogen fertilizers. In *Proc. 38th GCSAA Int. Turf. Conf.*, 5–10 Feb, Washington, DC, pp. 15–19.

Siviour, T.R. (1975) The differential reaction of *Agrostis* spp. and *Cynodon dactylon* to applications of thiabendazole formulations and the efficacy of these and other selected fungicides for control of dollar spot and fusarium patch of bentgrass turf. *J. Sports Turf Res. Inst.*, **51,** 52–61.

Smith, J.D. (1953) Fungi and turf diseases 3. Fusarium patch disease. *J. Sports Turf Res. Inst.*, **8**(29), 253–8.

Smith, J.D. (1954a) A disease of *Poa annua. J. Sports Turf Res. Inst.*, **8**(29), 344–53.

Smith, J.D. (1954b) Dollar spot disease – fungicide trials, 1954. *J. Sports Turf. Res. Inst.*, **8**(29), 439–44.

Smith, J.D. (1955) Fungi and turf diseases; 5. Dollar spot disease. *J. Sports Turf. Res. Inst.*, **9**(31), 35–9.

Smith, J.D. (1956) Dollar spot trials. 1956. *J. Sports Turf Res. Inst.*, **9**(32), 233–4.

Smith, J.D. (1957a) The control of certain diseases of sport turf grasses in the British Isles. MSc. Dissertation. University of Durham.

Smith, J.D. (1957b) The effects of dollar spot disease on the botanical composition of turf of sea marsh fescue. *J. Sports Turf Res. Inst.*, **9**(33), 322–3.

Smith, J.D. (1957c) Dollar spot-Fungicide trial. 1956. *J. Sports Turf Res. Inst.*, **9**(33), 353–4.

Smith, J.D. (1957d) The effect of cadmium chloride with sulphate of ammonia application on earthworm casts. *J. Sports Turf Res. Inst.*, **9**(33), 358–9.

Sprague, R. (1950) Diseases of cereals and grasses in North America. Ronald Press, New York, 538 pp.

Turgeon, A.J. and Meyer, W.A. (1974) Effect of mowing height and fertilization level on disease incidence in five Kentucky bluegrasses. *Plant Dis. Rept.*, **58,** 514–16.

Vaughn, J.R. and Klomparens, W. (1952) Drugs on the green. *Golf Course Rept.*, **20**(2), 5–7.

Warren, C.G., Sanders, P.L. and Cole, H. Jr (1974) *Sclerotinia homoeocarpa* tolerance to benzimidazole-configuration fungicides. *Phytopathology*, **64,** 1139–42.

Warren, C.G., Sanders, P.L. and Cole, H. Jr (1977) Relative fitness of benzimidazole and cadmium tolerant populations of *Sclerotinia homoeocarpa* in the absence and presence of fungicides. *Phytopathology*, **67,** 704–8.

Whetzel, H.H. (1946) The cypericolous and juncicolous species of *Sclerotinia. Farlowia*, **2,** 385–437.

Wilkinson, J.F., Martin, D.P. and Larsen, P.O. (1975) Kentucky bluegrass and perennial ryegrass cultivar susceptibility to Sclerotinia dollar spot. *Plant Dis. Rept.*, **59,** 935–8.

Wilson, C.G. and Grau, F.W. (1952) National cooperative fungicide trials. *Golf Course Rept.*, **20**(2), 8–13.

20 *Downy mildew or yellow tuft disease*

A disease of bentgrass (*Agrostis* spp.) turf, commonly referred to as 'yellow tufts' has been known in the northeastern US and Europe since the early 1920s. Monteith and Dahl (1932) published a description of this unsightly, although rarely very damaging disease, but they did not name a causal agent. During the ensuing forty years, various causal relationships were suggested (Tarjan and Ferguson, 1951a;b, 1955), but they were never verified. The disease received little attention until the early 1970s when severe yellow tuft symptoms occurred in fields of Kentucky bluegrass (*Poa pratensis* L.) sod in the northeastern US. Investigation of the problem resulted in the implication of *Sclerophthora macrospora* as the possible causal agent (Jackson, Mueller and Fenstermacher, 1974; Mueller, Jackson and Fenstermacher, 1974). Pathogenicity tests later confirmed this common downy mildew fungus as the incitant of yellow tuft disease on cool-season turf grasses (Jackson and Dernoeden, 1978, 1980). Downy mildew was first reported on *Stenotaphrum secundatum* (Walte) Kuntze in Texas and Florida during the spring of 1969 (Jones and Amador, 1969). Further records of the disease on this host in other southern states of the US have been documented since (Bruton, Toler and Blasingame, 1981; Dale and Toler, 1972; Holcomb *et al.*, 1973).

The fungus is an obligate parasite on a wide range of gramineous hosts and has an extensive global distribution. Disease symptoms have been reported on turf grasses in North America, Australasia and Europe but probably they are of much wider occurrence. Yellow tuft impairs surface trueness and detracts from the visual appearance of closely mown *Agrostis* spp. turf, especially on golf and bowling greens. On commercial *P. pratensis* sod, yellow tuft symptoms may be severe enough to incur economic losses by rendering the product temporarily unsaleable. Control measures are warranted on severely affected turf but, although unsightly and recurrent otherwise symptoms are transient and seldom cause any permanent damage to turf of cool-season grasses (Jackson, Mueller and Fenstermacher, 1974). Susceptible cultivars of *S. secundatum* exhibit a significant reduction in plant vigour when infected with downy mildew, justifying efforts to control the disease (Grisham, Toler and Bruton, 1985).

20.1 SYMPTOMS

Early symptoms of downy mildew are often hard to discern. Leaf blades may be thickened or broadened slightly and if unmown, the infected plants may show some degree of stunting. In regularly mown turf this characteristic is masked, and even heavily infected plants may appear normal in colour and texture for long periods of time.

Advanced symptoms on bentgrasses (*Agrostis spp.*) and red fescue (*Festuca rubra* L.) turf appear as small yellow spots, 1–3 cm in diameter; on bluegrass (*Poa pratensis* L.) and perennial ryegrass (*Lolium perenne* L.), the spots are larger in the range 3–10 cm diameter (Fig. 31, colour plate section). Each spot is composed of a dense cluster of yellow shoots due to proliferation of axillary buds at crowns or at the nodes and terminals of creeping stems. Individual shoots making up the tufts form few adventitious roots and the tufts are easily detached from the turf (Fig. 32, colour plate section). Prominent symptoms usually appear in late spring and again in the autumn especially if cool, wet weather conditions prevail. Whole tufts may wither and die during dry periods, but commonly a proportion of the many tillers comprising the tuft will survive the stress situation. Further depradation may occur due to infection by *Ustilago striiformis* and/or *Helminthosporium* spp. By June of the following year, autumn seedings may show well-developed yellow tuft symptoms, noticeable first in low-lying areas subject to previous flooding. The disease spreads outward from the initial infection sites and once estalished in a turf, yellow tuft will recur indefinitely with varying severity.

On *Stenotaphrum secundatum* turf, infection by *S.*

macrospora does not result in tiller proliferation and tuft formation. The disease is characterized on this host by white, raised, linear streaks running parallel to the venation of leaf blades and sheaths. These symptoms appear in the spring and remain throughout the summer. Leaves become yellow and necrosis of the leaf tips may occur. In addition, leaf distortion and reduction of internode length are frequently observed. In some situations the disease is not appreciably debilitating to the grass but, where surface water is plentiful, severe cases of the disease can occur to retard growth.

The abundant, pearly-white, turgid sporangia are most obvious in the early morning while leaf surfaces are moist and the relative humidity is high. These typical signs of a downy mildew infection are short lived if the leaves dry out. As drying occurs the sporangia collapse to a dirty-white residue.

20.2 THE CAUSAL FUNGUS

Sclerophthora macrospora (Sacc.) Thirum. (Shaw and Naras, *Bull. Torr. Bot. Club.*, **80**, 299–307, 1953). The synonomy of the fungus, according to Ullstrup (1970), is as follows:

Sclerospora macrospora Sacc. (1890)
Sclerospora kreigeriana Magnus. (1985)
Sclerospora oryzae Brizi. (1919)
Nozemia macrospora (Sacc.) Tasugi. (1931)
Phytophthora macrospora (Sacc.) Tanaka. (1940)

The mycelium is coenocytic, multinucleate, intercellular, without haustoria, with a variable diameter 3–60 μm. Sporangia are limoniform, apically poroid (Fig. 33, colour plate section), 37.8–60.8 μm × 70–100.8 μm; sporangiophores are determinate, 4.2–25.2 μm emerging through stomata from robust, lobed sporangiophoric pads. Zoospores are ovoid to pyriform, 14–20 × 5–10 μm, laterally biflagellate with anterior whiplash and posterior tinsel flagella; encysted zoospores are spherical 11.2–15.3 μm diam. Germination occurs by a single germ tube 5–60 μm long with protoplasm concentrated in a swollen proto-appressorium. Oogonia are globose to roughly spherical, 60–90 μm diam. with irregularly pitted walls about 7 μm thick; antheridia are amphigynous. Oospores are globose to roughly spherical, 50–75 μm in diam. with a smooth wall 5–8 μm thick, germinating to produce a single sporangium.

After clearing and suitable staining of the tissues (Roth, 1967; Ullstrup, 1952), microscopic examination of turf grasses showing symptoms of downy mildew invariably reveals the systemic mycelium of *Sclerophthora macrospora* within the crowns, stems and leaves. A few axillary buds may escape the colonizing hyphae emanating from the crown tissue and may produce an occasional healthy tiller free of mycelium. Mycelium has not been observed to progress far into roots.

Narrow-diameter primary hyphae form a complex network within the crown tissue, ramifying in leaf and shoot meristems. Primary hyphae closely associated with vascular bundles carry the fungus up the leaf sheaths. Localized branching and proliferation of these hyphae into tissue between the bundles occurs, increasingly so in the upper sheath. Massive development of multi-diameter secondary mycelium occurs in the lamina again associated initially with the bundles. Hyphae then branch out into the mesophyll tissue and may bridge to adjacent bundles, often via the small transverse bundles. Hyphae reaching the substomatal cavities on upper portions of sheaths and on both surfaces of the lamina form distinctive, lobed, irregularly thickened pads from which the sporangiophores eventually develop. These sporangiophoric pads are very numerous in heavily infected leaves.

The intercellular hyphae mould closely to the cell walls, filling all the interstices. Fine hyphae and the many finger-like projections from the lobed and convoluted larger hyphae press between the cells. Minute protruberances occur sparsely on some hyphae, especially those adjacent to the vascular bundles. Although cell walls appear invaginated by these structures, actual penetration of the wall has not been observed (Jackson, 1980).

Sporangial production may be demonstrated readily by placing infected leaves or plants in a moist chamber at 15–20°C for 12–24 hours. Detached leaves submerged in distilled water or in soil water at 15°C produce copious numbers of sporangia after four hours. Sporangia mature rapidly, each releasing up to 50 or more motile zoospores. Direct germination of sporangia has not been observed but elongation of the apical region frequently occurs, often to form a secondary sporangium. Germination of imperfectly cleaved zoospores or zoospores encysted within the sporangium may result in sporangia with several germ tubes issuing from the apical region.

Zoospores swim vigorously in a rolling, spiral motion. Temperature affects swimming time; at 5°C spores may remain active for more than 24 hours, above 20°C spores encyst within an hour. At encystment the flagella are absorbed and the spores round up. After a brief resting period, a single germ tube is produced with the spore protoplasm concentrated near the advancing tip.

Zoospores are remarkably chemotactic (Akai and Fukutomi, 1964; Dernoeden and Jackson, 1978, 1981a) and respond to low concentrations of sugars (fructose, glucose, maltose, sucrose), hydrolysed casein, yeast extract and several individual amino acids. Guttation fluid and glutamine are very active stimulants. Imbibed seeds and seeds in the early stages of germination are extremely attractive to the zoospores, which congregate in large numbers at the region of the mesocotyl and encyst there. Rapid germination of the spores ensues.

The actual penetration process has not been observed in turf grasses but was described in detail on *Oryza sativa* L. (rice). Germ tubes penetrated directly into juvenile plumule tissues of rice plants in the early stages of seed germination and the mycelium developed intercellularly (Akai and Fukutomi, 1964). Germinating seed of several festucoid grasses exposed to suspensions of *S. macrospora* zoospores produced seedlings containing typical mycelium of the fungus two weeks after inoculation, and such plants developed yellow tuft symptoms (Jackson and Dernoeden, 1980). Toler and Bruton (1983) reported that healthy *S. secundatum* could be inoculated successfully if plants were flooded with water and naturally infected leaves were floated on the surface to allow sporangia and subsequent zoospore production. The shortest exposure time required for optimum infection was 24 hours at 15°C.

Sexual reproduction in *Sclerophthora macrospora* is accomplished by means of oospores. Oogonia, each with an attendent antheridium, are initiated by those hyphae associated closely with the vascular bundles of the lamina. Distortion of the vascular tissue occurs as the globose to roughly spherical oogonia expand. Germination of these resting spores has been observed confirming the development of a single sporangium from the oospore and subsequent release of zoospores (Dernoeden and Jackson, 1981b); Semeniuk and Mankin, 1964). To date, oospore formation has not been reported in *S. secundatum* (Bruton, Toler and Blasingame, 1981).

20.3 HOST RANGE

S. macrospora attacks a wide range of grasses including most, if not all, cool-season turf grasses (Jackson, 1980; Semeniuk and Mankin, 1964), and also *Stenotaphrum secundatum* (Walte) Kuntze (Bruton, Toler and Blasingame, 1981; Dale and Toler, 1972; Grisham, Toler and Bruton, 1985; Holcomb *et al.*, 1973; Jones and Amador, 1969). *Agrostis* spp. are common hosts, but more frequent reports of affected *Poa pratensis* L, *Festuca rubra* L. and *Lolium perenne* L. turfs are occurring. The weed grasses, *Digitaria* spp. and *Agropyron repens* L. are very susceptible to invasion.

20.4 DISEASE DEVELOPMENT

Disease caused by *S. macrospora* in various crop plants is most serious and widespread in low-lying, wet poorly drained soils which are subject to frequent flooding and water logging (Roth, 1967, Safeeulla, 1976; Summers, 1952; Ullstrup, 1955, 1970). Downy mildew of turf grasses is no exception to this general observation (Jackson, Mueller and Fenstermacher 1974; Toler and Bruton, 1983). Invariably the appearance of symptoms in a new stand of turf can be linked with flooding or prolonged saturation of the soil at the time of sward establishment.

Soil saturation may be required to initiate germination of oospores. These long-viable resting spores and potential incitants of primary infection can be carried in debris from turf grasses and collateral hosts or borne on the seed (Safeeulla, 1976; Ullstrup, 1970). In general, oospores have shown a poor capacity for germination when tested under laboratory conditions (Safeeulla, 1976; Semeniuk and Mankin, 1964). Dernoeden and Jackson (1981b), demonstrated germination percentages of 16–23% by subjecting one-year-old oospores, freed from leaves of crabgrass (*Digitaria* spp.) to various physical and chemical treatments. By comparison, in distilled water controls the germination was 8.5%. Regardless of treatment, oospores which germinated did so within 14 days and over a wide optimum temperature range of 20–28°C. Semeniuk (1976) and Safeeulla (1976) both report successful infection of grasses and cereals utilizing oospores as inoculum but the significance of these structures as primary inoculum in turf is unknown.

Resting mycelium in seed, or zoospores from adjacent susceptible grasses (collateral hosts) may also serve as primary inoculum (Safeeulla, 1976; Ullstrup, 1970). Seed transmission has been demonstrated by Ullstrup (1976) in corn but he argues that this means of transmission is of little significance in nature. Mycelium of the fungus has been noted in all parts of *Poa pratensis* inflorescences, but the seed produced was infertile. Infected collateral hosts have been proposed as a very important primary inoculum source (Semeniuk and Mankin, 1964). Field observations in *P. pratensis* sod in growing areas support this proposal. The initial symptoms in sod fields usually occur in low sites at the edge of the field where surface flooding inundates the headland vegetation containing infected grasses, and spreads out over the adjacent establishing stand of turf grass. Infected volunteer plants from the previous sod crop may also provide primary inoculum in new seedings (Jackson, Mueller and Fenstermacher, 1974).

Given suitable conditions of cool temperatures, high soil moisture and high relative humidity, infected plants produce successive crops of sporangia. In Rhode Island, sporangia have been observed in the field on many turf, agricultural and weed grasses, over the period from early May to late November but particularly in the cooler spring and autumn seasons (Jackson, 1980a,b). The precise conditions for field infection have not been established but in other crop plants prolonged and abundant moisture is a major requirement (Shaw and Safeeulla, 1970), and soil saturation for 24–48 hours appears to provide sufficient time for infection to occur in grasses and corn (Semeniuk and Mankin, 1964; Ullstrup, 1970). The indications from laboratory tests with some festucoid turf species is that the time required for infection may be considerably shorter.

Infected excised leaves of *Poa pratensis* inundated with water at 15–20°C generated sporangia and actively swimming zoospores in as little as 4–6 hours. Newly germinating seeds immersed in the zoospore suspensions could be infected in 6 hours exposure (Dernoeden and Jackson, 1980a). Healthy *S. secundatum* plants placed in water at 15°C on which infected leaves were then floated, required 24 h immersion for optimum infection (Toler and Bruton, 1983).

Inoculation studies using zoospores have shown seedlings of *Lolium perenne* to be far more susceptible to infection than mature plants. Furthermore germinating grass seeds liberated substances attractive to *S. macrospora* zoospores which congregated in large numbers in the mesocotyl region. Positive chemotaxis to both seed exudates and also leaf surface fluids is suggested as a mechanism that aids the pathogen in exploiting the host (Dernoeden and Jackson, 1980a, 1981a). Although actual penetration of the host has not been confirmed histologically on turf grasses, it appears that meristematic tissues are the vulnerable regions. The region of the mesocotyl on young seedlings; the shoot apices, axillary buds and possibly leaf intercalary meristems on mature plants are the suggested infection sites. Easy access of zoospores to the exposed mescotyl tissues of seedlings as opposed to the sheathed meristems of mature plants, probably accounts for the marked vulnerability of seedlings. Soil saturation suffices for seedling infection, but flooding of the sward and cool temperatures to prolong motility of zoospores is probably necessary for zoospores to reach infection sites on mature turf stands.

Once the process of infection in complete the intracellular mycelium grows into the meristematic regions of the shoot meristem and axillary buds, and rapidly colonizes developing leaf primordia. Growth of the mycelium then keeps pace with the leaf growth and at the same time massive colonization of the crown region occurs, ensuring that any further leaf primordia and new axillary buds are infiltrated by mycelium. Primary hyphae in leaf laminae soon generate the robust secondary hyphae on which the sub-stomatal sporangiophoric pads are formed (Dernoeden and Jackson, 1980a). Sporangial production has been noted on zoospore-inoculated rye (*Secale cereale* L.) and ryegrass (*Lolium perenne*) seedlings three weeks after exposure to the inoculum (Jackson and Dernoeden, 1980).

Infected plants may survive at least two years and show recurrent symptoms during cool moist periods. As with stripe smut, infected plants may occasionally give rise to healthy tillers if an axillary bud escapes systemic invasion from mycelium present in the crown. This feature and the ephemeral nature of the symptoms may account for the apparent disappearance of the disease from turf without any major thinning of the sward (Dernoeden and Jackson 1980a,b).

Oospores have been found in *Poa pratensis* and *Agrostis* spp. on several occasions, sporadically over the growing season, but most commonly in the autumn. *Digitaria* spp. (crabgrasses) frequently develop yellow tuft symptoms in mid-summer and the fungus may sporulate abundantly on these grasses. Since no host specificity has been established for this fungus, crabgrass may provide a large amount of asexual inoculum at a time when sporulation by infected turf grass species is much reduced. Prodigious numbers of oospores are produced in crabgrass plants if early frosts do not cause premature killing. Hence crabgrasses may be an important source of both current and persistent inoculum.

The effects of soil pH and soil nutrient status on the activities of this fungus in turf has not been investigated fully. Jackson and Dernoeden (1980) inoculated seedings of various grasses growing in soils of pH 5.3 and 7.6 but the different pH values had little effect on the percentage infection. No association between yellow tuft incidence and soil pH has been noted in the field.

Reduced downy mildew severity has been reported for several grain crops when nitrogenous fertilizers were applied (Deshmukh, Mayee and Kulkarni, 1978; Singh *et al.*, 1970). In contrast, nitrogen fertilizer application to severely diseased Kentucky bluegrass (*Poa pratensis*) turf in commercial sod fields exacerbated yellow tuft symptoms, especially if the nitrogen source was urea (Jackson, unpublished). On an *Agrostis canina*/*Festuca rubra* mixed turf, Dernoeden and Jackson (1980b) found yellow tuft symptoms less noticeable when nitrogenous fertilizers were applied but there was no reduction in disease incidence, the effect was one of masking the symptoms. Greenhouse trials using rye (*Secale cereale*) as the test plant demonstrated that nutrients applied to the growing medium may enhance the sporulation capacity of the fungus.

20.5 CONTROL

20.5.1 Cultural

Soil saturation is a prime requirement for the production, dissemination and infection by the propagules of *S. macrospora*. Good soil drainage and any other water-management measures which keep the periods of soil saturation to a minimum are of obvious benefit. This applies particularly to new seedings of turf grasses.

Late autumn seeding in low-lying areas prone to flooding should be avoided and late-spring sowing of such areas will lessen the risk of severe yellow tuft disease. On sod fields, collateral hosts should be eradicated from the headlands, where feasible, and harvested areas should be ploughed as soon as possible to eradicate any volunteer growth from the previous sod crop. Excessive use of nitrogenous fertilizers should be avoided on infected turf. Iron sulphate at 10–20 lb (4.5–9 kg) acre may be utilized to mask symptoms in sodfields (Jackson, 1980b). Superior resistance between and within species of cool-season turf grasses to downy mildew has not yet been demonstrated. Toler Bruton and Grisham, (1983) evaluated a number of *S. secundatum* accessions and cultivars for resistance to *S. macrospora*. In field trials, the percentage of infected leaves among the candidate grasses ranged from 0 to 95. However, results from artificial inoculation trials with downy mildew on these grasses in the greenhouse were not consistent with the field results.

20.5.2 Fungicides

The control of downy mildew diseases in the Poaceae by fungicides, in general, has proved unreliable and uneconomic (Exconde, 1970, Frederiksen *et al.*, 1970; Renfro, 1975). Evaluation of available fungicides for yellow tuft control met with a similar lack of success until the introduction of the fungicide metalaxyl. Bruton and Toler (1979; 1980) reported the successful control of *S. macrospora* on *Stenotaphrum secundatum* in greenhouse trials. The effectiveness of this material when applied as a soil drench to individual infected plants of *Poa pratensis*, and its efficacy in field trials was demonstrated in Rhode Island during 1979. Three applications of the material to a mixed *Agrostis canina*/*Festuca rubra* turf showing severe yellow tuft symptoms resulted in the complete elimination of the symptoms (Dernoeden and Jackson, 1980b).

20.6 REFERENCES

Akai, S. and Fukutomi, M. (1964) Mechanism of the infection of plumules of rice plants by *Sclerophthora macrospora*. Studies on the downy mildews of rice plants 11. *Spec. Res. Rep. Dis. Insect Forecasting*, **17**, 47–54. (Japanese with English summary).

Bruton, B.D. and Toler, R.W. (1979) Efficacy of Ciba-Geigy experimental fungicide. CGA-48988 in control of St Augustinegrass downy mildew caused by *Sclerophthora macrospora Phytopathology*. **69,** 1023. (Abstr.)

Bruton, B.D. and Toler, R.W. (1980) Fungicidal control of *Sclerophthora macrospora* the downy mildew fungus on St Augustinegrass. (Abstr). In *Advances in Turfgrass Pathology* (eds P.O. Larson and B.G. Joyner), Harcourt, Brace, Jovanovich, Duluth, Ma., p. 192.

Bruton, B. D., Toler, R.W. and Blasingame, D.L. (1981) Downy mildew on St Augustinegrass in Mississippi. *Plant Dis.*, **65**, 925.

Dale, J.L. and Toler, R.W. (1972) Downy mildew on St Augustinegrass in Arkansas. *Plant Dis. Rept.*, **56**, 658.

Dernoeden, P.H. and Jackson, N. (1978) Zoospore chemotaxis in *Sclerophthora macrospora*. *Phytopathol. News*, **12**, 234. (Abstr.)

Dernoeden, P.H. and Jackson, N. (1980a) Infection and mycelial colonization of gramineous hosts by *Sclerophthora macrospora*. *Phytopathology*, **70**, 1009–13.

Dernoeden, P.H. and Jackson, N. (1980b) Managing yellow tuft disease. *J. Sports Turf Res. Inst.*, **56**, 9–17.

Dernoeden, P.H. and Jackson, N. (1981a) Zoospore chemotaxis in *Sclerophthora macrospora*. In *Proc. IVth Int. Turfgrass Res. Conf.* (ed. R.W. Sheard), University of Guelph and Int. Turfgrass Soc., pp. 443–7.

Dernoeden, P.H. and Jackson, N. (1981b) Enhanced germination of *Sclerophthora macrospora* oospores in response to various chemical and physical treatments. *Trans. Br. Mycol. Soc.*, **76**, 337–41.

Deshmukh, S.S., Mayee, C.D. and Kulkarni, B.S. (1978) Reduction of downy mildew of pearl millet with fertilizer management. *Phytopathology*, **68**, 1350–3.

Exconde, O.R. (1970) Phillipine corn downy mildew. *Indian Phytopathol.*, **23**, 275–84.

Frederiksen, R.A., Bockholt, A.J., Rosenow, D.T. and Reyes, L. (1970) Problems and progress of sorghum downy mildew in the United States. *Indian Phytopathol.*, **23**, 321–38.

Grisham, M.P., Toler, R.W. and Bruton, B.D (1985) Effect of *Sclerophthora macrospora* on growth and development of St. Augustinegrass. *Plant Dis.*, **69**, 289–91.

Holcomb, G.E., Derrick, K.S., Carver, R.B. and Toler, R.W. (1973) Downy mildew and rust found on St. Augustinegrass in Louisiana. *Plant Dis. Rept.*, **57**, 16.

Jackson, N. (1980a) Yellow tuft. In *Advances in Turfgrass Pathology* (eds P. O. Larsen and B.G. Joyner), Harcourt, Brace, Jovanovich, Duluth, MN, pp. 135–7.

Jackson, N. (1980b) Yellow tuft disease of turf grasses: A review of recent studies conducted in Rhode Island. In

Proc. III Int. Turfgrass Res. Conf. (ed. J.B. Beard), American Soc. Agron., Crop Sci. Soc. of America and Int. Turfgrass Soc., Madison, Wisconsin. pp. 265–70.

Jackson, N. and Dernoeden, P.H. (1978) *Sclerophthora macrospora* the causal agent of yellow tuft disease in turf grasses. *Phytopathol. News,* **12,** 236. (Abstr.)

Jackson, N. and Dernoeden, P.H. (1980) *Sclerophthora macrospora*: The incitant of yellow tuft disease of turf grasses. *Plant Dis.*, **64**, 915–16.

Jackson, N., Mueller, W.C. and Fenstermacher, J.M. (1974) The association of downy mildew with yellow tuft disease of turf grasses. *J. Sports Turf Res. Inst.*, **50,** 52–4.

Jones, B.L. and Amador, J. (1969) Downy mildew, a new disease of St. Augustinegrass. *Plant Dis. Rept.*, **53**, 852–4.

Monteith, J. and Dahl, A.S. (1932) Turf diseases and their control. *Bull. US Golf Assoc. Green Sect.*, **12,** 87–186.

Mueller, W.C., Jackson, N. and Fenstermacher, J.M. (1974) Occurrence of *Sclerophthora macrospora* in turf grass affected with yellow tuft. *Plant Dis. Rept.*, **58**, 848–50.

Renfro, B.L. (1975) Downy mildew disease of pearl millet. In *Proc. Consultant Group Meet., Downy Mildew, Ergot, Pearl Millet.* ICRISAT, Hyderabad, India, pp. 77–83.

Roth, G. (1967) *Sclerophthora macrospora* (Sacc.) Thirum. *et al.* (Syn. *Sclerospora macrospora* Sacc.) on sugar cane in South Africa. *Z. Pflanzenkr. Pflanzenpathol. Pflanzenschutz,* **74,** 83–100.

Safeeulla, K.M. (1976) *Biology and Control of the Downy Mildews of Pearl Millet, Sorghum and Finger Millet.* Wesley Press, Mysore, India, 304 pp.

Semeniuk, G. (1976) *Sclerophthora macrospora* infection of three annual grasses by oospore and asexual inocula. *Plant Dis. Rept.*, **60**, 745–8.

Semeniuk, G. and Mankin, C.J. (1964) Occurrence and development of *Sclerophthora macrospora* on cereals and grasses in South Dakota. *Phytopathology*, **54**, 409–16.

Semeniuk, G. and Mankin, C.J. (1966) Additional hosts of *Sclerophthora macrospora* in South Dakota. *Phytopathology*, **56**, 351.

Shaw, C.G. and Safeeulla K.M. (1970) Round table discussion No. 1. The pathogen–emphasizing taxonomy, morphology and life cycles. *Indian Phytopathol.*, **23**, 399–412.

Singh, R.S., Chaube, H.S., Singh, N. and Asnani, V.L. (1970) Observations on the effect of host nutrition and seed, soil and foliar treatments on the incidence of downy mildews. 1. A preliminary report. *Indian Phytopathol.*, **23**, 209–15.

Summers, T.E. (1952) Downy mildew (*Sclerospora macrospora*) of oats in the south. *Plant Dis. Rept.*, **36**, 347–8.

Tarjan, A.C. and Ferguson, M.H. (1951a) Observations of nematodes in yellow tuft of bentgrasses. *US Golf Assoc. J. Turf Mgmt*, **4**, 28–30.

Tarjan, A.C. and Ferguson, M.H. (1951b) Association of certain nematodes with yellow tuft of bentgrass. *Phytopathology*, **41**, 566. (Abstr.)

Tarjan, A.C. and Hart, S.W. (1955) Occurrence of yellow tuft of bentgrass in Rhode Island. *Plant Dis. Rept.*, **39,** 185.

Thirumalachar, M.J., Shaw, C.G. and Narasimhan, M.J. (1953) The sporangial phase of the downy mildew on *Eleusine coracana* with a discussion of the identity of *Sclerospora macrospora* Sacc. *Bull. Torrey Bot. Club*, **80**, 299–307.

Toler, R.W. and Bruton, B.D. (1983) Downy mildew of St. Augustinegrass. *Grounds Maitenance*, **18**(2), 98.

Toler, R.W. and Bruton, B.D. and Grisham, M.P. (1983) Evaluation of St. Augustinegrass accessions and cultivars for resistance to *Sclerophthora macrospora. Plant Dis.*, **67**, 1008–10.

Ullstrup, A.J. (1952) Observations on crazy top of corn. *Phytopathology*, **42**, 675–80.

Ullstrup, A.J. (1955) Crazy top of some wild grasses and the occurrence of the sporangial stage of the pathogen. *Plant Dis. Rept,* **39,** 839–41.

Ullstrup, A.J. (1970) Crazy top of mazie. *Indian Phytopathol.*, **23**, 250–61.

21 *Grey leaf spot*

Leaf spots and blights caused by *Pyricularia* spp. occur on many different grasses and on other plants. Rice 'blast' caused by *P. oryzae* is a common and often severe disease of that crop in many parts of the world. In amenity turf, grey leaf spot caused by *P. grisea* often becomes epidemic, especially on turf of *Stenotaphrum secundatum* in moist summers in the south and southeast of the United States. However, many other turf grass species may become infected (see Section 21.3) in tropical to warm-temperate regions of Africa, Asia, Australasia, the Caribbean, Europe, North and South America. In N. America, the pathogen has been recorded as far north as the southern Canadian prairies and southern Ontario (Bain, Patel and Patel, 1972; Carver, Rush, and Lindberg, 1972; Conners, 1967; Freeman, 1967; Malca and Owen, 1957; Sprague, 1950; Subramanian, 1968; Trevathan, 1982a; Wellman, 1977; Yamanaka, 1982).

21.1 SYMPTOMS

Distinctive lesions develop on most or all of the aerial parts of susceptible grass species. On *Stenotaphram secundatum* (St Augustinegrass), nearly round or elongate elliptical spots with bluish-grey to ashen-grey centres and purple or reddish to dark-brown margins, that in turn are bordered by a ring of chlorotic tissue, occur most profusely on leaf laminae and sheaths (Fig. 34, colour plate section). Abundant conidiophores of the fungus may give the spots a grey velvety appearance, but failure or subsidence of sporulation results in tan-coloured, dried and wrinkled spots. Under optimum environmental conditions the enlargement of the minute, water-soaked, brown, then greyish-green initial lesions is rapid. Fully developed lesions range mostly from 0.2 to 0.7 cm, occasionally reaching 2.0 cm in length. Lesions may be concentrated along the midrib, but they can occur at the base, tip and margins of the blade. Thirty or more lesions can develop on a single leaf blade resulting in chlorosis and leaf death. As increasing numbers of leaves become withered and brown, stands of *S. secundatum* turf look scorched with symptoms similar to severe drought stress.

Leaf blight epidemics, with transient symptoms similar to those described have occurred in Louisiana and Mississippi on forage stands of *Lolium multiflorum* since 1971. Infection of young seedlings may result in melting-out (Bain, Patel and Patel, 1972; Carver, Rush and Lindberg, 1972; Trevathan, 1982b). On *Cynodon dactylon* leaf spot symptoms and nodal rot occur.

21.2 THE CAUSAL FUNGUS

Teleomorph – *Magnaporthe grisea* (Hebert) Barr, 1977 synonyms – *Magnaporthe grisea* (Hebert) Yaegashi and Udagawa, 1978; *Ceratosphaeria grisea* Hebert, 1971; anamorph – *Pyricularia grisea* (Cooke) Sacc. (*Michelia*, **2**, 20, 1880).

The causal fungi of rice blast, *P. oryzae*, and grey leaf spot, *P. grisea* are morphologically very similar and the need for separate taxa is very doubtful (Sprague, 1950; Asuyama, 1965) although Ellis (1971) retained them. Asuyama (1965) proposed that these two *Pyricularia* spp. (and perhaps others) should be considered as one species with eleven specialized forms, based on their pathogenicity to various cereals and grasses. More recently, Yamanaka (1982) has suggested that there is one species, *P. grisea*, with the form species *grisea*, *oryzae* and *elusines* on different groups of grasses. The demonstration of compatible matings between isolates of the two species and the formation of a common teleomorphic state confirmed the synonomy (Hebert, 1975; Yaegashi and Udagawa, 1978a,b). *Pyricularia grisea* has precedence over *P. oryzae* according to the rules of nomenclature (Asuyama, 1965) but *P. oryzae* has a voluminous literature and remains in common usage as the name for the rice blast incitant.

In *P. grisea*, the faint grey or olive, sparsely septate, mostly simple, geniculate condiophores, $<150\,\mu m \times 2.5–4.5\,\mu m$, emerge in small groups through stomata. Solitary, dry, hyaline to fuscous, mostly 2-septate conidia, $17–28 \times 6–9\,\mu m$, are borne sympodially on denticles and have a conical or slightly beaked apical cell, and are broadest at the basal septum, usually with a protruberant scar.

In *M. grisea*, perithecia have resulted from mating compatible isolates of *P. grisea* with *P. grisea* or *P. oryzae* in culture (Herbert, 1971; Yaegashi and Udagawa, 1978). Perithecia were often gregarious, glabrous, dark brown to black, partially immersed, with a spherical or sub-spherical base 80–260 μm diam. × 500–1200 μm long and a protruding, cylindrical neck <1100 μm long and 55–160 μm diam. Asci were unitunicate, eight-spored, hyaline, cylindrical-clavate, 55–110 × 8–15 μm. Paraphyses were hyaline soon deliquescent. The hyaline, fusiform, 3-septate, ascospores, 18–23 × 5–7 μm, had rounded ends and slight septal constrictions. (For detailed descriptions see Hebert, 1971 and Yaegashi and Udagawa, 1978.)

Both *P. oryzae* and *P. grisea* grow well on 'natural' culture media and sporulation of isolates of *P. grisea* from *S. secundatum* was abundant on 2% rice polish or V-8 juice agar, but colony morphology and growth rate on some 'natural' media may vary considerably (Malca and Owen, 1957; Ou, 1972; Trevathan, 1982a).

21.3 HOST RANGE

Although *Stenotaphrum secundatum* (Walt.) Kunze is the turf grass suscept of major importance (Freeman, 1967; Haygood and Spencer, 1979; Malca and Owen, 1957), other warm-season species, e.g. *Cynodon dactylon* (L.) Pers., *Eremochloa ophiuroides* (Munro) Hack., *Eragrostis, Paspalum* and *Pennisetum* spp. are also hosts of *P. grisea*. Cool-season turf grasses in the genera *Agrostis, Anthoxanthum, Festuca* and *Lolium* are also susceptible. Although *L. multiflorum* cultivars are generally susceptible to *P. grisea* (Trevathan, 1982b), *L. perenne* has been found to be field resistant (Narita, Iwata and Yamanuki, 1956).

21.4 DISEASE DEVELOPMENT

In *P. grisea*, primary inoculum is derived from spores produced on overwintered debris and leaf lesions, and probably from seed-borne propagules since *P. oryaze* is reported to be seed-transmitted in grasses (Richardson, 1979). Wild grasses and both symptomatic and asymptomatic dicotyledonous plants may serve as additional reservoirs of inoculum (Trevathan, 1982a). In their detailed studies of grey leaf spot on St Augustinegrass in Florida, Malca and Owen (1957) found that the splash- or wind-dispersed conidia initiate primary infections on the new leaves following spring rains which then trigger renewed sporulation. Three factors were deemed necessary for the disease to reach epidemic proportions; (1) the presence of actively growing susceptible tissues, (2) rainy weather or prolonged periods of high humidity, and (3) temperatures in the range 21°–32°C. When these requirements were met, few St Augustinegrass lawns in Florida remained free of disease by mid-June. Unfertilized turf with little active growth had little disease. Dry periods of ten days or more, together with high temperatures, appeared to diminish the observable symptoms in established outbreaks of grey leaf spot. With the recurrence of more favourable weather, lesions soon developed on the healthy new leaves as spore inoculum was dispersed from previously infected lower leaves.

Nitrogenous fertilizers have a marked influence on host susceptibility to *P. grisea* and exacerbate rice blast (see Ou, 1972). Grey leaf spot on St Augustinegrass responds in a similar manner, and increased severity is correlated with the amount of fertilizer applied. Varying the rate from 1 to 8 lbs of N/1000 ft^2 (0.5–4 kg/100 m^2) resulted in a 65–125% increase in the number of lesions per 100 leaves over that found in plots receiving no nitrogen (Freeman, 1964). The mechanism whereby nitrogenous fertilizers affect susceptibility has not been elucidated, but Freeman (1967) noted that damage by *P. grisea* is especially evident on newly sprigged, rapidly growing grass. The stimulation of new susceptible growth in response to nitrogenous fertilization may partially explain the increased disease incidence. Promotion of leaf exudates stimulatory to spore germination, appressorium formation and penetration of leaf tissues could also be involved (Kapoor and Singh, 1977).

On rice (Ou, 1972) the fungus penetrates tissues directly or through stomata. Free water on the leaf surface is required for conidial germination and relative humidity near saturation is necessary for successful infection. Temperatures in the range 24–28°C with 16–24 hours of continuous wetting are the conditions that maximize infection and also provide the minimum incubation period of 5–6 days. Development of grey leaf spot on St Augustinegrass follows much the same pattern. Malca and Owen (1957) showed that spores of *P. grisea* formed appressoria on leaves incubated at 22–24°C in a moist chamber within 16 hours. Penetration was accomplished within 24 hours, predominantly directly through the cuticle, and active parasitism ensued. Hyphae moved inter- and intracellularly to colonize mesophyll cells, causing their collapse, and within four to five days sunken lesions, typical of grey leaf spot were apparent. Conidial inoculation of *L. multiflorum* was successful when plants were misted for three days at c. 25°C. Symptom development was most obvious five days after the plants were inoculated (Trevathan, 1982a).

21.5 CONTROL

P. grisea is widely recognized as of extremely variable pathogenicity on species and cultivars of plants. Breeding for grey leaf spot resistance in turf grass has not been attempted although differences in susceptibiliy within species have been noted. Freeman (1967) reported that the yellow–green or Roselawn types of *S. secundatum* exhibit an inherent degree of resistance to grey leaf spot not found in the blue–green or bitter blue types. However, the highly resistant strains selected to date lack the desirable agronomic features for lawn turf.

Cultural measures to reduce incidence and severity of the disease include moderation in the use of nitrogen fertilizer and timing of irrigation to ensure the shortest period of leaf wetness (Freeman, 1967). Removal of clippings is also effective in reducing disease (Parris, 1971). Treatment with certain herbicides and heavy traffic or drought may exacerbate grey leaf spot damage (Freeman, 1982). These stress situations should be avoided.

Fungicides are usually required to give complete disease control in *S. secundatum* turf. Thiram, chlorothanil and cyloheximide and thiram in combination afford good control when applied at 10–14 day intervals (Freeman 1967, 1982).

21.6 REFERENCES

Asuyama, H. (1965) Morphology, taxonomy, host range and life cycle of *Pyricularia oryzae*. In *The Rice Blast Disease. Proc Symp Int. Rice. Res. Inst.*, July 1963. Johns Hopkins University Press, Baltimore, MD, pp. 9–22.

Bain, D.C. Patel, B.N. and Patel, M.V. (1972) Blast of ryegrass in Mississippi. *Plant Dis. Rept.*, **56,** 210.

Barr, M.E. (1977) *Magnaporthe*, *Telimenella* and *Hyponectria* (Physosporellaceae). *Mycologia*, **69,** 952–66.

Carver, R.B., Rush, M.C. and Lindberg, G.D. (1972) An epiphytotic of ryegrass blast in Louisiana. *Plant Dis. Rept.*, **56,** 157–9.

Conners, I.L. (1967) An annotated index of plant diseases in Canada (and fungi recorded on plants in Alaska, Canada and Greenland). Res. Br. Can. Dept. Agric., Publ. 1251. Queen's Printer, Ottawa, 381 pp.

Ellis, M.B. (1971) *Dematiaceous Hyphomycetes*. Commonwealth Mycological Institute. Kew, Surrey, England, 608 pp.

Freeman, T.E. (1964) Influence of nitrogen on severity of *Pyricularia grisea* infection of St. Augustinegrass. *Phytopathology*, **54,** 1187–9.

Freeman, T.E. (1967) Diseases of southern turf grasses. *Fla. Agric. Exp. Stn. Tech. Bull.*, **731.** 31 pp.

Freeman, T.E. (1982) Lawn diseases in the sunbelt. *Am. Lawn Applicator*, **3**(1), 4–11.

Haygood, R.A. and Spencer, J.A. (1979) Geographical and seasonal distribution of turf grass diseases in Mississippi. *Plant Dis. Rept.*, **63,** 852–85.

Hebert, T.T. (1971) The perfect stage of *Pyricularia grisea. Phytopathology*, **61,** 83–7.

Hebert, T.T. (1975) *Proc. Sem. Horizontal Resistance to the Blast Disease of Rice.* Centro Internacional de Agricultura Tropical, Cali., Colombia, pp. 161–4.

Kapoor, A.S. and Singh, B.M. (1977) Overwintering of *Pyricularia oryzae* in Himachal Pradesh. *Indian Phytopathol.*, **30,** 213–16.

Malca, M.I. and Owen, J.H. (1957) The gray leaf spot disease of St. Augustinegrass. *Plant Dis. Rept.*, **41**, 871–5.

Narita, S., Iwata T. and Yamanuki, S. (1956) Studies on the host range of *Pyricularia oryzae* Cav., Report 1. *Hokkaido Pref. Agric. Expt. Sta. Rept.*, **7.,** 1–33.

Ou, S.H. (1972) *Rice Diseases.* Commonwealth Mycological Institute., Kew, Surrey, England, 368 pp.

Parris, G.K. (1971) Practical control of two lawn grass diseases without fungicides by removal of mower clippings. *Plant Dis. Rept.*, **55,** 778–9.

Richardson, M.J. (1979) *An Annotated List of Seed-borne Disease*, 3rd edn. Commonwealth Mycological Institute Kew, Surrey, 320 pp.

Sprague, R. (1950) *Diseases of Cereals and Grasses in North America.* Ronald Press, New York, 538 pp.

Subramanian, C.V. (1968) *Pyricularia oryzae. Descriptions of Pathological fungi and Bacteria,* **169.** Commonwealth Mycological Institute, Kew, Surrey, 2 pp.

Trevathan, L.E. (1982a) Pathogenicity on ryegrass and cultural variability of Mississippi isolates of *Pyricularia grisea*. *Plant Dis.*, **66**, 592–4.

Trevathan, L.E. (1982b) Response of ryegrass plant introductions to artificial inoculation with *Pyricularia grisea* under greenhouse conditions. *Plant Dis.* **66**, 696–7.

Wellman, F.L. (1977) *Dictionary of Tropical American Crops and Their Disease*. Scarecrow Press Metuchen, NJ 473 pp.

Yaegashi, H. and Hebert, T.T. (1976) Perithecial development and nuclear behavior in *Pyricularia*. *Phytopathology*, **66**, 122–6.

Yaegashi, H. and Udagawa, S. (1978a) The taxonomical identify of the perfect state of *Pyricularia grisea* and its allies. *Can. J. Bot.*, **56** 180–3.

Yaegashi, H. and Udagawa, S. (1978b) Additional note: the perfect state of *Pyricularia grisea* and its allies. *Can. J. Bot.*, **56**, 2184.

Yamanaka, S. (1982) A consideration of classification of *Pyricularia* spp. isolated from various gramineous plants in Japan. *Ann. Phytopathol. Soc. Japan*, **48**, 245–8 (Japanese).

22 *Powdery mildew*

Throughout the temperate zones, grasses and cereals are subject to powdery mildew, caused by the obligate plant parasitic fungus *Erysiphe graminis* DC. Species commonly used as turf grasses are susceptible to powdery mildew and the disease may be prevalent and sometimes destructive on seedling turf and also on mature stands for forage or seed production (Hardison, 1953; Shildrick, 1976, 1978, Smith, 1978). Damage to established turf is minimal in most instances but the disease may be a major problem on *Poa pratensis* turf when the latter is growing under low light intensity in cool, humid, cloudy climates (Gaskin and Britton, 1962; Beard, 1965; Hardison, 1953; Smith, 1978). In the United States increasing problems with powdery mildew have been linked with the widespread use of the very susceptible *Poa pratensis* cultivar Merion. The problem has been intensified by the high rates of nitrogenous fertilizers employed in the maintenance of this turf (Britton, 1969).

22.1 SYMPTOMS

Signs of the fungus in the form of small superficial patches of white to light-grey mycelium on the leaves and sheaths are the usual first indication of a powdery mildew invasion. The patches rapidly increase in size and number, mainly on the upper leaf surfaces, to form dense wefts of mycelium on which are borne numerous conidiophores and attendant chains of powdery conidia. Heavily colonized leaves assume a grey–white colour and the grey – white powdery appearance becomes the dominant feature of severely mildewed turf (Fig. 35, colour plate section).

As the pathogenic activities of the fungus intensify the host plants develop symptoms. Chlorotic lesions extend beneath the wefts of mycelium and gradually the whole leaf yellows. Host tissues beneath the mycelium become brown and necrotic and leaves may shrivel and die. Leaf necrosis and death of plants results in severe thinning of swards where mildew infections are heavy and prolonged.

22.2 THE CAUSAL FUNGUS

Erysiphe graminis DC ex Merat 1821

Various names have been suggested for the teleomorph (e.g. *Blumeria*) and for the anamorph (e.g. *Oidium*, *Acrosporium*) during the long history of the fungus. Yarwood (1978) has justified the retention of *E. graminis* as the most appropriate taxon through an extensive review of the taxonomic history of the fungus.

The primary, superficial septate mycelium forms patches mainly on upper leaf surfaces, at first white, becoming darker with age. From this mycelium, digitate haustoria penetrate leaf epidermal cells. Short conidiophores with a swollen basal cell arise from the primary mycelium and these bear long chains of ellipsoid conidia, 25–33 × 14–17 μm. The secondary, persistent, mycelium is woolly at first, becoming felted and bristly. Cleistothecia are embedded in the felt, scattered or in clusters, darkening to brown with age. These are globose or depressed, with simple appendages, 135–280 μm diam. and contain 8–25, pedicillate asci, 70–108 × 25–40 μm, each with 8 (or rarely 4) ovoid ascospores, which often fail to mature on the living plant (Brooks, 1953; Kapoor, 1967; Sampson and Western, 1954).

Confined to the host surface, the septate mycelium of this obligate parasite gains sustenance from the host by means of numerous haustoria formed almost exclusively within the epidermal cells. These modified hyphal branches, adapted for nutrient absorption, arise from appressoria and have a central elliptical body bearing two groups of finger-like lobes. Although they form within the cell lumen after direct penetration of the cell wall, an invagination of the host plasma membrane separates the haustorium from the cytoplasm. The processes at this specialized interface which govern the movement of solutes between the host and the fungus have not been interpreted fully, but a tentative scheme has been offered by Bushnell and Gay (1978).

22.3 HOST RANGE

The disease occurs generally on the cereals and grasses with the exception of those in the tribes Maydeae, Anropogoneae, Zoysieae, Paniceae and Oryzeae (Dickson, 1956). The host range of this fungus is large (Hardison, 1944, 1945a,b; Sampson and Western, 1954; Sprague, 1950. It includes all the common cool-season turf grasses and also *Cynodon dactylon* (L.) Pers (bermuda grass).

22.4 DISEASE DEVELOPMENT

The fungus can overwinter as cleistothecia on debris from the previous season's growth and as vegetative mycelium on living host tissue. Direct information on the relative importance of these means of overwintering in mown turf is lacking, but in powdery mildew of cereals it is concluded that the role played by the cleistothecia in overwintering is small. Where the winters are cold, dormant mycelium is proposed as the survival means whereas under less-severe winter conditions (not continuously below freezing) the mycelium can be active and produce conidia during mild periods (Jenkyn and Bainbridge, 1978). Conidia of powdery mildew generally were considered short lived and unsuitable for long-range dispersal. However, Butt (1978) has concluded that this popular view should be revised since conidia may survive for long periods if temperatures remain well below freezing (Hermansen, 1966; Johnston, 1974) and viable mildew spores have been collected at altitudes up to 1500 metres (Hermansen, 1968). Wind transport of spores over long distances is strongly suspected as the source of primary inoculum in some outbreaks of cereal powdery mildew (Hermansen, Trop and Prahm, 1975).

Although ascopores cannot be discounted entirely, it is presumed that conidia are the predominant spore form of primary inoculum in the turf environment. A free moisture requirement for the germination of ascospores is documented (Cherewick, 1944), but the moisture relations for conidial germination have been a subject of some controversy. Free surface moisture is accepted as being detrimental to conidial germination and subsequent mycelial development (Schnathorst, 1965) but the reported relative humidity (RH) values necessary for conidial germination vary widely. Cherewick (1944) showed that germination, infection and mildew development on wheat and barley could proceed normally at very low relative humidity but Grainger (1947) and Clayton (1942) observed that values only marginally below 100% RH precluded conidial germination. Manners and Hossain (1963), also working with cereal mildews, found optimum germination to occur at 100% RH in the absence of free water; however, some conidia germinated in dry air. Schnathorst (1965) designated *E. graminis* as belonging to a group of powdery mildews which germinate best at low moisture stress (high RH), but have a proportion of conidia capable of germinating under high moisture stress (low RH). *E. graminis* conidia, in common with conidia of all powdery mildews, have a high water content, a feature which enables the spores to germinate and infect in the absence of free surface moisture (Yarwood, 1978). Conidia can germinate over a wide temperature range of 0.5–30°C (Cherewick, 1944) and the germ tubes formed can elongate at temperatures between 2 and 30°C. Yarwood *et al.* (1954) combined the results obtained by several investigators and cited 17°C as the average optimum for *E. graminis* development. Manners and Hossain (1963) found 20°C, to be the optimum for germination at 100% RH but showed that the optimum temperature falls as moisture stress increases. Jenkyn and Bainbridge (1978) quote an optimum range of 15–20°C for infection and growth of the fungus with an upper limit of about 30°C.

The literature relating to the development of the mildew colony on the host has been reviewed by Bushnell and Gay (1978). Again the cereal mildews, particularly powdery mildew of barley, are the source of the information. A conidium germinates and forms an appressorium 9–10 hours after arrival on the leaf surface. Penetration of the wall of a host epidermal cell occurs within 1–12 hours and formation of a haustorium begins inside that cell. About 18 hours after inoculation, as the haustorium is starting to produce finger-like lobes, secondary hyphae are initiated from the appressorium which grow on the leaf surface and branch repeatedly. Secondary haustoria are produced 50–70 hours after inoculation and the mildew colony continues to grow, producing new batches of haustoria each night. On the fourth day, conidiophores and conidia start forming and by seven days up to 1000 haustoria and conidiophores may be produced. An individual epidermal cell of barley may have as many as 80 haustoria. One haustorium is matched by 1–1.3 conidiophores each of which produce 8–10 conidia per day over a 4–5 day period. Hirata (1971) states that a colony of barley powdery mildew requires two or three weeks to be fully grown and makes between 4000 and 5000 haustoria (and attendant conidiophores) in that period.

In contrast to other powdery mildews, production of conidia in *E. graminis* is not diurnal but continuous (Hammett and Manners, 1973). Maximum production is favoured by a moderately cool moist environment of about 20°C and 100% relative humidity. Spore release is passive, mechanical disturbance of infected leaves and air turbulence result in the spores becoming air-borne for widespread, local dissemination. Since mature conidia are available for dissemination from the primary mildew colonies within four or five days of their establishment, secondary infections soon establish and the disease can rapidly become severe.

Cleistothecia are reported to form on the mats of mycelium borne on dead and dying leaves in the late autumn and early winter (Britton, 1969; Couch, 1973). The extent to which these fruiting bodies are produced in turf and their importance in overwintering is not documented.

In reviewing the many variable reports relating to host nutrition and powdery mildew in cereals, Jenkyn and Bainbridge (1978) conclude that nitrogen is the element associated with the most consistent effect. Application of nitrogen frequently increases the susceptibility of the cereal host to mildew. Although other elements, including potassium, may tend to increase resistance, the effect is deemed of significance only when application is made to plants that are deficient in that element. This could explain the claim by Lowig (1935) that potassium application increased resistance of turf grasses to powdery mildew. The same author found nitrogen abundance or excess to favour the disease in turf grasses, an observation endorsed by Britton (1969).

Severe outbreaks of powdery mildew are most common, especially on *Poa pratensis*, when a susceptible turf is grown in a shaded location (Beard, 1965; Gaskin and Britton, 1962; Smith, 1978 and others). In such situations where mildew infections are heavy and prolonged, reduced growth of the grass plants, leaf necrosis and, ultimately, the death of many plants results in severe thinning of turf. Variation in light intensity may have a direct influence on the activity of *E. graminis* (Schnathorst, 1965) but the effect of shade in promoting mildew on turf is more likely to involve the predisposition of the host to infection and a tempering of the macro- and microclimate in favour of the pathogen.

22.5 CONTROL

22.5.1 Cultural

As summarized by Couch (1973) the conditions favouring powdery mildew on turf include (a) reduced air circulation, (b) high atmospheric humidity but an absence of free water on the surface of the leaves, (c) low light intensity, (d) an air temperature of 18°C. Any measures to improve air flow and reduce shade will, therefore, lessen the severity of powdery mildew.

Susceptible *P. pratensis* cultivars should not be used where improvements in light intensity cannot be made. More shade-tolerant *Festuca rubra* may be substituted, but regardless of the sward composition the height of cut should be raised, the mowing frequent, and excessive use of nitrogen fertilizers avoided. Where powdery mildew is a recurrent problem the only effective recourse is to adopt a preventive fungicide programme.

22.5.2 Resistant varieties

E. graminis shows marked specialization as to the genera, species and cultivars of grasses it will colonize. The existence of distinct pathogenic *formae speciales* was first shown by Marchal (1902) but the initial concept of *formae* restricted in infection to a single host genus was soon modified as the great genetic diversity existing within these groupings became established (Hiura, 1978). The fungus is heterothallic (Powers and Moseman, 1956) and hybridization between the different *formae* may occur (Hiura-1962) enabling the fungus to generate new physiological races.

Species and cultivar differences in susceptibility of turf grasses to powdery mildew have been reported by (Braverman, 1967, 1986; Braverman and Dolan, 1972; Hanson, 1965; Hardison, 1945a; Muhle and Frauenstein, 1970; Smith, 1978; Wood, 1970 and others). Most of the published information relates to *Poa pratensis*, in which striking cultivar differences in susceptibility have been demonstrated (Nelson, 1982). Some cultivars, for example Merion, contract severe mildew wherever the cultivar is grown. Others have shown inconsistent ratings in field plot evaluations depending on the disease pressure during the period of the trial (Braverman and Dolan, 1972) and on the geographic location of the trial (Braverman and Dolan, 1972; Smith, 1978; Wood, 1970). The distribution and prevalence of different physiological races of the pathogen influences the susceptibility of *P. pratensis* cultivars in a particular location (Hardison, 1945a) and emphasizes the need for regional trials to determine the turf grass cultivars most suited to the local area.

22.5.3 Fungicides

Howard, Rowell and Keil (1951) recommended the application of sulphur dusts or wettable sulphur sprays for the control of powdery mildew on turf grasses. A survey of turf fungicide practice conducted in 1966 showed that sulphur continued in use at that time but dinocap and cycloheximide were the fungicides most commonly recommended in the US (Gould, 1966). These two fungicides are still widely used.

An increasing number of chemicals are known which have good fungicidal activity against powdery mildews and their use has been reviewed by Bent (1978). Benomyl, thiophanate-methyl and ethyl, ethirimol, fenarimol, triadimefon, tridemorph and triforine represent a few of the more recently developed systemic materials which control powdery mildews on grasses.

Tolerance to the benzimidazole group of fungicides by a biotype of *E. graminis* occurring on *Poa pratensis* Merion turf in Michigan has been reported (Vargas, 1972) and biotypes of powdery mildew occurring on barley have shown varying tolerance to ethirimol, tridemorph and triforine (Bent, 1978; Hollomon, 1980; Jenkyn and Bainbridge, 1978; Walmsley-Woodward, Laws and Whittington, 1979; Wolfe and Schwarzbach, 1978). This would suggest that continued reliance on any one systemic fungicide for mildew control in turf is ill-advised and alternate spraying with unrelated fungicides should be adopted. In any event, chemicals specific for mildew control should be supplemented with a fungicide that has activity against '*Helminthosporium*' diseases. The low light-intensity conditions conducive to powdery mildew may also predispose turf to severe melting-out disease even when resistant cultivars are employed (Vargas and Beard, 1980).

22.6 REFERENCES

Beard, J.B. (1965) Factors in the adaptation of turf grass to shade. *Agron. J.*, **57**, 457–9.

Bent, K.J. (1978) Chemical control of powdery mildews. In *The Powdery Mildews* (ed. D.M. Spencer), Academic Press, London, New York and San Francisco, pp. 258–82.

Braverman, S.W. (1967) Disease resistance in cool-season forage, range and turf grasses. *Bot. Rev.*, **33**, 329–78.

Braverman, S.W. (1986) Disease resistance in cool-season, forage, range and turf grasses. *Bot. Rev.*, **52** (1), 1–112.

Braverman, S.W. and Dolan, D.D. (1972) Field resistance to *Erysiphe graminis* on *Poa* and its relation to bluegrass improvement, *Phytopathology*, **35**, 62–71.

Britton, M.P. (1969) Turf grass diseases. In *Turfgrass Science, Am. Soc. Agron.*, No. 14 (eds A.A. Hanson and F.V. Juska), American Society of Agronony, Madison, WI, pp. 309–10.

Brooks, F.T. (1953) *Plant diseases*. 2nd edn, Oxford University Press, London, New York, Toronto, 457pp.

Bushnell, W.R. and and Gay, J. (1978) Accumulations of solutes in relation to the structure and function of haustoria in powdery mildews. In *The Powdery Mildews* (ed. D.M. Spencer), Academic Press, London, New York and San Francisco, pp. 183–231.

Butt, D.J. (1978) Epidemiology of powdery mildews. In *The Powdery Mildews* (eds. D.M. Spencer), Academic Press, London, New York and San Francisco, pp. 51–81.

Cherewick, W.J. (1944) Studies on the biology of *Erysiphe graminis* DC. *Can. J. Res.*, **22**, 52–86.

Clayton, C.N. (1942) The germination of fungus spores in relation to controlled humidity. *Phytopathology*, **32**, 921–43.

Couch, H.B. (1973) *Diseases of Turfgrasses*. Kreiger, Huntington, New York, 348pp.

Dickson, J.C. (1956) *Diseases of Field Crops*. McGraw-Hill, New York, 517 pp.

Gaskin, T.A. and Britton, M.P. (1962) The effect of powdery mildew on the growth of Kentucky bluegrass. *Plant Dis. Rept.*, **46**, 724–5.

Gould, C.J. (1966) Use of fungicides in controlling turf grass diseases. *Golf Superintendent*, **34** (9,10), **35** (1).

Grainger, J. (1947) The ecology of *Erysiphe graminis* DC. *Trans. Br. Mycol. Soc.*, **31**, 54–65.

Hammett, K.R.W. and Manners, H.G. (1973) Conidium liberation in *Erysiphe graminis*. II. Conidial chain and pustule structure. *Trans. Br. Mycol. Soc.*, **61**, 121–33.

Hanson, A.A (1965) *Grass Varieties in the United States*. USDA, Agricultural Handbook No. 170, 102 pp.

Hardison, J.R. (1944) Specialization of pathogenicity in *Erysiphe graminis* on wild and cultivated grasses. *Phytopathology*, **34**, 1–20.

Hardison, J.R. (1945a) Specialization of pathogenicity in *Erysiphe graminis* on *Poa* and its relation to bluegrass improvement. *Phytopathology*, **35**, 62–71.

Hardison, J.R. (1945b) Specialization in *Erysiphe graminis* for pathogenicity on wild and cultivated grasses outside the tribe Hordeae. *Phytopathology*, **35**, 394–405.

Hardison, J.R. (1953) Leaf diseases of range grasses. In *Plant Diseases: The Yearbook of Agriculture*, 1953. USDA, Washington, pp. 253–8.

Hermansen, J.E. (1966) Use of detached leaves to test viability of stored conidia of barley mildew (*Erysiphe graminis* DC.) *Arsskr. K. Vet-Landbohojsk.*, 1966 pp. 61–7.

Hermansen, J.E. (1968). Studies on the spread and survival of cereal rust and mildew diseases in Denmark. *Friesia*, **8,** 1–206.

Hermansen, J.E., Trop, U. and Prahm, L. (1975) Evidence of distant dispersal of live spores of *Erysiphe graminis* f. sp. hordei. *Arsskr. K. Vet-Lanbohojsk,* 1975, pp. 17–30.

Hirata, K. (1971) In *Morphological and Biochemical Events in Plant–parasite Interaction* (eds S. Akai and S. Ouchi), Phytopathological Society of Japan, Tokyo, pp. 207–28.

Hiura, U. (1962) Hybridization between varieties of *Erysiphe graminis, Phytopatchology*, **52,** 664–6.

Hiura, U. (1978) Genetic basis of *formae speciales* in *Erysiphe graminis* DC. In *The Powdery Mildews* (ed. D.M. Spencer), Academic Press, London, New York and San Francisco, pp. 101–28.

Hollomon, D.W. (1980) Resistance of barley powdery mildew to fungicides. *Agric. Devel. Advis. Serv. Quart. Rev.*, **39,** 226–33.

Howard, F.L. Rowell, J.B. and Keil, H.L. (1951) Fungus diseases of turf grasses. *Rhode Is. Expt. Sta. Bull.*, **308,** 56 pp.

Jenkyn, J.F. and Bainbridge, A. (1978) Biology and pathology of cereal powdery mildew. In *The Powdery Mildews* (ed. D.M. Spencer), Academic Press, London, New York and San Francisco. pp. 283–321.

Johnston, H.W. (1974) Overwintering of *Erysiphe graminis* f. sp. *tritici* in maritime grown winter wheat. *Can. Plant Dis. Surv.*, **54,** 71–3.

Kapoor, J.N. (1967) *Erysiphe graminis*. C.M.I. Desc. Path. Fungi and Bacteria. No. 153.

Lowig, E. (1935) On the influence of potassium salts, particularly their anions as well as silicic acid and nitrogen on the mildew resistance of cereals and fodder plants. *Landw. J.*, **81**, 273–335. (German)

Manners, J.G. and Hossain, S.M.M. (1963) Effects of temperature and humidity on conidial germination in *Erysiphe graminis*. *Trans. Br. Mycol. Soc*,, **46**, 225–34.

Marchal, E. (1902) On the specialization of parasitism in *Erisiphe graminis*. *Compt. Rend. Acad. Sci.*, **135**, 210–12. (French).

Muhle, E. von and Frauenstein, K. (1970) Researches on physiological specialization of *Erysiphe graminis*. The important reactions of cultivated fodder grasses against races of powdery mildew in the German Democratic Republic. *Theor. Appl. Genet.*, **40**, 56–8. (German).

Nelson, S.H. (1982) Susceptibility of *Poa* spp. to powdery mildew. *J. Sports Turf Res. Inst.*, **58**, 73–5.

Powers, H.R. and Moseman, J.G. (1956). Heterothallism in *Erysiphe graminis tritici*. *Phytopathology*, **46**, 23.

Sampson, K. and Western, J.G. (1954) *Diseases of British Grasses and Herbage Legumes*. Cambridge University Press, London, 118 pp.

Schnathorst, W.C. (1965) Environmental relationships in the powdery mildews. *Ann. Rev. Phytopathol.*, **3**, 343–66.

Shildrick, J.P. (1976) Evaluation of red fescue cultivars. 1973–6: Part 1. Row plots. *J. Sports Turf Res. Inst.*, **52**, 14–25.

Shildrick, J.P. (1978) Preliminary trials of cultivars of smooth-stalked meadow-grass, 1975–78: Part 1. Row Plots (Trial Y2). *J. Sports Turf. Res. Inst.*, **54**, 53–78.

Smith, J.D. (1978) Powdery mildew on *Poa pratensis* cultivars and selections. *J. Sports Turf Res. Inst.*, 54, 48–53.

Sprague, R. (1950) *Diseases of Cereals and Grasses in North America*. Ronald Press, New York, 538 pp.

Vargas, J.M. Jr (1972) A benzimidazole-resistant strain of *Erysiphe graminis*. *Phytopathology*, **62**, 795. (Abstr.)

Vargas, J.M. Jr and Beard, J.B. (1980) Shade environment-disease relationships of Kentucky bluegrass cultivars. In *Proc. IV Int. Turf. Res. Conf.* (ed. R.W. Sheard), University of Guelph, Guelph, pp. 391–5.

Walmsely-Woodward D.J., Laws, F.A. and Whittington, W.J. (1979) Studies on the tolerance of *Erysiphe graminis* f. sp. *hordei* to systemic fungicides. *Ann. App. Biol.*, **92**, 199–209.

Wolfe, M.S. and Schwarzbach, E. (1978) Barley powdery mildew: evolution in Erupoe. In *The Powdery Mildews* (ed. D.M. Spencer), Academic Press, London, New York and San Francisco, pp. 129–57.

Wood, J.M. (1970) Shade grasses. *Proc. Mass. Ann. Turfgrass Conf. 1970*, pp. 26–35.

Yarwood, C.E. (1978) History and taxonomy of powdery mildews. In *The Powdery Mildews* (ed. D.M. Spencer), Academic Press, London, New York and San Francisco. pp. 1–37.

Yarwood, C.E., Sidky, S. Cohen, M. and Santilli, V. (1954) Temperature relations of powdery mildews. *Hilgardia*, **22**, 603–22.

23 *Pythium foliar blights (spot blight, grease spot, cottony blight)*

The genus *Pythium* Pringsheim embraces a large number of species that are ubiquitous in soil and aquatic environments, world-wide in distribution and have extremely broad and diverse host ranges (Tsao, 1974). Many are pathogenic on members of the Poaceae causing diseases of mild to devastating severity during the establishment period and in mature stands. The manifold depredations of *Pythium* species on cereals and grasses were categorized by Sprague (1950) but ongoing research indicates that there is still much to learn about the aetiology of *Pythium*-incited diseases of turf grasses. More precise information is needed regarding the identity, the preponderance and the relative importance of the individual species and species complexes which are variously involved in seed decay, damping-off, root necrosis and foliage blights of turf grasses. In addition, the environmental conditions most conducive to the pathogenic activities of some *Pythium* species have not been fully elucidated (see also the Section 5.4).

23.1 SYMPTOMS

The common names 'spot blight' (Monteith and Dahl, 1932) or grease spot' (Sharvelle and Likes, 1940) and 'cottony blight' (Fig. 36, colour plate section) and the blight syndrome of turf grasses. Since the symptoms on the infected plants are the same, these names essentially reflect differences in the amount of aerial mycelium that is readily visible on the diseased turf. Profuse production of aerial mycelium is a feature of 'cottony blight' (Fig. 36, colour plate section) and the name was coined by Wells and Robinson (1954) to describe a seedling disease of ryegrasses in the southern US caused by *P. aphanidermatum*. According to Freeman (1980a,b), in such a context, cottony blight may be regarded as an exaggerated form of the damping off/seedling blight syndrome. It was demonstrated later that *P. aphanidermatum* could cause cottony blight on a wide range of turf grasses in both the seedling and the mature stage (Freeman, 1963; Freeman and Horn, 1963) and recently the number of *Pythium* species implicated in cottony blight of mature turf has risen to six. All six *Pythium* species may produce similar signs and symptoms and the identity of the causal agent can be determined with certainty only by microscopic examination of the isolated fungus after growing the latter on special media which promote the production of both sporangia and oospores (Schmitthenner, 1980).

On closely mown turf, pythium blight appears as small circular to irregularly shaped spots of diseased plants which are watersoaked and dark coloured. Occurring usually in clusters and ranging in size from 1 to 10 cm in diameter, the spots fade to a light brown or reddish brown as the leaves dry out and wither. The margins of infection centres may coalesce rapidly to involve large irregular areas of killed turf. Often this progression is in the form of broad streaks and serpentine patterns due to the local dispersal of propagules in surface water flow when irrigation or rainfall is excessive. Blighted individual leaves at first are watersoaked, becoming dark coloured and slimy, soon collasping to give the spots a characteristic border of blackened matted blades intertwined with mycelium (Figs 37, 38, colour plate section). In the early morning or in other periods when ample free moisture and high atmospheric humidity conditions prevail, a copious cottony mycelial growth may invest the diseased tissues especially on higher mown turf. In general, the disease development on higher cut turf mirrors closely that found on low cut areas but the diseased patches may be somewhat larger. Incompletely blighted leaves near the margins of affected patches may exhibit individual lesions. These girdling lesions are uniformly straw-coloured and while resembling dollar spot lesions they lack the reddish, brown margin characteristic of the latter.

23.2 THE CAUSAL FUNGI

Six species of *Pythium* have been implicated in the 'cottony blight' syndrome affecting mature turf (Muse, Schmitthenner and Partyka, 1974; Saladini, 1980; Saladini, Schmitthenner and Larsen, 1983; Schmitthenner, 1980). Five of these species *P. graminicola*, *P. aphanidermatum*, *P. torulosum*, *P. vanterpoolii* and *P. myriotylum* apparently are closely related and belong in the *P. graminicola* complex of widely distributed, highly prevalent pathogens, primarily of grasses. The sixth, *P. ultimum* identifies and dominates a separate complex which is also widely distributed but with an extensive host range that includes grasses (Hendrix and Papa, 1974).

P. ultimum, P. graminicola, P. aphanidermatum, and tentatively *P. myriotylum*, join with *P. irregulare*, *P. arrhenomanes* and possibly others to form a list of *Pythium* species which incite disease in seedling turf (Freeman, 1980) (see pages 32–40.)

In the USA, *P. aphanidermatum* has been reported as the predominant pathogen in the southeastern region causing particularly severe problems in swards of all cool-season species used to overseed dormant turf composed of warm-season grasses (Freeman and Horn, 1963; Gould, 1964). However, a survey of turf grass soils conducted in 1970 throughout the south and southwest documented a rich *Pythium* flora with more than 20 species isolated from 95 of the 150 soils sampled (Hendrix, Campbell and Moncrief, 1970). It is of interest to note that *P. aphanidermatum* ranked a low fourth in order of frequency, mainly associated with *Agrostis* spp. and appearing in only one-third of the samples assayed. The species isolated most consistently was *P. irregulare-debaryanum* followed by *P. torulosum*. *P. ultimum* was listed eighth outranking *P. myriotylum* which was of low frequency. Taking into consideration the common practice of not keying out *Pythium* species beyond the generic level when diagnosing disease problems (Hendrix and Campbell, 1973), a strong possibility exists that species other than *P. aphanidermatum* may also be contributing to the pythium blight syndrome in the southern US (Freeman, 1980b).

A multi-species involvement for pythium blight certainly obtains in the northern United States (Salandini, Schmitthenner and Larsen, 1983). Monteith (1933) made the initial observations on pythium blight incited by *P. aphanidermatum* (as *P. butleri*) and reported the disease to occur under conditions of high temperature and abundant moisture on bentgrass putting greens throughout the middle western and eastern states but less commonly in the northern states (Monteith and Dahl, 1932). The importance of pythium blight as a disease of established turf in the cooler northern regions, particularly that incited by *P. ultimum*, was stressed later by Moore, Couch and Bloom (1963). Three more species, *P. graminicola*, *P. torulosum* and *P. vanterpoolii* were identified in the early 1970s as the cause of a cool-temperature foliar blighting of creeping bentgrass on golf greens in Ohio (Muse, Schmitthenner and Partyka, 1974). As part of a research programme on pythium blight initiated by Ohio State University in 1974, 54 golf courses in the state were surveyed over a two year period and a total of 18 *Pythium* species were recorded from diseased or healthy turf (Saladini, 1978). Pythium blight was seen only under conditions of high temperatures and relative humidity and such diseased turf yielded *P. aphanidermatum* as the most frequent species (Saladini, Schmitthenner and Larsen, 1983). *P. graminicola* (alone or with *P. aphanidermatum*) was also isolated from blighted turf but at a lower frequency. Two species isolated from healthy plants, *P. vanterpoolii* and *P. torulosum* respectively proved to be highly pathogenic and mildly pathogenic on foliage when turf grasses were inoculated with fungi at cool temperatures. Surprisingly, *P. ultimum* was not isolated during this survey but in pathogenicity studies this species (at both high and low temperatures) and also *P. myriotylum* (at high temperatures) readily caused foliage blighting of turf grasses. Comparative tests with the six *Pythium* species to measure their potential both as foliage and as root pathogens revealed a wide diversity of response which was species related and temperature dependent. Three of the group, *P. graminicola*, *P. aphanidermatum* and *P. myriotylum* showed good potential as root pathogens of turf grasses. However, the significance of the root rot and decline which conceivably may occur either alone or in conjunction with foliage blighting awaits clarification (Saladini, 1980). Golf greens on old courses renovated with a growing medium high in sand content seem to be especially susceptible to *Pythium*-induced root problems (Hodges, 1982; Hodges and Coleman, 1985).

Brief descriptions of *P. aphanidermatum*, *P. graminicolum*, *P. myriotylum*, *P. torulosum*, *P. ultimum* and *P. vanterpoolii* are given in Table 5.1 (p. 32–37).

23.3 HOST RANGE

Most cool-season turf grasses, with the exception of *Festuca arundinacea* Schreb. and *Poa trivialis* L. are susceptible to foliage infection by *P. ultimum* (Moore and Couch, 1961). All the cool-season species and one warm-season grass, *Cynodon dactylon* (L.) Pers., are susceptible to foliage infection by *P. aphanidermatum* (Freeman and Horn, 1963). Muse, Schmitthenner and Partyka (1974) demonstrated that *P. graminicola, P. vanterpoolii* and *P. torulosum* all caused foliar blighting of *Agrostis* spp. at 27°C, the severity of the disease decreasing in the order the pathogens are listed. *P. graminicola* also attacked cultivars of *Poa pratensis* L., *Festuca rubra* L. and *Lolium perenne* L. at 27°C but the other two species were not so disposed. McCarter and Littrell (1968) established the pathogenicity of *P. myriotylum* on *Lolium multiflorum* Lam. in greenhouse tests and this finding was confirmed by Freeman (1980b) who also showed *L. perenne* to be a susceptible host for this fungus. Saladini (1980), also reported this fungus as the cause of a foliage blight on various turf grasses (species not listed) in Ohio. Extensive sampling of Ohio golf courses that included turf from greens, tees and fairways (species not listed) indicated *P. aphanidermatum, P. graminicola* and *P. torulosum* as the fungi most commonly associated with foliar blight symptoms (Saladini, Schmitthenner and Larsen, 1983). (See also Table 5.1 on pages 32–37).

23.4 DISEASE DEVELOPMENT

Investigation into the biology of *Pythium* species as they occur in the turf environment has not kept pace with the prolific and productive research endeavours conducted in the past two decades on the same fungi occurring in other crop systems. It seems reasonable to assume that basic information generated by these studies would relate also to turf and, with this mind the relevant published data have been summarized by Schmitthenner (1980).

In general, all *Pythium* mycelia and zoospores and the sporangia of the lobulate sporangial species (including *P. aphanidermatum, P. graminicola*, and *P. myriotylum*, *P. torulosum* and *P. vanterpoolii*) tend to be shortlived in nature. Dormant propagules formed during pathogenic and/or saprobic colonization of aerial and/or root tissues probably constitute the chief means of survival of *Pythium* species in turf.

Oospores (Figs 23.1. and 23.2) are usually the most important and persistent of the propagules formed, but sporangia (Fig. 23.3) (sometimes called conidia or chlamydospores) of the spherical sporangial species (e.g. *P. ultimum*) may equal or surpass oospores in importance as survival structures. Sporangia are subject to soil fungistasis and exhibit an exogenous dormancy which may be overcome by nutrient concentrations of about 5 ppm. Oospores of many *Pythium* species, including *P. graminicola* and *P. ultimum*, exhibit a constitutive dormancy. This property diminishes with time as degradation of the colonized host tissues occurs and the oospores assume an exogenously dormant state. Such oospores are capable of germination under suitable environmental conditions and nutrient concentrations of around 100 ppm. Although not considered constitutively dormant, oospores of *P. aphanidermatum* require aging and liberation from host tissues before they too will germinate maximally. Again exogenous nutrients are necessary and germination is best at high temperatures (Schmitthenner, 1980).

Propagules of *Pythium* species present in the upper soil level and the thatch layer may germinate over a range of environmental conditions to generate myce-

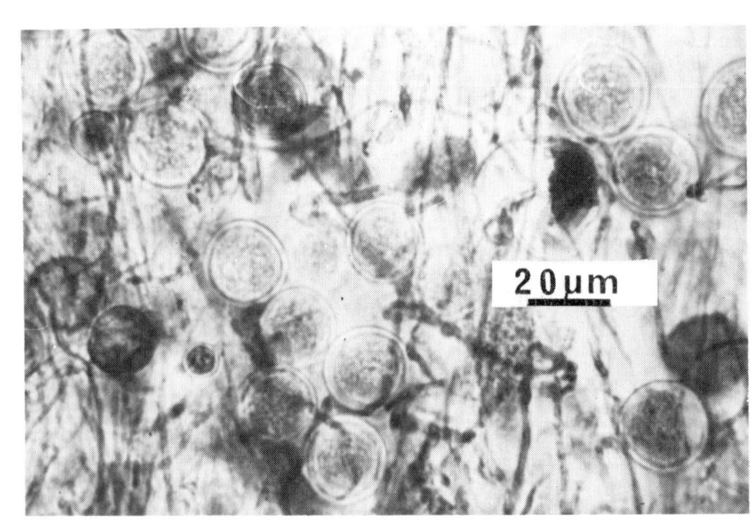

Figure 23.1 Oospores of a *Pythium* sp.

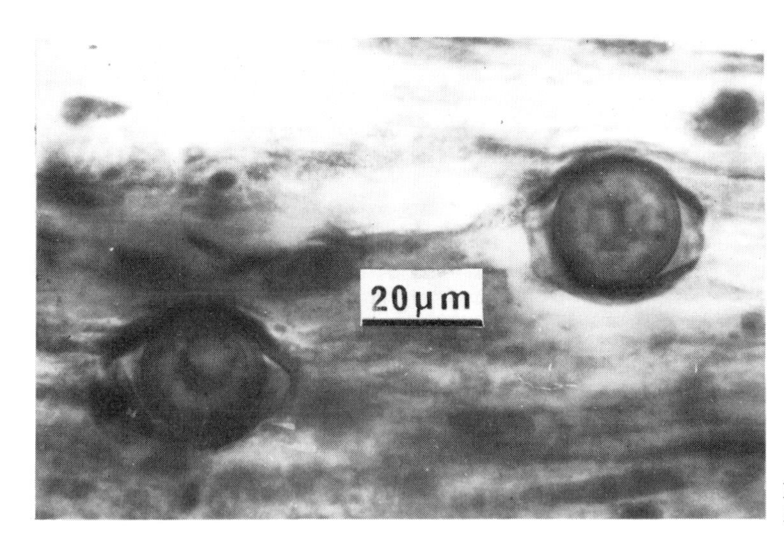

Figure 23.2 Intercalary oospores of a *Pythium* sp.

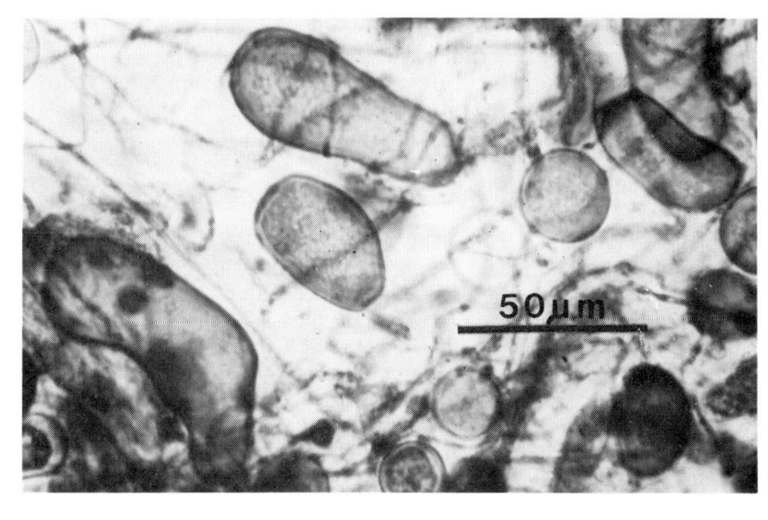

Figure 23.3 *Pythium afertile* Sporangia.

lium directly, or indirectly by means of zoospores. Extensive and rapid mycelial growth may serve to spread the pathogens locally from plant to plant. Wider dissemination may occur with the transport of infected plant material and infested soil or propagules may be dispersed in lateral movement of surface water from irrigation or flash flooding.

Direct penetration of the root hairs of *A. stolonifera* by *P. aphanidermatum* was described by Kraft *et al.* (1967). Encysted zoospores formed appressoria in contact with the cell walls and penetration and death of root hair cells was achieved in less than an hour. Multiple infections resulted in extensive colonization and root necrosis within 24 hours. Little information is available on the means whereby *Pythium* species initiate infection of the aerial parts of turf grasses but direct penetration from zoospores or vegetative mycelium is the most likely means. However, Endo and Colt (1974), in reviewing infection by *Pythium* species point to the paucity of detailed information about most aspects of *Pythium* diseases and warn that it is dangerous to generalize from the limited studies available.

Air and soil temperatures and air and soil water content play major roles in determining the form, extent and severity of the various *Pythium*-incited diseases in turf. Although leaf blighting by some *Pythium* species, e.g. *P. torulosum* and *P. vanterpoolii,* may occur during cool, wet weather with temperatures in the range 13–18°C (Saladini, 1978, 1980), most serious pythium blight outbreaks are associated with warm to hot, rainy weather that ensures prolonged leaf wetness, persisting high humidity and temperatures that do not fall below 20 °C. Disease severity increases markedly as air temperatures rise. *P. aphanidermatum*, *P. graminicola*, *P. myriotylum* and *P. ultimum* are most actively pathogenic on foliage between 30 and 35 °C (Freeman, 1960, 1980a; Hall, 1980; Monteith, 1933; Moore, Couch and Bloom, 1963; Saladini 1978, 1980; Saladini, Schmitthenner and Larsen, 1983). Nutter, Cole and Schein (1983) found that the appearance of warm temperature pythium blight in Pennsylvania was always preceded by a warm day and a warm, moist night. Over three seasons they monitored weather conditions and blight incidence on several golf courses and established that blight could be predicted by selected temperature and relative humidity criteria. The associated weather variables were (1) a maximum daily temperature higher than 30 °C, (2) a following period of at least 14 h when the relative humidity exceeded 90%, and the minimum temperature did not fall below 20 °C. A forecasting day ran from noon to noon. Preventive fungicides (see below) applied soon after the above-mentioned criteria were met substantially reduced any damage from pythium blight.

Hydrogen ion concentration of the soil solution may influence diseases caused by *Pythium* species, generally by indirect means through adverse changes in plant vigour and a consequent increase in predisposition to infection (Hendrix and Campbell, 1973). Using a drip culture technique with container-grown Highland bentgrass (*Agrostis castellano* Bois and Reuter), Moore, Couch and Bloom (1963) investigated the impact of various environmental factors on pythium blight incited by mycelial inoculum of *P. ultimum*. At pH 5.6, plants grown under low balanced fertility (0.1 H) were less prone to disease than those grown under normal balanced fertility (1.0 H). An increase to pH 9.0 did not alter susceptibility at the aforementioned fertility levels but when combined with high balanced fertility (3.0 H) there was a significant increase in disease proneness. Low soil moisture content also increased the susceptibility of the bentgrass to invasion by *P. ultimum*. Population density of this pathogen is able to increase in soil at metric potentials as low as −5 to −8 bars (Lifshitz and Hancock, 1983) suggesting that *P. ultimum* may cause turf disease at times of moisture stress. Field data are not available to confirm this contention. Moore, Couch and Bloom (1963) concluded that of all the environmental factors studied, variations in calcium nutrition, particularly a calcium deficiency, had the most pronounced effect on proneness of Highland bentgrass to pythium blight. This observation was supported by a further investigation demonstrating that the activity of pectolytic enzymes associated with the leaf blighting by *P. ultimum* was greater in leaves of *Agrostis* plants grown under low calcium nutrition (Moore and Couch, 1968).

23.5 CONTROL

Given favourable conditions *Pythium* foliar blights hold the dubious honour of being the fastest developing and most devastating of the fungal diseases affecting turf. As a consequence, where recurring problems with the diseases are common, emphasis should be placed on preventive measures. Cultural practices aimed at minimizing the period of leaf wetness (efficient soil and surface drainage, judicious watering early in the day, provision for free air circulation over the turf), combined with thatch removal and a balanced nutritional programme may lessen incidence of the disease, but reliable protection of turf is afforded only by implementing a preventive fungicide programme. Forecasting the periods of high risk should aid considerably in this endeavour, (see above).

23.5.1 Fungicides

Inorganic mercury compounds were recommended initially for the control of pythium blight (Monteith and Dahl, 1932). In the 1950s captan, dichlone, cycloheximide, several carbamates, organic cadmium and organic mercury compounds were employed with varying success, to be superseded by fenaminosulph, introduced early in the next decade (Wells, 1962). The latter is a specific and effective fungistat against *Pythium* species and related phycomycetes but is somewhat unstable in sunlight (Hills and Leach, 1962), necessitating frequent application as a foliar spray to afford adequate protection. More persistent control was obtained by the use of chloroneb (Littrell, Gay and Wells, 1969; Freeman and Horn, 1967; Wells, 1969) or by using ethazole (Freeman and Meyers, 1969).

Fenaminosulph, chloroneb and ethazole are among the fungicides still employed to control the disease (Freeman, 1980a; Raabe, 1974). All are primarily protectant in function requiring repeated and frequent spraying to prevent the appearance of pythium blight. The recent development of systemic fungicides, which are persistent and highly effective against *Pythium* species, marked a major advance in the struggle to control pythium blight. Several materials, including pyroxyclor, propamocarb and metalaxyl have given excellent results in laboratory and field evaluations (Ashbaugh and Larsen, 1984; Couch, 1981; McCoy 1975; Sanders 1982; Sanders, Warren and Cole, 1976; Sanders, Houser and Cole, 1983; Sanders, Nutter and Cole, 1980; Sanders *et al.*, 1978; Tillman and Ferguson, 1980; Windham and Spencer, 1981). Toxicological problems prompted the withdrawal of pyroxyclor but marketing and successful use of the other materials are proceeding. Fosetyl-aluminium provides a novel approach to controlling the disease. Greenhouse and field trials have demonstrated that when *Lolium perenne* plants inoculated with *P. aphanidermatum* were treated with this chemical, excellent symptom suppression occurred. No direct fungitoxicity was observed by fosetyl-aluminium to the eight pathogenic species of *Pythium* utilized in the tests and it was concluded that the chemical control may result from elicitation of antifungal responses in the host (Sanders, Houser and Cole, 1983).

Effective fungicides should be alternated or combined in any pythium blight programme since field tolerance and laboratory-induced tolerance by some phycomycetes (including some *Pythium* species) to metalaxyl has already been reported (Bruin and Edgington, 1982; Sanders, 1984; Sanders *et al.*, 1985). An integrated approach to controlling pythium blight using low rates of fungicide (metalaxyl) in conjunction with the mycoparasite *Trichoderma hamatum* is being explored (Rasmussen-Dykes and Brown, 1982).

23.6 REFERENCES

Ashbaugh, F.M. and Larsen, P.O. (1984) Comparison of fungicides for control of pythium blight on *Festuca rubra*. *Phytopathology*, **74**, 812. (Abstr.)

Bruin, G.C.A. and Edgington, L.V. (1982) Induction of fungal resistance to metalaxyl by ultraviolet irradiation. *Phytopathology*, **72**, 476–80.

Burr, T.J. and Stanghellini, M.E. (1973) Propagule nature and density of *Pythium aphanidermatum* in field soil. *Phytopathology*, **63**, 1499–501.

Couch, H.B. (1981) Comparative effectiveness of systemic and non-systemic fungicides in the control of Pythium blight of bentgrass. *Phytopathology*, **71**, 765. (Abstr.)

Endo, R.M. (1961) Turfgrass diseases in southern California. *Plant Dis. Rept.*, **45**, 869–73.

Endo, R.M. and Colt, W.M. (1974) Anatomy, cytology and physiology of infection by *Pythium*. In *Proc. Am. Phytopathol. Soc. Minnesota*, **1**, 215–23.

Freeman, T.E. (1960) Effects of temperature on cottony blight of ryegrass. *Phytopathology*, **50**, 575. (Abstr.)

Freeman, T.E. (1963) Age of ryegrass in relation to damage by *Pythium aphanidermatum*. *Plant Dis. Rept.*, **47**, 844.

Freeman, T.E. (1980a) Pythium blight on overseeded grasses. *Golf Course Mmt*, **48**, 20–2.

Freeman, T.E. (1980b) Seedling diseases of turf grasses. In *Advances in Turfgrass Pathology*. (eds P.O. Larsen and B.G. Joyner), Harcourt, Brace, Jovanovich, Duluth, MN, pp. 41–4.

Freeman, T.E. and Horn, G.C. (1963) Reaction of turf grasses to attack by *Pythium aphanidermatum* (Edson) Fitzpatrick. *Plant Dis. Rept*, **47**, 425–7.

Freeman, T.E. and Horn, G.C. (1967) Pythium fungicides. *Golf Superintendent*, **35**(2), 58,60.

Freeman, T.E. and Meyers, H.G. (1969) Control of Pythium blight. *Golf Superintendent*, **37**(5), 24–5, 44–5.

Gould C.J. (1964) Turf grass disease problems in North America. *Golf Course Rept.*, **32**(5), 36,38, 40–43, 46, 48, 50, 52, 54.

Hall, T.J. (1980) Survival of *Pythium aphanidermatum* in golf course turfs. *Plant Dis.*, **64**, 1100–3.

Hendrix, F.F., Jr and Campbell, W.A. (1973) Pythiums as pathogens. *Ann. Rev. Phytopathol.*, **11**, 77–98.

Hendrix, F.F., Jr and Papa, K.E. (1974) Taxonomy and genetics of *Pythium*. *Proc. Am. Phytopathol. Soc.*, **1**, 200–7.

Hendrix, F.F., Jr Campbell, W.A. and Moncrief, J.B. (1970) *Pythium* species associated with golf turf grasses in the south and southeast. *Plant Dis. Rept.*, **54**, 419–21.

Hills, F.J. and Leach, L.D. (1962) Photochemical decomposition and biological activity of *p*-dimethyl amino benzene diazo sodium sulfonate (Dexon). *Phytopathology*, **52**, 51–6.

Hodges, C.F. (1982) Root dysfunction of creeping bentgrass. *Grounds Maintenance*, **17**, 32–4.

Hodges, C.F. and Coleman, L.W. (1985) *Pythium*-induced root dysfunction of secondary roots of *Agrostis palustris*. *Plant Dis.*, **69**, 336–40.

Kraft, J.M., Endo, R.M. and Erwin, D.C. (1967) Infection of primary roots of bentgrass by zoospores of *Pythium aphanidermatum*. *Phytopathology*, **57**(1), 86–90.

Krywienczyk, J. and Dorworth, C.E. (1980) Serological relationships of some fungi of the genus *Pythium*. *Can. J. Bot.*, **58**, 1412–17.

Lifshitz, R. and Hancock, J.G. (1983) Saprophytic development of *Pythium ultimum* in soil as a function of water metric potential and temperature. *Phytopathology*, **73**, 257–61.

Littrell, R.H., Gay, J.D. and Wells, H.D. (1969) Chloroneb fungicide for control of *Pythium aphanidermatum* on several crop plants. *Plant Dis. Rept.*, **53**, 913–15.

McCarter, S.M. and Littrell, R.H. (1968) Pathogenicity of *Pythium myriotylum* to several grass and vegetable crops. *Plant Dis. Rept.*, **52**, 179–83.

McCoy, R.E. (1975) Pythium control in ryegrass in laboratory and greenhouse tests. *Proc. Florida Turfgrass Mgmt. Conf.*, **23**, 107–11.

Middleton, J.T. (1943) The taxonomy, host range and geographic distribution of the genus *Pythium*. *Mem. Torrey Bot. Club*, **20**, 1–171.

Monteith, J. Jr (1933) A *Pythium* disease of turf. *Phytopathology*, **23**, 23–4. (Abstr.)

Monteith, J. Jr and Dahl, A.S. (1932) Turf diseases and their control. *US Golf Assoc. Green Sect. Bull.*, **12**(4), 85–187.

Moore, L.D. and Couch, H.B. (1961) *Pythium ultimum* and *Helminthosporium vagans* as foliar pathogens of Gramineae. *Plant Dis. Rept.*, **45,** 616–19.

Moore, L.D. and Couch, H.B. (1963) Influence of calcium nutrition on pectolytic and cellulytic enzyme activity of extracts of Highland bentgrass foliage blighted by *Pythium ultimum. Phytopathology*, **58,** 833–8.

Moore, L.D., Couch, H.B. and Bloom, J.R. (1963) Influence of environment on diseases of turf grasses. III. Effect of nutrition, pH, soil temperature, air temperature and soil moisture on Pythium blight of Highland bentgrass. *Phytopathology*, **53,** 53–7.

Muse, R.R., Schmitthenner A.F. and Partyka, R.E. (1974) *Pythium* spp. associated with foliar blighting of creeping bentgrass. *Phytopathology*, **64,** 252–3.

Nutter, F.W., Cole, H. Jr and Schein, R.D. (1983) Disease forecasting system for warm weather Pythium blight of turf grass. *Plant Dis.*, **67,** 1126–8.

Raabe, R.D. (1974) Pythium diseases in turf grass. In *Grass Pests and Golf Course Management. Proc. Calif. Golf Course Superintendents Inst.*, Univ. of California, Davis, 136 pp.

Rasmussen-Dykes, C. and Brown, W.M. Jr (1982) Integrated control of Pythium blight on turf using metalaxyl and *Trichoderma hamatum. Phytopathology*, **72,** 974. (Abstr.)

Saladini, J.L. (1978) Pythium blight. *Golf Superintendent*, **46**(9), 34–5.

Saladini, J.L. (1980) Cool versus warm season Pythium blight and other related *Pythium* problems. In *Advances in Turfgrass Pathology* (eds P.O. Larsen and B.G. Joyner), Harcourt, Brace and Jovanovich Duluth, MN, pp. 37–9.

Saladini, J.L., Schmitthenner, A.F. and Larsen, P.O. (1983) Prevalence of *Pythium* species associated with cottony-blighted and healthy turfgrasses in Ohio. *Plant Dis.*, **67,** 517–19.

Sanders, P.L. (1982) New fungicides against *Pythium. Plant Dis.*, **66,** 265.

Sanders, P.L. (1984) Failure of metalaxyl to control pythium blight on turfgrass in Pennsylvania. *Plant Dis.*, **68,** 776–7.

Sanders, P.L., Warren, C.G. and Cole, H. Jr (1976) Control of Pythium blight on Penncross bentgrass with pyroxychlor. *Phytopathology*, **66,** 1033–7.

Sanders, P.L., Burpee, L.L., Cole, H. Jr and Duich, J.M. (1978) Control of Pythium blight of turf grass with CGA-48988 (metalaxyl). *Plant Dis. Rept.*, **62,** 663–7.

Sanders, P.L., Nutter, F.W. and Cole, H. Jr (1980) Control of Pythium blight of turf grass with SN66752 (propamocarb). *Phytopathology*, **70,** 468. (Abstr.)

Sanders, P.L., Houser, W.J. and Cole, H. Jr (1983) Control of *Pythium* spp. and Pythium blight of turf grass with fosetyl aluminum. *Plant Dis.*, **67,** 1382–3.

Sanders, P.L., Houser, W.J., Parish, P.J. and Cole, H. Jr (1985) Reduced rate fungicide mixtures to delay fungicide resistance and to control selected turf grass diseases. *Plant Dis.*, **69,** 939–43.

Schmitthenner, A.F. (1980) *Pythium* species. Isolation, biology and identification. In *Advances in Turfgrass Pathology* (eds P.O. Larsen and B.G. Joyner), Harcourt, Brace, Jovanovich, Duluth, MN, pp. 33–6.

Sharvelle, E.G. and Likes, D.F. (1949) 'Grease spot' suggested as name for new disease. *Midwest Turf*, **3**(4), 3.

Sprague, R. (1950) *Diseases of Cereal and Grasses in North America*. Ronald Press, New York, 538 pp.

Tillman, R.W. and Ferguson, M.W. (1980) Toxicity of pyroxychlor to *Pythium aphanidermatum. Phytopathology*, **70,** 441–4.

Tsao, P.H. (1974) Introductory remarks. Symposium on the genus *Pythium. Proc. Am. Phytopathol. Soc.*, **1,** 200.

Wells, H.D. (1962) Cottony blight disease. *Golf Course Rept.*, **30**(5), 33–6.

Wells, H.D. (1969) Chloroneb, a foliage fungicide for control of cottony blight of ryegrass. *Plant Dis. Rept.*, **53,** 528–9.

Wells, H.D. and Robinson, B.P. (1954) Cottony blight of ryegrass caused by *Pythium aphanidermatum. Phytopathology*, **44,** 509–10. (Abstr.)

Windham, A.S. and Spencer, J.A. (1981) Identity, pathogencity and control of *Pythium* spp. from selected turf plots and golf greens. *Phytopatology*, **71,** 913. (Abstr.)

24 *Red thread and pink patch diseases*

The long-held concept of a single incitant for red thread disease, also known as pink patch or corticium disease, was refuted recently with the discovery that several macroscopically similar basidiomycetous fungi may generate red thread/pink patch symptoms in turf (Cahill, O'Neill and Dernoeden, 1982; Kaplan and Jackson, 1982, 1983; O'Neill, 1983; Stalpers and Loerakker, 1982). *Corticium fuciforme* (Dennis) Wakef., designated by Wakefield in 1916 as the causal agent is no longer an acceptable taxon and the fungi involved are currently assigned to two newly erected genera, *Laetisaria* and *Limonomyces* (Burdsall 1979; Stalpers and Loerakker, 1982). Retention of the common name red thread disease for symptoms where *Laetisaria fuciformis* is the incitant, and pink patch disease where *Limonomyces* species are implicated, has been proposed (Kaplan and Jackson, 1983).

Components of this fungal complex were first observed in 1854 causing disease on ryegrass pasture in Australia (Berkeley, 1873), later occurring on turf and agricultural grasses in England (Cooke, 1880) but remaining unknown in the United States until 1932 (Erwin, 1941). Red thread disease is now recognized to be of widespread distribution on sports turf and agricultural grasses in the cooler humid regions of Australia and New Zealand, North America, northern Europe and the United Kingdom (Bahuon, 1986; Bennett, 1935; Britton, 1969; Couch, 1973, 1983; Erwin, 1941; Gould, 1964; Hims, Dickinson and Fletcher, 1984; Julich, 1976; McAlpine, 1906; Smith, 1965; Sprague, 1950). Pink patch disease, has been recorded to date in the Netherlands (Stalphers and Loerakker, 1982), the United Kingdom, France and the eastern United States (Cahill, O'Neill and Dernoeden, 1982; Kaplan and Jackson 1982, 1983; O'Neill, 1983). It probably has a distribution similar to that of red thread disease. The disease was frequently noted on *Lolium perenne* L. in a survey of amenity turf south of Paris in October 1985 (Smith, unpublished) causing considerable injury. The diseases are most common on turf of low fertility during periods of cool moist weather.

24.1 SYMPTOMS

24.1.1 Red thread disease

Water-soaked lesions develop on the leaves when hyphae from pink, stranded investing mycelia penetrate the stomata or enter through mowing wounds. As the lesions enlarge to involve most or all of the laminae, such blighted foliage assumes a tan to light-straw or bleached appearance and the leaves shrivel, usually from the tip down. Affected plants occur in small patches initially 2–5 cm or so in diameter, not unlike dollar spot and copper spot symptoms, but as the patches increase in size their margins take on an irregular outline. Given weather and soil conditions favourable for vigorous leaf production, an attack may be transient, but when grass growth is slow the fungus may kill back leaves and shoots in patches up to 35 cm or more in diameter.

The most noticeable feature on infected turf is the presence of pink to red, gelatinous, stranded hyphae investing the blighted leaves and extending from them in distinctive simple or branched outgrowths. These threadlike, needlelike or antlerlike stromata,* formed from conglutinated parallel hyphae, extend up to 1 cm or more, becoming brittle and darkening in colour to deep red on drying (Fig. 39, colour plate section). Under conditions of ample moisture an abundant weblike mycelium may bind the plants together in patches that assume an overall reddish colour and are slimy to the touch. Close scrutiny may reveal the pink, mealy, effuse basidiocarps investing the necrotic blades and extending onto some of the stromata. With

*Interpretations differ as to whether these structures should be termed stromata or sclerotia.

limited moisture, mycelial development is reduced but prominent cottony aggregates (flocks) of pink hyphae may form sporadically on the affected turf that on drying fragment to form arthroconidia. *Clamp connections are not present on the mycelium of this fungus.*

24.1.2 Pink patch disease

The overall symptom pattern of pink patch disease closely resembles that of red thread disease. Since both ailments can occur concurrently in a turf stand and since mixed infections are possible within the same patch, the past failure to separate the two diseases using macroscopic features is understandable.

Stalpers and Loerakker (1982) first described the symptoms caused by *Limonomyces roseipellis*, confined, in Holland, mainly to perennial ryegrass turf. During wet, warm conditions invasive mycelium, free growing or from adjoining, previously attacked leaves, forms a thin pink, later thicker and ceraceous membrane, largely following the margin of the host leaves. Entry is followed by the development of a chlorotic lesion that enlarges to encompass the entire width of the blade resulting in necrosis of the terminal part. On turf that is mown frequently and the clippings removed, patches seldom reach 60 cm in diameter. A low proportion of the grass leaves are damaged and continued growth allows the patches to retain much of their green colour. Under these conditions basidiocarps rarely occur.

In relatively long, infrequently mown turf, damage may be more severe as all parts of the plant above ground are killed and pink, effuse basidiocarps are formed in abundance on dead leaves and stems. The diseased plants assume an overall pinkish tinge and an infection may progress to affect large areas of a turf stand. Arthroconidial flocks and stroma are absent but small, red, effuse ceraceous bodies may occur on necrotic tissues. *Clamp connections are present throughout the mycelium of the incitant fungus.*

Pink patch signs and symptoms (Fig. 40, colour plate section) similar to those described above have been recognized during cool moist weather in the United Kingdom and in North America. The disease has been observed on turf composed of bluegrasses, bentgrasses, fescues and perennial ryegrass (Cahill, O'Neill and Dernoeden, 1982; Kaplan and Jackson, 1983; O'Neill, 1983, personal communication; Dernoeden, 1983, personal communication) and on bermuda grass (Filer, 1966; Freeman, 1982, personal communication). In most cases, sporadic mild patch symptoms typify the disease. Damage is largely superficial; lesion development follows direct penetration by hyphae from pale pink, stranded, leaf-investing mycelia. Older senescing laminae, sheaths and dead leaves become bleached and matted together but colonization of actively growing leaves is usually limited. On dormant or slow-growing turf, more severe damage can occur as tillers are killed back to the crown and some plants succumb to the infection. A pink colouration develops at the severed tips of the stubble and a thin pink hymenial layer is often discernible encrusting the upper portion. Senescent sheath material on adjacent living plants may also support basidiocarps in appreciable quantities.

24.2 THE CAUSAL FUNGI

24.2.1 Red thread disease

Laetisaria fuciformis (McAlp.) Burdsall (1979); synonym: *Corticium fuciforme* (Berk.) Wakef. (1916). For detailed treatment of the synonomy and descriptions of this fungus see Burdsall (1979), Julich (1976), Smith (1965) and Stalpers and Loerakker (1982).

Sterile mycelium varies in colour from almost hyaline to bright pink, orange or red, simple or branched, or in flocks, threads or, branched like antlers, or as needle-like stromata up to 25 mm or more in length, often at leaf tips or connecting leaf blades (Fig. 39, colour plate section). The brittle flocks consist of masses of hyaline, irregularly-shaped arthroconidia (Fig. 24.1) mostly $10–47 \times 5–17\,\mu m$.

Monomitic-type hyphae, multinucleate, without clamp connections, and sometimes stranded, give rise to the hyaline to pale pink, resupinate hymenium which bears basidia with typically four sterigmata and pip-shaped, apiculate basidiospores, $12.5–20 \times 5–6\,\mu m$ (For more detailed descriptions see Burdsall (1979) and Stalpers and Loerakker (1982).)

The fungus may be cultured readily on many artificial media but the rate of growth, the habit and the colour intensity of the cultures can vary widely depending on the medium used (Erwin, 1941) and on the source of the inoculum (Bennett, 1935; Smith, 1954). If taken from the mycelial web found on infected plants the culture tends to produce aerial mycelium but if taken from stranded mycelium or outgrowths the culture tends to be stranded with little aerial mycelium. Stalpers and Loerakker (1982) described two distinct strains of *L. fuciformis* that occur on 2% malt agar at room temperature again mainly depending on the source of inoculum. A fast-growing strain (70 mm radial increase in 12–14 days) with stranded hyphae was obtained from stromata or basidiocarps, while slow-growing, floccose cultures (10–25 mm in 14 days) were derived

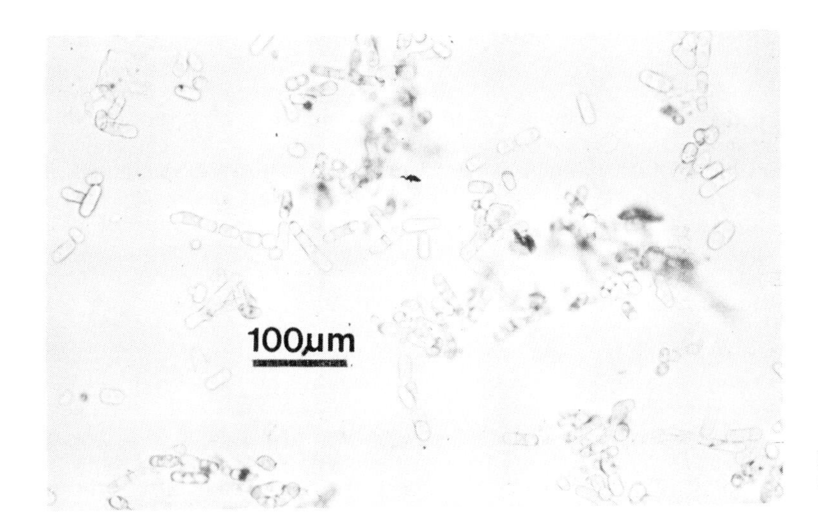

Figure 24.1 *Laetisaria fuciformis* arthrospores.

Figure 24.2 *Laetisaria fuciformis* basidia and basidiospores.

from brittle flocks of arthrospores. Reversion of one form to the other could occur.

Optimum growth for isolates of *L. fuciformis* may occur over the range 15.5 to 25 °C (Bennett, 1935) but most reports place the optimum nearer 20 °C (Erwin, 1941; Kaplan and Jackson, 1983; Stalpers and Loerakker, 1982). Endo (1963) cited 21°C as ideal for growth in culture and found this temperature most effective for inducing foliage infection on artificially inoculated bentgrass seedlings. Kaplan and Jackson (1983) successfully inoculated perennial ryegrass and red fescue plants in moist chambers maintained at 15 °C. Growth ceases at temperatures between 28° and 30 °C, lower than the maximum for *L. roseipellis*, but the minimum temperature for growth (approximately 4 °C) is similar for both the red thread and the pink patch fungus (Bennett, 1935; Erwin, 1941; Kaplan and Jackson, 1983; Stalpers and Loerakker, 1982). Arsvoll (1975) reported '*Corticium fuciforme*' in Norway actively growing at 3 °C and capable of growth at 1 °C. A recent study by Bahuon (1986) gives the growth temperatures for *L. fuciformis* as 14 to 18°C optimum and 30 °C maximum. She found the fungus gained entry into the grass plant through wounds or stomata and invasion of tissues was intercellular.

In culture, growth of *L. fuciformis* can occur over the range pH 3.5–7.5 (Bennett, 1935; Erwin, 1941) encompassing the range of soil pH usually encountered in managed turf.

24.2.2 Pink patch disease

(a) *Limonomyces roseipellis* Stalpers and Loerakker (1982); synonyms: *Corticium fuciforme* (Berk.) Wakefield (1916); *Athelia fuciformis* Wakef. (Burdsall, 1979).

The mycelium is film-like, hyaline, waxy, attached firmly to the grass leaves when moist, becoming detached when dry, usually pink to red, occasionally sterile, with pale to red tufts, joining leaf blades together (Fig. 40, colour plate section). *The hyphae of L. roseipellis, unlike those of L. fuciforme are binucleate and clamp connections are present. There are no arthroconidia as in L. fuciforme.*

Monomitic-type hyphae give rise to resupinate basidiocarps, waxy when fresh, pruinose, when dry, pale cream to red. The hyaline, basidiospores are borne on four, stout sterigmata and are 8–14 × 4.5–6.5 µm, distinctly pip-shaped. (For a detailed description of the fungus see Stalpers and Loerakker, 1982.)

L. roseipellis can be cultured on many artificial media. Initial radial growth of hypae is appressed and submerged, but as the colony approaches the perimeter of the Petri dish, a zone of pink, cottony, aerial mycelium develops, often growing up the sides and onto the Petri dish lid. Submerged, red ceraceous bodies, irregular in shape and 10 mm or more across, may develop in older cultures. The growth rate of the fungus in culture exceeds that of *L. fuciformis*, optimum growth occurs between 20° and 23°C but growth ceases at temperatures beyond 31°C and below 4°C (Cahill, O'Neill and Dernoeden, 1982; Kaplan and Jackson, 1983; Stalpers and Loerakker, 1982).

(b) Other basidiomycetous fungi with pink-clamped mycelium including *Limonomyces culmigenus* Stalpers and Loerakker (1982); synonyms: *Exobasidiellum culmigenum*. Webster and Reid (Reid, 1969); *Galzinia culmigena* (Webster and Reid), Johri and Bandoni (Bandoni and Johri, 1975).

24.3 HOST RANGE

L. fuciformis is pathogenic on most of the common cool-season turf and agricultural grasses (Bennett, 1935; Erwin, 1941; Julich, 1976; Smith, 1954; Stalpers and Loerakker, 1982). *Lolium perenne* L. and *Fustuca rubra* L. are particularly vulnerable to attack but widespread infection of other turf species may occur when cool moist weather conditions obtain. Couch (1983) found considerable variation in susceptibility in 14 cultivars of *Lolium perenne*.

L. roseipellis has been recorded on *Lolium perenne, Festuca rubra, Poa pratensis* L., *Poa annua* L., *Agrostis* spp. and *Cynodon dactylon* (L.) Pers. (Couch, 1983); Hims, Dickinson and Fletcher, 1984; Kaplan and Jackson, 1983; O'Neill, 1983; Stalpers and Loerakker, 1982). The host range of this fungus (and related pink, clamped basidiomycetes) may extend as more discriminating diagnostic procedures are used to determine the causal agents in red thread/pink patch symptom complexes.

24.4 DISEASE DEVELOPMENT

McAlpine (1906) reported that *L. fuciformis* may be seed-borne and that the presence of the hymenium on seeds of *L. perenne* in Australia was not uncommon. Both *L. fuciformis* and *L. roseipellis* readily produce basidiocarps on affected turf but the significance of basidiospores in the incidence of either disease remains obscure. Their role in local and long-distance dispersal probably matches in importance any vegetative means of dissemination. The latter is well documented for *L. fuciformis*. McAlpine (1906) observed that the pink stromata functioned in the local spread of red thread disease. In dry conditions stromata shrink, becoming brittle, easily detached, readily fragmented and dispersed. The stromata are resistant structures; Bennett (1935) showed that exposure to high (32°C) or low (−20°C) temperatures would not kill them and Libbey (1938) found that, if dry, they could survive for two years. Erwin (1941) reported that they would remain viable for at least eighteen months when stored at 25°C. In culture they may germinate from any point but generally do so from the ends. Small pink, ceraceous bodies are produced by *L. roseipellis* which may function as resting structures but probably these do not function as importantly in vegetative propagation as do the stromata of *L. fuciformis* (Stalpers and Loerakker, 1982). Both fungi may survive for long periods as resting mycelium in infected plant remains and can be dispersed through transport of such debris. Arthrospores developed in pink cottony tufts (flocks) also may function in local and long distance dispersal of *L. fuciformis*.

Under conditions of moderate temperature and ample moisture, resting propagules of either fungus may generate mycelia that grow up to invest and infect the foliage. Growth progresses radially from an infection centre as the hyphal strands bridge from leaf to leaf, leading to visible patch symptoms. Bennett (1935) considered that the limiting factor for growth of *L. fuciformis* during the optimum temperature period in summer was the amount of moisture. In dry periods the disease makes little progress but symptoms of red thread may become more conspicuous.

In regions where moderate summer temperatures and mild winters prevail, and there is sufficient moisture, red thread disease may occur year round, as for example, in the Pacific northwest, west of the Cascade Mountains (Gould, 1966). More commonly the disease appears in the spring, increases into the early summer, subsides with the higher summer temperatures and recurs in late summer and autumn. Often major epidemics occur late in the year, the fungus making good growth at the expense of near dormant or dormant turf grasses. *L. perenne* appears particularly vulnerable to these late infections. Activity of the fungus may continue throughout the winter on various turf grasses under mild, wet conditions or when deep snow persists on unfrozen turf. Similar environmental conditions appear conducive to pink patch disease with *L. perenne, P. pratensis* and *Agrostis* spp. the common hosts. A. major outbreak of pink patch disease occurred on *C. dactylon* in the southern United States in January 1982 (Freeman, 1982, personal communication) during a two-week period of cool, overcast weather. Filer (1966) recorded what was probably a mixed infection on the same host during late autumn and winter of 1965/1966 in Mississippi.

Severe cases of both diseases are often associated with turf of low fertility, particularly with respect to nitrogen. Application of nitrogenous fertilizers will alleviate red thread symptoms Bahuon, 1986; Cahill *et al.*, 1983; Erwin, 1941; Goss and Gould, 1971; Gould, Miller and Goss, 1967; Smith, 1954). The influence of potassium and phosphorus was also investigated (Cahill *et al.*, 1983; Goss and Gould, 1971) but it was concluded that these elements were secondary to nitrogen in reducing the incidence of red thread disease. Muse and Couch (1965) showed in growth-chamber experiments that calcium-deficient red fescue plants grown in silica sand were more susceptible to red thread. However, this finding has not been substantiated in field studies nor has the application of lime been shown to reduce the disease (Bennett, 1935; Cahill *et al.*, 1983; Erwin, 1941). In their studies, Muse and Couch (1965) also found that, surprisingly, nitrogen did not influence the susceptibility of *F. rubra* to the red thread fungus. They suggested that the observed field benefits of high nitrogen fertility are due to enhanced recovery in growth of infected plants. Results of field experiments by Goss and Gould (1971) and Cahill *et al.* (1983) support this theory. On the other hand, Hims, Dickinson and Fletcher (1984) found no significant benefit of increasing soil nitrogen in effectively eliminating red thread disease (in reclaimed heathland pasture) compared with fungicide treatments.

24.5 THE CONTROL OF RED THREAD AND PINK PATCH DISEASES

24.5.1 Cultural

Both diseases usually appear on turf in a low state of fertility particularly with regard to nitrogen. By increasing the vigour of the turf through the judicious application of fertilizers, particularly nitrogenous fertilizers, it is possible to minimize the effects and favour recovery from attacks of red thread and pink patch diseases.

Eliminating symptoms by manipulating nitrogenous fertilizer usage is not always desirable or indeed feasible. Low maintenance fine-leaved *Festuca* spp. turf, though very susceptible to both diseases may deteriorate rapidly under high fertility conditions due to invasion by undesirable competing grasses. Similarly, either applications of nitrogen late in the growing season, when sod temperatures are low, may not stimulate turf growth sufficiently to surmount a disease attack, or the benefits gained against red thread and pink patch may be outweighed in the predisposition of the stimulated turf to other cool-season diseases, such as the snow moulds. It is better to rely on timely application of a suitable fungicide coupled with judicious nitrogenous fertilizer use.

24.5.2 Fungicides

Bennett (1935) reported that malachite green/Bordeaux mixture was effective in the control of red thread disease in plot work and on bowling greens. Smith (1953) found that three applications of this material applied at ten-day intervals reduced an infection from 28% of the area infected to zero. Four applications of fungicides based on mercurous/mercuric chloride mixtures or fungicides based on thiram, cadmium and a phenyl mercury compound respectively reduced heavy infections to zero in four weeks (Smith, 1953). One application of a wettable powder based on mercurous/mercuric chlorides gave persistent control over a five-week period and the efficiency of control increased with the dosage of the material (Smith, 1954). Erwin (1941) found that stromata of *L. fuciformis* were killed after 72 hours exposure to mercury vapour without injury to grass blades and demonstrated the effectiveness of some mercurial fungicides against red thread in field trials. Smith (1957) found that a mercurous/mercuric chloride fungicide, cadmium chloride with urea, malachite green/Bordeaux mixture or cycloheximide applied at approximately monthly intervals functioned as efficient preventives.

In a 1966 survey of turf fungicide use in the USA, cadmium and mercury fungicides received the best rating for control of red thread disease (Gould, 1966–67). Cadmium compounds afforded the longest residual control (Gould, Miller and Goss, 1967). As restrictions on the use of these heavy metal fungicides became operative their availability declined and alternative fungicides were evaluated. Under severe disease pressure over the period of late autumn to spring, anilazine and thiram gave good control in trials reported by Gould, Goss and Vassey (1975). Chlorothalonil was an equally effective control material but gradually thinned the turf after repeated (eleven) applications. Results of a second trial reported in this same study showed that iprodione, triadimefon and benomyl were very effective in controlling red thread. Poor control was obtained with methyl and ethyl thiophanate. Quintozene and triadimefon, though effective, altered turf colour slightly when applied repeatedly. Cycloheximide mixtures, and fenarimol, also effective in most instances, caused injury in April on frost-stressed turf.

Handoll (1967) and Woolhouse (1969,1970, 1971, 1972, 1974) in a series of trials, demonstrated reliable control with anilazine, chlorothalonil, inorganic mercury compounds and quintozene. Benomyl performed satisfactorily in some years only. Efficacy of the fungicides was better in preventive rather than curative treatments and their performance was enhanced by the application of a nitrogenous fertilizer. Results of fungicide trials reported by Ashbaugh and Larsen (1983), Burpee *et al.* (1978), Dernoeden, O'Neill and Murray (1983), Nutter *et al.* (1979) and Sanders *et al.* (1978), indicated superior control of red thread by anilazine, benodanil, iprodione, triadimefon, propiconazole and vinclozolin, but cadmium fungicides included in some of the trials were comparable or superior to the newer systemic materials. Some of these fungicides showed non-target effects on the disease, increasing its severity (Dernoeden, Murray and O'Neill, 1985).

The results of tests made before the recognition of two distinct diseases have suggested that the causal fungi do not always respond similarly to fungicides. Kaplan and Jackson (1983) showed that iprodione and triadimefon were highly effective in suppressing growth of both fungi *in vitro*. Benomyl suppressed the growth of *L. fuciformis* but was much less efficient against *L. roseipellis*. This may explain the erratic performance of benomyl in reported field trials against 'corticium disease' and points to the need for confirmatory screening of fungicides and more care in diagnosis so that the most-effective chemical control for the two diseases can be adopted.

24.6 REFERENCES

Årsvoll, K. (1975) Fungi causing winter damage on cultivated grasses in Norway. *Meld Norg. LandbrHgosk.*, **54** (9), 49 pp.

Ashbaugh, F.M. and Larsen, P.O. (1983) Evaluation of fungicides for curative control of red thread (1982). *Fungicide Nematicide Tests*, **38**, 200.

Bahuon, A. (1986) Etude des conditions de propagation de *Corticium fuciforme* (Berk) Wakef.: mise an point d'une methode d'infection artificielle de *Lolium perenne* et *Festuca rubra* sp. Doctorate thesis, Univ. Poitiers, 121 pp.

Bandoni, R.J. and Johri, B.N. (1975) Observations on the genus *Exobasidiellum*. *Can. J. Bot*, **53**, 2561–4.

Bennett, F.T. (1935) Corticium disease of turf. *J. Board Greenkeeping Res.*, **4** 32–9.

Berkeley, M.J. (1873) Australian fungi. *J. Linn. Soc.*, **12**. 175.

Britton, M.P. (1969) Turfgrass diseases. In *Turfgrass Science* (eds A.A. Hansen and F.V. Juska). American Society of Agronomy, Madison, Wisconsin, pp. 288–335.

Burdsall, H.H. (1979) *Laetisaria* (Aphyllophorales, Corticiaceae), a new genus for the teleomorph of *Isaria fuciformis*. *Trans. Br. Mycol. Soc.*, **72**, 419–22.

Burpee, L.L., Sanders, P.L. Cole, H. Jr and Duich, J.M. (1978) Fungicide suppression of red thread in low nutrient status red fescue turf under home (lawn management, 1977). *Fungicide Nematicide Tests*, **33**, 147.

Cahill, J.V., O'Neill, N.R. and Dernoeden, P.H. (1982) Variations among isolates of *Corticium fuciforme* causing red thread disease of turf grasses. *Phytopathology*, **72**, 932. (Abstr.).

Cahill, J.V., Murray, J.J., O'Neill,N.R. and Dernoeden, P.H. (1983) Interrelationships between fertility and red thread fungal disease of turf grass. *Plant Dis.*, **67**, 1080–3.

Cooke, M.C. (1880) *Grevillea*, **9**, 94.

Couch, H.B. (1973) *Diseases of Turfgrasses*, 2nd edn, Kreiger, Huntington, New York, 348 pp.

Couch, H.B. (1983) Recent insights on the nature and control of corticium red thread. *USGA Green Sect. Rec.* **21** (6), 8–11.

Dernoeden, P.H., O'Neill, N.R. and Murray, J.J. (1983) Preventive control of red thread with fungicides. *Fungicide Nematicide Tests*, **38**, 201.

Dernoeden, P.H., Murray, J.J. and O'Neill, N.R. (1985) Non target effects of fungicides on turf grass growth and enhancement of red thread. *Proc. 5th. Int. Turf. Res. Conf.*, pp. 580–93.

Endo, R.M. (1963) Influence of temperature on rate of growth of five fungus pathogens of turf grasses and on rate of disease spread. *Phytopathology*, **53**, 857–61.

Erwin, L.E. (1941) Pathogenicity and control of *Corticium fuciforme*. *Rhode Island Agric. Expt. Sta. Bull.*, **278**. 34 pp.

Filer, T.H. Jr (1966) Red thread found on bermudagrass. *Plant Dis. Rept.*, **50**, 525–6.

Goss, R.L. and Gould, C.J. (1971) Interrelationships between fertility levels and corticium red thread disease of turf grasses. *J. Sports Turf Res. Inst.*, **47**, 48–53.

Gould, C.J. (1964) Turf grass disease problems in North America. *Golf Course Rept.*, **32** (5), 36, 38, 40–3, 46, 48, 50, 52, 54.

Gould, C.J. (1966–67) Use of fungicides in controlling turf grass diseases. *Golf Superintendent*, **34** (9)(10), **35**(1).

Gould, C.J., Miller, V.L. and Goss, R.L. (1967) Fungicidal control of red thread disease of turf grass in western Washington. *Plant Dis. Rept.*, **51**, 215–19.

Gould, C.J., Goss, R.L. and Vassey, W.E. (1975) Fungicidal tests of Corticum red thread in the Pacific Northwest. *J. Sports Turf Res. Inst.*, **51**, 62–6.

Handoll, C. (1967) Turf disease notes 1967. *J. Sports Turf Res. Inst.*, **43**, 28–33.

Hims, M.J., Dickinson, C.H. and Fletcher, J.T. (1984) Control of red thread, a disease of grasses caused by *Laetisaria fuciformis*. *Plant Pathol.*, **33**, 513–16.

Julich, W. (1976) Studies in resupinate basidiomycetes. IV. *Persoonia*, **8**, 431–42.

Kaplan, J.D. and Jackson, N. (1982) Variations in growth and pathogenicity of fungi associated with red thread disease of turf grass, *Phytopathology*, **72**, 262. (Abstr.)

Kaplan J.D. and Jackson, N. (1983) Red thread and pink patch diseases of turf grasses. *Plant Dis.*, **67**, 159–62.

Libbey, R.P. (1938) Corticium disease. *J. Bd Greenkpg Res.*, **5** (19), 269–70.

McAlpine, D. (1906) A new hymenomycete the so-called *Isaria fuciformis* Berk. *Ann. Mycol.*, **4**, 541–51.
Muse, R.R. and Couch, H.B. (1965) Influence of environment on diseases of turf grasses. IV. Effect of nutrition and soil moisture on Corticium red thread of creeping red fescue. *Phytopathology*, **55**, 507–10.
Nutter, F.W. Jr, Sanders, P.L., Loughner, D. and Cole, H. Jr (1979) Fungicide suppression of red thread in low nutrient status red fescue turf. *Fungicide Nematicide Tests*, **34**, 148.
O'Neill, N.R. (1983) Differentiation of basidiomycetes associated with red thread and pink patch diseases in grasses. *Phytopathology*, **73**, 1096. (Abstr.)
Reid, D.A. (1969) New and interesting British plant diseases. *Trans. Br. Mycol. Soc.*, **52**. 19–38.
Sanders, P.L., Burpee, L.K. Cole, H. Jr and Duich, J.M. (1978) Uptake, translocation and efficacy of triadimefon in control of turf grass pathogens. *Phytopathology*, **68**, 1482–7.
Smith, J.D. (1953) Corticium disease. *J. Sports Turf Res. Inst.*, **8**, 253–8.
Smith, J.D. (1954) Fungi and turfgrasses. IV. Corticium disease. *J. Sports Turf Res. Inst.*, **8**, 365–77.
Smith, J.D. (1957) Corticium disease – Fungicide trial, 1957. *J. Sports Turf Res. Inst.*, **9**, 367–8.
Smith, J.D. (1965) *Fungal Diseases of Turf Grasses.* 2nd edn (revised by N. Jackson and J.D. Smith), Sports Turf Research Institute, Bingley, UK, 97 pp.
Sprague, R. (1950) *Diseases of Cereals and Grasses in North America.* Ronald Press, New York, 538 pp.
Stalpers, J.A. and Loerakker, W.M. (1982) *Laetisaria* and *Limonomyces* species (Corticiaceae) causing pink diseases in turf grasses. *Can J. Bot.*, **60**, 529–37.
Wakefield, E.M. (1916) Notes on British Thelephoraceae. *Trans. Br. Mycol. Soc.*, **5**, 474–81.
Woolhouse, A.R. (1969) Fungicide trials (1969). *J. Sports Turf Res. Inst.*, **45**, 55.
Woolhouse, A.R. (1970) Fungicide trials (1970). *J. Sports Turf Res. Inst.*, **46**, 5–13.
Woolhouse, A.R. (1971) Fungicide trials (1971). *J. Sports Turf Res. Inst.*, **47**, 41–7.
Woolhouse, A.R. (1972) Fungicide trials (1972). *J. Sports Turf Res. Inst.*, **48**, 28–35.
Woolhouse, A.R. (1974) Fungicide trials (1974). *J. Sports Turf Res. Inst.*, **50**, 55–8.

25 *Rhizoctonia foliar diseases of established turf*

Until recently *Rhizoctonia solani* Kühn was considered the sole representative of the form-genus *Rhizoctonia* to function with any prominence as a turf grass pathogen. This ubiquitous soil-borne fungus was recognized early in the present century as the cause of disease in established and seedling turf and has been the subject of numerous studies. A substantial body of knowledge has accumulated pertaining mainly to the summer foliage disease, brown patch, less to damping-off diseases (see seedling diseases, pp. 40–42). Current developments suggest that for both disease complexes the literature includes some questionable information since increasingly it is apparent that other *Rhizoctonia* species, and *Rhizoctonia*-like fungi, may participate as turf grass pathogens.

In the species concept developed for *R. solani* by Parmeter and Whitney (1970), criteria deemed particularly important include, (1) a multinucleate condition of the cells in the young mycelium; (2) the presence of dolipore septa and (3) *Thanatephorus cucumeris* (Frank) Donk as the teleomorph. Even without reference to the teleomorph, fungi closely resembling *R. solani* may be characterized confidently as distinct if they possess predominantly binucleate cells in the young mycelium (Parmeter, Whitney and Platt, 1967; Burpee, 1980a; Burpee *et al.*, 1980a,b). Scrutiny of the *R. solani*-like fungi associated with both typical and atypical brown patch disease symptoms has indicated that some isolates are indeed binucleate (Dale, 1978; Hurd and Grisham, 1983; Martin and Lucas, 1984; Sanders, Burpee and Cole, 1978) and that some fit the species concept of *Rhizoctonia cerealis* Van der Hoeven (Burpee, 1980b). *R. cerealis* was confirmed in 1980 as the cause of 'cool weather brown patch' and the common name 'yellow patch' was coined to accommodate this disease syndrome (Burpee, 1980b). The multinucleate species *R. oryzae* Ryker and Gooch and *R. zeae* Voorhees are pathogenic on members of the Poaceae. Both species, and other as yet unidentified *Rhizoctonia* and *Rhizoctonia*-like fungi have been implicated as turf grass pathogens, but their significance remains largely undetermined (Burpee, 1980a; Christensen, 1979; Lucas, 1982; Martin, Campbell and Lucas, 1983, 1984; Martin and Lucas, 1984; Sprague, 1950; Wehlburg *et al.*, 1975).

25.1 BROWN PATCH

One of the first-documented fungal problems of cultivated turf, brown patch, also referred to as large brown patch or summer blight, has long been recognized as a major foliage disease (Piper and Coe, 1919; Monteith and Dahl, 1932). This widely occurring turf ailment severely damages both cool- and warm-season turf grasses particularly in locations where extended periods of high temperature, ample moisture and high relative humidity prevail. *R. solani* is a common causal agent but other *Rhizoctonia* species (e.g. *R. zeae*) may produce the typical symptoms (Lucas, 1982; Martin, Campbell and Lucas, 1983; Martin and Lucas, 1984). *R. cerealis* is the probable disease incitant where (a) weather conditions other than prolonged high temperatures and high humidity occur and (b) where atypical symptoms are reported. However, Traquair and Smith (1981) documented the occurrence of brown patch disease in western Canada both in cool spring months and during the summer. *R. solani* was verified as the causal agent in both cases.

25.1.1 Symptoms

On closely mown turf the leaves of infected grass plants become water soaked, blacken, then wither to a light-brown colour (Fig. 41, colour plate section) in irregular or roughly circular patches from a few centimetres to a metre in diameter. On higher cut turf withering and collapse of the affected plants often results in large brown or straw-coloured patches, which may appear sunken in relation to the surrounding healthy turf. When the air is moisture-saturated, especially in the early morning, the patches on close-mown turf are typically bordered by dark greyish, purple to black 'smoke rings' 1–5 cm wide (Fig. 42, colour plate section). The rings are composed of wilted, recently infected leaves, webbed with mycelium and constitute a useful, but ephemeral, diagnostic feature that disappears as the turf dries. Leaf spot symptoms, occurring concurrently or in the absence of the patch symptom, may be apparent in high cut swards of *Zoysia* sp., *Stenotaphrum secundatum* (Walt.) Kuntze, *Festuca arundinucea* Schreb., *Poa pratensis* L. and *Lolium perenne* L. The tan to straw-coloured leaf spots with dark borders (Fig. 43, colour plate section) develop from small water-soaked lesions to encompass large portions of the leaf lamina.

Given favourable environmental conditions of hot humid weather and free moisture on the grass leaves, development of the disease is rapid as patches enlarge and coalesce to involve large areas of turf within a period of 12–24 hours. If conditions favourable to the disease are prolonged the plants may be killed. More commonly, damage is restricted to the leaves which are partially or completely destroyed. Crowns and roots survive to generate a sparse recovery growth shortly after the activity of the pathogen subsides. Sclerotia (1–3 mm long) are often visible in sheath axils and in the thatch layer adjacent to infected plants (Fig. 25.1).

R. solani is intrinsically a highly variable fungus both in cultural and pathogenic characters (Parmeter and Whitney, 1970; and others). It is not surprising, therefore, that published descriptions of brown patch symptoms and the conditions favouring them may differ (Traquair and Smith, 1981). Isolates within *R. solani* may induce drastically different disease symptoms on cool- and warm-season grasses (Hurd and Grisham, 1983; Martin and Lucas, 1984). The situation is complicated further with the increasing realization that fungi previously implicated as *R. solani* in fact may be morphologically similar but quite distinct species (Burpee, 1980a,b; Joyner and Partyka, 1980; Lucas, 1982; Martin, Campbell and Lucas, 1983, 1984; Martin and Lucas, 1984).

25.1.2 The causal fungi

(a) Anamorph: *Rhizoctonia solani* Kuhn
Synonym: *Corticium solani* (Prill. and Delacr.) Bourd. and Galz
Teleomorph: *Thanatephorous cucumeris* (Frank) Donk, (Reinwardtia, 3,376, 1956)
Synonym: *Pellicularia filamentosa* (Pat.) Rogers.

Since it produces no spores, *R. solani* can be identified only from its mycelial and sclerotial states. In culture on potato dextrose agar the mycelium of different isolates (anastomosis groups, Sherwood, 1969) is variable in colour, from white turning buff, grey, and usually maturing to some shade of brown and from sparsely aerial to mealy or felted (Burpee, 1980a; Sherwood, 1969; Mordue, 1974). Branching of the main, or runner hyphae is mostly right-angled, with constrictions where they originate (Fig. 25.2) and septate above the junction. Young vegetative hyphae show multinucleate cells and dolipore septa. (Parmeter, Sherwood and Platt, 1969). There are no clamp connections. The diameter of the runner hyphae, according to Burpee, (1980a) and Sherwood, (1969) is 4–13 μm. Older hyphae are very variable in

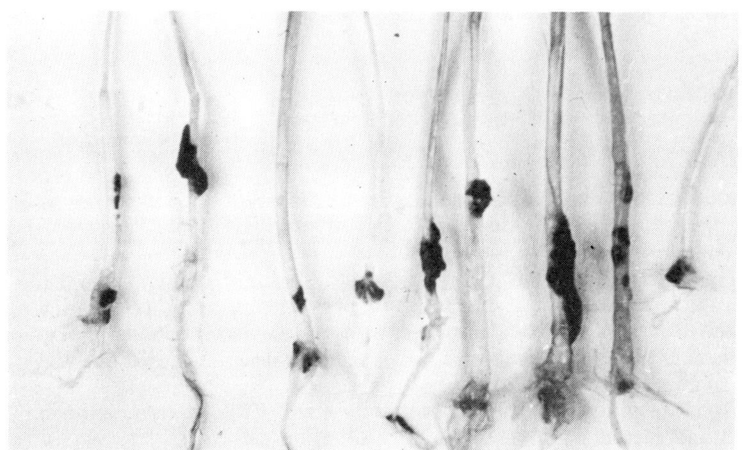

Figure 25.1 Sclerotia of *Rhizoctonia solani* on Astoria *Agrostis tenuis*. (M.C. Shurtleff).

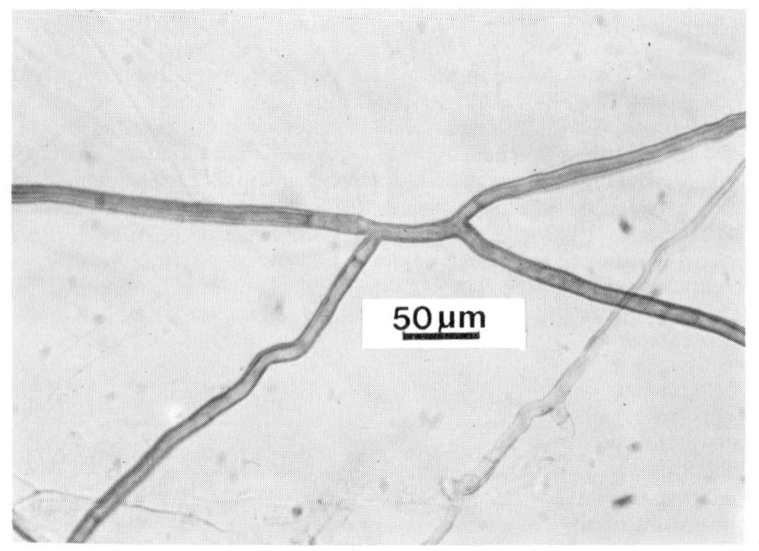

Figure 25.2 Mycelium of *Rhizoctonia solani.*

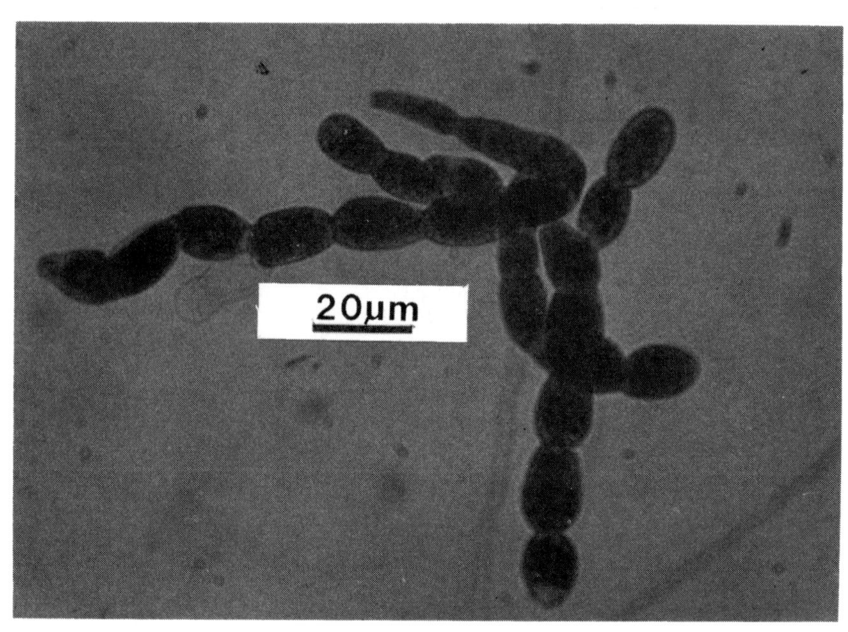

Figure 25.3 Moniloid cells of *Rhizoctonia solani.*

form and are often of much greater diameter than runner hyphae, with cruciform and moniloid cells (Mordue, 1974) (Fig. 25.3).

Sclerotia are very variable in size, colour and abundance, according to isolate, from >1 mm to <10 mm diam., dirty white or pale buff, brown to almost black, spherical to irregular, simple, aggregated or crust-like, with a fuzzy or hard, pitted surface, never differentiated into rind and medulla (Parmeter, Sherwood and Platt, 1969; Parmeter and Whitney, 1970).

The formation and structure of hyphal infection cushions, infection pegs and appressoria in relation to entry into the host tissues has been described in Section 5.5 under seedling diseases caused by *Rhizoctonia* spp.

The teleomorphic state, *T. cucumeris*, is not uncommon in turf situations, occurring on plants of

Festuca arundinacea, Lolium perenne, Poa pratensis, Digitaria spp. and broad-leaved weeds (*Taraxacum officinale, Plantago* spp. and *Trifolium repens*). It appears as effuse, white patches, readily removable, usually on the lower side of leaves. The upper portion of the plant tissue covered by the fungus may remain green and apparently unaffected (Frank, 1883; Luttrell, 1962; Jackson, unpublished).

The hymenium of *T. cucumeris* is formed on an arachnoid, membranous, grey or greyish brown sheet or collar of mycelium on the upper parts of roots and lower parts of plant stems. The investing hyphae consist of thick-walled, extensively septate cells on which the barrel-shaped or sub-cylindrical basidia, 6–9 μm wide are produced each usually with four sterigmata bearing hyaline, elliptic–oblong basidiospores, 8–14 × 4–8 μm (Frank, 1883; Mordue, 1974; Talbot, 1974).

Pathogenic isolates of *R. solani* from turf are typically quite fast growing (c.9 cm in 3 days), on a variety of natural and synthetic media with optimum growth around 28 °C (Dahl, 1933a; Dickinson, 1930; Endo, 1963; Monteith and Dahl, 1932; Sanders, Burpee and Cole, 1978; Sherwood, 1970). Biotypes of *R. solani* with lower temperature optima (c. 20 °C) have been reported, but in most instances their identity was not confirmed by nuclear examination, or production of the perfect state. Smith and Traquair (1982) demonstrated conclusively that low-temperature adapted strains of *R. solani* occur in Canada, but they did not determine the anastomosis grouping (Ogoshi, 1972; Parmeter, Sherwood and Platt, 1969). Isolates from turf that induced foliar blight are usually constituents of anastomosis groups AG1 and AG4 (Burpee, 1980a). However, *R. solani* AG2, AG5 and isolates of unknown anastomosis grouping have also been recovered from diseased turf grasses in North Carolina (Martin and Lucas, 1984). *R. solani*, AG2, has been identified as the cause of brown patch on *Stenotaphrum secundatum* in Texas (Hurd and Grisham, 1983). The latter can be pathogenic on corn roots over a wide temperature range (Sumner and Bell, 1982).

(b) Anamorph: *Rhizoctonia zeae* Voorhees (*Phytopathology*, **28**, 233–46, 1938).

Teleomorph: *Waitea circinata* Warcup and Talbot* (*Trans. Br. Mycol. Soc.* **45**, 495–518, 1962).

The hyphae of *R. zeae* are of similar diameter to those of *R. solani* and are also constricted at right-angled branches, hyaline when young, to pink and later reddish-brown when mature, later becoming multiseptate. Small sclerotia, often abundantly produced, < 1 mm diam. are produced in culture and on the host. The high optimum temperature for growth of 33°C has been confirmed by many workers (Martin, Campbell and Lucas, 1983; Luttrell, 1954; Ryker and Gooch, 1938; Ullstrup, 1963) with lower and upper limits of 11–14°C and 40–42°C, and sclerotia have a wide temperature tolerance range of 50°C and 0°C. Tu, Roberts and Kimbrough (1969) indicated that the cells of *R. zeae* are binucleate, but it is now generally agreed that they are multinucleate as in *R. solani* (Martin, Campbell and Lucas, 1983; Sumner and Bell, 1982). For detailed descriptions of the morphology of *R. zeae* and *Waitea circinata* see Voorhees (1938) and Warcup and Talbot (1962) respectively.

*According to Oniki *et al.* (1985), *R. zeae* and *R. oryzae* share the same teleomorph.

(c) Anamorph: *Rhizoctonia oryzae* Ryker and Gooch (1938)

Teleomorph: *Waitea circinata* Warcup and Talbot (*Trans. Br. Mycol. Soc.*, **45**, 495–518, 1962; Oniki *et al.*, 1985).

The main runner hyphae of *R. oryzae*, are white (Ryker and Gooch, 1938) to cream-buff (Burpee, 1980a) of similar diameter to those of *R. solani* and *R. zeae*, but branch at an acute angle, with only a slight constriction above the branch and they have a septum a short distance above the branch. Later branching becomes general, at right angles to the main hyphae and thick-walled barrel-shaped cells form salmon-coloured pseudosclerotia. Sclerotial masses are not produced on the host. An isolate from rice cultured by Burpee (1980a) produced no sclerotia and the young hyphae were multinucleate. Mycelial growth at 32°C was rapid, 9 cm in 2 days (Ryker and Gooch, 1938).

(d) Other *Rhizoctonia*-like fungi

Binucleate and multinucleate isolates of fungi resembling but not identical with, *R. cerealis*, *R. solani* or *R. zeae* have been isolated from diseased turf, exhibiting typical brown patch symptoms. The significance of these findings has not been established (Christensen, 1979; Hurd and Grisham, 1983; Lucas, 1982; Martin and Lucas, 1984).

25.1.3 Host range

(a) *Rhizoctonia solani*

The fungus attacks a broad range of gramineous plants and brown patch disease can occur with varying severity on all the cool- and warm-season grasses employed for turf use (Britton, 1969; Couch, 1973; Freeman, 1967; Shurtleff, 1953b). The disease is of general occurrence where warm temperatures and high humidity prevail in the United States, Canada, Europe, Africa, Australasia and Japan (Howard Rowell and Keil, 1951; Sprague, 1950). *Agrostis spp. L.* and *Poa annua L.*, under close mowing and intensive management are especially vulnerable and may suffer serious damage if environmental conditions favourable to the pathogen are prolonged. Moderate to severe damage may occur on turf composed of *Poa pratensis L.*, *Lolium perenne L. Festuca arundinacea Schreb.*, *Stenotaphrum secundatum* (Walt.) Kuntze and *Zoysia* spp. (Willd.). In the United States, increasing problems with brown patch disease have been associated with the introduction of fine-leaved, lawn-type cultivars of both *F. arundinacea* and *L. perenne.*

(b) *Rhizoctonia zeae*

Since the fungus was first described, *R. zeae* has been confirmed in the United States as the cause of a stalk, foliage or ear disease of pearl millet, rye and corn and as a root disease of corn (Sumner and Bell, 1982). Recently it has been associated with brown patch disease of *Festuca arundinacea* Schreb. turf in North Carolina (Lucas, 1982; Martin, Campbell and Lucas, 1983; Martin and Lucas, 1984).

(c) *Rhizoctonia oryzae*

Although primarily a pathogen of rice, the fungus has been isolated from diseased turf grasses in New Zealand (Christensen, 1979) and it has been reported on a few occasions in the United States as the cause of disease on *Stenotaphrum secundatum* (Walt.) Kuntze (Sprague, 1950; Wehlburg *et al.*, 1975). Presently, *R. oryzae* is not recognized as an important turf grass pathogen in the United States (Freeman, 1983 personal communication).

25.1.4 Disease development

R. solani is an ubiquitous inhabitant of cultivated soils and is capable of long survival in the form of saprobic mycelium or as sclerotia. The fungus may also be seedborne (Leach and Pierpoint, 1958). In established turf the fungus can persist as sclerotia, saprobic or dormant mycelia present in the plant debris and also as a mild root necrosis pathogen (Britton, 1969; Pitt, 1964). The significance of the teleomorph in turf is not known. Sclerotia dispersed in the thatch, supplementing those present in the upper soil layer, apparently serve as a principal means of perennation. These resting structures (1–2 mm in diameter) can often be observed on or between leaf sheaths near the crowns of previously infected grass plants (Fig. 25.1). Sclerotia are resistant to inclement environmental conditions for long periods and may remain viable after several successive germinations (Shurtleff, 1955). Other *Rhizoctonia* species are presumed to function similarly in the turf environment.

Dickinson (1930), contended that resumption of growth from sclerotia was promoted by a brief period of chilling, down to a germination optimum in the range 18–20 °C. Unless followed subsequently by an increase in temperature he demonstrated that the short hyphae initiating vegetative growth would not progress further and a drop in temperature to below 16 °C caused their demise. A rise in temperature above 23 °C and up to 32 °C (optimum 27–29 °C) induced rapid growth and aggressive parasitic activity of the fungus. Free moisture favoured germination of the sclerotia and the development of mycelia but it was not considered of vital significance. Dahl (1933a) cited a similar range for active growth of the fungus (optimum; about 28 °C) but he was unable to confirm any need for chilling as a stimulant to renewal of vegetative growth. Sclerotial germination was observed over the range 8–40 °C with an optimum around 28 °C. In relating rainfall data and irrigation practice to brown patch incidence, Dahl (1933a,b) concluded that ample water was a necessary factor for the brown patch fungus to attack turf grasses. More precise information on moisture requirements was provided by Shurtleff (1953a, 1955), who obtained germination of sclerotia from 15.5–37 °C, but only at sustained relative humidities of 98% or above. He concluded that high humidity with a dew or a film of moisture on the foliage, coupled with temperatures for several hours between 21–32 °C, were major requirements in determining disease onset. Although brown patch may occur under conditions of low atmospheric humidity (Dickinson, 1930), there is general consensus that activity of the disease is most intense on moist foliage in hot, humid weather (air temperature 28 °C or thereabouts) and when minimum air temperatures are sustained above 20–21 °C for a few days.

By utilizing stored food reserves, hyphae grow out radially from the resting structures in the soil or thatch, forming an ever-enlarging, roughly circular colony. Saprobic growth towards the plants and subsequent mycelial investment of the leaf surfaces is favoured by free moisture. The process is stimulated markedly by nutrients present in leaf exudates. Hyphae bridge from leaf to leaf via the guttation droplets and the hyphal aggregates formed therein can initiate infection of the leaf tissues by entering through stomata or through the cut end of leaves

(Dickinson, 1930; Rowell, 1951). The fungus may also penetrate directly from infection cushions or lobate appressoria adhering to the plant surfaces (Dodman and Flentje, 1970; Hurd and Grisham, 1983). Inter- and intracellular hyphae ramify the infected tissues leading rapidly to their collapse and discolouration. During prolonged brown patch attacks sclerotia may form on the grass leaves, commonly attched to or within the outermost leaf sheaths, on the crowns, or at the bases of the crowns where the roots attach. Some of the sclerotia may be distributed by greens mowers and, along with infected clippings (in which the fungus can survive for up to four months), they serve as inoculum in the local dissemination of the pathogen (Shurtleff, 1955).

R. solani grows satisfactorily over a broad soil pH range and certainly quite well over the range of pH deemed acceptable for the successful culture of turf grasses. Whilst not influencing brown patch disease directly, Bloom and Couch (1960), published data showing that hydrogen-ion concentration and certain fertility levels may interact to affect disease response. Disease severity was much greater at high nitrogen levels, a commonly observed and widely documented phenomenon (Dahl, 1933c; Dickinson, 1930; Hearn, 1943; Howard, Rowell and Keil, 1951; Shurtleff, 1955; and others), but they proposed that this increase in brown patch severity is not related directly to any increase in plant vigour or succulence. Rather, their data indicated it is due to an imbalance in nutrient levels involving nitrogen and other elements which interact with pH to change in some way the proneness of plants to infection.

The increase in brown patch severity following the application of soluble nitrogen fertilizers to turf may result from alterations in leaf exudate composition favourable to the saprobic and parasitic activities of the pathogen (Britton, 1969). Guttation fluid is probably necessary for rapid saprobic growth of *R. solani* and its subsequent invasion of the leaves, predominantly through the cut ends. Observations in Rhode Island showed that guttation droplets persisted for overnight periods of 14 hours or more, allowing severe disease incidence during the hot humid conditions referred to as 'brown patch weather'. High relative humidity at night-time, prolonged well into the morning by cloud cover or fog, ensured the slow drying of the turf and the continued activity of the pathogen (Rowell, 1951).

25.1.5 Control

(a) Cultural control

Since free water on the leaf surfaces plays an important role in the epidemiology of brown patch disease, any management stategy instrumental in reducing the period of leaf wetness can be expected to help in preventing serious outbreaks of the disease. These cultural practices include the provision of good surface and subsurface drainage to enable rapid removal of surplus water from rainfall or irrigation, watering turf early in the day to allow drying of the leaf surfaces before evening; removing dew and associated leaf exudates in the early morning by irrigating, poling (switching) or dragging a mat or hose over the turf; and pruning or selectively removing any barriers to free air flow across the turf (Dickinson, 1930; Dahl, 1933a,b; Rowell, 1951; Shurtleff, 1955).

Surface-active agents (wetting agents) may prevent the accumulation of guttative droplets on the leaf tips (Keil, 1944) or hydrated lime may be used *with caution* as a desiccant to remove leaf surface moisture and so limit brown patch development. Broadcast application of hydrated lime is made at the onset of symptoms at a rate of 4.9 kg/100 m^2 to dry turf. The material is worked into the turf surface by drag mat or poling, and irrigation withheld for a period of 24 hours. Applications may be repeated at about three week intervals (Couch, 1973).

Regardless of how nitrogen is involved in the promotion of brown patch disease, the practical observation that abundant soil nitrogen does indeed increase the incidence and severity prompts caution in the use of this nutrient on high risk turf. Balanced nutrition, based on needs as determined by soil testing and local recommendations for species involved, should be related to the service or aesthetic demands being placed on the turf, and coordinated with a fungicide programme.

(b) Chemical control

Cultural practices may delay or reduce the incidence of brown patch but, in general, recurrent seasonal problems with the disease are common. This is especially the case on intensively managed, low cut turf (e.g. *Agrostis* spp.) or on high cut turf composed of very susceptible species (e.g. *L. perene*). A preventative fungicide programme should be implemented commencing when night air temperatures in the range of 19–21 °C are anticipated. Repeat applications at 5–7-day intervals are recommended if daytime temperatures are 28 °C and above, high night temperatures are sustained and conditions of high humidity prevail. For curative treatment, or in periods of intense disease risk, the interval between applications may be reduced to 3 days. Conversely the interval may be extended to 7–10 days if night temperatures fall below 18 °C.

Formulations of inorganic or organic mercury compounds figured prominently as the most effective fungicides in the early recommendations for control of brown patch (Howard, Rowell and Keil, 1951; Monteith and Dahl, 1932). Their efficacy remained unchallenged, but for environmental pollution reasons the use of mercurial fungicides in brown patch control is forbidden in most jurisdictions. Early substitutes for mercury compounds which gave reliable control included thiram and quintozene (Freeman, 1967; Howard, Rowell and Keil, 1951). Other organic contact fungicides, which were developed later and afford moderate to good protection, include anilazine, chlorothalonil, and mancozeb. Cycloheximide combined with thiram and quintozene is more effective than cycloheximide alone against brown patch but occasional problems with phytotoxicity may arise when cycloheximide, quintozene or the two combined are used during high temperature conditions.

Systemic fungicides also control the disease with varying efficiency and include benomyl, carboxin, fenarimol and propiconazole, iprodione, methyl and ethyl thiophanate, thiabendazole and triadimefon (Sanders *et al.*, 1978a,b; Smiley, Craven and Thompson, 1978).

It is becoming increasingly apparent that control of rhizoctonia leaf blights requires correct identification of the specific pathogens involved since differential isolate and species responses to fungicides are common (Martin and Lucas, 1984; Martin, Campbell and Lucas, 1984).

25.2 YELLOW PATCH

The existence of cool-temperature *Rhizoctonia* diseases was implied by Hearn (1943), Holt (1963) and Zummo and Plakidas (1958) and in reporting brown patch on some warm-season grasses as occurring during the cool periods of spring and autumn. Broadfoot (1936) reported a *Rhizoctonia* snow mould pathogenic on *Festuca rubra* spp. *commutata* in Alberta and Sprague (1950) made reference to *R. solani* as a common pathogen on a variety of grass hosts during humid winter months in Oregon. Madison, Peterson and Hodges (1960) observed similar low-temperature damage to *Agrostis* spp. in California.

Endo (1961, 1963) corroborated this Californian report and demonstrated that typical symptoms of brown patch could be incited by both the warm- and cool-temperature *Rhizoctonia* strains on both warm- and cool-season grasses. When these observations were made, *R. solani* was considered to be the causal agent. In some instances this may indeed have been the case since *R. solani* has been confirmed by Traquair and Smith (1981) as the cause of a cool-temperature disease on turf in the Canadian prairie provinces, similar in symptom appearance to snow mould damage or cold injury.

Madison (1966, 1971), however, concluded that the classic brown patch symptoms occur only infrequently in the dry south western United States, and described as more common a *Rhizoctonia* disease which developed during mild weather. It seems likely that the 1966 paper by Madison was the first published description of what is now designated yellow patch disease (Figs. 44 and 45, colour plate sections), caused by *R. cerealis* (Burpee, 1980b).

25.2.1 Symptoms

Madison (1966) observed the fungus growing out radially from infection centres, progressing initially without visible damage to the grass. With warming temperatures, the pathogenic activity of the fungus increased and the grass was damaged in the form of brown rings. In contrast to typical brown patch, the disease resulted primarily from a root or crown infection, little aerial mycelium was produced, the leaves probably died from desiccation rather than fungal invasion, and a brown ring was produced rather than a brown patch. Other California turf grass pathologists endorsed the separation of these *Rhizoctonia*-induced symptoms, recognizing as distinct a cool-season and a warm-season disease (McCain, Endo and Raabe, 1971).

Similar cool-temperature *Rhizoctonia* diseases have been reported since from Europe and other regions of the United States. Large-scale damage to *Agrostis stolonifera* turf in Sweden by a *Rhizoctonia* active at temperatures below 20 °C was reported by Dahlsson (1975). Joyner, Partyka and Larsen (1977), documented the widespread occurrence of a *Rhizoctonia* disease on *Poa pratensis* swards in the mid-western USA during the autumn of 1975 and 1976. Dale (1978), described atypical brown patch symptoms in autumn stands of *Zoysia japonica* growing in Arkansas. In all the American reports, the authors drew attention to a ring of frogeye symptom. Irregular yellow rings and foliar blight symptoms occurring commonly in autumn of 1974, 1975 and 1976 on turf composed of *Poa pratensis* or *Agrostis* sp. in Pennsylvania and New York State, consistently yielded isolates of 'cool weather *Rhizoctonia*' (Sanders, Burpee and Cole, 1978).

After isolating the same incitant from similarly diseased *Cynodon dactylon* turf in Bermuda, Burpee (1980b) proposed the name 'yellow patch' to accommodate this distinctive turf ailment and established that the binucleate *Rhizoctonia* spp. inciting the disease matched the species concept of *R. cerealis* Van der Hoeven. Designation of the anamorph was determined on the basis of cultural morphology and anastomosis reactions of the binucleate isolates. Recently, the associated teleomorph was induced to form in culture and named *Ceratobasidium cereale* (Murray and Burpee, 1984).

Overall, symptoms of yellow patch disease take the form of separate and merging, uniformly diseased areas in circles, streaks and ribbons; or commonly, the patches comprise concentric rings that have a bullseye or target appearance (Fig. 44, colour plate section).

25.2.2 The causal fungus

Anamorph: *Rhizoctonia cerealis* van der Hoeven (Boerema and van der Hoeven, 1977)
Teleomorph: *Ceratobasidium cereale* (Murray and Burpee, 1984)

R. cerealis may be distinguished from *R. solani* by its slower growth rate, the lower optimum growth temperature, c. 20–23 °C, in culture. Runner hyphae are narrower (3.8–6.2 μm, Boerema and Van der Hoeven, 1977; 2–6 μm, Burpee, 1980a), although these generally branch at right angles and the branches are constricted and are septate above the point of origin in both species. Hyphae of *R. cerealis* are predominantly binucleate and binucleate isolates of *Rhizoctonia* spp. which have been persuaded to fruit also belong to *Ceratobasidium* (Burpee, *et al.*, 1978; Burpee, 1980a). Sclerotia of *R. cerealis* are small, 0.3–1.4 μm diam. and white to pale brown in culture. Pigmentation in the mycelium is also much lighter than in *R. solani.*

For a description of the teleomorph, C. cereale, refer to the paper by Murray and Burpee (1984).

Burpee *et al.*, (1980a) established that group 1 (CAG1) isolates were associated with gramineous hosts. Isolates assigned to this group anastomosed with the type culture of *R. cerealis* (Burpee, 1980b) indicating a common anastomosis group CAGl for all graminicolus *R. cerealis* isolates.

25.2.3 Host range

R. cerealis is one of the causes of sharp eyespot of cereals in the Netherlands (Boerema and Van der Hoeven, 1977; Van der Hoeven and Bollen, 1980), Germany (Reinecke and Fehrmann, 1979), South Africa (Scott, Visser and Ruzenacht, 1979), Japan (Murray and Burpee, 1984), and the United States (Lipps and Herr, 1982). Studies in England on sharp eyespot of cereals reported by Pitt (1964a) probably involved *R. cerealis* as the causal agent. As the incitant of yellow patch in the United States (and other as yet unconfirmed locations in temperate regions) the fungus has been reported on *Agrostis* spp. L., *Festuca arundinacea* Schreb, *Poa annua* L., *Poa pratensis* L., *Lolium perenne* L., *Cynodon dactylon* (L.) Pers. and *Zoysia japonica* Steud. (Burpee, 1980b; Burpee *et al.*, 1980b; Dale, 1978; Joyner, Partyka and Larsen, 1977; Martin and Lucas, 1984; Parmeter, Whitney and Platt, 1967; Sanders Burpee and Cole, 1978).

25.2.4 Disease development

The development of yellow patch disease from resting propagules is similar to that described for *R. solani* (page 270). Isolates of the fungus may be pathogenic at temperatures ranging from 10 to 27 °C (Burpee *et al.*, 1980b; Sanders, Burpee and Cole, 1978), but field symptoms are most common during cool, moist conditions in spring and autumn with temperatures between 10° and 20°C (Burpee, 1980b). Madison (1966) considered the infection as confined primarily to the roots and crown but in cool, moist weather (or if infected turf samples are incubated in a moist chamber), hyphae of *R. cerealis* may be observed growing from the thatch layer, up the leaf sheaths, and onto the adaxial surfaces of the blades. Aggregates of white, later fawn to tan, moniloid mycelium form infection cushions and the leaves involved soon yellow. No 'smoke ring' forms as in brown patch disease. Sclerotia (bulbils) 0.25–2 mm diameter are initiated close to infected plants in the thatch from mycelial aggregates that are at first white, changing to purplish-brown then brown as the structures mature.

Crown infections may eventually kill plants but in most instances the damage is confined to the outer leaves and sheaths. Although chlorotic in appearance the plants remain viable and recover slowly as increasing temperatures favour grass growth and suppress pathogenic activity of the fungus.

25.2.5 Control

The influence of turf management practices on the incidence and severity of yellow patch disease has not been investigated fully. Apparently the fertility level of the turf does not directly affect the disease but a balanced fertilization programme encourages rapid recovery of diseased turf once fungal activity ceases and weather conditions improve (Shurtleff, 1983). Any management practice aimed at reducing the duration of leaf wetness and/or dispersing guttation droplets may be expected to reduce the pathogenic activities of this, and other, *Rhizoctonia* spp. on turf grasses (Smiley, 1983).

Information on chemical control measures for this disease is sparse. Control of yellow patch has been observed when inorganic mercury fungicides were applied in late autumn as snow mould preventives (Jackson, 1982, personal observation). Sanders, Burpee and Cole (1978) reported that of six fungicides evaluated for activity *in vitro*, none completely inhibited the growth of all eighteen test isolates of *R. cerealis*. Chlorothalonil, chloroneb and iprodione showed the greatest overall activity in reducing fungal growth. Field trials are needed to confirm these laboratory results and determine the most effective chemical control treatment.

25.3 REFERENCES

Bloom, J.R. and Couch, H.B. (1960) Influence of environment on diseases of turf grasses. 1. Effect of nutrition, pH, and soil moisture on Rhizoctonia brown patch. *Phytopathology*, **50**, 532–5.

Boerema, G.H. and Van der Hoeven, A.A. (1977) Checklist for scientific names of common parasitic fungi. Series 26. Fungi of field crops: Cereals and grasses. *Neth. J. Plant Pathol.*, **83**, 165–204.

Britton, M.P. (1969) Turfgrass diseases. In *Turfgrass Science*, (eds A.A. Hanson and F.V. Juska), American Society of Agronomy, Madison, Wis., pp. 288–335.

Broadfoot, W.C. (1936) Experiments on the chemical control of snowmold of turf in Alberta. *Sci. Agric.*, **16**, 615–18.

Burpee, L.L. (1980a) Identification of *Rhizoctonia* species associated with turfgrass. In *Advances in Turfgrass Pathology* (eds P.O. Larsen and B.G. Joyner), Harcourt Brace Jovanovich, Duluth, Minn., pp. 25–8.

Burpee, L.L. (1980b) *Rhizoctonia cerealis* causes Yellow patch of Turfgrasses. *Plant Dis.*, **64**, 1114–16.

Burpee, L.L., Sanders, P.L., Cole, H. Jr and Kim, S.H. (1978) A staining technique for nuclei of *Rhizoctonia solani* and related fungi. *Mycologia*, **70**, 1281–3.

Burpee, L.L., Sanders, P.L., Cole, H. Jr and Sherwood, R.T. (1980a) Anastomosis groups among isolates of *Ceratobasidium cornigerum* and related fungi. *Mycologia*, **72**, 689–701.

Burpee, L.L., Sanders, P.L., Cole, H. Jr and Sherwood, R.T. (1980b) Pathogenicity of *Ceratobasidium cornigerum* and related fungi representing five anastomosis groups. *Phytopathology*, **70**, 843–6.

Christensen, M.J. (1979) *Rhizoctonia* species associated with diseased turfgrasses in New Zealand. *NZ J. Agric. Res.*, **22**, 627–9.

Couch, H.B. (1973) *Diseases of Turfgrasses*. Krieger Huntington, New York, 348 pp.

Dahl, A.S. (1933a) Effect of temperature and moisture on occurrence of brown patch. *US Golf Assoc. Green Sect. Bull.*, **13**, 53–61.

Dahl, A.S. (1933b) Effect of watering putting greens on occurrence of brown patch. *US Golf Assoc. Green Sect. Bull.*, **13**, 62–6.

Dahl, A.S. (1933c) Relationship between fertilizing and drainage on the occurrence of brown patch. *US Golf Assoc. Green Sect. Bull.*, **13**, 136–9.

Dahlsson, S.O. (1975) *Rhizoctonia solani* on Swedish golf greens. *Weibulls Gras. Tips. (Hort. Abs.* **46**, 10522).

Dale, J.L. (1978) Atypical symptoms of *Rhizoctonia* infection on *Zoysia. Plant Dis. Rept.*, **62**, 645–7.

Dickinson, L.S. (1930) The effect of air temperature on pathogenicity of *Rhizoctonia solani* parasitizing grasses on putting green turf. *Phytopathology*, **20**, 597–608.

Dodman, R.L. and Flentje, N.T. (1970) The mechanism and physiology of plant penetration by *Rhizoctonia solani*. In *Rhizoctonia solani, Biology and Pathology* (ed. J.R. Parmeter, Jr), University of California Press, Berkeley, Los Angeles and London, pp. 149–60.

Endo, R.M. (1961) Turfgrass diseases in southern California. *Plant Dis. Rept.*, **45**, 869–73.

Endo, R.M. (1963) Influence of temperature on rate of growth of five fungus pathogens of turfgrasses and on rate of disease spread. *Phytopathology*, **53**, 857–61.

Frank, B. (1883) On a new and little known plant disease. *Ber. Deut. Bot. Ges.*, **1**, 62–3. (German). (Cited by Talbot, 1970).

Freeman, T.E. (1967) Diseases of southern turfgrasses. *Florida Agric. Exp. Sta. Tech. Bull.*, **713**, 31 pp.

Hearn, J.H. (1943) *Rhizoctonia solani* Kühn and the brown patch disease of grass. *Proc. Texas Acad. Sci.*, **26**, 41–2.

Holt, E.C. (1963) Control of large brown patch on St. Augustinegrass. *Golf Course Rept.*, **31**, 48–50.

Howard, F.L., Rowell, J.B. and Keil, H.L. (1951) Fungus diseases of turfgrasses. *Univ. RI. Agric. Exp. Sta. Bull.*, **308**, 56pp.

Hurd, B. and Grisham, M.P. (1983) Characterization and pathogenicity of *Rhizoctonia* spp. associated with brown patch of St. Augustinegrass. *Phytopathology*, **73**, 1661–5.

Joyner, B.G. and Partyka, R.E. (1980) Rhizoctonia brown patch: Symptoms, diagnosis and distribution. In *Advances in Turfgrass Pathology* (eds P.O. Larsen and B.G. Joyner), Harcourt Brace Jovanovich, Duluth, Minn., pp. 21–3.

Joyner, B.G., Partyka, R.E. and Larsen, P.O. (1977) Rhizoctonia brown patch of Kentucky bluegrass. Plant Dis. Rept., **61**, 749–52.

Keil, H.L. (1944) New fungicide developments for turf. *Greenkeepers Rept.*, **12**, 5–6.

Leach, C.M. and Pierpoint, M. (1958) *Rhizoctonia solani* may be transmitted with seed of *Agrostis tenuis. Plant Dis. Rept.*, **42**, 240.

Lipps, P.E. and Herr, L.J. (1982) Etiology of *Rhizoctonia cerealis* in sharp eyespot of wheat. *Phytopathology*, **72**, 1574–7.

Lucas, L.T. (1982) Brown patch in the transition zone. *Am. Lawn Applic.*, **3**(4), 10–12.

Luttrell, E.S. (1954) Diseases of pearl millet in Georgia. *Plant Dis. Rept.*, **38**, 507–14.

Luttrell, E.S. (1962) Rhizoctonia blight of tall fescue. *Plant Dis. Rept.*, **46**, 661–4.

Madison, J.H. (1966) Brown patch of turfgrass caused by *Rhizoctonia solani* Kühn. *California Turfgrass Culture*, **16**(2), 9–13.

Madison, J.H. (1971) *Practical Turfgrass Management*. Van Nostrand, New York, 466 pp.

Madison, J.H., Peterson, L.J. and Hodges, T.K. (1960) Pink snowmold on bentgrass as affected by irrigation and fertilizer. *Agron. J.*, **52,** 591–2.
Martin, S.B., Campbell, C.L. and Lucas, L.T. (1983) Horizontal distribution and characterization of *Rhizoctonia* spp. in tall fescue turf. *Phytopathology*, **73,** 1064–8.
Martin, S.B. Campbell, C.L. and Lucas, L.T. (1984) Response of rhizoctonia blights of tall fescue to selected fungicides in the greenhouse. *Phytopathology*, **74,** 782–5.
McCain, A.H., Endo, R.M. and Raabe, R.D. (1971) Diseases of Turfgrass. *Calif. Agric. Expt. Sta. Man.*, **41,** 42–5.
Martin, S.B. and Lucas, L.T. (1984) Characterization and pathogenicity of *Rhizoctonia* spp. and binucleate *Rhizoctonia*-like fungi from turfgrasses in North Carolina. *Phytopathology*, **74,** 170–1.
Monteith, J. Jr and Dahl, A.S. (1932) Turf diseases and their control. *Bull.* US *Golf Assoc. Green Sect.*, **12,** 85–187.
Mordue, J.E.M. (1974) *Thanatephorus cucumeris. CMI Descriptions of Pathogenic Fungi and Bacteria.* no. 406.
Murray, D.I.L. and Burpee, L.L. (1984) *Ceratobasidium cereale* sp. nov., the teleomorph of *Rhizoctonia cerealis. Trans. Br. Mycol. Soc*, **82,** 170–2.
Musa, A.A., Dale, J.L. and Jones, J.P. (1982) Host range and pathogenicity of stem- (*Rhizoctonia cerealis*) and root attacking-(*R. solani*) isolates in Arkansas. *Phytopathology* **72,** 975. (Abstr.)
Ogoshi, A. (1972) Some characters of hyphal anastomosis groups in *Rhizoctonia solani* Kuhn. *Ann. Phytopathol. Soc. Jpn*, **38,** 123–9.
Oniki, M., Ogoshi, A., Araki, T. Sakai, R. and Tanaka, S. (1985) The perfect state of *Rhizoctonia oryzae* and *R. zeae* and the anastomosis groups of *Waitea* circinata. *Trans. Mycol. Soc. Jpn* **26,** 189–98.
Parmeter, J.R. Jr, Sherwood, R.T. and Platt, W.D. (1969) Anastomosis groupings among isolates of *Thanatephorus cucumeris. Phytopathology,* **59,** 1270–8.
Parmeter, J.R. and Whitney, H.S. (1970) Taxonomy and nomenclature of the imperfect state. In *Rhizoctonia solani, Biology and Pathology* (ed. J.R. Parmeter, Jr) University of California Press, Los Angeles, pp. 7–31.
Parmeter, J.R. Jr, Whitney, H.S. and Platt, W.D. (1967) Affinities of some *Rhizoctonia* species that resemble mycelium of *Thanatephorus cucumeris. Phytopathology*, **57**, 218–23.
Piper, C.V. and Coe, H.S. (1919) *Rhizoctonia* in lawns and pastures. *Phytopathology*, **9**, 89–92.
Pitt, D. (1964) Studies on sharp eyespot disease of cereals. I. Disease symptoms and pathogenicity of isolates of *R. solani* Kühn and the influence of soil factors and temperature on disease development. *Ann. Appl Biol.*, **54,** 77–89.
Reinecke, P. and Fehrmann, H. (1979) Infection experiments with *Rhizoctonia cerealis* van der Hoeven on cereals. *J. Plant Dis. Prot.*, **86**, 241–6.
Rowell, J.B. (1951) Observations on the pathogenicity of *Rhizoctonia solani* on bentgrasses. *Plant Dis. Rept.*, **35**, 240–2.
Ryker, T.C. and Gooch, F.S. (1938) Rhizoctonia-sheath spot of rice. *Phytopathology*, **28**, 233–46.
Sanders, P.L., Burpee, L.L. and Cole, H. Jr (1978) Preliminary studies on binucleate turfgrass pathogens that resemble *Rhizoctonia solani. Phytopathology*, **68**, 145–8.
Sanders, P.L., Burpee, L.L., Cole, H. Jr and Duich, J.M. (1978a) Uptake, translocation and efficacy of triadimefon in control of turfgrass pathogens. *Phytopathology,* **68,** 1482–7.
Sanders, P.L. Burpee, L.L., Cole H. Jr and Duich, J.M. (1978b) Control of fungal pathogens of turfgrass with the experimental iprodione fungicide RP 26019. *Plant Dis. Rept*, **62**, 549–3.
Scott, D.B., Visser, C.P.N. and Ruzenacht, E.M.C. (1979) Crater disease of summer wheat in African drylands. *Plant Dis. Rept.*, **63**, 836–40.
Sherwood, R.T (1969) Morphology and physiology in four anastomosis groups of *Thanatephorus cucumeris. Phytopathology*, **59**, 1924–9.
Sherwood, R.T. (1970) Physiology of *Rhizoctonia solani.* In *Rhizoctonia solani, Biology and Pathology,* (ed. J.R. Parmeter, Jr.) University of California Press, Los Angeles pp. 69–92.
Shurtleff, M.C. (1953a) Factors that influence *Rhizoctonia* to incite turf brown patch. *Phytopathology*, **43**, 484. (Abstr.)
Shurtleff, M.C. (1935b) Susceptibility of lawn grasses to brown patch. *Phytopathology*, **43,** 110. (Abstr.)
Shurtleff, M.C. (1955) Control of turf brown patch. *URI Agric. Expt. Sta. Bull.*, **328**, 25 pp.
Shurtleff, M.C. (1983) *Nigrospora* or *Rhizoctonia*? *Am. Lawn Applic.*, **4** (1) 10–13.
Smiley, R.W. (1983) *Compendium of Turfgrass Diseases.* American Phytopathology Society, St Paul, Minn., 102 pp.
Smiley, R.W., Craven, M.M. and Thompson, D.C. (1978) New fungicides for controlling brown patch. *Fungicide Nematicide Tests*, **33**, 136.
Sprague, R. (1950) *Diseases of Cereals and Grasses in North America.* Ronald Press, New York, 538 pp.
Sumner, D.R. and Bell, D.K. (1982) Root diseases induced in corn by *Rhizoctonia solani* and *Rhizoctonia zeae. Phytopathology*, **72**, 86–91.
Talbot, P.H.B (1970) Taxonomy and nomenclature of the perfect state. In *Rhizoctonia solani*, Biology and Pathology (ed. J.R. Parmeter, Jr), University of California Press, Berkeley, pp. 20–31.
Traquair, J.A. and Smith, J.D. (1981) Spring and summer brown patch of turfgrass caused by *Rhizoctonia solani* in Western Canada. *Can J. Plant Pathol.*, **3**, 207–10.
Tu, C.C., Roberts, D.A. and Kimbrough, J.W. (1969) Hyphal fusion, nuclear condition and perfect stages of three species of *Rhizoctonia. Mycologia*, **61**, 775–83.
Ullstrup, A.J. (1963) A note on the geographic distribution of *Rhizoctonia zeae. Mycologia*, **55**, 682–3.
Van der Hoeven, E.P. and Bollen, G.J. (1980) Effect of benomyl on soil fungi associated with rye. 1. Effect on the incidence of sharp eyespot caused by *Rhizoctonia cerealis. Neth. J. Plant Pathol.*, **86**, 163–80.
Voorhees, R.K. (1934) Sclerotial rot of corn caused by *Rhizoctonia zeae* n. sp. *Phytopathology*, **24**, 1290–303.
Wehlburg, C.S., Alfieri, S.A. Jr, Langdon, K.R. and Kimbrough, J.W. (1975) *Index of Plant Diseases in Florida.* Bull. 11. Div. of Plant Industry, Gainsville, Fla.
Zummo, N. and Plakidas, A.G. (1958) Brown patch of St. Augustine. *Plant Dis. Rept.*, **42**, 1141–6.

26 *Southern blight or southern sclerotium blight*

Sclerotium rolfsii Sacc. is a widely distributed soil inhabitant reported from the tropics and the warmer parts of the temperate zones throughout the world (Aycock, 1966; West 1947, 1961). The fungus is an unspecialized parasite pathogenic in varying degree to numerous plant species including many ornamentals, vegetables and crop plants. Disease attributable to *S. rolfsii* has been documented on a wide range of hosts in most southern states of the United States and commonly designated southern blight. Both dicotyledenous, and to a lesser extent, monocotyledenous families are represented in the host listings (Weber, 1931) but turf grasses, in particular cool-season species, are quite recent additions. Lucas first reported the occurrence of southern blight on *Agrostis stolonifera* turf in North Carolina during 1975 (Lucas, 1976). Further instances of the disease have since been reported in North Carolina (Lucas, 1982), California (Ohr, Humphrey and Henry, 1977; Punja, Grogan and Unruh, 1982b; Punja and Unruh, 1982), and Maryland (O'Neill, 1980). No explanation is available as to why this long-recognized and widely researched pathogen on other components of the flora in the region (Aycock, 1966) should emerge now as a potentially serious turf grass problem.

Southern sclerotium blight and sclerotium blight have since been proposed as more appropriate names for this turf disease (Punja, 1982; Endo, Ohr and Wilbur, 1982).

26.1 SYMPTOMS

The disease becomes apparent during the spring and summer as yellow to reddish-brown rings or crescents of dying grass with apparently healthy grass in the centre of the patch (Fig. 46, colour plate section). During hot, humid weather, rapid radial expansion of the diseased area occurs and the initial rings, up to 20 cm in diameter, may double in size each week to reach diameters in excess of one metre. With ample free moisture present, wefts of white mycelium occur on plant debris and dying plants at the periphery of the ring and frequently extend well beyond the leading edge of the diseased turf. Mycelium is readily visible only during periods of intense disease activity and disappears during dry weather conditions. Sclerotia, initially white in colour, turning light to dark brown, roughly spherical and 1–3 mm in diameter form on the mycelium, on the diseased grass plants and in the upper layers of any thatch present.

Recolonization of the dead area from the central island of living plants may occur during the summer, but at a rate much slower than the patches are expanding at the outer margin. Patches initiated late in the summer often exhibit no green centre or the turf surviving there may be chlorotic and thin. As the season progresses and temperatures moderate, activity of the fungus subsides and the affected patches heal in slowly.

26.2 THE CAUSAL FUNGUS

Teleomorph – *Athelia rolfsii* (Curzi) Tu and Kimbrough (1978). The synonomy of the fungus according to Punja, Grogan and Adams (1982a) is as follows: *Corticium centrifugum* (Lev.) Bres. (Goto, 1930); *Corticium rolfsii* (Sacc.) Curzi (1931); *Pellicularia rolfsii* (Sacc.) West (1947); *Botryobasidium rolfsii* (Sacc.) Venkatar. (1950)
Anamorph – *Sclerotium rolfsii* Sacc. (1911)

In the anamorph, *S. rolfsii*, the mycelium in culture on potato dextrose agar is white, and rapid growing, with fan-shaped or stranded extensions. Main hyphae are coarse, up to 9 μm diam, with clamp connections, cells mostly up to 250 μm long; secondary branches are narrower, <2 μm diam. some originating from below a clamp at a septum and growing at a narrow angle with a main hypha, others widely branched usually without clamps. Globose sclerotia, usually 1–3 mm diam., are produced in the mycelium; they have a distinct dark rind and a pale medulla. (For a more detailed description see Mordue, 1974.)

Fruiting of the fungus has been documented infrequently from nature and on hosts other than turf grasses. There are reports of the fungus fruiting in culture but no clear description of the conditions promoting basidiocarp production was available until recently. Punja, Grogan and Adams (1982a) developed a procedure utilizing potato – dextrose agar amended with 2% activated charcoal to fruit an impressive number of *S. rolfsii* cultures. Of 76 isolates from various hosts and geographic locations, 47 were induced to form fertile basidiocarps. Most consistent fruiting in culture occurred with isolates obtained in California from *Poa annua Agrostis* spp. golf greens.

Although isolates of *S. rolfsii* from turf grasses have been induced to develop the teleomorph, *Athelia rolfsii* in culture, fruiting has has not been reported on turf grasses. (For descriptions of the basidiocarp and spores on plant material (other than grasses) and in culture see Barrett, 1934; Goto, 1930; Punja, Grogan and Adams, 1982a; West, 1947.) The importance of the teleomorph in turf grass disease is not known.

S. rolfsii can be cultured on media containing a wide assortment of carbon and nitrogen sources but requires an external source of thiamine and biotin. The fungus will grow over a pH range of 1.4–8.8 with optimum growth between pH 3.0 and 6.5. Growth occurs over a temperature range of 8–40°C with a growth optimum at 25–35°C (Mordue, 1974).

26.3 HOST RANGE

The reported turf grass hosts of *S. rolfsii* include *Agrostis stolonifera, Poa annua* L., *Lolium perenne* L., *Poa pratensis* L. (Ohr, Humphrey and Henry, 1977; O'Neill, 1980; Lucas, 1982), *Lolium multiflorum* Lam. (Wells, 1959) and *Cynodon dactylon* (L) Pers. (Sprague, 1950; Punja, Grogan and Unruh, 1982b).

26.4 DISEASE DEVELOPMENT

Survival of *S. rolfsii* is ensured through the abundant production of sclerotia. Vegetative hyphae may survive a year or more in plant tissues maintained in cool dry air but are killed by a 24 hour exposure to –2°C; sclerotia can tolerate temperatures down to – 10°C for at least 48 hours but succumb to persistent cold temperatures (Epps, Patterson and Freeman, 1951; Higgins, 1927, Watkins, 1961). The various factors that influence germination of the sclerotia have been investigated extensively. The literature was reviewed and additional information on the subject was presented by Punja and Grogan (1981a, b, 1982). Their research confirmed the stimulatory effects of drying, mechanical rupture of the rind, low pH, exposure to volatile compounds released from dried and remoistened plant tissues, but also demonstrated two distinct forms of sclerotial germination. Hyphal germination, in which individual hyphal strands grow from the surface of the sclerotium, can occur repeatly. However, the emerging hyphae require a food base before infection of susceptible host tissue can take place. Eruptive germination, not previously described for this fungus, involves the depletion of all food reserves within the sclerotium to accommodate a single eruptive emergence of a plug (or plugs) of hyphae. Such aggregates of hyphae can infect host tissues in the absence of exogenous nutrients, their growth extending over several centimetres to do so, or the mycelium may form secondary sclerotia.

In agricultural crops a combination of factors that results in a warm, moist, well-aerated condition, in or on the upper inch or so of soil favours the growth and pathogenic activity of the fungus. Day time temperatures in the range 30–35°C and night-time temperatures holding at 24°C or above with intermittent rain showers constitute the optimum weather conditions. Frequently the disease is most severe when wet weather follows a protracted dry period, probably the result of enhanced sclerotial germination following drying (Boyle, 1961). Drought stress has not been associated particularly with the disease in turf situations, but since most sclerotia occur very close to the turf surface (Endo, Ohr and Wilbur, 1982), short drying periods of 1.5 to 2 hours, not deleterious to the turf, but sufficient to stimulate eruptive germination of sclerotia (Punja and Grogan, 1981a), must occur routinely. Easy accessibility to the host and/or plant debris would ensure that either method of sclerotial germination could serve to initiate disease in turf, given suitable environmental conditions. Punja and Unruh (1982) cite temperatures usually above 24°C and the presence of 'adequate' moisture as necessary for infection of crowns and roots to occur. Hot humid conditions (30–35°C and 90–100% relative humidity) favour abundant mycelium proliferation, sclerotium formation and the rapid development of southern blight symptoms in turf (Lucas, 1976).

S. rolfsii mycelium invests plant organs and penetrates directly or through wounds. Colonization is both inter- and intracellular resulting in a rapid degradation of tissues that soon leads to the chlorotic, later reddish-brown appearance of the diseased grass plants. Sclerotia then form on the lower leaves of the dying plants and in and on the thatch. The fungus may spread locally by extension of mycelial strands growing saprophytically in the thatch layer but the transport of sclerotia and infected clippings on feet and turf-maintenance equipment are considered the primary means of wider dissemination (Endo, Ohr and Wilbur, 1982; Punja, 1982). Although isolates of *S. rolfsii* from turf have been induced to form the sexual spore state in laboratory culture (Punja, Grogan and Adams, 1982a), basidiospores are not at present considered important in the spread of the disease (Punja, 1982; Punja, Grogan and Unruh, 1982b).

26.5 CONTROL

Cultural, biological and chemical control practices have been established for southern blight on various crop plants but for the newly occurring problem on turf grasses appropriate information is not available. Responding to the 1977 outbreak of the disease in northern California, Punja, Grogan and Unruh (1982b, c) conducted a series of laboratory and field trials directed towards finding effective chemical control measures. Few of the 22 chemicals, in a list that included many of the common turf fungicides, were active in suppressing germination of *S. rolfsii* sclerotia *in vitro*. With the exception of mancozeb, those that were effective in the preliminary laboratory screening proved satisfactory in field trials conducted in 1980. Good control was afforded by dichloran (DCNA) and cycloheximide in combination, captan, carboxin, and quintozene (PCNB) with fertilizer. Plots receiving ammonium bicarbonate or ammonium sulphate also had significantly less diseased turf than the untreated control plots. Materials were applied bi-monthly over 3.5 months in the 1980 season starting on 5 May. In further field trials commencing on 1 May, 1981, the effectiveness of bi-weekly applications of the materials named above was confirmed over a 4-month period and furmecyclox was added to the list. Carboxin applied at reduced rates in combination with reduced amounts of either captan or ammonium bicarbonate provided better disease control than carboxin or captan applied alone at higher rates.

Satisfactory control of the disease was not obtained by application of either calcium nitrate or hydrated lime and inoculation of turf with cultures of *Trichoderma* spp. over a two-year period did not reduce disease severity in either year. The fungicides that were effective in northern California against southern blight proved equally effective in field trials in southern California during 1980 and 1981. In addition, captafol was reported to afford good protection (Endo, Ohr and Wilbur, 1982). Excellent control of the disease was obtained on severely affected *P. pratensis* turf in North Carolina with one application of a high rate of triadimefon (Lucas, 1982).

Reliance solely on the application of materials containing nitrogen to control southern blight is not advised since, although effective, their repeated use may influence turf quality adversely or predispose the turf to other disease problems. Satisfactory southern blight control is obtained, however, when the materials are used in combination with selected fungicides in a preventive spray programme (Punja, 1982; Punja Grogan and Unruh, 1982b, c).

26.6 REFERENCES

Aycock, R. (1966) Stem rot and other diseases caused by *Sclerotium rolfsii*. *N.C. Agric. Exp. Sta. Tech. Bull.*, **174**, 202 pp.

Barrett, J.T. (1934) Observations on the basidial stage of *Sclerotium rolfsii*. *Phytopathology*, **24**, 1137–8.

Boyle, L.W. (1961) The ecology of *Sclerotium rolfsii* with emphasis on the role of saprophytic media. *Phytopathology*, **51**, 117–19.

Curzi, M. (1931) Some cases of 'foot canker' caused by *Sclerotium* observed in Italy. *Rendic R. Accad. Lincei*, **14**, Ser. VI, 5–6, 233–6.

Endo, R.M., Ohr, H.D. and Wilbur, W.D. (1982) Sclerotium blight of cool season turfgrasses in southern California and its control. *Calif. Turfgrass Culture*, **32**, 6.

Epps, W.M., Patterson J.C. and Freeman, I.E. (1951) Physiology and parasitism of *Sclerotium rolfsii*. *Phytopathology*, **41**, 245–56.

Goto, K. (1930) On the perfect stage of *Sclerotium rolfsii* Sacc. produced on culture media. Preliminary report. *J. Soc. Tropic Agric.*, **2**, 165–75.

Higgins, B.B. (1927) Physiology and parasitism of *Sclerotium rolfsii* Sacc. *Phytopathology*, **17**, 417–48.

Lucas, L.T. (1976) *Sclerotium rolfsii* on bentgrass greens in North Carolina. *Plant Dis. Rept.*, **60**, 820–2.

Lucas, L.T. (1982) Southern blight. *Grounds Maintenance* **17**, 40–2.

Mordue, J.E.M. (1974) *Corticium rolfsii*. *CMI Descriptions of Pathogenic Fungi and Bacteria*, no. 410.

Ohr, O.H., Humphrey, W.A. and Henry, M.J. (1977) *Sclerotium rolfsii* on turf in California. In (ed. V. Gibeault), *Proc. Turf. Landscape Inst*, Anaheim, CA., Riverside Publication, University of California, pp. 12–13.

O'Neill, N.R. (1980) Southern blight of cool season grasses. *Phytopathology*, **70**, 691. (Abstr.)

Punja, Z.K. (1982) Southern sclerotium blight on golf greens in northern California and its control. *Calif. Turfgrass Culture*, **32**, 4–5.

Punja, Z.K. and Grogan, R.G. (1981a) Eruptive germination of sclerotia of *Sclerotium rolfsii*. *Phytopathology*, **71**, 1092–9.

Punja, Z.K. and Grogan, R.G. (1981b) Mycelial growth and infection without a food base by eruptively germinating sclerotia of *Sclerotium rolfsii*. *Phytopathology*, **71**, 1099–103.

Punja, Z.K. and Grogan, R.G. (1982) Effects of inorganic salts, carbonate-bicarbonate anions, ammonia, and the modifying influence pH on sclerotial germination of *Sclerotium rolfsii*. *Phytopathology*, **72**, 635–9.

Punja, Z.K., Grogan, R.G. and Adams, G.C. Jr (1982a) Influence of nutrition, environment, and the isolate, on basidiocarp formation, development and structure in *Athelia (Sclerotium) rolfsii*. *Mycologia*, **74**, 917–26.

Punja, Z.K., Grogan, R.G. and Unruh, T. (1982b) Chemical control of *Sclerotium rolfsii* on golf greens in northern

California. *Plant Dis.*, **66**, 108–11.

Punja, Z.K., Grogan, R.G. and Unruh, T. (1982c) Comparative control of *Sclerotium rolfsii* on golf greens in northern California with fungicides, inorganic salts and *Trichoderma spp. Plant Dis.*, **66**, 1125–8.

Punja, Z.K. and Unruh, T. (1982) Southern blight. *Grounds Maintenance*, **17**, 46–8.

Saccardo, P.A. (1911) Notae mycologicae. *Ann. Mycol.*, **9**, 249–57.

Sprague, R. (1950) *Diseases of Cereals and Grasses in North America.* Ronald Press, New York, 538 pp.

Tu, C.C. and Kimbrough, J.W. (1978) Systematics and phylogeny of fungi in the *Rhizoctonia* complex. *Bot. Gaz.* (Crawfordsville), **139**, 454–66.

Venkatarayan, S.V. (1950) Notes on some species of *Corticium* and *Pellicularia. Indian Phytopathol.*, **3**, 81–6.

Watkins, G.M. (1961) Physiology of *Sclerotium rolfsii* with emphasis on parasitism. *Phytopathology*, **51**, 110–13.

Weber, G.F. (1931) Blight of carrots caused by *Sclerotium rolfsii*, with geographic distribution and host range of the fungus. *Phytopathology*, **21**, 1129–40.

Wells, H.D. (1959) Annual ryegrass, *Lolium multiflorum*, host for *Sclerotium rolfsii. Plant Dis. Rept.*, **43**, 834.

West, E. (1947) *Sclerotium rolfsii* Sacc. and its perfect stage on climbing fig. *Phytopathology*, **37**, 67–9.

West, E. (1961) *Sclerotium rolfsii*, history, taxonomy, host range and distribution. *Phytopathology*, **51**, 108–9.

Part nine

Leaf spots, leaf blights and leaf streaks

27 *Leaf spots, leaf blights, crown rots and seedling blights of turf grasses caused by Drechslera, 'Helminthosporium' and Curvularia species*

27.1 INTRODUCTION

'*Helminthosporium*' and *Curvularia* diseases of the Poaceae have a world-wide distribution from boreal to tropical regions (Cagas, 1973; Del Vescovo, 1962; Drechsler, 1923; Ellis, 1966, 1971; Endo, 1961; Gornostai, 1970; Kenneth, 1958; Latch, 1966; Mäkelä, 1971; Misra, Singh and Prakesh, 1971; Nishihara, 1967; Putterill, 1954; Smedegard-Petersen and Hermansen, 1972; Sprague, 1950; Subramanian, 1954). Many have been reported seed-borne (Chidambaram, Mathur and Neergaard, 1973; Mäkelä, 1972a, 1981; McKenzie, 1978; Richardson, 1979). Some of the '*Helminthosporium*' diseases are among the most serious of turf grasses. *Curvularia* spp. often occur in complexes with other pathogens and their role in causing disease is still not well understood. Where other pathogens are often confined to roots, stems, leaves or inflorescences, many of the '*Helminthosporium*' and *Curvularia* diseases affect aerial and underground organs of seedling and mature plants.

27.2 GENERAL TAXONOMY OF THE PATHOGENS

Before the teleomorphs were known in these fungi, their classification was based mainly on conidial (anamorphic) characters. Nevertheless, these were used, in many cases, to predict their teleomorphic states. Several of the species have more than one synonym for both teleomorph and anamorph, but by an evolutionary process, evoking considerable controversy and not a little confusion, most have been fitted into the teleomorph/anamorph combinations:

Pyrenophora–Drechslera
Cochliobolus–Bipolaris and *Curvularia*
Setosphaeria–Exserohilum

The names '*Helminthosporium*' and *Curvularia* for the conidial states of the graminicolous species of these pathogens have been gradually disappearing from the literature as anamorphic and teleomorphic affinities have been determined. Since we agree with Luttrell (1977a) and Shoemaker (1981) that we are now dealing with a group of pathogens whose classification is based on teleomorphic states (Alcorn, 1981; Sivanesan, 1984), the name of the latter will be used for the fungus where this is known and is practicable. There are some difficulties in correlating the teleomorphs of *Curvularia* with *Cochliobolus*, but the need for the separate genus *Pseudocochliobolus* (Tsuda, Ueyama and Nisada, 1977) for these, does not appear to be warranted (Alcorn, 1983a,b; Luttrell, 1977a,b). The salient features of this classification are given in Table 27.1.

Drechslera spp. are essentially restricted to festucoid grasses to which all the cool-season turf grass species belong, whereas *Cochliobolus* and *Exserohilum* spp. are pathogenic on panicoids which include the warm-season genera *Pennisetum, Paspalum* and *Stenotaphrum* and on the eragrostoids *Eragrostis, Cynodon* and *Zoysia*, but also on dicotyledonous plants. An exception to this is *Cochliobolus sativus*, pathogenic on many of the cool-season festucoids. *Drechslera* spp. are occasionally found on seeds of grasses outside the festucoids, but essentially as saprobes. *Drechslera gigantea* is misplaced as a *Drechslera*. Although its conidia are cylindrical, they are much longer than the remaining species of the gcnus, havc an unusual, small apiculum at the point of attachment to the conidiophore, an unusual mode of germination and the production of a *Dendriophion* microconidial stage. It has a very wide host range (plurivorous), including cool- and warm-season grasses, and bananas (Hagan, 1980; Meredith, 1963).

Where only the anamorphs are available, conidium and conidiophore size and morphology are the main means by which species of '*Helminthosporium*' and *Curvularia* may be identified. However, cultural characters such as pigment formation, colony appearance, occurrence of protothecia or sclerotia may provide valuable additional diagnostic data and these have been used extensively by several workers, to whom reference should be made (Ammon, 1963; Braverman and Graham, 1960; Ellis, 1966; Groves and Skolko, 1945; Shoemaker, 1962; Subramanian, 1954). Characteristics may vary considerably under different cultural conditions. in particular of light and temperature. Spore morphology on natural substrates sometimes differs greatly from that in artificial culture. Keys for the identification of *Bipolaris, Drechslera* and *Curvularia* are given by Ellis (1971), Luttrell (1951a) and Shoemaker (1962).

In the following sections diseases are dealt with in alphabetical order of pathogens.

Table 27.1

Anamorph	*Teleomorph*
In *Drechslera*, conidia develop porogenously (at an apical pore of the conidiophore), are cylindrical or nearly so when mature and are often pale in colour. The attachment scar is included in the contour of the basal cell. Germination is by the lateral production of one or more germ tubes from each cell.	*Pyrenophora* has pale elongate–oblong ascospores with transverse and longitudinal septa, surrounded by a thin, gelatinous sheath. Ascocarps are large, globoid, sclerotioid and covered with hairs or conidiophores, beakless or with a small beak.
In *Bipolaris*, conidia develop porogenously, germinating by one germ tube from each end; they are fusoid, straight or curved, often dark-coloured, without conspicuously swollen cells, long and many septate; the hilum is contained within the cell wall of the basal cell or at the tip of a conical protrusion of that cell wall.	*Cochliobolus* has elongate, cylindrical asci containing filiform, transversely septate ascospores, spirally fasciculate. Ascocarps are globoid with a long, cylindrical beak.
In *Curvularia*, conidia develop acropleurogenously (at the end and sides of the conidiophore). The *lunata* group of species has 3-septate conidia, these are strongly curved and have one cell disproportionately large. The *maculans* group has 3-septate spores where the two central cells are larger and darker than the two end cells and the conidia are only slightly curved. The *geniculata* group have spores with more than three septa, the third cell from the base is swollen to one side and the conidium is curved. The hilum in some species is protuberant. Some of the species have *Cochliobolus* teleomorphs.	
In *Exserohilum*, conidia develop porogenously, germination is bipolar; conidia are sub-cylindrical to fusoid or broadly obclavate and beaked, olive to brown; the hilum protrudes from the basal cell.	*Setosphaeria* has cylindrical or cylindro-clavate asci with phragmosporous ascospores which are hyaline, fusoid, 2–6-septate, enclosed in a thin, gelatinous sheath, projecting beyond the spore. Ascocarps are black, globose, with rigid, dark-brown hairs surrounding the ostiole.

27.3 BIPOLARIS SPECIES

27.3.1 *Bipolaris mediocre* (Putterill) (Shoemaker, 1959)

B. mediocre is associated with spots on discoloured leaf sheaths and blades of *Pennisetum clandestinum* Hochst, a turf forming species native to east central Africa. Conidia are brownish-yellow, 40–108 × 13–18 μm, 2–10 septate, sometimes curved, but usually straight, tapering slightly to the rounded ends from the middle of the spore of slightly below.

27.3.2 *Bipolaris micropa* (Drechs.) (Shoemaker, 1959)

B. micropa has an unnamed *Cochliobolus* teleomorph (Luttrell, 1958) and causes a leaf blight on *Paspalum boscianum* Flügge and *P. notatum* Flügge. It is seed-borne on the latter species. Conidia are sooty to light brown, borne acropleurogenously, straight or slightly curved, with a smooth contour, tapering gently from the middle or just below to both ends. The apex is rounded, and the basal portion of conidium wall is protruding (not just the hilum), 37–53 × 12–14 μm, 4–7 septate.

27.3.3 *Bipolaris stenospila* (Drechs.) (Shoemaker, 1959)

(*Cochliobolus stenospilus* An anomorph) was described by Matsumoto and Yamamoto (1936), but without Latin description, therefore invalid). Although primarily a disease of sugar cane, with a wide distribution, some lines of *Cynodon dactylon* are very susceptible, causing leaf spotting and blotching. Conidia usually curved, cylindrical to fusiform, dark olivaceous to golden brown with a thick peripheral wall and smooth outline, 7–135 × 14–22 μm 6–14 septate, scar not conspicuous.

27.3.4 *Bipolaris tetramera* (Kinney) (Shoemaker, 1959)

B. tetramera causes a root rot of wheat which is seed-borne on many grass species. It has been confused with *Cochliobolus spicifer* (q.v.). Conidia are acropleurogenous on conidiophores; light to dark brown, ellipsoid to almost cylindrical, symmetrical, slightly tapering to rounded ends, 20–38 × 9–14 μm, usually 3-septate.

27.4 COCHLIOBOLUS SPECIES

27.4.1 *Cochliobolus australiensis* (Tsuda and Ueyama) (Alcorn, 1983)

C. australiensis (*Bipolaris australiensis* (M.B. Ellis) Tsuda and Ueyama, 1981) is pathogenic on *Chloris* and *Pennisetum* spp., turf-forming grasses in Australia, Kenya and India. The conidia are straight, ellipsoidal or oblong, rounded at ends, mostly 3-septate, 13–40 × 6–11 μm.

27.4.2 *Cochliobolus hawaiiensis* Alcorn (1978)

C. hawaiiensis (*Bipolaris hawaiiensis* (M.B. Ellis) Tsuda and Ueyama) is pathogenic on *Cynodon dactylon* and on *Chloris gayana* Kunth., a tropical turf-forming species, but is widely distributed on other tropical species. The conidia are pale to mid-brown, straight, ellipsoidal oblong or cylindrical, rounded at ends, 12–37 × 5–11 μm, and 2-7 septate.

27.4.3 *Cochliobolus peregianus* Alcorn (1978)

C. peregianus (*Bipolaris peregianus* Alcorn) is associated with oblong–elliptical, dark-purple lesions up to 2 × 1 mm on leaves of *Cynodon dactylon.* The conidia are mid-olivaceous, brown to reddish-brown, fusoid, very curved, with a smooth outline and hemiellipsoidal basal and apical cells, 60–88 × 11–17 μm; the hilum is not conspicuous, but sometimes slightly protruding.

27.4.4 Leaf spot and mould of bermuda grass caused by *Cochliobolus cynodontis*

The disease is common on *Cynodon dactylon* L.: the fungus is plurivorous, cosmopolitan and seed-borne (Chidambaram, Mathur and Neergaard, 1973; Drechsler, 1923; Ellis, 1971; Kenneth, 1958; Marignoni, 1909; Misra, Singh and Prakesh, 1971; Nattrass, 1939; Nelson, 1964b; Nishihara, 1967; Putterill, 1954; Ricci, Geraldi and Ito, 1981; Riley, 1960; Sprague, 1950; Winter, Mathur and Neergaard, 1974). While it is often associated with senescing leaves of the grass (Endo, 1961; Rogerson, 1956; Sprague, 1950) and is less pathogenic than *Cochliobolus stenospilus* (Freeman, 1957) and probably other '*Helminthosporium*' species such as *Cochliobolus spicifer* (Gudauskas, 1962), it may be highly pathogenic on some *Cynodon* spp. and cultivars (Slana, 1977) especially in moist, warm weather and is pathogenic on other grasses (Nelson, 1964b).

(a) Symptoms

The fungus may cause olive to dark-brown spots on leaf blades and leaf sheaths of *Cynodon* spp., but more often blotches of indeterminate shape develop on leaves which may turn sooty with conidia and conidiophores of the fungus or they may just bleach, wither and dry up. Patches of turf up to a metre or so in diameter may be affected. Crowns and roots may be lesioned also.

(b) The fungus

Teleomorph – *Cochliobolus cynodontis* Nelson (1964b); anamorph – *Bipolaris cynodontis* (Marig.) Shoemaker (1959): (For a description of the teleomorph see Nelson (1964b).)

Conidiophores are dark-brown, usually occurring singly or in pairs, usually curved and geniculate with up to 8 septa, 40–150 μm × 4–7 μm. Conidia are usually curved, tapering to rounded ends, subhyaline to sooty with a smooth contour and with a thin peripheral wall; up to 9 septa, 30–80 × 4–7 μm (Fig. 27.1). Germination is bipolar.

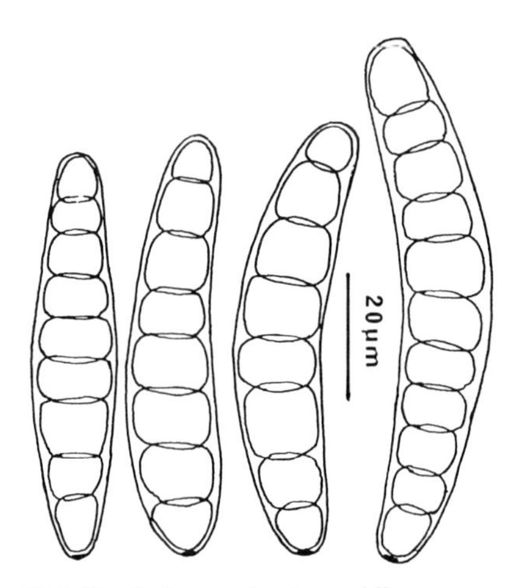

Figure 27.1 *Bipolaris cynodontis*–conidia.

(c) Disease development

In mild winter climates the disease may be visible during all seasons, but the fungus survives unfavourable conditions in cold periods and hot, dry weather as mycelium in the crowns or in plant debris, becoming active and pathogenic in cool, moist periods in autumn and again in spring.

(d) Control

Balanced fertilization improves the resistance of *C. dactylon* to the pathogen (Eichorn, 1976; Matocha and Smith, 1980). For general management practice and fungicidal treatment for this 'cool-season' leaf spot disease see *D. poae* (p. 299). Tifgreen, Turftex 10 and the Ormond cultivars of *C. dactylon* were reported resistant in field and greenhouse tests (Slana, 1977).

27.4.5 Leaf spot, leaf blight, seedling, crown and root rot of turf grasses caused by *Cochliobolus sativus*

The cosmopolitan, plurivorous fungus *C. sativus* (CMI Map 332; Conners, 1967; Ellis, 1971; USDA 1960), especially important as a pathogen on the Poaceae generally, injures turf grasses and is seed-borne (Chidambaram, Mathur and Neergaard, 1973; Richardson, 1979; McKenzie, 1978; Mäkelä, 1971). It has been shown, by inoculation, to be pathogenic on a wide range of seedling or mature turf grasses (Berkenkamp, 1971; Christensen, 1922; Endo, 1961; Hodges and Watschke, 1975; Kline and Nelson, 1963; Klomparens, 1953; Mower, 1962). Severe turf disease, caused by this fungus, sometimes in complex with other species, has been reported from southern Canada (Bolton and Cordukes, 1979; Lebeau, 1964; Noviello, 1963; Platford, Bernier and Ferguson, 1972; Smith and Evans 1985; Vanterpool, 1948), the northern states of the USA from the Great Plains to West Virginia (Bean and Wilcoxson, 1964; Hodges, 1980a; Klomparens, 1953; Vargas and Kelly, 1981; Weihing, Jensen and Hamilton, 1957) and southern California (Endo, 1961). It is often encountered, sometimes as a major turf grass pathogen, elsewhere in North America, and is a potential turf grass pathogen from cool-temperate to warm-temperate climates.

In cooler climates, such as those of northern and western Europe, only occasional damage to grass species has been reported (Andersen, 1955; Del Vescova, 1963; Mäkelä, 1975). In Britain, *C. sativus* caused severe seedling blight on a newly sown lawn of *Lolium perenne* L. (Smith, 1953) and damage and leaf blight and death of mature *Festuca rubra* L. in a bowling green of sea-marsh turf (Smith, 1955a). However, since the fungus has recently been isolated from several new wild grass hosts in Britain, including turf grass species, it may be more prevalent than previously supposed (Whittle, 1977).

(a) Symptoms

Damage to seedling grasses may be so severe that they do not emerge (pre-emergence blight), or if they do the coleoptile or primary leaves blacken and wither (post-emergence blight). Leaves and roots of seedlings and young plants may be lesioned and killed. On mature plants of *Poa pratensis*, leaf spots may be abundant, giving a speckled appearance and the whole leaf blade may become chlorotic. At first, the spots on the leaf blade are almost circular, dark brown or purple and solid, but as they increase in size their centres become light in colour with dark spot margins which may be surrounded with a yellowish halo. They are almost indistinguishable from those caused by *Dechslera poae*. (Correct diagnosis requires incubation in a moist chamber and microscopical examination. *C. sativus* will sporulate more freely than *D. poae*.) Leaf spots may coalesce and the whole leaf may be blighted, bleached, withered and may drop. The fungus invades crown and root tissues, blackening them. Irregular-shaped patches of brown, damaged turf develop.

The symptoms of disease caused by *C. sativus* on *Agrostis stolonifera* L. were described by Klomparens (1953), mainly from Michigan, but also from Ohio, Illinois, Wisconsin and Texas. A smoky-blue cast developed on the turf in patches 30–100 cm in diameter which was followed by yellowing and complete killing of the grass plants. The edges of the patches were definite, irregular, the damaged areas watersoaked in appearance and matted. Since roots were rotted at an early stage of attack there was little recovery in severely damaged patches. The leaves of individual plants show early symptoms as oval spots 3–6 mm × 1.5–3 mm approx. starting as yellow flecks and progressing through the spot stage to a watersoaked blotch, at the yellow patch stage. On *Poa annua* L. and *A. stolonifera* in southern California, the brown foot rot stage common to most grass hosts developed, but few leaf spots, only a general chlorosis, turning brown later. In Britain, spotting of leaves of the sea-marsh fescue, *Festuca rubra* L. ssp. *litoralis*, was not seen but they were blackened and blighted (Smith, 1955b).

(b) The fungus

Teleomorph – *Cochliobolus sativus* (Ito and Kuribay) Drechsl. ex Dast. (For synonyms see Drechsler (1934), Shoemaker (1955), Kuribayashi (1929) and Alcorn (1983a).) Anamorph – *Bipolaris sorokiniana* (Sacc.) Shoem. (see Shoemaker 1959).

Ascocarps develop on dead host tissue or sterilized grain in culture media (Harding and Tinline, 1983; Kuribayashi, 1929; Shoemaker, 1955; Tinline, 1951). They are erumpent early with a black pseudoparenchymatous wall, globose or sub-globose, 370–530 × 340–470 μm with well-developed, subconical or cylindrical beaks, 80–110 μm wide and 90–150 μm high.

Asci are numerous, hyaline, thin-walled, bitunicate, cylindrical or long fusiform, straight or curved with a short stalk and rounded apex, 110–220 × 32–45 μm, usually with 4–8 ascospores (Kuribayashi, 1929), sometimes fewer in crosses (Shoemaker, 1955). Pseudoparaphyses and present, downward directed.

Ascospores are scolecosporous, light olivaceous, coiled in a tight helix (Tinline, 1951) within the ascus when mature, 6–14 septate, each cell is multinucleate, 160–360 × 6–9 μm, extruded when ascus top ruptures. They occasionally germinate *in situ*. Ascospore production occurs regularly at 20 °C only (Kuribayashi, 1929; Shoemaker, 1955; Tinline, 1951).

Spermagonia are formed in culture regularly on sterilized maize seed at 24 °C on all isolates, up to 2.5 mm diam. with spermatia produced centrally in slime on phialides, mostly 2 μm diam. with a large single nucleus. Spermatia are functional (Shoemaker, 1955).

Conidiophores are brown, emerging singly or in groups of twos and threes from the stomata or between the epidermal cells after the death of the host tissues, 6–8 × 100–150 μm, up to 8-septate with well-defined geniculations and scars (Sprague, 1950).

Conidia are dark-brown when mature, with a thick, brittle, peripheral wall; curved, tapering evenly towards each rounded end or irregularly boomerang-shaped, bent elliptical-ovate or even pip-shaped, rarely bilobed or triangular, 3–10 septate, 15–20 × 60–120 μm with a conspicuous hilum which projects very slightly outside the basal cell contour. Dimensions of conidia may vary considerably with cultural conditions and host (Ammon, 1963; Berkenkamp, 1971; Smith, 1955a). Spore germination is bipolar. Spores are produced in the dark or light, freely on inoculated sterile grass straws, on end, partly immersed in sterile water (Tinline, 1951). Optimum temperature for conidial formation is 28°C (Shoemaker, 1955). Conidia in culture or on decayed material tend to be straight, short, few-celled and even pip-shaped (Fig. 1, colour plate section; Fig. 27.2).

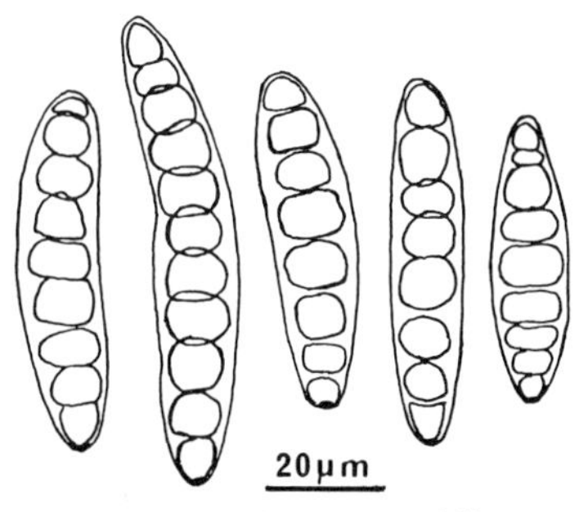

Figure 27.2 *Bipolaris sorokiniana*–conidia.

(c) Hosts

Sprague (1950) lists more than 100 grass hosts from Canada and the United States. *Agrostis stolonifera* L. is an additional host from the USA (Klomparens, 1953). *Cynosurus cristatus* L., *Festuca rubra* L., *Phleum pratense* L. and *Lolium perenne* L. have been reported as natural suscepts from Britain (Smith, 1953, 1955a; Whittle, 1977), while Mäkelä (1971) recorded *C. sativus* several times on *Agrostis tenuis* L. and *Phleum pratense* L. in Finland and Andersen (1955) noted *Festuca pratensis* Huds. and *L. perenne* plants as hosts in Denmark. *Elymus junceus* Fisch., used as a turf grass in the Canadian prairies, and other *Elymus* species are natural suscepts (Berkenkamp, Folkins and Meeres, 1973). The fungus has frequently been isolated from *Poa annua* in golf turf in eastern Ontario and western Quebec where it was associated with melting-out symptoms (Bolton and Cordukes, 1979). In complex with *Colletotrichum graminicola* (Ces.) Wils. Vargas and Kelly (1981) implicate it in *P. annua* (HAS) decline in Michigan.

The range of plants of turf grass species which have been shown susceptible when inoculated is also very wide, including some which have not been shown as natural hosts (Berkenkamp, 1971; Christensen, 1922; Kline and Nelson, 1963). The range of turf grass and non-grass species on which *C. sativus* has been shown seed-borne is also very wide, although many of these species are not necessarily suscepts.

(d) Disease development

C. sativus is an unspecialized, primary pathogen of grasses (Berkenkamp, 1971; Butler, 1953; Padwick and Henry, 1933; Smith, 1955a) causing seedling blight, (Fig 1, colour plate section) alone or in complexes with other organisms with which it may compete (Madsen and Hodges, 1980a). It shows poor competitive saprobic activity and sporulation and spore germination are inhibited in soil (Chinn and Tinline, 1963; Simmonds, Sallans and Ledingham, 1950). Inoculum may be seed-borne or soil-borne on colonized plant debris (Simmonds, Sallans and Ledingham, 1950) or on infected plants in the sward, on which the fungus sporulates. In those grass seedlings which manage to emerge, the coleoptile may show a dark rot, which passes to the cortex, sub-crown and adventitious roots. The lesioned leaves of seedlings and young plants which survive support the growth of conidia which are splashed or carried by the wind and infect other aerial tissues.

At 25 °C and 100% relative humidity conidia will start to germinate on *P. pratensis* leaves in 30–45 min. Substantial infection of leaves requires 8–10 hours under these conditions (Weihing, Jensen and Hamilton, 1957). Penetration of leaves is through a cell or stomata with or without the formation of an appressorium (Mower, 1962; Robinson and Hodges, 1977). In *Agrostis stolonifera* penetration of leaves is nearly always preceded by appressorium formation and most penetrations are direct. Macroscopical symptoms may be visible, as minute spots after 24 hours (Mower, 1962).

After initial leaf spotting in spring, foliage symptoms on *P. pratensis* increase in severity and frequency accompanied by a destructive leaf blight and crown rot. Atmospheric temperatures were shown to govern the kind of foliage damage which developed in controlled environment studies:

1. Slight leaf spotting, but no leaf blight at 20 °C.
2. Considerable spotting, with slight leaf blight at 25 °C.
3. Severe leaf blight, but slight leaf spotting at 30 °C.
4. Severe blight and heavy plant destruction at 35 °C.

These conditions simulated the seasonal progression of the temperature during the progress of an epidemic, especially in the north central states of the USA (Hodges and Blaine, 1974). Under alternating dry and moist conditions the disease was much less active and spotting rather than blighting took place (Weihing, Jensen and Hamilton, 1957). The pathogen is a common cause of severe seedling blight or *Agrostis stolonifera* in hot summer weather, particularly on sand-based greens in Saskatchewan (Smith, unpublished). Leaf spot usually occurs during the cooler, moist spring and autumn weather, ceasing as midsummer temperatures rise when infections of sheaths, crowns and roots develop leading to melting-out.

Generally, increasing dosages of nitrogen increase the susceptibility of *P. pratensis* cultivars (Cheeseman, Roberts and Tiffany, 1965; Couch and Moore, 1971; Roberts and Cheeseman, 1964) to *C. sativus* leaf spot, as is the case with that caused by *D. poae* (Lukens, 1970). However, other management factors, such as irrigation (Hodges and Blaine, 1974), mowing height, as with *D. poae* leaf spot (Drechsler, 1930; Halisky, Funk and Engel, 1966), thatch accumulation (Healey and Britton, 1968), and post-emergence herbicides (Hodges, 1980a) may also alter cultivar susceptibility.

A relationship between shading and increased nitrogen applications and reduced sugar content of leaves (Alberda, 1957; Jones, Griffiths and Walters, 1961) led to the concept that the leaf spot of *P. pratensis* caused by *D. poae* was a 'low sugar' disease (Lukens, 1970). However, a correlation between foliage sugar content and susceptibility to *C. sativus* leaf spot was not established (Couch and Moore, 1971). Ammonium sulphate, ammonium nitrate and calcium nitrate applications to soil increased conidial germination, germ-tube length and branching on leaf surfaces of leaves of *Poa pratensis*. None of the nitrogen sources significantly changed frequency of lesioning but all increased lesion size. Ammonium sulphate increased disease severity, and this and ammonium nitrate decreased total sugars in leaf tissue. Decreases in fructose, glucose and sucrose were negatively correlated with increased pathogen growth on leaf surfaces, but not correlated with leaf disease. It was suggested that increases in glutamic acid and proline amino acids resulting from nitrogen fertilization might contribute more directly to severity of leaf disease than the decrease in soluble sugars (Robinson and Hodges, 1977).

Leaf blight of *P. pratensis* increases following periods of drought stress. Drying and wetting cycles appear to trigger the saprophytic and parasitic activities of *C. sativus*. *P. pratensis* litter kept moist inhibited conidial germination and mycelial growth: when dried the volatile inhibitor disappeared. Immediately after remoistening, the litter was not inhibitory and released greatly increased amounts of carbohydrates and proteins, that encouraged rapid conidial germination, mycelial growth and infection by *C. sativus*. The cycle of inhibition and stimulation is repeated as the litter is wetted and dries out (Colbaugh and Endo, 1974). Fewer lesions were found on the leaf blades of water-stressed than non-stressed and more on the leaf sheaths of stressed than non-stressed plants. A water-saturated atmosphere was needed to induce sporulation and no sporulation of *C. sativus* occurred at 97% relative humidity or lower (Atalla and Cappellini, 1977).

Changes in leaf spot symptoms caused by *C. sativus* on *P. pratensis* have been related by Nilsen, Hodges and Madsen, (1979) Nilsen, Madsen and Hodges (1978) and Hodges (1980a) to changes in host physiology associated with sequential senescence of understorey leaves in the plant canopy. In spring and early summer leaf spots are small with or without chlorotic halos, but in late autumn and early winter they take the form of enlarged lesions joined by chlorotic streaks or complete straw-coloured chlorosis. As day-length shortens, symptoms become more severe on each older infected leaf (Hodges, 1980b). These leaf spot symptoms can be induced by the manipulation of photoperiod and light quality. This more severe lesion type symptom also occurs in response to *P. pratensis* exposed to the herbicides 2,4,5-T, MCPP and dicamba at concentrations of 10^{-9} and 10^{-6} M before inoculation with conidia of *C. sativus*. It has been suggested that the stimulation of *C. sativus* leaf spot on *P. pratensis* by these herbicides is mainly the result of physiological disturbances in this herbicide-tolerant host that is exploited by the pathogen (Hodges, 1980a).

Christensen (1926) considered that there were numerous physiological forms of *C. sativus* distinguished by particular cultural characters which also differed from each other pathogenically. Virulence varied from high to very low. Kline and Nelson (1963) referred differences in pathogenic capacities to the occurrence of pathogenic biotypes rather than specialized physiological forms. The pathogenicity of 28 isolates of the fungus from 15 species of wild and cultivated grasses from Alaska, Canada, United States, Mexico and Paraguay did not appear to be related to source host species of geographic origin. Mating types of these isolates were randomly distributed in all geographic areas. Isolates obtained from diverse hosts and locations were in general pathogenic to widely different cultivated and wild grasses, indicating that the populations of *C. sativus* examined were not highly specialized pathogenically. A later study confirmed that the qualitative pathogenicity to a particular species was simply inherited and dependent on one or two genes. Seven different genes conditioning lesioning in six grasses were identified, so crosses between isolates differing in pathogenicity to grasses could result in many pathotypes (Kline and Nelson, 1971). Five isolates from grasses in central Alberta examined for pathogenicity on foliage and roots of five cereals and foliage of 24 grasses showed significant differences in plant resistance and in isolate pathogencity. All grasses were infected by at least three of the isolates. This indicated that the isolates were not pathogenic races and not specific in their reaction to the grasses (Berkenkamp, 1971).

(e) Resistant turf grass species and cultivars

Few turf grass species show much resistance to *C. sativus*. *Poa trivialis* L., *Festuca ovina* L. and *F. rubra* L. were the least susceptible to foliage infection (Berkenkamp, 1971) and *Phleum pratense*, occasionally used in turf, was also reported as resistant (Berkenkamp, 1976; Padwick and Henry, 1933).

Cultivars of *P. pratensis* vary greatly in susceptibility in response to nitrogen nutrition, although susceptibility is usually least when it is low. Susceptibility needs to be evaluated under a wide range of nutritional levels. Couch and Moore (1971) found Anheuser resistant and Delta very susceptible at all nitrogen levels. It is also necessary to evaluate the resistance of *P. pratensis* cultivars to *C. sativus* on mature turf since the incidence of leaf spot increases with successive years of growth (Hodges and Blaine,

1974; Hodges, 1980b). Twelve of 25 cultivars tested in Iowa in midsummer 1973 had less than 30% of leaves infected (Hodges and Blaine, 1974). In a further test only 5 of 32 cultivars tested had less than 30% leaf infection after 4 years, comprising Adelphi, Olymprisp, Merion, Rugby and S. Dakota Certified (Hodges and Blaine, 1974). Subsequently, S. Dakota proved more susceptible (Hodges, 1980b). These increases in disease severity might have been due to increases in the inoculum potential or in virulence of *C. sativus*. Changes of the latter kind do not appear to have been documented (Dr H. Harding, personal communication). The resistance of cultivars of *P. pratensis* to *C. sativus* may vary considerably in different regions. In Virginia (Muse *et al.*, 1972) and in the Canadian prairies, Merion cv is regarded as susceptible to melting out, caused by *C. sativus* (Noviello, 1963; Traquair and Smith, 1984), although it is moderately resistant to *D. poae* melting out (see p. 299). Muse *et al.* (1972) explained the more severe spotting of Merion, compared with common *P. pratensis* on the basis of the higher pectolytic and cellulolytic enzyme activity in infected Merion than in the common sort.

In *Poa annua*, 20 clones originating across Canada varied from very susceptible to very resistant to *C. sativus*. Six of these showed greater resistance to the pathogen than did Penncross, Seaside and Highland cvs of *Agrostis* spp. and Bristol, Merion and Nugget cvs of *P. pratensis* (Bolton and Cordukes, 1979).

(f) Control

(i) GENERAL. The strategies adopted for the control of *C. sativus* leaf and crown diseases of *P. pratensis* and that for *D. poae* leaf spot and melting out are similar in many ways, but the decision on which cultivar(s) to use is critical. The choice of the latter should be based on regional knowledge of which pathogen is likely to be of most importance. Where summers are very hot, such as in the northern part of the continental interior of North America *C. sativus* is likely to be more important than *D. poae* and more reliance must be placed on the use of fungicides since the resistance of available cultivars to *C. sativus* is inadequate. The effects of winter diseases must also be considered (Traquair and Smith, 1984). Regional test data on cultivars resistance frequently refers to *Helminthosporium* sp. or spp. In such a case the best agronomic performer should be selected until specific resistance of cultivars in mature turf is known. Regional recommendations should be consulted.

Fertilize to give adequate growth response, but avoid over-stimulation of grasses with soluble nitrogen in spring and autumn. Obtain as precise control as possible over fertility with either a slow-release nitrogen fertilizer or split-applications of conventional materials.

Raise the height of cut on lawns and similar turfs above 5 cm, especially when spotting becomes apparent and continue mowing into autumn. Use a sharp mower, since shredding of leaves with a blunt one affords a suitable substrate for the pathogen. Where feasible, pick up the clippings. When composted and mixed with soil for top dressing they will no longer provide a source of inoculum. Limit thatch build up to less than 2.5 cm by regular dethatching in spring and/or autumn and by top dressing. Phytotoxicity to turf grasses may result from the application of some herbicides applied for the control of broad-leaved and grassy weeds, especially during droughty weather: such applications should be avoided since they may increase turf grass susceptibility.

Irrigate as needed to give deep penetration of water, to 15 cm if possible rather than frequently and shallowly and avoid evening applications which increase the duration of surface wetness.

(ii) CHEMICAL. Several studies have been reported on the effectiveness of fungicides against *C. sativus* on excised leaves or potted grass plants and these have given useful information on the mechanism of action and fungicide persistence under specified environmental conditions (Danneberger and Vargas, 1982; Hagan and Larsen, 1979; Joyner and Couch, 1976; Nicholson, Gray and Sinclair, 1971; Sanders *et al.*, 1978; Weihing, Jensen and Hamilton, 1957). However, in the field more than one '*Helminthosporium*' or other turf grass pathogen may be involved, e.g. *C. sativus, D. poae* and *Fusarium* spp. on *P. pratensis*. While many fungicides are effective against more than one species they may not all be adequate in controlling all the different phases of the disease, e.g. seedling blight, leaf spot, crown or root rot or against non-target species. Published data on control of disease with fungicides sometimes does not specify the '*Helminthosporium*' species involved. However, effective control of leaf spot and/or crown rot of *P. pratensis* has been reported with several contact and some systemic fungicides including anilazine, chlorothalonil, dicyclidine, iprodione, mancozeb, maneb, quintozene, thiram and vinclozolin alone or in combination with other materials. (Reports on the effectiveness of these and other experimental fungicides may be found in *Fungicide and Nematicide Test* published annually by the American Phytopathological Society.)

Autumn and spring applications of fungicides, which may be required for the control of '*Helminthosporium*' diseases of the cool, moist season kind, e.g. those caused by *D. poae, Pyrenophora lolii, Pyrenophora dictyoides* and *Cochiobolus cynodontis* (see p. 299), will also assist in the prevention of the build up of the crown and root phases of these diseases and of those caused by *C. sativus* during the hot, dry summer months. Contact fungicides are usually applied at 7–14-day intervals whereas systemic materials may be effective for longer periods (Danneberger and Vargas, 1982). When the crown rot phase of *C. sativus* is severe in summer fungicides, additional soluble nitrogen and adequate irrigation may be needed to assist in recovery.

27.5 CURVULARIA SPECIES

27.5.1 Disease of turf grasses caused by *Curvularia* species alone and in disease complexes

Wernham and Kirby (1941) reported the thinning out of plants of *Agrostis stolonifera* L. in turf of Metropolitan bentgrass in the eastern United States, with which '*Helminthosporium*' or *Curvularia* species, with high optimum temperature requirements for growth, were associated. Howard (1953) described a condition–*Curvularia* blight–'fading out' in Rhode Island, where putting greens went 'off colour' following rain or heavy watering during hot weather. Bean (1964) showed that *C. pallescens*, isolated frequently with two '*Helminthosporium*' species from *Poa pratensis* L. in the Washington, D.C. area was pathogenic on that species, as were the two other fungi. Bell (1967) found that *C. geniculata*, *C. inaequalis*, *C. lunata* and *C. pallescens* would retard the growth of *Zoysia japonica* Steudel for 6 weeks, but only *C. pallescens* would do this for an extended period. Hodges (1972) concluded that *C. geniculata* was a turf saprobe after failing to show its pathogenicity to *Agrostis stolonifera* and *P. pratensis*. Endo (1961) found that *C. lunata* would kill seedlings of *A. stolonifera* at high, but not low temperatures; *C. geniculata* and *C. trifolii* were much less pathogenic. Brown, Cole and Nelson (1972) demonstrated that 80% of isolates of *C. geniculata*, *C. intermedia*, *C. lunata*, *C. maculans* and *C. protuberata* were pathogenic to some extent on cultivars of *Poa pratensis, Festuca rubra* and *A. stolonifera* at high, but not low temperatures. Madsen and Hodges (1980a) established that *C. geniculata* was a potentially important pathogen of germinating seed of *F. rubra*, interacting synergistically with *Cochliobolus sativus* to reduce seedling emergence. *C. geniculata* alone reduced seedling emergence, but had no effect on rate of emergence or on seedling mortality. In New Zealand, Falloon (1975) found that *Poa annua* L. and *A. tenuis* Sibth. more susceptible to *C. trifolii* than *A. stolonifera* Penncross. *Festuca rubra* var. *commutata* Gaud. was of low susceptibility. Isolates differed in their virulence, but required high temperatures to cause severe disease.

Although the role of *Curvularia* spp. in causing turf grass disease especially in complexes is still debatable, there seems no reason to doubt that in particular circumstances they are potentially important pathogenically, alone or in complexes.

Details of the *Curvularia* spp. found on turf grasses are shown in Table 27.2 and Fig. 27.3.

(a) Symptoms

Some *Curvularia* spp. cause pre-emergence disease by rotting the seed or killing the germinating seedling (Groves and Skolko, 1945; Sprague, 1950). Crown and root lesioning also occurs (Bell, 1967; Rogerson, 1956). Curvularia blight or 'fading-out' of fine bentgrass turf appears as generalized irregularly shaped patches of yellow or green, dappled grass occurring on unshaded portions of turf. Invaded leaves shrivel, turn tan to brown and eventually are covered with brown mould, due to conidial production by the fungus. Bentgrasses, bluegrasses and fescues on domestic lawns are also attacked (Howard, 1953; Pirone, Dodge and Rickett, 1960). Symptoms in inoculated grasses are similar to those seen in turf. In *Festuca rubra* and *Poa pratensis* there is a leaf tip dieback with a progressive yellowing down the leaf blade. Then a brown or grey colouration develops and

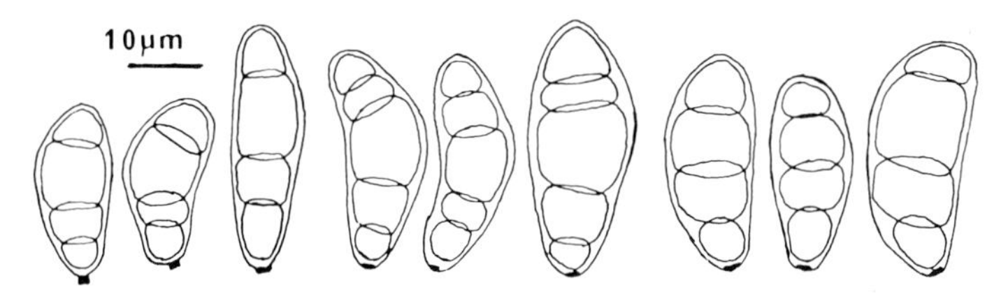

Fig. 27.3 Conidia of *Curvularia* spp. Left *C. trifolii*, centre *C. geniculata*, right-*C. intermedia*.

Table 27.2 Some *Curvularia* species on turf grasses

Teleomorph (*Cochliobolus*)	Anamorph (*Curvularia*)	Conidium		Sclerotia in culture	Occurrence
		Type*	Size (μm) (Ellis, 1971)		
Not known	*C. eragrostidis* (P. Henn) J.A. Meyer (Ellis, 1971)	M	22–33 × 10–18	S	*Eragrostis* spp.
C. geniculatus Nelson (1964a)	*C. geniculata* (Tracy and Earle) Boedijn (1933)	G	26–48 × 8–13	—	Common in tropics. On roots and seeds of native and exotic grasses in N. America. Pathogenic on *Festuca rubra* and *Zoysia japonica* Steudel
C. intermedius Nelson (1960)	*C. intermedia* Boedijn	M	27–40 × 13–20	—	*Agrostis stolonifera* L., *Cynodon dactylon* L., *F. rubra, Lolium perenne* L *Poa pratensis* L.
Not known	*C. inaequalis* (Shear) Boedijn	G	24–45 × 9–16	S	*Festuca elatior* L., *F. rubra*, *Poa pratensis*, *Zoysia japonica*
C. lunatus Nelson and Haasis (1964)	*C. lunata* (Wakker) Boedijn	L	20–32 × 9–15	—	Cosmopolitan *A. stolonifera*, *Z. japonica*
Not known	*C. pallescens* Boedijn	L	17–32 × 7–12	—	Common in tropics *Poa pratensis*
Not known	*C. protuberata* Nelson and Hodges	G†	27–35 × 10–14	S	*A. stolonifera*, *F. rubra*, *Phleum pratense* L., *Poa pratensis*
Not known	*C. trifolii* (Kauffm.) Boedijn	L†	28–38† × 12–16	S	Many grasses, including *A. stolonifera*, *A. tenuis* Sibth., *F rubra* ssp. *commutata* Gaud. *P. annua* L.

* *maculans* (M) type conidia – mostly 3-septate, 2 central cells larger and darker, spore only slightly curved.
lunata (L) type conidia – mostly 3-septate, only one cell (third from base) disproportionately larger, spore strongly curved.
geniculata (G) type conidia – mostly 4-septate, third cell from base disproportionately large.
† hilum protuberant.

finally the leaf shrivels. Sometimes a red or brown border may mark off the diseased portion and on these species, but not on bentgrass, a tan-centred leaf spot with red or brown border may occur. Crown and sheath blighting may result in irregular diseased patches (Brown, Cole and Nelson, 1972). On *Cynodon dactylon*, *Curvularia* spp. cause root and crown rots and a leaf blight (Brown, 1957; Putterill, 1954, Smith, 1955a).

(b) Disease development

The pathogenicity of some *Curvularia* spp. on aerial tissues of turf grasses, under controlled environmental conditions, particularly at higher temperatures (24–35°C) accords well with the original field observations on symptom expression which is most noticable in hot summer weather (Brown, Cole and Nelson, 1972; Falloon, 1976; Howard, 1953; Wernham and Kirby, 1941). Falloon (1975) showed that there was a positive correlation between the growth rate in culture (optimum 30°C) and the virulence of *C. trifolii*. Variation in disease development at different temperatures in different *Curvularia* spp. is apparent since Falloon (1975), with *C. trifolii*, noticed slight disease at 15 and 20°C. Brown, Cole and Nelson (1972) could not detect disease at 20–24°C with *C. lunata*, *C. geniculata*, *C. intermedia* and *C. protuberata* although it was obvious at 24–31°C and 29–35°C. Since they noticed good growth of fungal isolates at 20–24°C they suggested that the higher temperatures may dispose grass plants to attack.

Because *Curvularia* species are ubiquitous in turf grass litter, which they decompose, this is the main source of inoculum for aerial, and crown infections. *Curvularia* spp. have been recorded from soil but not with high frequencies. Many species are seed-borne on grasses (Benoit and Mathur, 1970; Groves and Skolko, 1945; Richardson, 1979; Sprague, 1950) and this is the most likely source of seedling and root infections. It was shown that before *C. geniculata* could reduce seedling emergence in *Festuca rubra*, seed must be inoculated with conidia (Madsen and Hodges, 1980a); soil infestation did not reduce emergence. *Curvularia* spp. seem poor competitors with other soil micro-organisms (Bell, 1967; Madsen and Hodges, 1980a). Isolates show marked differences in virulence on the same cultivar and a particular isolate can show great differences on different species and cultivars (Brown, Cole and Nelson, 1972). Applications of ammonium sulphate and calcium nitrate equivalent to 43 kg N/ha increased the pathogenicity of *C. geniculata* and *C. sativus*, alone or in combination, to seedlings of *F. rubra* (Madsen and Hodges, 1980b). There is great uncertainty about the role of *Curvularia* spp. in disease complexes other than with *C. sativus*.

(c) Control

There is little reliable information on the resistance of turf grass species and cultivars to disease in the field caused by *Curvularia* spp. alone or in complex with '*Helminthosporium*' spp.

'Fading-out' associated with *Curvularia* spp. occurs at higher temperatures than disease caused by *C. sativus* alone, but similar management practices and fungicidal treatment applicable to this hot-weather disease should be undertaken. If the affected turf is in an area with good 'air drainage' and sun exposure, leaf and turf surface temperatures and disease severity may be reduced by light sprinkling or syringing before the heat of the day. This should not be attempted on shaded turf or late in the day as the increased humidity would favour the disease.

27.6 *DRECHSLERA SPECIES*

27.6.1 *Drechslera biseptata* (Sacc. and Roum.) Richardson and Fraser (1968)

D. biseptata is pathogenic on *Anthoxanthum odoratum* L., *Deschampsia caespitosa* (L.) Beauv., *Dactylis glomerata* L. and *Festuca* spp.; it is a seed-borne mycotoxin producer. Conidiophores may arise in groups from dark-brown stromata or solitarily from hyphae. Conidia are straight, obovate, elliptical or clavata, yellowish-brown to mid-brown, smooth or rough, hilum dark-brown, 20–42 × 11–19 μm, 2–3-septate (Fig. 27.4).

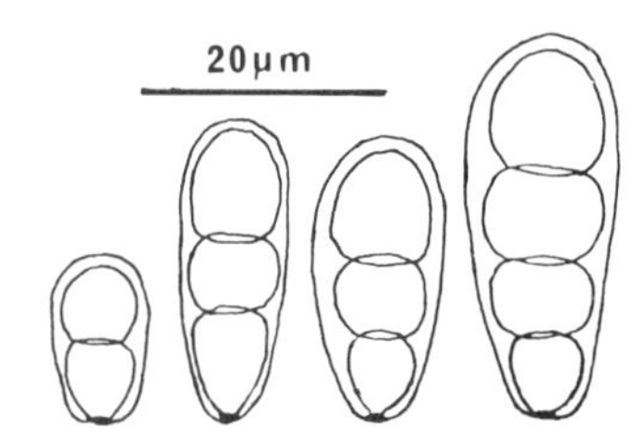

Figure 27.4 *Drechslera biseptata*–conidia.

27.6.2 *Drechslera dematioidea* (Bubak and Wroblewski) Subramanain and Jain (1966)

This is weakly pathogenic on *Agrostis* spp., *Cynodon* spp. and on *Anthoxanthum odoratum*, *Festuca duriuscula* L. and *Dactylis glomerata* L. and is seed-borne. Conidia are yellow–brown, clavate, broader at tip, tapering to base with a wide, dark hilum; basal cell lighter in colour, 25–41 × 9–15 μm, 3–5 septate.

27.6.3 *Drechslera triseptata* (Drechsl.) Subramanian and Jain (1966)

This is weakly pathogenic on withering leaves of *Agrostis* spp., *Holcus lanatus* L., *Dactylis glomerata* L. and *Phleum pratense* L. Conidia are dark brown to olivaceous, ellipsoidal or tapering to base, 35–50 × 15–21 μm, 2–3 septate.

27.6.4 Leaf blight and crown rot of turf grasses caused by *Drechslera catenaria*

Natural suscepts from North America and Europe in the cool-season turf grasses are notably, *Agrostis stolonifera* L., *Lolium perenne* L., *F. arundinacea* Schreb., and *Poa trivialis* L. (Andersen, 1959; Labruyere, 1977; Larsen and Hagan, 1981; Mäkelä, 1975; Shoemaker, 1962; J.D. Smith and R.A. Shoemaker, 1972, unpublished). The fungus is also seed-borne on *A. tenuis* Sibth., *Festuca pratensis* Huds., *Poa pratensis* L., *P. trivialis*, *Festuca rubra* L. and *Phleum pratense* L. (Andersen, 1959; Chidambaram, Mathur and Neergaard, 1973; Neergaard, 1958; Richardson, 1979; de Tempe, 1968). It may be seed-borne on other grass hosts since it is often not distinguished from *P. dictyoides* in seed health tests (Labruyere, 1977). On *L. perenne* the pathogen is less prevalent than *D. siccans* (Wilkins, 1973), but it may incite severe disease on *Agrostis stolonifera* in turf in Ohio (Larsen and Hagan, 1981; Spilker and Larsen, 1980). In the mid-west of the United States, a disease on 'Toronto' *A. stolonifera* with similar symptoms had previously been diagnosed as caused by *D. erythrospila* (see p. 304).

(a) Symptoms

On *Lolium perenne* and *Festuca arundinacea* disease symptoms comprise chlorosis and withering of leaf tips and leaf margins and a reticulate necrosis, similar to the net blotch symptoms caused by *P. dictyoides* with chlorotic streaking.

On *Agrostis stolonifera* 'Toronto' the pathogen causes a severe leaf blight and crown rot (Larsen and Hagan, 1981; Spilker and Larsen, 1980). Symptoms appear in early spring as a leaf tip dieback and reddish leaf blade lesions which may eventually involve the whole leaf and lead to crown necrosis and plant death. Small patches of infected plants appear as reddish-brown sunken areas on golf greens. Patches may coalesce until several square metres are affected.

(b) The fungus

Teleomorph – Not known; anamorph – *Drechslera catenaria* (Drechs.) Ito (1930).

Conidiophores are yellowish-brown, single, simple, erect mostly 50–150 μm, but up to 210 μm long with globose bases up to 12 μm wide, geniculate and scarcely tapering to truncate apices. Conidia are light yellow-brown with hemi-ellipsoidal basal cells, mostly 85–105 μm, but up to 129 μm long and 15–18 μm wide, 4–6 septate. The apical cell bears a scar or secondary conidiophore (Shoemaker, 1962). Conidial dimensions are variable: from Danish material 41–220 × 12–26 μm (Andersen, 1959); from the United States 30–200 × 14–18 μm (Drechsler, 1923) or 40–170 μm long (Spilker and Larsen, 1980) and from British Columbia on *Poa trivialis* (36) 57–101 (168) × 10 – 17 μm (J.D. Smith and R.A. Shoemaker, 1972, unpublished). Conidia are usually widest at the first septum or second cell from the base, tapering to approximately half that diameter at the apex. Small spores are usually straight, but longer ones are usually twisted (Fig. 27.5).

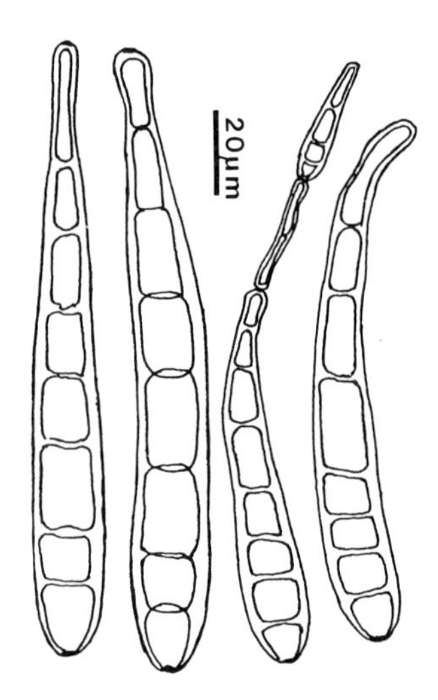

Figure 27.5 *Drechslera catenaria*–conidia.

D. catenaria conidia have a similar obclavate shape to those of *D. phlei*, but *D. catenaria* conidia have a smooth contour, not swollen at the second cell from the base and are generally longer and wider than those of *D. phlei* and more tapering than those of *P. dictyoides*. *D. noblea* which also has conidia tapering from the base to the apex has thinner-walled and lighter-coloured first or second cells from the base. In *D. catenaria* secondary conidia are produced more freely and secondary conidiophores are longer than in

D. phlei. For recent revisions in the taxonomy of *D. catenaria* see *Pyrenophora dictyoides* (p. 302).

(c) Disease development

As with *D. siccans*, the level of contamination of seed of *Lolium* spp. with *D. catenaria*/*P. dictyoides* complex had little effect on subsequent disease development. Although some organo-mercury fungicides and imazilil almost completely eliminated these fungi from seed they did not retard disease development (Anon., 1969; Labruyere, 1977). Incubation under high humidity for 72 hours was needed for adequate infection and no lesions developed until 6 days afterwards–much longer than for *D. siccans*. Resistance shown by the development of necrotic spots and traces only of reticulate necrosis was shown by some genotypes and the cultivar 'S.24' of *L. perenne*. Resistance to *D. catenaria* seemed to depend on the limitation of hyphal growth after infection, perhaps due to genotypic control over phytoalexin formation (Wilkins, 1973). The *D. catenaria*/*P. dictyoides* complex may have a greater competitive ability against saprophytic fungi on senescent and dead leaves since it was found that they became established soon after seedling emergence on senescent and dead leaves (Labruyere, 1977).

Inoculation of an isolate of *D. catenaria* from the Toronto cv of *A. stolonifera* produced typical lesions on Toronto, Penncross and Seaside cultivars of that species, but only small, restricted, reddish-brown spots on the Kentucky 31 cv of *Festuca arundinacea*. It failed to infect Delta *Poa pratensis* (Larsen and Hagan, 1981; Spilker and Larsen, 1980).

(d) Control

Excellent control of the leaf blight and crown rot on Toronto *A. stolonifera* throughout the growing season was obtained with two applications of iprodione at 61 g a.i./100 m^2 made in spring (Larsen and Hagan, 1981).

As the disease appears to be most severe on 'Toronto' *A. stolonifera*, where it is prevalent the use of this cultivar should be avoided. Management practices similar to those employed for the control of *D. poae* should be used (see pp. 298–299).

27.6.5 Leaf mould of bentgrasses caused by *Drechslera fugax*

(a) Symptoms and occurrence

Doubtfully pathogenic, but of common occurrence, especially on withered leaves of *Agrostis* spp. in Europe and North America (Drechsler, 1923; Ellis, 1971; Shoemaker, 1962; Smith, 1965).

(b) The fungus

Teleomorph–not known; anamorph–*Drechslera fugax* (Wallr.) Shoemaker in Hughes (1958).

Conidiophores are dark, reddish-brown, thick-walled, 45–260 μm long with a swollen base 9–18 μm and apex 7–14 μm, often arising from spherical stromata in leaves. Conidia are dark, yellowish-brown and variable in shape. Although the shape is generally cylindrical, most mature spores have the widest point just below the middle, tapering slightly to both ends. Young spores may be broadest near the apex. The distal end of the spore may have a prolongation. Conidia measure 45–172 μm long and 12–24 μm wide and are usually 4–8 septate (Drechsler, 1923; Shoemaker, 1962).

27.6.6 Zonate eyespot of bermuda grass and bentgrasses caused by *Drechslera gigantea*

Although this disease was first described on *Cynodon dactylon*. L. from Texas (Heald and Wolf, 1911), and has been regarded principally as a disease of grasses in the southern states of the USA, Central America and northern South America (Drechsler, 1923, 1928, 1929; Ellis, 1971; Shoemaker, 1962; Sprague, 1950) it has been found on turf grasses as far north as Rhode Island in the USA (Jackson and Fenstermacher, 1973). Many grass species are natural suscepts, but it may cause very severe leaf spotting and defoliation in *Cynodon dactylon* in turf and in some cultivars of *Agrostis stolonifera* L., *A. canina* L. and *A. tenuis* Sibth. (Drechsler, 1923, 1928, 1929; Heald and Wolf, 1911; Jackson and Fenstermacher, 1973; Monteith and Dahl, 1932).

(a) Symptoms

On *C. dactylon*, eyespots with yellow or bleached centres and brown margins develop in spring often becoming confluent and involving most of the leaf blades. Leaves may be completely killed and patches of turf affected. On *Agrostis canina* the first symptoms are of small, yellow leaf-blade lesions, but these quickly give place to complete yellowing then withering and browning of the leaves. Irregularly shaped patches of blighted grass, 5–23 cm in diameter may develop. *A. tenuis* is similarly affected. (Fig. 47, colour plate section; Fig. 27.6).

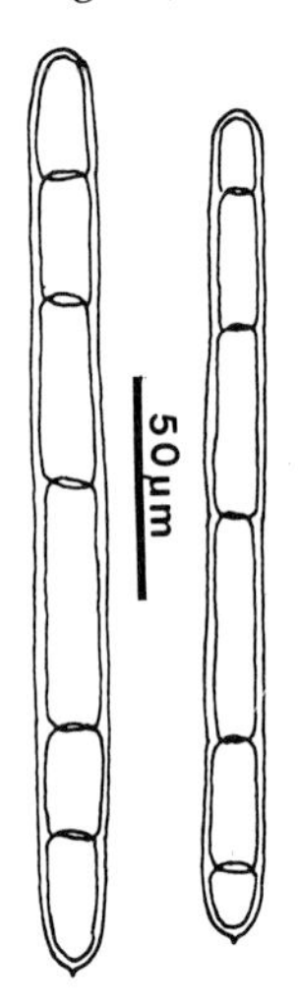

Figure 27.6 *Drechslera gigantea*–conidia.

(b) The fungus

Teleomorph–not known; anamorph–*Drechslera gigantea* (Heald and Wolf) Ito (1930). (For synonomy see Shoemaker, 1962.)

Conidiophores are dark brown with a lighter-coloured tip, 75–300 µm long × 9–15 µm wide at the apex with saucer-shaped depressions where the conidia are formed and 1–5-septate. Conidia are often sparsely produced, sub-hyaline to pale yellow with thin walls, cylindrical with hemispherical ends, much longer than most other members of the genus, mostly 220–320 µm long × 20–26 µm wide (Shoemaker, 1962) and from *Agrostis* spp. 150–420 µm long × 17–21 µm wide (Jackson and Fenstermacher, 1973). There are up to five septa. There is a small apiculum on the conidium at the point of attachment to the conidiophore, unlike any other *Drechslera* sp.

(c) Control

See *D. poae* (pp. 298–299).

27.6.7 Leaf spot of ryegrasses caused by *Drechslera noblea*

(a) Symptoms and occurrence

Moderately pathogenic on *Lolium multiflorum* Lam. and causes oval, brownish-black leaf lesions, leaf yellowing and dieback and pathogenic on *L. perenne* L. × *L. multiflorum* in New Zealand. Seed-borne on *Lolium* spp. generally in New Zealand and on *L. multiflorum* in Britain (Lam, 1982; Matthews, 1971; McKenzie, 1978; McKenzie and Matthews, 1977; Morrison, 1982; Richardson, 1979).

(b) The fungus

Teleomorph – not known; anamorph–*Drechslera noblea* McKenzie and Matthews (1977).

Conidiophores are solitary, unbranched, hyaline to dark brown, sometimes swollen to 9–25 µm at base, with several geniculations at apex, up to 450 µm long × 6–11 µm wide. Conidia are solitary, pale straw coloured, straight, smooth, obclavate with first or second cell from base thinner-walled and paler, inflated to 11–15 µm wide, tapering uniformly to 6–10 µm at apex, 3–8-septate, 55–125 µm long, usually germinating from inflated cell and apex, but not from the basal cell. The first septum is produced near mid-length. Conidia are produced closely together at apex of conidiophore. Secondary conidia may be produced on incubation (McKenzie and Matthews, 1977; Morrison, 1982).

Conidia are similar in shape to those of *P. dictyoides, D. catenaria* and *D. phlei* (and *D. andersenii*). However, in *D. noblea* the conidia are produced at the tip of long, dark conidiophores, the first or second cell above the basal one is inflated and germination usually takes place from the inflated cell.

27.6.8 Leaf streak of timothy caused by *Drechslera phlei*

(a) Symptoms and occurrence

Brown, necrotic streaks and blotches develop on leaves of *Phleum pratense* L., *P. bertolonii* DC and occasionally on other species, sometimes becoming severe in autumn or on seed crops, rarely causing significant damage to timothy in turf; seed-borne.

(b) The fungus

Teleomorph – not known; anamorph – *Drechslera phlei* (Graham) Shoemaker (1959, 1962).

Conidiophores are light yellow – brown, single, straight and unbranched, mostly 50–130 µm long, base enlarged, 11–14 µm wide, apex truncate, 6–8 µm wide. Conidia are hyaline to light brown, obclavate with hemi-spherical base; second cell from base abruptly enlarged, spores mostly 80–100 µm long × 14–16 µm wide with 4–6 septa. Secondary conidia produced on incubation in moist atmosphere. (Ammon, 1963; Drechsler, 1923; Graham, 1955; Mäkelä, 1971; Richardson, 1979; Scharif, 1961; Shoemaker, 1959, 1962; Smith, 1970a,b) (Fig. 27.7).

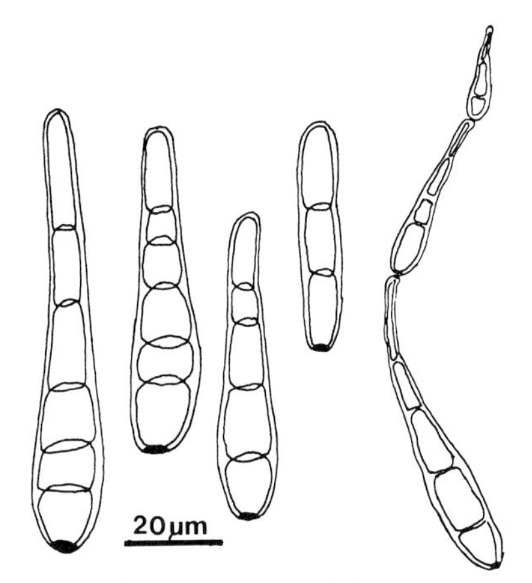

Figure 27.7 *Drechslera phlei*–conidia.

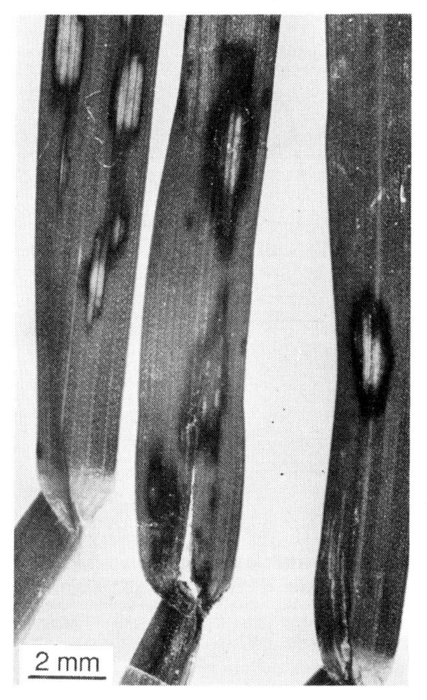

Figure 27.8 *Drechslera poae* leaf spot on *Poa pratensis*.

27.6.9 Leaf spot and foot rot ('melting-out') of *Poa pratensis*

The leaf spot phase of this disease was first reported on *Poa trivialis* L. from Bohemia (Czechoslovakia) by Baudys (1915–16) and noted generally prevalent on *Poa pratensis* L., its major host, by Drechsler (1922) around Madison, Wisconsin in the United States. The disease is now known to be coextensive with the latter species, one of the principal turf grasses in many cool regions (Carr, 1971; Halisky and Funk, 1966; Jensen, 1970; de Leeuw and Voss, 1970; Mäkelä, 1972a,b; Mühle, 1971; Nishihara, 1973; Smedegard-Petersen and Hermansen, 1972; Smith, 1965; Teutberg, 1978). In the British Isles, the disease is important only on *Poa pratensis* in amenity turf. Plants of *P. trivialis* and *Poa annua* L. which may develop from impurities in the commercial seed of *P. pratensis*, are usually only slightly infected (Smith, 1965).

(a) Symptoms

The disease is characterized by a leaf spotting phase followed by a more generalized infection and foot rot. In milder climates leaf spots may be seen at most times of the year, but are most obvious in spring, autumn and early winter. Individual spots are circular to elongate, 0.5–3 × 1–8 mm on leaves and leaf sheaths. The margins of the spots are purplish-black to reddish-brown with lighter, sometimes bleached centres. In close-mown turf the disease starts as small discoloured patches and the leaf and stem lesions may be more clearly seen by parting the grass to the sole. In susceptible cultivars the leaf spot symptoms are accompanied by rhizome lesioning which shows as brown necrotic tissues. Severe leaf-dropping and foliage dieback may develop, referred to as 'melting-out' (Fig. 48, colour plate section). Root infection may also occur. Other pathogens may be involved in both the leaf spotting and foot rot stages, their relative importance being related to climatic conditions.

(b) The causal fungus

Anamorph – *Drechslera poae* (Baudys) Shoemaker (1962; see this for synonomy); Teleomorph – not known.

Conidiophores of the fungus emerge from the stomata or between the epidermal cells of the host, singly or less frequently in pairs soon after the death of the host cells; they are stout and straight or slightly curved, usually simple, but occasionally branched, geniculate below the apex, dark yellowish-brown with a slightly enlarged base, 11–15 µm wide tapering slightly to the apex which may be slightly swollen at the tip with 1–11 septa, usually 55–165 µm, but occasionally as long as 280 µm (Drechsler, 1923; Shoemaker, 1962; Sprague, 1950).

Conidia are dark brown with a dark-brown basal scar contained within the cell contour, almost cylindrical with broadly hemispherical ends, most measuring 65–90 × 18–20 µm (Fig. 27.9) with 5–6 septa.

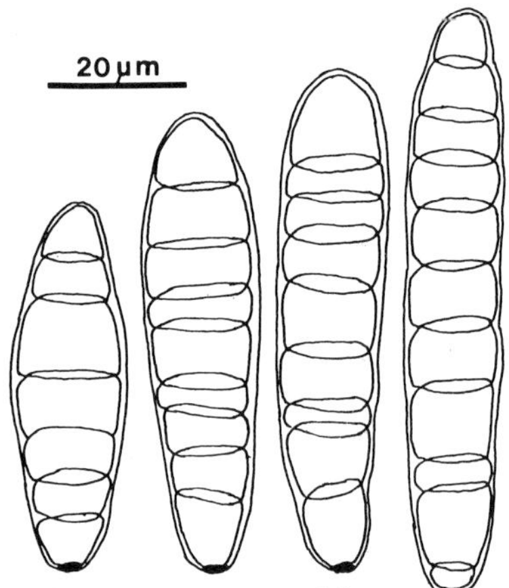

Figure 27.9 *Drechslera poae*–conidia.

Often the septa do not coincide with constrictions in the peripheral wall (Shoemaker, 1962; Sprague, 1950). In some British collections the apical cell was prolonged, lighter in colour than the remainder, the prolongation being either straight or curved. The distal segment was narrower than in normal spores of *D. poae*, but not tapering abruptly as in *D. siccans* (Drechsl.) Shoem. The spore length was increased to 167 µm (Smith, 1955b). Sporulation is usually poor on younger lesions, but spores may be found on withered, lesioned leaf tissues and on older agar cultures or on cultures exposed to long wave UV light (black light) (Schlosser, 1970). Abundant spores for inoculation tests may be produced on washed cultures (Halisky and Funk, 1966). Spore germination is from any or all cells of the conidium (Drechsler, 1923), unlike that in *Bipolaris sorokiniana*, which may attack plants concurrently with *D. poae* (see p. 287).

(c) Other hosts

Several other *Poa* species are natural suscepts (Sprague, 1950, 1956). The fungus was also found on *Koeleria cristata* (L.) Pers. (Shoemaker, 1962). Mäkelä (1971) was able to infect *Festuca pratensis* Huds., *Lolium perenne* L., *L. multiflorum* Lam. and *Phleum pratense* L. Moore and Couch (1961) found 30 species of grasses from 18 genera susceptible to *D. poae* and Couch (1973) lists 34 susceptible species.

(d) Disease development

The disease may progress from the leaf sheaths and leaves to the inflorescences; glumes and florets become infected (Gray and Guthrie, 1977; Lukens, 1967; Smedegard-Petersen, 1970). The pathogen is seed-borne on *P. pratensis* and *P. trivialis* (Richardson, 1979). Most infected seeds of *P. pratensis* produced infected seedlings (Gray and Guthrie, 1977).

Sporulation of the fungus and the subsequent leaf spotting phase of the disease is favoured by moist, overcast, cool weather (Halisky and Funk, 1966; Lukens, 1970). In maritime, or cool, humid temperate climates spore production and new infections may develop during autumn, winter and spring. In the northeastern United States a peak of leaf destruction occurs in late May and early June, although highly susceptible cultivars of *P. pratensis* may be completely brown, as a result of the disease, by early March (Funk, Halisky and Babinsky, 1975). Their recovery in spring depends mainly on nutrient reserves and favourable environmental conditions (see below). In this region, the disease is of less importance in bright, warm late summer weather. On the other hand, in northern England (at The Sports Turf Research Institute) the peak periods for leaf spotting on *P. pratensis* are in spring and autumn and severe melting-out does not occur unless susceptible cultivars are subjected to very close mowing (Smith, 1965).

In the eastern United Sates Nutter, Cole and Schein (1982) found that spore dispersal began with leaf elongation, at about the time of the first mowing, with peak conidial release in mid-May. Only a few conidia were caught in spore traps from mid-June to the end of August. Incidence of leaf spot symptoms coincided with late-April to late-May peaks in spore trap counts. It was suggested that spore release and dispersal played no role in disease development after mid-June, although existing lesions may continue development, tiller death and melting-out severity being related to the leaf spot incidence. Mowing caused dislodgement of spores and appeared to be a major mechanism for spore dispersal, but dispersal of conidia seemed to be limited largely to heights below 20 cm. Hagan and Larsen (1985) found leaf litter clippings, not thatch, to be the primary source of conidia.

Although mowing was shown (Nutter, Cole and Schein, 1982) to be of major importance in the dispersal of conidia of *D. poae* in amenity turf, other means are likely involved, e.g. the 'rain tap and puff' mechanism (Hirst and Stedman, 1963) which is particularly effective in dispersing dry spores of fungi. Strong air currents will dislodge conidia of various species of *Bipolaris* and *Drechslera*, including *D. poae*, from their conidiophores (Kenneth, 1964). Sprinkler irrigation may also contribute to an increased population of the pathogen on leaves of *P. pratensis* by the splashing action of the water droplets

(Gregory, 1961) dispersing the conidia (Gray and Guthrie, 1977).

Close mowing greatly increases disease severity, particularly with susceptible cultivars. Monteith and Dahl (1932) found that by raising the mowing height above 13 mm and supplying nitrogenous and phosphatic fertilizers, both leaf spot and foot rot symptoms, caused by *D. poae* could be reduced. Mowing above 5 cm height and more frequent applications of lower dosages of nitrogenous fertilizer were recommended by Juska and Hanson (1963) to reduce the serious effects of the leaf spot. However, in New Jersey, it was found that *P. pratensis* turf maintained at high fertility suffered less leaf spot damage in early spring than under moderately low fertility. Conversely, root rot damage in May and June was greatest on high fertility plots. The most extensive crabgrass (*Digitaria* sp.) invasion was related to the most severe melting-out in the turf maintained at high fertility and mown at 19 mm (Funk, Halisky and Babinsky, 1975; Halisky, Funk and Engel, 1966).

It has been shown that the incidence of melting-out and leaf sugar content in *P. pratensis* were negatively correlated (Lukens, 1970). Mowing at 25 mm or shading the turf increased disease severity and resulted in a lower leaf sugar content than in turf mown at 50 mm or left unshaded (Lukens, 1970). However, in *Cochliobolus sativus* (see p. 288) which also causes an important leaf spot, crown and root rot of turf grasses, including *P. pratensis*, Couch and Moore (1971) were unable to show a correlation between susceptibility to infection and foliage sugar content.

(e) Disease control

The selection and use of resistant *P. pratensis* cultivars and the appropriate cultural treatment of the turf are the keys to the control of the leaf spot and the foot rot stages of the disease. Cultivars that will produce high quality turf and that are resistant to *D. poae* (and other '*Helminthosporium*' leaf and crown diseases) are commercially available. Regional listings should be consulted because different spectra of pathogens are characteristic of different climates. The Merion cultivar, one of those most extensively used has good resistance to *D. poae*, but, unfortunately it lacks the necessary resistance in different regions to the leaf, crown and root disease caused by *C. sativus*, and to stripe smut (*Ustilago striiformis*), stem rust (*Puccinia graminis*) and LTB snow mould (*Coprinus psychromorbidus*). Among the more recently available *P. pratensis* cultivars reported to have good resistance are Adelphi, Baron, Bonnieblue, Bristol, Enmundi, Fylking, Galaxy, Glade, Kimono, Majestic, Nugget, Pac, Parade, Pennstar, Ram 2, Sodco, Sydsport, Touchdown and Victa (Anon., 1979; Bourgoin, 1976; Bourgoin *et al.*, 1974; Funk, Halisky and Babinsky, 1975; STRI, 1980; Smith and Evans, 1985; Vargas, 1981; Watkins and Shearman, 1976).

While many of the newer cultivars, particularly dwarf types, will tolerate shorter mowing than their forebears without suffering severe damage from *D. poae* during periods of active leaf growth, during danger periods for the root rot phase, the risk of severe damage will be reduced by raising the mowing height above 20 mm. Fertilization in autumn should be adequate to encourage spring growth recovery and, when the grass is dormant, nitrogen application lessens the risk of leaf spot problems in spring. Where the root rot phase is an annual occurrence, split applications of nitrogen should be made in spring. Where severe melting-out is of regular occurrence fungicide applications should be used to control the leaf spot phase. Irrigate to avoid drought stress in summer.

There are available very effective fungicides for the control of leaf spot caused by *D. poae* and other '*Helminthosporium*' species which include: anilazine, actidione with thiram, captan, chlorothalonil, iprodione, mancozeb, maneb with zinc sulphate and quintozene (Funk, Halisky and Babinsky, 1975; Jackson, 1982; Shurtleff, 1981; Smith, 1980). More complete listings of effective materials are given by Vargas (1981) and Watkins, Shearman and Bruneau (1979). Where the leaf spot phase is not very severe two or three spring applications of some of these chemicals at 3-week intervals may adequately control the disease (Funk, Halisky and Babinsky, 1975). When the leaf spot phase is very active from autumn to spring it may be necessary to apply a fungicide every 10 to 14 days during autumn and spring (Jackson, 1982) in Rhode Island. In Nebraska, spring and autumn applications may be needed to control the leaf spot phase (Watkins, Shearman and Bruneau, 1979).

27.6.10 Leaf spot, leaf blight and foot rot of *Lolium perenne* and other turf grasses caused by *Drechslera siccans*

The pathogen causes disease which is widely distributed in temperate regions of North America, Europe and Australasia. Turf grass species in *Lolium* and *Festuca* are natural suscepts (Ammon, 1963; Ellis and Holliday, 1976; Mäkelä, 1971; Mühle, 1971; Sampson and Western, 1940; Teuteberg, 1977), but other species are susceptible and the fungus is seed-borne on several grass species often with *P. dictyoides*, *D. catenaria* and *D. noblea* (see pp. 294, 296 and 302; Andersen, 1955; Lam, 1982; Mäkelä, 1971;

Mckenzie, 1978; Richardson, 1979). Mown swards of *Lolium perenne* may become severely infected in autumn (Bonis and Heard, 1980) and the disease occasionally causes problems in amenity turf in Britain (Smith, 1965). Sprague (1950) found it common on uncut lawns in winter and early spring in western Oregon.

(a) Symptoms

The pathogen has been reported to cause a seedling blight of *Lolium perenne* in Ireland (O'Rourke, 1976). Leaves of this species become infected before they emerge from the sheath and eyespots with greyish-white centres develop (Wilkins, 1973). Whole tillers may be killed under wet conditions. In Scotland *D. siccans* may cause a foot rot as well as leaf spot of *L. perenne* (Sampson and Western, 1954). Multiple infections of leaves of host species may occur, resulting in large numbers of small, chocolate-brown, interveinal spots which may be so numerous that the whole leaf becomes chlorotic and is killed. This is typical of autumn outbreaks of the disease (Frauenstein, 1968). Small lesions may extend in size into larger brown blotches or longitudinal streaks, a centimetre or more in length. If lesions extend across a leaf blade the portion above withers. Net blotch markings are absent (Fig. 49, colour plate section).

(b) The fungus

Teleomorph – *Pyenophora lolii* Dovaston (1948)? (many authorities do not accept this as the perfect state); anamorph – *Drechslera siccans* (Drechs.) Shoemaker (1962).

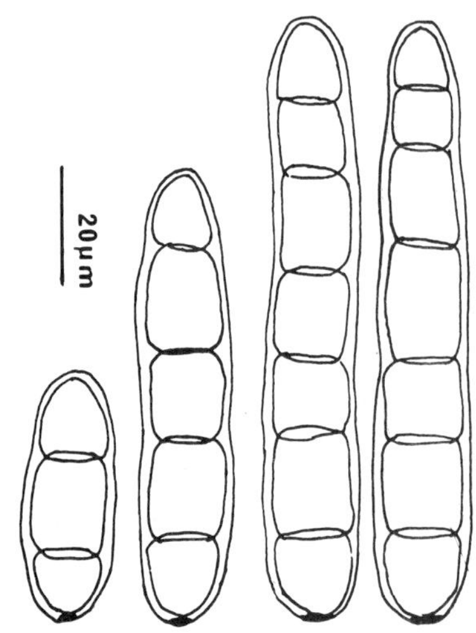

Figure 27.10 *Drechslera siccans*–conidia.

Conidiophores are chestnut brown, single or in groups, simple erect, slightly curved, long and narrow, up to 270 µm long often 8–9 µm at the apex with a globose base usually 12–15 µm wide. Scars near the apex are close together and geniculations are not abrupt. Conidia are subhyaline to light olive when young, Golden-brown when mature, mostly cylindrical, apex and base broadly hemi-elliptical. Scar broad and obvious but not protruding, mostly 5–7-septate, cell length and breadth almost equal. Most conidia measure 90–110 × 16–18 µm (Ammon, 1963; Drechsler, 1923; Shoemaker, 1962). *D. siccans*, *D. dictyoides* and *D. catenaria* often occur together on *L. perenne*, but conidia of *D. siccans* are cylindrical, those of *D. dictyoides* are obclavate while those of *D. catenaria* are usually much longer and secondary conidia develop on conidiophores which arise from the apices of the primary conidia (see pp. 294 and 302).

(c) Disease development

Although the disease may be transmitted by infested or infected seed, Labruyere (1977) demonstrated that the susceptibility or resistance of cultivars of *L. perenne* to *D. siccans* and to *P. dictyoides* and *D. catenaria* had a greater effect on disease incidence than seed contamination level. Considerable differences in susceptibility were shown in forage cultivars of *Lolium* spp. The systemic fungicide imazilil was ineffective in improving emergence or health of emerged plants although it freed seedlings from *Drechslera* spp. in germination tests and retarded *D.siccans* development. This was taken to indicate the survival of internally seed-borne inoculum. Spread of infection in the growing crop was considered to be by rain splash and wind transport of conidia as in other '*Helminthosporium*' diseases.

Wilkins (1973) found that a range of genotypes in *L. perenne*, *L. multiflorum*, *Festuca arundinacea* and *F. pratensis* and hybrids of *L. multiflorum* × *F. arundinacea* were equally susceptible to *D. siccans* in greenhouse infection studies. Symptoms were similar in *Lolium* and *Festuca* spp. although the latter are not usually infected in the field. An incubation period under high humidity of 48 hours was adequate for initial infection. Since high humidity was found to assist in the subsequent spread of the pathogen, wet and dry cycles of incubation of 48 hours were used. Spotting by *D. siccans* became apparent during or soon after the first incubation period. With plants inoculated in the field, sufficient variation in resistance was noted between some of the species or genotypes of these to indicate that selection within a range of cultivars of *Lolium* and *Festuca* spp. in the field might result in raising the general level of resistance to *D. siccans*. Resistance to the fungus appeared partly dependent on restriction in conidial production.

As with '*Helminthosporium*' diseases of other turf grasses conflicting results have been obtained with applications of nitrogenous fertilizer on the incidence of the *D. siccans*, *D. catenaria* and *P. dictyoides* complex on *L. perenne*. For Broom *et al.* (1975), N applications did not affect the incidence of *D. siccans* and *P. dictyoides*. Cook (1975) found *D. siccans* damage was more severe at high than at low N dosage when disease level was low, but at a higher disease level the plane of nitrogen nutrition had little effect on disease severity. Bonis and Heard (1980) and Lam and Lewis (1982) found that damage caused by *P. dictyoides* (the '*D. andersenii*' of Scharif (1963)) and *D. siccans* was increased by N applications. Potassium applications had no apparent effect on either pathogen, However, *D. siccans* was rarely found on high N (600 kg N/ha), but occurred with increasing frequency on plots with N and most often on plots with no N. N level had little effect on *P. dictyoides*.

(d) Control

Although usually considered of minor importance in amenity turf, the incidence and effects of disease caused by *D. siccans* have not been closely examined. However, mown agricultural grass may be damaged severely enough for control with resistant cultivars, cultural practices and fungicides to be investigated.

There is little information on the effect of fungicides on disease caused by *D. siccans* on amenity turf. Smith (1965) reported phenyl mercuric acetate to be effective. Bonis and Heard (1980) controlled a mixed infection of *P. dictyoides* and *D. siccans* in mown *L. perenne* turf with chlorothalonil sprays. Broom *et al.* (1975) found that when the fungicide carboxin was used to control rust on *L. perenne*, caused by *Puccinia coronata* Cda. a higher leaf infection by *D. siccans* and *P. dictyoides* resulted. Where significant amounts of the disease are found in amenity turf control measures suggested for *D. poae* on *Poa* spp. (see p. 299) which develops under similar cool, moist environmental conditions may be used. In the absence of cultivars of *L. perenne*, with resistance to *D. siccans*, those with persistence under close mowing (STRI, 1982) should be used.

27.7 *PYRENOPHORA* SPECIES

27.7.1 Net blotch and leaf blight of *Festuca* and *Lolium* species caused by *Pyrenophora dictyoides*

Although this disease is widely distributed on *Festuca* and *Lolium* species in North America, Europe and Australasia (Ellis and Waller, 1976) it seems to reach epidemic proportions only occasionally. First noted as common on meadow fescue, *Festuca pratensis* Huds. around Washington, DC, by Drechsler (1923), it was later considered by Kreitlow, Sherwin and Lefebre (1950) to be among the most prevalent of the leaf diseases of the latter species and of tall fescue, *F. arundinacea* Schreb. in the eastern United States. Epidemics of the disease on *F. arundinacea* Schreb. in Georgia and on *F. rubra* L. in Pennsylvania were reported (Luttrell, 1951b; Couch and Cole, 1957) in winter and early spring and in a prolonged period of wet weather in summer, respectively.

Net blotch is widespread on *Lolium perenne* L., *L. multiflorum* Lam. and *Festuca pratensis* in Fenno-Scandia (Mäkelä, 1971, 1972b), and on *F. pratensis* and *F. rubra* in E. Germany (Frauenstein, 1968). In West Germany although isolated frequently from *L. perenne*, *P. dictyoides* was less common than *D. siccans* (Teuteberg, 1977, and p. 298). In southern and south-western Britain and in Wales, the anamorph of *P. dictyoides*, referred to as '*Drechslera andersenii*' in some cases, was the most common *Drechslera* sp. on *Lolium perenne* (Bonis and Heard, 1980; Lam and Lewis, 1982; Latch and McKenzie, 1977) associated with leaf spot and net blotch symptoms. *P. dictyoides* causes herbage loss and winter yellowing on *L. perenne* in New Zealand and occurs throughout the year on ryegrasses and fescues in Ireland (Latch, 1966; O'Rourke, 1976). Many forage and wild grasses as well as turf grasses are natural suscepts of *P. dictyoides* (sensu lato) and the fungus is seed-borne in these (Braverman and Graham, 1960; Frauenstein, 1968; Graham, 1955; Kenneth, 1958; Labruyere, 1977; Lam, 1982; Mäkelä, 1971, 1972b, 1975, 1981; Matthews, 1971; McKenzie, 1978; Paul and Parberry, 1968; Richardson, 1979; Shoemaker, 1962; Sprague, 1950).

(a) Symptoms

In the broad-leaved *Festuca* spp. and in *Lolium* spp. the symptoms appear first on the leaf blades as small brown spots with chlorotic margins; these spots may then coalesce and develop into lesions with net-like markings – reticulate necrosis or net blotch. Leaf tips and margins may wither progressively until much leafage is killed – leaf blight. In the fine-leaved *Festuca* spp. brown leaf spots develop and leaves may be blighted, but the net blotch symptoms are not usually prominent. Severe crown infections may develop and lead to the death of groups of plants in the turf.

(b) The fungus

Teleomorph – *Pyrenophora dictyoides* Paul and Parberry (1968); anamorph – *Drechslera dictyoides* (Drechs.) Shoemaker (1962).

Graham (1955) described a variety of *Helminthosporium dictyoides* Drechs. on *Phleum pratense* L. as var. *phlei.* Scharif (1961) examined this in detail and renamed it *H. phlei* (Graham) which in turn was changed to *Drechslera phlei* (Graham) by Shoemaker (1962). Braverman and Graham (1960) distinguished between *D. dictyoides* on *Festuca pratensis* which they called f. sp. *dictyoides* and that on *Lolium* spp. which they named f. sp. *perenne.* They distinguished these *formae speciales* culturally and by pathogenicity tests. Isolates from fescues were pathogenic only on fescues, whereas those from ryegrasses and an intergeneric hybrid were pathogenic on *Lolium* spp. and to a lesser extent on the *Festuca* spp. Mäkelä (1971) was able to infect *L. perenne* with isolates from *F. pratensis* and Latch (1966) established infection on *F. arundinacea*, *F. pratensis* and *F. rubra* L. *commutata* Gaud. with the fungus from *Lolium* sp. Since in the literature there is some confusion concerning the correct names, the name of the teleomorph, *P. dictyoides* (sensu lato) will be used here. '*Drechslera andersenii*' sometimes used for the *forma specialis* on *Lolium* spp. was an invalid name. Lam (1985) has now prepared a valid Latin description for *D. andersenil.* She suggests that the *D. catenaria* described by Andersen (1959), Labruyere (1977) and Wilkins (1973), because of its tendency to produce elongated and secondary conidia, is *D. andersenii* Form 2. D. catenaria isolations from *Agrostis*, *Phalaris* or *Cinna* produce a persistent, yellow pigment on sucrose–proline agar (Shoemaker, 1962). *D. dictyoides* produces a pink pigment and *D. andersenii*, no pigment.

P. dictyoides frequently occurs in complexes with other species in the same genus, including *D. siccans* (Drechs.) Shoem. *(Pyrenophora lolii* Dovaston?), *D. catenaria* (Drechs.) Ito, which is morphologically similar (Labruyere, 1977; Mäkelä, 1972b; Welling and Heard, 1978; Wilkins, 1973). *Drechslera poae* (Baudys) Shoem. was misidentified as *D. dictyoides* on *Poa pratensis* L. on which the latter species had not been previously reported, by several workers (Morrison, 1980).

For a description of the teleomorph see Paul and Parberry (1968).

Conidiophores are dark-brown to olive-brown, simple, erect, emerging singly or in groups of 2–6, uniform in diameter, 6–8 μm wide, but with a globose base about twice as wide as the stalk. There are few scars at geniculations and the conidiophore tapers very gradually to a rounded, sometimes inflated apex 6–9 μm wide. Conidiophore length is mostly 50–130 μm (Shoemaker, 1962) up to 250 μm (Ellis, 1971).

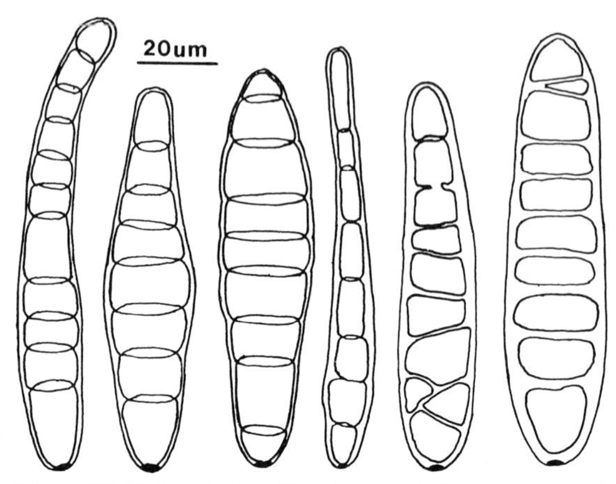

Figure 27.11 *Drechslera dictyoides*–conidia.

Conidia are typically straight, light yellowish-brown with a darker hilum, often widest near or just above the basal septum (Fig. 27.11), mostly 15–19 μm wide, tapering evenly to the apex which may be elongated. The apical septum is usually 10–15 μm wide. The conidium base is broad and hemispherical and the apex hemi-elliptical if not elongated. Conidium length 52–277 μm, mostly 72–145 (Shoemaker, 1962) mainly from *Festuca* spp., 19–99 μm from *Festuca rubra*, 20–255 μm from *F. pratensis* and 54–124 μm from *Lolium perenne* (Mäkelä, 1971). Drechsler's (1923) original description gave conidium dimensions as 14–17 μm wide at the basal segment and length 23–115 μm, usually 50–70 μm, 1–7-septate, usually 3–5 (Fig. 27.11). According to Ellis (1971) conidia may become catenate in culture or in moist chambers, but Shoemaker (1962) notes that conidia elongate, but do not form secondary spores under moist conditions. Sporulation is light-dependent and in culture is improved by incubation under near ultra-violet light (Schlosser, 1970; MacVicar and Childers, 1955).

(c) Disease development

Initial infections with *P. dictyoides* in *Festuca* and *Lolium* spp. may develop from infected seed (Labruyere, 1977; Lam, 1979; Wells and Allison, 1952). However, the pathogen may not show up in routine germination tests or significantly affect germination capacity. Fungicides applied to contaminated seed may eradicate superficially borne inoculum but do not eliminate internal infection, which in some cases, may be deep-seated inside the caryopsis (Anon, 1969; Lam, 1982; Labruyere, 1977). Although leaves emerging from coleoptiles produced by infected seeds did not show disease symptoms (de Tempe, 1968) cultures from leaf tips of some of the first emerged leaves showed that a *Drechslera* sp. was present (Lam, 1982). Level of seed contamination in *Lolium* spp. could not be correlated with disease severity in the subsequent crop (Labruyere, 1977; Lam, 1982; Tribe and Herriot, 1970). Lam (1982) found that high levels of infection with *Drechslera* spp. (mainly *P. dictyoides*), developed in fields sown with apparently uncontaminated seed, but the same *Drechslera* sp. occurred on the seed in the same proportions as on the sward foliage. This suggested that seed-borne inoculum could be important in the development of infection, although weather or cultural pratices may control disease development independently (see also *D. siccans* and *D. catenaria* – pages 293 and 300).

In regions with mild, open winters, such as New Zealand, the British Isles and maritime parts of Western Europe, the fungus persists as mycelium in infected living leaves and in attached senescent and dead tissue on which it may sporulate throughout the year. Where winters are severe and most aerial tissues are killed, the mycelium overwinters in the crowns and in leaf debris. Conidia produced on these substrates in spring are carried to new leaves by rain splash and wind. Under suitably moist conditions the disease may then spread from initial infection foci on leaves to start epidemics of the leaf spot phase in summer and autumn. Under drier conditions the leaf spotting phase is suppressed, but crown infections develop, as in *D. poae* on *Poa pratensis* (see p. 298).

There is little information on the resistance of *L. perenne* turf grass cultivars to *P. dictyoides*. Incidence of *Drechslera* spp. was lower in the seed of samples of the later flowering S.23, Melle or Endura than the earlier S.24, all mainly forage types (Lam, 1982). The higher resistance of S.23 cultivar compared with S.24 was noted by O'Rourke (1976). A general relationship exists between later maturing and flowering varieties of *Bromus inermis* Leyss and resistance to *Pyrenophora bromi* (Died.) Drechs. (Smith, 1969). On the other hand, Frandsen, Honne and Juléñ (1981) found no positive correlation between lateness and reduced incidence of *P. dictyoides* leaf spot in *Festuca pratensis*. However, there appeared to be a significant heritable variation in resistance among clones in this species and additive and non-additive genetic effects. Field attacks were noted to cause almost complete destruction of *Lolium perenne* polycross progenies in the third crop of the second harvest year (after inoculum build-up). In a parallel study plots mown weekly were not infected.

Kreitlow, Sherwin and Lefebre (1950) found that although none of the strains of *F. arundinacea* or *F. pratensis* tested were resistant to *P. dictyoides*, individual plants were resistant. Resistance was also found in clones of inbred lines of *F. arundinacea* and in *Lolium perenne* × *F. pratensis* crossed and backcrossed to *F. arundinacea*. However, there was generally a poor correlation between resistance in the field and in greenhouse tests, as with *D. siccans* (Henson and Buckner, 1957; Sherwood, Zeiders and Beg, 1973 and p. 299).

Resistance to *P. dictyoides* leaf spot was found in twelve. *F. rubra* plants of superior turf-forming ability and this resistance was incorporated into the turf grass cultivar Durlawn (Hanson, 1972; MacVicar and Childers, 1959).

(d) Control

Management practices similar to those suggested for the control of leaf spot and foot rot of *Poa pratensis* caused by *D. poae* with regard to use of resistant, or at least, known persistent cultivars of the turf grass species (Hanson, 1972), mowing height and fertilization should be employed. Fungicides effective against *D. poae* and other '*Helminthosporium*' diseases may be expected to control *P. dictyoides*. Maneb was effective against the pathogen on *Lolium perenne* in France (Courtillot and Herve, 1970) and chlorothalonil, applied at 3–4 week intervals was effective against *P. dictyoides* on *L. perenne* in England (Bonis and Heard, 1980). Joyner and Couch (1976) reduced the severity of the disease on K-31 *Festuca arundinacea* with the systemic fungicides benomyl, thiophanate-methyl and thiabendazole, but they were not as effective as contact fungicides.

27.7.2 Red leaf spot and leaf blight of bentgrasses caused by *Pyrenophora erythrospila*

The disease was first described on *Agrostis alba* L. and *A. perennans* (Walt.) Tuckerm. by Drechsler (1925) and on *A. stolonifera* Huds. and *A. tenuis* Sibth. from some of the eastern and mid-western states of the USA (Drechsler, 1935). Sprague (1950) referred to the fungus as 'obscure' yet apparently common and noted that *A. tenuis* and *A. canina* seemed less susceptible than *A. alba* and *A. stolonifera*. In Victoria, Australia the pathogen is common wherever *A. tenuis* and *A. stolonifera* occur and a fungus similar to *P. erythrospila* is often isolated from *Anthoxanthum odoratum* L. (Paul, 1972). The fungus has been isolated from withered leaves of *A. tenuis* in turf in Britain (Smith, 1965).

(a) Symptoms

Circular or eyespot lesions, with or without light-coloured centres and with reddish-brown borders develop on leaves of *Agrostis* spp., but especially on *A. alba*. These spots may extend rapidly, and become surrounded with light, water-soaked halos in warm, wet weather. In drier, hot weather infected leaves wither and desiccate and patches of the disease may take on a dull cast and appear drought-stricken (Fig. 50, colour plate section). In Australia, the eyespot lesions are rarely seen on *A. tenuis* and *A. stolonifera*, and instead extensive irregular blotches may occupy much of the leaf blade surface (Paul, 1972). In Britain, leaves supporting the fungus growth may show a dry, reddish-brown appearance (Smith, 1965). Symptoms might be confused with those caused by *D. catenaria* (Larsen, 1979), but the conidia of the two species are quite distinct morphologically (see *D. catenaria* leaf blight – p. 294).

(b) The fungus

Teleomorph – *Pyrenophora erythrospila* Paul (1972); anamorph – *Drechslera erythrospila* (Drechs.) Shoemaker (1959, 1962). For a description of the teleomorph see Paul (1972).

Conidiophores are medium to dark brown with an enlarged, globose base, narrow, usually only 6–8 μm wide at the apex, rarely branched, with geniculations tending to be clustered with the conidia, towards the apex, up to 275 μm long. Conidia are light greyish-brown, usually straight and nearly cylindrical with a hemispherical apex and slightly papillate base. The basal cell is slightly longer than others. Conidia mostly 50–70 μm with 4–8 septa, but often fewer (Smith, 1965) (Fig. 27.12). In Australian collections from *Agrostis* spp. conidiophores measured 110–195 μm × 6–8 μm and conidia 32–66 × 10–14 μm with 4 to 7 septa (Paul, 1972).

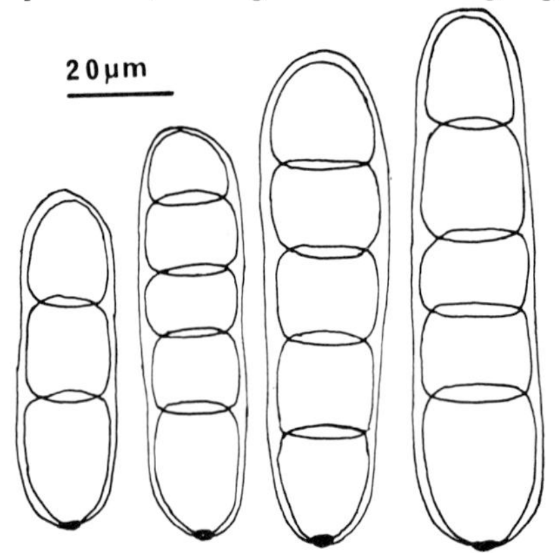

Figure 27.12 *Drechslera erythrospila*–conidia.

D. fugax may occur on *Agrostis* leaves with *D. erythrospila* (Shoemaker, 1962; Smith, 1965). The conidia and conidiophores of *D. fugax* are darker and wider than those of *D. erythrospila* (see p. 295).

(c) Disease development and control

This is a warm, wet-weather disease which usually is not prominent until the onset of summer. Subsequent dry weather results in leaf desiccation even when there is abundant soil moisture. Leaves, crowns and roots are infected (Meyer and Turgeon, 1975). Blighting is favoured by excessive nitrogen applications as with several other '*Helminthosporium*' diseases (Muse, 1974). Weekly applications of chlorothalonil or alternate applications of chlorothalonil and anilazine at 137 g a.i. and 61 g a.i./100 m^2 gave good control of the disease and turf recovered best from severe attacks where fungicides were used combined with spring fertilization (Meyer and Turgeon, 1955). Residual inoculum for new outbreaks occurs as mycelium in crowns and plant debris. The disease is probably seed transmitted (Drechsler, 1935).

27.7.3 Leaf spot and leaf blight caused by *Pyrenophora tritici-repentis*

(a) Symptoms and occurrence

P. tritici-repentis Causes a leaf spot, leaf blight and dieback of grasses including turf grasses in *Agrostis, Agropyron, Cynodon* and *Elymus* in Europe, America, Asia, Australia and Africa and has the widest host range of any of the *Drechslera* spp. It is seed-borne (Ellis, 1971; Ellis and Waller, 1976b; Hosford, 1972; Morrall and Howard, 1975; Richardson, 1979; Shoemaker, 1957, 1962).

(b) The fungus

Teleomorph – *Pyrenophora tritici-repentis* (Died.) Drechsler (1923) (Wehmeyer, 1949); anamorph – *Drechslera tritici-repens* (Died.) (Shoemaker, 1959, 1962). For a description of the teleomorph see Drechsler (1923) and Shoemaker (1962). Ascocarps develop on overwintered straw.

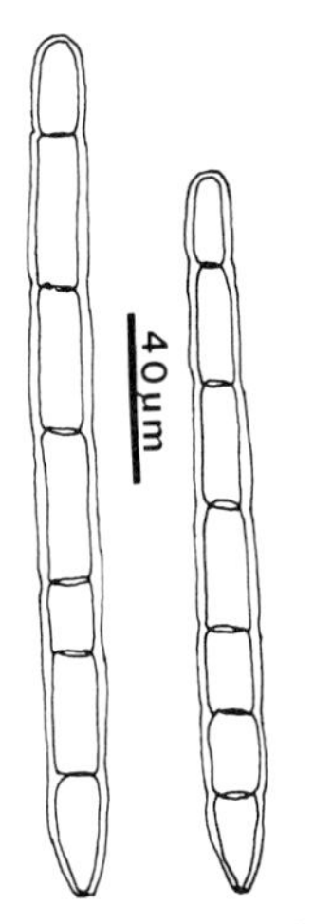

Figure 27.13 *Drechslera tritici-repentis*–conidia.

Conidiophores are light yellow–brown, solitary or in groups of 2 or 3, 11–15 µm wide at the base, 7–9 µm above, 50–250 µm long, apex not enlarged. Conidia light yellow–brown, straight to slightly bent, hemispherical at apex, conical at base, almost cylindrical, but sometimes with slight constrictions at septa, 45–175 µm × 12–21 µm, basal cell sometimes inflated, hilum minute (Drechsler, 1923; Shoemaker, 1962) (Fig. 27.13).

27.8 *SETOSPHAERIA ROSTRATA*

27.8.1 *Setosphaeria rostrata* Leonard (*Mycologia*, 68, 402–411 (1976) (teleomorph).

(*Exserohilum rostratum* Leonard and Suggs (anamorph)); causes a seedling blight on *Cynodon dactylon* (L.) Pers.; seed-borne on *Poa pratensis* L., *Festuca arundinacea* Schreb. and *Agrostis stolonifera* L., a wide range of other grasses is susceptible. Conidia are acrogenous, becoming pseusdopleurogenous, elliptical or narrowly obclavate-rostrate, brown or olivaceous, thick-walled except for a small sub-hyaline region at the apex and a similar region surrounding the hilum which protrudes as a darkened cylinder or truncate cone from the end of the basal cell. Conidia are usually straight, 1–15-septate, basal septum darker and thicker than intermediate septa, 15–190 × 7–29 µm (Fig. 27.14). They germinate from sub-hyaline regions of end cells; germ tubes grow parallel to the conidial axis.

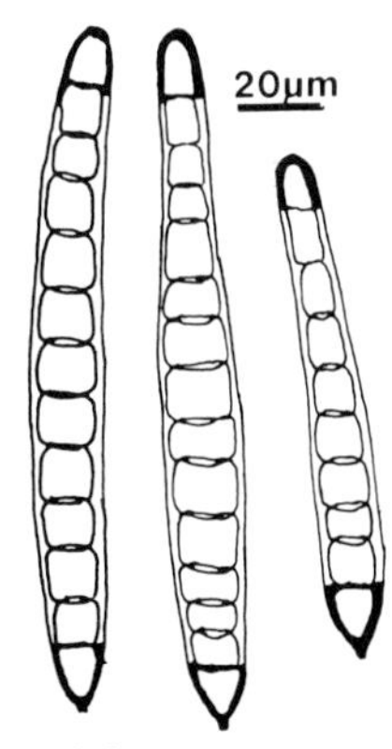

Figure 27.14 *Exserohilun rostratum*–conidia.

27.9 REFERENCES

Alberda, T. (1957) The effects of cutting, light intensity and night tempeature on growth and soluble carbohydrate content of *Lolium perenne* L. *Plant and Soil*, **8**, 199–230.

Alcorn, J.L. (1978) *Trans. Br. Mycol. Soc.*, **70**, 61–5.

Alcorn, J.L. (1981) Ascus structure and function in *Cochliobolus* species. *Mycotaxon*, **13**, 349–60.

Alcorn, J.L. (1982) New *Cochliobolus* and *Bipolaris* species. *Mycotaxon*, **15**, 1–19.

Alcorn, J.L. (1983a) On the genera *Cochliobolus* and *Pseudocochliobolus*. *Mycotaxon*, **16**, 353–79.

Alcorn, J.L. (1983b) Generic concepts in *Drechslera*, *Bipolaris* and *Exserohilum*. *Mycotaxon*, **17**, 1–86.

Ammon, H.U. (1963) On some species of the genera *Pyrenophora* Fries and *Cochliobolus* Drechsler with *Helminthosporium* as the secondary stage. *Phytopathol. Z.*, **47**, 244–300 (German, En.).

Andersen, H. (1955) Species of *Helminthosporium* on cereals and grasses in Denmark. *Friesia*, **5**, 80–9.

Andersen, H. (1959) *Helminthosporium catenarium* on grasses in Denmark. *Planteavl.*, **63**, 710–36. (Danish, En.).

Anon. (1969) Fungus eradication from perennial ryegrass. *Statens Forsogvirksomhed i Plantekultur. Lyngby, Denmark Med.*, **901**, 2 pp. (Danish).

Anon. (1979) *Beschriebende Sortenliste für Rasengräser, 1979*. A. Strothe Verlag, Hanover, 198 pp.

Atalla, N.S. and Cappellini, R.A. (1977) Effect of water stress on *Helminthosporium sativum* infection and development in Kentucky bluegrass. *Proc. Am. Phytopathol. Soc.*, **4**, 182 (Abstr.)

Baudys, E. (1915–1916) A contribution to knowledge of *Micromyces* in Bohemia. *Lotos*, **63**, 103–112; **64**, 11–29, 42–64, 80–85. (German) (Cited by Drechsler, C. 1923 and Shoemaker, R.A. 1962 q.v.)

Bean, G.A. (1964) Prevalence of *Curvularia pallescens* and *Helminthosporium* spp. pathogenic on bluegrass in Washington D.C. area. *Phytopathology*, **54**, 888.

Bean, G.A. and Wilcoxson, R.D. (1964) Pathogenicity of three species of *Helminthosporium* on roots of bluegrass. *Phytopathology*, **54**, 1084–5.

Bell, A.A. (1967) Fungi associated with root and crown rots of *Zoysia japonica*. *Plant Dis. Rept.*, **51**, 11–14.

Benoit, M.A. and Mathur, S.B. (1970) Identification of species of *Curvularia* or on rice seed. *Proc. Int. Seed Test. Assoc.*, **35**, 99–119.

Berkenkamp, B. (1971) Host range of Alberta isolates of spot blotch (*Bipolaris sorokiniana*) from forage grasses. *Phytoprotection*, **52,** 52–7.

Berkenkamp, B. (1976) Diseases of timothy in Alberta. *Timothy Plus*, **5,** 1 pp.

Berkenkamp, B., Folkins, I.P. and Meeres, J. (1973) Diseases of *Elymus* and other grasses in Alberta, 1972. *Can. Plant Dis. Surv.*, **53,** 36–8.

Boedijn, K.B. (1933) On some phragmosporous Dematiaceae. *Bull. Jard. Bot. Buitenzorg. Ser. 3*, **13,** 120–34.

Bolton, A.T. and Cordukes, W.E. (1979) Differences in susceptibility to leaf spot caused by *Bipolaris sorokiniana* among strains of *Poa annua. Can. J. Plant Sic.*, **59,** 1113–16.

Bonis, A. and Heard, A.J. (1980) Fungal diseases of grass. In *Ann. Rept. Grassld. Res. Inst.*, Hurley, pp. 46–7.

Bourgoin, B. (1976) Behaviour of the principal turf grass species in French climatic conditions. *Pepinieristes Horticulteurs Maraichers*, **171,** 33–44 (French).

Bourgoin, B., Billot, C., Kerguelen, M., Hentgen, A. and Mansat, P. (1974) Behaviour of turfgrass species in France, In *Proc. 2nd Int. Turfgrass Res. Conf.* (ed. E.C. Roberts), American Society of Agronomy, Crop Science Society of America, pp. 35–40.

Braverman, S.W. and Graham, J.H. (1960) *Helminthosporium dictyoides* and related species on forage grasses. *Phytopathology*, **50,** 691–5.

Broom, E.W., Heard, A.J., Griffiths, E. and Valentine, M. (1975) Effect of fungicides and fertilizer on yield and diseases of grass swards. *Plant Pathol.*, **24,** 144–9.

Brown, G.E., Cole, H. and Nelson, R.R. (1972) Pathogenicity of *Curvularia* spp. to turf grass. *Plant Dis. Rept.*, **56,** 59–63.

Brown, L.R. (1957) Outbreaks and new records: *FAO Plant Prot. Bull.*, **6,** 26.

Butler, F.C. (1953) Saprophytic behaviour of some root rot fungi. 1. Saprophytic colonization of wheat straws. *Ann. Appl. Biol.*, **40,** 284–97.

Cagas, B. (1973) Species of the genus *Helminthosporium*– important parasites of forage grasses. *Ochrana Rostlin*, **9**(2), 141–2.

Carr, A.J.H. (1971) Grasses. In *Diseases of Crop Plants* (ed. J.H. Western), Macmillan, London, pp. 286–307.

Cheeseman, J.H., Roberts, E.C. and Tiffany, L.H. (1965) Effects of nitrogen level and osmotic pressure of the nutrient solution on the incidence of *Helminthosoporium sativum* infection in Merion bluegrass. *Agron. J.*, **57,** 599–602.

Chidambaram, P., Mathur, S.B. and Neergaard, P. (1973) Identification of seed-borne *Drechslera* species. *Friesia*, **10,** 165–207.

Chinn, S.H.F. and Tinline, R.D. (1963) Spore germinability in soil as an inherent character in *Cochliobolus sativus. Phytopathology*, **54,** 349–52.

Christensen, J.J. (1922) Studies on the parasitism of *Helminthosporium sativum. Minn. Agric. Expt. Sta. Tech. Bull.*, **11,** 42 pp.

Christensen, J.J. (1926) Physiologic specialization and parasitism of *Helminthosporium sativum. Univ. Minn. Agr. Exp. Sta. Tech. Bull.*, **37,** 101 pp.

Colbaugh, P.F. and Endo, R.M. (1974) Drought stress: an important factor stimulating the development of *Helminthosporium sativum* on Kentucky bluegrass. In *Proc. 2nd Int. Turfgrass Res. Conf.* (ed. E.C. Roberts,), American Society of Agronomy, Madison, Wis., pp. 328–34.

Commonwealth Mycological Institute (1974) *Cochliobolus sativus* Distribution Map 332.

Conners, I.L. (1967) An annotated index of plant diseases in Canada. *Res. Br. Can. Agric. Publ. 1251*, Queen's Printer, Ottawa, 381 pp.

Cook, F.G. (1975) Production loss estimation in *Drechslera* infection in ryegrass. *Proc. Fed. Br. Plant Pathologists. Ann. Appl. Biol.*, **81,** 251–6.

Couch, H.B. (1973) *Diseases of Turfgrasses.* Krieger, New York, 348 pp.

Couch, H.B. and Cole, H. Jr (1957) Chemical control of melting out of Kentucky bluegrass. *Plant Dis. Rept.*, 41, 205–8.

Couch, H.B. and Moore, L.D. (1971) Influence of nutrition and total non-structural carbohydrate content on *Helminthosporium sativum*-incited leafspot of Kentucky bluegrass. *Phytopathology*, **61,** 888. (Abstr.)

Courtillot, M. and Hervé, J.-J. (1970) Treatment trials against crown rust of oats and ryegrass. *Phytiat.-Phytopharm.*, **19,** 123–31.

Danneberger, T.K. and Vargas, J.M. (1982) Systemic activity of iprodione in *Poa annua* and post infection activity of *Drechslera sorokiniana* leaf spot management. *Plant Dis.*, **66,** 914–15.

Del Vescovo, M. (1962) A contribution to the knowledge of some Helminthosporioses of wild and cultivated grasses in Apulialucania. *Ann. Fac. Agric. Univ. Bari.*, **16,** 137–59 (Italian).

Dovaston, H.F. (1948) A new species of *Pyrenophora* from Italian ryegrass. *Trans. Br. Mycol. Soc.*, **31,** 249–53.

Drechsler, C. (1922) A new leaf spot of Kentucky bluegrass caused by an undescribed species of *Helminthosporium. Phytopathology*, **12,** 35. (Abstr.)

Drechsler, C. (1923) Some graminicolous species of *Helminthosporium*: 1. *J. Agric. Res.*, **24,** 641–740.

Drechsler, C. (1925) A leaf spot of redtop caused by an apparently undescribed species of *Helminthosporium. Phytopathology*, **15,** 51–52. (Abstr.).

Drechsler, C. (1928) Zonate eyespot of grasses caused by *Helminthosporium giganteum. J. Agric. Res.*, **37,** 473–92.

Drechsler, C. (1929) Occurrence of the zonate eyespot fungus *Helminthosporium giganteum* on some additional grasses. *J. Agric. Res.*, **39,** 129–36.

Drechsler, C. (1930) Leaf spot and foot rot of Kentucky bluegrass caused by *Helminthosporium vagans. J. Agric. Res.*, **40,** 447–56.

Drechsler, C. (1934) Phytopathological and taxonomic aspects of *Ophiobolus, Pyrenophora, Helminthosporium* and a new genus *Cochliobolus. Phytopathology*, **24,** 953–83.

Drechsler, C. (1935) A leaf spot of bentgrasses caused by *Helminthosporium erythrospilum* n.sp. *Phytopathology*, **25,** 344–61.

Eichorn, M.M. Jr (1976) Potash helps Coastal Bermudagrass resist disease while improving yield. *Better Crops with Plant Food*, **60,** 29–31.

Ellis, M.B. (1966) Dematiaceous hyphomycetes, VII, *Curvularia, Brachysporium*, etc. *Mycol. Pap.* **106,** Commonwealth Mycological Institute, Kew, p. 57 pp.
Ellis, M.B. (1971) Dematiaceous hyphomycetes. Commonwealth Mycological Institute Kew, 608 pp.
Ellis, M.B. and Holliday, P. (1976) *Drechslera siccans*. CMI descriptions of pathogenic fungi and bacteria No. 492. Commonwealth Mycological Institute, Kew.
Ellis, M.B. and Waller, J.M. (1976a) *Pyrenophora dictyoides*. CMI Descriptions of pathogenic fungi and bacteria. Commonwealth Mycological Institute, Kew.
Ellis, M.B. and Waller, J.M. (1976b) *Pyrenophora tritici-repentis*: (Conidial state: *Drechslera tritici-repentis*) CMI Descriptions of Pathogenic Fungi and Bacteria No. 494. Commonwealth Mycological Institute, Kew.
Endo, R.M. (1961) Turfgrass disease in southern California. *Plant Dis. Rept.*, **45,** 869–73.
Falloon, R.E. (1975) *Curvularia trifolii* as a high-temperature turf grass pathogen. *NZ J. Agric. Res.*, **19,** 243–8.
Frandsen, K.J., Honne, B.I. and Julén, G. (1981) Observations on the inheritance of resistance to *Drechslera dictyoides* (*Helminthosporium dictyoides* Drechsl.) in a population of meadow fescue (*Festuca pratensis* Huds.). *Acta Agric. Scand.*, **31,** 91–9.
Frauenstein, K. (1968) Observations on the occurrence of leaf spot diseases on forage grasses. *Nachrichtenblatt. Deut. Pflanzenschutzdienst* (*Berlin*) *N.F.*, **22,** 4–14, (German, En.).
Freeman, T.E. (1957) A new *Helminthosporium* disease of Bermudagrass. *Plant Dis. Rept.*, **41,** 389–91.
Funk, C.R., Halisky, P.M. and Babinsky, P.L. (1975) *Helminthosporium* leaf spot and crown rot of Kentucky bluegrass. *Greenmaster*, **11,** 10.
Gornostai, V.I. (1970) *Drechslera* spp. on grasses in the Primorsk region. *Mikol. i Fitopathol.*, **4**(1), 69–73.
Graham, J.H. (1955) *Helminthosporium* leaf streak of timothy. *Phytopathology*, **45,** 227–8.
Gray, P.M. and Guthrie, J.W. (1977) The influence of sprinkler irrigation and post-harvest residue removal practices on the seed borne population of *Drechslera poae* on *Poa pratensis* 'Merion'. *Plant Dis. Rept.*, **61,** 90–3.
Gregory, P.H. (1961) *The Microbiology of the Atmosphere*. Leonard Hill, London.
Groves, J.W. and Skolko, A.J. (1945) Notes on seed-borne fungi. 3. *Curvularia. Can. J. Res. C.*, **23,** 94–104.
Gudauskas, R.T. (1962) Stem and crown necrosis of Coastal Bermudagrass caused by *Helminthosporium spiciferum*. *Plant Dis. Rept.*, **46,** 498–500.
Hagan, A. (1980) Isolation, identification and taxonomy of *Drechslera* and *Bipolaris* species on turfgrasses. In *Advances in Turfgrass Pathology* (eds B.G. Joyner and P.O. Larsen), Harcourt Brace Jovanovich, Duluth, Minn. pp. 89–96.
Hagan, A. and Larsen, P.O. (1979) Effect of fungicides on conidium germination, germ tube elongation and appressorium formation by *Bipolaris sorokiniana* on Kentucky bluegrass. *Plant Dis. Rept.*, **63,** 474–8.
Hagan, A.K. and Larsen, P.O. (1985) Source and dispersal of conidia of *Drechslera poae* in Kentucky bluegrass turf. *Plant Dis.*, **69**(1), 21–4.
Halisky, P.M. and Funk, C.R. (1966) Environmental factors affecting growth and sporulation of *Helminthosporium vagans* and its pathogenicity to *Poa pratensis. Phytopathology*, **56,** 1294–6.
Halisky, P.M., Funk, C.R. and Engel, R.E. (1966) Melting-out of Kentucky bluegrass varieties by *Helminthosporium vagans* as influenced by turf management practices. *Plant Dis. Rept.*, **50,** 703–6.
Hanson, A.A. (1972) Grass varieties in the United States. *US Dept. Agric. Handbk.*, **170,** 124 pp.
Harding, H. and Tinline, R.D. (1983) The existence of differentially fertile strains in two populations of *Cochliobolus sativus. Can. J. Plant Pathol.*, **5,** 17–20.
Heald, F. de F. and Wolf, F.A. (1911) New species of Texas fungi. *Mycologia*, **3,** 5–22.
Healy, M.J. and Britton, M.P. (1968) Infection and development of *Helminthosporium sorokinianum* in *Agrostis palustris. Phytopathology*, **58,** 273–6.
Henson, L. and Buckner, R.C. (1957) Resistance to *Helminthosporium dictyoides* in inbred lines of *Festuca arundinacea. Phytopathology*, **47,** 523. (Abstr.)
Hirst, J.M. and Stedman, O.J. (1963) Dry liberation of fungus spores by raindrops. *J. Gen. Microbiol.*, **33,** 335–44.
Hodges, C.F. (1972) Interaction of culture age and temperature on germination and growth of *Curvularia geniculata* and on virulence. *Can. J. Bot.*, **50,** 2093–6.
Hodges, C.F. (1980a), Postemergent herbicides and pathogenesis by *Drechslera sorokiniana* on leaves of *Poa pratensis*. In *Advances in Turfgrass Pathology* (eds B.G. Joyner and P.O. Larsen), Harcourt, Brace Jovanovich, Duluth., Minn. pp. 101–12.
Hodges, C.F. (1980b) Control factors of leaf spot and their effect on symptoms. (Leaf spot symposium–18th Nebraska Turf grass Conference) *Weeds, Trees Turf* (March, 1980), pp. 52–3, 62.
Hodges, C.F. and Blaine, W.M. (1974) Midsummer occurrence of *Helminthosporium* leaf spot on selected cultivars of *Poa pratensis. Plant Dis. Rept.*, **58,** 448–50.
Hodges, C.F. and Watschke, G.A. (1975) Pathogenicity of soil-borne *Bipolaris sorokiniana* on seeds and roots of three perennial grasses. *Phytopathology*, **65,** 398–400.
Hosford, R.M. Jr (1972) Propagules of *Pyrenophora trichostoma. Phytopathology,* **62,** 627–9.
Howard, F.L. (1953) *Helminthosporium–Curvularia* blights on turf and their cure. *Golf Course Rept.*, **21,** 5–9.
Hughes, S.J. (1958) Revisiones hyphomycetum aliquot cum appendice de nominibus rejiciendis. *Can. J. Bot.*, **36,** 727–836.
Ito, S. (1930) On some new ascigerous stages of the species *Helminthosporium* parasitic on cereals. *Proc. Imp. Acad. Tokyo*, **6,** 352–5.
Jackson, N. (1983) Rhode Island turfgrass disease control recommendations–1982. *R.I. Agric. Expt. Sta.*, 5 pp.
Jackson, N. and Fenstermacher, J.M. (1973) A patch disese of two bentgrasses caused by *Drechslera gigantea. Plant Dis. Rept.*, **57,** 84–5.
Jensen, A. (1970) Turf grass diseases and their importance in Scandinavia. *Rasen-Turf-Gazon*, **1,** 3, 69–70. (German, Fr., En.).
Jones, D.I.H., ap. Griffiths, G. and Walters, R.J.K. (1961)

Effect of nitrogen fertilizer on the water-soluble carbohydrate content of perennial ryegrass and cocksfoot. *J. Br. Grassld. Soc.*, **16,** 272–5.
Joyner, B.G. and Couch, H.B. (1976) Relation of dosage rates, nutrition, air temperature and suscept genotype to side effects of systemic fungicides on turfgrasses. *Phytopathology*, **66,** 806–10.
Juska, F.V. and Hanson, A.A. (1963) The management of Kentucky bluegrass on extensive turfgrass areas. *Park Maintenance*, **16,** 22–7.
Kenneth, R. (1958) Contribution to the knowledge of the *Helminthosporium* flora on Gramineae in Israel. *Bull. Res. Council of Israel*, **60,** 191–210.
Kenneth, R. (1964) Conidial release in some Helminthosporia. *Nature*, **202,** 1025–6.
Kline, D.M. and Nelson, R.R. (1963) Pathogenicity of isolates of *Cochliobolus sativus* from cultivated and wild gramineous hosts of the Western Hemisphere to species of the Gramineae. *Plant Dis. Rept.*, **47,** 890–4.
Kline, D.M. and Nelson, R.R. (1971) The inheritance of factors in *Cochliobolus sativus* conditioning lesion induction on gamineous hosts. *Phytopathology*, **61,** 1052–4.
Klomparens, W. (1953) A study of *Helminthosporium sativum* P.K. & B. as an unreported parasite of *Agrostis palustris* Huds. Ph.D Thesis, Univ. Mich. 77.
Kreitlow, K.W., Sherwin, H. and Lefebre, C.L. (1950) Susceptibility of tall and meadow fescues to *Helminthosporium* infection. *Plant Dis. Rept.*, **34,** 189–90.
Kuribayashi, K. (1929) The ascigerous state of *Helminthosporium sativum. Trans. Sapporo Nat. Hist. Soc*, **10,** 138–45.
Labruyere, R.E. (1977) Contamination of ryegrass seed with *Drechslera* species and its effect on disease incidence in the ensuing crop. *Neth. J. Plant Pathol.*, **83,** 205–15.
Lam, A. (1979) Drechslera leaf spot. *Grassland Res. Inst., Hurley. Ann. Rept.*, p. 64.
Lam, A. (1982) Presence of *Drechslera* species in certified ryegrass seed lots. *Grass Forage Sci.*, **37**, 47–52.
Lam, A. (1985) *Drechslera andersenii* sp. nov. and other *Drechslera* spp. on ryegrass in England and Wales. *Trans. Br. Mycol. Soc.*, **85,** 592–602.
Lam, A. and Lewis, G.C. (1982) Effects of nitrogen and potassium fertilizer application on *Drechslera* spp. and *Puccinia coronata* on perennial ryegrass foliage. *Plant Pathol.*, **31,** 123–31.
Larsen, P.O. (1979) Leaf blight and crown rot of Toronto creeping bentgrass. *Greenmaster*, **15,** 15–16.
Larsen, P.O. and Hagan, A.K. (1981) Leaf blight and crown rot of creeping bentgrass, a new disease caused by *Drechslera catenaria. Plant Dis.*, **65,** 79–81.
Latch, G.C.M. (1966) Fungus diseases of ryegrasses in New Zealand. 1. Foliage diseases. *NZ J. Agric. Res.*, **9,** 394–409.
Latch, G.C.M. and McKenzie, E.H.C. (1977) Fungal flora of ryegrass swards in Wales. *Trans. Br. Mycol. Soc.*, **68,** 181–4.
Lebeau, J.B. (1964) Lawns and turf. *Can. Plant. Dis. Surv.*, **44**(1), 32.
de Leeuw, W.P. and Vos, H. (1970) Diseases and pests on turf grasses in the Netherlands. *Rasen-Turf-Gazon*, **1**(3), 65–68, 84. (German, Fr., En.).
Lukens, R.J. (1967) Infection of inflorescence of *Poa pratensis* by *Helminthosporium vagans. Plant Dis. Rept.*, **51,** 752.
Lukens, R.J. (1970) Melting out of Kentucky bluegrass, a low sugar disease. *Phytopathology*, **60,** 1276–80.
Luttrell, E.S. (1951a) A key to the species of *Helminthosporium* in the United States. *Plant Dis. Rept.*, Suppl. 201, 59–67.
Luttrell, E.S. (1951b) Diseases of tall fescue in Georgia. *Plant Dis. Rept.*, **35,** 83–5.
Luttrell, E.S. (1958) The perfect stage of *Helminthosporium turcicum. Phytopathology*, **48,** 281–7.
Luttrell, E.S. (1977a) Biosystematics of *Helminthosporium*: Impact on agriculture. In *Biosystematics in Agriculture. Beltsville Symposium in Agricultural Research*, May 8–11, Beltsville, Mol, pp. 193–209.
Luttrell, E.S. (1977b) Correlation between conidial and ascigerous state characters in *Pyrenophora, Cochliobolus* and *Setosphaeria. Rev. Mycologie*, **4**(1) 271–9.
MacVicar, R.M. and Childers, W.R. (1955) In *Progress Rept. of the Forage Crops Div. Cent. Expt. Farm, Ottawa, Ont., 1949–1953*, pp. 21–5.
Madsen, J.P. and Hodges, C.F. (1980a) Pathogenicity of some select soil-borne dematiaceous hyphomycetes on germinating seeds of *Festuca rubra. Phytopathology*, **70,** 21–5.
Madsen, J.P. and Hodges, C.F. (1980b) Nitrogen effects on pathogenicity of *Drechslera sorokiniana* and *Curvularia geniculata* on germinating seed of *Festuca rubra. Phytopathology*, **70,** 1033–6.
Mäkelä, K. (1971) Some graminicolous species of *Helminthosporium* in Finland. *Karstenia*, **12,** 5–35.
Mäkelä, K. (1972a) Disease damage to the foliage of cultivated grasses in Finland. *Acta Agralia Fenn.*, **124**(1), 1–56.
Mäkelä, K. (1972b) Seed-borne fungi on cultivated grasses in Finland. *Acta Agralia Fenn.*, **124**(2), 1–44.
Mäkelä, K. (1975) Leaf spot diseases of grass. *Nordisk Jordbruks Forskning*, **57,** 511–13. (Swedish).
Mäkelä, K. (1981) On the seed-borne microfungi on wild grasses in Finland. *Acta Agralia Fenn.*, **20,** 132–55.
Marignoni, G.B. (1909) Micromycetes of Schio. First contribution to the fungus flora of the Province of Vicenza. *Illus. Schio*, **8,** 1–44.
Matocha, J.E. and Smith, L. (1980) Influence of potassium on *Helminthosporium cynodontis* and dry matter yields of 'Coastal' Bermudagrass. *Agron. J.*, **72,** 565–7.
Matsumoto, and Yamamoto (1936) *J. Plant Prot.* (Japan), **23,** 9–14, 107–15.
Matthews, D. (1971) A survey of certified ryegrass seed for the presence of *Drechslera* species and *Fusarium nivale* (Fr.) Ces. *NZ J. Agric. Res.*, **14,** 219–26.
McKenzie, E.C.H. and Matthews, D. (1977) *D. noblea* sp. nov. on ryegrass. *Trans. Br. Mycol. Soc.*, **68,** 309–11.
McKenzie, E.H.C. (1978) Occurrence of *Drechslera* and *Curvularia* on grass seed in New Zealand. *NZ J. Agric. Res.*, **21,** 283–6.
Meredith, D.S. (1963) 'Eyespot', a foliar disease of bananas caused by *Drechslera gigantea* Heald and Wolf. 1. Preliminary studies on pathogenicity and spore dispersal. *Ann. Appl. Biol.*, **51,** 29–40.

Meyer, W.A. and Turgeon, A.J. (1975) Control of red leaf spot on 'Toronto' creeping bentgrass. *Plant Dis. Rept.*, **59,** 642–5.

Misra, A.P. and Gupta, V. (1971) A comparative study of graminicolous species of *Helminthosporium. Proc. Nat. Acad. Sci., India*, **B 41**(1), 33–42.

Misra, A.P., Singh, R.A. and Prakesh, O. (1971) Two new leafspot diseases of Bermudagrass incited by Helminthosporia in India. *Sic. Cult.*, **32,** 95–6.

Monteith, J. Jr and Dahl, A.S. (1932) Turf diseases and their control. *US Golf Assoc. Green Sect. Bull.*, **12,** 87–187.

Moore, L.D. and Couch, H.B. (1961) *Pythium ultimum* and *Helminthosporium vagans* as foliar pathogens of the Gramineae. *Plant Dis. Rept.*, **45,** 616–19.

Morrall, R.A.A. and Howard, R.J. (1975) The epidemiology of leaf spot disease in a native prairie. II Airborne spore populations of *Pyrenophora tritici-repentis. Can. J. Bot.*, **53,** 2345–53.

Morrison, R.H. (1980) Leaf spot of Kentucky bluegrass in Minnesota; misidentification of *Drechslera poae* as *D. dictyoides*. In *Advances in Turfgrass Pathology* (eds B.G. Joyner and P.O. Larsen), Harcourt Brace Jovanovich, Duluth, Minn. pp. 113–19.

Morrison, R.H. (1982) *Drechslera noblea* on *Lolium multiflorum* in North America. *Mycologia*, **74,** 391–7.

Mower, R.G. (1962) Histological studies of suscept-pathogen relationships of *Helminthosporium sativum* P.K. and B., *Helminthosporium vagans* Drechs. and *Curvularia lunata* (Wakk.) Boed. on leaves on Merion and of common Kentucky bluegrass (*Poa pratensis*). Ph.D. Thesis, Cornell University, 150 pp.

Mühle, E. (1971) *Diseases and Pests of Forage Grasses*. S. Hirzel Verlag, Leipzig, 422 pp. (German).

Muse, R.R. (1974) Influence of nutrition on the development of *Helminthosporium* red leaf spot on Seaside bentgrass, *Agrostis palustris. Physiol. Plant Pathol.*, **4,** 99–105.

Muse, R.R., Couch, H.B., Moore, L.D. and Muse, B.D. (1972) Pectolytic and cellulolytic enzymes associated with *Helminthosporium* leaf spot on Kentucky bluegrass. *Can. J. Microbiol.*, **18,** 1091–8.

Nattrass, R.M. (1939) Annual Report of the Senior Plant Pathologist. *Rept. Dept. Agric. Kenya, 1938*, **2,** 42–9. (*R.A.M.*, (1939) **19,** 71).

Neergaard, P. (1957) Eighth and ninth annual reports relating to pathological seed testing–1 June 1955 to 31 May 1957. Statens Plantettilsyn, Copenhagen 15 pp. (Danish). (*Rev. Appl. Mycol.*, **38,** 54–5).

Nelson, R.R. (1960) *Cochliobolus intermedius* the perfect stage of *Curvularia intermedia. Mycologia*, **52,** 775–8.

Nelson, R.R. (1964a) The perfect stage of *Curvularia geniculata. Mycologia*, **56,** 777–9.

Nelson, R.R. (1964b) The perfect stage of *Helminthosporium cynodontis. Mycologia*, **56,** 64–9.

Nelson, R.R. and Haasis, F.A. (1964) The perfect stage of *Curvularia lunata. Mycologia*, **56,** 316–17.

Nicholson, J.F., Gray, G.G. and Sinclair, J.B. (1971) Excised leaf blades for fungicidal evaluation against *Helminthosporium sorokinianum. Plant Dis. Rept.*, **55,** 959–60.

Nilsen, K.N., Hodges, C.F. and Madsen, J.P. (1979) Pathogenesis of *Drechslera sorokiniana* on progressively older leaves of *Poa pratensis* as influenced by photoperiod and light quality. *Physiol. Plant Pathol.*, **15,** 171–6.

Nilsen, K.N., Madsen, J.P. and Hodges, C.F. (1978) Enhanced *Drechslera sorokiniana* leaf spot expression on *Poa pratensis* in response to photoperiod and blue biassed light. *Physiol. Plant Pathol.*, **14,** 57–9.

Nishihara, N. (1967) *Helminthosporium* leafspot of Bermudagrass, *Cynodon dactylon* in Japan. *Ann. Phytopathol. Soc. Japan*, **33,** 276–9.

Nishihara, N. (1973) Helminthosporium leafspot of Kentucky bluegrass. *Bull. Nat. Grassld. Res. Inst. Japan*, **2,** 41–5.

Noviello, C. (1963) Cultivated grasses: Lawns and turf. *Can. Plant Dis. Surv.*, **44**(1), 32–3.

Nutter, F.W., Cole, H. Jr and Schein, R.D. (1982) Conidial sampling of *Drechslera poae* to determine role of mowing in spore dispersal. *Plant Dis.*, **66,** 721–3.

O'Rourke, C.J. (1976) *Diseases of Grasses and Forage Legumes in Ireland*. An Foras Taluntais (The Agricultural Institute), Carlow, 115 pp.

Padwick, G.W. and Henry, A.W. (1983) The relation of species of *Agropyron* and certain other grasses to the foot rot problem of wheat in Alberta. *Can. J. Res. Sect.*, **C 8,** 349–363.

Paul, A.R. (1972) *Pyrenophora erythrospila* sp. nov., the perfect stage of *Drechslera erythrospila. Trans. Br. Mycol. Soc.*, **59,** 97–102.

Paul, A.R. and Parberry, D.G. (1968) *Pyrenophora dictyoides* sp. nov., the perfect state of *Helminthosporium dictyoides. Trans. Br. Mycol. Soc.*, **51,** 707–10.

Pirone, P.P., Dodge, B.O. and Rickett, H.W. (1960) Fading-out. In *Diseases and Pests of Ornamental Plants*. Ronald Press, New York, 538 pp.

Platford, R.G., Bernier, C.C. and Ferguson, A.C. (1972) Lawn and turf diseases in the vicinity of Winnipeg. *Can. Plant Dis. Surv.*, **52,** 108–9.

Putterill, K.M. (1954) Some graminicolous species of *Helminthosporium* and *Curvularia* occurring in South Africa. *Bothalia*, **6,** 347–78.

Ricci, A. Jr., Geraldi, M.A. and Ito, M.F. (1981) Occurrence of *Bipolaris cynodontis* (Marig.) Shoemaker (*Helminthosporium cynodontis* Marighioni) on *Cynodon dactylon* (L.) Pers. (Portuguese). *Summa Phytopathol.*, **7,** 44–8.

Richardson, M.J. (1979) *An Annotated List of Seed-borne Diseases*. Commonwealth Mycological Institute, Int. Seed Test. Assoc., 320 pp.

Richardson and Fraser (1968) New combinations in *Drechslera. Trans. Br. Mycol. Soc.*, **51,** 147–8.

Riley, E.A. (1960) A revised list of plant diseases in Tanganyika Territory. *Commonw. Mycol. Inst. Mycol. Pap.*, **75,** 42 pp.

Roberts, E.C. and Cheeseman, J.H. (1964) Effect of nitrogen and moisture stress on leaf spot infection in Merion bluegrass. *Golf Course Rept.*, **32,** 18–20.

Robinson, P.W. and Hodges, C.F. (1977) Effect of nitrogen fertilization and free amino acid and soluble sugar content of *Poa pratensis* on infection and disease severity by *Drechslera sorokiniana. Phytopathology*, **67,** 1239–44.

Rogerson, C. (1958) Diseases of grasses in Kansas, 1953–55. *Plant Dis. Rept.*, **40**, 338–87.

Rogerson, C.T. (1956) Diseases of grasses in Kansas: 1953–55. *Plant Dis. Rept.*, **40**, 388–97.

Sampson, K. and Western, J.H. (1940) Two diseases of grasses caused by *Helminthosporium* not previously recorded in Britain. *Trans. Br. Mycol. Soc.*, **24**, 255–63.

Sampson, K. and Western, J.H. (1954) *Diseases of British Grasses and Herbage Legumes*. 2nd edn, Br. Mycol. Soc., Cambridge University Press, 118 pp.

Sanders, P.L., Burpee, L.L., Cole, H. Jr and Duich, J.M. (1978) Control of fungal pathogens with the experimental iprodione fungicide. *Plant Dis. Rept.*, **62**, 549–53.

Scharif, G. (1961) Studies on graminicolous species of *Helminthosporium*. 1. *H. phlei* (Graham) comb. nov. *Trans. Br. Mycol. Soc.*, **44**, 217–29.

Scharif, G. (1963) *Continue of Studies on Graminicolous Species of Helminthosporium*. Ministry of Economy's Press, Tehran, 97 pp. (and 50 pp. in Persian).

Schlösser, U.G. (1970) On the improvement and maintenance of sporulation in a culture collection of parasitic fungi from the Gramineae through long-wave ultra-violet light. *Phytopathol. Z.*, **68**, 171–80.

Sherwood, R.T., Zeiders, K.E. and Beg, C.C. (1973) Selecting resistance to *Helminthosporium dictyoides* in *Lolium-Festuca* derivatives. *Plant Dis. Rept.*, **57**, 563–6.

Shoemaker, R.A. (1955) Biology, cytology and taxonomy of *Cochliobolus sativus*. *Can. J. Bot.*, **33**, 562–76.

Shoemaker, R.A. (1957) *Helminthosporium* on western grasses. *37th Ann. Rept. Can. Plant Dis. Surv.*, pp. 24–5.

Shoemaker, R.A. (1959) Nomenclature of *Drechslera* and *Bipolaris*, grass parasites segregated from '*Helminthosporium*'. *Can. J. Bot.*, **37**, 880–7.

Shoemaker, R.A. (1962) *Drechslera* Ito. *Can. J. Bot.*, **40**, 809–36.

Shoemaker, R.A. (1981) Changes in taxonomy and nomenclature of important genera of plant pathogens. *Ann. Rev. Plant Pathol.*, **19**, 297–307.

Shurtleff, M.C. (1981) New turfgrass fungicides. In *Proc. 22nd Ill. Turfgrass Conf.* (ed. W. Fermanian), Coop. Ext. Serv. Coll. Agr. Univ. Illinois, Urbana Champaign, pp. 62–3.

Simmonds, P.M., Sallans, B.J. and Ledingham, R.J. (1950) The occurrence of *Helminthosporium sativum* in relation to primary infections in common root rot of wheat. *Sci. Agric.*, **30**, 407–17.

Sivanesan, A. (1984) *The Bitunicate Ascomycetes and Their Anamorphs*. Strauss and Kramer, Lichtenstein.

Slana, L.J. (1977) Reactions of Bermuda grasses (*Cynodon* spp.) to *Helminthosporium cynodontis* Marignoni. *Proc. Am. Phytopathol. Soc.*, **4**, 219 (Abstr.).

Smedegard-Petersen, V. (1970) *Drechslera poae and Rhynchospörium orthosporum recorded as pathogens on grasses in Denmark* Royal Veterinary Agricultural University of Copenhagen. (Preprint Yrbk. 1971) 10 pp.

Smedegard-Petersen, V. and Hermansen, J.E. (1972) Leaf spot diseases on graminicolous species at a locality in Greenland. *Friesia*, **10**, 25–9.

Smith, J.D. (1953) New record. In Turf Disease Notes, 1953. *J. Sports Turf Res. Inst.*, **8**(29), 260.

Smith, J.D. (1955a) *Curvularia lunata* from *Cynodon*. In Turf disease notes, 1955. *J. Sports Turf Res. Inst.*, **31**, 73.

Smith, J.D. (1955b) *Helminthosporium* spp. on turfgrasses. In Turf disease notes, 1955. *J. Sports Turf Res. Inst.*, **31**, 60–75.

Smith, J.D. (1965) *Fungal Diseases of Turf Grasses*. 2nd edn (revised N. Jackson and J.D. Smith), The Sports Turf Research Institute 97 pp.

Smith, J.D. (1969) Field resistance to *Pyrenophora bromi* in *Bromus inermis* and *Bromus* spp. *Can. Plant Dis. Surv.*, **49**, 8–13.

Smith, J.D. (1970a) Resistance of timothy cultivars to *Heterosporium phlei, Drechslera phlei*, and frost injury. *Can. Plant Dis. Surv.*, **50**, 95–7.

Smith, J.D. (1970b) Seed-borne *Drechslera phlei* on *Phleum* spp. *Can. J. Plant Sci.*, **50**, 746–7.

Smith, J.D. (1980) *Major Diseases of Turf grasses in Western Canada*. University of Saskachewan Saskatoon, 14 pp.

Smith, J.D. and Evans, I.R. (1985) *Major diseases of turf grasses in western Canada. Epidemiology and control.* Agdex 273/636-6. Alberta Agriculture, Edmonton.

Spilker, D.A. and Larsen, P.O. (1980) Leaf blight and crown rot, new disease on creeping bentgrass. In *Advances in Turfgrass Pathology* (eds B.G. Joyner and P.O. Larsen), Harcourt Brace Jovanovich, Duluth, Minn. 147–9.

STRI (1980) *Turfgrass Seed, 1980.* Sports Turf Research Institute, Bingley, 105 pp.

STRI (1982) *Turfgrass seed, 1983.* The Sports Turf Research Institute, Bingley, 12 pp.

Sprague, R. (1950) *Diseases of Cereals and Grasses in North America.* Ronald Press, New York, 538 pp.

Sprague, R. (1956) Some leafspot fungi on western Gramineae. 10. *Mycologia,* **48,** 741–6.

Subramanian, C.V. (1954) Fungi Imperfecti from Madras–V. *Proc. Indian Acad. Sci. B.*, **38,** 27–39.

Subramanian, C.V. and Jain (1966) A revision of some graminicolous Helminthospora. *Curr. Sci.*, **35,** 352–5.

Tempe, J. de (1968). The detection of *Helminthosporium* and *Fusarium* spp. in ryegrass and meadow fescue seed samples. *Proc. Int. Seed Test. Assoc.*, **33,** 541–5.

Teuteberg, A. (1977) A contribution to the occurrence of leaf spot fungi on *Lolium perenne* L. and *Lolium multiflorum* Lam. in the Federal Republic of Germany. *Nachrichtenbl. des. Deutsh. Pflanzenschutzdienst*, **29,** 121–3. (German, En.).

Teuteberg, A. (1978) *Drechslera poae* (Baudys) Shoem. as a pathogen on *Poa pratensis* L. *Rasen-Turf-Gazon*, **9**(2), 36–8.

Tinline, R.D. (1951) Studies on the perfect stage of *Helminthosporium sativum. Can. J. Bot.*, **29,** 467–78.

Traquair, J.A. and Smith, J.D. (1984) Turf grass problems in the Prairie Provinces. *Commun. Br. Agric. Can. Ottawa*, Publ. 176E, 25 pp.

Tribe A.J. and Herriot, J.B.D. (1970); Fungicidal treatment for herbage seed. *Edin. School Agric. Exp. Work Rept. for 1969*, p. 63.

Tsuda, M. and Ueyama, A. (1981) *Pseudocochliobolus australiensis*, the ascigerous stage of *Bipolaris australiensis. Mycologia,* **73,** 88–96.

Tsuda, M., Ueyama, A. and Nishihara, N. (1977) *Pseudocochliobolus nisikadoi*, the perfect state of *Helminthos-*

porium coicis. Mycologia, **69,** 1109–20.

United States Department of Agriculture (1960) Index of plant diseases in the United States. *Agric. Handbk.*, **165.** Crops Res. Div. USDA 531 pp.

Vanterpool, T.C. (1948) Cultivated grasses: Lawns and turf. *Can. Plant Dis. Surv.*, **28,** 35.

Vargas, J.M. (1981) *Management of Turfgrass Diseases.* Burgess, Minneapolis 204 pp.

Vargas, J.M. and Kelly, K.J. (1981) HAS decline of annual bluegrass (anthracnose). *Mich. State Univ. Coop. Ext. Serv. Bull.*, 1451, 2 pp.

Watkins, J.E. and Shearman, R.C. (1976) *Melting out disease of lawns.* Nebraska Guide. Coop. Ext. Service. Institute of Agriculture Natural Resources. University of Nebraska. G76–307. 3 pp.

Watkins, J.E., Shearman, R.C. and Bruneau, A. (1979) *Melting-out Disease of Lawns.* NebGuide, Coop. Ext. Serv. Inst. Agr. Nat. Resources, Univ. Neb. Lincoln, Neb. 3 pp.

Wehmeyer, L.E. (1949) Studies in the genus Pleospora. Mycologia, 41 (5), 565–93.

Weihing, J.L., Jensen, S.G. and Hamilton, R.L. (1957) *Helminthosporium sativum* a destructive pathogen of bluegrass. *Phytopathology*, **47,** 744–6.

Welling, B. and Heard, A.J. (1978) Descriptions of grass diseases No. 1. Net blotch *Drechslera dictyoides* (Drechsler) Shoemaker. *Weibulls Gräs-tips*, **21,** 3–4.

Wells, H.D. and Allison, J.L. (1952) Tall fescue seed can carry net blotch fungus. *Phytopathology*, **42,** 22–3 (Abstr.).

Wernham, C.C. and Kirby, R.S. (1941) A new turf disease. *Phytopathology*, **31,** 24.

Whittle, A.M. (1977) *Cochliobolus sativus* on barley in Scotland. *Plant Pathol.*, **26,** 67–74.

Wilkins, P.W. (1973) Infection of *Lolium* and *Festuca* spp. by *Drechslera siccans* and *Drechslera catenaria. Euphytica*, **22,** 106–13.

Winter, W.E., Mathur, S.B. and Neergaard, P. (1974) Seedborne organisms of Argentina: A Survey. *Plant Dis. Rept.*, **58,** 507–11.

28 *Some leaf spots associated with pycnidial fungi*

28.1 *ASCOCHYTA*

Ascochyta leaf spots and leaf blights occur commonly on grasses in many parts of the world. Rarely do they cause extensive injury to mown amenity turf, in fact, the pathogenicity of many found on turf grass species is uncertain, since little work has been done on them as agents of turf grass disease.

Leaf spot symptoms caused by the various species vary greatly according to the host and pathogen, from discrete, dark-bordered leaf spots, circular to longitudinal or striate in form, to leaf-girdling lesions, vague yellow laminar or leaf-sheath blotches or leaftip blights with immersed or erumpent pycnidia. The latter vary in colour from yellow to reddish-brown, dark-brown or black. Some of the lesions are found in living leaves, but the fungi also occur on fading or dead leaves or leaf sheaths (Fig. 52, colour plate section). Sometimes floral parts are affected; seed transmission has been noted in some species and seedling infection is likely.

There are about 20 graminicolous *Ascochyta* spp. according to the recent classification, of which perhaps five or six may occasionally cause severe or moderately severe disease on turf grass species (Buchanan, 1984a,b; Punithalingam, 1979; Sprague, 1950). The taxonomy of several of the species is still uncertain and their diagnosis difficult (Boerema and Bollen, 1975; Punithalingam, 1979; Sprague, 1950; Sutton, 1980). In Punithalingam's (1979) classification used here, some of the older (and more familiar) species have been excluded, some other genera and species of *Ascochyta* have been placed in synonymy, departing from a common previous practice of distinguishing species according to host. Indeed some of the species have wide host ranges and several species may occur on one host.

Ascochyta in the sense used by Punithalingam (1979) includes pycnidial species of the Sphaeropsidales with hyaline and pale-straw or yellow–brown, cylindrical, oblong to ellipsoidal or boat-shaped conidia, mostly nearly medianly uniseptate (occasionally unicellular to rarely 3-septate) conidia. The conidia may split along thc scptum into two cells. There are *Didymella* or *Didymosphaeria* teleomorphs in some graminicolous *Ascochyta* spp.

Ascochyta species in the *Apiocarpella* Group (I) have conidia mostly unequally uniseptate. In the *Eu-Ascochyta* Group (II) conidia are predominantly hyaline and mostly medianly or near medianly uniseptate. In the *Ascochytella* Group (III) conidia are mostly medianly or near medianly uniseptate and pale yellow to straw yellow.

28.1.1 *Ascochyta* spp. on turf grass hosts, location, pathogenicity and group (from Punithalingam 1979)

A. agrostis Polosova on *Agrostis alba* L. and *A. tenuis* Sibth.; USSR, GB, Norway and S. America; pathogenic on several other grass spp. in laboratory tests (II).

A. anthoxanthi Kalymbetov on *Lolium perenne* L.; USSR and GB; doubtfully pathogenic (II).

A. avenae (Petrak) Sprague and Johnson on *Lolium perenne*; GB, Europe and N. America; pathogenicity uncertain (III).

A. controversa Punith. on glumes of *L. perenne*; Italy; pathogenicity uncertain (II).

A. cynosuricola Punith. *Cynosurus cristatus* L.; GB rare; pathogenicity uncertain (II).

A. desmazieresii Cav. on *L. perenne*; USA, British Isles; widely distributed, locally important pathogen of *L. perenne* (II).

A. digraphidis Polosova on *A. alba* and *L. perenne*; USSR, USA; probably widely distributed, pathogenicity uncertain (II).

A. festucae Punith. on fading leaves of *A. tenuis*, *F. ovina* L., *F. rubra* L. and *L. perenne*; Europe and British Isles; doubtfully pathogenic (II).

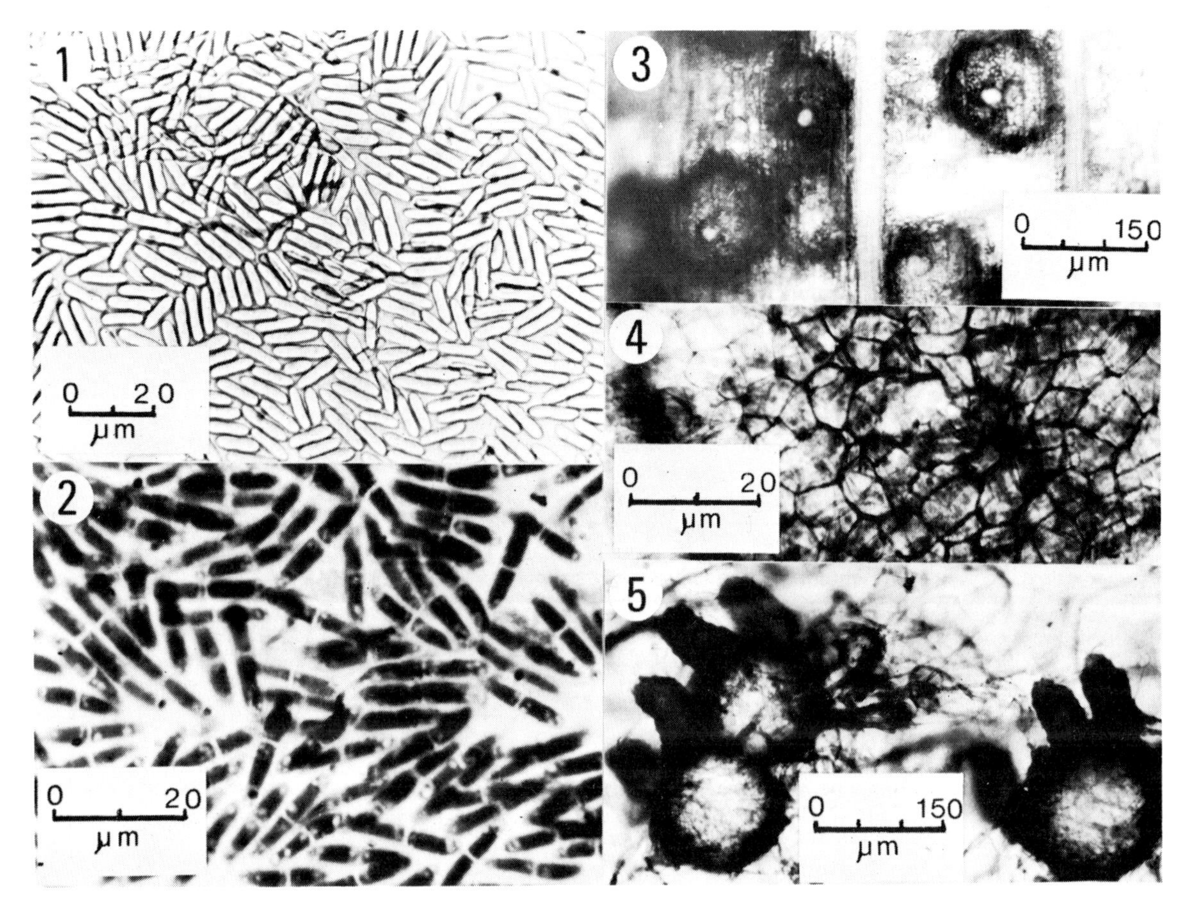

Figure 28.1 *Ascochyta phleina.* 1. Conidia (unstained). 2. Conidia stained to show septation. 3 and 5. Pycnidia from culture. 4. Pycnidium wall.

A. festucae-erectae P. Henn. on *L. perenne*, *F. rubra*; Antarctica, British Isles and W. Germany; pathogenic (II and III).

A. graminea (Sacc.) Sprague and Johnson on *Cynodon dactylon* (L.) Pers.; USA, Mexico, Ethiopia, Europe; doubtfully pathogenic (II).

A. hordei Hara and vars. on *Agropyron cristatum* Gaertn and *Festuca arundinacea* Schreb.; USA; widely distributed but pathogenicity uncertain (III).

A. hordeicola Punith. on *Poa pratensis* L.; mildly pathogenic (II).

A. leptospora (Trail) Hara and vars. on *A. stolonifera* L. *A. tenuis*, *Chloris gayana* Kunth., *F. rubra*, *Holcus lanatus* L., *L. perenne*, *Cynosurus cristatus*, *Phleum pratense*; wide host ranges and distribution, variable pathogenicity (II).

A. missouriensis Sprague and Johnson on *L. perenne*; Wales; pathogenicity uncertain (II and III).

A. paspali (Syd.) Punith. on *Paspalum dilatatum* Poir.; USA, S. America and New Zealand (III). It is endophytic pathogenic and seed-borne (Buchanan, 1984a,b).

A. phleina Sprague on *F. arundinacea*, *Phleum pratense*, *Poa pratensis*; widely distributed in N. America; moderately pathogenic (II) (see Fig. 28.1).

A. rhodesii Punith. on *L. perenne*; GB only (II).

Didymella graminicola Punith. has an *Ascochyta* anamorph on seeds of *Lolium* spp.

Didymella phleina Punith. and K. Årsvoll has the anamorph *Ascochyta phyllachoroides* Sacc. and Malbr. forma *melicae* Fautrey on *Phleum pratense* and *Lolium* sp. in Norway; pathogenic on seedlings of *P. pratense* (III).

Didymosphaeria loliina Punith. has an *Ascochyta* anamorph derived from an isolate from seeds of *Lolium* (II).

For diagnostic features of these species see Punithalingam (1979).

Many of the species are probably saprobes and others need mechanically damaged or senescent tissues to provide infection courts from which to invade tissues of higher vitality. *Ascochyta* spp. are debris-borne as mycelium or pycnidia and their spores are rain-splash and wind dispersed. Usually special control measures are not required, but in epidemics clippings should be collected after mowing and suitable fungicides applied. For general management and fungicides, see *D. poae* leaf spot (p. 299).

28.2 *PHYLLOSTICTA*

Phyllosticta leaf spots, mostly called *Macrophoma* or *Sphaeropsis* in the literature, are of world-wide distribution, with about 50 definite fungal species, not all separable morphologically. Many names in *Phyllosticta* are referable to *Phoma, Phomopsis, Asteromella*, etc. with which their pycnospores may be confused. They are well represented on monocotyledonous species. Most of those on grasses are pathogenic, but most have been poorly studied. Where known, they have *Guignardia* as teleomorphs (Grove, 1935; Sprague, 1950; von Arx, 1970; Van der Aa, 1973).

Pathogenic species often produce a discrete spot with a dark margin, the centre-bearing rows of pycnidia between the leaf veins. *Pycnidia* are separate, immersed, then erumpent, ostiolate, thick- or thin-walled, up to 200 μm diam. *Conidia* are hyaline, single-celled, oval or sub-cylindrical with rounded ends and a delicate cell wall, usually with an appendage, distinguishable with phase optics, becoming slimy, variable in size, but usually large.

28.3 *PSEUDOSEPTORIA*

According to Sutton (1980) graminicolous *Selenophoma* spp. should be referred to *Pseudoseptoria* Speg. (Sphaeropsidales) because of their anneliform method of conidiogenesis. The species which may be found on turf grasses are:

1. *P. donacis* (Pass.) Sutton (syn. *Selenophoma donacis* (Pass.) Sprague and A.G. Johnson) (Punithalingam and Waller, 1973). Pycnidia erumpent, brown, globose ostiolate, 40–150 μm. Conidia mostly 20–23 × 2.5–4.0 μm. Bissett (1982c) considers this species quite distinct from *P. stomaticola* (below) and relatively restricted to hosts in the Arundinaceae and coarse grasses such as *Phragmites* and *Molinia*, so it is less likely to be found on turf grasses (Fig. 51, colour plate section).
2. *P. stomaticola* (Bauml.) Sutton (syn. *Selenophoma donacis* (Pass.) Sprague and A.G. Johnson var. *stomaticola* (Bauml.) Sprague and A.G. Johnson). Pycnidia are immersed to slightly erumpent in linear rows between the leaf veins (Fig. 28.3) golden-brown to black, similar in size to those of *P. donacis*. Conidia are hyaline, falcate to lunate, usually aseptate, mostly 16–18.5 × 2 μm. This is a very variable species (Bissett, 1982c; Jørstad, 1967; Mäkelä, 1977b; Sprague, 1950). (Fig. 28.2)

Figure 28.2 *Pseudoseptoria stomaticola*–conidia.

3. *P. everhartii* (Sacc. and Syd.) Sutton (syn. *S. everhartii* Sprague and A.G. Johnson and probably of *P. stomaticola* also (J. Bissett, personal communication)). Pycnidia smaller in than *P. stomaticola*, up to 100 μm diam. with smaller, sharply curved conidia, 10–15 × 1–2.5 μm.
4. *P. obtusa* (Sprague and A.G. Johnson) Sutton (syn. *S. obtusa* Sprague and A.G. Johnson) and (Sprague and Johnson, 1950). Pycnidia black 40–150 μm. Conidia are curved, sausage-shaped with rounded to sharp ends, 13–17 × 2.5–4.2 μm.

P. donacis, with the longest spores tends to be on the coarser, larger grasses; *P. stomaticola*, with intermediate length spores is found mostly on pasture grasses, and *P. everhartii*, with the shortest spores, on the smallest grasses (Sprague, 1950). Spore germination takes place from either or both ends of the spore (Fig. 28.4).

Disease development in *Pseudoseptoria* leaf spot is similar to that for other pycnidial graminicolous leaf pathogens such as *Septoria* and *Mycosphaerella*. *P. donacis* and *P. stomaticola* are seed-borne on grasses (Richardson, 1979). Abundant spores of *P. stomaticola* were found in seed washings of *Festuca rubra* L. from northern Alberta, but as with *P. bromigena* (Smith, 1970 and unpublished), the relationship between spore infestation and subsequent leaf spotting was not established. *Pseudoseptoria* leaf spots appear early in the growing season on mature plants and the fungus probably survives as mycelium or pycnidia on overwintering, living tissues. Control measures for these diseases on turf grasses have not been developed.

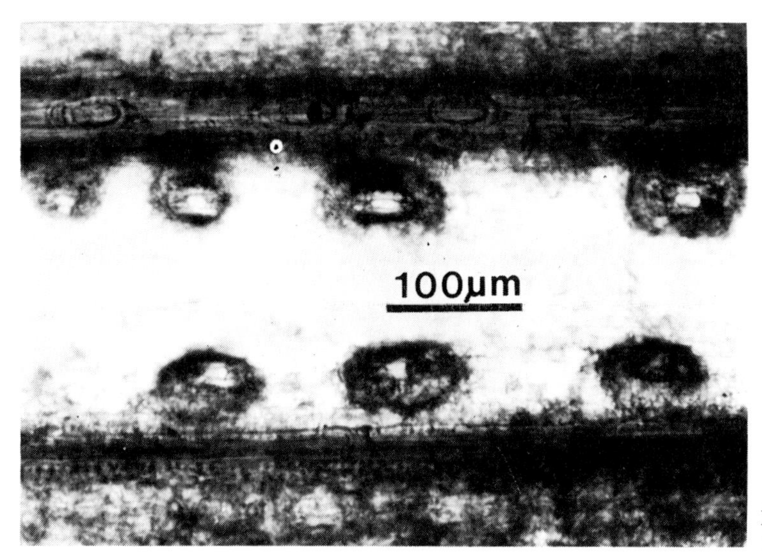

Figure 28.3 *Pseudoseptoria bromigena*– rows of substomatal pycnidia.

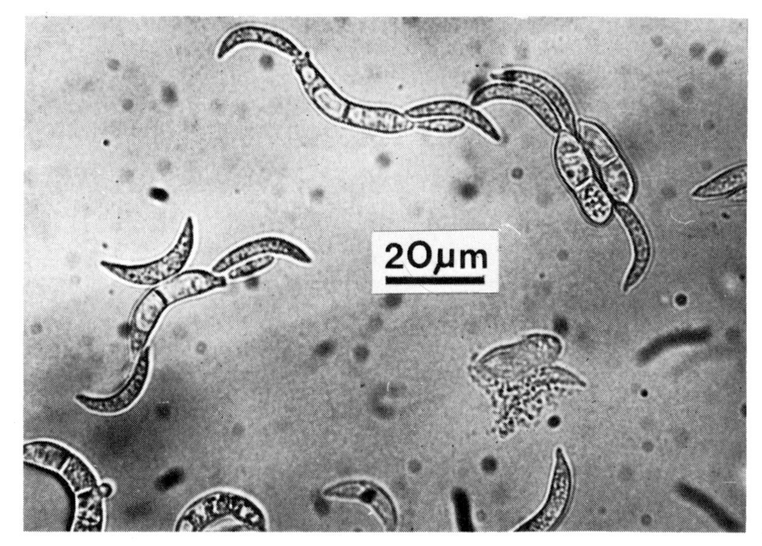

Figure 28.4 Germinating conidia of *Pseudoseptoria bromigena.*

28.4 *SELENOPHOMA*

See *Pseudoseptoria* (p. 314).

28.5 *SEPTORIA*

In leaf spots caused by *Septoria* spp. inflorescences, culms, leaf blades and sheaths, and sometimes roots and seedlings of turf grasses may be lesioned. Some of the pathogens are seed-borne. Actively growing tissues are damaged by some species, but many of the large number of graminicolous species reported are saprophytes or weak pathogens on dead, moribund or senescing tissues. Unmown or infrequently mown turf or swards in moist, shaded situations at the edges of woodland or near hedges are occasionally damaged during periods of cool, wet weather, particularly in sping and autumn.

The taxonomy of many of the *Septoria* spp. (Sphaeropsidales) on the Poaceae is confused, synonymy is frequent and there is no recent overview of graminicolous species such as that for *Ascochyta*

(Punithalingam, 1979). While N. American species were thoroughly examined by Sprague (1944, 1950, 1955), several of these appear absent from Europe (Cejp and Jechova, 1967; Grove, 1935; Jørstad, 1967; Mäkelä, 1977b). Cherepanova and Nguen van Tchank (1980) recently published a key to 29 *Septoria* spp. on Poaceae in the Leningrad region of the USSR. There is apparently little information on *Septoria* spp. on warm-season turf grass species. Host records and results of infection experiments, which would allow a better appreciation of the pathogenicity of species are sometimes unreliable because of uncertain diagnosis (see *Stagonospora*) and the existence of specialized forms and varieties. Some of the *Septoria* spp. likely to be found on turf grass species (mainly cool-season) are briefly described below.

28.5.1 Leaf spot of *Agrostis*, *Agropyron* and *Holcus* spp. – *Septoria calmagrostidis* (Lib.) Sacc.

S. calmagrostidis occurs on senescing and dead leaves of the grasses in Europe, N. America, N. Africa, Asia on pale or brown, elongated leaf spots. In the Pacific northwest of the USA it is found on *A. stolonifera* L., while *S. triseti* is mainly on *A. tenuis* Sibth. (Sprague, 1950).

Pycnidia are dark-brown, obscure, flattened, subepidermal, 50–180 μm. Conidia are filiform, straight, wavy or curved, mostly 3-septate, but septa are obscure (Jørstad, 1967), distinctly 0–5-septate, but mostly 3-septate (Sprague, 1950), 18–80 μm × 0.7–2 (Jørstad, 1967) or 25–73 × 1.0–2.0 μm (Sprague, 1950). Can be distinguished from *S. triseti* by its longer spores and larger pycnidia.

28.5.2 Speckled leaf blotch of *Agropyron* and *Elymus* spp. – *Septoria elymi* Ell. and Ev.

S. elymi on narrow brown leaf spots or on dry leaves of *Agropyron* spp. in Canada (Bissett, 1983b) and Fennoscandia (Jørstad, 1967; Mäkelä, 1977b), and according to Sprague (1950) associated with pale grey to tan or fuscous lesions on *Agropyron* or *Elymus* spp. widely in the USA on grasses in shaded areas near woodlands.

Pycnidia are brown, flattened, ostiolate, thin or thick-walled, 60–200 μm. Conidia are 0–3-septate, filiform or rod-shaped, hyaline, 25–50 × 1.2–2.1 μm (Sprague, 1950), but very variable, usually 11–43 × 1–1.5 μm (Jørstad, 1967) and 20–46 × 1–2 μm (Mäkelä, 1977b). Jørstad (1976) distinguishes between *S. elymi* and *S. elymicola*, Died. but Sprague (1944) considers them conspecific.

28.5.3 Leaf spot of *Cynodon* – *Septoria cynodontis* Fuckel.

This species occurs on *Cynodon dactylon* in USA, southern Europe, Iran.

Pycnidia are minute, in rows in a thin, black stroma. Conidia are filiform, hyaline, 50–65 × 1.5–2.0 μm.

28.5.4 Leaf spot or leaf blotch of *Poa* spp.

Septoria macropoda Pass. and vars. *grandis* Sprague and *septulata* (Gonz. and Frag.) Sprague cause brown to pale leaf spots, usually on faded leaves.

S. macropoda occurs mainly on *Poa annua* L. from N. America, China, Germany and Morocco. It is the most common form in Norway and Finland (Jørstad, 1967; Mäkelä, 1977b). Var. *septulata* was described from Spain on *Poa pratensis* L., also its commonest host in the USA. Var. *grandis* seems to be native to the rangelands and prairies of N. America (Sprague, 1950).

	S. macropoda	var. *septulata*	var. *grandis*
Pycnidia	60–160 μm sub-epidermal, flattened	60–140 μm, brown obscure	40–90 μm dark-brown
Conidia	30–40 × 1.0–1.5 μm filiform	40–60 × 1.3–1.7 filiform, needle-like	40–60 × 1.4–2.4 μm filiform, but stouter

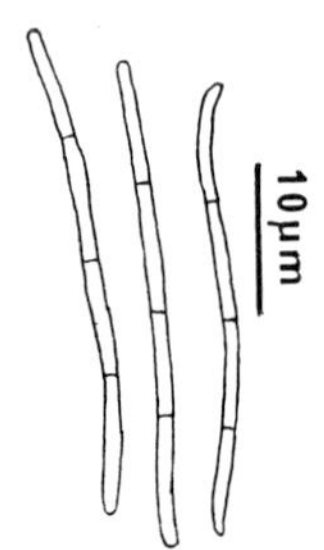

Figure 28.5 *Septoria macropoda*–conidia.

28.5.5 Leaf blotch of *Festuca* spp. – *Septoria tenella* Cke. and Ell.

This occurs on several turf grass species causing brown lesions in living leaves, culms and leaf sheaths in North America, Europe and Asia.

Pycnidia are dark-brown, elliptical in section, 80–190 μm. Conidia are very variable in shape and size from short, rod-shaped to filiform, straight or wavy, 5–70 × 0.8–2.1, often 25–45 × 1.0–1.5 μm (Sprague, 1950), 26–72 × 1.0–1.5 μm (Jørstad, 1967), to 17–60 × 0.5–2.0 (Mäkelä, 1977b), often without septa or indistinctly septate. Jørstad (1967) separated *S. festucae* Died. from *S. tenella*, but Sprague (1950) considered it conspecific.

28.5.6 Leaf blotch of *Agrostis* spp. – *Septoria triseti* Speg. em. Sprague

This occurs mainly on *A. tenuis*, causing a vague, linear or irregular spotting, but mainly of leaf tips; infrequent and sporadic distribution in N. America, Argentina and Finland.

Pycnidia brownish-black, sub-stomatal, 40–100 (mostly 40–80) μm diam. Conidia filiform to narrow fusiform, straight, bent or curved, 0–1-septate, 16–43 × 0.8–2 μm.

28.5.7 Leaf spot of *Lolium* spp.

Septoria tritici Rob. Desm. var. *lolicola* Sprague and Johnson (*S. tritici* Rob. ex Desm. has a teleomorph *Mycosphaerella graminicola* (Fuckel) Sand.) (Bissett, 1983a; Sprague, 1944). It occurs on senescing leaves of *L. perenne* causing vague circular to oval lesions particularly along the leaf edges, coalescing, with prominent pycnidia in rows between the leaf veins, in N. America, W. Europe, New Zealand and Brazil.

Pycnidia are light to dark-brown, 80–190 μm. Conidia are hyaline, 0–5-septate (mostly 1–3), curved, tapering from an enlarged base, 14–85 × 1–2.8 μm, stouter and shorter in summer than in winter (Sprague, 1950). Latch (1966) noted that the New Zealand isolates had atypical conidia and that it was a minor disease and O'Rourke (1976) that in Ireland the disease rarely occurred during active growth periods.

28.6 *STAGONOSPORA*

Stagonospora leaf spots occur on many grasses, including some turf grass species. Although some are pathogenic, causing highly coloured leaf spots, many are saprobic on grass culms. Some *Stagonospora* spp. are close to *Septoria*, others approach *Ascochyta* and *Hendersonia* (a name now rejected in favour of *Stagonospora*) in morphology. They are cosmopolitan in distribution. Some have teleomorphs in *Leptosphaeria*.

Pycnidia are globose or flattened, immersed or half-projecting, ostiolate, thin- or thick-walled, often darker round the ostiole, mostly up to 200 μm diam. approx. Conidia are multiseptate, usually coarse (broadened) and guttulate, cylindrical to fusiform, hyaline and obscurely septate when young, sometimes becoming yellow or yellow–brown when mature, grass species 15–70 × 2.5–10 μm.

28.6.1 Some *Stagonospora* spp. on turf grass species

(a) S. arenaria

S. arenaria (Sacc.) Sacc. (syn. *Septoria elymina* Prands. (1943)) is plurivorous on grasses (Grove, 1935), but reported on *Agrostis alba* from Japan (Nishihara, 1969), *Agropyron cristatum* and *Agrostis alba* L. and *A. stolonifera* from the USA (Sprague, 1950, 1955). It causes dark brown to purple stripes or spots on living leaves. Pycnidia occur in groups, subepidermal at first, subglobose, brown thin-walled, 130–200 μm. Conidia are fusoid to subcylindrical, 1–7-septate, straight to slightly curved, 30–35 (–40) × 3.5–4.0(–4.3) μm, mostly 3-septate (Castellani and Germano, 1977).

Septoria agropyrina Lobik (Sprague, 1950) is probably conspecific with *S. arenaria* (Jørstad, 1967). *S. agropyrina* is associated with other leaf-spotting fungi on *Agropyron* and *Elymus* in the great plains of North America, in northwestern USA and Canada. It is seed-borne on *A. cristatum*, but is usually saprophytic on leaf tips of dead and dying leaves (Sprague, 1950).

(b) Speckled blotch of *Agropyron, Agrostis, Festuca* and *Phleum* spp. – *Leptosphaeria avenaria* Weber *f.sp. triticea*. Anamorph – *Stagonospora avenae* (Frank) Bissett (1983a).

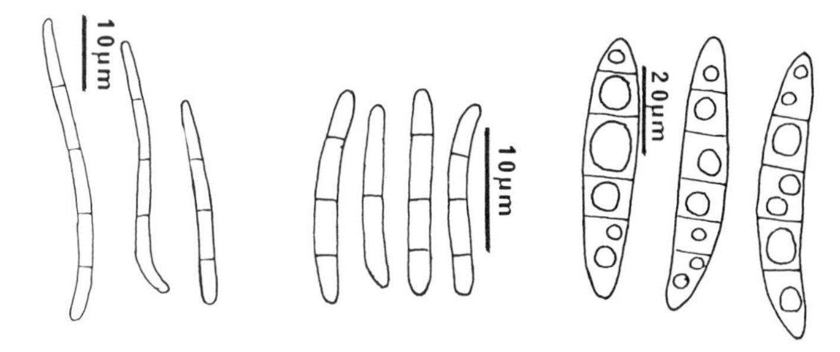

Figure 28.6 Conidia of *Stagonospora* spp. Left *S. avenae*, centre *S. nodorum*, right *S. subseriata*.

This occurs on senescing leaves of grasses on yellowish-brown blotch lesions which may turn grey and coalesce, in Canada, USA, Finland.

Ascocarps are brownish-black, 100–220 diam. Asci are clavate, straight or slightly curved, hyaline, 40–80 × 8–11 μm, 8-spored. Ascospores are biseriate, fusoid, 3-septate, constricted at septa, penultimate distal cell slightly swollen, 16–28 × 4–6 μm. Size differences do not allow a distinction between f.sp. *triticea* and f.sp *avenae* which is confined to oats. Ascocarps occur on overwintering tissues.

Pycnidia occur on leaf lesions, golden-brown to black, subepidermal, globose, 85–140 diam. Conidia and straight or curved, 17–46 × 2.6–4.4 μm, mostly 3-septate (Fig. 28.6).

S. phyllachoroides Pass. which occurs on *Agropyron*, *Agrostis*, *Festuca* and *Holcus* spp. in Norway, *S. secalis* Prill. and Delacroix and *S. loligena* Sprague are very similar and appear to belong to the same group as *S. avenae* (Bissett, 1982a; Jørstad, 1967; Mäkelä, 1977a; Sprague, 1950).

(c) Stagonospora nodorum

Leaf blotch of *Agropyron*, *Agrostis*, *Festuca*, *Phleum* and *Poa* spp. is caused by *Leptosphaeria nodorum* Müller (anamorph *Stagonospora nodorum* (Berk.) Cast. and Germano, syn. *Septoria nodorum* Berk.) (Fig. 28.6). For synonymy see Bissett (1982b). It occurs mainly on senescing and withering leaves of many grass species.

Ascocarps are mid-brown to black, 150–200 μm diam. Asci are clavate, cylindrical or curved, short-stalked, 8-spored, wall bitunicate. Ascospores are fusoid, hyaline to pale brown, 3-septate, constricted at septa, penultimate distal cell swollen, 19.5–22.5 × 4 μm (Sutton and Waterston, 1966). Ascocarps develop on overwintering stubbles.

Pycnidia are golden-brown, later black, variable in shape, thin-walled, 70–230 μm diam. Conidia are 13–28 × 2.8–4.6 μm, short cylindrical, 0–3-septate. Seed-borne in cereals; may cause seedling disease (Fig. 28.6).

S. holcina Unam. on *H. lanatus* may be conspecific with *S. nodorum* (Jørstad, 1967), as may *Septoria oudemansii* Sacc. which occurs on dry or withered leaves, especially leaf tips of *Poa* spp. and occasionally causes purple or straw-coloured leaf spots. The latter has a wide but sporadic distribution in North America and Europe.

Pycnidia are yellowish-brown, subglobose, 70–180 μm diam. Conidia are cylindrical, 0–3-septate, mostly 1-septate, 12–32 × 1–3 μm.

(d) Stagonospora paspali

S. paspali Atkinson causes a leaf spot on *Paspalum* spp., reported common along the Atlantic and Gulf coasts from N. Carolina to Texas in the United States (Sprague, 1950) and Australia (Castellani and Germano, 1977). Pycnidia are flattened-globose, 85–150 × 75–130 μm. Conidia are hyaline, 2–3-septate, oblong-elliptical, ends obtuse-pointed 24–28 × 7–9 μm.

(e) Stagonospora smolandica

S. smolandica Eliass. is weakly pathogenic on leaves of *Agrostis tenuis* Sibth. in Sweden and Norway (Jørstad, 1967) and occurs on the latter species and on *Festuca rubra* and *F. pratensis* in Finland, but is not common. *Pycnidia* 80–175 μm. Conidia are hyaline, usually with four large guttules, straight with rounded ends, mostly 3-septate, 14–20 × 3 μm (Mäkelä, 1977a).

(f) Stagonospora subseriata

S. subseriata (Desm.) Sacc. is doubtfully pathogenic on any of the turf grass species on which it has been found, e.g. *Phleum pratense*, *Poa pratensis* (Sprague, 1950), *Festuca* spp. (Grove, 1935) and *F. ovina* (Mäkelä, 1977a). Pycnidia are subglobose to oblong, black, thin-walled, 100–200 μm. Conidia are hyaline, broad, oblong fusoid, apex rounded with large guttules, mostly 3–4-septate, 23–32 × 4–8 μm (mean 6.1 μm) (Mäkelä, 1977a) (Fig. 28.6).

28.7 *'HENDERSONIA'* (SEE *STAGONOSPORA* FOR SYNONYMY)

The fungus described as *Hendersonia culmicola* Sacc. and reported associated with indistinct leaf spots with brown margins on a similar range of grass species, may be synonymous with *H. crastophila* (Jørstad, 1967–see *Wojnowicia graminia*, below). Its spores resemble those of the conidial stages of *Leptosphaeria pontiformis* (Fuck.) Sacc. and *L. eustomoides* Sacc. (Jørstad, 1967; Webster and Hudson, 1957).

'*H. culmicola*' is common in Britain and also occurs in Europe, Fennoscandia and North America (Årsvoll, 1975; Grove, 1937; Jørstad, 1967; Mäkelä, 1977a; Sprague, 1950). Pycnidia are black, 60–220 μm diam. Conidia are hyaline when young, later yellow–brown, straight or curved with a rounded, blunt or pointed apex, usually 3- to 4-, but up to 7-septate, 12–42 × 1.9–4.5 μm (Fig. 28.7). Årsvoll (1975) and Sprague (1950) considered *H. simplex* Schroet. synonymous with *H. culmicola*, but Jørstad (1967) and Mäkelä (1977a) retained the species which has considerably smaller conidia, range 11–35 × 2–4.5 μm with 1–3 septa. *H. culmicola* has a wide host range on almost the same species as the other two fungi.

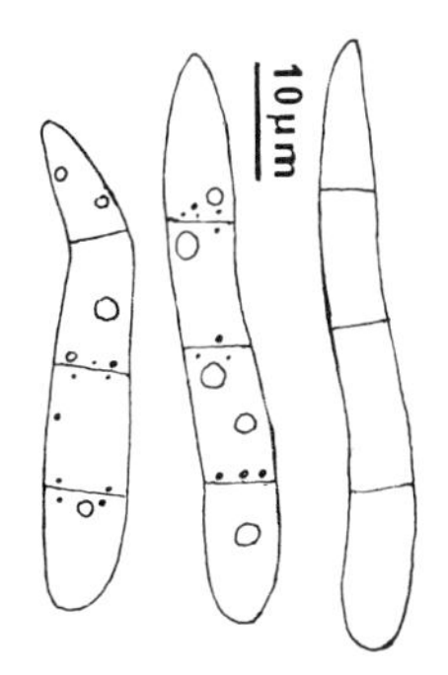

Figure 28.7 *Hendersonia culmicola*–conidia.

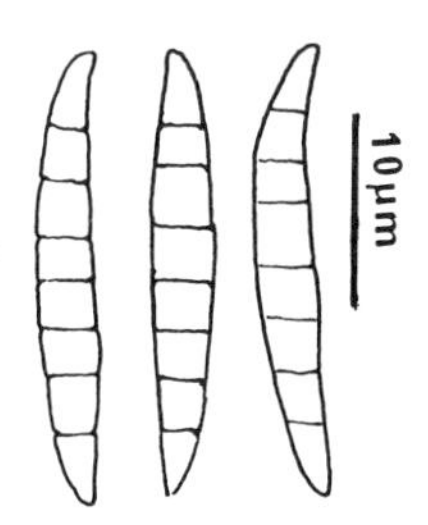

Figure 28.8 *Wojnowicia graminis*–conidia.

28.8 *PHAEOSEPTORIA*

Phaeoseptoria spp. are associated with leaf spots on turf grass species and all are saprobes or very weak pathogens. Some *Phaeoseptoria* spp. were included in *Hendersonia* (*nomen rejiciendum*–see above), but species in the latter genus should be referred to *Stagonospora* or *Wojnowicia* (q.v. for synonymy). *Phaeoseptoria* spp. have long, pale-brown conidia and dark, firm-walled pycnidia, up to 280 μm diam.

P. festucae Sprague, although described on *Festuca rubra* occurs on a wide range of turf grass species in North America, Europe and Fennoscandia (Grove, 1937; Jørstad, 1967; Mäkelä, 1977a; Sprague, 1943, 1950, 1962). It is saprobic. Pycnidia are brown or black, 50–280 μm diam. Conidia are obclavate-filiform, mostly wavy with a pointed apex, hyaline to pale-brown, range 36–120 × 2.4–7.5 μm with 3–17 septa, mostly 9-septate. Sprague (1962) considered *P. festucae* an aggregate species with three varieties showing a large range of spore sizes.

P. poae Sprague originally described from *Poa canbyi* (Scribn.) Piper in the United States by Sprague (1948) was reported to be common on *P. annua* and *P. pratensis*, frequent on *Festuca rubra*, and also on *Agrostis tenuis*, *Festuca pratensis* L., *Lolium perenne* L. and *Phleum pratense* L. in Finland (Mäkelä, 1977a) and on *F. arundinacea* Schreb. in Norway (Jørstad, 1967). Pycnidia are black, firm-walled, 110–240 μm diam. Conidia are hyaline to pale yellow, filiform to clavate, straight or wavy, tapering to a sharp apex, size range 21–91 × 1.5–5 μm.

28.9 *WOJNOWICIA*

Wojnowicia Sacc. comprises about five species of uncertain synonymy.

W. graminis (McAlp.) Saac. and Sacc. (synonyms *Hendersonia crastophila* Sacc. and *Phaeoseptoria multiseptata* Frands.?) occurs frequently on *Agropyron*, *Agrostis*, *Festuca* and *Poa* spp. in North America, Australia, and is widely distributed in Europe (Grove, 1937; Jørstad, 1967; Mäkelä, 1977a, 1979; Sprague, 1950; Webster, 1955). It is occasionally found on brown spots on living leaves and on crowns of perennial grasses associated with other foot rot pathogens. *Pycnidia* are brown, globose, immersed, with a short neck, 90–360 μm diam. (Sprague, 1950) or 60–180 μm (Jørstad, 1967), or 50–170, mostly 125 μm (Mäkelä, 1979). *Conidia*, pale brown in the mass, hyaline when young, later yellow to yellowish-brown, cylindrical, straight or curved and flattened on one side, tapering to a blunt apex, often 7- but up to 12-septate, very variable in size according to host and region; mostly 28–40 × 3.8–4.4 μm (Sprague, 1950), range 19–65 × 2.5–6.0 (Jørstad, 1967) 18–78 × 2–8 μm (Mäkelä, 1979) (Fig. 28.8).

28.10 DISEASE DEVELOPMENT AND CONTROL OF PYCNIDIAL LEAF SPOTS

Some of the species are seed-borne (Richardson, 1979), but in turf the main sources of infection are mycelium, pycnidia and conidia in lesioned leaves and plant debris. Under favourable conditions, inoculum may remain viable for a complete growing season at least. Spores are released from the pycnidia only under conditions of high humidity and are dispersed by rain splash, wind and leaf contact. Lower leaf infection usually occurs first, spreading to successively higher leaves. Wet weather is also necessary for infection. Although infection by *S. tritici* on wheat, for example, can occur in 24 h at 22–24°C, probably several days of cool, humid weather are needed for severe infection. Infection takes place through stomata or directly through cell walls. Probably high levels of nitrogen predispose turf grasses to the pathogenic species of *Septoria*, *Pseudoseptoria* and *Stagonospora*, but senescing or dying leaves resulting from earlier droughts, infrequent mowing, herbicide injury or inadequate nutrition provide suitable infection courts for the weaker pathogens or saprobes. The role of the teleomorphs in the infection cycle is probably minimal (Shipton *et al.*, 1971).

Some of the fungicides effective against *S. nodorum* and *S. tritici* on cereals, such as benomyl plus maneb, carbendazim plus maneb, thiophanate methyl plus maneb, tridemorph plus metiram, tridemorph plus carbendazim and ethirimol plus captafol (Marsh, 1977) may be expected to give some control of *Septoria* diseases of turf grasses even if they are not officially registered for such diseases or crops.

28.11 REFERENCES

Årsvoll, K. (1975) Fungi causing winter damage on cultivated grasses in Norway. *Meld. Norges Landbrhøgsk.*, **54**(9), 47 pp.

Bissett, J. (1982a) *Stagonospora avenae.* Fungi Canadenses 239, 2 pp.

Bissett, J. (1982b) *Stagonospora nodorum.* Fungi Canadenses 240, 2 pp.

Bissett, J. (1982c) *Pseudoseptoria stomaticola.* Fungi Canadenses, 238, 2 pp.

Bissett, J. (1983a) *Septoria tritici.* Fungi Canadenses 244, 2 pp.

Bissett, J. (1983b) *Septoria elymi.* Fungi Canadenses 242, 2 pp.

Boerema, G.H. and Bollen, G.J. (1975) Conidiogenesis and conidial septation as differentiating criterial between *Phoma* and *Ascochyta. Persoonia*, **8,** 111–44.

Buchanan, P.K. (1984a) *Ascochyta paspali*, a fungal parasite on *Paspalum dilatatum. N.Z. J. Bot.*, **22,** 515–23.

Buchanan, P.K. (1984b) Systemic growth of *Ascochyta paspali* in paspalum. *NZ J. Agric. Res.*, **27,** 451–7.

Castellani, E. and Germano, E.G. (1977) Graminicolous *Stagonospora* species. *Ann. Fac. Agric. Univ. Torino*, **10,** 132 pp.

Cejp, K. and Jechova, V. (1967) Contribution to the knowledge of some Czechoslovakian species of the genus *Septoria* Fries. Acta Musei Nationalis Pragae. *23B* (4), 101–23. (German, Czech).

Cherepanova, N.P. and Nguen van Tchank (1980) Key for the identification of *Septoria* Sacc. spp. on Gramineae in the Leningrad region. *Novosti Sistematici Nizshikh Rastenii*, **17,** 100–2. (Russian).

Grove, W.B. (1935) *British Stem- and Leaf-fungi* (*Coelomycetes*) I. Sphaeropsidales. Cambridge University Press, 488 pp.

Grove, W.B. (1937) *British Stem- and Leaf-fungi* (*Coelomycetes*) II. Sphaeropsidales. Cambridge University Press, 407 pp.

Jørstad, L. (1967) *Septoria and Septoroid Fungi on Gramineae in Norway.* Norske Videnskaps-Akad., Oslo. Universitetsforlaget. 63 pp.

Latch, G.C.M. (1966) Fungus diseases of ryegrass in New Zealand. 1. Foliage diseases. *NZ J. Agric. Res.*, **9,** 394–409.

Mäkelä, K. (1977a) *Hendersonia, Phaeoseptoria* and *Stagonospora* in Finland. *Ann. Agric. Fenn.*, **16,** 238–55.

Mäkelä, K. (1977b) *Septoria* and *Selenophoma* species on Gramineae in Finland. *Ann. Agric. Fenn.*, **16,** 256–76.

Mäkelä, K. (1979) *Wojnowicia graminis* on Gramineae. *Karstenia*, **19,** 54–7.

Marsh, R.W. (ed.) (1977) *Systemic Fungicides*, 2nd. edn, Longmans, Edinburgh, 400 pp.

Nishihara, N. (1969) Leaf rot of redtop (*Agrostis alba* L.) caused by *Stagonospora intermixa* (Cooke) Sacc. *Bull. Nat. Inst. Anim. Ind.*, **20,** 79–84.

O'Rourke, C.J. (1976) *Diseases of Grasses and Herbage Legumes in Ireland.* An Foras Taluntais. 115 pp.

Punithalingam, E. (1979) Graminicolous *Ascochyta* species. *Mycol. Pap.* **12,** Commonw. Mycol. Inst. Kew, 214 pp.

Punithalingam, E. and Waller, J. (1973) *Selenophoma donacis. CMI Descriptions of Pathogenic Bacteria and Fungi*, **400,** 2 pp. Commonw. Mycol. Inst. Kew.

Richardson, M.J. (1979) *An annotated list of seed-borne diseases.* Commonwealth Mycological Institute and International Seed Test Association 320 pp.

Sampson and Western, J.H. (1954) *Diseases of British Grasses and Herbage Legumes.* Cambridge University Press, 118 pp.

Shipton, W.A., Boyd, W.R.J., Rosielle, A.A. and Shearer, B.I. (1971) The common *Septoria* diseases of wheat. *Bot. Rev.*, **37,** 231–62.

Smith, J.D. (1970) Viability of stored bromegrass seed and seed-borne spores of a leaf spot pathogen. *Phytopathology*, **60,** 1470–1.

Sprague, R. (1943) The genus *Phaeoseptoria* on grasses in the Western Hemisphere. *Mycologia*, **35,** 483–91.

Sprague, R. (1944) *Septoria diseases of Gramineae in the United States.* Oregon State Coll. Press, Corvallis, Ore., 151 pp.

Sprague, R. (1948) Some leafspot fungi on western Gramineae–2. *Mycologia*, **40,** 177–93.

Sprague, R. (1950) *Diseases of Grasses and Cereals in North America*, Ronald Press, New York, 538 pp.

Sprague, R. (1955) Some leafspot fungi on western Gramineae – 9. *Mycologia*, **47,** 835–45.

Sprague, R. (1962) Some leafspot fungi on western Gramineae – 15. *Mycologia*, **54,** 44–61.

Sprague, R. and Johnson, A.G. (1950) *Species of Selenophoma on North American Grasses.* Oregon State College Press, Corvallis, Ore. 43 pp.

Sutton, B.C. (1980) *The Coelomycetes.* Commonwealth Mycological Institute Kew, 696 pp.

Sutton, B.C. and Waterston, J.M. (1966) *Leptosphaeria nodorum. CMI Descriptions of Pathogenic Fungi and Bacteria.* **86,** 2 pp. Commonwealth Mycological Institute, Kew.

van der Aa, H.A. (1973) *Studies in Phyllosticta* 1. *Studies in Mycology* 5. Centraalbureau voor Schimmelcultures, Baarn.

von Arx, J.A. (1970) *The Genera of Fungi Sporulating in Pure Culture.* J. Cramer, Lehre 288 pp.

Webster, J. (1955) Graminicolous Pyrenomycetes – 5. *Trans. Br. Mycol. Soc.*, **38,** 347–65.

Webster, J. and Hudson, H.J. (1957) Graminicolous pyrenomycetes – 6. *Trans. Br. Mycol. Soc.*, **40,** 509–22.

29 *Some other leaf spots, leaf blights and leaf streaks*

29.1 LEAF SPOTS CAUSED BY *CERCOSPORA* AND *PSEUDOCERCOSPORELLA*

Of the 2000 or so species of the genus *Cercospora* on plants, many of which are synonymous, about 30 have been found on members of the Poaceae (Chupp, 1954; Deighton, 1973; Ellis, 1971, 1976; Sprague, 1950). Many of the fungi originally described as *Cercospora* spp., and named according to the host on which they were associated, have now been accommodated in genera such as *Cercoseptoria, Mycovellosiella* and *Pseudocercosporella* (Deighton, 1973; Ellis, 1976). In the latter genus, *P. herpotrichoides* (Fron) Deighton is well known as the cause of a major eyespot disease of the culms of cereals and also occurs on wild and cultivated grasses (Booth and Waller, 1973; Cunningham, 1981; Sampson and Western, 1954; Sprague, 1936, 1950). Some *Cercospora* spp. are pathogens causing distinct spots on leaves and leaf sheaths, but others are associated with vaguely defined lesions on the aerial parts and may be secondary invaders. Where heavy infection occurs, particularly in wet weather, leaf death and defoliation may occur causing turf thinning. This is the case with the leaf spot on St Augustinegrass, *Stenotaphrum secundatum* (Walt.) Kunze in tropical regions (see below). Some *Cercospora* spp. produce the photosensitizing phytotoxin, cercosporin, implicated in the necrosis, chlorosis and water-soaking of plant cells (Daub, 1982). Fewer than ten *Cercospora* spp. are likely to be encountered on amenity turf grasses. Most of the studies on these have been made in the United States. Brief descriptions of some leaf, culm and inflorescence diseases caused by *Cercospora* and *Pseudocercosporella* spp. are given below.

29.1.1 Causal fungi, hosts, symptoms, distribution

Cercospora agrostidis Atkinson, occurs on *Agrostis* spp. *Leaf spots* broadly elliptical, 3–4 mm long with pale-brown centre and broad, reddish-brown margin. Stromata are rudimentary. Conidia are hyaline, tapering smoothly, straight or slightly curved, 1–7-septate, 10–60 × 1.5–3 μm. Location: Alabama, N. Dakota and Idaho (Chupp, 1953; Sprague, 1950).

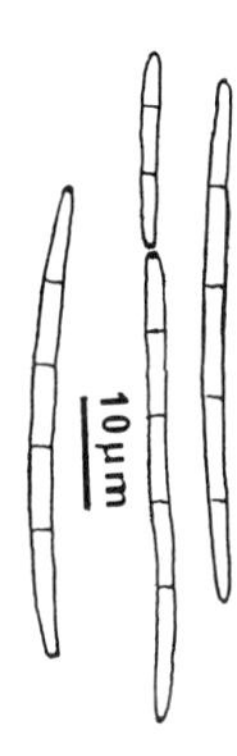

Figure 29.1 *Cercospora fusimaculans*–conidia.

C. boutelouae (Chupp and Greene (syn. *C. apii* Fres. ?), occurs on *Bouteloua* spp. Leaf spots are elliptical or elongated, 0.5–4 mm long, pale brown to dark with no margin or with a chlorotic halo. Stromata are absent or comprise few dark cells. Conidia are hyaline to pale brown, obclavate, straight or mildly curved, septate, obconic/truncate base, subobtuse tip, 20–80 × 3.5–5 μm. Location: Wisconsin and Oklahoma (Chupp, 1953; Sprague, 1950).

C. elymi Rostrup occurs on *Elymus arenarius* L. *Leaf lesions* are long narrow streaks pale to dark brown. Stromata are dark brown, globular or irregular. Conidia are hyaline, cylindrical, straight, mostly 3-septate; the base is subtruncate–long obconic, tip blunt, rounded, 15–50 × 2–4 μm. Location: Iceland (Chupp, 1953).

C. festucae Hardison (syn. *C. apii* Fres.) occurs on *Festuca* spp. Leaf spots are oval to elongate, 0.5–4 mm long with grey centre and purple border. Stromata are absent or a few brown cells. Conidia are acicular, curved or wavy, indistinctly multiseptate, with base truncate, tip acute 40–300 × 2–4 μm. Location: Kentucky and Georgia (Hardison, 1945; Luttrell, 1951).

C. fusimaculans Atkinson, has been reported on many warm-season grasses but notably *Stenotaphrum secundatum*, in the southern United States, Central America, West Indies, Africa, India, Japan and Asia (Chupp, 1953). Leaf spots on *S. secundatum* (Freeman, 1959) are dark brown to purple on sheaths and leaf blades uniformly when young and with a tan centre later, usually discrete, but coalescing when severe 1–3 mm long × 0.5–1 mm wide. Stromata are small, filling the stomatal opening. Conidia are hyaline to subhyaline, cylindrical to slightly tapering, straight, sometimes catenulate, indistinctly 2–7-septate, base truncate, tip blunt, 33–60 × 1.8–3 μm (Chupp, 1953; Freeman, 1959; McCoy, 1973) (Fig. 29.1).

C. seminalis occurs on *Buchloe* and *Cynodon* spp. and causes a leaf and glume spot (false smut). In the inflorescence the fungus penetrates the seed which is replaced by a mass of olive-brown to black mass of conidia. Conidia are produced acrogenously, subhyaline to dark brown, obclavate to cylindrical, multiseptate, base truncate, tip blunt, 20–160 × 5–7 μm (Chupp, 1953, Sprague, 1950; Weiss 1945). Location: S. Dakota, Nebraska, Kansas, Colorado, Wisconsin and Texas. This fungus is uncertainly placed as a *Cercospora*. It is seed-borne.

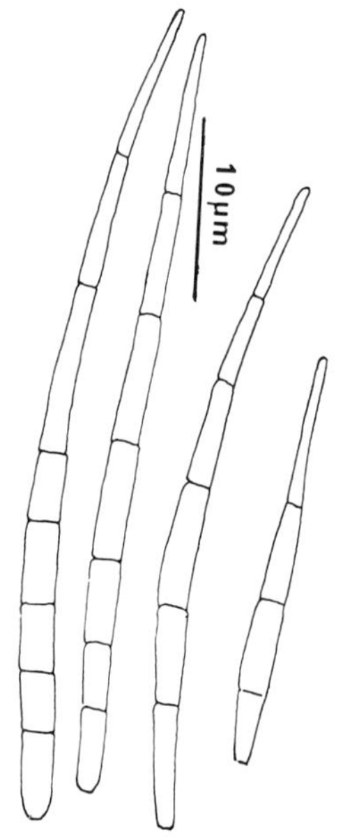

Figure 29.2 *Pseudocercosporella herpotrichoides*–conidia.

Pseudocercosporella herpotrichoides (Fron) Deighton (syn. *Cercosporella herpotrichoides* Fron) causes an eyespot of basal leaf sheaths and later brown to black-charred lesions on culms at ground level in cereals, wild and cultivated grasses (Booth and Waller, 1973; Cunningham, 1965, 1981; Sprague, 1936, 1950). Sheath spots are typically fawn or white with a dark margin, but may be obscure in some species. The mycelium penetrates the sheath causing a light lesion which later becomes black and charred. Stromata are black and pustular in the lesion centre. Mycelium is grey, in stem cavity of split straw. Conidia are hyaline, acicular, obclavate, straight to slightly curved, smooth with a truncate, unthickened hilum, usually 3–7-septate, 30–80 μm (mostly 40–60) × 1.5–3.5 μm (Fig. 29.2). Location: world-wide in temperate regions.

29.1.2 Disease development

Diseases of turf grasses caused by *Cercospora* spp. have been little studied. Warm, humid weather favours the development of leaf spot of *S. secundatum* caused by *C. fusimaculans* (Freeman, 1967). The disease has been shown to be severe under low fertility conditions (McCoy, 1973).

Cercospora, Cercosporella and *Pseudocercosporella* spp. survive on infected plant debris, such as stubble, infected leaves and litter. Living, infected secondary hosts may also be concerned (Cunningham, 1981; Sprague, 1936). *C. seminalis* is seed-borne. *P. herpotrichoides* produces abundant spores on plant debris during wet weather in autumn winter and spring at low temperatures. Spore germination occurs as low as 0 °C, and in the field, soil temperatures of 5–10 °C and abundant moisture are most favourable for disease development. Pathotypes of *P. herpotrichoides* on cereals and grasses have been found (Cunningham, 1981).

Spore dispersal in *P. herpotrichoides* from plant debris is mainly by rain splash (Pitt and Bainbridge, 1983; Rowe and Powelson, 1973). Lawrence and Meredith (1970) concluded that while rain may be an important factor affecting release and dispersal of spores of *C. beticola*, on beet etc. wind plays an important role in their dissemination. Most spores were trapped on warm, dry days preceded by a period of rain or overnight dew. No studies have been found relating to spore dispersal of *Cercospora* spp. on grasses. The presence or role of the toxin, cercosporin in pathogenesis in grasses has not been demonstrated.

29.1.3 Control

Except in *C. fusimaculans* on *S. secundatum*, *Cercospora* or *Pseudocercosporella* leaf spots generally cause little concern in amenity turf, and no specific control methods have been developed. The disease caused by *Cercospora fusimaculans* on *S. secundatum* can be effectively controlled by improving fertility with nitrogenous fertilizer and combined with applications of systemic or contact fungicides such as thiabendazole, thiophanate methyl, anilazine and chlorothalonil (McCoy, 1973). Measures should be undertaken to reduce duration of leaf-wetness periods. Yellow–green types of *S. secundatum* are more frequently damaged than blue–green ones.

29.2 *CERCOSPORIDIUM GRAMINIS* – BROWN STRIPE OR LEAF STREAK

This is a very common foliar disease of grasses found in Europe, Asia, North and South America and Australasia. Its host range includes all the cool-season, common turf grass species. It is rarely of any significance in close-mown turf, but it may cause severe, late-season leaf symptoms in unmown turf and in turf grass seed crops, including fine-leaved *Festuca* spp. and *Phleum pratense* (Berkenkamp, Folkins and Meres, 1972; Braverman, 1958; Brummer, 1937; Latch and Wenham, 1958; Mäkelä, 1972a; Melchers, 1925; Orlob, 1960; O'Rourke, 1976; Sampson and Western, 1954; Smith, unpublished; Sprague, 1950; Vicar and Childers, 1955).

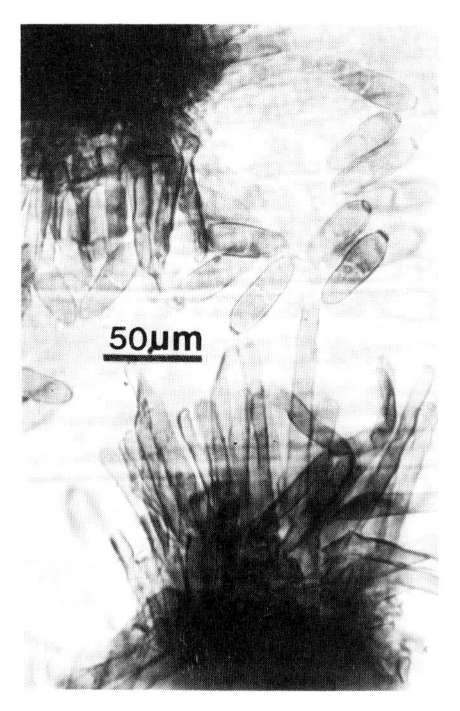

Figure 29.3 *Cercosporidium graminis*–conidiophores and conidia.

29.2.1 Symptoms

On broad-leaved grasses, the first symptoms are of water-soaked spots on the leaf blades and sheaths. These spots have grey centres and brown or dark-purple margins and lie mainly between the leaf veins. They may elongate to become streaks with golden-yellow, bronze, purple or chocolate margins as the leaves are killed and wither. Parallel rows of black dots, the groups of close-packed conidiophores, develop on the lesions.

In the fine-leaved *Festuca* spp. the leaf blades may become partitioned by the lesions into green, unaffected and lesioned sections. The dark, central portion of lesions, which may completely girdle the leaf lamina, are separated from the green sections of the leaf by orange or brown bands (Fig. 53, colour plate section).

29.2.2 The fungus

Cercosporidium graminis (Fuckel) Deighton (1967) (*synonym Scolecotrichum graminis* Fuckel) is a Hyphomycete which produces groups of fasciculate, olive-brown conidiophores on dark-brown stromata arranged in rows on the lesions. Conidiophores are unbranched, irregular or slightly geniculate with swellings below the apices; they vary in size, probably according to humidity at time of formation rather than because of host species (Braverman, 1958; Brummer, 1937; Ellis, 1971; Mäkelä, 1972a; Sprague, 1950) from $30–150 \times 5–12\,\mu m$. Conidia are olive-brown, 1–3-, but usually 1-septate, elongate bottle-shaped with a broadly rounded base and an obvious hilum, variable in size as for the conidiophores, $16–56 \times 4–12\,\mu m$ (Fig. 29.3).

The occurrence of an ascomycete teleomorph has been reported, but not confirmed (Sprague, 1950). Some pathogenic specialization on host species has been found, but the full extent of this has not been studied (Braverman, 1958; Graham, Zeiders and Braverman, 1963) and the fungus may develop and sporulate abundantly on many hosts on which it is only slightly pathogenic when tissues senesce during moist autumns and early winters (Sprague, 1950).

29.2.3 Disease development and control

The pathogen overwinters as stromata in leaf and sheath tissues and these erupt through the epidermis in spring and the conidiophores produce conidia which are the source of new infections. The optimum temperature for spore production (in culture) is c. 20 °C and when cold weather returns sporulation ceases and stromata are formed. The disease may be found on leaves of different ages throughout the growing season. Although older leaves are more severely attacked, seedlings are also susceptible (Braverman, 1958; Graham, Zelders and Braverman, 1963; Vicar and Childers, 1955). Compared with many other leaf spots, disease development is slow. Very little disease resistance has been noted in *Phleum pratense* (Vicar and Childers, 1955). Control methods for the pathogen in turf have not been worked out, but in seed crops effective control of inoculum results from burning of stubble.

29.3 *CLADOSPORIUM PHLEI*–TIMOTHY EYESPOT

This foliar leaf spot has been reported very common on *Phleum* spp. in all types of grassland, particularly in the cool climates of northern North America, northwestern and northern Europe, northern Asia and New Zealand (Årsvoll, 1975; Berkenkamp, 1976; Creelman, 1956; Ellis, 1976; Frauenstein, 1968; Gregory, 1919; Horsfall, 1930; Jacques, 1941; Kreitlow *et al.*, 1953; Leeuw and Voss, 1970; Mäkelä and Sundheim, 1982; O'Rourke, 1976; Sakuma and Narita, 1961; Sampson and Western, 1954; Smedegard-Petersen and Hermansen, 1971; Smith, 1970; Sprague, 1950; Sundheim and Aarvold, 1969; Tsutomu and Takeshi, 1963; Wilkins and Carr, 1971). Most damage occurs in seed production fields (Sundheim and Aarvold, 1969; Smith, 1970), but sometimes heavy forage losses occur (Creelman, 1956; Horsfall, 1930; Sakuma and Narita, 1961). In western Canada crop loss is usually slight (Berkenkamp, 1976) and damage on amenity turf is light. However, this foliage eyespot is considered the most important disease of timothy in Europe (Sundheim and Aarvold, 1969; Wilkins and Carr, 1971).

29.3.1 Symptoms

Leaf blades only are infected. Individual leaf spots are small, elongated, parallel to the veins, up to 1 mm wide and 4 mm long on upper and lower leaf surfaces. Small spots are brown or purplish-brown in colour; larger ones have straw-coloured centres and dark purple or purplish-brown borders (Fig. 54, colour plate section). Heavy lesioning results in chlorosis of the tissue between the spots and leaves may wither progressively from the tip to the base.

29.3.2 The fungus

Cladosporium phlei (Gregory) de Vries (1952); synonym *Heterosporium phlei* Gregory (1919).

Conidiophores are olive-brown or brown up to 250 μm long × 5–10 μm wide (Årsvoll, 1975). Conidia are produced acropleurogenously in simple or short-branched chains; they are dark-brown, ovoid to oblong with blunt ends, sometimes slightly dumbell-shaped, densely verrucose, 5–12 × 14–36 μm and usually 1–3-septate (Årsvoll, 1975; Sprague, 1950; Sundheim and Aarvold, 1969). Conidia are sparingly produced on new lesions, but are more common on heavily lesioned leaf tips (Creelman, 1956; Gregory, 1919); they develop abundantly on infected leaves on incubation for 48 h in moist chambers (Sundheim and Aarvold, 1969) and on potato – dextrose agar cultures left at room temperature in the light. Årsvoll (1975) induced sporulation by exposing cultures to near u.v. illumination. Spores will germinate from 3 to 33°C, optimum 24°C and may be produced under the high humidity conditions under a snow cover (Sprague, 1950; Tsutomu and Takeshi, 1963). The minimum temperature for mycelial growth *in vitro* is −6 °C (Årsvoll, 1975).

29.3.3 Host range

The host range of *C. phlei* is confined to *Phleum pratense* L., *P. alpinum* L. and *P. bertolonii* DC. (Gregory, 1919; Sundheim and Aarvold, 1969; Smith, 1970; Sprague, 1950). Other species of *Cladosporium* and *Heterosporium* e.g. *C. herbarum* (Pers.) Link ex F.S. Gray, *C. cladosporoides* (Fresen.) de Vries and *H. avenae* Oud. (?) occur on grasses as saprobes or weak parasites. Sawada (1958) described another species, *C. phlei-pratensis* on timothy from Japan.

29.3.4 Disease development

The pathogen is adapted to a wide range of climatic conditions. Although the disease is of greatest significance during the growing season, and is favoured by cool to warm, wet weather, in temperate climates it may be found from spring to winter (O'Rourke, 1976; Sprague, 1950, 1955; Sundheim and Aarvold, 1969). In Ireland the disease is most noticeable on heavy or peaty soils and in Alaska it is more severe where soil potassium is deficient (Laughlin, 1965; O'Rourke, 1976). In Norway it is found to the latitudinal and altitudinal limits of cultivation of the hosts (Sundheim and Aarvold, 1969) and in the western prairies in the United States, Sprague (1950, 1955) reported its occurrence to a limited extend in shaded canyons.

Gregory (1919) showed that overnight incubation of inoculated plants in a moist chamber was sufficient to obtain leaf spot development in *P. pratense*. The first symptoms appeared in 8 days. Sundheim and Aarvold (1969) obtained maximum infection following incubation in a moist chamber in plants of *P. pratense* inoculated with a spore suspension at 10^6 spores/ml. Spots with typical tan centres and purple margins appeared in 8 days. They suggested that with its long conidiophores the fungus was adapted to wind dispersal and that a few days of humid weather would be needed for spore production in the field. The fungus can overwinter on green leaves as indicated by the presence of old lesions on them in early spring.

Gregory (1919) and Sundheim and Aarvold (1919) were unable to obtain infection in grasses other than *P. pratense* with *H. phlei*. Tsutomu and Takeshi (1963) and Smith (1970) found that although no cultivars of *P. pratense* or *P. bertolonii* showed complete resistance some were significantly less susceptible than others, e.g. Essex was considerably less susceptible than S.50. However, Sundheim and Aarvold (1969) found no significant differences in susceptibility in three Norwegian *P. pratense* cultivars and contradictory ratings have been noted for the same cultivar in European tests (O'Rourke, 1976). Resistant lines have been developed from the Canadian cultivar Bounty (Smith, 1970) and from Aberystwyth cultivars S.50 and S.51 (Wilkins and Carr, 1971).

29.3.5 Control

The disease is not usually a problem in amenity turf and specific control measures have not been developed. The use of a less-susceptible cultivar, reduction in irrigation when leaf-spotting and leaf death is occurring and applications of a foliar fungicide are suggested. Horsfall (1930) obtained effective control with applications of colloidal sulphur. Where potassium is likely to be deficient, e.g. on light soils, an adequate level of this nutrient should be maintained (Laughlin, 1965; Sakuma and Narita, 1961) and a satisfactory N, P and K soil nutrient status achieved, since this also may reduce disease severity (Sakuma and Narita, 1961). Although in seed crops effective control may be expected from the burning of crop debris, severe plant thinning has resulted from this sanitation technique (Smith, 1970).

29.4 *HADROTRICHUM* LEAF SPOT

This disease has been noted causing considerable injury locally in Central Europe, is occasionally found in Fennoscandia and North America, and is not uncommon in Britain, but it is rarely destructive. The principal hosts are *Agrostis stolonifera* L. and *A. tenuis* Sibth., but *Lolium perenne* L., *Agropyron* and *Poa* spp. are also spotted. The disease is probably unimportant except on neglected or unmown amenity turf (Conners, 1967; Guyot, 1932; Jørstad, 1945; Mäkelä, 1976; Siemaszko, 1933; Sampson and Western, 1954; Sprague, 1950).

29.4.1 Symptoms

Leaf spots on living leaves of *Agrostis* spp. are dark olive-green or brown and could be mistaken for telia of a rust. The spot colour is due mainly to the dark mycelial pad, close-packed conidiophores and conidia.

29.4.2 The fungi

On *Agrostis*, *Hadrotrichum virescens* Sacc. and Roum (Hyphomycetes) is regarded as the causal agent, but there are unresolved problems of synonomy. Possibly *H. phragmitis* Fckl. (Lindau, 1907) has priority and *H. microsporum* Sacc. and Malbr. var. *microsporum* Karst. is the same fungus. *Scirrhia agrostidis* (Fckl.) Wint. (syn. *Phyllachora helvetica* Fckl.) is the suggested, but unproven teleomorph.

Conidiophores produced from a stromatic pad in the leaf, are close-packed, unbranched, basally septate, 20–40 × 4–8 μm, bearing solitary, terminal spores produced in succession, leaving scars below the apex of the conidiophores. Conidia are spherical to slightly ovate, echinulate, dark olive-green, 8–16 μm diam. (Guyot, 1932; Hughes, 1953; Lindau, 1907; Lind, 1913; Mäkelä, 1976; Pape, 1928; Parberry, 1967; Sampson and Western, 1954; Sprague, 1950) (Fig. 29.4).

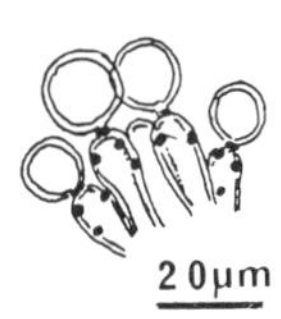

Figure 29.4 *Hadrotichum virescens*–conidia and conidiophores.

29.4.3 Disease development and control

The disease has been noted from spring to early winter, but the modes of infection and overwintering and methods of control have not been examined.

29.5 *LEPTOSPHAERULINA* LEAF BLIGHT

This leaf blight has been found on many different plant hosts, including amenity grasses such as *Agros-*

tis, *Bromus*, *Festuca*, *Lolium* and *Poa* species in Australia, Canada, Europe and the United States (Dunlap, 1944; Graham and Luttrell, 1961; McAlpine, 1902; Müller, 1951; Ormrod, Hughes and Shoemaker, 1970; Smith, unpublished; Sprague, 1950; Wehmeyer, 1955). It is often noted on senescing leaves or leaves damaged by other agencies. In the Canadian prairies, it can be found commonly on leaves of cultivars of *Poa pratensis* which senesce early, from late summer to winter (Smith, unpublished). The virulence of the fungus and the significance of the disease is uncertain.

29.5.1 Symptoms

In *P. pratensis* it is usually the leaf tips which first show yellow or brown discolouration and this advances down the leaf blade preceded by water-soaking of the leaf tissues. This blighting may extend into the leaf sheaths. Small patches or larger areas of turf may be affected. Later, small brown ascocarps of the fungus form in dead leaf tissue still attached to the plant or in clippings deposited at the sole of the turf.

29.5.2 The causal fungus

Teleomorph – *Leptosphaerulina australis* McAlpine (1902); no anamorph; synonyms – *Pleospora gaumannii* Müller (1951); *Pseudoplea gaumannii* (Müller) Wehmeyer (1955).

Graham and Luttrell (1961) give the earlier synonomy. Dunlap (1944) implicates a *Pleospora* sp. which may be the same fungus, fruiting on dead leaves of *P. pratensis* and on dying leaves of *Agrostis* sp. in a lawn and golf green respectively in Texas.

Ascocarps are pale brown, immersed, spherical with membranous, parenchymatous walls with a short collar of darker cells, opening by a broad pore, 40–170 μm (in culture, 69–188 μm). Asci are few, saccate, thick-walled, bitunicate, 50–90 × 30–45 μm in culture, 65–91 × 35–55 μm), with eight ascospores. Ascospores are bluntly conical at the base and rounded at the tip, hyaline, turning to pale brown when mature, muriform with 2–6 transverse septa (usually 4) and 1–3 longitudinal septa, 25–41 × 10–15 μm (29–41 × 11–15 μm in culture (Graham and Luttrell, 1961) (Fig. 55, colour plate section). The second cell or pair of cells above the base is usually the widest point of the spore. The fungus makes rapid growth and produces fertile ascocarps which are necessary to distinguish the fungus from *Pleospora herbarum* (Fr.) Rabenh. (teleomorph of *Stemphylium botryosum* Wallr.) (Booth and Pirozynski, 1967b) where the ascospores have two lines of longitudinal septa or from pycnidial leaf spotting fungi (e.g. *Ascochyta*, *Phyllosticta* and *Stagonospora* spp. – see pp. 314 and 317). The plurivorous *L. trifolii* (Booth and Pirozynski, 1967a) may also develop on grasses.

29.5.3 Disease development and control

L. australis is generally regarded as a saprobe or a very weak, unspecialized pathogen and practically nothing is known about its survival and invasion of host tissues. Ascocarps can be found in leaves in the litter of turf until winter and the fungus is presumed to survive as ascocarps or mycelium. In France, *L. australis* has been frequently isolated from leaves of *Poa* spp. (Dr G. Raynal, personal communication). Infection of leaves is probably ascosporic as in *L. trifolii* (Booth and Pirozynski, 1967a). No specific control measures have been developed, but it is unlikely that well-managed turf will be greatly damaged. In the Canadian prairies it seems to have been overlooked as one of the grass-leaf inhabitants competing with and influencing the primary, prehibernal attacks of snow mould fungi (see LTB snow mould, p. 104).

29.6 *MASTIGOSPORIUM* SPP. – LEAF FLECK OF BENTGRASSES AND TIMOTHY

This leaf spot disease is common and may cause considerable injury to *Agrostis* and *Phleum* spp., and rarely other turf grass species (Lacey, 1967), particularly in cool or cold humid climates such as are found in western and northern Europe, Fennoscandia, the USSR, north western and eastern North America and Japan (Andersen and Gjaerum, 1979; Bollard, 1950; Conners, 1967; Gunnerbeck, 1971; Mäkelä, 1970a, 1972a; O'Rourke, 1976; Sampson and Western, 1954; Schlösser, 1970; Smedegard-Petersen and Hermansen, 1971; Sprague, 1950). It is not important on close-mown turf grass, but may develop in longer swards in cool or cold, wet weather.

29.6.1 Symptoms

On *Agrostis tenuis*, in particular, but also on *A. stolonifera* L. and *A. canina* L., oval or elliptical leaf flecks develop which may increase in size to about 2 mm in length with purple or deep-red margins, darkest along the veins. The centres are light-coloured and may bear shining clusters of slimy, white conidia.

On *Phleum pratense* L. leaf lesions are very similar to those on *Agrostis* spp. and could be confused with those caused by *Cladosporium phlei* (see p. 324). Flecking or spotting may be sufficiently heavy to cause defoliation, particularly of lower leaves. Spots may be seen soon after snow melt and become frequent again in humid weather in autumn.

29.6.2 The fungi

Mastigosporium spp. (Hyphomycetes) are highly specialized pathogens found only on grasses. Host records are confused because of the uncertain diagnosis of pathogen and host species. Four species of the pathogen are currently recognized (Boerema and Verhoeven, 1977).

In *Mastigosporium* spp. the hyaline, narrow, intercellular mycelium produces individual conidiophores which pierce the outer cell wall and cuticle. The short conidiophores (annellophores) are produced in groups and each bears, usually, one conidium, less commonly more, which has a basal frilled hilum. There are no known teleomorphs.

M. rubricosum (Dearn and Barth.) Nannf. occurs on *Agrostis* spp. Conidia are ellipsoidal, hyaline, with a rounded, but elongated apical cell, normally 3-septate, constricted at the septa, 29–60 × 9–18, mostly 15–16 μm.

M. kitzebergense Schlösser occurs on *Phleum pratense.* Conidia are cylindrical, hyaline with a rounded apical cell, usually 3-rarely 1-septate, 29(22–39) × 10(8.4–13.4) μm. Conidia form only in the light.

For synonymy see Boerema and Verhoeven (1977), Gunnerbeck (1971) and Schlösser (1970).

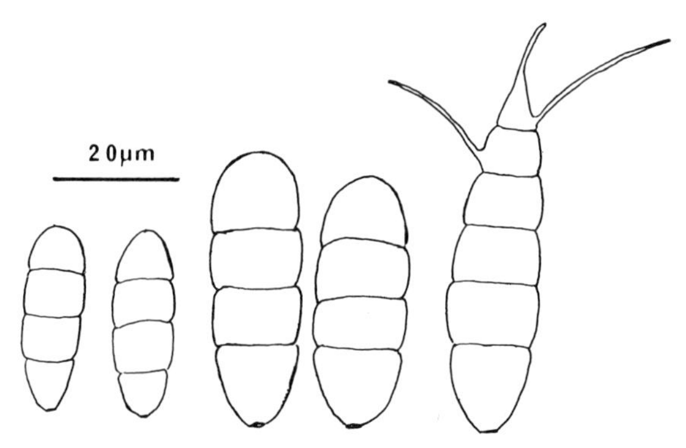

Figure 29.5 Conidia of *Mastigosporium* spp. Left *M. kitzbergensis*, centre *M. rubricosum*, right *Mastigosporium album* from *Alopecurus pratensis*.

29.6.3 Disease development

In cold climates, some *Mastigosporium* spp. produce sporiferous sclerotial or stromatic resting stages on some hosts, and they develop in some isolates in culture, but not on *Phleum pratense* and rarely on

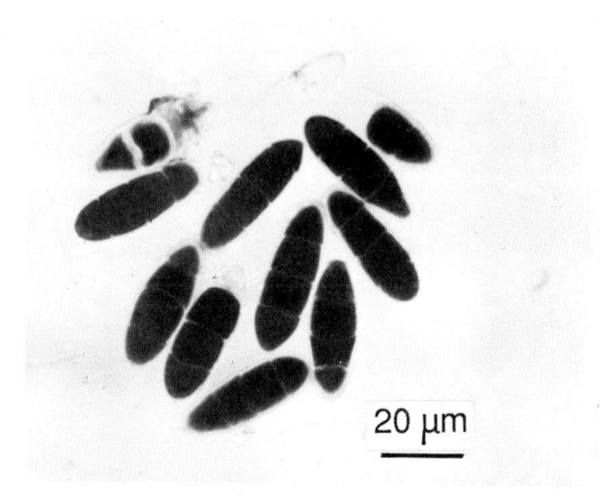

Figure 29.6 *Mastigosporium kitzbergense*–conidia (stained) from *Agrostis* sp.

Agrostis tenuis. Viable spores can be found at snow melt, having probably developed on overwintering mycelium in the leaves under a snow cover (Mäkelä, 1970b). Spore dispersal is by rain or dew. Wet weather or high humidity and cool temperatures favour spore germination (15–20°C) and the development of epidemics (Andersen and Gjaerum, 1979; Sampson and Western, 1954). Infection by germinating spores is direct through the epidermal cells. *Mastigosporium* leaf spots are most obvious in Britain during autumn and spring (Sampson and Western, 1954), from midsummer to autumn in Fennoscandia (Andersen and Gjaerum, 1979) and during rainy and foggy weather in autumn and open winter weather in northwestern North America (Sprague, 1950). Control methods for turf diseases caused by the pathogens have not been developed.

29.7 *PHYLLACHORA* SPP.–BLACK LEAF SPOT OR TAR SPOT OF TURF GRASSES

Black leaf spot disease which affects practically all genera of turf grasses used in turf has also a world-wide distribution (Koponen and Mäkelä, 1976; Mühle, 1953; Orton, 1944; Parberry, 1967; Sampson and Western, 1954, Sprague, 1950). While the symptoms produced may be occasionally spectacular in turf grasses, especially if they are unmown, it is doubtful whether much injury is caused.

Affected turf grass appears chlorotic or mottled green; infected leaves show scattered or congregated black stromata-like structures on upper and/or lower leaf surfaces often surrounded by a zone of chlorotic tissue which may disappear as the colonies of the

Table 29.1 *Phyllachora* species on turf grass genera and their distribution (From Parberry, 1967)

Species	Turf grass genus	Distribution
P. acuminata Starb.	*Paspalum*	S. America, common
P. bonariensis Speg.	*Axonopus, Brachiaria*	Australia, Central and South America, Japan, Java, Nepal, Taiwan, West Indies.
P. bulbosa D. Parberry	*Zoysia*	Australia
P. cynodontis (Sacc.) Niessl.	*Bouteloua, Buchloe, Chloris, Cynodon*	Almost world-wide on major host, *Cynodon dactylon*
P. dactylidis. Delacr.	*Festuca*	Europe
P. elusines P. Henn.	*Eragrostis*	Australia, China, Africa, Taiwan, Southern USA
P. eragrostidis Chard.	*Eragrostis*	South America, southern USA
P. fuscescens Speg.	*Agrostis*	Argentina, New Zealand
P. graminis (Pers. ex Fries) Nke. in Fkl.	*Agropyron, Agrostis Festuca, Phleum, Poa*	Widespread in Europe, Canada, northern USA, Scandinavia, Japan, Persia, Taiwan, not in southern Asia, but from Gold Coast and S. Africa
P. helvetica Fckl.	*Agrostis*	Germany, Switzerland and Europe?
P. infuscans Wint.	*Paspalum*	South America
P. koondrookensis D. Parberry	*Chloris*	Australia
P. minutissima (Welv. and Curr.) Sm.	*Pennisetum*	Africa, Australia, N., S. and Central America, W. Indies
P. paspalicola P. Henn.	*Paspalum*	Australia, Central and South America, Japan, New Guinea, Papua, Philippines, S. Africa, W. Africa, W. Indies
P. pennisetina Syd.	*Pennisetum*	China, Taiwan
P. punctum (Schw.) Ort. and Stev.	*Paspalum, Stenotaphrum*	Australia, Brazil, Canada, India, S. Africa, USA
P. guadraspora Tehon	*Eremochloa*	Australia, S. America, West Indies
P. stenospora (B. and Br.) Sacc.	*Brachiaria*	Ceylon, Philippines
P. sylvatica Sacc. and Speg.	*Festuca, Poa*	Australia, Britain, Central, South and North America, Europe and Scandinavia

fungus mature. 'Green islands' of tissue may occur as the leaves senesce. Leaf sheaths, but rarely grass culms, may be infected.

Of the 300 or so 'species' of *Phyllachora* described as the cause of the disease on grasses, many are synonymous and now only about 100 species are recognized (Parberry, 1967). The genus is now usually included in the order Sphaeriales.

The colonies of graminicolous *Phyllachora* spp. usually develop in the mesophyll of the leaf, but the dark-coloured clypei, which are the only visible parts of the fungus colony, usually develop in the epidermis of the host on one or both sides of the leaf, often at the same time as perithecia and spermogonia are being formed. The clypei may be shiny-surfaced, giving the name 'tar spot' to the disease.

Colonies may contain variable numbers of perithecia depending on fungus and host species. When not restricted, perithecia are globose, and are paraphysate and ostiolate, ostioles opening on either or both leaf surfaces, containing variable numbers of saccate, clavate, elliptical or cylindrical and stalked asci. The latter are single-walled with a crown-like, angular thickening around the apical pore. Ascospores are aseptate, hyaline or pale yellow, varying greatly in size and shape: the latter in each species is constant and useful in diagnosis. There are usually eight, but a few species have four ascospores per ascus. Many species produce spermagonia with scolecosporous spermatia. None of the *Phyllachora* spp. on turf grasses listed in Table 29.1 have known conidial states.

Characters other than ascus and ascospore size, such as spore shape and arrangement, ascus stalk length and spermatial characters are needed to separate the species satisfactorily (Parberry and Langdon, 1963). For these, reference should be made to the keys and descriptions in Parberry's (1967) monograph.

29.7.1 Disease development

Phyllachora spp. are obligate parasites and the life cycles of graminicolous species are not completely known. In Britain the colonies of *P. graminis* develop in autumn and winter on *Agrostis* spp. (Sampson and Western, 1954). Probably graminicolous species require living tissues in which to (overwinter and) complete their development (Orton, 1956; Parberry, 1967). The role of spermatia in fertilization is uncertain; in *P. punctum* they were not infective (Orton, 1956). In *P. sylvatica* ascospores are produced in abundance, oozing in pink droplets from the ostiole, some germinating to produce appressoria on and in the perithecia, but further growth in culture after germination did not occur (Sampson and Western, 1954). In *P. punctum* the ascospores are forcibly discharged from the perithecium (Orton, 1956), but the infection process has not been observed.

29.7.2 Control

The development of methods to control black leaf spot in turf is scarcely justifiable and methods for this have not been reported. In seed crops of turf grass species, especially of fine-leaved *Festuca* spp. which may become heavily infected (Sampson and Western, 1954), flame sanitation seems applicable.

29.8 *PHYSODERMA* LEAF SPOT AND LEAF STREAK

These are uncommon diseases of species of turf grasses in the genera *Agropyron, Agrostis, Axonopus, Festuca, Paspalum, Phleum* and *Poa* (Childers, 1948; Karling, 1950; Sampson and Western, 1954; Sparrow and Griffin, 1964; Sparrow, Griffin and Johns, 1961; Stevenson, 1946; Thirumalachar and Dickson, 1947). Many grasses were susceptible to a *Physoderma* sp. from *Agropyron repens* when artificially inoculated (Sparrow and Griffin, 1964).

29.8.1 Symptoms

These vary according to the host species and fungus isolate (species?) involved (Sparrow, Griffin and Johns, 1961; Sparrow and Griffin, 1964). Small, yellow elliptical spots or streaks develop on leaves, leaf sheaths and/or culms and tillers may be stunted with shortened internodes and erect leaves to give a similar appearance to yellow tuft disease (p. 238). The spots or streaks change colour to purplish and then brown and may be erumpent, finally turning ash grey. In severe infections leaves may become shredded and basal culm rot may develop. Sori of yellow–brown sporangia in the interveinal tissues are apparent on microscopical examination.

29.8.2 Causal fungi and hosts

Several different obligately parasitic species of *Physoderma*, a genus of the Chytridiales, have been reported concerned with these diseases, although their taxonomy is still confused (Sparrow, 1965 and see Table 5.2, p. 44).

Physoderma maydis Miyabe (syn. *P. zeae-maydis* Miyabe) is the cause of a serious disease of *Zea mays* L. in humid, hot climates (Walker, 1983). It will only infect *Zea mays* and *Euchlaena mexicana* (teosinte).

P. graminis (Busgen) de Wild. (syn. *Cladochytrium graminis* Busgen) is an uncommon pathogen on *Agropyron repens* L. in Europe and North America (Childers, 1948; Gopalkrishnan, 1951; Sampson and Western, 1954; Thirumalachar and Dickson, 1947), but has also been noted on a wide range of grasses, including amenity turf grass species from Europe (Karling, 1950). Whether these cases were of *P. graminis* or other *Physoderma* spp. is uncertain (Sampson and Western, 1954; Sparrow, 1965). Sparrow and Griffin (1964) showed that a *Physoderma* sp. from *A. repens* was pathogenic on species from many tribes of the Poaceae, but did not consider that this implied they were field hosts.

In *P. graminis* the rhizomycelium is intracellular, occasionally with thin-walled zoosporangia. Epibiotic sporangia are ephemeral, abundant in spring, oval to ellipsoidal, perhaps lobulate, $20-55 \times 20-40\,\mu m$. Zoospores are elongate, ellipsoidal, $2-2.5 \times 4-6.5\,\mu m$, monoflagellete, $10-20\,\mu m$ long. Resting sporangia filling the host cells, are thick-walled, flat on one surface, and $20-34 \times 20-40\,\mu m$ (Karling, 1950).

P. agrostidis Lagerheim with ellipsoidal resting

sporagia 13–21 × 17–25 μm was reported from leaves of *A. stolonifera* from Sweden and Germany (Cook, 1934; Karling, 1950; Sampson and Western, 1954). Germination of the sporangia has not been seen and the identity of the species is doubtful (see Table 5.2).

P. gerhartii Schroeter (syn. *Cladochytrium gerhartii* (Schroeter) Fischer) was found parasitizing leaves and sheaths of aquatic or hydrophilic grasses in Europe. The resting sporangia are subspherical, ellipsoidal to angular or irregular, depending on host-cell shape these were slightly smaller, 12–30 × 16–32 μm, than those of *P. graminis*. Germination was not seen, but a lid on the resting sporagium like that of *P. graminis* is probably present (Karling, 1950).

P. paspali Stevenson was described from *Paspalum plicatum* from Puerto Rico (Stevenson, 1946). It may also be parasitic on *P. millegrana* from Brazil. Resting sporangia measure 15–24 × 18–33 μm.

29.8.3 Disease development in *P. graminis*

Under moist conditions, a blunt, cone-shaped endospore with a blunt papilla pushes up a saucer-shaped epispore lid of the resting sporangium. The endospore bursts and releases the zoospores. These swim in water films and may encyst, usually on the coleotile or first leaf, or produce germ-tubes while free in the water. Either endobiotic or epibiotic development may then occur.

Endobiotic (resting sporangium) development starts by the movement of most of the contents of the encysted zoospore into the host cell through a fine tube. This mass of naked cytoplasm becomes the primary turbinate organ. From the base of this structure fine rhizoids are produced on which further turbinate organs are produced which move from cell to cell in the host. These eventually become the resting sporangia which are liberated by the breakdown of the host tissue. They survive in host tissue fragments or in the soil.

Epibiotic development takes place mainly on the grass coleoptile. The encysted zoospore penetrates the cell wall of the host and inside produces a whorl of short rhizoids. A thin-walled sporangium develops externally, becomes vacuolated with equal-sized 'rings' which indicate the location of zoospore formation. Zoospores are discharged by the rupture of a blunt sporangial papilla (Sparrow, Griffin and Johns, 1964; Thirumalachar and Dickson, 1947).

Dispersal of resting sporangia in broken pieces of infected leaves, as has been shown to occur in *P. maydis* (Tisdale, 1919) is likely in *P. graminis*. According to Karling (1950) infection in *P. graminis* is systemic, commencing at the leaf base and moving to the leaf tip. Gopalkrishnan (1951) found *P. graminis* in the vascular tissues of the rhizomes and roots of *A. repens* and he was able to infect meristematic tissues. However, the *Physoderma* sp. from *A. repens* studied by Sparrow Griffin and Johns (1961) was not systemic.

29.8.4 Control

No specific control methods have been developed for *Physoderma* diseases in grasses. Since the pathogens are dependent on warm temperatures and on free water on plant surfaces for spore germination, zoospore migration and infection, adequate drainage and careful management of irrigation during hot weather will reduce risk of disease. Systemic fungicides such as benomyl, carbendazim and oxycarboxin which Lal and Chakravarti (1977) found effective in control of *P. maydis* on *Zea mays*, or those active against Phycomycetes, may also be effective against *Physoderma* diseases of turf grasses.

29.9 *RAMULARIA* AND *RAMULASPERA* LEAF SPOTS

Ramularia leaf spot is widely distributed, but usually uncommon, on a large number of turf grasses including *Agropyron, Agrostis, Anthoxanthum, Bromus, Elymus, Festuca, Lolium* and *Poa* species, mainly in temperate climates in Europe, Fennoscandia, Asia, Canada, New Zealand and the United States. The *Ramulaspera* leaf spot has been found on *Holcus lanatus* in continental Europe, Wales and New Zealand. Latch (1964) and Mäkelä (1976) found the *Ramularia* leaf spot frequently on *Poa annua* in Finland and New Zealand respectively and Latch (1964) noted that it was one of the common components of complexes of leaf spots capable of considerable leaf destruction. The importance of both diseases in close-mown turf is likely to be small (Griffiths, 1957; Latch, 1964, 1966; Mäkelä, 1976; Orlob, 1960; Sprague, 1955).

29.9.1 Symptoms

Ramularia leaf spot symptoms vary according to the host species, but are of the circular or elliptical eyespot type. Those on *Agrostis* have light-coloured centres, but on *Lolium* and *Festuca* spp. this is not usually the case. Otherwise the spot centre is dark red or brown and the spot is surrounded by a yellow halo. With multiple spotting, halos may coalesce and premature leaf death occurs (Fig. 29.7).

Figure 29.7 Stitch spot on *Bromus inermis* caused by *Ramularia pusilla*.

Ramulaspera leaf spots are similar in appearance to those caused by *Ramularia* but the spots on *H. lanatus* are elongated up to 5 × 2 mm.

29.9.2 The fungi

(1) *Ramularia pusilla* (Ung.) (syns. *Ovularia pusilla* (Ung.) Sacc. and Sacc. emend. Sprague, *O. pulchella* (Ces.) Sacc. and *O. hordei* (Cav.) Sprague). (2) *Ramulaspera holci-lanati* (Cav.) Lindroth.

Sprague (1955) grouped all the *Ovularia* spp. on Gramineae in one species, *O. pusilla*, but Latch (1964) pointed out the *Ramularia* has priority over *Ovularia* and also maintained that *Ramulaspera holci-lanati* was a separate species, although closely related to *R. pusilla*.

In *R. pusilla* conidiophores arise directly from a small stroma in the leaf, usually simple, unbranched, 1–3-septate, indeterminate, bent towards the tip with up to six scars, range 22–170 × 1–5 μm. Conidia are few, borne terminally in succession (acropleurogenously), clear hyaline, ellipsoidal to ovate, aseptate, rarely 1-septate, smooth or finely echinulate with a prominent scar, variable in size according to host, range 7–27 × 2–10 μm (Griffiths, 1957; Latch, 1964; Mäkelä, 1976; Sprague, 1955) (Figs 29.8 and 29.9).

In *R. holci-lanati* conidiophores are few to a group, rising directly from the mycelium in the leaf, sparingly branched, septate, 70–235 × 2.3–3.0 μm. Conidia are few, hyaline, occasionally 1-septate, borne terminally, finely echinulate with a prominent hilum, 13–24 × 5–11.5 μm.

Figure 29.8 Conidiophores and conidia of *Ramularia pusilla*.

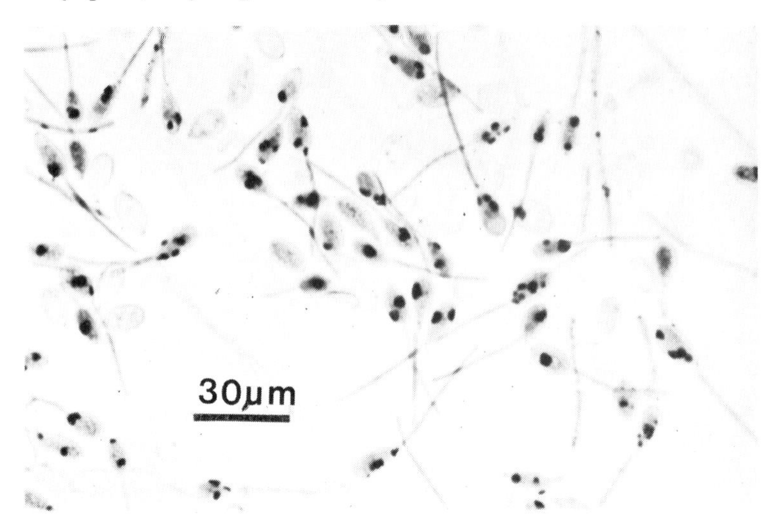

Figure 29.9 Conidia of *Ramularia pusilla*, stained giemsa, germinating on culture media showing basal germination.

29.9.3 Disease development and control

The fungi probably overwinter as stromata (*R. pusilla*) or mycelium (*R. holci-lanati*) in leaves or leaf debris. Conidia germinate in a bipolar or basal fashion and penetration of leaves is through stomata. In *R. pusilla* 20–24°C is optimum for disease development and lesions of both fungi are most noticeable in spring and autumn. Leaf flecking develops in *Lolium perenne* and *Poa annua* 4 and 5 days respectively after inoculation. In *R. pusilla* some host spcificity is shown in *Lolium/Festuca* and *Agrostis* isolates for these hosts. *R. holci-lanati* isolates are most pathogenic on that host but will infect *Lolium* and *Festuca* spp.

No control methods have been reported.

29.10 *RHYNCHOSPORIUM* SPP. – LEAF BLOTCH OR SCALD

While this disease has a wide distribution on species of cool-season turf grass genera (Latch, 1966; Mäkelä, 1972c; Sampson and Western, 1954; Sprague, 1950), it is of little importance in short-mown turf. In northern latitudes it is typically a disease of grasses in spring and early summer, but will develop during cool, humid weather at other times of the year (Caldwell, 1937; Latch, 1966; Mäkelä, 1972c). Two different species of *Rhynchosporium* (Hyphomycetes) cause the disease on different hosts.

29.10.1 Symptoms and the fungi

Rhynchosporium orthosporum Caldwell is the more important on turf grass species, attacking *Agrostis, Festuca, Lolium, Phleum* and *Poa* spp. (Latch, 1954; Mäkelä, 1972c; Sampson and Western, 1954). Initial symptoms are water-soaked lesions mainly on leaf laminae, of irregular shape, up to 25 mm long. These may develop into blotches which coalesce and girdle the leaf, the tips of which may be killed. Spot colour varies with the host species and within the species from reddish-brown to russet or greyish-brown. Spots sharply bordered with brown, grey in the centre may be found later in the season.

Rhynchosporium secalis (Oud.) J.J. Davis, the cause of barley scald and leaf spotting of some grass species is occasionally found on *Agrostis* spp. and *L. perenne.* Most evidence points to the existence of physiological races of both fungi on different grass species (Shipton, Boyd and Ali, 1974; Wilkins, 1973). Both fungal species may occasionally be found on the same host. Leaf symptoms are often similar for both fungi and may be confused, in grasses, for those

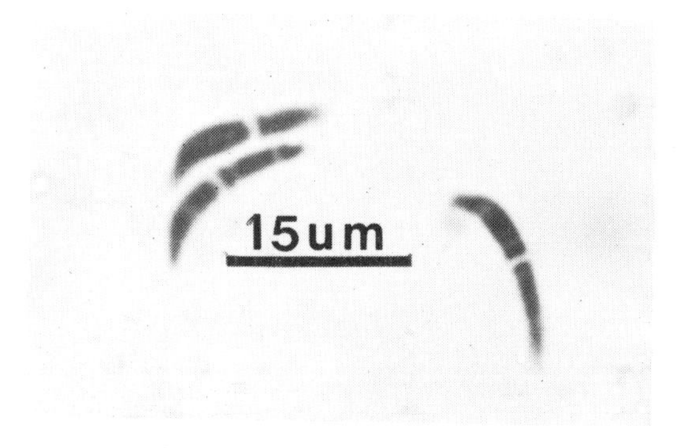

Figure 29.10 *Rhynchosporium secalis*–conidia (stained).

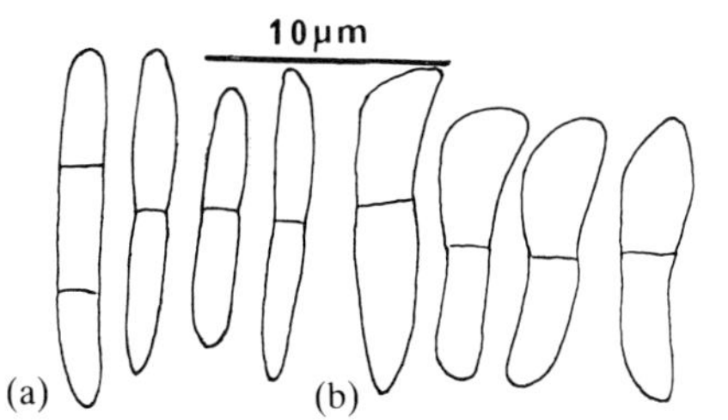

Figure 29.11 (a) *Rhynchosporium orthosporum*, (b) *R. secalis*.

caused by some species of '*Helminthosporium*' (Fig. 56, colour plate section).

The shape of the cylindrical, symmetrical conidia of *R. orthosporum* which are hyaline, usually medianly 1-septate, measuring 8–25 μm × 1.5–4.0 μm distinguishes them from the beaked or curved spores of *R. secalis*, 12–20 μm × 2.2–4.8 μm in size (Figs 29.10 and 29.11). Spores are sometimes difficult to find on typical lesions during dry weather and lesions require rewetting before further spore crops are produced (Skoropad, 1962). Under cool, moist conditions even lightly infected leaves may produce abundant spores. The superficial, hyaline mycelium is sparse and the thallus is confined mainly to the tissues below the cuticle through which a stroma erupts which later bears the sporiferous hyphae.

29.10.2 Disease development and control

Initial infection of leaves may take place in autumn and lesions may be found on overwintering leaves after snow melt (Mäkelä, 1972c; Stedman, 1980). Dispersal of spores occurs during rain in water drops and as dry spores (Stedman, 1980). After spore germination and formation of a small appressorium, mycelial penetration is direct through the epidermal cells of either leaf surface. The fungus overwinters as a flat dark stroma in the leaf. *R. secalis* is seed-borne.

To reduce susceptibility excessive levels of nitrogen should be avoided and balanced fertilizers used. Triadimefon fungicide, alone or with carbendazim, is effective against barley scald (Marsh, 1977) and may give control of the disease in turf also, if this should be necessary.

29.11 *SPERMOSPORA* LEAF SPOT, SCALD AND LEAF BLIGHT

These diseases are widely distributed on several important cool-season turf grass species in regions with cool, moist climates in northern and western Europe, North America, Asia and New Zealand (Deighton, 1968; Latch, 1966; Mäkelä, 1972a,b; MacGarvie and O'Rourke, 1969; O'Rourke, 1976; Smedegard-Petersen and Hermansen, 1971; Sprague, 1948, 1950, 1962). Although they may be encountered in amenity turf they are usually of no great agronomic significance, but *Agrostis* spp. in longer turf are occasionally severely infected. Several different highly pathogenic *Spermospora* spp. (Hyphomycetes) are involved.

29.11.1 Symptoms

On *Agrostis* spp., *S. cilata* causes olive-grey spots with dark-red borders; the spots may become diffuse to give scald-type lesions which may cover large portions of the leaf blade (Fig. 57, colour plate section).

On *Lolium* spp., *S. lolii* attacks may result in numerous, oval spots up to 4 mm long with greyish-brown or reddish-brown borders with off-white centres, bearing small groups of white conidia usually on the upper leaf surface. Scald symptoms may develop also. On *Festuca arundinacea* lesions are at first chocolate-brown, then reddish-brown and like those on *Lolium* spp. are surrounded with chlorotic halos of leaf tissue.

On *Poa* spp., *S. poagena* leaf lesions are at first small, circular, light brown with a straw-coloured centre and a halo of chlorotic tissue, but these may extend considerably, finally fading to dull-brown and then straw-coloured.

On *Holcus lanatus*, *S. holci* spots are tan-coloured with buff or brown borders.

29.11.2 The fungi

The mycelium of *Spermospora* spp. is internal, hyaline, septate, branched and narrow except in the epidermis where swollen, vesicular cells develop, which give rise to narrow hyphae which penetrate the epidermis and bear the conidia.

In *S. cilata* (Sprague) Deighton (syn. *S. subulata* f. *cilata* Sprague) conidia are 2- to 4-septate, hyaline, fusiform or sperm-like with the apical cell sharply tapered to a fine, straight or curved beak, 34–48 × 4.5–5.5 μm (including the beak). There may be a small, lateral process above the hilum (Fig. 29.12).

In *S. lolii* MacGarvie and O'Rourke, conidia are hyaline, similar in shape to those of *S. cilata*, but 2–6-septate and 40–70 × 3.5–4.8 μm, including the beak and with a basal lateral appendage.

In *S. poagena* (Sprague) MacGarvie and O'Rourke (syn. *Cercosporiella poagena* Sprague), conidia are hyaline, 3–7-septate, usually almost straight, broadly filiform, tapering gradually from blunt bases to

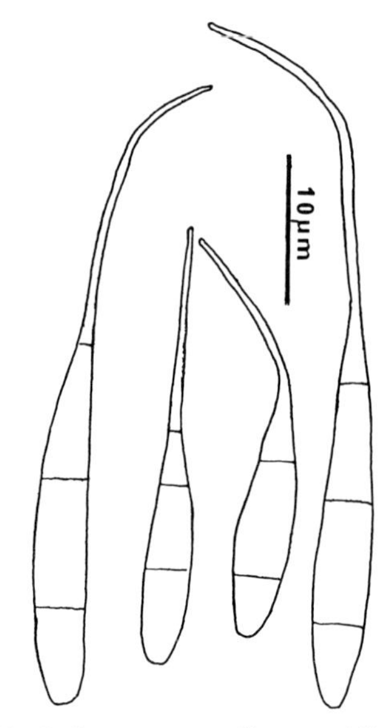

Figure 29.12 *Spermospora ciliata*–conidia.

pointed tips, 45–95 × 3.4–4.5 µm, without a lateral basal appendage.

In *S. holci* (Sprague) MacGarvie and O'Rourke (syn. *Cercosporiella holci* Sprague), conidia are narrowly fusiform or broadly filiform, straight to slightly curved, tapering to a rounded tip from a rounded base, 40–105 × 1.5–3.0 µm, without a basal appendage.

S. lolii may produce 0–2-septate, simple microspores or triradiate structures resembling those of *Volucrispora* spp. in culture (MacGarvie and O'Rourke, 1969).

29.11.3 Disease development and control

In the wet winter regions of western Europe and northwestern North America the disease may persist throughout the year, declining in summer from peak incidence in (late winter) spring and autumn (Booth and Waller 1973; Sprague, 1950).

Frequent mowing appears to reduce disease severity by removing inoculum. No specific controls have been developed.

29.12 *CHEILARIA AGROSTIS*–CHAR SPOT

This is a commonly recorded disease of grasses including the cool-season turf grass species, *Agrostis canina, A. stolonifera, A. tenuis, Agropyron cristatum, Holcus lanatus, Lolium perenne, Phleum pratense* and *Poa pratensis* (Gibson and Sutton, 1969; Latch, 1966; Mäkelä and Koponen, 1976; O'Rourke, 1976; Sampson and Western, 1954; Smith, 1969; Sprague, 1950) in Scandinavia, Western Europe, Canada, USA and New Zealand. While it probably causes little injury under dry conditions its effect may be more severe on irrigated areas (Smith, 1969).

29.12.1 Symptoms

Lesions are mainly on the upper leaf surface, at first tawny with a yellow margin, the centre dirty-white with abundant conidia under moist conditions, elliptical, later becoming elongated, up to 15 µm long depending on the host and darkening with the gradual formation of brown or black stromata, variable in size (Fig. 58, colour plate section). Lower, senescing leaves and leaf sheaths are most frequently attacked.

29.12.2 The fungus

Cheilaria agrostis Lib. (synonym *Septogloeum oxysporum* Bomm. Rouss. and Sacc. (Melanconiales). This is the anamorph: perithecia of various fungi have been found associated with it (Mäkelä and Koponen, 1976; Sprague, 1950) but have not been proven as the teleomorph. The black stromata are often not sporiferous. The conidia are mostly hyaline, thin-walled without guttules, smooth, fusiform, flattened on one side, 0–3 septate, but mostly 2 septate, 15–34 × 3–6 µm (Fig. 29.13). They are borne on enteroblastic, phialidic conidiogenous cells in shallow acervuli, replacing the epidermal cells of the host, or in pycnidia (Mäkelä and Koponen, 1976; Sprague, 1950).

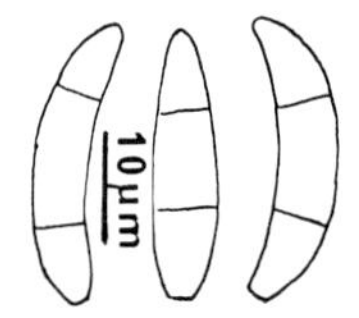

Figure 29.13 *Cheilaria agrostis*–conidia.

29.12.3 Disease development

The disease is favoured by cool, moist conditions and the conidia are presumed to be splash dispersed. In mild climates (e.g. Ireland, O'Rourke, 1976) the disease persists on old leaves over winter, but the role of the stromata in overwintering and control methods have not been investigated.

29.13 REFERENCES

Andersen, I.L. and Gjaerum, H.B. (1979) Descriptions of grass diseases No. 9. Mastigosporium leaf spot. *Mastigosporium kitzebergense* Schlösser, *M. muticum* (Sacc.) Gunnerb., *M. rubricosum* (Dearn. and Barth.) Nannf. *Weibulls Gräs-tips*, **22,** 2 pp.

Årsvoll, K. (1975) Fungi causing winter damage on cultivated grasses in Norway. *Meld. Norg. LandbrHøgsk.*, **54** (9), 49 pp.

Berkenkamp, B. (1976) Diseases of timothy in Alberta. Timothy Plus, *Timothy Newsletter*, **5,** lp.

Berkenkamp, B., Folkins, L.P. and Meeres, J. (1972) Diseases of *Elymus* and other grasses in Alberta, 1972. *Plant Dis. Surv.*, **53.**

Boerema, G.H. and Verhoven, A.A. (1977) Check-list for scientific names of common parasitic fungi. Series 2b. Fungi on field crops, cereals and grasses. *Neth. J. Plant Pathol.*, **83,** 165–204.

Bollard, E.G. (1950) Studies in the genus *Mastigosporium.* 1. General account of the species and their host ranges. *Trans. Br. Mycol. Soc.*, **33,** 250–64.

Booth, C. and Pirozynski, K.A. (1967a). *Leptosphaerulina trifolii. CMI Descriptions of Pathogenic Fungi and Bacteria*, **146,** Commonwealth Mycological Institute, Kew, 2pp.

Booth, C. and Pirozynski, K.A. (1967b) *Pleospora herbarum CMI Descriptions of Pathogenic Fungi and Bacteria.* **150,** Commonwealth Mycological Institute, Kew, 2pp.

Booth, C. and Waller, J.M. (1973) *Pseudocercosporella herpotrichoides. CMI Descriptions of Pathogenic Fungi and Bacteria*, **386,** Commonwealth Mycological Institute, Kew, 2 pp.

Braverman, S.W. (1958) Leaf streak of orchardgrass, timothy and tall oatgrass incited by *Scolecotrichum graminis. Phytopathology*, **48,** 141–3.

Brummer, V. (1937) Observations on outbreaks of fungal disease on timothy in Finland. *J. Sci. Agric. Soc. Finland*, **9,** 165–80.

Caldwell, R.M. (1937) Rhynchosporium scald on barley, rye and other grasses. *J. Agric. Res.*, **55,** 175–98.

Childers, W.R. (1948) Record of the occurrence of *Physoderma graminis* in Canada. *Science*, **108,** 484.

Chupp, C. (1953) *A monograph of the Fungus Genus Cercospora*, Cornell University, Ithaca, New York 667 pp.

Conners, I.I. (1967) *An Annotated Index of Plant Diseases in Canada.* Queen's Printer, Ottawa, 381 pp.

Cook, W.R. Ivimey (1934) Some observations on the genus *Cladochytrium* with special reference to *C. caespitis* Griffon and Manblanc. *Ann. Bot.*, **48,** 177–85.

Creelman, D.W. (1956) The unusual occurrence of three leaf-spotting fungi on grasses in Nova Scotia. *Plant Dis. Rept.*, **40,** 510–12.

Cunningham, P.C. (1965) *Cercosporella herpotrichoides* Fron. on gramineous hosts in Ireland. *Nature, Lond.*, **207,** 1414–15.

Cunningham, P.C. (1981) Occurrence, role and pathogenic traits of a distinct pathotype of *Pseudocercosporella herpotrichoides. Trans. Br. Mycol. Soc.*, **76**(1), 3–15.

Daub, M.E. (1982) Cercosporin, a photosensitizing toxin from *Cercospora* species. *Plant Physiol.*, **69**(6), 1361–4.

Deighton, F.C. (1967) Studies on *Cercospora* and allied genera. II. *Passalora, Cercosporidium* and some species of *Fusicladium* on *Euphorbia. Mycol. Pap.*, **112,** Commonwealth Mycological Institute, Kew, 80 pp.

Deighton, F.C. (1968) *Spermospora. Trans. Br. Mycol. Soc.*, **51,** 41–9.

Deighton, F.C. (1973) Studies on *Cercospora* and allied genera. 4. *Cercosporella* Sacc, *Pseudocercosporella* gen. nov. and *Pseudocercosporidium* gen. nov. Commonwealth Mycological Institute, Kew, *Mycol. Pap.*, **133,** 62 pp.

Dunlap, A.A. (1944) *Pleospora* on lawn grass in Texas. *Plant Dis. Rept.* **28** (4–5), 168.

Ellis. M.B. (1971) *Dematiaceous Hyphomycetes.* Commonwealth Mycological Institute, Kew, 608 pp.

Ellis, M.B. (1976) *More Dematiaceous Hyphomycetes.* Commonwealth Mycological Institute, Kew 507 pp.

Frauenstein, K. (1968) Studies on the leaf spots of fodder grasses. *Nachrichtenblatt Dt. Pflanzenshutzdienst. Berlin. N.F.*, **22,** 4–14.

Freeman, T.E. (1959) A leafspot of St. Augustinegrass caused by *Cercospora fusimaculans. Phytopathology*, **49,** 160–1.

Freeman, T.E. (1967) Diseases of southern turfgrasses. *Fla Agric. Expt. Stat., Tech. Bull.*, **713,** 31 pp.

Gibson, I.A.S. and Sutton, B.C. (1969) *Cheilaria agrostis. CMI descriptions of pathogenic fungi and bacteria*, **488,** Commonwealth Mycological Institute, Kew, 2 pp.

Gopalkrishnan, K.S. (1951) Development and parasitism of *Physoderma graminis* (Bus) Fischer on *Agropyron repens. Phytopathology*, **41,** 1065–76.

Graham, J.H. and Luttrell, E.S. (1961) Species of *Leptosphaerulina* on forage plants. *Phytopathology*, **51,** 680–93.

Graham, J.H., Zeiders, K.E. and Braverman, S.W. (1963) Sporulation and pathogenicity of *Scolecotrichum graminis* from orchardgrass and tall oatgrass. *Plant Dis. Rept.*, **47,** 255–6.

Gregory, C.T. (1919) *Heterosporium* leafspot of timothy. *Phytopathology*, **9,** 576–80.

Griffiths, E. (1957) Occurrence of *Ramulaspera holci-lanati* in Britain. *Trans. Br. Mycol. Soc.*, **40,** 232–6.

Gunnerbeck, E. (1971) Studies on foliicolous Deuteromycetes. 1. the genus *Mastigosporium* in Sweden. *Svensk Bot. Tidskr.*, **65,** 39–52.

Guyot, A.L. (1932) Observations on the comparative geographical distribution of some plant species and of certain of their natural parasites (2nd note). On some fungi parasitic on Gramineae. *Rev. Pathol. Veg. et Ent. Agric.*, **19,** 36047 (*Rev. Appl. Mycol.*, **11,** 721–2 (1932)).

Hardison, J.R. (1945) A leaf spot of tall fescue caused by a new species of *Cercospora. Mycologia*, **37,** 492–4.

Horsfall, J.G. (1930) A study of meadow-crop diseases in New York. *Cornell Agric. Expt. Sta. Mem.*, **130,** 1–139.

Hughes, S.J. (1953) Some foliicolous Hyphomycetes. *Can. J. Bot.*, **31,** 560–76.

Jacques, J.E. (1941) Studies in the genus *Heterosporium. Contrib. Inst. Bot. Univ. Montreal*, **39,** 46 pp.

Jørstad, I. (1945) Parasitic fungi on cultivated and useful plants in Norway. *Meld. Stat. Pl. Patol. Inst.*, **1,** 142 pp. Oslo. (Norwegian).

Karling, J.S. (1950) The genus *Physoderma* (Chytridiales). *Lloydia*, **13**, 29–71.

Koponen, H. and Mäkelä, K. (1976) *Phyllachora graminis, P. silvatica, Epichloe typhina* and *Acrospermum graminum* on grasses in Finland. *Karstenia*, **15**, 46–55.

Kreitlow, K.W., Graham, J.H. and Garber, R.J. (1953) Diseases of forage grasses and legumes in the Northeastern States. *Bull. Pa. Agric. Expt. Sta.*, **573**, 42 pp.

Lacey, J. (1967) Mastigosporium leaf fleck of perennial ryegrass. *Plant Pathol.*, **16**, 48.

Lal, B.B. and Chakravarti, B.P. (1977) Root and collar inoculation and control of brown spot of maize by post-infection spray and soil application of systemic fungicides. *Plant Dis. Rept.*, **61**(4), 334–6.

Latch, G.C.M. (1964) *Ramularia pusilla* Ung. and *Ramulaspera holci-lanati* (Cav.) Lind. in New Zealand. *NZ J. Agric. Res.*, **7**, 405–16.

Latch, G.C.M. (1966) Fungous diseases of ryegrasses in New Zealand. I. Foliage diseases. *NZ J. Agric. Res.*, **9**, 394–409.

Latch, G.C.M. and Wenham, H.T. (1958) Fungal leaf-spot diseases of cocksfoot (*Dactylis glomerata* L.) in the Manawatu. I. Leaf streak caused by *Scolecotrichum graminis* Fckl. *NZ J. Agric. Res.*, **1**, 182–8.

Laughlin, W.M. (1965) Effect of fall and spring application of four rates of potassium on yield and composition of timothy in Alaska. *Agron. J.*, **57**, 555–8.

Lawrence, J.S. and Meredith, D.S. (1970) Wind dispersal of conidia of *Cercospora beticola. Phytopathology*, **60**, 1076–8.

Leeuw, W.P. de, and Voss, H. (1970) Diseases and pests of turf grasses in the Netherlands. *Rasen-Turf-Gazon*, **1**, 65–9 (German, Fr., En.).

Lind, J. (1913) *Danish Fungi, as represented in the Herbarium of E. Rostrup*, Copenhagen, 648 pp.

Lindau, G. (1907) The fungi. Fungi Imperfecti. In *Rabenhorst's Kryptogamenflora Deutschland, Osterr.* 2nd, edn, Schweitz, vol. 1, pp. 1–852.

Luttrell, E.S. (1951) Diseases of tall fescue in Georgia. *Plant Dis. Rept.*, **35**(2), 83–5.

MacGarvie, Q.D. and O'Rourke, C.J. (1969) New species of *Spermospora* and *Cercosporella* affecting grasses and other hosts in Ireland. *Ir. J. Agric. Res.*, **8**, 151–67.

Mäkelä, K. (1970a) The genus *Mastigosporium* Riess in Finland. *Karstenia*, **11**, 5–22.

Mäkelä, K. (1970b) Resting stage of *Mastigosporium* Riess Genus in Finland. *Acta Agric. Scand.*, **20**, 219–24.

Mäkelä, K. (1972a) Disease damage to the foliage of cultivated grasses in Finland. *Aeta Agric. Fenn.*, **124**, 1–56.

Mäkelä, K. (1972b) *Spermospora ciliata* (Sprague) Deighton and *Scolecotrichum graminis* on Finnish grasses. *Karstenia*, **13**, 9–15.

Mäkelä, K. (1972c) *Rhynchosporium* species on Finnish grasses. *Karstenia*, **13**, 23–31.

Mäkelä, K. (1976) *Ovularia pusilla, Hadrotrichum virescens, Deightonella arundinacea* and *Discosia artocreas* on grasses in Finland. *Karstenia*, **15**, 38–45.

Mäkelä, K. and Koponen, H. (1976) *Telimenella gangraena* and *Septogloeum oxysporum* on grasses in Finland. *Karstenia*, **15**, 56–63.

Mäkelä, K. and Sundheim, L. (1982) *Cladosporium* eyespot. *Weibull's Gräs-tips*, **23**, 2 pp.

Marsh, R.W. (ed.) (1977) *Systemic Fungicides*, 2nd edn, Longmans, Edinburgh, 400 pp.

McAlpine, D. (1902) *Fungus Diseases of Stone Fruit Trees in Australia and Their Treatment.* Govt. Printer, Melbourne, Australia, 134 pp.

McCoy, R.E. (1973) Relation of fertility level and fungicide application to incidence of *Cercospora fusimaculans* on St. Augustinegrass. *Plant Dis. Rept.*, **57**, 33–5.

Melchers, L.E. (1925) Diseases of cereal and forage crops in the United States in 1924. *Plant Dis. Rept.*, Suppl. **40**, 106–91.

Mühle, E. (1953) *The Diseases and Pests of Fodder Grasses Grown for Seed.* S. Hirzel Verlag, Leipzig, 167 pp (German).

Müller, E. (1951) On the development of *Pleospora gaumannii* nov. spec. *Ber. Schweitz Bot. Ges.*, **61**, 165–74 (German).

O'Rourke, C.J. (1976) *Diseases of Grasses and Forage Legumes in Ireland.* An Foras Taluntais. 115 pp.

Orlob, G.B. (1960) Observations on the occurrence of grass and forage diseases in New Brunswick. *Can. Plant. Dis. Surv.*, **40**, 78–86.

Ormrod, D.J., Hughes, E.C. and Shoemaker, R.A. (1970) Newly recorded fungi from colonial bentgrass in coastal British Columbia. *Can. Plant Dis. Surv.*, **50**(3), 111–2.

Orton, C.R. (1944) Graminicolous species of *Phyllachora* in North America. *Mycologia*, **36**, 28–53.

Orton, C.R. (1956) The morphology and life history of *Phyllachora punctum. Phytopathology*, **46**, 441–4.

Pape, H. (1928) Diseases and pests of fodder and meadow crops and their control. *Deutsche Landw. Presse*, **40**, (6), 83; (18), 270; (25), 378; (28), 418. (*Rev. Appl. Mycol.*, **7**, 641–2 (1928)).

Parberry, D.G. and Langdon, R.F.N. (1963) Studies on graminicolous species of *Phyllachora* Fckl. IV. Evaluation of the criteria of spcies. *Aust. J. Bot.*, **12**, 265–81.

Parberry, D.G. (1967) Studies on graminicolous species of *Phyllachora* Nke. in Fckl. V. A taxonomic monograph. *Aust. J. Bot.*, **15**, 271–375.

Pitt, D.L. and Bainbridge, A. (1983) Dispersal of *Pseudocercosporella herpotrichiodes* spores from infected wheat straw. *Phytopathol. Z.*, **106**, 214–25.

Rowe, R.C. and Powelson, R.L. (1973) Epidemiology of cercosporella foot rot of wheat: disease spread. *Phytopathology*, **63**, 984–8.

Sakuma, T. and Narita, T. (1961) *Heterosporium* leaf spot of timothy and its causal fungus, *Heterosporium phlei* Gregory. *Bull. Hokkadio Agric. Expt. Sta.*, **96**, 96–100 (Japanese).

Sampson, K. and Western, J.H. (1954) *Diseases of British Grasses and Herbage Legumes.* Cambridge University Press, 118 pp.

Sawada, K. (1958) Researches on fungi in the Tohuku district in Japan. 4. Fungi Imperfecti. *Bull. For. Exp. Sta. Meguro*, **105**, 35–140.

Schlösser, V.G. (1970) *Mastigosporium kitzebergense* spec. nov., a parasitic fungus on *Phleum pratense. Phytopathol. Z.*, **67**, 248–58 (German, En.).

Shipton, W.A., Boyd, W.J.R. and Ali, S.M. (1974) Scald of

barley. *Rev. Plant Pathol Commonw. Mycol. Inst.*, **53,** 839–61.

Siemaszko, W. (1933) Some notes on plant diseases in Poland. *Rev. Pathol. Veg. et Ent. Agric.*, **20,** 139–47. (*Rev. Appl. Mycol.*, **12,** 550 (1933).)

Skororpad, W.P. (1962) Effect of alternate wetting and drying on sporulation and survival of *Rhynchosporium secalis. Phytopathology*, **52,** 752 (Abstr.).

Smedegärd-Petersen, V. and Hermansen, J.E. (1971) Leaf spot diseases on graminicolus species at a locality in Greenland. *Freesia*, **10,** 25–9.

Smith, J.D. (1969) Char spot on wheatgrasses. *Can. Plant Dis. Surv.*, **49,** 140–1.

Smith, J.D. (1970) Resistance of timothy cultivars to *Heterosporium phlei, Drechslera phlei* and frost injury. *Can. Plant Dis. Surv.*, **50,** 95–7.

Sparrow, F.K. (1965) The occurrence of *Physoderma* in Hawaii with notes on other Hawaiian Phycomycetes. *Mycopathol. Mycol. Applic.*, **25,** 119–43.

Sparrow, F.K. and Griffin, J.E. (1964) Observations on chytridiaceous parasites of phanerogams. 15. Host range and species concept in *Physoderma. Archiv Microbiol.*, **49,** 103–11.

Sparrow, F.K., Griffin, J.E. and Johns, R.M. (1961) Observations on chytridiaceous parasites of phanerogams. 11. A *Physoderma* on *Agropyron repens. Am J. Bot.*, **48**(9), 850–8.

Sprague, R. (1936) Relative susceptibility of certain species of Gramineae to *Cercosporella herpotrichoides. J. Agric. Res.*, **53,** 659–7.

Sprague, R. (1948) Some leaf spot fungi on western Gramineae. II. *Mycologia*, **40,** 117–93.

Sprague, R. (1950) *Diseases of Grasses and Cereals in North America.* Ronald Press, New York, 538 pp.

Sprague, R. (1955) Some leaf spot fungi on western Gramineae 8. *Mycologia*, **47,** 249–62.

Sprague, R. (1962) Some leaf spot fungi on western Gramineae. 16. *Mycologia*, **54,** 593–610.

Stedman, O.J. (1980) Observations on the production and dispersal of spores, and infection by *Rhynchosporium secalis. Ann. Appl. Biol.*, **95,** 163–75.

Stevenson, J.A. (1946) Fungi novi denominati 2. *Mycologia*, **38,** 524–33.

Sundheim, L. and Aardvold, O. (1969) Inoculation experiments with timothy eyespot fungus *Cladosporium phlei. Meld. Norg. LandbrHøgsk.*, **48**(26), 10 pp.

Thirumalachar, M.J. and Dickson, J.G. (1947) A *Physoderma* disease of quackgrass. *Phytopathology*, **37,** 885–8.

Tisdale, W.H. (1919) *Physoderma* disease of corn. *J. Agric. Res.*, **16,** 137–54.

Tsutomu, S. and Takeshi, N. (1963) Timothy leafspot caused by *Heterosporium phlei* Gregory. *J. Hokkaido Nat. Agric. Expt. Sta.*, **7,** 77–80 (Japanese).

Vicar, R.M. and Childers, W.R. (1955) In *Progress Report*, Forage Crop Div., Central Expt. Farm, Ottawa, Canada 1949–1953 61 pp.

Vries, G.A. de. (1967) Contribution to the knowledge of the genus *Cladosporium* (Thesis, Baarn Netherlands, 1952). Reprinted in *Bibliotheca Mycologica*, **3,** 1–121.

Walker, J.C. (1983) *Physoderma maydis. CMI Descriptions of Pathogenic Fungi and Bacteria.* Commonwealth Mycological Institute, Kew, 753, 2 pp.

Wehmeyer, L.E. (1955) The development of the ascocarp in *Pseudoplea gaumannii. Mycologia*, **47,** 163–76.

Weiss, F. (1945) Check list revision. *Plant Dis. Rept.*, **29,** 34–9.

Wilkins, P. (1973) Infection of *Lolium multiflorum* with *Rhynchosporium* species. *Plant Pathol.*, **22,** 107–11.

Wilkins, P.W. and Carr, A.J.H. (1971) *Heterosporium* leaf spot of timothy. In *Plant Pathol. Rept. Welsh Plant Breeding Sta.*, 1970, p. 41.

Part ten
Fairy rings

30 *Fairy rings*

The rings, ribbons or arcs of stimulated plant growth or of the fruiting bodies of the larger fungi which often occur in floors of woodlands, agricultural or amenity grassland in most parts of the world are commonly called 'fairy rings' or 'witches rings'. There are many myths or superstitions concerning rings and their origin which have been dealt with in considerable detail by Ramsbottom (1927, 1953) to whom the interested reader is referred. These rings are the result of the activities of soil-inhabiting fungi, most commonly Basidiomycetes which may elicit growth responses in the vegetation in which they are growing. In amenity turf, fairy rings usually are most obvious where soil nutrients are in only moderate or are in short supply (Gilbert, 1875; Gould, Austenson and Miller, 1958; Gould, Miller and Polley, 1955; Mathur, 1970; Smith, 1957) or in droughty conditions (Halisky and Peterson, 1970; Lebeau and Hawn, 1961; Smith, 1978a). The stimulated grass growth which may be symptomatic of the activities of some fairy ring fungi results from the breakdown of organic nitrogen compounds as fungus mycelium or other organic material is decomposed and the nitrogen made available is then taken up by the green plants (Norstadt, Frey and Wilhite, 1968; Shantz and Piemeisel, 1917; Smith, 1955, 1957; Weaver, 1975).

The mycelium of the fungus in the soil of a fairy ring grows in a centrifugal fashion as a fungal isolate would do in pure culture on a plate of agar media, because of the apical growth habit of the fungi. In its progression through the soil it may form circles or part circles of stimulated grass growth. The fungus dies out in the centre of the ring after the soil nutrients it metabolizes have been used up or because of the accumulation of staling substances or from these causes in combination with the drying out of the soil (see p. 344). Actively growing hyphae are found at the periphery of the ring and fruiting bodies, if present, are often found closer to the centre of the ring. Rings growing on slopes are usually incomplete downslope, without mycelium in the gap (Smith, 1975).

Shantz and Piemeisel (1917), in the USA, classified fairy ring fungi into three types according to their effect on grassland. This classification holds good elsewhere.

- Type 1 Those where the grass is killed or badly damaged.
- Type 2 Those where the grass growth is stimulated only.
- Type 3 Those where there is no damage to the turf, but the fruiting bodies of the fungus are found in rings.

Bayliss (1911) distinguished three distinct zones in the rings caused by the common fairy ring fungus, *Marasmius oreades* (Bolt ex Fr.) Fr (a Type 1 ring former). Shantz and Piemeisel (1917) also recognized these zones in rings caused by *Agaricus tabularis* Pk. These zones are:

- An inner zone where there is luxuriant grass growth.
- A middle zone where the grass may be droughted or dead.
- An outer zone where there is some stimulated grass growth (Fig. 30.1).

30.1 CAUSAL FUNGI

Shantz and Piemeisel (1917) and Ainsworth, James and Hawksworth (1971) both state that many fairy ring species, mostly Basidiomycetes, have been recorded associated with fairy rings. Other listings are given by Smith (1957), Waterhouse (1957), Halisky and Peterson (1970) and Gregory (1982). Some of the commonest causes of marked fairy rings are given in Table 30.1. Other cap fungi grow in rings, arcs or troops without causing any noticeable effect on the vegetation and yet others occur in an apparently random fashion as solitary specimens. Waterhouse (1957) has suggested that some of the larger fungi can be regarded as truly lawn fungi, others are typically found in other grassy situations such as in pastures, or on heaths or downs with a short turf while others can be classed as being quite out of place in grass at all and occur there fortuitously. Some of these may fruit on buried debris in the lawn, on tree roots spreading into it or on twigs carried on to the turf (Figs 29 and 32, colour plate section; Fig. 30.1).

Table 30.1 Larger fungi from amenity turf in Great Britain (names taken from Wakefield and Dennis (1950); many were named by Dr R.W. Dennis*).

Species	Habitat	Ring type
Marasmius oreades	Greens, fairways, lawns, pitches	1–2
Psalliota arvensis	Golf fairways (as Type 1)	2–1
Collybia butyracea	Golf fairways, seaside	2
Lycoperdon perlatum	Bowling greens, lawns, golf courses	2
Lycoperdon hiemale	Bowling greens, lawns, golf courses	2
Lycoperdon spadiceum	As above, but much less common	2
Psalliota campestris	Turf of all types	2
Scleroderma verrucosum	Golf fairways	2
Tricholoma personatum	Golf fairways	2
Unidentified Basidio-mycete–no fruiting body	Golf and bowling greens, lawns	2
Hygrophorus psittacinus		2–3
The fungi given below occasionally occur in rings or in troops, but they do not usually produce any noticeable effect on turf.		
Clavaria corniculata	Golf greens and fairways	
Clitocybe aurantiaca	Turf bank on a golf course	
Clitocybe rivulosa	Tennis turf	
Coprinus comatus	On turf areas made on rubbish tips	
Coprinus atramentarius	On turf areas made on rubbish tips	
Crepidotus sp.	At the side of golf holes	
Hygrophorus niveus	Coarse and fine amenity turf	
Hygrophorus coccineus (Scarlet and yellow forms)	Coarse and fine amenity turf	
Hygrophorus reai	Coarse and fine amenity turf	
Hygrophorus pratensis	Coarse and fine amenity turf	
Mycena flavo-alba	Coarse and fine amenity turf	
Nolanea staurospora	Coarse and fine amenity turf	
Panaeolus campanulatus	Coarse and fine amenity turf	
Peziza coccinea	On bowling greens usually near edges, perhaps associated with wood surrounds	
Psilocybe semilanceata	All types of turf in autumn	
Psilocybe foenisecii	All types of turf in summer	
Tricholoma carneum	Coarser turf	

*Identification of specimens from mown turf is often made difficult by their damaged condition. Often they are smaller in size than specimens from unmown turf or pastures.

Tahama (1980) in Japan has recorded four types of symptoms in *Zoysia* and *Agrostis* turf in which *Lycoperdon perlatum* or *Lepista sordida* were growing: no green ring; green ring; green ring; green and necrotic ring and necrotic ring. A *Curvularia* sp. was associated with the necrosis. He has suggested that the rings are not caused by the *Lycoperdon* or *Lepista* spp., but they simply induce the conditions under which the rings form.

Other types of rings or circular patches on lawn turf with which fungi (usually non-fruiting or shyly fruiting Basidiomycetes) are associated have been described as 'superficial fairy rings' (Smith, 1955) because the fungus is usually confined to the grass shoot bases and the top of the thatch layer (see Section 30.6).

30.2 THE RATE OF GROWTH OF FAIRY RINGS

Fairy rings are fugitive phenomena (Smith, 1955), especially those of Types 2 and 3. In some years they are clearly visible: they may then not appear for a year or two and then appear again in a slightly different location. In some cases they die out completely. This may make observations on their rate of spread through the soil difficult. Rates of growth for different species of fairy ring fungi have been given by several observers, Thomas (1905), Bayliss (1911), Shantz and Piemeisel (1917), Smith (1957), Ingold (1974), Wolf, (1971). Dickinson (1979) made measurements on approximately 30 *Marasmius oreades* rings over a 3-year period in Norfolk, England and found the annual radial increase to be between 115 and 135 mm. Rings up to 800 m in diameter have been noted in rangeland in Southern Alberta (Lebeau and Hawn, 1961).

30.3 THE EFFECTS OF FAIRY RINGS ON THE SOIL

In rings due to *M. oreades*, Moliard (1910) found most ammonium nitrogen under the dead grass zone and noted very little difference between the amount present outside and inside the ring. In rings of both *C. cyathiformis* and *Agaricus tabularis*, Shantz and Piemeisel (1917) found more ammonium nitrogen in soil which had an abundant mycelium than elsewhere; most nitrates were found in the zone containing the dying mycelium of the fungi. Bayliss-Elliott (1926) found that in *M. oreades* ring soil, sampled in August, there was most ammonium nitrogen in the area of the inner stimulated zone and very nearly the same in the bare zone. In three *M. oreades* rings sampled in June, before the different zones showed their full expression, Smith (1957) found more ammonium and nitrate nitrogen in the soil of the bare zone, which contained most mycelium, than in the other zones (Table 30.2). In all three rings there was approximately the same amount of ammonium nitrogen in the outer and inner stimulated grass zones and more nitrate nitrogen in the inner stimulated zone than in the outer one. However, the amount of total nitrogen was approximately the same in zones of the three rings. In *M. oreades* rings the inner stimulated zone usually shows more clearly than the outer one. This is probably because the nitrate nitrogen is more readily available than that in the ammonium form. Norstadt, Frey and Wilhite (1968) found that the amounts depended on fungal activity which changed with sampling position in the fairy rings. Normal soil outside rings had 1–37 and 1–7 ppm of ammonium and nitrate nitrogen respectively but where the fungus was most active the ammonium nitrogen ranged from 60 to 262 ppm and the nitrate nitrogen ranged from 8 to 30 ppm. Nitrification was not stimulated as much as ammonification, although the larger quantities of nitrate were associated with the larger amounts of ammonium nitrogen.

The fungus may liberate nitrogen from dead roots killed by its advance in the stimulated zone as suggested by Bayliss-Elliott (1926). However, the stimulation of herbage growth may also result from nitrogen liberated by the degradation of humus by the fungus as suggested by Norstadt, Frey and Wilhite (1968) and Mathur (1970). Smith (1957) found in the outer stimulated zone of the three rings sampled greater amounts of ammonium and nitrate nitrogen than in the non-ring soil beyond it which would account for the stimulated grass growth.

Granules of uric acid were collected under the leaf litter in the collision zones of two *M. oreades* fairy rings in Ontario, Canada. Miles and Mathur (1972) suggested that the uric acid might have been produced by (a) the initially oxidative reaction between HCN from the bare zone and NH_4OH from the outer

Table 30.2 Soil analyses of three *Marasmius oreades* fairy rings (From Smith, 1957)

	Loss on ignition (%)	Total nitrogen (ppm)	Ammonia nitrogen (ppm)	Nitrate nitrogen (ppm)
Ring No. 1	Sampled 9 Apr. 1952		Sample depth 22.5 cm	
Non-ring soil	8.8	3116	7.6	16.9
Outer stimulated zone	9.1	3664	17.1	24.5
Bare zone	9.1	3847	234.2	106.4
Inner stimulated zone	8.8	3007	20.1	64.4
Ring No. 2	Sampled 19 June 1952		Sample depth 15 cm	
Non-ring soil	10.3	3660	8.6	3.7
Outer stimulatd zone	11.9	2883	20.5	5.1
Bare zone	10.7	3852	260.0	23.0
Inner stimulated zone	9.4	3440	21.2	11.8
Ring No. 3	Sampled 30 June 1952		Sample depth 10 cm (stony soil)	
Non-ring soil	15.6	4412	7.2	2.4
Outer stimulated zone	15.6	4257	27.0	4.6
Bare zone	14.1	4400	199.2	11.3
Inner stimulated zone	14.5	4394	23.4	8.5

stimulated zone, (b) inhibition of purine catabolism at the uric acid stage or (c) a secondary infection of a soil organism such as *Aspergillus oryzae* which could produce uric acid.

30.3.1 Carbon and humus

Lawes, Gilbert and Warrington (1883) showed that *M. oreades* rings depleted the organic matter in soils growing through them by an average of approximately 16%, but Shantz and Piemeisel (1917) found no significant differences in the organic carbon content in the different zones of *Agaricus tabularis* rings. Table 30.2 shows only slight differences in loss on ignition between soil samples from the different zones of rings caused by the latter species. Norstadt, Frey and Wilhite (1968) found a highly significant correlation between nitrogen and carbon dioxide production in *M. oreades* ring soil. Much more carbon dioxide was released from incubated mycelium-infested soil than from uninfested soil. Mathur (1970) demonstrated that *M. oreades* could utilize chernozem and podzol humic acid in unaerated cultures by degrading the humus fraction, but this was inhibited by increased aeration. He suggested that increased aeration of rings should be studied as a possible control measure.

30.4 FAIRY RINGS CAUSED BY *MARASMIUS OREADES* (BOLT EX FR.) FR.

M. oreades is the most common cause of the severe type of fairy ring (Type 1) in all types of amenity turf in the British Isles (Smith, 1957), New Zealand (Cunningham, 1934), Australia (Black, 1937), United States (Gould, Austenson and Miller, 1958), Canada (Lebeau and Hawn, 1961; Smith, 1972), West Germany (Roediger, 1978) and eastern Norway (Smith, 1975). The rings are very common on golf course fairways over a wide range of climate. They appear to be less common on the more intensively irrigated turf of golf greens. While *M. oreades* rings may be very disfiguring on fine turf, it is doubtful whether they greatly affect the turf trueness if it is well managed. It is only on the lighter soils, e.g. of seaside golf links, and in dry summers that the bare areas often become very conspicuous.

30.4.1 Symptoms of *M. oreades* rings

Typically there are three zones: one inner and one outer zone of stimulated grass growth with a zone between where there is little or no active growth or where the soil may be bare. The outer stimulated zone is usually narrower than the inner one (Fig. 61, colour plate section). In intensively fertilized and irrigated turf no bare zone may be visible. Young rings appear as solid circular patches of stimulated grass which can sometimes be confused with that resulting from fertilization with animal excreta. Fig. 30.1 shows a section and plan of a *M. oreades* ring.

Fruiting bodies of the fungus (Fig. 62, colour plate section) are usually produced in the bare zone or at the junction of the bare and the outer zone. In young rings where distinct zones have not yet developed the fruits may often be found in the single stimulated grass zone. The pileus (cap) of the sporophore is 25–65 mm across (dwarf specimens occur in mown turf), sometimes water-soaked in appearance, reddish-tan when wet, becoming buff-coloured or pale tan when dry, convex, then flattened with a slight boss, fleshy in texture, the edges sometimes becoming wavy when old. The stipe or stalk is 35–100 mm long, pale buff, tough and rigid, attached centrally, smooth with a matt surface 30–50 mm thick. The gills on which the spores are produced are free with shorter intermediate ones, pallid, pale cream or whitish, thick in the centre and thin at the edges of the cap. The spores are white, pip-shaped, 10.5–9.5 μm $\times$ 6.0–5.5 μm. (Description from Wakefield and Dennis, 1950, modified.)

The fruits are tough and do not readily decay. After drying they will revive in moist weather. They are edible and dry well (but ensure that they are of *M. oreades* if to be eaten). They may be produced from spring to late autumn. Rings may not produce fruits for years particularly on golf greens which are mown throughout the year. Prominent rings occasionally disappear without apparent cause.

Rings may attain a large diameter and the distance from inside the inner to outside the outer zone may be up to 1.5 m. The highest concentration of the white mycelium, which has a strong musty smell, is found in the soil of the bare zone where it may be 30–50 cm deep in light soil. In heavy soils it may not penetrate much deeper than the thatch (Fig. 59, colour plate section). It may be hard to find mycelium in winter. There is less mycelium in the soil under the inner and outer stimulated zones. When the fungus is growing actively it may be found on the surface of the soil at the sole of the turf in the form of a discontinuous sheet in the outer stimulated zone. Where the soil is permeated by mycelium, particularly in the bare zone it is most difficult to wet even if a surfactant is used. In soil with abundant mycelium it will fill cavities between soil particles and be found in and on plant roots. The bare zone is often not visible in winter, but grass on the stimulated zones may be attacked by *Microdochium nivale*, symptoms of which often persist until spring. When two rings collide, at the junction the rings disappear (Parker-Rhodes, 1955) (Fig. 30.2).

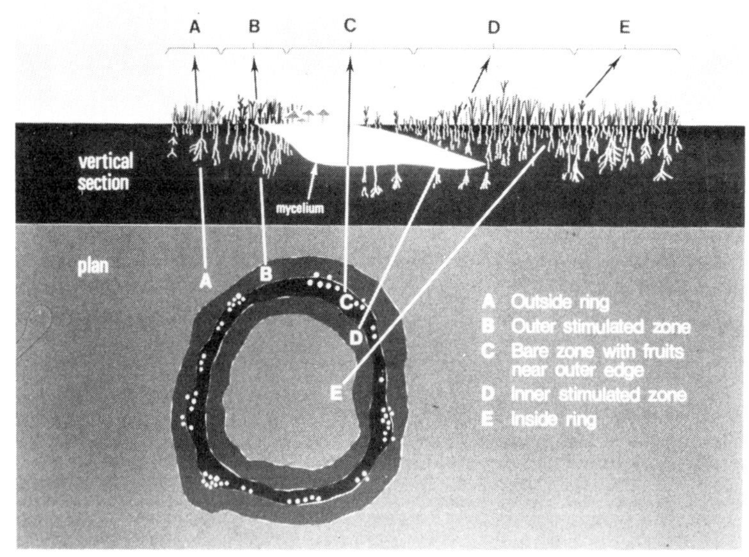

Figure 30.1 Vertical section and plan of a fairy ring caused by *Marasmius oreades*.

Rarely is one ring found inside another, but occasionally a small ring will be found inside an incomplete larger one (Smith, 1975).

30.4.2 Pathogenicity and hydrogen cyanide production by *M. oreades*

Bayliss (1911) showed that the fungus produced an excretion toxic to grass roots and that roots were invaded by the fungus. Lebeau and Hawn (1963a) and Filer (1964, 1965, 1966) demonstrated that the toxin, hydrogen cyanide (HCN) is produced by the mycelium of some isolates of the fungus in culture and under natural conditions. Filer (1964, 1965, 1966) showed that *M. oreades* could parasitize grass seedlings and inhibit root growth and that the amount of HCN produced by the mycelium was sufficient to damage higher plants. He concluded that interactions of HCN toxicosis, parasitism and drought conditions produced in the soil by the fungus are the chief factors which cause the death of grass in the bare zone of a *M. oreades* fairy ring. The relative importance of each of these factors is in doubt. Evans (1967) who studied these mechanisms concluded that the soil conditions in the mycelial zone, particularly drought, accelerated the decline of infected plants. Traquair and McKeen (1986) have demonstrated that cell-free filtrates from vegetative cyanogenic and acyanogenic isolates of *M. oreades* will cause discolouration and necrosis of the root tips of winter wheat seedlings.

30.4.3 The relation of the fungus to soil conditions

(a) Soil reaction

M. oreades rings have been noted on amenity turf with pH values between 5.1 and 7.4. Of 17 rings, eight occurred in soils having a pH between 6.0 and 6.5. The pH of the soil in the bare zone is usually lower than that of the non-ring soil (Smith, 1955). The optimum pH for the growth in diameter of an isolate of the fungus on buffered culture medium was 6.0; growth was slightly better at pH 6.8 than at 5.0 and considerably better at pH 5.0 than at pH 7.4 and 4.7. No growth took place at pH 3.8 or lower. Warcup (1951) found few larger Basidiomycetes (including *M. oreades* and other ring fungi) from grassland soils having pH values of 8.0, 4.4 and 3.8. Most occurred on soils with pH values of 7.0 and 6.4.

(b) Nutrient status

Many soils used for amenity turfs, especially golf fairways are infertile if compared with agricultural grassland and may be purposely maintained in that condition to limit herbage production and reduce mowing. However, *M. oreades* rings have been noted on amenity turf with phosphate and potash analyses in the range very low to high.

(c) Soil physical condition

In Britain, *M. oreades* rings are more noticeable on light, sandy or gravelly soils in the east than in the

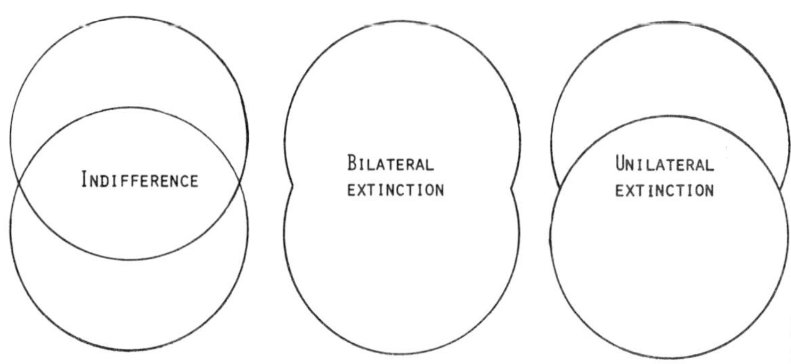

Figure 30.2 Types of reaction when fairy rings meet. M. oreades reaction in centre.

west or the remainder of the country. Most cases occur on the fairways of seaside golf courses and on light land in valley bottoms. However, rings may be found on all soil types. In the prairies of Canada, where average rainfall ranges from 250 to 380 mm/yr ring damage appears to be more frequent on the light brown and dark brown rather than on the heavier textured and inherently more fertile black soils. In Norway *M. oreades* rings are common on light soils round the Oslo Fjord but rare on the west coast with its much higher rainfall. It is difficult to distinguish between the effects of low soil moisture, light texture and low fertility in the incidence and severity of *M. oreades* rings. Probably the factors are cumulative.

30.4.4 The microflora of *M. oreades* rings

Most studies on fairy rings have been concerned with the biochemical changes taking place in the soil rather than in its microbiology, so the picture of the interaction of the ring fungus and the other microflora is very incomplete.

Warcup (1951) in Suffolk, England found that the mycelial zones of *M. oreades, Psalliota arvensis* and *Tricholoma nudum*, all Type 1 ring formers, had a lower population both in numbers and species than the normal soil around. *Arachniotus*, *Chaetomium*, *Gymnoascus* and *Penicillium* spp. were isolated more frequently from mycelial zones than from non-ring soil. In Yorkshire, Smith (1957) isolated *Rhizopus*, *Penicillium* and *Gymnoascus* spp. from the different zones of a *M. oreades* ring. *Penicillium* spp. were more common in the bare zone than in the others. Fungi were more frequent in the top 80 mm of soil than in that from below. Bacterial counts were made from two sampling depths (25–75 and 75–150 mm) in the soil of all zones of a *M. oreades* ring twice in August. On both occasions the highest count was obtained from the upper sample of the inner stimulated zone. This suggested greater bacterial activity associated with the breakdown of *M. oreades* mycelium in this zone and the release of nitrogenous compounds which stimulated grass growth. However, high bacterial counts in soil from both sample depths indicated that it is in this zone generally that the *M. oreades* mycelium is dead or dying. There was a reversal of this bacterial count in the outer stimulated zone which can be explained by postulating that the actively growing fairy ring fungus in colonizing fresh soil produces materials antagonistic to soil bacteria. Non-ring soil showed the expected normal gradient of high numbers in the top sample and low numbers in the lower. Smith and Rupps (1978) data for fungal and bacterial numbers in soil of *M. oreades* rings in spring in Saskatchewan gave equivocal results. However, they suggested that lower fungal and raised bacterial counts from bare zones of some of the rings resulted from the suppression of other fungi by *M. oreades*. Increased bacterial activity was related to the start of *M. oreades* mycelium decomposition. The superimposition on a normal soil microflora of the heavily colonized mycelial zone of *M. oreades* which advances at different depths centripetally creates a soil ecological problem of considerable complexity. Much more data are needed before there is reliable information on the microbiological changes taking place in rings in different zones and seasons.

Many of the fungi isolated from ring soil were found to be strongly to moderately antagonistic to *M. oreades* in pure culture. These included *Acremonium* sp., *Aspergillus ustus* (Bainer) Thom and Church, *Actinomyces* sp., *Alternaria alternata* (Fr.) Keissler, *Bipolaris sorokiniana* (Sacc. in Sorok) Shoem., *Cladosporium herbarum* Link, *Corynascus sepedonicum* (Emmons) v. Arx, *Dactylaria* sp., *Fusarium oxysporum* Schlecht. var *redolens* (Wollenw.) Gordon, *F. solani* (Mart). Sacc., *F. stilboides* Wollenw., *Penicillium canescens* Sapp., *P. notatum* Westling, *Preussia fleischhakii* (Allersw.) Cain and *Ulocladium atrum* (Smith, 1980b).

30.4.5 Soil exhaustion and self-inhibition in *M. oreades*

M. oreades rings show bilateral extinction when they collide (Wollaston, 1807) (Fig. 63, colour plate section; Fig. 30.2). This is the usual occurrence when fairy rings caused by the same fungus meet. Two reasons have been given for this:

1. It results from the exhaustion of soil nutrients. Wollaston (1807) referred to this as 'the exhaustion of a particular pabulum occasioned by each, obstructs the progress of the other and both are starved.' Although there are differences in chemical composition of the nitrogen in ring and non-ring soil, there is little difference in total nitrogen inside and outside rings. Although total carbon may be depleted it also recovers rapidly. Visible changes take place in the thatch as the fungus utilizes it, e.g. it turns reddish brown. The thatch contributes a considerable portion of the carbon and humus in the soil. Mathur (1970) suggested that the unusual ablity of *M. oreades* to degrade chernozem humic and podzol fulvic acid would deplete the humus content of the soil temporarily and allow it to dominate the soil microflora.
2. The fungus produces a self-inhibitory staling product. Coville (1897) suggested that decomposition products of fungal growth washed downslope were harmful to the growth of the fungus lower down. Bayliss (1911) produced some evidence that a toxin produced by the fungus was self-inhibitory. Smith and Rupps (1978) showed that dikaryotic isolates of *M. oreades* from the same or different rings would inhibit each other in culture, although they were unable to obtain an antibiotic by chemical extraction of fruits.

30.5 SUPPRESSION AND ELIMINATION OF FAIRY RINGS

The fugitive nature of fairy rings has an important bearing on the reliability of reports on the effectiveness of control measures.

Among the materials which have been tried for the control of fairy rings have been:

1. Stable manure applied to the grass in front of rings caused by *M. oreades* or *Clitocybe gigantea*. According to Bayliss-Elliott (1926) if this was applied in patches the ring died out in patches.
2. Ferrous sulphate, in solution in concentrations between 2.6 and 25%. This material was the main contituent of many so-called 'fairy ring killers' from the beginning of the twentieth century (Smith, 1957).
3. Bordeaux mixture (6:4:24) for control of puffball rings (McAlpine, 1898).
4. Potassium permanganate at 6.7% applied as a drench (Black, 1937).
5. Compounds of iron, sulphur, cadmium, magnesium and lime and various fungicides tried by Gould, Miller and Poley (1955) did not show sufficient promise in experiment to justify their recommendation as controls.
6. Howard, Rowell and Keil (1951) found that organo-cadmium fungicides appeared to reduce mushroom fruiting on turf in Rhode Island. Smith (1957) tested a wide range of fungicides and soil amendments on both *M. oreades* and *Lycoperdon* sp. rings between 1951 and 1957. None of these, other than formaldehyde showed promise of eliminating rings with certainty.

M. oreades fairy rings have developed where ripe fruits were placed in small heaps in marked positions on lawn turf at Saskatoon, but only in a few places was this successful (Smith, unpublished). It seems probable that the collection of fruits would reduce the amount of available spore inoculum which might start new rings. Mercury fungicides such as phenyl mercuric acetate, to which *M. oreades* is sensitive (Smith, 1955) has been shown to reduce its fruiting (Gould, Miller and Polley, 1955). Erratic success has resulted from attempts to start *M. oreades* rings on turf plots with plugs of infected soil as inoculum. Initial growth of the fungus from the plug is usually satisfactory, but then it stops as if unable to overcome antagonistic factors in its new environment. Although there is little experimental evidence on which to base management practices to discourage the start and progress of fairy rings, observations suggest that rings are less common or at least less deleterious to amenity turf which is well irrigated. *M. oreades* rings occur over the pH range at which amenity turf is usually maintained. Depression of the pH with acid-tending materials or raising it with lime has had little effect on *M. oreades* rings (Smith, 1957). Fertilization effects have not been fully examined, but rings caused by various fungi transgressed from plot to plot on fertilizer treatments without regard to nitrogen source at the Sports Turf Research Institute. Gould, Miller and Polley (1955) noted that fertilization with organic fertilizers such as chicken manure was associated with heavy mushroom infestation. Beard, Vargas and Rieke (1973) attributed the stimulation in growth of Type 1 rings caused by *Tricholoma sordidum* to increased nitrogen fertility levels. This is a factor to be considered when using nitrogen applications to 'mask' the effect of rings. However, Lebeau and Hawn (1963b) concluded from results obtained in Southern Alberta that nitrogenous fertilizers do not improve on the therapeutic effects of aeration and persistent watering. Their recommendation for the control of *M. oreades* fairy rings is based solely on these two factors (see below).

It is difficult to wet soils with suspensions or solutions of fungicides if the soil contains much mycelium of *M. oreades* even if penetration is assisted by pricking or spiking and the use of surfactants. *M. oreades* is moderately sensitive to fungicides. In laboratory disc assays its growth was completely inhibited by ethyl mercuric phosphate at 10 ppm, phenyl mercuric acetate at 100 ppm and mercuric chloride at 1000 ppm. Volatile materials such as pentachloronitrobenzene, tetrachloronitrobenzene, orthodichlorobenzene and formaldehyde at low concentrations will kill the fungus on artificial media (Smith 1955, 1957). The problem of fungicidal control of fairy rings with dense soil mycelium is not of finding a suitable fungicide, but of positioning it in a concentration long enough for it to exert its fungicidal or fungistatic activity without turf damage.

Systemic fungicides effective against Basidiomycetes have been reported effective in suppressing ring symptoms for varying periods. Dahlsson (1977) found that carboxin at 7.5 g a.i./m^2 and oxycarboxin at 11.25 g a.i./m^2applied in water to spiked rings suppressed or almost completely suppressed ring symptoms for 2 years. Benodanil at 7.5 g a.i./m^2 partially suppressed symptoms. Roediger (1978) found that benodanil at 2.5 g a.i./m^2 suppressed ring symptoms for one year. A 5% granular formulation of carboxin applied at 70 g/m^2 prevented further development of fruits of *M. oreades* when applied to moist soil (Hansen and Schadegg, 1980). A 75% wettable powder of this material at 5 g/m^2 was less effective. Quintozene emulsion (24% a.i.) at 10 g/m^2 had little effect on cap production. Karlson (1975) found that dosages of a granular fungicide based on carboxin (concentration not given) between 2.25 and 7.25 g a.i./m^2 reduced cap production of *M. oreades* to very low levels one year after application and there was no difference in colour between the ring herbage and the uninfested turf. Effective suppression of fruiting was obtained by Heimes and Loëcher (1980) with soil drenches of benodanil 2.5, 3.0 or 5.0 g a.i./m^3. Only in one trial were there any fruits showing on treated plots one year after application. Old rings had practically disappeared. Some of these materials are registered for fairy ring control in Europe, but not yet in North America*. While these results on fruit and ring suppression are promising, experience suggests that because of the fugitive nature of fairy rings, several years must elapse before treatment can be regarded as effective in eliminating the fungus from soil.

* Promising results in ring elimination and sporophore inhibition in *M. oreades* were obtained with applications of oxycarboxin, flutinol and triforine fungicides applied as soil drenches in field tests in Saskatchewan. However, repeated applications of the drenches over two years (1986–7) were necessary. Two applications of triadimefon completely suppressed sporophore production (Smith, in press).

Lebeau and Hawn (1963b) developed a method for the control of fairy rings caused by *M. oreades* which involves complete wetting of the infested soil, which is no small task, as stressed by the authors, because of the impervious nature of the soil on the bare ring. Aeration (thorough spiking) of the fairy ring followed by persistent watering each day for at least a month was advocated. The extremely wet soil conditions so produced may lead to the complete disappearance of the ring. Lebeau and Hawn (1963b) suggested that increased bacterial and fungal activity in the soil resulting from the persistent very wet conditions led to the destruction of the *M. oreades* mycelium or overcame its hydrophobic properties. Penetration of water into the ring soil may be improved by the use of a solution of a 1:10000 (0.01%) turf wetting agent (such as Aqua-gro) or less desirably, a mild liquid dish-washing detergent, diluted similarly, applied to the spiked area until the holes will take no more. Plain water from a hose or hydrogun is then applied daily. Forks and hydrogun should be sterilized with a 5% solution of domestic hypochlorite bleach (bleach containing 5% chlorine is diluted 20 times) and then washed with plain water before being used on clean areas of turf. If this ring treatment is done thoroughly rings may be suppressed and eradicated, but if they are inadequately treated they may fragment and become individual centres of infestation. The method is as laborious as complete excavation of the mycelium-infested soil, fumigation of the excavation with a volatile fungicide and replacement with clean soil. It is more laborious than the method given below.

If there are only a few small rings of *M. oreades* to be eradicated, a method developed by Smith (1955, 1957, 1978b) based on that of Cunningham (1934) is to be preferred. Originally, the sterilizing agent was formaldehyde, but other volatile materials such as chloropicrin and methyl bromide may be used (by professionals). Soil treatment (partial sterilization with formaldehyde) not only eliminates *M. oreades* and other ring fungi, but also results in the development of an antagonistic microflora (Smith, 1957). Disturbance of the soil before application of the fumigant is a necessary part of the treatment, facilitating fungicidal action (Smith, 1957). The soil is treated *in situ* which solves the problem of its disposal. For domestic use, a fungicide solution, sufficient for 10 m^2 of turf area consists of:

2.5 litres of 40% formaldehyde solution (formalin)
200 ml of surfactant (turf wetting agent or dish washing detergent)

50 litres of water
Mix this in a suitable vessel such as a plastic garbage can in the open air, being careful not to spill it on vegetation.
*Avoid inhaling the vapour.**

During warm weather in late spring, summer or early autumn, when the fungus is active, strip the turf with a sharp, flat spade or turfing iron from 300 mm outside the outer green ring to inside the inner green ring. Transport this turf away from the site on a polythene sheet without spilling any on the clean lawn. Compost this material, but do not use this for lawn top dressing. With a digging fork break up the soil of the ring as deeply as possible, starting at the outside of the ring, then sterilize tools used with the prepared fungicide. Apply the fungicide with a watering can equipped with a rose, taking care not to spill it on surrounding turf which would be damaged. Cover treated soil with weighed down-polythene sheets to seal in the vapour for 7 to 10 days, then remove the covers and stir the soil carefully with a digging fork. Leave it exposed for 2 to 3 weeks to eliminate the fumigant. Level the surface with soil from adjacent flower borders or vegetable plot. Imported grassland soil should not be used. The site may then be seeded.

The advance of fairy rings may be halted by digging temporary flower borders or constructing paths across the progression. Their mycelium will rarely transgress such barriers although that of *M. oreades* will cross 'cordons sanitaires' of fungicide laid in the turf (Smith, 1957).

A promising biological control method for ring eradication suited to large areas of amenity turf on lawns, playing fields and golf course fairways, heavily infested with *M. oreades* rings has been developed (Smith, 1978a, b, 1980a). No fungicides are used. It is based on (a) the observation that when two *M. oreades* rings collide they eliminate each other by mutual antagonism (Fig. 6.3, colour plate section), (b) that the soil contains fungi and probably other organisms which are antagonistic to *M. oreades* (Smith, 1978; Smith and Rupps, 1978) and (c) when soil infested with the fungus is disturbed it has to overcome the antagonism of the soil microflora before its typical ring form can develop again. Alternative techniques have been developed to exploit these principles.

*Formaldehyde has been shown to induce cancer in experimental animals and should be considered a cancer risk to humans (NIOSH, 1981). The chance of being exposed to high concentrations of the gas is considered slight if all mixing of solutions to be used in fairy ring treatment is done in the open air and exposure to the vapour is reduced to a minimum. However, it would be safer to leave the treatment to a suitably equipped professional pesticide applicator.

If there are deep-rooted, persistent, perennial grasses present as weeds in the infested turf use a non-residual total herbicide such as glyphosate, to kill all grassy vegetation and a herbicide for broad-leaved weeds also if necessary. Strip off the dead turf and compost it on a non-turf site. If such a location is not available or where fresh soil or turf is not to be imported, kill the turf in late summer with the herbicide, and when it is dead surface-cultivate it with a heavy-duty rotary cultivator in autumn and leave rain or snow melt water to soften the soil below. The soil and or soil and turf must then be repeatedly cultivated in different directions as deeply as possible to ensure intimate mixing of infested and normal soil. Between cultivations any dry areas which were the sites of rings should be flooded with water. The lawn is then levelled, firmed, fertilized and sown or turfed. Sowing is to be preferred since no grassland soil is then imported. Although fairy rings are rarely seen in young sod on sod farms latent infection may be present.

Whatever method of eradication is used, the risk of reinfection will be greatly reduced by good turf management, including adequate fertilization and irrigation to ensure deep penetration of water. Infrequent watering to 150 mm depth is to be preferred to frequent shallow irrigation which keeps the turf surface wet for longer periods, a factor known to increase turf disease susceptibility generally.

30.5.1 Control of Type 1 rings other than *M. oreades* and Type 2 rings

There are few published reports of the effectiveness of methods for the control of Type 1 rings other than those caused by *M. oreades*. Those with a dense mycelial zone like *M. oreades* are likely to prove as refractory and may require similar treatment, e.g. complete excavation or soil disturbance with fumigation. The reaction of ring-forming species other than *M. oreades* to fungicides active towards Basidiomycetes (e.g. carboxin, oxycarboxin and benodanil) is not known.

Type 2 fairy rings caused by *Agaricus* (Fig. 60, colour plate section), *Lycoperdon* and *Scleroderma* spp. may be abundant particularly on golf course fairways during wet periods. Their fruits, puffballs, may be very numerous and may be confused with white golf balls. These species and the common field mushroom often have inconspicuous mycelial soil zones, but can develop extesive ribbons and arcs of stimulated grass growth which can be unsightly. On greens these are usually well masked by the use of a little extra nitrogenous fertilizer. On fairways, it is rarely feasible to attempt their eradication. Because of their fugitive nature, it is difficult to determine the effectiveness of fungicides, and large amounts of fungicide are needed. However, formaldehyde fumigation is effective, using half-strength solution applied to the pricked ring without turf removal. The killed turf will need replacement afterwards. Probably other fungicides, such as mercuric chloride and malachite green would also be effective (Smith, 1955, 1957).

30.6 SUPERFICIAL FAIRY RINGS

Most of the activity of fairy ring fungi takes place in the thatch and the mineral soil just below it, although their mycelium may go 300 mm deep in light soil. However, there are fairy ring fungi, mostly Basidiomycetes, confined almost entirely to the litter layer, the thatch and the top of the mineral soil. Their mycelial zone is often visible in the sole of the turf. These are the fungi which cause 'superficial fairy rings' (Jackson, 1972; Smith, 1955, 1959). They are common in turf in many parts of the world. Most of them are saprobes and probably cause little damage, others produce unsightly rings and some may be implicated in the 'dry patch' condition (Anon, 1962), particularly on sand golf greens.

Superficial fairy rings may be categorized by their appearance and their effects on turf:

Type A – produce sparse to copious mycelium, with or without fruiting bodies, which develops on shoot bases and in the thatch with very little apparent effect on grass growth.

Type B – produce stimulated grass growth and/or turf grass discolouration develops; thatch degradation is apparent but the grass is not severely injured.

Type C – produce severe grass injury; turf grass growth may or may not also be stimulated.

30.6.1 Fungi associated, symptoms and occurrence

In Type A rings, mycelium of Basidiomycetes are found closely investing grass sheath bases (Fig. 64, colour plate section), rhizomes and stolons of many grass species, but apparently cause no host damage. However, the snow mould fungus, *Coprinus psychromorbidus* may fruit on dead sheaths of *Agrostis stolonifera*. Possibly some of these fungi are saprobic stages of grass pathogens.

A mesophilic Basidiomycete with abundant clamp connections and stranded mycelium investing mainly senescent, lower turf grass leaves, produced rings of faintly pink, downy mycelium, 80–300 mm in diameter in the sole of the turf on experimental plots of *Agrostis* sp. at the Sports Turf Research Institute (Fig. 30.3). A hymenial layer with small irregularly club-shaped basidia 7.5–28.0 μm × 4.0–7.5 μm was occasionally produced at grass leaf tips and on sand grains but no mature basidiospores were found. Several cases were seen. The fungus had no deleterious effect on turf which it had colonized (Smith, 1955). Fungicides based on phenyl mercury acetate eliminated these rings. A similar, non-sporing fungus was found in autumn 1971 (Jackson, 1972). A low-temperature tolerant Basidiomycete was found causing superficial rings in northeastern Scotland and such rings are common at snow melt in Saskatchewan (Smith, 1959, unpublished).

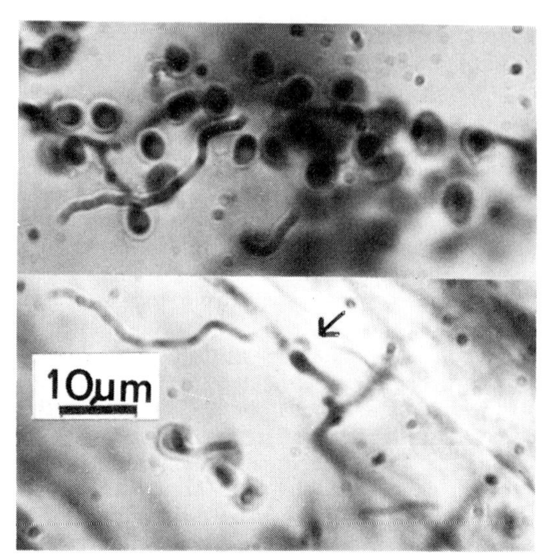

Figure 30.3 Conidia and mycelium with clamp connection from a superficial fairy ring.

Smith, Stynes and Moone (1970) reported Type B rings due to a sterile Basidiomycete. They took the form of white, circular patches which caused some turf disfigurement. The patches were formed on *Agrostis* turf following treatment with benomyl. In culture, growth of the fungus was stimulated by 0.5 and 1.0 ppm benomyl. Spotting and discolouration of dormant, overwintering bermuda grass following methyl bromide fumigation prior to establishment, was caused by a sterile Basidiomycete. This organism was not stimulated by benomyl (Dale, 1972). During autumn, winter and early spring 1971/72, circular patches of 150–380 mm in diameter with white mycelium felted on lower leaves and in the top 6 mm of the thatch were noticed on golf greens and fairways and on experimental turf of *Agrostis* spp, *Festuca rubra* and *Poa annua* at the Sports Turf Research

Institute in Yorkshire, England (Jackson, 1972). In the infested thatch, young roots developing at the nodes of creeping stems were stunted and discoloured. Felted mats of surface litter and algae contained copious crystals of calcium oxalate and carried on their undersurfaces the hymenial layers of the fungus. Hyphae were much branched, 1.3–3.3 μm in diameter with prominent clamps, sometimes swollen to 4.0–4.6 μm diameter. Basidia were variable, mostly short, cylindro-clavate, 5–14 μm × 1–5 μm with four slender, curved sterigmata, 2–3 μm long. Basidiospores were sub-globose to broadly oval, smooth, markedly apiculate, 2.0–3.0 × 2.5–3.5 μm usually with one oil droplet. It has been identified as *Trechispora alnicola* (Stalpers, 1983, personal communication). A Basidiomycete with conidia causes superficial rings with stimulated and discoloured zones on bent grass turf at Agassiz, British Columbia in autumn. It has also been found at Kingston, Rhode Island and Puyallup, Washington turf grass experiment grounds. Like the *Trechispora* sp. crystals of calcium oxalate are present in the inter-hyphal spaces (Smith, unpublished). Fairy ring polygons associated with thatch fungi have been noted on golf fairway turf in Alberta (Smith and Jackson, unpublished). *Marasmius siccus* colonizes senescent leaves and sheaths of *Poa pratensis* and *Lolium perenne* in Rhode Island. Affected areas of 500 mm or more in diameter appear off-colour and support large numbers of fruiting bodies (Jackson, unpublished).

Persistent, severe Type C rings were associated with *Coprinus kubickae* Pilat and Svrcek on *Agrostis stolonifera* 'Penncross' turf on a sand-base experimental golf green at Saskatoon (Redhead and Smith, 1981). The fungus caused no noticeable lesions on grass shoots but a growth of fine mycelium of the fungus developed in the turf base. The fungus fruited freely in culture on potato – malt agar. Benomyl had been used on the green to control snow moulds. Some turf grass pathogens such as *Microdochium nivale*, *Fusarium roseum* and *Rhizoctonia solani* may be associated with annular patches in turf. Patch centres may recover while the disease is progressing at the periphery. However, in these cases, unlike superficial fairy rings, the ring effect is produced mainly by an attack by the pathogen on the aerial parts of the grasses.

30.6.2 Significance and control of superficial fairy rings

The fungi associated with superficial fairy rings are essentially thatch inhabitants and many of the problems associated with the management of amenity turf can be related to the nature, depth and condition of the thatch layer (Adams and Saxon, 1979). The hydrophobic condition of the thatch and the soil immediately below it is 'referred to as dry patch' (Anon, 1962) or 'localized dry spots' (Beard, 1973). While these are usually regarded as problems on sand-based golf greens, they also occur on *Agrostis* turf formed on soil mixtures. In dry patches on sand greens, water penetration is greatly impeded and appears to be related to a cellular organic material coating individual coarse sand particles, rendering them water-repellant (Wilkinson and Miller, 1978). It is probable that this hydrophobic material is of fungal origin (Miller and Wilkinson, 1977; Parry, 1978 cited by Adams and Saxon, 1979) although the relationship of particular cap fungi to the colonization of soil particles on sand-based golf greens has not yet been demonstrated. Some of the superficial fairy ring fungi appear to be good candidates for further study as their fungal mycelium is often hydrophobic and they seem to be common on sand-based greens.

Although superficial fairy rings are litter and thatch inhabitants, little is known of the effect of different thatch management factors on their occurrence and persistence. Thatch management practices which restrict its accumulation and hence limit the substrate for these fungi, such as aeration, reduction of compaction in the mineral soil below the thatch, mechanical thatch removal and top dressing are likely to reduce ring incidence. Soil sterilization with methyl bromide, which probably had the effect of reducing competition from antagonists (Dale, 1972) and the use of a fungicide such as benomyl (Smith, Stynes and Moore, 1970) which may reduce competition from susceptible fungi, have both been implicated in the development of superficial Basidiomycete rings. The use of sand top-dressings on sand-based golf or bowling greens may lead to a low microbial population unable to decompose the organic matter produced by the grass or compete successfully with the superficial fairy ring fungi. Spot treatment with a wide spectrum fungicide such as phenyl mercury acetate has proved effective in controlling Type B rings.

30.7 REFERENCES

Adams, W.A. and Saxon, C. (1979) The occurrence and control of thatch in sports turf. *Rasen-Turf-Gazon*, **9**(3), 75–83.

Ainsworth, G.C., James, P.W. and Hawksworth, D.L. (1971) *Ainsworth and Bisbys' Dictionary of the Fungi.* 6th edn, Commonwealth Mycological Institute, Kew, 663 pp.

Anon. (1962) Dry patches. *J. Sports Turf Res. Inst.*, **10**(38), 467.

Bayliss, J.S. (1911) Observations on *Marasmius oreades* and *Clitocybe gigantea* as parasitic fungi causing fairy rings. *J. Econ. Biol.*, **6**, 111–32.

Bayliss-Elliott, J.S. (1926) Concerning fairy rings in pastures. *Ann. Appl. Biol.*, **13,** 227–88.
Beard, J.B. (1973) *Turfgrass Science and Culture.* Prentice-Hall, Englewood Cliffs, N.J., 658 pp.
Beard, J.B., Vargas, J.M. and Rieke, P.E. (1973) Influence of nitrogen fertility on *Tricholoma* fairy ring development in Merion Kentucky bluegrass (*Poa pratensis* L.). *Agron. J.*, **65,** 994–5.
Black, R.S. (1937) Fairy rings. *Aust. Greenkpr.*, **1,** 4, 12.
Coville, F.V. (1897) Observations on recent cases of mushroom poisoning in the district of Columbia. *US Dep. Agric. Div. Bot. Circ.*, **13,** 21 pp.
Cunningham, G.H. (1934) Control of fairy rings. *Greens Res. Committee NZ Golf Assoc. 2nd Ann. Rept.*, pp. 44–6.
Dahlsson, S-O. (1977) Control of fairy rings. *Weibulls Gras-tips*, **20,** 19–22.
Dale, J.L. (1972) A spotting and discoloration condition of dormant Bermudagrass. *Plant Dis. Rept.*, **56,** 355–7.
Dennis, R.W.G., Orton, P.D. and Hora, F.B. (1960) New check list of British agarics and boleti *Trans. Br. Mycol. Soc.* (*Supplement*), 225 pp.
Dickinson, C.H. (1979) Fairy rings in Norfolk. *Bull. Br. Mycol. Soc.*, **13,** 91–94.
Evans, E.J. (1967) A study of the fairy ring fungus *Marasmius oreades.* Ph.D. Thesis. Univ. Newcastle, England.
Filer, H.J. (1964) Parasitic and pathogenic aspects of *Marasmius oreades*, a fairy ring fungus. Ph.D. thesis. Wash. State Univ., 75 pp.
Filer, H.J. (1965) Damage to turf grasses caused by cyanogenetic compounds produced by *Marasmius oreades*, a fairy ring fungus. *Plant Dis. Rept.*, **49,** 571–4.
Filer, H.J. (1966) Effect on grass and cereal seedlings of hydrogen cyanide produced by mycelium and sporophores of *Marasmius oreades. Plant Dis. Rept.*, **50,** 264–6.
Gilbert, J.H. (1875) Note on the occurrence of fairy rings. *J. Linn. Soc. Bot.*, **15,** 17–24.
Gould, C.J., Austenson, H.M. and Miller, V.L. (1958). Fairy ring disease of lawns. *Wash. Agric. Exp. Sta. Circ.*, **330,** 5 pp.
Gould, C.J., Miller, V.L. and Polley, D. (1955) Fairy ring disease of lawns. *Golf Course Rept.*, **23**(8), 16–20.
Gregory, P.H. (1982) Fairy rings; free and tethered. *Bull. Br. Mycol. Soc.*, **16**(2), 161–3.
Hansen, T. and Schadegg, E. (1980) Experiments with insecticides, acaricides and fungicides in fruit and gardening crops 1978. *Tidsskriff for Planteavl.*, **84,** 55–74.
Halisky, P.M. and Peterson, J.L. (1970) Basidiomycetes associated with fairy rings in turf. *Bull. Torrey Bot. Club*, **97**(4), 225–7.
Heimes, R. and Loëcher, F. (1980) The possibility of controlling fairy ring and rust diseases in lawns with benodanil. In *Proc. 3rd Int. Turfgrass Res. Conf.* (ed. J.B. Beard), Am. Soc. Agron., Crop Sci. Soc. Am., Soil Sci. Soc. Am., Int. Turf grass Soc., pp. 283–91.
Howard, F.L., Rowell, J.B. and Keil, H.L. (1951) Fungus diseases of turf grasses *Agric. Expt. Sta. Univ. Rhode Is.*, 56 pp.
Ingold, C.T. (1974) Growth and death of a fairy ring. *Bull. Br. Mycol. Soc.*, **8**(2), 74.
Jackson, N. (1972) Superficial invasion of turf by two basidiomycetes. *J. Sports Turf Res. Inst.*, **48,** 20–3.
Karlson, K. (1975) Fairy rings in grassland controlled by new material. *Viola-Tradsgardsvarlden* (8 May), **19,** 13 (Swedish).
Lawes, J.B., Gilbert, J.H. and Warrington, R. (1883) Contribution to the chemistry of fairy rings. *J. Chem. Soc. Lond.*, **43,** 208–23.
Lebeau, J.B. and Hawn, E.J. (1961) Fairy rings in Alberta. *Can. Plant Dis. Surv.*, **41,** 317–20.
Lebeau, J.B. and Hawn, E.J. (1963a) Formation of hydrogen cyanide by the mycelial stage of the fairy ring fungus. *Phytopathology*, **53,** 1395–6.
Lebeau, J.B. and Hawn, E.J. (1963b) A simple method for the control of fairy ring caused by *Marasmius oreades. J. Sports Turf Res. Inst.*, **11**(39), 23–5.
Mathur, S.P. (1970) Degradation of soil humus by the fairy ring mushroom. *Plant Soil*, **33,** 717–20.
McAlpine, D. (1898) *Fairy Ring and the Fairy Ring Puffball.* Melbourne, 12 pp.
Miles, N. and Mathur, S.P. (1972) Seasonal incidence of anhydrous uric acid granules in the collision zone of two fairy rings. *Can. J. Soil Sci.*, **52,** 515–17.
Miller, O.K. Jr., Farr, D.F. (1975) An index of the common funngi of North America (Synonomy and common names). *Bibliotheca Mycologica,* **44,** J. Cramer, Vaduz, 206 pp.
Miller, R.H. and Wilkinson, J.F. (1977) Nature of organic coating on sand grains of non-wettable golf greens. *Soil Sci. Soc. Am. Proc.*, **41,** 1203–4.
Moliard, M. (1910) On the action of *Marasmius oreades* Fr. on the vegetation. *Bull. Soc. Bot. France*, **57,** 62–9 (French).
National Institute for Occupational Safety and Health (1981) Formaldehyde: Evidence of carcinogenicity. *US Dept. Health Human Services. Curr. Intelligence Bull.*, **34** (Apr. 15), 15 pp.
Norstadt, F.A., Frey, C.R. and Wilhite, F.M. (1968) *Marasmius oreades* (Bolt.) Fr. Effects on mineralizable nitrogen and carbon in soil. *Abstr. Am. Soc. Agron.*, **95.**
Parker-Rhodes, A.J. (1955) Fairy ring kinetics. *Trans. Br. Mycol. Soc.*, **38,** 59–72.
Parry, R. (1978) (Unpublished) The effect of pH on soil fungal and bacterial populations and the synthesis of hydophobic substances. BSc dissertation, University College of Wales, Aberystwyth (cited by Adams and Saxon, 1979).
Ramsbottom, J. (1927) Fairy rings. *J. Quekett Microscop. Club. Sr.* 2, **15**(92), 231–42.
Ramsbottom, J. (1953) *Mushrooms and Toadstools.* Collins, New Naturalist Series, 306 pp.
Redhead, S.A. and Smith, J.D. (1981) A North American isolate of *Coprinus kubickae* associated with a superficial fairy ring. *Can. J. Bot.*, **59**(3), 410–14.
Roediger, H. (1978) Witches ring caused by 'Nelken Schwindling' (*Marasmius oreades*). *Rasen-Turf-Gazon*, **3,** 60–2. (German, En., Fr.).
Shantz, H.L. and Piemeisel, R.L. (1917) Fungus fairy rings in eastern Colorado and their effect on vegetation. *J. Agric. Res.*, **11,** 191–245.

Smith, A.M., Stynes, B.A. and Moore, K.J. (1970) Benomyl stimulates the growth of a Basidiomycete on turf. *Plant Dis. Rept.*, **54,** 774–5.

Smith, J.D. (1955) Turf disease notes (1955) The control of fairy rings in sports turf. *J. Sports Turf Res. Inst.*, **9,** 62–8.

Smith, J.D. (1957) Fungi and turf diseases. 7: Fairy rings. *J. Sports Turf. Res. Inst.*, **33,** 324–52.

Smith, J.D. (1959) Turf diseases in the North of Scotland. *J. Sports Turf Res. Inst.*, **10**(35), 42–6.

Smith, J.D. (1972) Marasmius rings: Lawn age and incidence. *J. Sports Turf Res. Inst.*, **48,** 24–7.

Smith, J.D. (1975) Incomplete *Marasmius oreades* fairy rings. *J. Sports Turf Res. Inst.*, **51,** 41–5.

Smith, J.D. (1978a) Control of *Marasmius oreades* fairy rings: A review of methods and new approaches to their elimination. *J. Sports Turf Res. Inst.*, **54,** 106–14.

Smith, J.D. (1978b) Fairy ring biology and control. *Canadex. Can. Dept. Agric.* **273,** 630, 3 pp.

Smith, J.D. (1980a) Is biologic control of *Marasmius oreades* fairy rings possible? *Plant Dis.*, **64**(4), 348–54.

Smith J.D. (1980b) Fairy rings: Biology, antagonism and possible new control methods. In *Advances in Turfgrass Pathology* (eds P.O. Larsen and B.G. Joyner), Harcourt, Brace Jovanovich, New York, pp. 81–5.

Smith, J.D. (1987) Chemical control of *Marasmius oreades* fairy rings. Proc. 8th Ann. Mtg. Plant Path Soc. Alberta, 28–30 Oct. Lacombe Res. Sta. Agric. Can. (Abs.) (in press).

Smith, J.D. and Jackson, N. (1981) Superficial fairy rings in amenity turf. In *Proc. 4th Int. Turgrass Res. Conf. Ontario*, (ed. R.W. Sheard), Agric. Coll. and Int. Turf grass Soc.

Smith, J.D. and Rupps, R. (1978) Antagonism in *Marasmius oreades* fairy rings. *J. Sports Turf Res. Inst.*, **54,** 97–105.

Tahama, Y. (1980) On the ring of fairy rings on turf grasses. *J. Jap. Turfgrass Res. Assoc.*, **9**(2), 133–6. (Japanese, En. *R.A.P.P.* **60,** 12, 6517 (1981)).

Thomas, F. (1905) The rate of growth of a fungus ring of *Hydnum suavolens* Scop. *Ber. Deutsch. Bot. Gesell.*, **23**(9), 476–8.

Traquair, J.A. and McKeen, W.E. (1986) Fine structure of root tip cells of winter wheat exposed to toxic culture filtrates of *Coprinus psychromorbidus* and *Marasmius oreades*. *Can. J. Plant Pathol.*, **8,** 59–64.

Wakefield, E.M. and Dennis, R.W.G. (1950) *Common British Fungi*, P.R. Gawthorn, London.

Warcup, J.H. (1951) Studies in the growth of basidiomycetes in soil. *Ann. Bot. Land. NS*, **15,** 305–17.

Waterhouse, G.M. (1957) The larger fungi in lawns. *Trans. Lincolnsh. Naturalists Union*, **14**(2), 74–85.

Weaver, T. (1975) Fairy ring fungi as decomposers. *Proc. Montana Acad. Sci.*, **35,** 34–8.

Wilkinson, J.F. and Miller, R.H. (1978) Investigation and treatment of localised dry spots on sand golf greens. *Agron. J.*, **70**(2), 299–304.

Wolf, F.T. (1971) An unusual occurrence of 'fairy rings'. *Mycologia*, **63,** 671–2.

Wollaston, W.H. (1807) On fairy rings. *Phil. Trans. Roy. Soc. Land.*, **2,** 133–8.

Part eleven
Slime moulds and lichens

31 *Slime moulds*

The inconspicuous plasmodia of slime moulds may be encountered in turf. Later, the much more conspicuous patches of sometimes brightly coloured sporangia (individually 1–2 mm or so in diameter) or other spore masses of these organisms may be seen on the aerial parts of turf grasses and turf weeds or in the turf in damp or wet weather (Fig. 65, colour plate section), often from late summer to early winter. With the return of dry conditions the sporangia break down exposing the powdery spore masses and accessory sporangial structures.

31.1 CAUSAL ORGANISMS AND THEIR BIOLOGY

Slime moulds are usually classed as Myxomycetes. The relationship of this class of organisms to animals or plants is still in dispute since they have some characters of each. Mycologists often claim that they are 'primitive' fungi. They have a free-living, vegetative, jelly-like stage, the plasmodium, with a wide range of light or dark colours. The plasmodium has a non-cellular skin. It was described by Macbride (1922) as 'a simple mass of protoplasm, destitute of cell walls, protean in form and amoeboid in its movements'. The plasmodium is multinuclear, but whether it should be regarded as one multinucleate cell or a multicellular organism which has lost its cell walls is a matter for dispute (Martin, 1957). The inner part of the plasmodium shows reversible protoplasmic streaming and the whole structure can creep (flow) slowly over or up substrates, ooze through narrow gaps, perhaps dividing into fingers or branches in its progress. Plasmodia are holozoic and feed on living bacteria, fungi or dead organic matter (Martin, Alexopoulos and Farr, 1983).

The plasmodium may enter a resting condition when environmental conditions are unfavourable for further movement or development e.g. when temperatures are low or moisture is lacking (Macbride, 1922). In those species whose spore masses are found in patches on the aerial parts of turf grasses, under suitable environmental conditions, the plasmodia flow up the shoots and break up to form walled and sometimes stalked sporangia. Each of these sporangia contains many spherical elliptical or slightly angled, unicellular spores often in the range <5–20> μm. The sporangia mature and dry up, the peridial wall breaks and the spores are free to be dispersed leaving behind a network structure, the capillitium (Martin and Alexopoulos, 1969). Spore dispersal may be by rain, wind, animals and traffic of various sorts. The slightly elevated condition of sporangia probably facilitates spore dispersal. Under suitable environmental conditions, perhaps after a period of dormancy, the spores germinate in the turf to give 1 to 4 uninucleate amoebae (which may be flagellated) which obtain their nutrition by ingesting living organisms and organic food materials phagocytically. They may survive encysted for some time or they may fuse in pairs and develop into small plasmodia (Collins, 1979; Martin, Alexopoulos and Farr; 1983; Macbride, 1922).

In Britain, *Physarum* spp. with grey sporangia are commonly seen in turf: in Saskatchewan, a *Physarum* species with lilac-grey sporangia (Fig. 65, colour plate section) is usual in autumn and in northern France a member of this genus with bright-yellow sporangia was collected in October 1984 (Smith, 1985 and unpublished). When the sporangia mature and burst, dark-brown spores are exposed. The plasmodia of *Mucilago spongiosa* (Adanson) Morg. and *Didymium crustaceum* have been reported in turf in Britain (Smith, 1965).

31.2 EFFECTS ON TURF

The plasmodia probably have little effect on the turf grasses. The sporangia may give a spectacular appearance to the grass when they occur in patches 10 or more cm in diameter. Living grass tissues are not attacked, although the grass litter supplies the substrate for the development of the microorganisms which in turn supply much of the food material for the slime moulds. The sporangia or other sporing structures probably cause some grass shading, but rarely are symptoms of this noticeable when these structures are removed.

31.3 CONTROL

Where infestation with slime moulds is light, control measures are unnecessary, because they rapidly disappear when drier weather returns. Where an outbreak is heavy, sporangia or other sporing structures can be removed from the grass by mowing, switching, brushing and hosing down with water. Although their appearance is rather unpredictable, they are more likely to be a problem where thatching is heavy and the drainage is inadequate. Attention to these problems will reduce the likelihood of their development. Chemical control is rarely necessary and not usually recommended, but a light dressing of calcined ferrous sulphate will assist in their removal (Smith, 1965).

31.4 REFERENCES

Collins, O.R. (1979) Myxomycete systematics: Some recent developments and future research opportunities. *Bot. Rev.*, **45**(2), 145–201.

Macbride, T.H. (1922) *The North American Slime moulds.* 2nd edn, Macmillan, London, 347 pp.

Martin, G.W. (1957) Concerning the 'cellularity' and the acellularity of the Protozoa. *Science*, Ny, **125,** 155.

Martin, G.W. and Alexopoulos, C.J. (1969) *The Myxomycetes.* University of lowa Press, Iowa City, 560 pp.

Martin, G.W., Alexopoulos, C.J. and Farr, M.L. (1983) *The genera of Myxomycetes.* 2nd edn, University of Iowa Press, Iowa City, 102 pp and 41 plates.

Smith, J.D. (1965) *Fungal diseases of turfgrass.* 2nd edn, (Revised N. Jackson and J.D. Smith), The Sports Turf Research Institute Bingley, W. Yorks, England, 97 pp.

Smith, J.D. and Evans, I.R. (1985) *Major diseases of turf grasses in Western Canada.* Alberta Agriculture, Agdex 273/636–3, 15 pp.

32 *Lichens*

These are not fungi, but composite organisms of fungi, nearly always Ascomycetes, and usually a green or blue-green, unicellular or filamentous alga. The association is to their mutual benefit, since the alga gets its mineral salts and water from the fungus and the fungus benefits from the photosynthetic activities of the alga. The species of lichen fungi occur only in association with algae, but the algal species may grow independently. Of the different types of lichen structure the foliose or leafy lichens are the kind most commonly found in lawn turf.

In Britain, the most common leafy lichen, found in neglected lawns, of low fertility, on moist, or seasonally droughted, thatched turf or turf shaded by trees, is *Peltigera canina* Willd. This is a very variable species, also widespread through North America (Hale, 1979). The body or thallus consists of overlapping leafy structures, white and with raised veins and rhizines on the under surface, dull and brown or greenish-brown on the upper. Erect reproductive structures, or apothecia, are often present. The lichen may also reproduce from lobules of tissue which are torn or broken off from the edges of the thallus. It grows intermixed with turf grasses and turf weeds.

32.1 CONTROL

This is best effected by scarification to remove the lichen followed by operations designed to increase grass vigour, such as coring, thatch removal, surface cultivation, top dressing, liming if necessary, and application of the appropriate complete fertilizer (Smith, 1965).

32.2 REFERENCES

Hale, M.E. (1979) *How to know lichens*. 2nd edn, W.C. Brown, Dubuque, Iowa, 246 pp.

Smith, J.D. (1965) Fungal diseases of turfgrass. 2nd edn, (revised N. Jackson and J.D. Smith), The Sports Turf Research Institute Bingley, W. Yorks, England, 97 pp.

Part twelve
Summary

33 *Key to common turf diseases caused by fungi*

		Go to this number in left column	
1	Disease of seedling turf	2	
1a	Disease of established turf	3	
2	Grass seedlings fail to emerge	SEED ROT	(p. 29)
2a	Grass seedlings collapse and rot after emergence, mycelium often not apparent	DAMPING OFF	(p. 29)
3	Mycelium often observable on diseased turf	4	
3a	Mycelium generally not observable	25	
4	Mycelium occurring during the growing season	9	
4a	Mycelium occurring on winter dormant turf evident most often at the spring thaw. Low temperature snow moulds	5	
5	Mycelium white to grey, forming sclerotia on or in dead plant tissues	6	
5a	Mycelium white or grey-white, cottony, to sparse, sclerotia absent	8	
6	Sclerotia uniformly ovate to globose, initially white or pink turning black or brown, up to 2–5 mm in size. Patches straw coloured or bleached localized or extensive.	TYPHULA (GREY AND SPECKLED) SNOW MOULDS	(p. 75)
6a	Sclerotia irregular in shape	7	
7	Sclerotia globular to irregular and arched, elongate up to 8 mm × 4 mm, initially cream turning black. Patches with leaves bleached, wrinkled and threadlike.	SNOW SCALD OR SCLEROTINA SNOW MOULD	(p. 88)
7a	Sclerotia irregular, flattened, initially white to grey turning grey–brown to black, up to 1.5 mm. Patches bleached, localized or extensive, adjacent patches often not coalescing completely.	COPRINUS (SCLEROTIAL) SNOW MOULD	(p.67)
8	Mycelium often cottony white, dirty white to grey, no sclerotia present. Patches as in 7a.	COPRINUS (NON-SCLEROTIAL) SNOW MOULD	(p. 73)
8a	Mycelium dirty white to faintly pink, no sclerotia present. Patches roughly circular with bleached centre and tan to pink margins, localized or extensive.	MICRODOCHIUM (PINK) SNOW MOULD	(p. 59)
9	Mycelium grey or dark coloured	10	
9a	Mycelium hyaline, white, pink or red coloured	16	
10	Mycelium grey or dark coloured, sclerotia forming in plant debris	11	
10a	Mycelium grey or dark coloured, sclerotia absent	12	
11	Mycelium grey to fawn, sclerotia globular, tan to brown, up to 1.5 mm. Yellow to brown rings, circular patches and/or irregular streaks localized or extensive. Cool, wet-weather disease.	YELLOW PATCH	(p. 272)

11a	Mycelium grey to brown, sclerotia variable up to 5 mm or more, brown to black. Patches circular to irregular, brown, often extensive: margin of grey wilted grass forms an ephemeral 'smoke ring' round patches on low cut turf. Hot, wet and humid weather disease.	LARGE BROWN PATCH	(p. 266)
12	Mycelium grey, forming dark aggregates on blackened stem bases, dark, club shaped appressoria prominent. Cool-season basal rot of *Poa annua* and *Agrostis* (see 34).	ANTHRACNOSE AND BASAL ROT	(p. 219)
12a	Mycelium light to dark brown, appressoria absent	13	
13	Mycelium light brown, hyphopodia present	14	
13a	Mycelium dark brown, hyphopodia absent	15	
14	Mycelium as light-brown runner hyphae with hyphopodia, on roots, creeping stems and crowns, forming dark mats between leaf sheaths and stem bases. Patches straw to brown, circular or frog eyed, with bronzed-yellow margins, localized or extensive cool-season disease of *Agrostis* spp. predominantly.	TAKE-ALL PATCH	(p. 137)
14a	Mycelium as in 14. Patches similar to 24a. Localized or often extensive hot-season disease of *Poa* species predominantly.	SUMMER PATCH	(p. 133)
15	Mycelium as in 14 but dark brown, hyphopodia absent. Patches similar to 14. Localized or extensive cool-season disease of *Poa pratensis* predominantly.	NECROTIC RING SPOT	(p. 129)
15a	Mycelium as in 14. Patches circular or frog-eye shaped dead, grass completely bleached. Often extensive at spring greening on *Cynodon* spp.	SPRING DEAD SPOT	(p. 124)
16	Mycelium hyaline, white, pink or red, sclerotia formed in plant debris	17	
16a	Mycelium hyaline, white, pink or red, sclerotia absent	18	
17	Mycelium white, sclerotia numerous, roughly spherical, cream to brown, 1–2 mm diameter. Patches reddish brown, crescent or ring shaped, localized or extensive.	SOUTHERN BLIGHT	(p. 276)
17a	Mycelium hyaline to white, sclerotia black, minute, 0.1–0.2 mm in size, embedded in necrotic tissue. Patches small but numerous, coppery orange.	COPPER SPOT	(p. 227)
18	Mycelium white to buff or hyaline	19	
18a	Mycelium pink or red	22	
19	Mycelium white to buff, superficial, persistent and powdery. No distinct patches, generalized turf yellowing.	POWDERY MILDEW	(p. 247)
19a	Mycelium white or hyaline, ephemeral and not powdery	20	
20	Mycelium white and cottony. Distinct patches or irregular streaks with 'greasy' margins of dark water-soaked grass, collapsed plants form matted crusts. Often extensive.	PYTHIUM BLIGHTS	(p. 252)
20a	Mycelium hyaline, fine and cobweb-like	21	
21	Mycelium hyaline, fine and cobweb-like, non-sporulating. Patches small but numerous, eventually bleached, leaf lesions yellow to bleached with dark margins.	DOLLAR SPOT	(p. 230)
22	Mycelium pale pink to pink, in gelatinous strands drying to form brittle, antler or needlelike red stromata. Patches variable, irregular, straw-coloured, later pink to red, coalescing to involved large areas.	RED THREAD	(p. 259)
22a	Mycelium pale pink to pink, stromata absent	23	
23	Mycelium pale pink, stranded, matting leaves together. Patches circular to irregular, straw-coloured to pale pink, localized or extensive.	PINK PATCH	(p. 259)

23a	Mycelium pale pink, not stranded or gelatinous	24	
24	Mycelium white to faintly pink. Patches small, roughly circular with bleached centres and tan to pink margins. Often extensive during cool moist conditions and as a snow mould (see 6).	MICRODOCHIUM PATCH	(p. 59)
24a	Mycelium pink to red. Leaf lesions, when present reddish brown to dull tan often with bleached centres. Patches circular or irregular often frog-eye shaped, light tan to straw coloured. Often extensive during hot humid conditions on *Poa pratensis*.	FUSARIUM LEAF BLIGHT, AND/OR ROOT AND CROWN ROT	(p. 117)
25	Mycelium not immediately observable but present in thatch layer and soil, basidiocarps (mushrooms, toadstools and puffballs, etc.) present or absent	26	
25a	Mycelium generally not discernible	27	
26	Mycelium superficial in rings or arcs, around plant bases or in thatch, often without fruits (basidiocarps) and often without pronounced turf symptoms.	SUPERFICIAL FAIRY RINGS	(p. 350)
26a	Mycelium present in thatch and sometimes in soil below. May be associated with turf injury or growth stimulation of grass in rings or arcs, often with fruits (basidiocarps)	FAIRY RINGS	(p. 341)
27	Mycelium generally not observable but fruiting bodies present on the affected plants	28	
27a	Mycelium generally not observable, fruiting bodies absent	38	
28	Fruiting bodies present on apparently healthy plants	29	
28a	Fruiting bodies present on diseased leaves and stems	30	
29	Sporangia, dark coloured, crowded and encrusting leaves. Patches irregular in shape, grey or black. Localized.	SLIME MOULDS	(p.357)
29a	Sporangia, on leaves, pearly white and pyriform when turgid. Infected plants tiller profusely and may yellow. Localized or extensive.	DOWNY MILDEW (YELLOW TUFT)	(p.238)
30	Fruiting bodies brown or black, globose or flask-shaped structures on diseased plants	31	
30a	Fruiting bodies yellow, orange, pink, brown or black pustules on diseased plants	34	
31	Brown or black globose pycnidia, in straw-coloured or bleached leaf lesions. Sometimes without distinct patches, leaves yellowed or flecked.	LEAF BLIGHTS (PYCNIDIAL FUNGI)	(p. 312)
31a	Black perithecia or pseudothecia, globose or flask shaped, neck length variable in or on diseased tissues	32	
32	Perithecia globose, necks short, common in necrotic tissues especially in leaf laminae. Patches indefinite or lacking.	LEPTOSPHAERULINA LEAF BLIGHT	(p. 325)
32a	Perithecia or pseudothecia globose to flask shaped with distinct necks in or on leaf sheaths, crowns and roots. Patches very definite circular or frog-eye in shape	33	
33	Perithecia flask shaped with long often curved protruding necks, embedded in dying leaf sheath bases, crows and roots (see 14).	TAKE-ALL PATCH	(p. 137)
33a	Pseudothecia globose, necks variable in length usually short, gregarious on necrotic crowns and rhizomes, singly on roots (see 15).	NECROTIC RING SPOT	(p. 129)
34	Fruiting pustules linear, elongated, grey to black sori, leaves shredding and curling. No distinct patches turf ragged, thinned and off colour.	FLAG AND STRIPE SMUT	(p. 183)
34a	Fruiting pustules generally discrete, scattered on blighted leaves	35	

35	Pustules black, scab-like acervuli with setae. Symptoms variable, ranging from indistinct patches and thinning to extensive yellow-bronze to brown areas of dead turf, especially *Poa annua* (see 12).	ANTHRACNOSE	(p. 219)
35a	Pustules as sori or sporodochia	36	
36	Pustules as yellow, orange-brown or black, erumpent sori. No distinct patches, flecking or yellowing of turf.	RUSTS	(p. 157)
36a	Pustules as sporodochia	37	
37	Sporodochia cream to pink (see 8a, 24, 24a)	MICRODOCHIUM PATCH OR BLIGHT	(p. 59 or
37a	Sporodochia coppery orange (see 17a)	COPPER SPOT	117)
38	Irregular yellow-green brown patches in thinned turf, individual plants showing leaf spot, crown, and/or root rot symptoms. On *Poa pratensis*.	LEAF SPOT AND MELTING OUT	(p. 297)
38a	Leaf spots, leaf and crown blighting causing thinning, patching or deterioration of turf.	DRECHSLERA BIPOLARIS CURVULARIA PYRICULARIA AND OTHER LEAF SPOTS	(pp. 244,281 and 321)

34 *Summary of pathogens, hosts and controls*

In the following table we report information on the effectiveness of fungicides applied experimentally and in practice. *These data should be used as a guide to their use only* since the effectiveness of these chemicals against particular diseases may show considerable variation, depending on the prevailing conditions. Registration requirements often require both pathogen and plant species (or crop) to be specified and some chemicals may not be registered for use in particular jurisdictions even though they have been shown to be effective experimentally for the control of a particular disease. *Always observe the instructions given on the package relating to usage and precautions to be taken with the pesticide.*

Table 34.1 Summary of pathogens, hosts and controls

Disease	Pathogen	Suscepts	Cultural control	Resistant species and cultivars	Fungicidal control
Anthracnose/ foliage blight	*Colletotrichum graminicola (Glomerella graminicola)*	At high temperatures: Fine-leaved *Festuca* spp., *Poa* and *Agrostis* spp., *Lolium perenne*	Restrict irrigation during hot, humid weather to morning applications. Preventive fungicides are necessary.	Use less susceptible *F. rubra* cultivars. *Poa annua* selections differ markedly in susceptibility	See HAS syndrome (below)
Anthracnose/ basal rot	*C. graminicola*	At cool temperatures: mainly *Poa annua* but also: *A. stolonifera*, *A. canina*	Reduce traffic. Relieve soil compaction by aerifying or spiking. Replace *P. annua* with other turf grasses	Other species less severely affected than *P. annua*	Combinations of triadimefon or benomyl with chlorothalonil reduce severity, especially combined with nematicide.
'HAS' syndrome (Helminthosporium–Anthracnose–Senescence)	*Bipolaris sorokiniana (Cochliobolus sativus)* *C. graminicola* on senescing tissues	At high temperatures *Poa annua*	Provide cultural conditions to suit the agronomic character of the grass.	There are considerable differences in susceptibility between cultivars of commonly used cool-season turf grasses.	Chlorothalonil, mancozeb, maneb + zinc sulphate, benomyl, thiophanate-ethyl or methyl at 7–14 day intervals July–Sept. Triadimefon very effective.
Copper spot	*Gloeocercospora sorghi*	*Agrostis canina* *A. castellana* *A. stolonifera* *A. tenuis* *Cynodon dactylon*	Reduce acidity of soil with lime. High nitrogen fertility increases susceptibility of *A. canina*.	*A. canina* cultivars are much more susceptible than those of other *Agrostis* spp.	Chlorothalonil, amilazine, iprodione and cadmium fungicides; benomyl, triadimefon, thiabendazole, methyl and ethyl thiophanate give longest protection.
Damping off	*Pythium* spp. *Fusarium* spp. *'Helminthosporium'* spp.	All turf grasses	Pay careful attention to seed bed preparation. Balanced fertilizer		Use seed dressed with captan, thiram,

	Rhizoctonia spp.		and correct pH. Use correct seed rate and good quality seed. Do not seed if soil is too cold or too wet. Do not plant seed too deep.		carbendazim, drazoxolon, calcium peroxide, etridiazole, organo-mercurials.
Dollar spot	*Sclerotinia homoeocarpa*	In Britain on *Festuca rubra*, spp. *rubra*. In continental Europe also on *Agrostis* spp. Elsewhere on *Agrostis* spp. and many other cool and warm-season grasses.	Maintain vigorous growth in spring and summer with judicious nitrogen applications. Remove morning dew, but maintain soil moisture near field capacity. Permit free 'air drainage'.	In Britain avoid use of *Festuca rubra* cultivars of sea marsh origin. Elsewhere, resistance of cultivars of turf grass species influenced by locality, management, herbicide use.	Tolerance to anilazine, benzimidazoles, cadmium and iprodione fungicides has developed in some locations. Triadimefon, fenarimol, chlorothalonil and propiconazole give excellent control.
Downy mildew (Yellow tuft)	*Sclerophthora macrospora*	*Agrostis* spp. *Poa pratensis*, *Stenotaphrum secundatum* and many other grass species.	Maintain good soil drainage and keep soil saturation to a minimum especially in new seedings. Eradicate alternative hosts. Avoid excessive use of nitrogen	Resistant species and cultivars of turf grasses not yet available.	Metalaxyl soil drenches
Fairy rings	*Marasmius oreades*, *Lycoperdon*, *Agaricus* and *Lepiota* spp. and other Basidiomycetes.	All turf grasses	Reduce thatch accumulation. Replacement of infested soil or soil fumigation where practical. Repeated irrigation. Biological control for *M. oreades* rings.	Not applicable	Fungicidal control difficult. Some success reported for soil drenches with carboxin, benodanil, triadimefon. Fumigation with formaldehyde, metam-sodium, chloropicrin, methyl bromide.

Table 34.1 Continued

Disease	Pathogen	Suscepts	Cultural control	Resistant species and cultivars	Fungicidal control
Fusarium foliar blight, crown and root rot	*Fusarium culmorum* *Fusarium poae* and *Fusarium* spp.	Principally *Poa pratensis* cultivars in situations of full sun.	Judicious use of nitrogen in spring. Prevention and alleviation of drought-stress in hot, dry periods. Improvement of 'air drainage'. Removal of clippings. Prevention of thatch build-up. Preventive fungicide applications are necessary.	Resistance of *P. pratensis* cultivars related to their ability to tolerate stresses. Select cvs. well adapted to regional conditions. Use *Lolium perenne* in mixture with *P. pratensis*. Renovate affected bluegrass turf with *L. perenne*.	Tolerance to benzimidazole fungicides develops rapidly. Preventive applications of triadimefon, propiconazole or fenarimol.
Cercospora and *Pseudocercos-porella* leaf and culm spots	*Cercospora fusimaculans* *C. festucae* *Cercospora* spp. *Pseudocercos-porella herpotrichoides*	Many warm-season grasses, mainly on *Stenotaphrum secundatum* *Festuca* spp. Cool- and warm-season grasses Cereals, wild and cultivated grasses	In *C. fusimaculans* improve fertility with N fertilizer and use fungicide.	In *C. fusimaculans* yellow–green types of *S. secundatum* are less resistant that blue–green ones. Pathotypes of *P. herpotrichoides* occur.	Thiobendazole, thiophanate-methyl, anilazine, chlorothalonil
Necrotic ring spot	*Leptosphaeria korrae*	Primarily *Poa pratensis* turf 2–4 years after establishment, especially sodded lawns. Symptoms can occur in shaded areas.	Not fully determined. Reducing surface pH by applying ammonium sulphate may aid recovery. Maintain turf vigour	Not determined.	Benomyl, fenarimol, iprodione, propiconazole, preventive applications in early May to June.
Nigrospora blight	*Nigrospora sphaerica*	Common saprobe in turf. Pathogenic on stressed *Lolium perenne*,	Minimize drought stress by sound water management. Raise height of cut.	Differences in cultivar susceptibility have been noted.	Quintozene, iprodione, chlorothalonil.

		Festuca rubra, Poa pratensis and *Stenotaphorum secundatum* turf			
Powdery mildew	*Erysiphe graminis*	All commonly used turf grasses, but especially *Poa pratensis*.	Improve airflow over turf and reduce shading. Raise height of mowing, mow more frequently. Avoid excessive use of nitrogenous fertilizers. Substitute *Festuca rubra* for *Poa pratensis*. Use a preventive fungicide application where problem is recurrent.	*Festuca rubra* cultivars are generally much more resistant than many of *P. pratensis*. Resistant cultivars of the latter are available but inconsistently so, largely because of different physiological races of the pathogen.	Microfine sulphur dusts or wettable powders, dinocap. Systemic materials such as benomyl, thiophanates, ethirimol, fenarimol, propiconazole, triarimol, triadimefon, tridemorph and triforine show activity but fungus may show tolerance. Alternate unrelated materials.
Low-temperature diseases					
Microdochium patch and pink snow mould	*Microdochium nivale* (*Monographella nivalis*)	*Poa annua* and *Agrostis* spp. are particularly prone to the patch disease and the snow mould. Most cool-season turf grasses are susceptible under prolonged snow cover.	Improve 'air drainage'. Dissipate surface moisture. Control fertility, particularly avoiding nitrogen applications late in the growing season. Avoid sudden changes to alkaline turf surface. Exclude or eliminate *P. annua* if possible.	*Poa pratensis* and *Lolium perenne*, polystands of *Agrostis* and fine-leaved *Festuca* spp., or some of the less-susceptible stolonized *Agrostis* spp. may reduce liability to attack by the disease.	Inorganic and organic mercurials (where permitted), quintozene, chlorothalonil, maneb, mancozeb, benomyl, thiophanate methyl, cadmium compounds (where permitted), dichlorophen, triadimefon, iprodione, and thiobendazole.

Table 34.1 Continued

Disease	Pathogen	Suscepts	Cultural control	Resistant species and cultivars	Fungicidal control
Grey or speckled snow mould or typhula blight	*Typhula incarnata* and *T. ishikariensis* and varieties.	Most cool-season grasses are susceptible, but *Agrostis* and *Lolium* spp. are particularly prone.	Similar to microdochium patch and pink snow mould. Remove snow drifts in spring.	*Agrostis stolonifera* cvs. least resistant; *A. canina* cvs. usually intermediate; *A. tenuis* cvs. less susceptible. Resistance available in some *Poa pratensis* cvs.	*T. incarnata*. Inorganic mercury chlorides, phenyl mercuric acetate, cadmium succinate, chloroneb, quintozene, triadimefon, iprodione, chlorothalonil. *T. ishikariensis*. As for *T. incarnata*, but at higher dosage and/or more frequent applications.
Sclerotinia snow mould or snow scald	*Myriosclerotinia (Sclerotinia) borealis*	Most cool-season turf grasses, but particularly fine-leaved *Festuca* and *Agrostis* spp.	Although grass susceptibility increases with increasing N, adequate amounts, balanced with P are needed to ensure quick spring recovery. Maintain neutral turf surface pH. Speed spring snow drift removal or melting.	Cold-tolerant cultivars of northern origin are inherently more resistant than southern.	Quintozene, benomyl, methyl thiophanate, phenyl mercuric acetate + thiram, thiram. Inorganic mercury chlorides often ineffective.
Coprinus (cottony) snow mould. (Low-temperature basidiomycete or LTB and SLTB snow mould)	*Coprinus psychromorbidus*	Most cool-season turf grasses are susceptible, but *Poa annua* and most cvs. of fine-leaved *Festuca* and *Agrostis* spp. are very	Avoid heavy usage of nitrogenous fertilizer generally and adjust dosage to state of growth in summer. Do not apply N in growing season later than August, but N and P as dormant	Use cultivars of northern origin, especially those which go dormant early in autumn. *P. pratensis* cvs. vary greatly in susceptibility. No *Agrostis* cvs. will withstand heavy	Inorganic mercury chlorides are most effective. Quintozene, chloroneb, triadimefon, thiram, thiram + carbathiin + oxycarboxin.

		susceptible.	applications assist in spring recovery from disease. Exclude *P. annua* from turf	attacks without fungicidal protection. *Agropyron cristatum* is rarely damaged.	Usually, at least two fungicide applications are needed in autumn and early winter to give control.
Brown root rot	*Phoma sclerotiodes*	Most cool-season turf grasses which have suffered previous low-temperature injury may be attacked.	Not known	Use low-temperature resistant species and cultivars preferably of northern origin.	Not known
	Acremonium boreale (Nectria tuberculariformis)	Weakly pathogenic on unacclimated grasses. Antagonistic to other snow moulds.	Not needed		Not needed
String-of-pearls disease	*Sclerotium rhizoides*	*Poa pratensis*, *Agrostis* spp., fine-leaved *Festuca* spp.	Improve fertility and raise soil pH. Collect infected material	Not known	Inorganic mercury chlorides
Rusts	Mainly *Puccinia* and *Uromyces* spp.	All amenity turf grass spp.	Maintain grass vigour with judicious fertilization (especially N) and irrigation during stress. Remove inoculum by frequent mowing and clipping collection, but do not mow too closely.	Cultivars of *Poa pratensis* resistant to *P. graminis*, *P. striiformis*, *P. poae-nemoralis* and *P. poarum* are available, and in *Lolium perenne* and *Festuca arundinacea* to *P. coronata* and in *Zoysia* spp. resistant to *P. zoysiae*. For other species see text. Use blends of resistant cultivars of one species or species mixtures to 'space' susceptible component species.	Benodanil, carboxin, chlorothalonil, mancozeb, oxycarboxin, propiconazole, triadadimefon, zineb.

Table 34.1 Continued

Disease	Pathogen	Suscepts	Cultural control	Resistant species and cultivars	Fungicidal control
Southern blight	*Sclerotium rolfsii* (*Athelia rolfsii*)	*Agrostis stolonifera*, *Poa annua* *Lolium* spp. *Poa pratensis* *Cynodon dactylon*	Not fully determined. Ammonium sulphate or bicarbonate applied bimonthyl to biweekly commencing in spring with appropriate fungicide.	Not available.	Captafol, triadimefon. Captan, carboxin and quintozene + fertilizer.
Spring dead spot	In New South Wales, Australia and California. *Leptosphaeria korrae*. In eastern USA and Japan – cause not established.	*Cynodon dactylon;* *Cynodon* spp., *Axonopus compressus*, *Stenotaphrum secundatum*, *Pennisetum clandsetinum*, *Zoysia* spp.	Raise mowing height. Reduce N levels, thatch accumulation and soil compaction. Increase potassium usage to improve winter hardines in *Cynodon* spp.	Differences in *C. dactylon* and *Cynodon* spp. cultivars have been noted.	In Australia: nabam or thiram In USA: weekly applications of nabam from late summer. Benomyl or quintozene at high dosages.
Summer patch	*Magnaporthe poae* *Phialophora* sp.	Many grass species but causing severe damage mainly on *Poa pratensis* and *P. annua* turf.	Not fully determined.	Not determined.	Benomyl, fenarimol propiconazole, triadimefon.
Take-all patch	Most commoly *Gaeumannomyces graminis* var. *avenae* also vars. *tritici* and *graminis*	Many grass species but *Agrostis* spp. are very susceptible to *G. graminis* var. *avenae*.	Where disease is endemic: 1. Use species other that *Agrostis* in fine turf or use the latter in polystands with less-susceptible grass species. 2. Where liming is necessary, especially on *Agrosits* turf use coarsely ground calcium carbonate and/or reacidify turf	*Festuca rubra* and *Cynosurus cristatus* are resistant. *Poa pratensis* cultivars are often tolerant of the disease.	Phenyl mercuric acetate applied as a soil drench. Triadimefon, triadimenol, carbendazim, chlorothalonil and iprodione are possible alternatives. Reacidification is probably more reliable than fungicide use

			surface with sulphur, ammonium sulphate or phosphate. 3. Avoid suppression of antagonists with soil fumigants on non-target fungicides. 4. Use a balanced fertilizer programme.		treatment.
Yellow patch	Rizoctonia cerealis (*Ceratobasidium cereale*)	*Agrostis* spp., *Festuca arundinacea*, *Poa* spp. *Cynodon dactylon, Zoysia* spp.	Not known.	Not determined.	Mercury compounds where permitted. Iprodione, quintozene.
Pythium foliar blights	*Pythium aphanidermatum* and other *Pythium* spp.	Most cool-season turf grasses and *Cynodon dactylon* during hot, humid weather and high night temperatures (>20°C)	Minimize the period of leaf wetness, remove thatch and provide balanced nutritional programme.	Not determined.	Ethazole, chloroneb propamocarb, metalaxyl, fosetyl aluminium, as preventive treatments.
Red thread and Pink patch	*Laetisaria fuciformis*	All commonly used cool season turf grasses but especially *Lolium perenne* and fine-leaved *Festuca* spp.	Most common on turf of low vigour in cool/warm, wet weather conditions. Apply N fertilizer to encourage growth but, if outbreak is late in the growing season, omit N and use a fungicide. Both fungi may be active under snow with unfrozen ground.	Variation in resistance occurs in cultivars of both *Lolium perenne* and fine-leaved *Festuca* spp.	Cadmium compounds, benomyl, iprodione, chlorothalonil, anilazine, triadimefon, mancozeb, thiram, propiconazde, fenarimol. NB. fungicide response may vary with the causal fungus present. Correct diagnosis essential.
	Limonoyces roseipellis and *Limonomyces* spp.	*Lolium perenne, Poa* spp., *Agrostis* spp. *Cynodon dactylon*			

Table 34.1 Continued

Disease	Pathogen	Suscepts	Cultural control	Resistant species and cultivars	Fungicidal control
Rhizoctonia foliar blights (Brown patch)	*Rhizoctonia solani (Thanatephorus cucumeris) R. cerealis (Ceratobasidium cereale) R. oryzae* and *R. zeae (Waitea circinata)*	Most common cool- and warm-season turf grasses during periods of hot, humid weather and night temperatures above 20 °C	Minimize the period of leaf wetness. Moderate N fertilization balanced by P and K.	Species and cultivars differ in their susceptibility.	Mercury compounds where allowed, benomyl, thiophanate methyl and ethyl, chlorothalonil, iprodione, mancozeb, maneb, thiram, thiabendazole, triadimefon fenarimol.
'Helminthosporium' diseases					
Leaf spot, leaf and seedling blight, crown and root rot.	*Cochliobolus sativus (Bipolaris sorokiniana)*	Few turf grasses species show much resistance. Regional differences in cultivar reaction have been noted.	Use the least susceptible cultivar available. Fertilize to give adequate growth response. Raise height of mowing when spots appear. Avoid herbicide applications during droughts.	*Phleum pratense, Festuca ovina, F. rubra* and *Poa trivialis* reported least susceptible to foliage infection. Some less-susceptible cultivars of *Poa pratensis* are available.	Effective control of leaf spot or crown rot reported with anilazine, chlorothalonil, dicyclidine, iprodione mancozeb, maneb, quintozene, thiram and vinclozolin, alone or in combination.
Leaf spot-and foot-rot (melting out) of *Poa pratensis*	*Drechslera poae*	Mainly *Poa* spp. especially *Poa pratensis*	Use resistant cultivars. Adequate fertilization in autumn to encourage spring recovery. Split applications of nitrogen in spring where root-rot phase is a problem. Adequate irrigation to avoid drought stress in summer. Raise height of mowing.	There are several cultivars of *Poa pratensis* with good resistance to *D. poae* but may not have adequate resistance to *C. sativus* or other pathogens.	Fungicides which control *C. sativus* are effective against *D. poae*. For slight to moderate leaf in spring make 2–3 applications at 14 day intervals starting early. Where leaf spot is active autumn to spring apply fungicide at 7–14 day intervals.

Net blotch and leaf blight of fescues and ryegrass	*Pyrenophora dictyoides (Drechslera erythrospila)*	Mainly *Lolium perenne, Festuca arundinacea, F. rubra*	Use cultivars reported persistent if information on disease resistance not available. Similar management practices as for *D. poae*.	Resistant forage cultivars of *L. perenne* available. Little information on resistance of turf grass cultivars. Resistant *F. rubra* turf grasses available.	Maneb, mancozeb chlorothalonil applied at 3–4 week intervals.
Leaf spot, leaf blight and foot rot of ryegrass	*Drechslera siccans (Pyrenophora lolii)*	*Lolium perenne* less commonly *Festuca* spp.	Use cultivars reported persistant locally. Similar to *D. poae*. Use cultivars persistant under close mowing. Fertilize in spring to give adequate growth response.	Persistent cultivars of *L. perenne* are available.	Phenyl mercuric acetate, chlorothalonil. Carboxin may increase leaf infection.
Leaf blight and crown rot	*Drechslera catenaria*	*Agrostis* spp., *Lolium perenne, Festuca arundinacea, Poa trivialis*	Avoid highly susceptible cultivars. Fertilize in spring to give adequate growth response. Adequate irrigation to avoid drought stress.	Little information on resistance. Toronto, Penncross and Seaside (*A. stolonifera*) susceptible	Two applications of iprodione in spring reported to give excellent control.
Red leaf spot and leaf blight of bentgrasses	*Pyrenophora erythrospila (Drechslera erythrospila)*	*Agrostis* spp.	Avoid excessive nitrogen fertilization but fertilize in spring to give adequate growth response.	Toronto and Seaside *A. stolonifera* susceptible.	Weekly applications of chlorothalonil, or chlorothalonil and anilazine alternated.
Leaf spot and leaf mould of bermuda grass	*Cochliobolus cynodontis (Bipolaris cynodontis)*	*Cynodon dactylon* and *Cynodon* spp.	Balanced fertilization	Resistant cultivars of *C. dactylon* available.	See *D. poae*
Zonate eyespot of bermuda and bentgrasses	*Drechslera gigantea*	*Cynodon dactylon Agrostis stolonifera, A. canina, A. tenuis*		Not known	See *D. poae*

Table 34.1 Continued

Disease	Pathogen	Suscepts	Cultural control	Resistant species and cultivars	Fungicidal control
Pre-and post-emergence blight, Crown and root-lesioning. 'Fading out'	*Curvularia* spp. alone or in complex with *Helminthosporium* spp.	Many cool- and warm-season grasses including: *Festuca rubra, Poa pratensis, Lolium perenne, Agrostis* spp., *Zoysia* and *Cynodon* spp.	Similar to that employed for *C. sativus*. In open, sunny locations, reduce surface temperatures by syringing before the heat of the day.	Little information on cultivar resistance	Similar to that for *C. sativus*.
Leaf spots					
Pycnidial leaf spots	*Phyllosticta* spp., *Stagonospora* spp., *Phaeoseptoria* spp., *Wojnowicia* spp., *'Hendersonia'* spp., *Septoria* spp., *Pseudoseptoria* spp.	Most cool- and warm-season turf grass species. Several different pycnidial dial species may occur on the same host.	Usually special control measures are not necessary. Avoid excessive nitrogen fertilization, infrequent mowing, herbicide and dought injury, inadequate nutrition. During outbreaks collect clippings to reduce inoculum.	Little information available, but generally those species and cultivars which senesce early are more liable to injury.	No specific studies, but those fungicides effective against Septoria diseases of cereals may give control, e.g. benomyl + maneb, carbendazim + maneb, thiophanate methyl + maneb, tridemorph + carbendazim, ethirimol + captafol.
Grey leaf spot and leaf blight	*Piricularia grisea (Magnaporthe grisea)*	Mainly warm-season species. *Stenotaphrum secundatum* is the major turf grass suscept. *Cynodon Eremochloa, Erogrostis, Paspalum, Pennisetum, Agrostis, Festuca* and *Lolium* spp.	Moderation in use of nitrogen fertilizer. Time irrigation to ensure shortest period of leaf wetness. Avoid stress from herbicides, traffic and drought.	*Lolium perenne* is field resistant. Yellow–green types of *S. secundatum* are more resistant than blue–green or bitter-blue types.	Fungicides are usually required for control. Thiram, chlorothalonil at 10–14-day intervals.

Spermospora leaf spot, scald and leaf blight	*Spermospora ciliata, S. lolii, S. Poagena, S. holci*	*Agrostis spp., Lolium* and *Festuca* spp., *Poa* spp., *Holcus lanatus*	Reduce inoclum during outbreaks by frequent mowing and collection of clippings.	Species of pathogens confined to appropriate hosts.	Not determined.
Brown stripe or leaf streak	*Cercosporidium graminis*	Many hosts, but may cause significant damage to seed crops and fine-leaved *Festuca* spp. and *Phleum pratense.*	In seed crops burn crop debris.	Little resistance noted in *Phleum pratense* lines.	Not determined.
Blotch and char spot of grasses	*Cheilaria agrostidis*	*Agrostis* spp., *Agropyron cristatum, Holcus lanatus, Lolium perenne, Phleum pratense, Poa pratensis*	In outbreaks, reduce irrigation.	Not determined	Not determined.
Leptosphaerulina leaf spot	*Leptosphaerulina australis*	*Poa, Agrostis, Festuca,* and *Lolium* spp.	Select cultivars which do not senesce early.	Cultivars of northern origin more likely to be damaged than those which remain winter green.	Not determined.
Physoderma leaf spots	*Physoderma graminis, P. agrostidis, P. gerhartii, P. paspali*	*Agropyron repens, Agrostis stolonifera, Aquatic* or *hydrophilic* grasses *Paspalum* spp.	Improve drainage Manage irrigation in hot weather to reduce duration of plant surface wetness	Uncertain	Not determined.
Timothy eyespot	*Cladosporium phlei*	*Phleum pratense, P. bertolonii*	Not usually any concern in amenity turf but in seed production select less-susceptible cultivars. Maintain adequate levels of N, P + K and especially K on light-soils. Flame sanitation not recommended.	Resistant lines are available in *P. pratense* and *P. bertolonii.*	No present studies. Colloidal sulphur.

Table 34.1 Continued

Disease	Pathogen	Suscepts	Cultural control	Resistant species and cultivars	Fungicidal control
Leaf blotch or scald	*Rhynchosporium orthosporium R. secalis*	*R. orthosporium* on *Agrostis, Festuca Lolium, Phleum* and *Poa* spp. *R. secalis* on *Agrostis* and *Lolium* spp.	Avoid excessive nitrogen. Use balanced fertilizers.	Little information but some overlap of pathogens on particular host ranges.	No specific studies but fungicides effective against the disease on cereals, e.g. triadimefon alone or with carbendazim may give control of the disease in turf grasses.
Black leaf spot or char spot	*Phyllachora* spp.	A wide range of cool- and warm-season grasses.	The diseases are rarely severe enough to justify control in turf. Mow and collect clippings before autumn to remove inoculum. Flame sanitation in seed crops.	No information available.	Not determined.
Leaf fleck of bentgrasses and timothy	*Mastigiosporium rubricosim, M. kitzbergense*	*M. rubricosum* mainly on *Agrostis tenuis* but also on *A. stolonifera* and *A. canina. M. kitzbergense* on *Phleum pratense*	Not determined	Not determined	Not determined
Ramularia and Ramulaspora leafspots	*Ramularia pusilla*	Mainly *Agropyron, Agrostis, Lolium* and *Poa* spp.	Not determined.	Isolates of *R. pusilla* from *Lolium/Festuca* and *Agrostis* show some host-specificity: *R. holci-lanati* isolates from *H. lanatus* are more pathogenic on that host than on *Lolium* and *Festuca* spp.	Not determined.
	Ramulaspora holci-lanati.	Mainly *Holcus lanatus*			

Hadrotrichum leaf spot	*Hadrotrichum virescens*	*Agrostis stolonifera, A. tenuis, Lolium perenne, Agropyron* and *Poa* spp.	Improve fertility. Mow turf regularly.	Not determined.	Not determined.
Leaf smuts					
Leaf spot and leaf blister smut	*Entyloma dactylidis*	*Agrostis, Poa Holcus* and *Phleum* spp. in particular but many other species are suscepts.	Not known	Resistant cultivars occur in *Poa pratensis*	Not determined.
Stripe smut	*Ustilago striiformis* (but *U. trebouxii, U. macrospora,* and *U. cynodontis* may also infect leaves and sheaths.)	Most extensive and damaging in turf formed by species of *Poa* and *Agrostis*	Balanced fertilization and adequate irrigation in hot-weather favours survival of infected plants.	Blending of cultivars may lessen the severity of the disease. Resistant cultivars of *Poa pratensis* and *Agrostis palustris* are available but pathogenic specialization occurs at species and cultivar level	Tank-mix applications of quintozene with either benomyl, triarimol, fenarimol, or triadimefon in late autumn. Systemic fungicides applied in the growing season.
Sheath and culm smut	*Ustilago hypodytes*	*Agropyron cristatum* and many other grass species		Not known	Not developed, but fungicides used for stripe smut may be effective.
Flag smut	*Urocystis agropyri*	Mainly in *Poa pratensis* but turf forming *Agrostis* and *Poa* species also infected.		Physiological specialization known. Some resistance shown to stripe and flag smut by A34, Cougar and Nugget *Poa pratensis*.	Similar to stripe smut

Table 34.1 Continued

Disease	Pathogen	Suscepts	Cultural control	Resistant species and cultivars	Fungicidal control
Head smuts					
Covered smuts	*Tilletia decipiens* *T. pallida*	Cereals and grasses susceptible. May be important on *Agrostis* spp.	Stem bud infection by *T. controversa* in first seed production year can be reduced in spring-planted grasses by encouraging vigorous growth during first summer and autumn.	*A. tenuis*, *F. arudinacea*, *Phleum pratense*, *Poa pratensis* not infected by *T. controversa*.	Fungicide seed dressings as used on cereals, e.g. benomyl, thiobemdaxole, carboxin + maneb may not control because of soil-borne inoculum. Hexachlorbenzene and quintozene as soil fungicides to reduce soil-borne infection.
	T. caries *T. controversa* *T. foetida*	Mainly on *Poa* spp. and *Agropyron cristatum*. *T. foetida* on *Lolium perenne* also.			
Loose smuts	*T. lollii* *T. holci* *Ustilago bullata*	*Lolium perenne* *Holcus lanatus*, *H. mollis* and *Anthoxanthum odoratum*. Physiological races on different species. *Agropyron cristatum* and *Elymus junceus* likely hosts.	Of little importance	Resistant lines of some species are available.	Fungicidal seed dressings or soil drenches with thiram, carboxin and benomyl have given control.
	U. trebouxii *U. macrospora* *U. cynodontis* *U. affinis*	On *Poa pratensis* On *Agropyron cristatum* On *Cynodon dactylon* On *Stenotaphrum secundatum*	Use a smut-free source for vegetative plantings.		As for *U. bullata*
Turfgrass seed crop diseases					
Blind seed	*Gloeotinia granigena*	Many turf grass species are susceptible but *Lolium* spp. suffer most.	Store seed for 2 + years. Seed at 13 cm depth. Hot water or vapour heat treatment of seed. Annual burning	Resistance available in *Lolium perenne*, polygenically inherited	Sodium azide effective in inhibiting apothecial production, but

			of crop debris effective if done on district basis.		not used commercially.
Choke	*Epichloe typhina*	Most cool-season turf grasses are susceptible, but disease is of most significance in *Agrostis, Poa* and *Festuca* spp.	Roguing plants showing stromata usually not effective. Where disease is not seed-borne, plough and reseed.	Not available	Triforine. giberellic acid applied to breeding material allows some seed production.
Ergot	*Claviceps purpurea* *C. cynodontis,* *C. paspali,* *C. yanagawensis* *Claviceps* spp.	*Claviceps* species are grass-tribe specific on cool-season and warm-season turf grasses. *C. purpurea* on more than 200 spp.	Reduce available inoculum by deep ploughing, bury sclerotia minimum of 5–8 cm. Burn crop residues. Avoid seedling diseases and herbicide injury that may promote late tillering.	Not available. The disease is most severe in lines where seed setting is poor.	Sodium azide and other chemicals are available which suppress ascospore development. Economics of use doubtful.
Twist	*Dilophospora alopecuri*	Most cool-season turf grass species are susceptible. *Phleum pratense* frequently attacked in Europe.	Longer rotations. High cutting, burning of crop debris suggested, but detailed control measures not worked out.	Not known	Dry seed disinfectants are ineffective.
Stem eyespot of fescues	*Didymella festucae (Phleospora idahoensis)*	Fine-leaved *Festuca* spp.	Burning of crop debris, high seeding rates and high rates of N applied in autumn minimize effects of disease and improve yields of first crops after treatment.	Some lines of *F. rubra* show resistance but not available in commercial cultivars.	Effective control has been obtained with systemic fungicides experimentally but not yet in practice.
Silvertop	Physiological. Insects, mites, *Fusarium poae* (usually secondary)	Most cool-season turf grasses, but particularly *Agropyron cristatum, Festuca rubra, F. arundinacea, F. ovina* and *Poa pratensis.*	Field burning of crop debris.		Pesticide appropriate to insect or mite concerned, applied before inflorescences are exserted from sheath.

35 *Common and chemical names of commonly used turf fungicides*

Nomenclature:

Common names are those recommended by:

ISO (International Standardization Organization)
BSI (British Standards Institution)
ANSI (American National Standards Institute)

Chemical names are from:

CA = Chemical Abstracts – names under rules for the ninth and tenth collective index periods.
IUPAC = International Union of Pure and Applied Chemistry

Common or generic name	*Chemical name*
anilazine	4,6-dichloro-*N*-(2-chlorophenyl)-1, 3, 5-triazin-2-amine
benodanil	2-iodo-*N*-phenyl benzamide
benomyl	methyl 1-(butylcarbamoyl)-2-benzimidazolecarbamate
bordeaux mixture	copper sulphate plus calcium hydroxide
cadmium chloride	cadmium chloride
cadmium sebacate	cadmium sebacate
cadmium succinate	cadmium succinate
calomel	mercurous chloride
captan	*N*-[(trichloromethyl)thio]-4-cyclohexene-1, 2-dicarboximide
carbendazim	methyl benzimidazol-2-ylcarbamate
carboxin	5, 6-dihydro-2-methyl-*N*-phenyl-1, 4-oxathiin-3-carboxamide
chloranil	2, 3, 5, 6-tetrachloro-1, 4-benzoquinone
chloroneb	1,4-dichloro-2, 5-dimethoxybenzene
chlorothalonil	tetrachloroisophthalonitrile
corrosive sublimate	mercuric chloride
cycloheximide	3-[2-(3, 5-dimethyl-2-oxocyclohexyl)-2-hydroxyethyl-]glutarimide
dichlone	2, 3-dichloro-1, 4-naphthoquinone
dicloran or DCNA	2, 6-dichloro-4-nitroaniline
dichorophen	Di-(5-chloro-2-hydroxyphenyl)methane
dinocap	2-(1-methylheptyl)-4, 6-dinitrophenyl crotonate (a mixture of 1-methylheptyl, 1-ethylhexyl, and 1-propylpentyl isomers of the octyl 8-carbon chain)
etaconazole	1-[2-(2′, 4′-dichlorophenyl-4-ethyl-1, 3-dioxo-lan-n-2-yl-methyl]-1H-1, 2, 4-triazole

Common or generic name	*Chemical name*
ethirimol	5-butyl-2-ethylamino-4-hydroxy-6-methyl-pyrimidine
etridiazol (or ethazol)	5-ethoxy-3-(trichloromethyl)-1, 2, 4-thiadiazole
fenaminosulf	sodium [4-(dimethylamino)phenyl]diazene sulphonate
fenarimol	α-(2-chlorophenyl)-α-(4-chlorophenyl)-5-pyrimidinemethanol
formaldehyde	methanal
fosetyl aluminium	aluminium tris (-o-ethylphosphonate)
iprodione	3-(3, 5-dichlorophenyl)-*N*-(1-methylethyl)-2, 4-dioxo-1-imidazolidinecarboxamide
malachite green	basic cupric carbonates
mancozeb	manganese ethylenebis-(dithiocarbamate) complex with zinc salt
maneb	manganese ethylenebis-(dithiocarbamate)
maneb plus zinc ion	manganese ethylenebis-(dithiocarbamate) plus zinc sulphate
mercurous chloride (calomel)	mercurous chloride
mercuric chloride (corrosive sublimate)	mercuric chloride
oxycarboxin	5, 6-dihydro-2 methyl-*N*-phenyl-1, 4-oxathiin-3-carboxamide 4,-4, -dioxide
procymidone	*N*-(3, 5-dichlorophenyl)-1, 2-dimethyl-cyclopropane-1, 2-dicarboximide
propamocarb	propyl 3-(dimethylamino)propylcarbamate
propiconazole	1-[2-(2, 4-dichlorophenyl)4-propyl-1, 3-dioxo lan 2-ylmethyl]-1H-1, 2, 4-triazole
quintozene	pentachloronitrobenzene
thiabendazole	2 (thiazol-4-yl) benzimidazole
thiophanate-ethyl	1, 2-bis (3-ethoxycarbonyl-2-thioureido)-benzene
thiophanate-methyl	1, 2-bis (3-methoxycarbonyl-2-thioureido-benzene
triadimefon	1-(4-chlorophenoxy)-3, 3-dimethyl-1-(1H-1, 2, 4-triazol-1-yl)-2-butanone
tridemorph	2, 6-dimethyl-4-tridecyl-morpholine
triarimol	A(2, 4-dichlorophenyl)-a-phenyl-5-pyrimidinemethanol
triforine	*N*, *N′*-[1,4-piperazinediyl-bis (2, 2, 2-trichloroethylidene)]-bis-[formamide]
vinclozolin	3 (3, 5-dichlorophenyl)-5-ethenyl-5-methyl methyl-2, 4-oxazolidinedione
zineb	zinc ethylenebis(dithiocarbamate)

Common Combinations:

calomel plus corrosive sublimate; cycloheximide plus PCNB; cycloheximide plus thiram; thiophanate plus mancozeb; thiram plus cadmium chloride; thiram plus PMA; thiram plus thiophanate methyl.

Index

Page numbers in italics refer to illustrations

Acaulospora endomycorrhizae 147, 148
Acremonium
  boreale 94–5
  endophytes 151, 152
Acrostalagmus roseus 92
Aeration of turf 11–12, 348
Agaricus tabularis fairy rings 341
Agropyron
  leaf spot caused by *Septoria calmagrostidis* 316
  Speckled leaf blotch 316
Agropyron cristatum
  loose smut susceptibility 195
  LTB resistance 71
  Pythium root rot 40
  sheath smut attack 191–2
  silvertop in 207
  tolerance to mowing 8
Agrostis
  anthracnose infection 223
  basal rot caused by *Collectrichum graminicola* 221
  black leaf spot caused by *Phyllachora* 327
  for bowling and putting greens 15
  char spot 334
  choke infection 204
  copper spot 227–8
  dollar spot disease 231
  downy mildew 238, 240
  effect of methyl bromide 10
  emergence, seedling 30
  Fusarium blight susceptibility 117, 120
  Fusarium patch 62, 66
  grey leaf spot 245
  head smut attack 192
  leaf blight 32
  leaf blotch caused by *Septoria triseti* 317
  leaf fleck caused by *Mastigosporium* 326–7
  leaf mould caused by *Drechslera fugax* 295
  leaf spot caused by *Cercospora agrostidis* 321
  leaf spot caused by *Hadrotrichum* 325
  leaf spot caused by *Septoria calmagrostidis* 316
  LTB attack 67
  LTB resistance 70, 71
  microdochium patch 62,66
  pH of turf 9
  pink patch disease 262
  pink snow mould 62, 66
  red leaf spot and leaf blight caused by *Pyrenophora erythrospila* 303–4
  Sclerotinia snow mould 89, 90
  Sclerotium rhizoides infection 75
  smut 183
  snow mould susceptibility 81
  southern blight 227
  stripe smut susceptibility 185
  take-all turf patch susceptibility 139, 141, 143
  tolerance to mowing 8
  twist attack 211
  Typhula spp. susceptibility 81, 84
  zonate eyespot 295
Agrostis canina
  copper spot 228
  Gloeocercospora sorghi susceptibility and pH 4
  in putting green turf 16
  vegetative establishment of turf 7
Agrostis stolonifera
  Cochliobolus sativus diseases 287, 288
  crown rot 294
  Drechslera 16
  grey snow mould control 81–2
  leaf blight 294
Rhizoctonia solani susceptibility and pH 4
  stripe smut 185, 186, 187
  susceptibility to disease 16
  Typhula ishikariensis control 88
  vegetative establishment of turf 7
  volume of spray required 23
Agrostis tenuis
  calcium deficiency 6
  mycorrhizal effects 149
Alternaria alternata 67
Ammonium nitrate application
  Cochliobolus sativus diseases 288
  grey snow mould 87
  LTB 72
Ammonium phosphate application
  grey snow mould 87
  take-all turf patch 141
Ammonium sulphate application
  Cochliobolus sativus diseases 288
  dollar spot disease 234, 235
  LTB 72
  pink snow mould attack 65
  snow moulds 80
  in southern blight control 278
  take-all turf patch 141
Anilazine 20
  against dollar spot disease 235
  against *Microdochium nivale* 66, 67
  for brown patch control 271–2
  for *Cochliobolus sativus* disease control 290
  for copper spot control 228–9
  for red thread disease control 263, 264
Antagonists, beneficial and partial sterilization 11
Anthracnose 219, 368
  basal rot 220, 225
  control 225
  disease development 224–5
  foliage blight 220
  susceptible cultivars 225
Antibiotics 20
Argentine stem weevil 152

Ascochyta
 effect of mowing 8
 leaf spots and leaf blights 312–14
 phleina 313
 turf grass hosts 312–14
Ascomycetes 359
Asterocystis radicis 45
Athelia rolfsii 276
Atkinsonella endophytes 151, 152
Aureobasidium bolleyi 44
Axonopus compressus 126

Balansia endophytes 151, 152
Balansiopsis endophytes 151
Bandane 10
Barley, susceptibility to *Pythium* root rot 48
Basal rot and soil compaction 11
Basidiomycetes, fairy ring fungi 341, 350
Benodanil
 against rusts 175
 for fairy ring control 348
Benomyl
 against dollar spot disease 235
 against fusarium blight 122
 against *Microdochium nivale* 66, 67
 against spring dead spot 127
 against *Typhula incarnata* 81
 for blind seed disease control 203
 for choke control 205
 for copper spot control 228–9
 for grey snow mould control 87–8
 for powdery mildew control 249
 for red thread disease control 263, 264
 for sclerotinia snow mould control 92
 for seed dressing 50
 for stem eyespot of fescue control 210
 for stripe smut control 188
Bentgrass
 cultivars for putting greens 16
 microdochium patch 62
 pink snow mould 62
 see also Agrostis
Benzimidazole 21
 against *Microdochium nivale* 66
Berberis as alternate host for black rust 169
Bermuda grass
 leaf spot and mould 285–6
 rust 158
 spring dead spot 124, 126
 see also Cynodon
Biloxazol for stem eyespot of fescue control 210
Bipolaris 283–285
 cynodontis 286
 sorokiniana 220, 287
Black leaf spot 380
 caused by *Phyllachora* 327–9
Black rust 157, 159, 168–70
 disease development 169
 host resistance 169–70
Blights
 causal fungi 30
 caused by *Fusarium* 30–1, 38
 disease development 31, 38
 foliar 36
 post-emergence 30–1, 378
 pre-emergence 30, 378
 symptoms 30–1
Blind seed disease 201–3, 382
 control 203
 environmental conditions 202
 epidemiology 202–3
 symptoms 201
Blister or spot smut 183–4
Blotch 379
Bluegrass turf, yellow tuft disease 238
Bone meal 10
Boom sprayer, motorized 23, *24*
Bordeaux mixture 18
 for control of fairy rings 347
Bouteloua
 gracilis and mycorrhizal effects 148
 leaf spots caused by *Cercospora boutelouae* 321
Bowling green turf 8, 15
Brown
 basal rot 93
 fleck rust 159, 170–1
 patch 19
 rust 160, 171–2
 stripe 323–4, 379
Brown patch 266–72, 376
 Bordeaux mixture for 18
 Control 271–2
 disease development 270–1
 fungicide control 271–2
 symptoms 266–7
 thiram treatment 19
 on vegetatively propagated bentgrass clones 16
 and water sprinkling 12
Brown root rot 93–4, 373
 identification 99
Browning root rot 48
Buchloe, leaf spots caused by *Cercospora seminalis* 322
Burning crop debris for rust control 175
 see also Fire sanitation and Stubble burning

Cadmium compounds
 against dollar spot disease 235
 against *Microdochium nivale* 66
 as fungicides 19
 for copper spot control 228–9
 for fairy ring control 347
 for red thread disease control 263, 264·
Cadmium succinate against *Typhula incarnata* 81, 82
Calcined ferrous sulphate 6
Calcium
 arsenate 11
 deficiency 6
 and fusarium blight 121
 nitrate application and *Cochliobolus sativus* diseases 288
 peroxide seed dressing 50
 soil level and pythium foliar blight 256
 soil level and *Rhizoctonia solani* infection 43
Captafol for southern blight control 278
Captan 20
 for southern blight control 278
 seed dressing 50
Carbathiin for sclerotinia snow mould 92
Carbon
 dioxide and fungal growth 4
 dioxide sensitivity of *Rhizoctonia solani* 42
 disulphide as fumigant 24
 effect of fairy rings on soil level 344
Carboxin
 against LTB 73
 against rust 175
 for fairy ring control 348
 for southern blight control 278
 for stripe smut control 188
Cat's tail disease *see* Choke

Ceratobasidium cereale 41, 273
Cercospora leaf spots 322–3, 370
Cercosporella leaf spot 322
Cercosporidium graminis 323–4
Char spot 334, 379
Cheilaria agrostis 334
Cheshunt compound 50
Chewing's fescue 15
Chlordane 11, 141
Chloroneb 20, 384
  against LTB 72
  against *Microdochium nivale* 66
  against *Typhula incarnata* 81, 82
  for control of yellow patch 274
  for grey snow mould control 87–8
  for pythium foliar blight control 256–7
  for Sclerotinia snow mould 92
Chlorophenol mercury 19
Chloropicrin as fumigant 24
Chlorothalonil 20
  against dollar spot disease 235
  against *Microdochium nivale* 66
  against rust 175
  against *Typhula incarnata* 81, 82
  for brown patch control 271–2
  for *Cochliobolus sativus* disease control 290
  for copper spot control 228–9
  for *Drechslera siccans* control 301
  for grey leaf spot control 246
  for grey snow mould control 87–8
  for *Pyrenophora dictyoides* control 303
  for red thread disease control 263
  for sclerotinia snow mould control 92
  for yellow patch control 274
Choke 151, 152, 203–5, 383
  control 205
  disease development 204–5
  susceptible grass species 204
Cladochytrium 44
Cladosporium
  herbarum 67
  phlei 5, 324–5
Claviceps purpurea 205, 206–7
  control 207
  sclerotium *206*
  sphacelia stage 206
  susceptible grasses 207
Climate and selection of grass cultivars 15–16
Cochliobolus 3, 283, 284
  cynodontis 285–6
  nitrogen fertilizer and susceptibility to 5
Cochliobolus sativus 286–91
  as cereal pathogen 48
  disease control 290–1
  disease development 288–9
  disease symptoms 286–7
  and dry soil 12
  effect of post-emergence herbicides 10
  environmental conditions for development 288
  and excess nitrogen 5
  hosts 288
  resistant turf grass species 289–90
Colletotrichum basal rot 219
  control 225
Colletotrichum graminicola 4, 219, 222
  anthracnose symptoms 220–2
  host range 223–4
  pathogenicity tests 222–3
Colletotrichum on grass roots 43
Compost, partial sterilization 11
Copper
  accumulation from Bordeaux mixture 18, 19
  application and microdochium patch resistance 65
  deficiency 6
Copper spot 4, 227–9, 368
  cadmium fungicides 19
  control 228–9
  disease development 228
  symptoms 230
Coprinus
  kubickae fairy rings 351
  snow mould 372
  urticicola 67–8
Coprinus psychromorbidus 68–9, 73–4, 372
  complex 88, 95
  disease resistance *70*
  fairy rings 350
  and fertilizer application 5
  HCN production 97
  quintozene for control 20
Corticium fuciforme
  low-temperature-tolerant pathogen 92
  see also Laetisaria and *Limonomyces*
Corticium red thread disease 6
Cottony blight *see* Pythium foliar blight
Cottony snow mould 67–74, 372
  low temperature basidiomycete (LTB) 67–73
  control 72–3, 102–3
  distribution 102–3
  environmental conditions 71
  epidemiology 71
  identification 99–100
  pathogenesis 69–70
  symptoms 67–9
Covered smut 192–3, 382
Crabgrass *see Digitaria*
Cricket
  top dressing of tables 9
  type of turf 9, 15
Crotonylidene urea
  application and LTB 72
  fertilizers 10
Crown rot 117, 376, 377
  caused by *Cochliobolus sativus* 286–91
  caused by *Drechslera catenaria* 294–5
Crown rust 158, 162, 167–8
Culm smut 381
Curvularia
  classification 284
  complexes 291
  diseases 291–3
  on grass roots 43
  identification 284
  lunata involvement with anthracnose 220
  pallescens 291
  turf diseases 283
Cycloheximide 20
  against dollar spot disease 235
  against rust 175
  for brown patch control 271–2
  for grey leaf spot control 246
  for powdery mildew control 249
  for red thread disease control 263
  for southern blight control 278
Cylindrocarpon 43, 44
Cynodon
  leaf blight 32
  leaf spots caused by *Cercospora seminalis* 322
  pH of turf 9
  rust 168
  rust resistance 175
  use in southern Europe 15

Cynodon (cont'd.)
vegetative establishment of turf 7
Cynodon dactylon
copper spot 227–8
crown rot 293
Curvularia disease 293
cynodon rust resistance 168
grey leaf spot 244, 245
leaf blight 293
leaf spot and mould 285–6
loose smut attack 195
pink patch disease 262
pythium foliar blight 254
root rot 293
Septoria leaf spot 316
southern blight 277
spring dead spot 124, 126
yellow patch 272
zonate eyespot 295
Cynosurus cristatus
microdochium patch 59
pink snow mould 59
seedling emergence 29

2, 4–D 10
Dactylis glomerata
brown basal rot 93
choke disease 204–5
mycorrhizal effects 148
sclerotinia snow mould 89
Damping-off 4, 368–9
causes 30–1
nitrogen and phosphate deficiency 50
rapid seed germination 7
and soil moisture levels 7
soil nitrogen and phosphate levels 40
Dazomet partial sterilization of compost 11
Deschampsia flexuosa mycorrhizal effects 149
Dew 3
dispersal 11, 65
Dicamba 10
Dichloran for southern blight control 278
Dichlorophen against *Microdochium nivale* 66, 67
Dicyclidine for *C. sativus* disease control 290
Didymella festucae 209
Didymium crustaceum 357
Digitaria
brown patch 269
dollar spot disease 231
downy mildew infection 240, 241
Dilophospora alopecuri 211
Dinocap 20
for powdery mildew control 249
Disease resistance of cultivars 15–16
Ditylenchus graminophilus involvement in twist 211
Dollar spot 5, 230–6, 369
cadmium fungicides 19
control 235–6
development 232–4
fungicide resistance 21
host range 231
mercuric chloride treatment 19
in sod 8
and soil compaction 11
spread by mowing 8
symptoms 230–1
on vegetatively propagated bentgrass clones 16
Downy mildew 238–42, 369
control 242
disease development 240–2
effect of fertilizer application 241
humidity requirements 240–1
survival of infected plants 241
symptoms 238–9
temperature requirements 240–1
Drainage 3, 65
Drazoxolon seed dressing 50
Drechslera 3
catenaria 294–5
classification 284
dictyoides 302
diseases of grass 293–305
effect of mowing 8
erythrospila 304
fugax 295
gigantea 295–6
nitrogen fertilizer and susceptibility to 5
noblea 296
phlei 296
siccans 299–301
tritici-repentis 304–5
turf diseases 283
Drechslera poae 297–9
anthracnose 219
environmental conditions for growth 298
foot rot phase 8
spore dispersal by mowing 8
Dried blood 10
Drought
reduction of transpiration 12
susceptibility 12
Drought stress
and grey leaf spot 246
leaf blight of *Poa pratensis* 289
and rust 12, 175
and stripe smut 187, 188
of turf 3

Ectendomycorrhizae 147
Ectomycorrhizae 147
Elymus
leaf spots caused by *Cercospora elymi* 321
speckled leaf blotch 316
Endoconidium temulentum 201–2
Endomycorrhizae 147, 148
Endophytes 151–3
effect on cattle and sheep 151
eradication 153
fungicides 153
insect resistance 152
Entyloma 183
crastophilum 183, 184
dactylidis 183, 184
irregulare 183, 184
Epichloe
endophytes 151
typhina 204, 205
Eragrostis grey leaf spot 245
Eremochloa ophiuroides
fusarium blight susceptibility 120
grey leaf spot 245
vegetative establishment of turf 7
Ergot 205–7, 383
alkaloids 151, 205
control 207
poisoning 205
susceptible grasses 205, 207
Erysiphe graminis 247
conidia production 248
effect of mowing 8
host range 247–8
humidity required for germination 248
potassium deficiency and susceptibility 5
resistant grass varieties 249
temperature required for germination 248

tolerance to fungicides 250
Etaconazole against dollar spot disease 235
Ethazole
 -etridiazole 20
 for pythium foliar blight control 256–7
Ethirimol for powdery mildew control 250
Ethylene bisdithiocaramate (EBDC) 19
ETMT seed dressing 50
Etridiazole seed dressing 50
Evaporation, soil moisture 3
Evapotranspiration control 13
Exserohilum 283, 284
 rostratum 305

Fading-out 293
Fairy rings 25, 341–51, 369
 biological control 349
 causal fungi 341–2
 effect on soil 343–4
 elimination 347–50
 rate of growth 342–3
 superficial 350–1
 suppression 347–50
Fenaminosulph 20
 for pythium foliar blight control 256–7
Fenarimol 21
 against dollar spot disease 235
 against fusarium blight 122
 for powdery mildew control 250
 for stem eyespot of fescue control 210
 for stripe smut control 188
Ferrous sulphate 11
 for fairy ring control 347
 for microdochium patch disease 19
 for slime mould control 358
Fertilization of established turf 9–10
Fertilizer 4, 5
 application and Fusarium patch disease 65–6
 application and LTB 71, 72
 application and pink snow mould 65–6
 in cold and dry conditions 10
 in control of stripe smut 188
 controlled nitrogen release 10
 in seedling disease control 49
 effect on take-all turf patch 141, 144
 for sod 8
 and microdochium patch 65
 and pH control 9
 and pink snow mould 65
 prior to seeding 5
Fescue
 microdochium patch 62
 pink snow mould 62
 rust 158
 toxicity syndrome 151
Festuca
 brown stripe 323
 crown rust 167
 crown rust resistance 167–8
 dollar spot disease 231
 emergence 30
 endophytes and insect resistance 152
 endophytic parasites 151
 for bowling and putting greens 15
 fusarium blight susceptibility 120
 grey leaf spot 245
 leaf blotch caused by *Septoria tenella* 316–17
 leaf spots caused by *Cercospora festucae* 321–2
 LTB resistance 70, 71
 microdochium patch 62
 net blotch and leaf blight caused by *Pyrenophora dictyoides* 301–3
 ovina 149
 pH of turf 9
 pink snow mould 62
 sclerotinia snow mould susceptibility 90
 Sclerotium rhizoides infection 75
 seedling blight symptoms 30
 silvertop in 207
 stem eyespot 208
 tenuis 15
 Typhula ishikariensis susceptibility 84, 85
Festuca arundinacea
 brown patch 269
 crown rot 294
 leaf blight 294
 rust resistance 175
 tolerance to mowing 8
Festuca rubra
 anthracnose infection 223, 224
 calcium deficiency 6
 choke infection 204
 Cochliobolus sativus diseases 287, 288
 Curvularia disease 291
 dollar spot disease 234
 downy mildew 238, 240
 LTB attack 67
 microdochium patch resistance 66
 pink patch disease 262
 pythium foliar blight 254
 red thread disease 262
 take-all turf patch resistance 143
 tolerance to mowing 8
Fine-leaved fescue
 rust 158, 168
 susceptibility to anthracnose 225
Fire sanitation
 in silvertop control 208
 in stem eyespot of fescue control 210
 stubble and crop waste burning 175, 203
Fish meal 10
Flag smut 183, 189–91, 381
 disease development 189–91
 fungicides for control 191
 symptoms 189
Fly parasite of *Epichloe typhina* 205
Fog 3
Foliar
 blight 117
 diseases 4
Foot rot 376, 377
 cycloheximide for control 20
 effect of mowing 8
 of *Lolium perenne* 299–301
 of *Poa pratensis* 297–9
Football pitches, turf type 16
Formaldehyde
 as fumigant 24, 25
 for control of fairy rings 348–9, 350
Fosetyl-aluminium 21, 256–7
Frost scorch 74–5
 identification 100
Fumigants 24–5
Fumigation
 against mycorrhizal infection 149
 effect on take-all turf patch 142
Fungal endophytes see Endophytes

Fungicides 7, 384–5
against dollar spot disease 235–6
against microdochium patch disease 66
against pink snow mould 66
against rusts 175
against stripe smut 188–9
application 22–4
for anthracnose control 225
effect on take-all turf patch 141
for choke control 205
for *Cochliobolus sativus* disease control 290–1
for colletrichum basal rot control 255
for copper spot control 228–9
for downy mildew control 242
for flag smut control 191
for grey leaf spot control 246
for grey snow mould control 87–8
for mycorrhizal infection 149
for pink patch disease control 263–4
for powdery mildew control 249–50
for pythium foliar blight control 256–7
for red thread disease control 263–4
for sclerotinia snow mould 92
for stem eyespot of fescue control 210
for yellow patch control 274
formulations 22, 67
fumigants 24–5
granular formulations 22
historical development 18–21
in brown patch control 271–2
mode of action 21
phytotoxicity 24
resistance 21
seed dressing 49–50
spray formulations 22
Fusarium 3
anthracnose 219
avenaceum 92
biological control of ergot 207
crown rot 370
damping-off 4
effect on cereals 48
foliar blight 370
on grass roots 43
heat stressed turf 4
nivale see Microdochium nivale
oxysporum 31
pathogenicity 30
poae 117, 119–20
pre-and post-emergence seedling blights 30–1, 38
root rot 370
roseum 119
see also Fusarium culmorum
seed rot and seedling blight 30
in silvertop attack 208
tricinctum see Fusarium poae
Fusarium blight 117–22
causal fungi 119–20
control 121–2
disease development 120–1
fungicides against 118, 122
host range 120
nematode involvement 118
resistant grass varieties 121–2
symptoms 118–19
syndrome 118
temperature for 121
Fusarium culmorum
and dry soil 12
fusarium blight 117, 118, 119
root rot complexes 31
seed dressing against 50
seedling blight 30, 31, 38
Fusarium patch
see Microdochium patch, Pink snow mould
Fusarium species 119–20

Gaeumannomyces graminis 4
antagonists 142–3
control with chlordane 11
effect of methyl bromide 10
pH and growth 142
phosphorus deficiency and susceptibility to 5
and soil organic matter 5
and sulphur deficiency 6
take-all turf patch 137–9
Germination 30
capacity and disease control 49
delayed 49
Giberellic acid for choke control 205
Gigaspora, endomycorrhizae 147–8
Gloeocercospora sorghi
cadmium fungicide 19
copper spot 227
nitrogen fertilizer and susceptibility to 5
pH effects 4
Gloeosporium bolleyi 44
Gloeotinia granigena 201–2
nitrogen deficiency and susceptibility 5
Glomerella graminicola 222
Glomus, endomycorrhizae 147, 148, 149
Golf courses, turf type 9, 16
Grass
nutrient requirements 9
parasitic root fungi 43–9
removal of clippings 8
root pathogens 43–9
sward density 3
temperature change during rapid growth 4
transpiration 3
Grease spot *see* Pythium foliar blight
Grey leaf blight 377
Grey leaf spot 244–6, 378
control 246
disease development 245
humidity levels required 245
nitrogenous fertilizer application 245
temperature required 245
Grey snow mould 75–82, 372
control 80–2, 87–8, 101
diagnosis 84
distribution 101
epidemiology 80, 85–6
identification 99
plant nutrients 87
symptoms 84
suscepts 245
Griseofulvin 20
Guignardia 314
Guttation
drops 3, 271
increase with excess nitrogen 5

Hadrotrichum leaf spot 325, 381
HCN *see* Hydrogen cyanide
Head smuts 192–3
disease development 195
Heat stress
of turf 4
and wilting of *Poa annua* 219
'Helminthosporium'
association with anthracnose 219

captan for control 20
disease 5, 376–7
following downy mildew 238
identification 284
turf disease 283
on vegetatively propogated bentgrass clones 16
Helminthosporium-anthracnose-senescence (HAS) syndrome 220, 225, 368
Hendersonia culmicola 318
Herbicides 10–11
and grey leaf spot damage 246
Hockey, type of turf 9
Holcus
lanatus 149
leaf spot caused by *Septoria calmagrostidis* 316
Honeydew 205
Hoof and horn meal 10
Hose dragging 11, 65
Humidity
atmospheric 3
relative for *Rhizoctonia solani* growth 42
requirements for fungal infection 3
and temperature 4
turf microclimate 3
Humus, effect of fairy rings on 344
Hydrogen cyanide
production by *Coprinus psychromorbidus* 97
production by low temperature basidiomycete 69–70
production by *Marasmius oreades* 345
Hydrogen-ion concentration *see* pH

Idrella
bolleyi 44
on grass roots 43
Insecticides in silvertop control 208
Insects, damage to stems and silvertop attack 208
Iprodione 21
against dollar spot disease 235
against fusarium blight 122
against spring dead spot 127
against *Typhula incarnata* 81, 82
for *Cochliobolus sativus* disease control 290
for copper spot control 228–9
fusarium blight control 118
for grey snow mould control 88
leaf blight and crown rot control 295
red thread disease control 263, 264
tolerance to in microdochium patch 66
yellow patch control 274
Iron
deficiency 6
sulphate application for downy mildew 242
Irrigation 12–13
effects on stripe smut 187, 188
and LTB 71
and nutritional plane of turf 66
Isobutylidene fertilizers 10

Kentucky bluegrass
microdochium patch 63
pink snow mould 63
Key to turf diseases 363–6
Knapsack sprayers 23

Laetisaria fuciformis 4, 5, 259, 260–1
basidia and basidiospores *261*
control with malachite green and Bordeaux mixture 18–19
effect of scarification 11
host range 262
nitrogen deficiency and susceptibility 5
Lagena
parasitic root fungus 43
radicicola 45
Lawns, turf for 8
Lead arsenate 11
Leaf
blister smut 183–4, 381
galls 211
mould of bentgrass 295, 377
rot 74–5
rust 161
see also Brown rust
smut on vegetatively propogated bentgrass clones 16
Leaf blight 376, 377
ascochyta 312–14
of bentgrass caused by *Pyrenophora erythrospila* 303–4
caused by *Cochliobolus sativus* 286–91
caused by *Drechslera catenaria* 294–5
caused by *Leptosphaerulina* 325–6
caused by *Pyrenophora tritici-repentis* 304–5
caused by *Spermospora* 333–4
of *Festuca* and *Lolium* caused by *Pyrenophora dictyoides* 301–3
of *Lolium perenne* 299–301
Leaf blotch 380
caused by *Rhynchosporium* 332–3
caused by *Stagonospora nodorum* 318
of *Agrostis* caused by *Septoria triseti* 317
of *Festuca* caused by *Septoria tenella* 316–17
Leaf fleck 370
caused by *Mastigosporium* 326–7
Leaf scald 379
caused by *Rhynchosporim* 332–3
caused by *Spermospora* 333–4
Leaf spot 3, 370, 376, 377
ascochyta 312–14
caused by *Cercospora* 321–3
caused by *Cochliobolus sativus* 286–91
caused by *Hadrotrichum* 325
caused by *Physoderma* 329–30
caused by *Pseudocercospella* 321–3
caused by *Pseudoseptoria* 314
caused by *Pyrenophora tritici-repentis* 304–5
caused by *Ramularia* 330–2
caused by *Ramulaspera* 330–2
caused by *Spermospora* 333–4
control of by mowing 8
control of pycnidial 319–20
cycloheximide for control 20
disease development of pycnidial 319–20
of *Agrostis*, *Agropyron* and *Holcus* caused by *Septoria calmagrostidis* 316
of *Cynodon* caused by *Septoria cynodontis* 316
of *Lolium* caused by *Septoria tritici* 317

Leaf spot *(cont'd.)*
  of *Lolium perenne* 299–301
  of *Poa* caused by *Septoria macropoda* 316
  of *Poa pratensis* 297–9
  on vegetatively propogated bentgrass clones 16
  of ryegrass 296
  Phaeoseptoria 319
  phyllosticta 314
  pycnidial 312–20
  Septoria 315–17
  smut 183–4, 381
  Stagnospora 317–18
  and systemic fungicides 21
  and water sprinkling 12
  Wojnowicia 319
  zineb and maneb for control 19
Leaf streak 323–4
  caused by *Physoderma* 329–30
  of timothy 296
Leptosphaeria
  nabam control 19
  narmari 125
  patch 129–31
  spring dead spot 124
Leptosphaeria korrae
  effect of mowing 8
  in fusarium blight syndrome 118
  necrotic ring spot 129, *130*, 131
  nitrogen fertilizer and susceptibility to 5
  potassium deficiency and susceptibility 5
  spring dead spot 125–6
  summer patch 133
Leptosphaerulina leaf
  blight 325–6
  spot 379
Lichens 359
Light intensity and *Pythium* disease 39
Ligniera
  parasitic root fungus 43
  pilorum 45
Lime application 4, 9
  brown patch control 271
  effect on take-all turf patch 140, 141, 142, 143
  effect on *Typhula* infection 87
  and microdochium patch disease 64, 66
  and pink snow mould 66
Limonomyces roseipellis 4, 5, 259, 261–2
  host range 262
  nitrogen deficiency and susceptibility 5
Lolitrems 151, 152
Lolium
  crown rust 167
  grey leaf spot 245
  head smut susceptibility 193
  leaf spot caused by Drecslera noblea 296
  leaf spot caused by *Septoria tritici* 317
  multiflorum 244, 277
  net blotch and leaf blight caused by *Pyrenophora dictyoides* 301–3
  pythium foliar blight 254
  sclerotinia snow mould susceptibility 90
  Typhula ishikariensis susceptibility 84
Lolium perenne
  black rust susceptibility 170
  blind seed disease 201, *202*
  brown patch 269
  crown rot 294
  crown rust resistance 167
  downy mildew 240
  emergence 30
  endophytes and insect resistance 152
  endophytic parasites 151
  Fusarium infection 31
  leaf blight 294
  leaf spot and blight and foot rot caused by *Drechslera siccans* 299
  microdochium patch 62–3, 66
  mycorrhizal effects 148, 149
  pink patch disease 259, 260, 262
  pink snow mould 62–3
  red thread disease 262
  rust resistance 175
  seedling susceptibility to downy mildew 241
  southern blight 277
  tolerance to mowing 8
Lolium perenne
  pH of turf 9
  'S 23' 15
Loose smut 193–5, 382
  disease development 195
Low fertility disease 5
  and removal of grass clippings 8
  unfertilized turf 10
Low-maintenance turf 5
Low temperature basidiomycete (LTB) 67-73
  causal fungus 67–9
  control 71–3
  cultivar resistance 70–1
  environmental conditions 71
  epidemiology 71
  fertilizer application 72
  fungicidal control 72–3
  HCN production 69–70
  hosts 70–1
  pathogenesis 69–70
  and snow cover 71, 72
  Low-temperature-tolerant turf grass pathogens 92–5
  Lycoperdon fairy rings 349–50
  D-Lysergic acid diethylamine (LSD) 205

Macrophoma 314
Magnaporthe
  grisea 5, 244–5
  poae 118, 133–5
Magnesium deficiency 6
Malachite green 18
  against *Pythium* 50
  and Bordeaux mixture 263
Maleic hydrazide 11
Mancozeb 19–20
  against fusarium blight 122
  against rust 175
  against *Typhula incarnata* 81
  brown patch control 271–2
  for *Cochliobolus sativus* disease control 290
Maneb 19
  against fusarium blight 122
  against *Microdochium nivale* 66
  against rust 175
  for *Cochliobolus sativus* disease control 290
  for *Pyrenophora dictyoides* control 303
  for stem eyespot of fescue control 210
Manganese application and microdochium patch resistance 65
Manganese deficiency 6
Marasmius oreades
  effect on soil 343–4
  fairy rings 341, 344–7
  hydrogen cyanide production 345
  microflora of rings 346
  pathogenicity 345

rate of growth 342–3
relation to soil conditions 345–6
self-inhibition 347
sensitivity to fungicides 348
soil exhaustion 347
Marasmius siccus fairy rings 351
Mastigosporium leaf fleck of bentgrass and timothy 326–7
Meadowgrass
microdochium patch 63
pink snow mould 63
see also Poa
Melting-out 297–9, 376
of *Poa pratensis* 297–9
Mercuric chloride
against LTB 72
brown patch treatment 19
dollar spot treatment 19
Mercury fungicides 67
against fusarium blight 122
against LTB 72
against take-all turf patch 141
brown patch control 271–2
Drechslera siccans control 301
fairy ring control 347, 350, 351
for copper spot control 228–9
for *Sclerotium rhizoides* infection 75
for snow mould 81
grey snow mould control 87–8
red thread disease control 263
sclerotinia snow mould control 92
seed dressing 48, 50
yellow patch control 274
Metalaxyl 21
for downy mildew control 242
for pythium foliar blight control 256–7
Metham-sodium, partial sterilization of compost 11
Methyl bromide 10
as fumigant 24, 25
fumigation and take-all turf patch 141, 142
partial sterilization of compost 11
Methyl isocyanate 24
Microdochium
bolleyi 44
on grass roots 43
patch 59–67
Microdochium nivale 4
annular patches 351
attack on grass around fairy rings 344
bridging growth and dew 11
combination with *Fusarium culmorum* 38
complexes 62, 66, 95
disease association with worm casts 11
effect of liming 4
effect of quintozene 11
effect of temperature 63–4
epidemiology 63–5
fusarium patch 59, 60–1
host range 61–3
and iron deficiency 6
isolation 61
low-temperature-tolerant pathogen 92
nitrogen fertilizer and susceptibility to 5
on vegetatively propagated bentgrass clones 16
and organic top dressing 5
patch disease of unmown turf 8
pathogenicity 38, 61
pink snow mould 59, 60–1
potassium deficiency and susceptibility 5
seedling blight 30
soil pH 65
source of inoculum 63
and sulphur deficiency 6
Microdochium patch 59–67
causal fungus 60–1
control 65–7, 100
fungicides against 18–20, 66–7
Mites and silvertop attack 208
Moisture
control for Fusarium patch disease 65
control for pink snow mould 65
effect on *Microdochium nivale* 64
and LTB 72
requirement of snow moulds 96
tension 3
Monographella nivalis 30, 60, 61
Moorland turf 15
Mowing 8, 65
and dispersal of *Drechslera poae* 298–9
and fusarium blight 121
and rust disease control 175
Mucilago spongiosa 357
Mycorrhizae 147–9
control 149
development 148
effects on grasses 148–9
vesicular-arbuscular (VA) type 147, 148, 149
Myriogenospora endophytes 152
Myriosclerotinia borealis 89–92
biological control 92
complexes 95
control 11, 91–2
effect of scarification 11
epidemiology 90–1
host range 90
pathogenicity 90
phosphorus deficiency and susceptibility to 5
susceptibility 90
temperature and disease 95
Myxomycetes 357

Nabam 19, 127
Necrotic ring spot 129–31, 370
control 131
fungicides against 131
host range 131
symptoms 129
Nectria radicicola 44
Nematicides 149
Nematodes, involvement with fusarium blight 118
Net blotch 377
of *Festuca* and *Lolium* 301–3
Nickel sulphate against rust 175
Nigrospora blight 370
Nitrogen
deficiency 5
effect of fairy rings on soil level 343
excess 5
influence on host susceptibility in grey leaf spot 245
in rust disease management 174–5
protection against *Pythium* root disease 49
requirements of grass 9–10
use on sod 8
Nitrogen application and
Cochliobolus sativus diseases 288
copper spot 228
Curvularia diseases 293
effect on *Drechslera siccans* 301
fairy rings 347
fusarium blight 121
microdochium patch 65
pink snow mould 65
powdery mildew 247, 249

Nitrogen application *(cont'd.)*
- red thread disease 263
- stripe smut control 188
- susceptibility to *Typhula ishikariensis* 87

Nitrogen fertilizer and
- blind seed disease control 203
- downy mildew 241, 242
- LTB 71, 72
- Sclerotinia snow mould 91
- snow mould 80–1
- spring dead spot 127

Nitrogen levels and
- brown patch 271
- damping-off 40
- dollar spot disease 234, 235
- leaf spot development 319
- *Rhizoctonia solani* infection 43

Nutrient
- demand by grass 9
- retention 4
- soluble 5

Olpidium
- *agrostidis* 45
- *brassicae 43*, 45, 48
- parasitic root fungus 43
- *radicale* 46
- *radicicolum* 45

Oospore germination rate in *Pythium* 39
Ophiobolus patch *see* Take-all turf patch
Orange stripe rust 160, 171
Organo-mercury fungicides 66
- *see also* Mercury fungicides

Oryza sativa
- grey leaf spot penetration 245
- penetration process of *Sclerophthora macrospora* 240

Oxathiins 21
Oxycarbathiin for sclerotinia snow mould control 92
Oxycarboxin
- against LTB 73
- against rust 175
- for fairy ring control 348
- for stripe smut control 188

Paederia 173
Parasitic grass root fungi 43–9
Paspalum
- grey leaf spot 245
- leaf blight 285
- leaf spot 318

Patch disease
- liming 9
- on unmown turf 8

Pathogen
- antagonists 4, 92–7
- soil-borne 5

Peltigera canina 359
Pennisetum clandestinum
- leaf spot 285
- spring dead spot 126

Pennisetum grey leaf spot 245
Peramine 152
Perennial ryegrass, blind seed disease 201, *202*
Pesticides 10–11
- effect on mycorrhizal infection 149

pH
- effect on growth of pathogens 4
- effect on take-all turf patch 140, 142, 143
- of soil around fairy rings 347
- of soil and brown patch 271
- of soil and microdochium patch 64–5
- of soil and pythium foliar blight 256
- of turf 9
- and *Pythium* disease 39

Phaeoseptoria leaf spot 319
Phenyl mercury acetate 19, 81
Phialophora 47
- *graminicola* 46, *48*
- in fusarium blight syndrome 118
- summer patch 133, 134

Phleospora
- *graminearum* leaf spot 210
- *idahoensis* 209

Phleum
- brown stripe 323, 324
- emergence 30
- leaf fleck caused by *Mastigosporium* 326–7
- leaf streak 296
- sclerotinia snow mould susceptibility 90
- smut 183
- Timothy eyespot 324–5
- *Typhula ishikariensis* susceptibility to 84, 87

Phleum pratense
- black rust susceptibility 170
- stripe smut 186
- susceptibility of seedlings to pythium blight 38
- twist attack 211

Phoma on grass roots 43
Phoma sclerotiodes
- brown root rot 93–4
- complexes 95

Phosphate
- deficiency 5
- fertilizer 5
- levels and damping-off 40

Phosphorus
- application and microdochium patch resistance 65
- protection against *Pythium* root disease 49
- and *Typhula borealis* resistance 87

Phyllachora black leaf spot 327–9
Physarum 357
Physoderma 44
- leaf spot 329–30, 379
- leaf streak 329–30

Phytotoxicity of fungicides 22, 24–5
Phytotoxin 233
Pink patch 5, 259, 261–4, 375
- control 263–4
- fungicide control 264
- symptoms 260

Pink snow mould 59–67, 371
- causal fungus 60–1
- control 65–7, 100
- distribution 100
- epidemiology 63–5
- identification 98
- symptoms 60

Plant nutrients 5–6
Plantago lanceolata 149
Plugs, stolous, sprigs for turf establishment 7
Poa
- emergence 30
- leaf spot caused by *Septoria macropoda* 316
- smut 183
- take-all turf patch 139–40, 143

Poa annua
- anthracnose infection 219, 223
- basal rot 11
- basal rot caused by *Colletrichum graminicola* 11, 220–2
- *Cochliobolus sativus* diseases 287, 288
- grey snow mould control 81–2
- invasion of sod 8
- LTB attack 67, 70, 72

microdochium patch 61–2, 66
pink patch disease 262
pink snow mould 61–2, 66
resistant clones to *Cochliobolus sativus* 290
southern blight 277
susceptibility to disease 16
tolerance to mowing 8
twist attack 211
winter dessication of turf 13
Poa pratensis
black rust resistance 169–70
brown fleck rust resistance 170–1
brown patch 269
Cochliobolus sativus diseases 286–9
cultivars resistant to flag smut 191
Curvularia disease 291
downy mildew 240
flag smut 189, 190–1
foot rot 297–9
Fusarium blight 117, 118, 120
leaf blight 289
leaf blight caused by *Leptosphaerulina* 326
leaf spot 297–9
LTB resistance 70, 71
melting-out 297–9
Merion cultivar 185, 187, 188, 249
microdochium patch 63, 66
necrotic ring spot 129
pH of turf 9
pink patch disease 262
pink snow mould 63
powdery mildew 247, 249
resistant cultivars to *Cochliobolus sativus* 286–90
rust resistance 175
sclerotinia snow mould resistance 91
Sclerotium rhizoides infection 75
silvertop in 207
southern blight 277
spot smut susceptibility 183, 184
stripe smut 10, 12, 185–8
tolerance to mowing 8
Typhula ishikariensis susceptibility 84, 85
volume of spray required 23
Poa trivialis
crown rot 294
leaf blight 294
sclerotinia snow mould susceptibility 90
Poling 11
Polymyxa
graminis 47
parasitic root fungus 43
Polythene covers against cold injury and dessication 13, 72
Potassium
deficiency 5
fertilizer 5
leaching 5
soil level and *Rhizoctonia solani* infection 43
and *Typhula borealis* resistance 87
Potassium application and
microdochium patch 65
pink snow mould 65
powdery mildew 249
spring dead spot 127
Powdery mildew 247–50, 371
control 249–50
disease development 248–9
overwintering 248
symptoms 247
Pricking 12, 65
Prochloraz for choke control 205
Procymidone 21
Propamocarb 21
for pythium foliar blight control 256–7
Propiconazole 21
against fusarium blight 122
for choke control 205
Pseudocercosporella
herpotrichoides 322
leaf spots 312–3, 370
Pseudoseptoria 314, *315*
Puccinia
crandalii 157, 173
nitrogen deficiency and susceptibility 5
stenotaphri 160, 174
Puccinia coronata 157–9, 162–3, *165–6*, 167–8, 174–5, 373
Puccinia cynodontis 157–8, *166*, 168, 175
Puccinia festucae 158–9, *166*, 168, 174, 373
Puccinia graminis 157, 159, *162*, *164–5*, 167–70, 174–5, 373
Puccinia poae-nemoralis 157, 159, *165*, 170–1, 174–5, 373
Puccinia poarum 160, *166*, 171 174, 373
Puccinia recondita 160, *165*, 171–2, 174, 373
Puccinia striiformis 157, 161, *165*, 172–4
Puccinia zoysiae 161, 173
effect of mowing 8
Puffballs 349
Putting greens, turf type 15, 16
Pycnidial leaf spot 312–20
Pyrenophora 283, 284
dictyoides 301–3
erythrospila 303–4
lolii 300
tritici-repentis 304–5
Pyricularia
grisea 244–5
oryzae 244, 245
Pyroxyclor for pythium foliar blight control 257
Pythium 3
aphanidermatum 32, 252, 253
aristosporum 32, 40
arrhenomanes 32, 40, 48
blight and seed and root rot species 32–7
catenulatum 33
climatic conditions for root rot 40
control with malachite green and Bordeaux mixture 18–19
culture 38–9
damping-off 4
debaryanum 33
disease development 39
disease and excessive irrigation 12
dissotocum 33
graminicola 34, 40
on grass roots 43
hydrogen-ion concentration 39
identification 39
infection and sand top dressing 9
intermedium 34
irregulare 34
isolation 38–9
iwayamai 34
light intensity and infection 39
moisture requirements 39

Pythium (cont'd)
  myriotylum 35, 253
  oospores 31–9, 254
  root rots 38
  rostratum 35
  seed dressing against 50
  seed rot and seedling blight 30
  seedling diseases 38–40
  snow rot 93
  tardicrescens 36, 40
  taxonomy 38
  temperature requirements 39
  torulosum 36, 253
  ultimum 6, 36, 253
  vanterpoolii 36–7, 253
  volutum 40
  zineb and maneb for control 19
Pythium foliar blight 252–7, 375
  control 256–7
  disease development 254–6
  host range 253–4
  humidity required for outbreak 256
  soil conditions for 256
  symptoms 252–3
  temperature required for outbreak 256

Quintozene 11, 20
  against LTB 72
  against *Microdochium nivale* 66, 67
  against spring dead spot 127
  against *Typhula incarnata* 81, 82
  brown patch control 271–2
  for *Cochliobolus sativus* disease control 290
  grey snow mould control 87–8
  red thread disease control 263
  sclerotinia snow mould control 92
  southern blight control 278
  stripe smut control 188

Raindrop and spore dispersal 3
Ramularia leaf spot 330–2, 380
Ramulaspora leaf spot 330–2, 380
Ranunculus, alternate host for a *Uromyces* 173
Red leaf spot
  of bentgrass caused by *Pyrenophora erythrospila* 303–4, 377
Red thread 5, 259–61, 262–4, 375
  control with malachite green and Bordeaux mixture 18–19
  effect of mowing 8
  environmental conditions 262
  fungicide control 263–4
  maleic hydrazide treatment 10
  nitrogen application 263
  in sod 8
  symptoms 259–60
  zineb and maneb for control 19
Respiration, cellular in waterlogged soil 3
Rhamnus alternate host for crown rust 167
Rhizoctonia spp.
  as parasites 41
  cerealis 266, 273
  complexes 41
  disease development 41–3
  disease symptoms 41
  grass seedling disease 40–3
  low-temperature-tolerant pathogen 92
  on grass roots 43
  oryzae 266, 269, 270
  quintozene for control 20
  saprobes 41
  seed rot and seedling blight 30
  taxonomy 41
  zeae 266; 269, 270
Rhizoctonia foliar blight 376
Rhizoctonia foliar diseases 266–74
  cool-temperature 272
Rhizoctonia solani 4, 266, 267–9
  annular patches 351
  Bordeaux mixture for 18
  bridging growth and dew 11
  disease development 41–2
  effect on Graminae 48
  effect of mowing 8
  effect of scarification 11
  environmental conditions for sclerotial germination 270
  host range 269–70
  moniloid cells *268*
  mycelium *268*
  nitrogen fertilizer and susceptibility to 5
  pH effects 4
  and soil organic matter 5
  on vegetatively propagated bentgrass clones 16
Rhizophidium graminis 47
  parasitic root fungus 43
Rhynchosporium leaf blotch 332–3
Rice blast 244
Root
  respiration 4
  rot 117, 286–91, 376
  senescence 127
Rugby turf type 9, 15
Rusts 3, 5, 158, 373
  control 174
  diagnosis 174
  and drought stress 12
  effect of mowing 8
  fertilizer effects 174–5
  fungi 158, 162
  fungicides against 175
  life cycles 162
  resistant grass cultivars 175
  spore germination 11
  spore states 157
  symptoms 157
  zineb and maneb for control 19
Rye
  blind seed disease 201
  ergot of 205
Ryegrass
  microdochium patch 62–3
  pink snow mould 62–3
  see also *Lolium*

St Augustingrass
  rust 160, 174
  Sclerophthera macrospora resistance 242
  see also Stenotaphrum secundatum
Sand for top dressing 9
Scarifying 11, 65
Sclerocystis endomycorrhizae 147, 148
Scleroderma fairy rings 349
Sclerophthora macrospora 239–40
  host range 240
  nitrogen fertilizer and susceptibility to 5
Sclerotia of *Typhula* in unmown turf 8
Sclerotial low-temperature basidiomycete (SLTB) 73–4
Sclerotinia borealis see Myriosclerotinia borealis
Sclerotinia homoeocarpa 4, 5, 230, 231–2
  bridging growth and dew 11
  effect of mowing 8
  effect of scarification 11

fungicide resistance 21
nitrogen deficiency and susceptibility 5
on vegetatively propagated bentgrass clones 16
overwintering 233
potassium deficiency and susceptibility 5
toxin production 233
Sclerotinia snow mould 88–92, 372
control 91–2, 101
distribution 101
epidemiology 90–1
identification 98
symptoms 89
Sclerotium
rhizoides 74–5
rolfsii 276, 277
Sea-marsh turf 8, 15
Secale cereale blind seed disease 201
Seed
dressing 7, 48, 49–50
germination 7
rot 30
rot control 49–50
sowing and disease control 49
sowing times and blight 38
Seeding 7
Seedling
attack by low-temperature-tolerant fungi 48
blight 30, 376
disease control 49
HCN damage 70
post-emergence blight 30
pre-emergence blight 30
rot caused by *Cochliobolus sativus* 286–91
Selenophoma 314, 315
see also Pseudoseptoria
Septoria
cynodontis 316
effect of mowing 8
elymi 316
fungicide control 320
macropoda 316
tenella 316–17
triseti 317
tritici 317
Setosphaeria 283, 284
rostrata 305
Sharp eyespot 41
Sheath smut 191–2, 381
Silvertop 207–8, 383
Slime moulds 357–8
control 358
plasmodia 357
Snow
blight 93
rot 93
Snow cover
effect on *Microdochium nivale* 64
and grey snow mould 77
and LTB 71, 72
and *Myriosclerotinia borealis* 90, 91
Sclerotium rhizoides infection 75
and *Typhula* infection 76
Snow moulds 5, 59–114
antagonism 96–7
competition 95–7, 104
complexes 95–7, 104
fungicides for control (*see* individual diseases and 18–26)
on vegetatively propagated bentgrass clones 16
rhizoctonia 272
temperature for activity 95–6
Soccer turf type 9, 15
Sod growing 7–8
Sodium azide
for blind seed disease control 203
for ergot control 207
Sodium nitrate application and snow mould 80
Soil
aeration 4
aeration and fairy ring control 348
compaction 11
drainage 3, 4
effect of fairy rings 344
exhaustion with *Marasmius oreades* 347
moisture 3, 5, 7
moisture control and disease 49
moisture and incidence of *fusarium* seedling blight 31
moisture and *Pythium* germination 39
organic matter 4–5
pH 4
pH and microdochium patch 64–5
reaction 4
relation of *Marasmius oreades* to conditions 345
structure 4
structure and seed germination 7
temperature 30
temperature and germination 49
texture 4
texture and microdochium patch 65
water retention 5
waterlogged 3
Sorghum vulgare, copper spot disease 227
South German Mixed Bentgrass 15, 16
Southern blight 276–8, 374
control with ammonium bicarbonate 278
disease development 277–8
environmental conditions for growth 277
symptoms 276
Speckled
blotch 317–18
leaf blotch of *Agropyron* and *Elymus* caused by *Septoria elymi* 316
Speckled snow mould 75–6, 82–8, 372
control 102
distribution 102
identification 99
Spermospora leaf spot, scald and leaf blight 333–4, 379
Sphaeropsis 314
Spiking 65
Sporangia germination rate in *Pythium* 39
Spore dispersal
by mowing 8
by water 3
and irrigation 13
Sporulation, leaf spot pathogens 3
Spot
blight *see* Pythium foliar blight
smut 183–4
Spray
dilution rate 23
equipment 23
nozzles 23
optimum volume 23
Sprayers
mobile 23
motorized 23, *24*

Sprigs 7
Spring dead spot 19, 124–7, 374
 causal fungi 125–6
 control 127
 disease development 126–7
 fungicides 127
 host range 126
 symptoms 124
Sprinkling 12
Stagonospora
 arenaria 317
 effect of mowing 8
 leaf spots 317–18
 nodorum 318
 paspali 318
 smolandica 318
 subseriata 318
Stem
 eyespot of fescue 208–10, 383
 rust 159
 see also Black rust
Stenotaphrum leaf spots caused by *Cercospora fusimaculans* 322, 323
Stenotaphrum secundatum spring dead spot 126
 grey leaf spot 244, 245
 mode of infection with grey leaf spot 245
 resistance to grey leaf spot 246
 turf and downy mildew infection 238–9, 240
 vegetative establishment of turf 7
Stolons 7
Straw burning
 of blind seed disease infected fields 203
 see also Fire sanitation
String-of-pearls disease 74–5, 373
 identification 100
Stripe rust 161
Stripe smut 10, 183, 185–9, 381
 chemical control 188
 control 188–9
 disease development 186–8
 environmental conditions 187
 fertilizer application for control 188
 and irrigation 12
 symptoms 185
Stubble burning
 of blind seed disease infected fields 203
 see also Fire sanitation
Sulphur 4
 application and microdochium patch resistance 65
 coated urea fertilizers 10
 deficiency 6
 for powdery mildew control 249
Summer patch 117, 133–5, 374
 control 135
 disease development 135
 host range 134–5
 symptoms 133
Switching 11
Syringing 12
Systemic fungicides 20–1

2, 4, 5-T 10
Take-all turf patch 4, 137–44, 374
 antagonism 141–3
 control 143–4
 decline 141–3
 epidemiology 140–3
 fertilizers 144
 host range 139–40
 and irrigation 12
 liming and pH 4, 9, 144
 resistant grass species 143
 susceptibility 139–40
 symptoms 137
Tar spot caused by *Phyllachora* 327–9
Temperature 3–4
 and activity of *Microdochium nivale* 63–4
 extremes 4
 and Fusarium blight development 31, 121
 optimum for disease 3–4
 optimum for growth 3–4
 requirements of *Pythium* 39
 and *Rhizoctonia solani* disease 42
 and snow mould activity 95–6
Temporary covers for turf 13
Tetrachloroparabenzoquinone 50
Tetramethylthiuram disulphide (TMTD) 19
Thanatephorus cucumeris 41, 266, 267, 268–9
Thatch
 accumulation rate 12
 build-up and fusarium blight 121
 build-up and spring dead spot 126, 127
 management and superficial fairy rings 351
Thiabendazole 11
 against dollar spot disease 235
 against *Microdochium nivale* 66
 for copper spot control 228–9
Thiophanate
 against dollar spot disease 235
 for copper spot control 228–9
 for powdery mildew control 249
Thiophanate methyl
 against fusarium blight 122
 against *Microdochium nivale* 66, 67
 against *Typhula incarnata* 81
 for sclerotinia snow mould 92
 for stripe smut control 188
Thiram 19
 against dollar spot disease 235
 against LTB 73
 against rust 175
 against spring dead spot 127
 against *Typhula incarnata* 81, 82
 brown patch control 271–2
 Cochliobolus sativus disease control 290
 grey leaf spot control 246
 grey snow mould control 88
 red thread disease control 263
 sclerotinia snow mould control 92
 seed dressing 50
Thrips and silvertop attack 208
Tilletia head smuts 192
Timothy eyespot 324–5, 379
Tip blight 74–5
Top dressing 8–9
 partial sterilization 11
Topping 7
Toxicity, mammalian of fungicides 22
Transpiration
 of grass plants 3
 reduction of rate 12
Trechispora alnicola fairy rings 350–1
Triadimefon 21
 against dollar spot disease 235
 against fusarium blight 122
 against LTB 72
 against *Microdochium nivale* 66
 against rust 175
 against *Typhula incarnata* 82
 copper spot control 228–9
 fusarium blight control 118
 powdery mildew control 250
 red thread disease control 263, 264

southern blight control 278
stripe smut control 188
Triarimol
against dollar spot disease 235
powdery mildew control 250
stripe smut control 188
Tricholoma sordidum fairy rings 347
Tridemorph for powdery mildew control 250
Trifolium repens 149
Triforine for powdery mildew control 250
Turf
density and rate of disease spread 3
establishment by seeding 7
types 8, 15–16
vigour in microdochium patch disease 65–6
vigour in pink snow mould 65–6
Twist 210–11, 383
Typhula
blight 372
complexes 95
distribution 76
effect of scarification 11
incidence 76
patch disease of unmown turf 8
phacorrhiza 88
phosphorus deficiency and susceptibility to 5
rind patterns of sclerotia *78*
snow moulds 75–88
temperature ranges 79
on vegetatively propagated bentgrass clones 16
Typhula incarnata
barrage effect 97
control 80–2
epidemiology 80
grey snow mould 76–82
host range 79–80
isolation 79
low-temperature-tolerant pathogen 92
pathogenicity 79
susceptibility 79–80
temperature range 79
Typhula ishikariensis
basidiospores *84*
biological control 88
complex 88
control 87–8
diagnosis 84
effect of thiabendazole 11
epidemiology 85–6
host range 84–5
low-temperature-tolerant pathogen 92
speckled snow mould 82–8
susceptibility 84–5
symptoms 84
temperature range 79
varieties 83–4, 85

Urea application and snow mould 80
Urea-formaldehyde fertilizers 10
Urocystis agropyri 183
flag smut 189
soluble sugar and free amino acid levels of affected plants 185
teliospores *190*
Urocystis on vegetatively propagated bentgrass clones 16
Uromyces
dactylidis 161, *166*, 173, 174
leaf rust 173
Ustilago
affinis 195
bullata 193
hypodytes 191–2
macrospora 194–5
on vegetatively propagated bentgrass clones 16
trebouxii 194
Ustilago striiformis 183, 185–6
following downy mildew 238
mycelial colonization 186, 187
resistant grass cultivars 188
teliospores *186*

Vapam as fumigant 24
Vegetative establishment of turf 7
Velvet bentgrass turf, copper spot 228
Virus, plant 43

Waitea circinate 269
Water
and release and dispersal of spores 3
requirements of fungal infection 3
Watering can application of fungicides 22
Weed seeds in top dressings 9
White-ears *see* Silvertop
Whiteheads *see* Silvertop
Wojnowicia leaf spot 319
Worms 11

Yellow
rust 157
stripe rust 160–1, 172–3
tuft and water sprinkling 12
see also Downy mildew
Yellow patch 41, 272–4, 375
control 273–4
disease development 273
symptoms 272–3
Yellow patch 272

Zineb 19, 175
Zonate eyespot of bentgrass and Bermuda grass 295–6, 377
Zoospores
motile 3
Pythium 40
Zoysia
pH of turf 9
rust 161, 173
rust resistance 175
use in southern Europe 15
vegetative establishment of turf 7
Zoysia japonica
fusarium blight susceptibility 120
yellow patch 272